FUNDAMENTALS OF GAS SHALE RESERVOIRS

FUNDAMENTALS OF GAS SHALE RESERVOIRS

Edited by

REZA REZAEE
Department of Petroleum Engineering
Curtin University

Copyright © 2015 by John Wiley & Sons, Inc. All rights reserved

Published by John Wiley & Sons, Inc., Hoboken, New Jersey
Published simultaneously in Canada

No part of this publication may be reproduced, stored in a retrieval system, or transmitted in any form or by any means, electronic, mechanical, photocopying, recording, scanning, or otherwise, except as permitted under Section 107 or 108 of the 1976 United States Copyright Act, without either the prior written permission of the Publisher, or authorization through payment of the appropriate per-copy fee to the Copyright Clearance Center, Inc., 222 Rosewood Drive, Danvers, MA 01923, (978) 750-8400, fax (978) 750-4470, or on the web at www.copyright.com. Requests to the Publisher for permission should be addressed to the Permissions Department, John Wiley & Sons, Inc., 111 River Street, Hoboken, NJ 07030, (201) 748-6011, fax (201) 748-6008, or online at http://www.wiley.com/go/permissions.

Limit of Liability/Disclaimer of Warranty: While the publisher and author have used their best efforts in preparing this book, they make no representations or warranties with respect to the accuracy or completeness of the contents of this book and specifically disclaim any implied warranties of merchantability or fitness for a particular purpose. No warranty may be created or extended by sales representatives or written sales materials. The advice and strategies contained herein may not be suitable for your situation. You should consult with a professional where appropriate. Neither the publisher nor author shall be liable for any loss of profit or any other commercial damages, including but not limited to special, incidental, consequential, or other damages.

For general information on our other products and services or for technical support, please contact our Customer Care Department within the United States at (800) 762-2974, outside the United States at (317) 572-3993 or fax (317) 572-4002.

Wiley also publishes its books in a variety of electronic formats. Some content that appears in print may not be available in electronic formats. For more information about Wiley products, visit our web site at www.wiley.com.

Library of Congress Cataloging-in-Publication Data:

Fundamentals of gas shale reservoirs / edited by Reza Rezaee.
   pages cm
  Includes bibliographical references and index.
  ISBN 978-1-118-64579-6 (hardback)
1. Shale gas reservoirs. I. Rezaee, Reza.
  TN870.57.F86 2015
  553.2'85–dc23

                                 2015007792

Printed in the United States of America

10 9 8 7 6 5 4 3 2 1

CONTENTS

Contributors   xv

Preface   xvii

1 Gas Shale: Global Significance, Distribution, and Challenges   1

  1.1 Introduction, 1
  1.2 Shale Gas Overview, 1
      1.2.1 Shale Gas Geology, 2
      1.2.2 Characteristics of a Producing Shale Gas Play, 3
  1.3 The Significance of Shale Gas, 4
  1.4 Global Shale Gas Resources, 5
      1.4.1 Sources of Information, 5
      1.4.2 Resource Estimation Methodologies, 5
  1.5 Global Resource Data, 7
      1.5.1 China, 7
      1.5.2 The United States, 7
      1.5.3 Mexico, 7
      1.5.4 Southern South America, 7
      1.5.5 South Africa, 8
      1.5.6 Australia, 8
      1.5.7 Canada, 8
      1.5.8 North Africa, 8
      1.5.9 Poland, 9
      1.5.10 France, 9
      1.5.11 Russia, 9
      1.5.12 Scandinavia, 9
      1.5.13 Middle East, 9
      1.5.14 India, 9
      1.5.15 Pakistan, 10
      1.5.16 Northwest Africa, 10
      1.5.17 Eastern Europe (Outside of Poland), 10
      1.5.18 Germany and Surrounding Nations, 10
      1.5.19 The United Kingdom, 10

   1.5.20 Northern South America, 11
   1.5.21 Turkey, 11
 1.6 Data Assessment, 11
   1.6.1 Distribution, 11
   1.6.2 Basin Type, 11
   1.6.3 Depositional Environment, 12
   1.6.4 TOC Content, 12
   1.6.5 Clay Content, 13
 1.7 Industry Challenges, 13
   1.7.1 Environmental Challenges, 13
   1.7.2 Commercial/Economic, 14
 1.8 Discussion, 14
 1.9 Conclusions, 15
   Appendix A.1 Global Shale Gas Resource Data, 16

2 Organic Matter-Rich Shale Depositional Environments   21

 2.1 Introduction, 21
 2.2 Processes Behind the Deposition of Organic Matter-Rich Shale, 23
   2.2.1 Processes Behind the Transport and Deposition of Mud, 23
   2.2.2 Production, Destruction, and Dilution:
      The Many Roads to *Black* Shale, 23
 2.3 Stratigraphic Distribution of Organic Matter-Rich Shales, 25
 2.4 Geographic Distribution of Organic Matter-Rich Shales, 27
   2.4.1 Background, 27
   2.4.2 Controls on the Geographic Distribution of Black Shales, 30
 2.5 Organic Matter-Rich Shale Depositional Environments, 34
   2.5.1 Continental Depositional Environments, 34
   2.5.2 Paralic Depositional Environments, 36
   2.5.3 Shallow Marine Depositional Environments, 37
   2.5.4 Deep Marine Depositional Environments, 38
 2.6 Conclusion, 39

3 Geochemical Assessment of Unconventional Shale Gas Resource Systems   47

 3.1 Introduction, 47
 3.2 Objective and Background, 49
 3.3 Kerogen Quantity and Quality, 49
 3.4 Sample Type and Quality, 51
 3.5 Kerogen Type and Compositional Yields, 52
 3.6 Thermal Maturity, 54
 3.7 Organoporosity Development, 55
 3.8 Gas Contents, 57
 3.9 Expulsion–Retention of Petroleum, 57
 3.10 Secondary (Petroleum) Cracking, 58
 3.11 Upper Maturity Limit for Shale Gas, 58
 3.12 Gas Composition and Carbon Isotopes, 59
 3.13 Additional Geochemical Analyses for Shale Gas Resource System Evaluation, 61
 3.14 Oil and Condensate with Shale Gas, 63
 3.15 Major Shale Gas Resource Systems, 64
 3.16 Conclusions, 65

4 Sequence Stratigraphy of Unconventional Resource Shales   71

 4.1 Introduction, 71
 4.2 General Sequence Stratigraphic Model for Unconventional Resource Shales, 71

    4.3 Ages of Sea-Level Cycles, 72
    4.4 Water Depth of Mud Transport and Deposition, 73
    4.5 Criteria to Identify Sequences and Systems Tracts, 74
    4.6 Paleozoic Resource Shale Examples, 74
          4.6.1 Barnett Shale (Devonian), 74
          4.6.2 Woodford Shale (Late Devonian–Early Mississippian), 74
          4.6.3 Marcellus Shale (Devonian), 78
          4.6.4 New Albany Shale (Upper Devonian–Lower Mississippian), 78
    4.7 Mesozoic Resource Shale Examples, 80
          4.7.1 Montney Formation (Early Triassic), 80
          4.7.2 Haynesville/Bossier Shales (Late Jurassic), 80
          4.7.3 Eagle Ford Formation (Cretaceous), 80
          4.7.4 LaLuna Formation (Upper Cretaceous), 82
    4.8 Cenozoic Resource Shale Example, 83
    4.9 Conclusions, 84
    4.10 Applications, 84

5 **Pore Geometry in Gas Shale Reservoirs**   89
    5.1 Introduction, 89
          5.1.1 Gas Shales and Their Challenges, 89
          5.1.2 Pore Size Classification, 90
    5.2 Samples Characteristics, 90
          5.2.1 Sample Collection, 90
          5.2.2 Mineral Composition, 90
    5.3 Experimental Methodology, 91
          5.3.1 Capillary Pressure Profile, 91
          5.3.2 Nitrogen Adsorption (N_2), 92
          5.3.3 Low-Field NMR, 92
          5.3.4 Image Acquisition and Analysis, 93
    5.4 Advantages and Disadvantages of Experimental PSD Methods, 95
    5.5 Permeability Measurement, 95
    5.6 Results, 96
          5.6.1 Pore Size Distribution from MICP Experiments, 96
          5.6.2 Pore Size Distribution from Nitrogen Adsorption Experiments, 98
          5.6.3 NMR T_2 Relaxation Time, 98
          5.6.4 Scanning Electron Microscopy, 100
          5.6.5 Focused Ion Beam/Scanning Electron Microscopy, 100
          5.6.6 Capillary Pressure and Permeability, 102
    5.7 Discussion, 103
          5.7.1 Porosity and PSD Comparisons, 103
          5.7.2 Interchanging MICP with NMR Data, 103
          5.7.3 Pore-Body to Pore-Throat Size Ratio: Pore Geometry Complexity, 107
          5.7.4 Pore Throat Size and Permeability, 107
          5.7.5 Mineralogy, 108
    5.8 Conclusions, 112
          Appendix 5.A XRD Results, 114

6 **Petrophysical Evaluation of Gas Shale Reservoirs**   117
    6.1 Introduction, 117
    6.2 Key Properties for Gas Shale Evaluation, 117
          6.2.1 Pore System Characteristics, 117
          6.2.2 Organic Matter Characteristics, 118
          6.2.3 Permeability, 118

    6.2.4 Gas Storage Capacity, 119
    6.2.5 Shale Composition, 120
    6.2.6 Geomechanical Properties, 120
6.3 Petrophysical Measurements of Gas Shale Reservoirs, 121
    6.3.1 Pore Structure Evaluation Techniques, 121
    6.3.2 Fluid Saturation Measurement, 122
    6.3.3 Permeability Measurement, 123
    6.3.4 Adsorbed Gas Measurement, 124
6.4 Well Log Analysis of Gas Shale Reservoirs, 125
    6.4.1 Well Log Signatures of Gas Shale Formations, 125
    6.4.2 Well Log Interpretation of Gas Shale Formations, 128

7 Pore Pressure Prediction for Shale Formations Using well Log Data   139

7.1 Introduction, 139
    7.1.1 Normal Pressure, 139
    7.1.2 Overpressure, 139
7.2 Overpressure-Generating Mechanisms, 140
    7.2.1 Loading Mechanisms, 141
    7.2.2 Unloading Mechanisms (Fluid Expansion), 142
    7.2.3 World Examples of Overpressures, 143
    7.2.4 Overpressure Indicators from Drilling Data, 144
    7.2.5 Identification of Shale Intervals, 144
7.3 Overpressure Estimation Methods, 146
    7.3.1 Overview of the Compaction Theory, 146
    7.3.2 Eaton's Method, 147
    7.3.3 Effective Stress Method, 149
    7.3.4 Bowers's Method, 150
7.4 The Role of Tectonic Activities on Pore Pressure In Shales, 151
    7.4.1 Geology of the Study Area, 151
    7.4.2 Stress Field in the Perth Basin, 152
    7.4.3 Pore Pressure in Tectonically Active Regions (Uplifted Areas), 154
    7.4.4 Pore Pressure in Tectonically Stable Regions, 154
    7.4.5 Origins of Overpressure in Kockatea Shale, 156
7.5 Discussion, 160
    7.5.1 Significance of Pore Pressure Study, 163
    7.5.2 Overpressure Detection and Estimation, 163
    7.5.3 Pore Pressure and Compressional Tectonics, 163
    7.5.4 Overpressure-Generating Mechanisms, 164
    7.5.5 Overpressure Results Verifications, 164
7.6 Conclusions, 165

8 Geomechanics of Gas Shales   169

8.1 Introduction, 169
8.2 Mechanical Properties of Gas Shale Reservoirs, 170
    8.2.1 Gas Shale Reservoir Properties under Triaxial Loading, 170
    8.2.2 True-Triaxial Tests, 171
    8.2.3 Gas Shale Reservoir Properties under Ultrasonic Tests, 172
    8.2.4 Nanoindentation Tests on Gas Shale Plays, 173
    8.2.5 Scratch Tests, 174
8.3 Anisotropy, 175
    8.3.1 Anisotropy in Gas Shale Reservoirs, 175

8.4 Wellbore Instability in Gas Shale Reservoirs, 176
    8.4.1 Structurally Controlled Instability, 177
    8.4.2 Instability Due to Directional Dependency of Geomechanical Parameters, 178
    8.4.3 Time-Dependent Instability, 184

9 Rock Physics Analysis of Shale Reservoirs   191

9.1 Introduction, 191
9.2 Laboratory Measurements on Shales: Available Datasets, 192
9.3 Organic Matter Effects on Elastic Properties, 192
9.4 Partial Saturation Effects, 195
9.5 Maturity Effects, 197
9.6 Seismic Response of ORSs, 201
9.7 Conclusions, 203

10 Passive Seismic Methods for Unconventional Resource Development   207

10.1 Introduction, 207
10.2 Geomechanics and Natural Fracture Basics for Application to Hydraulic Fracturing, 209
    10.2.1 Basics of Earth Stress and Strain, 209
    10.2.2 Natural Fracture Basics and Interaction with Hydraulic Fractures, 211
10.3 Seismic Phenomena, 213
    10.3.1 MEQs and Their Magnitudes, 213
    10.3.2 Earthquake Focal Mechanisms, 213
    10.3.3 Other Types of Seismic Activity Produced by Hydraulic Fracturing, 216
10.4 Microseismic Downhole Monitoring, 216
    10.4.1 Downhole Monitoring Methodology, 216
    10.4.2 Advantages and Disadvantages of Downhole Monitoring, 220
10.5 Monitoring Passive Seismic Emissions with Surface and Shallow Buried Arrays, 222
    10.5.1 Recording, 222
    10.5.2 Seismic Emission Tomography, 223
    10.5.3 MEQ Methods, 229
    10.5.4 Imaging Cumulative Seismic Activity, 230
    10.5.5 Direct Imaging of Fracture Networks, 232
    10.5.6 Comparison of Downhole Hypocenters and Fracture Images, 232
    10.5.7 Summary, 233
10.6 Integrating, Interpreting, and Using Passive Seismic Data, 235
    10.6.1 General Considerations, 235
    10.6.2 Interpreting Reservoir Stress from Focal Mechanisms, 236
    10.6.3 Fracture Width, Height, SRV, and Tributary Drainage Volume, 240
    10.6.4 Using Passive Seismic Results for Frac, Well-Test, and Reservoir Simulation, 240
10.7 Conclusions, 241

11 Gas Transport Processes in Shale   245

11.1 Introduction, 245
11.2 Detection of Nanopores in Shale Samples, 247
11.3 Gas Flow in Micropores and Nanopores, 248
11.4 Gas Flow in a Network of Pores in Shale, 251
11.5 Gas Sorption in Shale, 252
11.6 Diffusion in Bulk Kerogen, 253
11.7 Measurement of Gas Molecular Diffusion into Kerogen, 255
11.8 Pulse-Decay Permeability Measurement Test, 256
    11.8.1 Pulse-Decay Pressure Analysis, 257
    11.8.2 Estimation of Permeability Parameters with the Pulse-Decay Experiment, 259

11.9 Crushed Sample Test, 260
    11.9.1 Porosity Measurement, 260
    11.9.2 Crushed Sample Pressure Analysis for Permeability Measurement, 261
    11.9.3 Crushed Sample Permeability Estimation with Early-Time Pressure Data, 262
    11.9.4 Crushed Sample Permeability Estimation with Late-Time Pressure Data, 262
11.10 Canister Desorption Test, 262
    11.10.1 Permeability Estimation with Early Time Cumulative Desorbed Gas Data, 263
    11.10.2 Permeability Estimation with Late-Time Cumulative Desorbed Gas Data, 264

12 A Review of the Critical Issues Surrounding the Simulation of Transport and Storage in Shale Reservoirs   267

12.1 Introduction, 267
12.2 Microgeometry of Organic-Rich Shale Reservoirs, 268
12.3 Gas Storage Mechanisms, 269
12.4 Fluid Transport, 270
12.5 Capillary Pressure, Relaxation to Equilibrium State, and Deposition of Stimulation Water, 273
12.6 Characterization of Fluid Behavior and Equations of State Valid for Nanoporous Media, 274
    12.6.1 Viscosity Corrections, 276
    12.6.2 Corrections for Interfacial Tension, 277
12.7 Upscaling Heterogeneous Shale-Gas Reservoirs into Large Homogenized Simulation Grid Blocks, 277
    12.7.1 Upscaling Fine Continuum Model of Shale to Lumped-Parameter Leaky Tank Model of Shale, 278
    12.7.2 Upscaling Finely Detailed Continuum Model of Shale to Coarse Continuum Model of Shale, 279
12.8 Final Remarks, 280

13 Performance Analysis of Unconventional Shale Reservoirs   283

13.1 Introduction, 283
13.2 Shale Reservoir Production, 283
13.3 Flow Rate Decline Analysis, 284
    13.3.1 Decline Curve Analysis in Unconventional Reservoirs, 285
    13.3.2 Flow Rate Transient Analysis (RTA) and its Relation to Rate Decline Analysis, 286
    13.3.3 Field Applications, 287
13.4 Flow Rate and Pressure Transient Analysis in Unconventional Reservoirs, 288
    13.4.1 Bilinear Flow Regime in Multistage Hydraulic Fracturing, 288
    13.4.2 Linear Flow Analysis for Reservoir Permeability, 289
    13.4.3 Field Applications, 290
    13.4.4 Type-Curve Matching, 290
13.5 Reservoir Modeling and Simulation, 292
    13.5.1 History Matching and Forecasting, 292
    13.5.2 Dual-Porosity Single-Phase Modeling, 293
    13.5.3 Dual-Porosity Multicomponent Gas Modeling, 294
13.6 Specialty Short-Term Tests, 295
    13.6.1 Mini-DST, 295
    13.6.2 Mini-Frac Test, 296
13.7 Enhanced Oil Recovery, 297
13.8 Conclusion, 298

14 Resource Estimation for Shale Gas Reservoirs — 301

14.1 Introduction, 301
- 14.1.1 Unique Properties of Shale, 301
- 14.1.2 Petroleum Resources Management System (PRMS), 301
- 14.1.3 Energy Information Administration's Classification System, 301
- 14.1.4 Reserves Estimate Methodology for Unconventional Gas Reservoirs, 302
- 14.1.5 Monte Carlo Probabilistic Approach, 302
- 14.1.6 Analytical Models, 303
- 14.1.7 Economic Analysis, 303
- 14.1.8 Region-Level World Shale Gas Resource Assessments, 304
- 14.1.9 Shale Gas OGIP Assessment in North America, 305
- 14.1.10 Recent Shale Gas Production and Activity Trends, 306
- 14.1.11 Drilling, Stimulation, and Completion Methods in Shale Gas Reservoirs, 308

14.2 Methodology, 309

14.3 Resource Evaluation of Shale Gas Plays, 310
- 14.3.1 Reservoir Model, 310
- 14.3.2 Well Spacing Determination, 310
- 14.3.3 Reservoir Parameters Sensitivity Analysis, 311
- 14.3.4 Reservoir Parameters, 312
- 14.3.5 Model Verification, 312
- 14.3.6 Resource Assessment, 313
- 14.3.7 Reserve Evaluation, 318

14.4 Discussion, 320

15 Molecular Simulation of Gas Adsorption in Minerals and Coal: Implications for Gas Occurrence in Shale Gas Reservoirs — 325

15.1 Introduction, 325
- 15.1.1 Molecular Dynamics Simulation, 325
- 15.1.2 Major Challenges in Shale Gas Research, 326
- 15.1.3 MS of Gas Adsorption, 326
- 15.1.4 Methodology and Workflow of Molecular Simulation, 327
- 15.1.5 Simulation Algorithms and Software, 327

15.2 MS of Gas Adsorption on Minerals, 327
- 15.2.1 MD Simulation of Gas Adsorption on Quartz, 328
- 15.2.2 Molecular Dynamic Simulation of Gas Adsorption on Wyoming-Type Montmorillonite, 330
- 15.2.3 MD Simulation of Gas Adsorption on Zeolite, 332
- 15.2.4 MD Simulation of Gas Adsorption on Coal, 334

15.3 Conclusions, 337

16 Wettability of Gas Shale Reservoirs — 341

16.1 Introduction, 341
16.2 Wettability, 341
16.3 Imbibition in Gas Shales, 342
16.4 Factors Influencing Water Imbibition in Shales, 343
- 16.4.1 Sample Expansion, 343
- 16.4.2 Depositional Lamination, 346
- 16.4.3 Chemical Osmosis, 346
- 16.4.4 Water Film and Salt Crystals, 348
- 16.4.5 Water Adsorption (Clay Swelling), 348
- 16.4.6 Connectivity of Hydrophobic and Hydrophilic Pore Networks, 349
- 16.4.7 Effect of Polymer and Surfactant, 351

16.5 Quantitative Interpretation of Imbibition Data, 352
  16.5.1 Scaling Imbibition Data, 352
  16.5.2 Modeling Imbibition Data, 352
16.6 Estimation of Brine Imbibition at the Field Scale, 354
16.7 Initial Water Saturation in Gas Shales, 356
16.8 Conclusions, 356

17 Gas Shale Challenges Over The Asset Life Cycle   361

17.1 Introduction, 361
17.2 The Asset Life Cycle, 361
  17.2.1 Exploration Phase Objectives—Recommended Practices, 361
  17.2.2 Appraisal Phase Objectives—Recommended Practices, 362
  17.2.3 Development Phase Objectives—Recommended Practices, 362
  17.2.4 Production Phase Objectives—Recommended Practices, 362
  17.2.5 Rejuvenation Phase Objectives—Recommended Practices, 362
17.3 Exploration Phase Discussion, 362
  17.3.1 Screening Study—Current Practice, 362
  17.3.2 Screening Study Recommended Practices, 363
  17.3.3 Reservoir Characterization—Current Practice, 363
  17.3.4 Reservoir Characterization—Recommended Practices, 363
  17.3.5 Determining Initial Economic Value and Reservoir Potential, 365
17.4 Appraisal Phase Discussion, 365
  17.4.1 Drill Appraisal Wells—Current Practice, 365
  17.4.2 Drill Appraisal Wells—Recommended Practices, 365
  17.4.3 Build Reservoir Models for Simulation—Current Practice, 365
  17.4.4 Build Reservoir Models for Simulation—Recommended Practices, 365
  17.4.5 Generate a Field Development Plan—Current Practice, 366
  17.4.6 Generate a Field Development Plan—Recommended Practices, 366
  17.4.7 Validate Economics of the Play or Pilot Project, 366
17.5 Development Phase Discussion, 367
  17.5.1 Implement the Field Development Plan, 367
  17.5.2 Surface Facilities, 367
  17.5.3 Design Wells and Optimize Drilling Costs—Current Practice, 367
  17.5.4 Design Wells and Optimize Drilling Costs—Recommended Practice, 368
  17.5.5 Refine and Optimize Hydraulic Fracturing and Wellbore Completion Design—Current Practices (Characterize the Lateral), 369
  17.5.6 Current Hydraulic Fracturing Practices, 369
  17.5.7 Hydraulic Fracturing—Recommended Practices, 370
  17.5.8 Characterize the Lateral, 372
  17.5.9 Current Wellbore Completion Practice, 373
  17.5.10 Wellbore Completion—Recommended Practices, 373
  17.5.11 Drilling Considerations for Completion Methods, 375
  17.5.12 Fracturing Considerations for Completion Method, 375
17.6 Production Phase Discussion, 375
  17.6.1 Monitor and Optimize Producing Rates—Current Practice, 375
  17.6.2 Monitor and Optimize Producing Rates—Recommended Practices, 375
  17.6.3 Manage the Water Cycle—Recommended Practices, 376
  17.6.4 Preventing Corrosion, Scaling, and Bacterial Contamination in Wells and Facilities, 376
  17.6.5 Protecting the Environment, 376
17.7 Rejuvenation Phase Discussion, 376
17.8 Conclusions—Recommended Practices, 377

18 Gas Shale Environmental Issues and Challenges   **381**

    18.1   Overview, 381
    18.2   Water Use, 381
    18.3   The Disposal and Reuse of Fracking Wastewater, 382
    18.4   Groundwater Contamination, 384
    18.5   Methane Emissions, 386
    18.6   Other Air Emissions, 387
    18.7   Social Impacts on Shale Gas Communities, 388
    18.8   Induced Seismicity: Wastewater Injection and Earthquakes, 388
    18.9   Regulatory Developments, 389
    18.10 Disclosure of Fracking Chemicals, 389
    18.11 At the Federal Government Level, 390
    18.12 Conclusion, 391

Index   **397**

CONTRIBUTORS

Ahmad, Abualksim, Dr., Department of Petroleum Engineering, Curtin University, Perth, WA, Australia

Al Hinai, Adnan, Dr., Department of Petroleum Engineering, Curtin University, Perth, WA, Australia

Civan, Faruk, Prof., Mewbourne School of Petroleum and Geological Engineering, The University of Oklahoma, Norman, OK, USA

Deepak, Devegowda, Dr., Mewbourne School of Petroleum and Geological Engineering, The University of Oklahoma, Norman, OK, USA

Dehghanpour, Hassan, Dr., School of Mining and Petroleum Engineering, University of Alberta, Edmonton, Alberta, Canada

Dewhurst, Dave N., Dr., CSIRO Energy Flagship, Perth, WA, Australia

Dong, Zhenzhen, Dr., PTS, Schlumberger, College Station, TX, USA

Eker, Ilkay, Dr., Colorado School of Mines, Golden, CO, USA

Ettehadtavakkol, Amin, Dr., Bob L. Herd Department of Petroleum Engineering, Texas Tech University, Lubbock, TX, USA

Fish-Yaner, Ashley, Global Geophysical Services, Inc., Denver, CO, USA

Golodoniuc, Pavel, Dr., CSIRO Mineral Resources Flagship and Department of Exploration Geophysics, Curtin University, Perth, WA, Australia

Gurevich, Boris, Prof., Department of Exploration Geophysics, Curtin University, and CSIRO Energy Flagship, Perth, WA, Australia

Habibi, Ali, Dr., School of Mining and Petroleum Engineering, University of Alberta, Edmonton, Alberta, Canada

Holditch, Stephen A., Prof., Petroleum Engineering Department, Texas A&M University, College Station, TX, USA

Jarvie, Daniel M., Worldwide Geochemistry, LLC, Humble, TX, USA

Javadpour, Farzam, Dr., Bureau of Economic Geology, Jackson School of Geosciences, The University of Texas at Austin, Austin, TX, USA

Kazemi, Hossein, Prof., Colorado School of Mines, Golden, CO, USA

Kennedy, Robert "Bobby," Petroleum Engineering, Unconventional Resources Team, Baker Hughes, Inc., Tomball, TX, USA

Kurtoglu, Basak, Dr., Marathon Oil Company, Houston, TX, USA

Labani, Mehdi, Dr., Department of Petroleum Engineering, Curtin University, Perth, WA, Australia

Lacazette, Alfred, Dr., Global Geophysical Services, Inc., Denver, CO, USA

Lebedev, Maxim, Assoc., Prof., Department of Exploration Geophysics, Curtin University, Perth, WA, Australia

Lee, W. John, Prof., UH Energy Research Park, University of Houston, Houston, TX, USA

Liu, Keyu, Dr., PetroChina Research Institute of Petroleum Exploration and Development, Beijing, China; and CSIRO Division of Earth Science and Resource Engineering, Perth, WA, Australia

Liu, Shaobo, Dr., PetroChina Research Institute of Petroleum Exploration and Development, Beijing, China

Pervikhina, Marina, Dr., CSIRO Energy Flagship, Perth, WA, Australia

Rasouli, Vamegh, Prof., Department of Petroleum Engineering, University of North Dakota, Grand Forks, ND, USA

Rezaee, Reza, Prof., Department of Petroleum Engineering, Curtin University, Perth, WA, Australia

Rothwell, Mark, HSEassist Pty Ltd, Perth, WA, Australia

Sicking, Charles, Dr., Global Geophysical Services, Inc., Dallas, TX, USA

Sigal, Richard F., Dr., Mewbourne School of Petroleum and Geological Engineering, The University of Oklahoma, Norman, OK, USA

Slatt, Roger M., Prof., Institute of Reservoir Characterization, School of Geology and Geophysics, Sarkeys Energy Center, University of Oklahoma, Norman, OK, USA

Thorn, Terence H., President, JKM 2E Consulting, Houston, TX, USA

Tian, Hua, Dr., PetroChina Research Institute of Petroleum Exploration and Development, Beijing, China

Tibi, Rigobert, Dr., Global Geophysical Services, Inc., Denver, CO, USA

Torcuk, Mehmet A., Dr., Colorado School of Mines, Golden, CO, USA

Trabucho-Alexandre, João, Prof., Department of Earth Sciences, University of Durham, Durham, UK; and Institute of Earth Sciences Utrecht, Utrecht University, Utrecht, The Netherlands

Xu, Mingxiang, Dr., School of Mining and Petroleum Engineering, University of Alberta, Edmonton, Alberta, Canada

Zhang, Shuichang, Dr., PetroChina Research Institute of Petroleum Exploration and Development, Beijing, China

PREFACE

The hydrocarbon source from conventional reservoirs is decreasing rapidly. At the same time, global energy consumption is growing so quickly that conventional reserves alone cannot solely satisfy the demand. Therefore, there is a pressing need for alternative sources of energy. As things currently stand from a technical viewpoint, the more expensive clean-sustainable energy sources cannot compete with the relatively cheap nonrenewable fossil fuels. Thus, the obvious immediate alternative energy source would be found in non-conventional oil and gas resources. These non-conventional resources come in many forms and include gas hydrate, tar sand, oil shale, shale oil, tight gas sand, coal bed methane, and of course, shale gas. Shale gas has for some time been the focus of gas exploration and production in the USA and in other countries. Based on a recent EIA report, there is an estimated 7299 trillion cubic feet (Tcf) of technically recoverable shale gas resource to be found in some 137 basins located in 41 countries.

Following notable successes in shale gas production in the USA, to the point where that country now produces more shale gas than gas from the conventional sources, other countries are pursuing the same course. Even so, in order to be successful in the exploration and the development of shale gas plays, a number of important factors have to be taken into account:

- A vast knowledge of the different aspects of shales, such as organic geochemistry, mineralogy, petrophysical properties, shale geomechanics, reservoir engineering and so on, is required in order to properly evaluate and map shale gas sweet spots in each sedimentary basin.
- Shale gas environmental issues together with challenges such as the high water demands and possible contamination risks posed by hydraulic fracturing fluids and waste have to be addressed.

The aim of this book is to provide some guidance on the major factors involved in evaluating shale gas plays. The book is structured as follows:

Chapter 1 introduces shale gas from the point of view of its global significance, distribution and inherent challenges.

Chapter 2 discusses the environments suitable for organic matter-rich shale deposition.

Chapter 3 assesses the organic geochemical properties of shale gas resource systems.

Chapter 4 highlights important points about the sequence stratigraphy of shales.

Chapter 5 discusses methods used for evaluating pore geometry in shales.

Chapter 6 details the steps required for the petrophysical analysis of shale gas plays.

Chapter 7 deals with pore pressure estimation of shales using conventional log data.

Chapter 8 covers shale gas geomechanics.

Chapter 9 discusses the rock physics of organic-rich shales.

Chapter 10 introduces passive seismic methods for non-conventional resource development.

Chapter 11 discusses gas transport processes in shale.

Chapter 12 reviews the critical issues surrounding the simulation of transport and storage in shale reservoirs.

Chapter 13 provides important information about the performance analysis of shale reservoirs.

Chapter 14 presents methodologies to determine original gas in place (OGIP), technically recoverable resources (TRR) and the recovery factor (RF) for shale reservoirs.

- Chapter 15 discusses molecular simulation of gas adsorption.
- Chapter 16 deals with the wettability of gas shale reservoirs.
- Chapter 17 summarises gas shale challenges expected to occur over the life cycle of the asset.
- Chapter 18 presents gas shale environmental issues and challenges.

The study of shale gas plays is advancing rapidly in many countries, and I hope this book will provide some useful fundamental information on the topic.

PROFESSOR REZA REZAEE
Curtin University, Department
of Petroleum Engineering
August 7, 2014

1

GAS SHALE: GLOBAL SIGNIFICANCE, DISTRIBUTION, AND CHALLENGES

REZA REZAEE[1] AND MARK ROTHWELL[2]

[1] *Department of Petroleum Engineering, Curtin University, Perth, WA, Australia*
[2] *HSEassist Pty Ltd, Perth, WA, Australia*

1.1 INTRODUCTION

The central geological properties of a shale gas play are generally assessed in terms of depositional environment, thickness, organic geochemistry, thermal maturity, mineralogy, and porosity. The key features of successful shale gas plays include high total organic carbon (TOC) content (>2%), thermally mature (Ro 1.1–1.5%), shallow for the given maturity, and a low clay content/high brittle mineral content. However, porosity, in situ stress regime, stress history, and mineralogy are also significant factors.

Technically recoverable (although not necessarily economically recoverable) gas shale is abundant across the globe. It is also located in a very wide range of geographical regions, and in many of the nations with the highest energy consumption. For certain nations, shale gas therefore has the potential to reduce energy prices and dependence on other nations, hence impact on both the political and economic outlook. However, the prospects for and significance of shale gas are greater where there is a lack of existing conventional gas production, where there is proximity to demand (i.e., population), and where some form of existing gas distribution infrastructure exists.

The definition of a "resource" can follow a number of classifications. However, in the context of this chapter, the class of "technically recoverable resources" (TRRs) has been adopted, which includes both economic and uneconomic resources.

The assessment of the global data included the identification of the shale depositional environment and basin type. A brief summary of the shale gas plays is presented for each country, which is followed by a statistical assessment of certain data subsets to illustrate where shale gas is located, the expected range of properties in terms of TOC, depth, age, and basin type.

There are a number of key challenges that the industry faces, including environmental issues and commercial challenges. The key issues relate to the management of the hydraulic fracturing process, the prediction and improvement of EUR/well, and the consideration of variable production costs in different regions.

1.2 SHALE GAS OVERVIEW

In very simple terms, shale gas refers to gas produced from fine-grained gas-prone sedimentary rocks (i.e., organic-rich shale) (Lakatos and Szabo, 2009). Shale gas is considered an "unconventional" gas resource, since conventionally gas is produced from granular, porous, and permeable formations (i.e., sandstone), within which gas can readily flow. Although shale gas is considered an unconventional hydrocarbon resource, the gas produced essentially serves the same market (Staff, 2010). The term "unconventional," therefore, only refers to the rock from which the natural gas produced in this particular case.

In a conventional gas play, gas shale[1] is often present, but it serves as the source rock rather than the reservoir.

---

[1] For the purposes of this document, "shale gas" refers to the hydrocarbon, whilst "gas shale" refers to the geologic material from which the gas is extracted in a shale gas play.

Fundamentals of Gas Shale Reservoirs, First Edition. Edited by Reza Rezaee.
© 2015 John Wiley & Sons, Inc. Published 2015 by John Wiley & Sons, Inc.

Shale, as a function of its traditionally low permeability, also often serves as a sealing lithology within the trapping mechanism of a conventional gas play, which prevents oil and gas accumulations from escaping vertically (Gluyas and Swarbrick, 2009).

Generic global hydrocarbon estimates have always somewhat reflected resources in-place within tight formations and shale. However, it is the relatively recent technological developments and higher gas prices that have now resulted in a vast resource being considered potentially economic, which had previously been considered uneconomic to develop (Ridley, 2011).

Sources indicate that shale is present in a very wide range of regions across the globe, with an estimated 688 shale deposits occurring in approximately 142 basins (Ridley, 2011).

1.2.1 Shale Gas Geology

Shale gas is a natural gas produced from organic-rich fine-grained low-permeability sedimentary rocks, such as shale, where the rock typically functions as both the "source rock" and the "reservoir rock," to use terms associated with conventional plays (US DOE, 2009). The relationship between conventional and unconventional gas is illustrated in Figure 1.1.

Gas shale is similar to traditional shale in terms of the range of environments of deposition. For example, Caineng et al. (2010) note that organic-rich shale can be divided as marine shale, marine–terrigenous coal bed carbonaceous shale, and lacustrine shale. The depositional setting directly controls key factors in shales, such as organic geochemistry, organic richness, and rock composition. According to Potter et al. (1980), the organic matter preserved in shales depends on the dissolved oxygen level in the water.

Shale gas organic geochemistry is a function of the depositional environment and is similar to conventional source rock geochemistry. Marine shale is typically associated with Type II kerogen (i.e., organic matter associated with a mixture of membraneous plant debris, phytoplankton, and bacterial microorganisms in marine sediments). Lacustrine shale is generally associated with Type I kerogen, due to the organic matter being associated with an algal source rich in lipids (typically only in lacustrine and lagoonal environments). Finally, terrestrial/coal bed shale is typically associated with Type III kerogen, due to the organic matter being associated with higher plant debris, as commonly found in coal-bed-forming environments such as delta tops (Gluyas and Swarbrick, 2009).

Target TOC (wt% kerogen) values are somewhat interrelated to the thickness and other factors that influence gas yield. However, for commercial shale gas production, Staff (2010) notes a target TOC of at least 3%, whilst Lu et al. (2012) states that a TOC of 2% is generally regarded as the lower limit of commercial production in the United States. That said, TOC varies considerably throughout any one shale gas play.

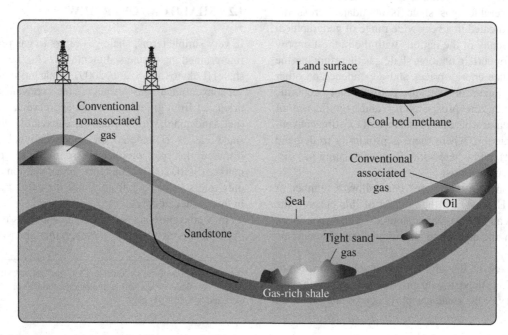

FIGURE 1.1 Schematic geological section illustrating the fundamental geological principles associated with conventional and unconventional hydrocarbons. Shale gas is designated as "gas-rich shale" (from EIA, 2010).

The thickness of economic gas shales is one of many considerations. However, as an example, in North America, the effective thicknesses of shale gas pay zones range from 6 m (Fayetteville) to 304 m (Marcellus) (Caineng et al., 2010). Caineng et al. (2010) note a guidance thickness for economic plays of 30–50 m, where development is continuous and the TOC (wt%) is greater than 2%.

TOC is only an indication of shale gas potential. The actual accumulation of gas from the organic compounds within the shale requires the organic matter to first generate the gas. The degree to which this has happened in a shale gas play is a function of the thermal maturity of the shale (Lu et al., 2012). Significant shale gas is typically only generated beyond vitrinite reflectance (Ro%) values of approximately 0.7% (Type III kerogen) to 1.1% (Types I and II kerogen), which corresponds to depths of between 3.5 and 4.2 km (Gluyas and Swarbrick, 2009). However, the most favorable situation is when vitrinite reflectance values range from 1.1 to 1.4 (Staff, 2010).

Mineralogy plays a central role when evaluating gas shale, due to its impact on the performance of fracture treatment (also known as hydraulic fracturing and "fracking"). In terms of mineralogy, brittle minerals (i.e., siliceous and calcareous minerals) are favorable for the development of extensive fractures throughout the formation in response to fracture treatment. Caineng et al. (2010) note that a brittle mineral content greater than 40% is considered necessary to enable sufficient fracture propagation. Alternatively, Lu et al. (2012) notes that within the main shale-gas-producing areas of the United States, the brittle mineral content is generally greater than 50% and the clay content is less than 50%. In more simplistic terms, high clay content results in a more ductile response to hydraulic fracturing, with the shale deforming instead of shattering. Mineralogy and brittle mineral content can be linked to the depositional environment. For instance, marine-deposited shales tend to have a lower clay content and, hence, a higher brittle mineral content (EIA, 2011a). It should be noted that the fracture susceptibility of shale is also influenced by the stress regime and the degree of overpressure in the formation, amongst other factors.

Petrophysical considerations are beyond the scope of this review. However, it is worth noting that porosity is an important petrophysical consideration, as it will influence the amount of free gas that can be accumulated within the shales. Staff (2010) note that it is preferable for porosity to be greater than 5%. However, for the main producing gas shales in the United States porosity range from 2 to 10% (Staff, 2010).

1.2.2 Characteristics of a Producing Shale Gas Play

The general geological features of a gas shale determine the general framework of a commercial-scale shale gas development. Some of these development features are outlined in the following text for the purposes of providing a clearer picture of what a shale gas development comprises, and how this differs to a conventional gas play.

Shale gas is currently only viable onshore since it would be cost prohibitive to drill the large quantity of wells required in an offshore environment due the higher cost per offshore well. For instance, the day rate for offshore drilling can be an order of magnitude higher than for onshore drilling.

Shale gas wells are generally drilled horizontally so as to maximize exposure to the "reservoir." However, vertical wells may also be drilled where the shale interval is very thick.

Extensive hydraulic fracturing (fracking) is undertaken within the shale gas reservoir to further increase the permeability and hence gas yield. Fracturing is generally undertaken in multiple stages, with the fracturing treatment of each individual section being undertaken separately, so as to maximize the control and effectiveness of the process. It is also not usually possible to maintain a downhole pressure sufficient to stimulate the entire length of a well's reservoir intersection in a single stimulation/treatment event (US DOE, 2009), and it would also probably result in the concentration of fractures in the most susceptible zones. Each treatment stage involves a series of substages which involve using different volumes and compositions of fluids, depending on the design (US DOE, 2009). For example, the sequence of substages may be as follows:

1. Test phase—validating the integrity of the well casings and cement,
2. Acid treatment—pumping acid mix into the borehole to clean walls of "damage,"
3. Slickwater pad—pumping water-based fracturing fluid mixed with a friction-reducing agent in the formation, which is essentially designed to improve the effectiveness of the subsequent substage,
4. Proppant stage—numerous sequential substages of injecting large volumes of fracture fluid mixed with fine-grained mesh sand (proppant) into the formation, with each subsequent substage gradually reducing the water-to-sand ratio, and increasing the sand particle size. The fracture fluid is typically 99.5% water and sand, with the remaining components being additives to improve performance.

A large number of wells are required to extract economic quantities of gas from shale. The approximate quantity of wells required to produce 1 Tcf (trillion cubic feet) of gas within various producing shale gas plays in the United States varies widely (Kennedy, 2010). The suggested typical quantity of wells per Tcf gas is 200–250. This equates to an estimated ultimate recovery per well (EUR/well) of 5 Bcf/well (billion cubic feet per well). However, other sources

(EIA, 2011a) indicate the average EUR/well is 1.02 Bcf/well. To compare this to a conventional development, the Gorgon/Jansz-lo field is estimated to have a EUR/well of 750 Bcf/well (Chevron, 2012). The principal reason why the development of shale plays remains economically risky is that the EUR/well is poorly constrained during the early stages of field development (Weijermars, 2013). As a function of the large quantity of wells, significant infield infrastructure is required to transport gas to processing facilities.

1.3 THE SIGNIFICANCE OF SHALE GAS

There is an estimated 6000 Tcf of TRR of shale gas within the countries listed by EIA (2011b) (Table 1.1). This compares to a global total proven gas reserve of approximately 6000 Tcf. Shale gas therefore has the potential to be a very significant source of natural gas, and has the potential to greatly increase the gas resource of many nations across the globe. This is best illustrated in

TABLE 1.1 Summary of production, consumption, reserves, and resources for various nations[a]

	2009 natural gas market (trillion cubic feet, dry basis)			Technically recoverable shale gas resources (trillion cubic feet)
	Production	Consumption	Imports (exports)	
Europe				
France	0.03	1.73	98%	180
Germany	0.51	3.27	84%	8
Netherlands	2.79	1.72	(62%)	17
Norway	3.65	0.16	(2156%)	83
United Kingdom	2.09	3.11	33%	20
Denmark	0.30	0.16	(91%)	23
Sweden	–	0.04	100%	41
Poland	0.21	0.58	64%	187
Turkey	0.03	1.24	98%	15
Ukraine	0.72	1.56	54%	42
Lithuania	–	0.10	100%	4
Others	0.48	0.95	50%	9
North America				
United States	20.6	22.8	10%	862
Canada	5.63	3.01	(87%)	388
Mexico	1.77	2.15	18%	681
Asia				
China	2.93	3.08	5%	1275
India	1.43	1.87	24%	63
Pakistan	1.36	1.36	–	51
Australia	1.67	1.09	(52%)	396
Africa				
South Africa	0.07	0.19	63%	485
Libya	0.56	0.21	(165%)	290
Tunisia	0.13	0.17	26%	18
Algeria	2.88	1.02	(183%)	231
Morocco	0.00	0.02	90%	11
Western Sahara	–			7
Mauritania	–			0
South America				
Venezuela	0.65	0.71	9%	11
Colombia	0.37	0.31	(21%)	19
Argentina	1.46	1.52	4%	774
Brazil	0.36	0.66	45%	226
Chile	0.05	0.10	52%	64
Uruguay	–	0.00	100%	21
Paraguay	–	–	–	62
Bolivia	0.45	0.10	(346%)	48
Total of above areas	53.1	55.0	(3%)	6622
Total world	106.5	106.7	0%	

[a]Modified from EIA (2011b).

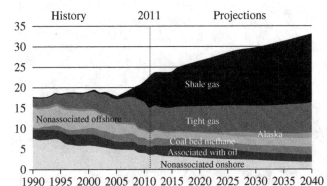

FIGURE 1.2 Historical and projected sources of natural gas in the United States (EIA, 2013).

Table 1.1, which defines the existing proven resources and the TRRs of shale gas for many major nations. Although proven and TRRs represent very different estimates, there is clearly still the potential for a shift in the distribution of gas away from the traditional central producing hubs of the Middle East and Russia, and toward more local domestic supply, with many consequential political impacts.

Using the United States as an example, where conventional gas has been in decline over recent decades (EIA, 2011a), it is shale gas that is forecast to provide the greatest contribution to domestic gas production. This forecast pattern is illustrated in Figure 1.2 in terms of projected sources of natural gas. Note that shale gas is estimated to represent over 50% of gas production by 2040. Similar outcomes may apply to other nations in the future; although for various reasons many countries face major challenges before the same success in shale gas can be enjoyed as in the United States (Stevens, 2012).

As outlined by Ridley (2011), the significance and future of shale gas will also be influenced by the interplay of a wide variety of other issues, including the following:

- Potentially falling gas prices, due to increased production
- Reduced production costs due to technological developments, and the associated competitiveness of gas produced from shale in comparison to other sources
- Increased demand for gas due to increased adoption of natural gas to produce energy and in new markets (i.e., natural gas-fuelled vehicles)
- The regulatory environment for shale gas development in each country

1.4 GLOBAL SHALE GAS RESOURCES

This section collates shale gas resource data from a variety of sources. It is structured as follows:

- Sources of information
- Resource estimation methodologies
- TRR data

As noted previously, shale gas is widespread within the world's sedimentary basins. For example, Figure 1.3 (from EIA, 2011b) illustrates that shale gas plays occur in all of the regions assessed within the study of concern. However, it is also known that Russia and the Middle East also have considerable shale gas resources, but are unlikely to develop them in the next decade due to the abundance of conventional gas resources.

1.4.1 Sources of Information

For assessing the global resources, this chapter has extracted data from EIA (2011a, b). This source was the primary source of data. However, it does not include data for the Russia or the Middle East. The other source is obtained from Rogner (1997). This source was used to provide resource estimates for Russia and the Middle East.

In addition to the above sources, two regional maps published by the Society of Petroleum Engineers were referenced, as they both include "shale resource" values. However, the values are identical to those presented by the EIA.

These sources provided data for the most significant developed nations globally. It is certain that many other nations will have shale gas resources, but they are currently lacking demand for local production and also lack infrastructure for distribution and export, and would therefore have difficulty attracting investment.

1.4.2 Resource Estimation Methodologies

The different sources of data quote a slightly different category of resource. The resource category framework presented by Dong is used as a baseline for comparing the differing resource estimation techniques associated with various sources.

The primary objective was to identify a TRR for each region, including a play-specific breakdown where available. This was relatively straightforward for the EIA sources since they quote something very similar to TRR. However, some assumptions were required to convert the values presented by Rogner (1997).

It should be noted that TRR includes both economic and uneconomic resources. As such, despite the large TRR values sometimes quoted, it may be uneconomic to produce gas from these resources.

6 GAS SHALE: GLOBAL SIGNIFICANCE, DISTRIBUTION, AND CHALLENGES

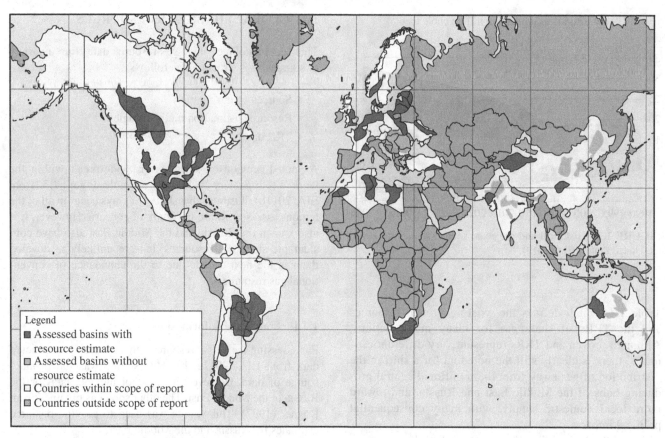

FIGURE 1.3 Map of 48 major shale basins in 32 countries (from EIA, 2011b).

1.4.2.1 EIA Global Resource Estimation Methodology
The resource estimates presented by the EIA in the global shale gas review were calculated using a basin-by-basin approach, using the following methodology:

1. Conducting preliminary geological and reservoir characterization of shale basins and formation(s)
2. Establishing the areal extent of the major gas shale formations (i.e., specific to certain shale formations within a basin)
3. Defining the prospective area for each gas shale formation
4. Estimating the total gas in-place (GIP)
5. Estimating the risked shale GIP, accounting for the following:
   - Play success probability factor
   - Prospective area success risk factor
6. Calculating the TRR of shale gas in terms of Tcf. On a "by region" average, this value was generally between 24 and 29% of GIP.

Naturally, the accuracy of the estimate is a function of the availability and quality of data, but this is generally reflected in the calculation of "risked GIP" and the subsequent calculation of TRR. The TRR values presented by the EIA effectively correlate with the TRR zone defined by Dong et al. (2013).

1.4.2.2 EIA USA Resource Estimation Methodology
The resource estimates provided by the EIA for individual US shale gas plays were assessed using a comparable method to that adopted for the global resources assessment. However, the main difference is that production data (i.e., well recovery data) was used to support the estimate. This reflects the fact that many US shale gas plays are in a production phase, with approximately 25,000 producing wells in 2007, (Vidas and Hugman, 2008), whilst the rest of the world is still largely in the exploration phase.

The resource estimates quoted represent a TRR for each shale gas play, although they do reduce the gas already produced. The TRR for this source effectively correlates with the TRR zone defined by Dong et al. (2013).

1.4.2.3 Rogner Resource Estimation Methodology
The Rogner study (1997) provides shale gas resource data for Russia and the Middle East. Rogner states that the estimates presented are very speculative as a result of the lack of data.

The estimation methodology involved applying knowledge about US gas shales to other shales in different regions. In simple terms, this involved assuming that all prospective shales contain 17.7 Tcf of gas for every Gt (gigatonne) of shale in-place. The value presented by Rogner is a GIP estimate, which does not conform to the definition of TRR used by the EIA and defined by Dong et al. (2013).

The Rogner GIP estimates were converted to TRR values by averaging the GIP:TRR ratios for global shale gas plays from other sources, then applying this average ratio to the Rogner GIP values. It was also necessary to adjust Rogner Middle East values to account for overlap with EIA sources.

1.5 GLOBAL RESOURCE DATA

The shale gas resource data is presented in Appendix A.1. The information is presented as a hierarchy in terms of region, country, basin, and shale play. A summary of each prospective country, and in some cases region, is presented further. This chapter is limited to the general geological reservoir characteristics and a brief summary of the status of exploration or production.

All quantitative reservoir properties and characteristics (i.e., TOC, depth, and thickness) are indicative nonweighted averages only, will vary greatly across any one play, and are not representative of the likelihood of commercial shale gas production. However, they do give an indication of the potential resource quality.

All information has been sourced from the EIA documentation (2011a, b, c), except where stated otherwise.

1.5.1 China

China has two major prospective basins, the Sichuan Basin and the Tarim Basin, with a combined estimated TRR of 1275 Tcf. This is the largest TRR of any single nation within this review and supports the opinion that China is widely regarded as having excellent potential for shale gas development.

The four target shales within both basins were deposited on a passive margin in a marine environment from Cambrian to Silurian times. They are thick (200–400 ft), dry gas mature (Ro of 2.0–2.5), and have moderate clay content. However, the shales are situated relatively deep at depths of 10,000–14,000 ft, and have only moderate organic content (2–3%). Geological complexity is high in certain parts of both basins, which is the reason why large parts of the basin have currently been disregarded in preparing TRR estimates (EIA, 2011b).

There is considerable exploration activity in China due to the potential significance in terms of domestic energy supply, less reliance on the Middle East, and high domestic demand for energy. Although there is currently no shale gas production, the Sichuan Basin has a well-developed network of natural gas pipelines, in addition to proximity to large cities with considerable energy demand. That said, the prospective areas do suffer from remoteness and often a lack of water (UPI, 2013).

1.5.2 The United States

The United States has numerous producing shale gas basins, many of which are very well understood due to production-related data. It also has the second largest TRR within this study.

A total of 16 basins comprising 20 shale gas plays are noted within the source study, with a cumulative TRR of 751 Tcf. All the prospective shales are of marine origin, with the majority associated with foreland basins (e.g., Appalachian Basin) and Devonian deposition. The majority are of favorable depth, with some as shallow as 3000 ft, although the national average is approximately 7500 ft. This shallow depth combined with competitive drilling costs often equates to relatively low cost production. Organic content is generally very favorable, with an extremely high average of some 6–7%, with some shale gas plays (i.e., Marcellus Shale) reporting average TOC of 12%. The United States also has considerable local experience in the drilling and hydraulic fracturing service industry.

EIA sources (2011a) have considerable information on each US shale gas play; as such, no further information is provided here.

1.5.3 Mexico

Mexico has the third largest TRR within this review, at approximately 681 Tcf. The shales are of marine origin and were deposited in rift basins during Jurassic and Cretaceous. The shale plays are favorable in terms of thickness (200–400 ft), low clay content, organic richness (3–5% average), and gas mature. However, the majority of resources occur quite deep at between 10,000 and 12,000 ft.

Mexico's most prospective resources, within the Eagle Ford Shale, are time comparable to those in the SE USA. However, Mexico's coastal shale zone is narrower, less continuous, and structurally much more complex than the equivalent in the United States (EIA, 2011b). However, due to the similarities, there is the potential for similar success.

There has only been very limited exploration activity in Mexico, with no wells as of 2011.

1.5.4 Southern South America

Southern South America is considered as one "zone" in this section because the key basins are very large and span many borders.

Of particular interest are the Parana-Chaco Basin (Paraguay, Brazil, Argentina, Bolivia) and the Neuquen Basin (Argentina), since they are associated with the majority of the 1195 Tcf TRR associated with this region. All shales within these two basins are of marine origin, and were deposited in a rift and back-arc basin, respectively.

The Parana-Chaco Basin shales are at a relatively shallow depth (7500 ft), are extremely thick (1000 ft), have low clay content, and have moderate TOC (2.5%). However, they are relatively low in terms of maturity (0.9% Ro).

The Neuquen Basin has two prospective shales, at depths of 8,000 and 12,000 ft. They are generally more mature, have higher TOC, and are more overpressured than the Parana-Chaco Basin.

There is also a sizeable TRR in the Austral-Magnallanes Basin on the border between Argentina and Chile, which has similar characteristics to the Parana-Chaco Basin shales, but with lower estimated TOC and higher clay content.

Active exploration is underway within the Neuquen Basin in Argentina. Argentina also has existing gas infrastructure and favorable policy to support unconventional gas production.

1.5.5 South Africa

South Africa has approximately 485 Tcf of shale gas (TRR) within the vast Karoo Basin, which extends across nearly two-thirds of the country. There are three prospective shales within this basin, all of which were deposited during the Permian in a marine environment associated with a foreland basin. The shales are relatively thick (ca. 100–150 ft), shallow (8000 ft), low in clay content, highly organic rich (6% within the Whitehill Formation), mature, and overpressured.

However, one notable downside is the presence of intruded volcanics (sills), which may impact on resource quality, limit the use of seismic, increase the risk of exploration, and elevate the CO_2 content. There is also no significant gas pipeline infrastructure within the Karoo basin, with existing gas supplies coming from Mozambique to the North.

Exploration activity is increasing in the region, with multinationals (i.e., Shell) holding large permits, and with drilling expected to commence sometime during 2015. However, there were wells drilled pre-1970, which indicated gas saturation and potential for flow through existing fractures.

1.5.6 Australia

Four prospective basins have been identified within Australia—the Cooper Basin in central Australia, the Maryborough Basin in Queensland, and the Perth and Canning Basins in Western Australia. The combined TRR for these basins is approximately 396 Tcf.

Each of the basins have quite different characteristics in terms of basin type and age, but all of the associated shales are of marine origin, with the exception of the Cooper Basin Permian shale that was deposited in a lacustrine environment.

The shallowest resources are within the Cooper Basin, at approximately 8,000 ft, with the other shales being at depths of between 10,000 ft (Perth Basin) and 12,000 ft (Canning Basin). All the shales have favorable characteristics, such as low clay content, thermal maturity, normal to overpressured, and high average TOC (>2.5%, typically around 3.5%).

Active exploration is underway within Australia, particularly within the Cooper Basin (Beach Petroleum) and the Canning Basin (Buru Energy). Although there is active gas production from conventional sources within the Cooper Basin, the shale is of the less favorable lacustrine origin, and there are reported higher CO_2 concentrations. The conditions within the Canning Basin seem more favorable, although the industry has to compete with high-domestic gas production from other conventional gas sources, relatively high production and labor costs, and a currently high Australian dollar.

1.5.7 Canada

Canada has approximately 388 Tcf of shale gas (TRR), the majority of which is within five subbasins within the vast Western Canadian Basin (WCB). The WCB is a modern foreland basin associated with the Rocky Mountains, although the prospective shales were deposited in a passive margin marine environment. Gas shale depths and thicknesses are relatively favorable, with the majority of resources at approximately 8000 ft, and with typical thickness of between 200 and 400 ft. The organic content is also generally good (>3.5%), clay content is low, thermal maturity is high, and the shales are often slightly overpressured.

The majority of WCB subbasins are very favorable for development due to proximity to significant conventional gas pipeline infrastructure. Exploration has been active for many years, with significant development phase work being undertaken. However, commercial scale production has not yet commenced.

There are some smaller prospective shale formations on the east coast, with the Appalachian Basin being the most significant and favorable, although the resource quality is less than west coast equivalents, with lower TOC (2%). However, there is also existing conventional gas infrastructure and some active exploration, hence good potential for the development of favorable areas. It also has proximity to US shale gas basins, which have good industry capability.

1.5.8 North Africa

North Africa has a considerable shale gas TRR of approximately 557 Tcf, with the majority being within Libya (290 Tcf) and Algeria (230 Tcf). There are two key basins: the Ghadames Basin (mainly Algerian) and the Sirt Basin

(Libya), with a combined TRR of approximately 504 Tcf. Both are intracratonic basins associated with marine shale deposition during the Devonian and Silurian.

Both basins have favorable characteristics, such as good thickness (100–200 ft), high TOC (3–5%, locally up to 17%), overpressured/normal pressure, medium clay content, and thermally mature. However, all the prospective shales are relatively deep, at depths of between 9,400 and 13,000 ft, with an average of approximately 11,000 ft.

There is already considerable exploration activity within the Ghadames Basin, but no production as of 2011. There is no reported exploration or production in the Sirt Basin.

1.5.9 Poland

Poland is the most active nation in Europe in pursuit of shale gas, due to both the relatively abundance of shale gas in comparison to other European nations—a favorable regulatory environment—and as a result of currently being a net importer of natural gas, the majority of which comes from Russia. Poland has an estimated TRR at approximately 187 Tcf.

There are three main prospective basins: the Baltic, the Lublin, and the Poladsie. In all three cases, the prospective shale formations are of marine origin, Silurian age, and were either rift or passive margin basin associated. Each target also has a moderate clay content and favorable thickness (i.e., 200–300 ft).

The Poladsie Basin has the most favorable organic content (6% TOC) and depth (8000 ft). However, the resource is relatively small (14 Tcf), and there is not much exploration activity in the basin to date to validate potential.

The Lublin Basin shale target is of intermediate depth, but has only a moderate organic content of 1.5%, and moderate maturity (wet-dry gas, Ro 1.35%).

The Baltic Basin has a large resource, with optimum maturity within the dry gas window, but with a deep pay zone (12,000 ft).

Exploration is active within the Baltic and Lublin Basins, which are also associated with small conventional oil and gas fields. To date, drilling results in the Baltic Basin seem to have been mixed with companies such as ExxonMobil, Talisman Energy, and Marathon Oil deciding to withdraw from shale gas operations in the area based on the results of drilling and testing operations (BBC, 2013).

1.5.10 France

France has an estimated TRR of 180 Tcf of shale gas, relatively evenly distributed between the Paris Basin and the South-East Basin.

All of the shales are of marine origin; they have low-to-medium clay content, good organic content (2.5–4%), good maturity (Ro ~1.5%), and moderate thickness (100–150 ft). However, the majority of the resource is relatively deep (85% of TRR is at depths of between 10,000 and 12,000 ft).

The Teres Noires Shale within the South-East Basin is relatively small at 28 Tcf, but it is very shallow (5000 ft), has low clay content, 3.5% average TOC, and reasonable maturity.

The Paris Basin target has similar characteristics, but the target is significantly deeper at nearly 11,000 ft, although average TOC is 4%.

There is currently a ban on hydraulic fracturing in France, and exploration permits are being revoked, despite the fact that France has considerable shale gas potential.

1.5.11 Russia

Russia has vast conventional oil and gas resources, and is a major exporter, hence is unlikely to produce shale gas in the near future. No detailed information was available for Russia from the sources considered. However, the estimated TRR of shale gas is approximately 162 Tcf.

1.5.12 Scandinavia

Scandinavia has an estimated TRR of 147 Tcf of shale gas within the Alum Basin. The prospective shale is of marine origin and Ordovician age. Although the basin and shale deposits are widespread, only one area is predicted to be within the gas window, although the TRR is still very large.

It is regarded as a promising shale gas target, due to very high organic richness (average TOC ~10%), shallow depth (3300 ft), low clay content, reasonable thickness (150 ft), and predicted maturity within the gas window.

Shell completed an exploration program in Southern Sweden. However, as of 2011, they decided to not proceed with the operation based on the results of drilling (Bloomberg, 2011). There is only limited activity associated with the Alum Basin within Denmark and Norway, although exploration wells are planned.

1.5.13 Middle East

The Middle East (excluding Turkey) has an estimated TRR of shale gas of approximately 138 Tcf. No detailed information is available from the sources considered regarding specific shale gas plays within the Middle East. Also, due to abundant conventional energy resources, the Middle East is not likely to proceed with shale gas development in the near future.

1.5.14 India

India has a moderate estimated TRR of 63 Tcf of shale gas, defined within four basins: the Cambay Basin, the Domodar Valley Basin, the Krishna-Godavari Basin, and the Cauvery Basin. The two former basins are associated with marine shales, whilst the latter are terrestrial shales prone to Type III kerogen. With the exception of the Cambay Basin shales, all of the shales have high clay content. However, the Cambay Basin shales are very deep (13,000 ft), only marginally mature (Ro 1.1%), and only have moderate organic richness (TOC 3%). That said, they are very thick (500 ft), hence the GIP concentration is relatively high.

EIA (2011a) noted that as of 2011, there was no previous or specific planned shale gas exploration activity, although the National Oil and Gas Companies have identified the shales in the Cambay Basin as a priority area. Sharma and Kulkarni (2010) note that there was an accidental shale gas strike in well DK#30 within the Cambay Basin, in which hydraulic fracturing was undertaken and which yielded 200 m^3/day.

1.5.15 Pakistan

Pakistan has a moderate estimated TRR of 51 Tcf of shale gas, associated with the Southern Indus Basin. The target shale is of marine origin and was deposited in a foreland basin. Although the net thickness is large (300–450 ft) and the clay content is low, the average organic content is only moderate (TOC 2%) and the target zones are deep (11,500 and 14,000 ft). The target zones are considered within the wet gas to dry gas window (Ro 1.15–1.25%).

There is no information regarding any shale gas exploration activity in Pakistan. Also, Pakistan's natural gas production and consumption are in equilibrium, with growing proven conventional reserves.

1.5.16 Northwest Africa

The nations of Morocco, Algeria, Western Sahara, and Mauritania share coverage of the prospective Tindouf Basin, which is the most significant basin for shale gas in the region. It has an estimated TRR of 50 Tcf of shale gas, whilst the only other identified basin for shale gas (Tadla Basin) has a TRR of 3 Tcf.

The target horizon in the Tindouf Basin is associated with a thin zone of "hot shale," limited to approximately 50-ft thick, as such the GIP concentration is very low. However, the shale does have good organic richness (average TOC of 5%), appropriate clay content, and good maturity. However, the limited vertical thickness and formation underpressure are likely to be the limiting factor.

1.5.17 Eastern Europe (Outside of Poland)

Outside of Poland, the shale gas potential of Eastern Europe has not been explored to the same extent. However, there are three main basins—which may have potential and which have TRR data—the Baltic Basin in Lithuania, the Lublin Basin, and the Dnieper-Donets Basins in Ukraine. All three associated prospective shales are of marine origin.

1.5.17.1 Baltic Basin (Lithuania)
The Baltic Basin in Lithuania has an estimated TRR of 23 Tcf of shale gas, and is associated with the same Silurian age marine shale target that is attracting attention in Poland, hence has similar characteristics. However, the shale is less mature within Lithuania (Ro 1.2%) but is also at a much shallower depth (6,700 ft, as opposed to 12,000 ft). There has been no significant exploration activity in Lithuania to date.

1.5.17.2 Lublin Basin (Ukraine)
The Lublin Basin in Ukraine is an extension of the Lublin Basin in Poland, and has an estimated TRR of 30 Tcf. The shale characteristics are similar, although the average TOC is estimated to be approximately 2.5% instead of 1.5%. However, all exploration interest in this basin to date has focused on Poland, not the Ukraine.

1.5.17.3 Dnieper-Donets Basin (Ukraine)
The Dnieper-Donets Basin in central Ukraine has an estimated TRR of 12 Tcf. The target shale is relatively thin (100-ft thick), deep (13,000 ft), and is within the wet to dry gas window (Ro 1.3%). There has been no significant shale gas exploration within this basin to date, although there is interest in Ukrainian shale gas.

1.5.18 Germany and Surrounding Nations

The North Sea-German Basin extends across northern Germany, Belgium, and the West Netherlands. There is an estimated TRR of 25 Tcf of shale gas, within three different prospective shale formations. All of the shales are of marine origin and were deposited in a rift basin during the Carboniferous, Jurassic, and Cretaceous. The shales are recognized source rocks in the region, but have only recently been identified as having shale gas potential.

All three shales are quite thin, at between 75 and 120 ft, and have medium clay content, good organic content, and maturity within the wet to dry gas window (Ro 1.25–2.5%). The Wealden shale (TRR of 2 Tcf) is the shallowest (6,500 ft), whilst the Posidonia and Namurian Shales are at depths of approximately 10,000 and 12,000 ft, respectively.

ExxonMobil has undertaken considerable shale gas exploration in Germany. However, in recent years, there has been legislative uncertainty surrounding hydraulic fracturing, with a temporary ban imposed during 2012, which has since been lifted (Bloomberg, 2012).

1.5.19 The United Kingdom

The United Kingdom has an estimated TRR of 20 Tcf of shale gas, within the Northern Petroleum System (19 Tcf TRR) and the Southern Petroleum System (1 Tcf TRR), both of which are marine-associated shales deposited in a passive margin during the Carboniferous and Jurassic, respectively.

The Northern Petroleum System (NPS) seems most favorable, with the target at shallow depths (4800 ft), having high organic content (average of 5.8%), reasonable average thickness (150 ft), and maturity in the wet to dry gas window (average Ro 1.4%). However, it is thought to be associated with high clay content.

The Southern Petroleum System is minor by comparison, and also much less favorable in terms of depth (13,500 ft), organic content (average of 2.4% TOC), and thermal maturity (Ro 1.15%).

Recent exploration activity seems to have validated the shale gas potential of the Northern Petroleum System (BBC, 2013), with Caudrilla Resources suggesting that there may be 20 Tcf of TRR based on drilling.

1.5.20 Northern South America

Northern South America has a total estimated TTR of 30 Tcf shale gas, with 11 Tcf within the Venezuelan Maracaibo Basin and 19 Tcf within the Colombian Catatumbo Subbasin. Of the three prospective shales identified, all are in the wet to dry gas window, have moderate thickness (~200 ft), have medium clay content, and are age equivalent to the Eagle Ford Shale play in the United States.

The Colombian La Luna Formation within the Catatumbo Basin seems to be the most favorable, with both high average organic content (4.5% TOC) and a relatively shallow depth (6600 ft). The other two prospective shales are either of relatively low organic content (average of 1.3%), or relatively deep (13,500 ft).

Both prospective basins are associated with significant conventional gas, which is considered a geologically complex region. Conventional exploration does suggest there is gas potential, although the shale gas potential has not yet been validated.

1.5.21 Turkey

Turkey has an estimated TRR of 15 Tcf within the Anatolian (9 Tcf) and Thrace Basins (6 Tcf). All the prospective shales are of marine origin. The shale gas characteristics seem reasonable in all the prospective shales. However, the Anatolian Basin shale occurs at the shallowest depths (8000 ft), whilst still having reasonable net thickness (150 ft), high organic richness (5.5%), and a degree of maturity (predominately wet gas, Ro 1.1%). The Thrace Basin has two target shales, one of which is very deep (14,000 ft), whilst the other has only moderate organic richness (2.5% TOC) and marginal maturity (Ro 1.1%).

The Anatolian Basin has active conventional oil production, but shale gas exploration is still only in the leasing stage, with no specific plans for exploration yet.

1.6 DATA ASSESSMENT

This section presents some statistics and discussion on the distribution, geological characteristics, and general features of the shale gas plays based on the data collated.

1.6.1 Distribution

The relative and absolute distribution of an estimated shale gas TRR across different regions and countries is depicted in Figures 1.4 and 1.5. However, the following should be considered in this context:

- Shale maps published by the SPE (2012, 2013) define many large shale basins that are not included in any of the source studies, and hence are not reflected in the values noted. For example, the Sao Francisco Basin in Brazil covers approximately 750,000 mi^2, hence could greatly impact the TRR estimate for South America.
- The significance that shale gas in any particular region is not simply a function of the absolute volume of resources. Factors, such as the availability of other domestic sources of natural gas and the current demand for energy, influence significance considerably.

1.6.2 Basin Type

Based on the data sources available, an assessment of the basin type has been made. The type noted refers to the architecture of the basin at the time of deposition of the prospective shale. The basin types allocated to each play are passive margin basin, foreland basin, rift basin,

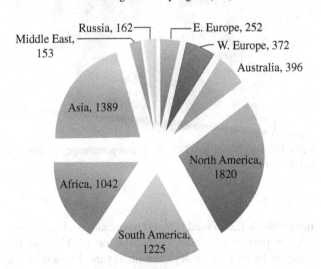

FIGURE 1.4 Chart illustrating the relative abundance of shale gas across different regions.

12 GAS SHALE: GLOBAL SIGNIFICANCE, DISTRIBUTION, AND CHALLENGES

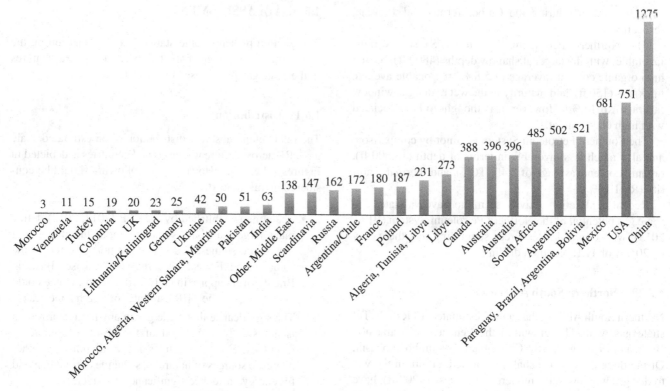

FIGURE 1.5 Estimated shale gas TRR (Tcf) across different countries.

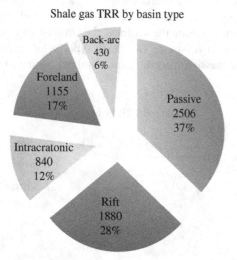

FIGURE 1.6 Shale gas TRR (Tcf) and percentage contribution to total for each basin type.

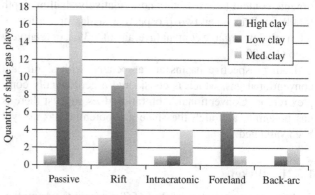

FIGURE 1.7 Distribution of shale gas plays by basin type.

1.6.3 Depositional Environment

Approximately 97% of the shales were deposited in a marine environment, and hence are likely associated with Type II kerogen (Gluyas and Swarbrick, 2009).

1.6.4 TOC Content

TOC data was available for all plays, with the exception of the "whole country" resource estimates for Russia and certain Middle East nations.

intracratonic (i.e., failed rift/sag) basin, and back-arc basin. The proportion and relative distribution of TRR on the basis of basin type is depicted on Figure 1.6, whilst the quantity of plays associated with each basin type is illustrated in Figure 1.7.

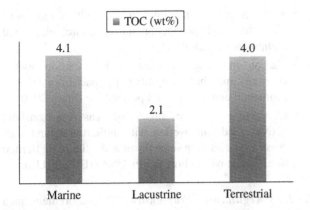

FIGURE 1.8 Average TOC by gas shale depositional environment.

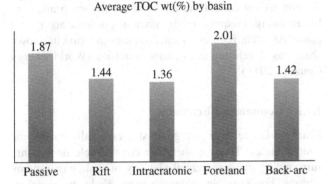

FIGURE 1.9 Average TOC by gas shale basin type at time of deposition.

Figures 1.8 and 1.9 illustrate the relationship among TOC, the inferred depositional environment, and basin type for each gas shale, respectively.

1.6.5 Clay Content

Qualitative clay content data was available for all non-US shale gas plays, with the exception of Russia and much of the Middle East. Values of high, medium, or low were extracted. Source documents acknowledge that there is considerable uncertainty regarding this assessment, and considerable use of analogous plays was made when selecting an average value.

It is generally accepted that there is a tendency for marine shales to have a lower clay content (EIA, 2011b) than lacustrine and terrestrial shales.

1.7 INDUSTRY CHALLENGES

The fundamental way in which gas is produced within a shale gas play presents some different technical and environmental challenges in comparison with conventional gas plays.

1.7.1 Environmental Challenges

As noted by the US Department of Energy (US DOE, 2009), "the key difference between a shale gas well and a conventional gas well is the reservoir stimulation (large-scale hydraulic fracturing) approach performed on shale gas wells." Also, on a play scale, the major difference between a shale play and a conventional gas play is the sheer quantity of wells required to produce the same quantity of gas.

The environmental concerns regarding shale gas production, in particular hydraulic fracturing, are so significant that certain countries, such as France and Switzerland, have imposed a ban (SPE, 2012).

As a result of these features, the challenges outlined below are likely to greatly influence the future of the industry.

1.7.1.1 Protecting Existing Water Resources In the United States, the regulatory framework places considerable emphasis on protecting groundwater (US DOE, 2009) and surface water, due to the potential for shallow fresh groundwater aquifers (or surface waters) to be contaminated by deeper saline water, gas, or fracturing fluids during the drilling and hydraulic fracturing process.

Literature (US DOE, 2009) suggests that a substantial amount of independent research has been carried out to assess the impact that shale gas operations have on shallow aquifers and surface water. The US Department of Energy (US DOE, 2009) have highlighted the importance of the following in managing environmental risks:

Drilling, casing, and cementing programs (US DOE, 2009) to isolate water-bearing zones from gas-bearing zones. These include consideration of factors such as preventing drilling mud entering the shallowest aquifers, corrosion of steel casing over time, testing to validate performance, and designing for redundancy. Studies have suggested that the current level of redundancy in the systems adopted in the United States means that "a number of independent events must occur at the same time and go undetected" for fluid from a pay zone to reach a shallow freshwater aquifer (Michie & Associates, 1988).

Fracture treatment design (US DOE, 2009) using robust, yet sophisticated, techniques to produce a controlled treatment within a specific target formation, reflecting the in situ reservoir conditions. This includes the implementation of microseismic-fracture-mapping techniques to map the development of fractures during treatment, and also fracture design refinement based on the outcome of monitoring.

Fracturing process (US DOE, 2009), including testing/certification of equipment and wells prior to fracture treatment to ensure each well is fit for treatment, and the implementation of staged treatments to ensure controlled fracturing of discrete intervals. Naturally, the hazard likelihood and hence risk that fracture treatments present to an aquifer is a function of the vertical distance above the gas shale. Thus literature recognizes the distinction between shale gas plays associated with deep pay zones and shallow pay zones. For example, as noted by Fisher (2010), "even in areas with the largest measured vertical fracture growth, such as the Marcellus, the tops of the hydraulic fractures are still thousands of feet below the deepest aquifers suitable for drinking water." However, there can be relatively close proximity between a gas pay zone and an aquifer. As an example, in the United States, the Antrim and New Albany shale gas plays are quite shallow and hence closer to groundwater aquifers than the likes of the Marcellus Shale (US DOE, 2009).

1.7.1.2 Sustainable Use of Groundwater Resources for Formation Fracturing Approximately three million gallons of freshwater is required on average for complete treatment of a shale gas well, although this value varies considerably. The water is also required over a relative short period of time; hence, there is significant demand on surface water, groundwater, and municipal sources (US DOE, 2009). Water resource management is therefore very important, in particular, in more arid areas.

1.7.1.3 Responsible Treatment and Disposal of Exploration and Production-Related Water Wells produce fracture treatment fluids mixed with formation fluid after pressure associated with treatment has been relieved from the well. The quality of this fluid ranges from fresh to saline, and the volume may range from 30 to 70% of the original volume pumped into the formation (US DOE, 2009). Environmental management of produced water is an important part of the overall environmental management plan, and successful management will directly influence the successful expansion of shale gas production (US DOE, 2009). Some of the methods being adopted and considered for disposal of produced water include the following:

- On-site injection into deep permeable and porous formations, when available in the play
- Transportation and disposal at remote injection sites
- On-site treatment
- Reuse of fluid for treatment of other wells
- Supplying the water to other users who may benefit (e.g., nearby mines, Queensland, Australia) (Staff, 2010)

1.7.1.4 Other Environmental Considerations Some other environmental considerations associated with shale gas developments include the following:

- Management of naturally occurring radioactive materials produced from the ground, which can be within drill cuttings and dissolved within produced water, and can precipitate out over time (US DOE, 2009).

- Management of air quality as a result of omissions associated with production infrastructure, plant, and equipment (US DOE, 2009).
- Carbon emissions management, in a carbon pricing economy, may have a significant impact from an environmental and commercial perspective (Staff, 2010).
- Competing land use, since shale gas is an onshore activity, and can overlap with agricultural land (i.e., Australia), and even some towns and cities (e.g., Barnett Shale wells within Fort Worth, USA) (EIA, 2011a).

1.7.1.5 Regulatory Framework The aforementioned environmental risks and issues are generally addressed at a regulatory level within the United States. However, literature suggests that other countries mentioned within this literature review do not yet have the regulatory framework for ensuring adequate environmental controls are put in place. As such, regulatory uncertainties are slowing down shale gas development in many countries (World Energy Council, 2011).

1.7.2 Commercial/Economic

Shale gas is a relative young industry, especially outside the United States. There is therefore considerable uncertainty surrounding the commercial viability of shale gas in many regions. For example, although many shale gas developments appear to be profitable within the United States, the economics are not necessarily comparable in other areas for the following reasons:

1. It has been suggested that it may cost as much as three times to drill a shale gas well in Poland compared to the United States (Pfeifer, 2012). This reflects the limited supply of rigs, with only 34 land rigs operating in all of Western Europe in 2010 (Stevens, 2012).
2. Commercial viability hinges on EUR/well, which is notoriously difficult to predict.

1.8 DISCUSSION

The geology of shale gas has much in common with source rock geology. However, geomechanical characteristics play a key role in shale gas plays. Geomechanical properties are somewhat influenced by mineralogy/clay content, and the tectonic stress history of the basin. As such, there would seem to be potential for such properties to be assessed during early exploration phase using basin history analysis, sequence stratigraphy, and facies association.

Although the geological characteristics presented for the various shale gas deposits are directly influencing the TRR volume presented, it remains unclear what the real factors are that influence the actual estimated ultimate recover (EUR) per well, which is a central aspect of assessing the economic viability of a shale gas play. Therefore, an appreciation of the methods for assessing likely well recovery is central to identifying economic shale gas plays. In the event that such methods are not currently reliable, then it seems it would be an area worthy of considerable research.

In addition to EUR/well, the cost of production has the potential to vary greatly from country to country and hence influence whether shale gas will be significance in each respective area. Although some of the published sources allude to the relative cost of production, a more quantitative approach would be beneficial for identifying opportunities-specific regions. It is expected that as the industry develops slowly in each respective region, that drilling services will become more competitive, and hence reduce the cost of production.

Due to the vast shale gas resource globally, there is considerable attention being drawn to the industry. However, a current concern for the global adoption of gas production from shales is the lack of non-US production examples on which to consider the economic viability in other regions. As such, shale gas remains quite speculative on a global scale. Gas prices, the growth of new markets for gas consumption (i.e., transport), and technology are likely to have a major impact on the future of shale gas in other countries.

1.9 CONCLUSIONS

- Key geological characteristics of a successful shale gas play include the following:
  - Organic rich, minimum TOC of 2%, although successful US shale gas plays have average TOC of 12%
  - Low clay content (<50%)/high brittle mineral content (>40%). Generally associated with marine shales
  - Thermally mature, R_o >1.1%, ideally 1.1–1.4% (Types II and III kerogen), >0.7% (Type I kerogen). Kerogen type is a function of depositional environment.
  - Thick (minimum of 100 ft).
  - Porosity >5%
- Technically recoverable (although not necessarily economically recoverable) shale gas resources are abundant across the globe. They are also located in a very wide range of geographical regions, and in many of the nations with the highest energy consumption.
- For certain nations, shale gas has the potential to reduce energy prices and reduce dependence on other nations, and hence impacts on both the political and economic outlook.
- The environmental concerns identified have the potential to halt development in many regions. Furthermore, should the industry fail to address the environmental issues at all levels, then there may not only be an impact on the environment but also on public perception and hence political support for the industry, and hence a favorable regulatory environment.
- Technological developments are a major reason why the production of gas from shale has become possible. As such, the future growth of the industry is likely to relate closely to technological developments that further improve well yield and the duration of well yield.
- The prospects for and significance of shale gas vary considerably by country, irrespective of the absolute TRR estimate and production costs. It seems industry success also requires high local demand for gas, a lack of existing large-scale conventional gas production, and an existing gas distribution network. The United States is a good example of the potential for shale gas under such favorable conditions.

APPENDIX A.1 GLOBAL SHALE GAS RESOURCE DATA

Region	Country	Basin	Shale Fm	TRR (Tcf)	GIP Conc. (Bcf/mi²)	Epoch/age	Basin type[a]	Dep. En.[b]	Depth (ft)	Net thick. (ft)	Clay content	TOC (avg wt%)	Ro[c]	PHI	Pressure	Source[d]
Africa	Algeria, Tunisia, Libya	Ghadames	Tannezuft	156	44	Silurian	Intracratonic	Marine	12,900	104	Medium	5.7	1.15	N/A	Overpressured	A
			Frasnian	75	65	M. Devonian	Intracratonic	Marine	9350	177	Medium	4.2	1.15	N/A	Overpressured	A
	Libya	Sirt	Sirt-Rachmat	162	61	U. Cretaceous	Intracratonic	Marine	10,000	200	Medium	2.8	1.1	N/A	Normal	A
			Etel	111	42	U. Cretaceous	Intracratonic	Marine	12,000	120	High	3.6	1.1	N/A	Normal	A
	Morocco, Algeria, W Sahara, Mauritania	Tindouf	L. Silurian	50	18	Silurian	Passive	Marine	9000	50	Medium	5	3.5	N/A	Underpressured	A
	Morocco	Tadla	L. Silurian	3	49	Silurian	Passive	Marine	6560	197	Medium	2	2.25	N/A	Underpressured	A
	South Africa	Karoo	Prince Albert	91	43	L. Permian	Foreland	Marine	8500	120	Low	2.5	3	N/A	Overpressured	A
			Whitehill	298	59	L. Permian	Foreland	Marine	8000	100	Low	6	3	N/A	Overpressured	A
			Collingham	96	36	L. Permian	Foreland	Marine	7800	80	Low	4	3	N/A	Overpressured	A
Asia	China	Sichuan	Longmaxi	343	80	Silurian	Passive	Marine	10,700	280	Medium	3	2.3	N/A	Normal	A
			Quongzhusi	349	57	Cambrian	Passive	Marine	11,500	195	Medium	3	2.5	N/A	Normal	A
		Tarim	O1/O2/O3	224	102	Ordovician	Passive	Marine	13,000	260	Medium	2	2	N/A	Normal	A
			Cambrian shales	359	141	Cambrian	Passive	Marine	14,000	404	Medium	2	2.5	N/A	Normal	A
	India	Cambay	Cambay shale	20	231	U. Creta.-Tertiary	Rift	Marine	13,000	500	Medium	3	1.1	N/A	Mod. Overpress.	A
		Domodar Valley	Barren measure	7	123	Permian-Triassic	Rift	Marine	4920	368	High	4.5	1.2	N/A	Mod. Overpress.	A
		Krishna-Godavari	Kommugudem	27	156	Permian	Rift	Terrestrial	11,500	300	High	6	1.6	N/A	Normal	A
		Cauvery	Andimadam	9	143	Cretaceous	Rift	Terrestrial	10,000	400	High	2	1.15	N/A	Normal	A
	Pakistan	Southern Indus	Sembar	20	100	E. Cretaceous	Foreland	Marine	14,000	300	Low	2	1.25	N/A	Normal	A
			Ranikot	31	157	Paleocene	Foreland	Marine	11,500	450	Low	2	1.15	N/A	Normal	A
Australia	Australia	Cooper	Roseneath, Epsilon-Murteree	85	105	Permian	Intracratonic	Lacustrine	8500	300	Low	2.5	2	N/A	Mod. Overpress.	A
		Maryborough	Godwood/Cherwell	23	110	Cretaceous	Back-arc	Marine	9500	250	Low	2	1.5	N/A	Slight. Overpress.	A
		Perth	Carynginia	29	107	U. Permian	Passive	Marine	10,700	250	Low	4	1.4	N/A	Normal	A
			Kockatea	30	110	L. Triassic	Passive	Marine	10,000	230	Low	5.6	1.3	N/A	Normal	A
		Canning	Goldwyer	229	106	M. Ordovician	Rift	Marine	12,000	250	Low	3	1.4	N/A	Normal	A
E. Europe	Poland	Baltic	Lower Silurian	129	145	L. Silurian	Passive	Marine	12,300	316	Medium	4	1.75	N/A	Overpressured	A
		Lublin	Lower Silurian	44	79	M. Silurian	Rift	Marine	10,005	228	Medium	1.5	1.35	N/A	Overpressured	A
		Podlasie	Lower Silurian	14	142	L. Silurian	Rift	Marine	8545	297	Medium	6	1.25	N/A	Overpressured	A
	Lithuania, Kaliningrad	Baltic	Lower Silurian	23	101	Silurian	Passive	Marine	6,724	284	Medium	4	1.2	N/A	Overpressured	A
	Ukraine	Dnieper-Donets	Rudov Bed	12	42	Carb.	Rift	Marine	13,120	102	Medium	4	1.3	N/A	Overpressured	A
		Lublin	Lower Silurian	30	79	Silurian	Rift	Marine	9,840	208	Medium	2.5	1.35	N/A	Overpressured	A
Middle East	Turkey	SE Anatolia	Dadas Shale	9	61	Devonian-Silurian	Passive	Marine	8,200	150	Medium	5.5	1.1	N/A	Normal	A
		Thrace	Hamitabat	4	128	M. Eocene	Passive	Marine	14,268	344	Medium	3.9	1.75	N/A	Normal	A
		Thrace	Mezardere	2	74	L. Oligocene	Passive	Marine	9,184	295	Medium	2.5	1.1	N/A	Normal	A
	Other, M. East	All remaining	All remaining	138	N/A	N/A	N/A	N/A	N/A	N/A	N/A	N/A	N/A	N/A	N/A	B

Region	Country	Basin	Formation			Age	Tectonic	Depositional	Depth						Pressure	
N. America	Canada	WCB: Horn River	Muskwa/Otter Park	132	152	Devonian	Passive	Marine	8,000	380	Low	3.5	3.8	N/A	Mod. Overpress.	A
			Evie/Klua	33	55	Devonian	Passive	Marine	8,500	144	Low	3.5	3.8	N/A	Mod. Overpress.	A
		WCB: Cordova	Muskwa/Otter Park	29	61	Devonian	Passive	Marine	6,000	207	Low	2	2.5	N/A	Normal	A
		WCB: Liard	Lower Besa River	31	161	Devonian	Passive	Marine	9,000	441	Low	3.5	3.8	N/A	Mod. Overpress.	A
		WCB: Deep	Montney Shale	49	99	Triassic	Passive	Marine	6,000	240	Low	3	1.5	N/A	Overpressured	A
			Doig Phosphate	20	67	Triassic	Passive	Marine	9,250	150	Low	5	1.1	N/A	Mod. Overpress.	A
		WCB: Colarado Group	2WS and Fish Scales	61	21	Cretaceous	Passive	Marine	6,900	105	Low	2.4	0.61	N/A	Underpressured	A
		Appalachian Fold Belt	Utica	31	134	Ordovician	Passive	Marine	8,000	400	Low	2	2	N/A	Slight. Overpress.	A
		Windsor	Horton Bluff	2	82	E. Carb.	Passive	Marine	4,000	300	Unknown	5	2	N/A	Normal	A
	Mexico	Burgos	Eagle Ford Shale	454	209	L. Cretaceous	Rift	Marine	10,380	400	Low	5	1.3	N/A	Normal	A
		Sabinas	Tithonian Shales	82	75	U. Jurassic	Rift	Marine	12,000	200	Low	3	1.3	N/A	Normal	A
			Eagle Ford Shale	44	113	L. Cretaceous	Rift	Marine	10,380	400	Low	4	1.3	N/A	Underpressured	A
			La Casita	11	58	L. Jurassic	Rift	Marine	11,500	240	Low	2	2.5	N/A	Underpressured	A
		Tampico	Pimienta	65	63	Jurassic	Rift	Marine	6,200	245	Low	3	1.3	N/A	Normal	A
		Tuxpan Platform	Tamaulipas	8	65	L. Cretaceous	Rift	Marine	7,900	225	Low	3	1.25	N/A	Normal	A
			Pimienta	8	72	Jurassic	Rift	Marine	8,500	245	Low	3	1.3	N/A	Normal	A
		Veracruz	Maltrata	9	29	U. Cretaceous	Rift	Marine	11,200	120	Medium	2	1.5	N/A	Normal	A
	United States	Appalachian	Marcellus	410	N/A	M. Devonian	Foreland	Marine	6,250	125	N/A	12	N/A	8	N/A	C
			Big Sandy	7	N/A	Devonian	Foreland	Marine	3,800	175	N/A	3.75	N/A	10	N/A	C
				14	N/A	Devonian	Foreland	Marine	3,000	371	N/A	N/A	N/A	7	N/A	C
			Greater Siltstone	8	N/A	Devonian	Foreland	Marine	3,911	623	N/A	N/A	N/A	5.8	N/A	C
		Illinois	New Albany	11	N/A	Devonian	Intracratonic	Marine	2,750	200	N/A	13	N/A	12	N/A	C
		Michigan	Antrim Shale	20	N/A	U. Devonian	Intracratonic	Marine	1,400	95	N/A	11	N/A	9	N/A	C
		N/A	Cincinnati Arch	1	N/A	N/A	Passive	N/A	N/A	N/A	N/A	N/A	N/A	N/A	N/A	C
		East Texas	Haynesville	75	N/A	L. Jurassic	Rift	Marine	12,000	250	N/A	2.25	N/A	8.5	N/A	C
		Texan Maverick	Eagle Ford	20	N/A	L. Cretaceous	Rift	Marine	7,000	200	N/A	4.25	N/A	9	N/A	C
		Black Warrior	Floyd–Neal and Conasauga	4	N/A	Cambrian	Passive	Marine	8,000	130	N/A	1.8	N/A	1.6	N/A	C
		Arkoma	Fayetteville	32	N/A	E. Carb.	Rift	Marine	4,000	110	N/A	6.9	N/A	5	N/A	C
		Ardmore / Arkoma	Woodford	22	N/A	Devonian	Rift	Marine	7,000	150	N/A	6.5	N/A	7	N/A	C
		Anadarko	Cana Woodford	6	N/A	E. Carb.	Intracratonic	Marine	13,500	200	N/A	6	N/A	7	N/A	C
		Fort Worth	Barnett	43	N/A	E. Carb.	Foreland	Marine	7,500	300	N/A	N/A	N/A	5	N/A	C
		Permian	Barnett Woodford	32	N/A	E. Carb.	Foreland	Marine	10,200	400	N/A	5.5	N/A	7	N/A	C
			Avalon & Bone Springs	2	N/A	Permian	Rift	Marine	8,750	1300	N/A	N/A	N/A	N/A	N/A	C
		Greater Green River	Hilliard-Baxter-Mancos	4	N/A	L. Cretaceous	Rift	Marine	14,750	3075	N/A	1.75	N/A	4.25	N/A	C
		San Juan	Lewis	12	N/A	U. Cretaceous	Rift	Marine	4,500	250	N/A	N/A	N/A	3.5	N/A	C
			Williston–Shallow, Niobraran	7	N/A	N/A	Passive	N/A	N/A	N/A	N/A	N/A	N/A	N/A	N/A	C
		Uinta	Mancos	21	N/A	U. Cretaceous	Passive	Marine	15,250	3000	N/A	14	N/A	3.5	N/A	C
Russia	Russia	All	All	162	N/A	N/A	Passive	N/A	N/A	N/A	N/A	N/A	N/A	N/A	N/A	B

(Continued)

Region	Country	Basin	Shale Fm	TRR (Tcf)	GIP Conc. (Bcf/mi²)	Epoch/age	Basin type[a]	Dep. En.[b]	Depth (ft)	Net thick. (ft)	Clay content	TOC (avg wt%)	Ro[c]	PHI	Pressure	Source[d]
S. America	Venezuela	Maracaibo	La Luna	11	93	L. Cretaceous	Passive	Marine	13,500	180	Medium	5.6	1.25	N/A	Normal	A
	Colombia	Catatumbo Sub-	La Luna	7	74	L. Cretaceous	Passive	Marine	6,600	180	Medium	4.5	1.05	N/A	Normal	A
			Capacho	12	106	L. Cretaceous	Passive	Marine	7,500	320	Medium	1.3	1.1	N/A	Normal	A
	Argentina	Neuquen	Los Molles	167	123	M. Jurassic	Back-arc	Marine	12,500	300	Medium	1.1	1.5	N/A	Overpressured	A
			Vaca Meurta	240	168	L. Jurassic	Back-arc	Marine	8,000	325	Medium	4	1.25	N/A	Overpressured	A
		San Jorge	Aguada Bandera	50	149	L. Jurassic	Rift	Lacustrine	12,000	400	Medium	2.2	2	N/A	Normal	A
			Pozo D-129	45	151	E. Cretaceous	Rift	Lacustrine	10,500	420	Medium	1.5	1.5	N/A	Normal	A
	Argentina, Chile	Austral-Magallanes	L. Inoceramus	84	86	E. Cretaceous	Passive	Marine	8,500	300	Medium	1.6	1.3	N/A	Slight Overpress.	A
			Magnas Verdes	88	72	E. Cretaceous	Passive	Marine	8,500	240	Medium	2	1.3	N/A	Slight Overpress.	A
	Paraguay, Brazil, Argentina, Bolovia	Parana-Chaco	San Alfredo	521	347	Devonian	Rift	Marine	7,500	1000	Low	2.5	0.9	N/A	Normal	A
W. Europe	France	Paris	PermianCarboniferous	76	47	Permian Carb.	Intracratonic	Marine	10,824	115	Medium	4	1.65	N/A	Normal	A
		South-East	Terres Noires	28	27	U. Jurassic	Foreland	Marine	4,920	100	Low	3.5	1.25	N/A	Normal	A
			Liassic	76	57	L. Jurassic	Foreland	Marine	12,300	158	Medium	2.5	1.45	N/A	Normal	A
	Germany	North Sea-German	Posidonia	7	33	Jurassic	Rift	Marine	9,840	100	Medium	5.7	1.5	N/A	Normal	A
			Namurian	16	54	Carb.	Rift	Marine	12,300	122	Medium	3.5	2.5	N/A	Overpressured	A
			Wealden	2	26	Cretaceous	Rift	Marine	6,560	75	Medium	4.5	1.25	N/A	Normal	A
	Scandinavia	Scandinavia Region	Alum	147	77	Ordovician	Passive	Marine	3,280	164	Low	10	1.85	N/A	Normal	A
	United Kingdom	N. Pet. System	Bowland	19	48	Carb.	Passive	Marine	4,800	148	High	5.8	1.4	N/A	Normal	A
	United Kingdom	S. Pet. System	Liassic	1	45	Jurassic	Passive	Marine	13,500	125	Medium	2.4	1.15	N/A	Normal	A

[a]Basin type at time of deposition. In some cases, this is a close call, for example, the transition from being a passive margin to a foreland basin during collisional events.
[b]Depositional Environment.
[c]Vitrinite reflectance (Ro).
[d]Letters A, B, and C correspond to the data sources listed in Section 4.1.

REFERENCES

1. BBC. 2013. North American firms quit shale gas fracking in Poland. BBC News. Available at http://www.bbc.co.uk/news/business-22459629. Accessed May 8, 2013.
2. Bloomberg. 2011. Shell ends shale gas search in Sweden; invests in China fields. Available at http://www.bloomberg.com/news/2011-07-28/shell-ends-shale-gas-search-in-sweden-invests-in-china-fields.html. Accessed December 1, 2014.
3. Bloomberg. 2012. German lawmakers reject ban on shale gas fracking in parliament. Bloomberg News. Available at http://www.bloomberg.com/news/2012-12-13/german-lawmakers-reject-ban-on-shale-gas-fracking-in-parliament.html. Accessed December 14, 2012.
4. Caineng Z, Dazhong D, Wang S, Jianzhong L, Xinjing L, Yuman W, Denghua L, Keming C. Geological characteristics and resource potential of shale gas in China. Petrol Explor Dev 2010;37 (6):641–653.
5. Chevron. 2012. The Gorgon project fact sheet. Available at http://www.chevronaustralia.com/Libraries/Chevron_Documents/May_2012_Gorgon_Project_Fact_Sheet.pdf.sflb.ashx. Accessed December 1, 2014.
6. Dong Z, Holditch SA, McVay DA. Resource evaluation for shale gas reservoirs. *Soc Petrol Eng Eco Manag* 2013. 16pp.
7. EIA. 2010. Schematic geology of natural gas resources. Available at http://www.eia.gov/oil_gas/natural_gas/special/ngresources/ngresources.html. Accessed April 19, 2013.
8. EIA. Review of emerging resources: US shale gas and shale oil plays. US Energy Information Administration; 2011a.
9. EIA. World shale gas resources: An initial assessment of 14 regions outside the United States. US Energy Information Administration; 2011b.
10. EIA. 2011c. Shale gas and the outlook for US natural gas markets and global gas resources. US Energy Information Administration. Available at http://www.eia.gov/pressroom/presentations/newell_06212011.pdf. Accessed December 1, 2014.
11. EIA. 2013. AEO2013 early release overview. US Energy Information Administration. Available at http://www.eia.gov/forecasts/aeo/er/pdf/0383er(2013).pdf. Accessed December 1, 2014.
12. Fisher K. Data confirm safety of well fracturing. The American Oil and Gas Reporter. 2010. 4 pp.
13. Gluyas J, Swarbrick R. *Petroleum Geoscience*. Malden: Blackwell Publishing; 2009.
14. Kennedy R. Shale gas challenges/technologies over the asset life cycle. US China Oil and Gas Industry Forum. Baker Hughes; September 2010.
15. Lakatos I, Szabo JL. Role of conventional and unconventional hydrocarbons in the 21st century: Comparison of resources, reserves, recovery factors and technologies. Society of Petroleum Engineers 2009;SPE-121775-MS.
16. Lu S, Huang W, Chen F, Li J, Wang M, Xue H, Wang W, Cai X. Classification and evaluation criteria of shale oil and gas resources: Discussion and application. Petrol Explor Dev 2012;39 (2):268–276.
17. Michie & Associates. Oil and gas water injection well corrosion. Prepared for the American Petroleum Institute; 1988.
18. Pfeifer S. Finds that form a bedrock of hope, *Financial Times*, April 22, 2012.
19. Potter PE, Maynard JB, Pryor WA. *Sedimentology of Shale*. New York: Springer Verlag; 1980. p 305.
20. Ridley M. The shale gas shock. The Global Warming Policy Foundation, Report 2, 2011.
21. Rogner HH. An assessment of world hydrocarbon resources. Ann Rev Energy Environ 1997;22:207–262.
22. Sharma S, Kulkarni P. Gas strike in shale reservoir in Dholka field in Cambay Basin. Oil and Gas Conference and Exhibition; Jan 20–22, 2010; Mumbai, India: Society of Petroleum Engineers; 2010.
23. SPE. European shale gas map. *J Petrol Technol* February 2012; Suppl.
24. SPE. South American shale map. *J Petrol Technol* March 2013; Suppl.
25. Staff. 2025 Unconventional gas outlook, the next wave. Volume 1. Draft for participant review. Core Energy Group; 2010.
26. Stevens P. *The shale gas revolution: Developments and changes*. Chantham House Briefing Paper. EERG BP 2012/04. Energy, Environment and Resources. London: Chantham; August 2012.
27. UPI. China's shale development faces obstacles. UPI Business News. Available at http://www.upi.com/Business_News/Energy-Resources/2013/05/01/Chinas-shale-development-faces-obstacles/UPI-51811367439864/. Accessed May 1, 2013.
28. US DOE. Modern shale gas development in the United States: A primer (prepared for United States Department of Energy, Office of Fossil Energy and National Energy Technology Laboratory). Oklahoma: Ground Water Protection Council; 2009.
29. Vidas H, Hugman B. *ICF International. Availability, economics, and production potential of North American unconventional natural gas supplies (prepared for the INGAA Foundation, Inc.)*. ICF International: Fairfax, VA; 2008.
30. Weijermars R. Economic appraisal of shale gas plays in Continental Europe. J Appl Energy 2013;106:100–115.
31. World Energy Council. *Survey of Energy Resources: Shale Gas—What's New?* London: World Energy Council; 2011.

2

ORGANIC MATTER-RICH SHALE DEPOSITIONAL ENVIRONMENTS

João Trabucho-Alexandre[1,2]

[1] Department of Earth Sciences, University of Durham, Durham, UK
[2] Institute of Earth Sciences Utrecht, Utrecht University, Utrecht, The Netherlands

2.1 INTRODUCTION

Shale is the most abundant rock type available at the surface of our planet and makes up about two-thirds of the stratigraphic record (Garrels and Mackenzie, 1969). The term "shale"[1] refers to all sedimentary rocks composed predominantly of mud[2] ($>4\phi$ or <0.0625 mm) particles (cf. Tourtelot, 1960, p. 342). Mud particles may be terrigenous, biogenous, or hydrogenous. Terrigenous—or siliciclastic—mud is always detrital, that is, produced by weathering and erosion of preexisting rocks, and comes into the depositional environment as individual particles and/or as aggregates. Biogenous mud is made by organisms, and includes a skeletal and an organic component. Skeletal mud may be calcareous or siliceous. Some benthic protozoans (agglutinated forams) build their tests by cementing terrigenous and other particles with an organic ligand. Some quartz silt aggregates in shales represent the collapsed or compacted tests of these organisms (Pike and Kemp, 1996; Schieber, 2009). When a shale is characterized by an organic matter content higher than the average marine shale (ca. 0.5%, Arthur, 1979), it is referred to as an organic matter-rich shale or, more commonly, a black shale.[3] Hydrogenous mud precipitates out of solution directly, either from seawater or from interstitial water during diagenesis, and it includes oxides and hydroxides, silicates, for example, zeolites and clay minerals, heavy metal sulfides, sulfates, carbonates, and phosphates. Clay minerals are therefore not only terrigenous, but they may also be hydrogenous, that is, formed *in situ*. Biogenous and hydrogenous mud may be detrital, that is, recycled from older deposits, in which case their origin is biochemical, but their texture will be clastic. The composition of the detrital fraction of a shale depends on the petrology of its source areas and on the intensity and effectiveness of chemical weathering.

---

[1] Although the original meaning of the word "shale" is "laminated clayey rock," our historical usage of the word has been that of "general class of fine-grained sedimentary rocks" (Tourtelot, 1960). There is no reason why we should restrict our usage of the word shale to laminated and/or fissile fine-grained sedimentary rocks. "Lamination" has a descriptive and a genetic definition with distinct sedimentologic implications (cf. McKee and Weir, 1953; Campbell, 1967), and fissility is a secondary property largely related to weathering (e.g., Ingram, 1953). If we define shale as a fissile fine-grained sedimentary rock, then "there are no shales in the subsurface, only potential shales" (Weaver, 1989, p. 6). Some rocks which are referred to in the literature as shales are actually metasedimentary rocks produced by regional low-grade metamorphism, and should therefore be called slates.
[2] Mud is the name given to particles or collections of particles smaller than sand, that is, smaller than 62.5 μm, that is, silt and clay, which typically occur together. Some authors talk about "mud and silt," perhaps making "mud" a synonym of "clay." Operationally, the mud/sand boundary may be defined based on sieve sizes around this value. Because mud is a term related to grain size, it has no connotations as to composition.

[3] Modern organic matter-lean muds may also be black when their iron sulfide content is high (e.g., Potter et al., 2005); however, they become light-colored on lithification as the sulfide changes into either marcasite or pyrite (Twenhofel, 1939). Black shale is the general term for any dark-colored, fine-grained, organic matter-rich sedimentary rock. In the words of Stow et al. (1996, p. 403): "[m]any black shales are hemipelagites; others, such as black cherts and organic matter-rich limestones, are pelagites; whereas still others are fine-grained turbidites." This *de facto* usage has also been noted by Arthur (1979) who states that "the term 'black shale' is used in a general sense to refer to relatively organic carbon-rich [...] mudstone and marlstone which may or may not be 'shale' in the classical sense."

Fundamentals of Gas Shale Reservoirs, First Edition. Edited by Reza Rezaee.
© 2015 John Wiley & Sons, Inc. Published 2015 by John Wiley & Sons, Inc.

22 ORGANIC MATTER-RICH SHALE DEPOSITIONAL ENVIRONMENTS

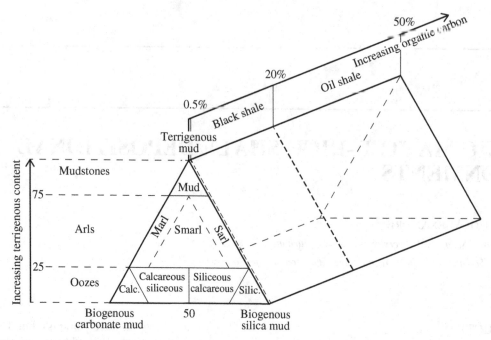

FIGURE 2.1 Classification of shales based on composition. This scheme is based on the classification scheme proposed by Hay et al. (1984, p.14). Shales are fine-grained sedimentary rocks with varying relative proportions of terrigenous mud and mud-sized biogenous components. Shales may also contain up to 25% of terrigenous or biogenous grains coarser than mud (>62.5 μm). A shale with an organic matter content higher than the average marine rock, that is, ca. 0.5%, is referred to as a black shale. Black shale is thus the general term for any dark-colored, fine-grained, organic matter-rich sedimentary rock.

Shales, particularly those deposited in marine environments, are usually combinations of mud from different sources (Fig. 2.1).

Although "black shale" is a useful term when referring to organic matter-rich sedimentary rocks in general, it is a collective noun that groups rocks of various types and origins (Trabucho-Alexandre et al., 2012b). In the petrologic hierarchy of sedimentary rocks, the term "black shale" is equivalent to such terms as "sandstone" or "limestone," rather than to more precise terms such as "lithic arenite" or "oöidal grainstone." However, shales are seldom subdivided based on their composition and texture (Fig. 2.1). For this reason, we are often not able to distinguish subtle compositional and textural differences that would otherwise allow us to subdivide what were once considered monotonous successions of shale and marl (cf. Lewan, 1978).

Shales do not lend themselves well to study in the field or in hand specimen. Sorby (1908, p. 196), who started the microscopic study of rocks, realized that the "examination of [...] rocks in a natural condition is enough to indicate that the structure of clays differs enormously, and indicates formation under very different conditions; but there is always some doubt as to their true structure, when not made into thin sections." The study of shales has also been hampered by the ingrained idea that shales always reflect deposition in quiet water. "Aside from this it is hard to say anything definitive about the environment of a shale since most environments have periods and places of quiet water deposition"[4] (cf. Potter et al., 2005, p. 75). Shales have been studied mostly for the unusual geochemical signals they carry, and the composition of shales is often only known in terms of organic matter and elemental content. However, we should not attempt to explain the origin of black shales by focusing solely on explanations for their high organic matter content, which typically constitutes but a few percent of the total rock volume.

It is now widely recognized that black shales show significant compositional and textural variability on a variety of scales (Aplin and Macquaker, 2011; Ghadeer and Macquaker, 2012; Lobza and Schieber, 1999; Macquaker et al., 2007, 2010b; O'Brien, 1996; Plint et al., 2012; Schieber, 1994, 1999; Schieber and Riciputi, 2004; Schieber et al., 2010; Trabucho-Alexandre et al., 2011, 2012a), which reflects the diverse and dynamic nature of the processes and environments behind their formation. The composition, textures (viz. grain size and fabric), structures, and fossil content of shales depend on the physical, chemical, and biological processes responsible for their deposition and on their depositional environments. Patterns of vertical and lateral variability, which can be observed on a variety of scales in shale successions, preserve a record of the evolving

---

[4] "SEPM Strata, Terminology List," accessed November 2, 2013. http://www.sepmstrata.org/TerminologyList.aspx?search=shale

depositional environments in which they formed. Indeed, shales can be deposited by a variety of processes in almost any environment (e.g., Schieber, 2011; Stow et al., 2001; Trabucho-Alexandre et al., 2012b).

2.2 PROCESSES BEHIND THE DEPOSITION OF ORGANIC MATTER-RICH SHALE

Shales are the end product of the processes that control the production, erosion, transport, deposition, and diagenesis of mud. The composition of shales is a product of the interaction of three key variables: sediment input, removal (or destruction), and mixing (or dilution). Diagenetic processes act on the sediment and result in changes to its composition and/or texture. Although organic enrichment of shales is always a function of the same basic variables, which combinations will yield organic matter-rich sediments depend on depositional environment.

2.2.1 Processes Behind the Transport and Deposition of Mud

Mud may be transported to its final resting place by gravitational settling, by advective processes, that is, mud transport resulting from net horizontal water movement, and by sediment gravity flows, that is, mud transport by density currents for which excess density is produced by the presence of suspended solids.

The deposition of particles smaller than about 10 µm is controlled by gravitational settling, that is, settling from suspension under the force of gravity toward the depositional interface. For particles larger than about 10 µm, depositional processes are dominated by shear stress at the depositional interface, and silt has a bedform succession similar to that of sand finer than ca. 80 µm (Mantz, 1978; Southard, 1971).

In freshwater, mud is mostly present as individual particles, because the excess negative charge present on the surface of fine mud particles keeps them from flocculating. In paralic and marine environments, aggregates are formed due to changes in the chemical environment, namely an increase in salinity, and due to the activity of organisms, and mud tends to be present as flocs, fecal pellets, pseudofeces, and other organominerallic aggregates (e.g., marine snow). Although salt flocculation is an important mechanism, particularly in environments where water masses of different salinities mix, biogenic aggregation is probably the most important process controlling the behavior of mud in paralic and shallow marine environments (Eisma, 1986; Pryor, 1975). Despite their lower density, the behavior of aggregates is comparable to silt- and sand-sized particles. Consequently, mud in paralic and marine environments settles relatively quickly through the water column and can be transported as bedload over a wide range of flow velocities (e.g., Richter, 1926; Schieber et al., 2007; Trusheim, 1929; van Straaten, 1951).

Mud can also accumulate in the presence of current velocities that exceed the threshold of mud erosion if suspended sediment concentrations exceed $1\,g\,l^{-1}$. Fluid mud, which is a highly concentrated aqueous suspension of mud in which settling is hindered by particle proximity, forms when the amount of mud entering the near-bed layer is greater than the dewatering rate of the high density suspension (McAnally et al., 2007). Fluid mud is a common feature of river, lake, estuarine, and shelf environments in which water is laden with fine-grained sediment. Along coastlines with abundant mud supply, fluid mud dampens waves (Wells and Coleman, 1978) and allows mud deposition in relatively high energy environments (Rine and Ginsburg, 1985). Mud drapes can therefore be formed over significant portions of the tidal cycle, rather than just at slack water; if fluid mud layers persist, mud can accumulate continuously over multiple tidal cycles (MacKay and Dalrymple, 2011). Large volumes of fluid mud can be transported downslope advectively by high energy events across low-gradient shallow marine environments as wave-enhanced sediment gravity flows (e.g., Macquaker et al., 2010b).

The fabric of freshly deposited mud that resulted from the gravitational settling of individual mud particles has a stable subparallel structure with comparatively little water; whereas, the fabric of aggregated mud deposited in the same way is open with a water content in excess of 90% by volume (Hedberg, 1936; Migniot, 1968; Terwindt and Breusers, 1972). The density and shear strength of aggregated mud deposits are therefore lower. However, if aggregated mud is transported as bedload to its final resting place, the deposits are denser, less porous, and contain less water than the deposits produced by gravitational settling of aggregated mud (J. Schieber, personal communication). The fabric of settling mud particles and of freshly deposited mud is difficult to observe directly, and burial of mud tends to obscure initial sedimentary fabrics (Allen, 1985, p. 144, fig. 8.5) unless there is early cementation of the sediment. Flocs are crushed and rearranged by the accumulating overburden (Migniot, 1968), while water loss and compactional processes normally destroy the pelletal character of fecal pellet mud (Pryor, 1975). For this reason, fine-grained sediments sampled from recent or fossil deposits and analyzed in the laboratory may show a textural composition quite different from the original or *in situ* material (e.g., de Boer, 1998).

2.2.2 Production, Destruction, and Dilution: The Many Roads to *Black* Shale

Production is the synthesis of organic compounds from nutrients, carbon dioxide, and water by terrestrial and aquatic organisms through photo- and chemosynthesis, that is, using (sun) light or the oxidation of inorganic molecules as an energy source, respectively. Primary organic production from photo- and/or chemosynthesis is the first and foremost prerequisite to generate an organic matter-rich sediment. In its broader sense, production also refers to the biomineralization processes by which aquatic organisms produce their skeletons. The relationship between organic productivity and biomineral productivity is typically nonlinear; a possible reason for this may be the effect of dissolution.

Organic matter in continental environments is terrigenous, that is, produced by land-dwelling organisms, whereas in marine sediments organic matter may be either of marine or terrestrial origin. On land, almost all primary production since the Devonian is by vascular plants. Land-derived organic matter, highly degraded and nitrogen poor, is brought into the ocean by rivers in dissolved and particulate forms. Most terrestrial particulate organic matter, for example, pollen, plant debris, and charcoal, is deposited in nearshore environments, whereas the dissolved component escapes removal and is carried out into the ocean. The bulk of dissolved organic carbon in seawater is marine. Land-derived dissolved organic matter entering the open ocean must therefore be extensively oxidized back to CO_2 (Emerson and Hedges, 1988; Hedges and Keil, 1995; Hedges et al., 1997). Marine organic matter is produced largely by phytoplankton, for example, cyanobacteria, diatoms, and dinoflagellates, in the photic zone. Productivity on the continental margin is favored by a combination of fluvial, eolian, and offshore nutrient supplies. Nutrients carried by rivers to the ocean are consumed quickly within and immediately off river mouths (Piper and Calvert, 2009). Nutrients supplied from the base of the thermocline by mixing and by upwelling are the main source of nutrients in highly productive areas of the ocean, and fuel about three-quarters of the new production in the ocean (Eppley and Peterson, 1979). Although coastal regions have higher rates of photosynthesis than the open ocean, most (ca. 80%) of the total photosynthetic production occurs in the open ocean (Emerson and Hedges, 1988), which accounts for about 90% of the total sea surface. However, export production, that is, the amount of organic matter that is not remineralized before it leaves the photic zone and sinks to the seafloor, is lower in the open ocean. At present, for example, most export production is concentrated along the relatively shallow continental margins (Laws et al., 2000; Walsh, 1991), where up to 90% of organic carbon burial takes place (Berner, 1982; Hedges and Keil, 1995).

Although primary productivity is important (e.g., Pedersen and Calvert, 1990), it is not sufficient by itself. In the modern Southern Ocean, for example, areas associated with oceanic divergence are characterized by high primary productivity; yet, sediments below these fertile surface waters are organic matter lean (Demaison, 1991). This is because the water column is well oxygenated, largely due to very low water temperatures, and because silica-secreting zoöplankton produce large amounts of skeletal debris, viz. frustules, which results in significant dilution of organic matter. To generate an organic matter rich sediment, the destruction of organic matter must be minimized. Destruction refers to the remineralization of organic matter by organisms (mainly bacteria) and oxidation in the water column. These processes can continue at the sediment–water interface and to some depth within the sediment column. Destruction also includes the dissolution of skeletal material in the water column. Dissolution of calcareous skeletal material increases with water depth and with an increase in supply of organic matter (Emerson and Archer, 1990), which lowers the pH of sediment interstitial waters unless sulfate-reducing conditions in the sediment prevail (Morse and Mackenzie, 1990). Dissolution of siliceous skeletal material occurs throughout the water column, but is more intense in the warmer surface layers of the ocean and shortly after deposition (Berger, 1974). A minimization of the destruction of organic matter can be achieved by lowering dissolved oxygen content in the water column, by making the export path and/or transit times shorter, and/or by reducing sediment exposure time to bottom water after reaching the sediment–water interface.

Oxygen levels in seawater depend on how much oxygen seawater can hold and on oxygen supply and demand. Oxygen levels are lower in warm climates due to the reduced solubility of oxygen in warmer water. This is the case in a geographic sense, that is, sea surface water at lower latitudes contains less oxygen, and in a geologic sense, that is, seawater during hot/greenhouse intervals contained less oxygen than at present. Dysoxia, and—depending on the frequency, intensity, and depth of mixing—anoxia, is favored in basins whose physiography (e.g., oxbow lakes and silled marine basins) and/or water column thermohaline structure (e.g., lakes) result in the stagnation of (part of) its water column. Dysoxia develops in response to runoff of nutrient-rich water from rivers to lakes and oceans, and upwelling of nutrients and consequent enhanced surface productivity in lakes (overturning) and oceans. Oxygen depletion is more dynamic than commonly assumed and depends on the interaction between lake/ocean circulation, biological activity, and nutrient distribution (Meyer and Kump, 2008). Biochemical processes are ultimately responsible for the consumption of oxygen, but ocean circulation is responsible for the distribution of dysoxic and anoxic water masses in the ocean (Wyrtki, 1962). Oxygen depletion may be either local or regional, and it may be seasonal or permanent (e.g., Lake Tanganyika). It has been suggested that the preservation of organic matter in mid-Cretaceous marine sediments was favored by decreased oxygen supply to deep water as a consequence of sluggish ocean circulation (e.g., Bralower and Thierstein, 1984; Erbacher et al., 2001). In a stagnant ocean, the supply of nutrients to the photic zone would not be sufficient to sustain the elevated primary productivity required to support high oxygen demand in deep water (e.g., Hotinski et al., 2001). Whereas a sluggish ocean would result in a significantly decreased oxygen supply to deep water, that reduction would be balanced by a reduction in oxygen demand (Meyer and Kump, 2008). Other studies have suggested that a more vigorous circulation resulted in a more productive mid-Cretaceous ocean (Hay and Floegel, 2012; Southam et al., 1982; Topper et al., 2011; Trabucho-Alexandre et al., 2010; Wilson and Norris, 2001).

The Black Sea is often used as a model for ancient sluggish or stagnant oceans. However, evidence suggests that this enclosed basin is not properly described as stagnant. Radiocarbon dating indicates a mean residence time of 935 years for deep Black Sea water, whereas mass balance calculations indicate a shorter residence time of 475 years (Östlund, 1974). Brewer and Spencer (1974) calculated a present-day upward advective velocity of $0.5\,m\,a^{-1}$ in the interior of the Black Sea (in Degens and Ross, 1974). These results suggest that the rates of vertical exchange in the Black Sea are of the same order of magnitude as those in the modern open ocean, and that euxinia, which refers to the presence of free hydrogen sulfide in the water column, in the Black Sea represents a dynamic balance (Southam et al., 1982).

In the open ocean, the fraction of the organic matter produced in surface waters that reaches the seafloor is inversely proportional to water depth (Hedges and Keil, 1995; Müller and Suess, 1979; Suess, 1980). All other variables being equal, the preservation of organic matter is favored where the seafloor is relatively shallow, namely, on the continental shelf and upper slope and on the top and flanks of seamounts. Long transit times through mildly oxidizing water (e.g., ca. $3\,ml\,l^{-1}\,O_2$ in the deep modern North Pacific, Southam et al., 1982) and slow sedimentation rates (ca. $2\,m\,Myr^{-1}$) are sufficient to result in the deposition of organic matter lean, red/brown pelagic clay on the deep ocean floor. A significant part of the vertical flux of particulate organic matter from the photic zone is in the form of organomineralic aggregates. Because they are larger than their constituent mud particles, these aggregates settle much faster through the water column, and transit times to the seafloor are within weeks. The preservation of organic matter is thus greatly favored.

Dilution is a consequence of the mixing of siliciclastic, skeletal, and organic material, because the composition of a sediment is a zero-sum game; an increase in one component must be accompanied by a relative decrease in the others. The input of siliciclastic material, which is a key control in the composition of a shale, may be included in dilution. Up to a certain point, an increase in the input of siliciclastic and/or skeletal material, that is, an increase in sediment accumulation rates which leads to relatively rapid burial, favors the preservation of organic matter. Indeed, the preservation of organic matter, particularly in oxidizing environments, is favored by sedimentation processes that deliver large quantities of sediment to the seafloor, including metabolizable organic material, in a short period of time (e.g., Degens et al., 1986; Ghadeer and Macquaker, 2012; Macquaker et al., 2010a). Moreover, large fluxes of metabolizable organic matter favor processes of natural vulcanization, which lead to the creation of resistant geobiopolymers (Lallier-Vergès et al., 1997; Sinninghe Damsté et al., 1989). However, too much of any component will "mask" others, especially if they are present in low absolute amounts in the sediment, as is typically the case for organic matter. Dilution of organic matter by inorganic material, terrigenous and/or biogenous (skeletal), is an important control in the accumulation of organic matter in sediments (Bohacs et al., 2005; Tyson, 2001). High dilution rates can result in organic matter-lean sediments even under regions of high surface productivity. In deltaic settings, for example, organic carbon contents are feeble where elevated sedimentation rates of terrigenous sediment dilute the organic component of sediments (Dow, 1978).

Along continental margins, the calcite compensation depth (CCD) is raised due to higher primary productivity and consequent respiration of organic matter in sediments, which releases metabolic CO_2 and thus increases carbonate dissolution (Berger, 1974; Seibold and Berger, 1996). As a result, dilution of organic matter by calcareous skeletal debris is minimized. On the other hand, dilution can also be too low. The fraction of organic carbon that reaches the basin floor is a function of water depth and of bulk sedimentation rate (Müller and Suess, 1979), but more than 90% of the organic matter that does reach the seafloor is nonetheless remineralized (Emerson and Hedges, 1988). Where sedimentation rates are low, the preservation of organic matter is reduced because the sediments are kept within the mixed sediment layer for too long, where they are exposed to active microbial reworking and oxidants in pore waters, as well as erosion and transport (Bohacs et al., 2005, and references therein). In condensed sequences in the Mesozoic of Alabama, United States, for example, organic matter was not preserved probably due to low sedimentation rates (Mancini et al., 1993). In conclusion, there is not just a single combination of variables that will yield organic matter-rich sediments, but optimum organic enrichment occurs where production is maximized, destruction minimized, and dilution optimized (Bohacs et al., 2000; Tyson, 2001).

2.3 STRATIGRAPHIC DISTRIBUTION OF ORGANIC MATTER-RICH SHALES

Although shales are a ubiquitous component of the stratigraphic record, the distribution of black shales in the Phanerozoic is predominantly limited to six stratigraphic intervals (Fig. 2.2), which together represent about one-third of Phanerozoic time (e.g., Bois et al., 1982; Klemme and Ulmishek, 1991; North, 1979; Tissot, 1979). The petroleum source rocks in these

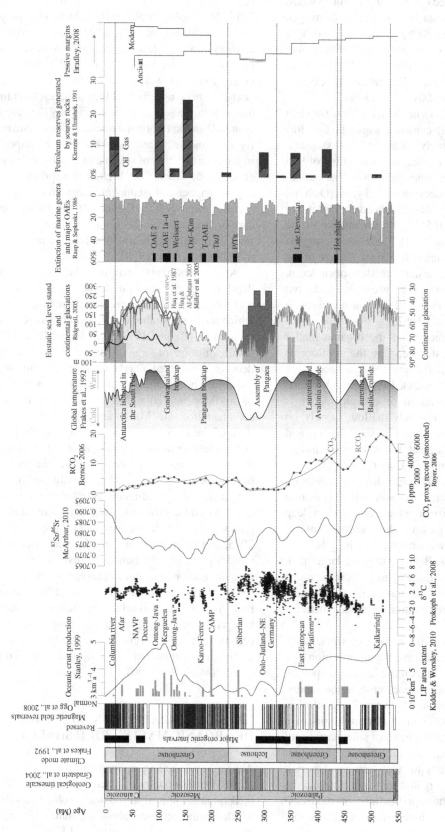

FIGURE 2.2 Phanerozoic patterns of various indicators of global change. From left to right: Phanerozoic geological timescale (Gradstein et al., 2004), climate mode (Frakes et al., 1992), major orogenic intervals, magnetic field reversals (Gradstein et al., 2004), oceanic crust production (Stanley, 1999), and large igneous province areal extent (Kidder and Worsley, 2010), carbon isotope curve (Prokoph et al., 2008), strontium isotope curve (McArthur, 2010), RCO$_2$ and CO$_2$ proxy record (Berner, 2006; Royer, 2006), global temperature, eustatic sea-level stand (Haq and Al-Qahtani, 2005; Haq et al., 1987; Miller et al., 2005), continental glaciations (Ridgwell, 2005), extinction of marine genera and major oceanic anoxic events (OAEs) (Raup and Sepkoski, 1986), petroleum reserves generated by source rocks (Klemme and Ulmishek, 1991), and passive margin extent (Bradley, 2008).

intervals have provided more than 90% of the world's known conventional hydrocarbon reserves (Klemme and Ulmishek, 1991; Tissot, 1979). Overmature oil-prone source rocks[5] deposited during these six stratigraphic intervals also function as reservoirs in the increasingly explored and exploited unconventional shale gas plays.

Correlations between geologic anomalies and black shale deposition (Condie, 2004; Kerr, 1998; Larson, 1991; Sheridan, 1987; Sinton and Duncan, 1997) may explain the stratigraphic clustering of black shales in specific stratigraphic intervals of the Phanerozoic (Klemme and Ulmishek, 1991; Tissot, 1979; Trabucho-Alexandre et al., 2012b). Phanerozoic intervals characterized by enhanced tectonic activity, namely, supercontinent breakup, ocean basin formation, and large igneous province emplacement, are associated with greenhouse climates, eustatic highstands, vigorous ocean circulation, and abundant nutrients in seawater (Fig. 2.2). Warm and humid greenhouse climates support abundant life, such as highly productive tropical rain forests on land, reef communities on the shelf, whose area is greatly expanded during eustatic highstands, and abundant plankton in the ocean. Abundant nutrients in seawater are a product of intense seafloor spreading activity and large igneous province emplacement, increased circulation of deep, nutrient-rich water, and an enhanced hydrologic cycle (e.g., Larson, 1991; Sinton and Duncan, 1997; Trabucho-Alexandre et al., 2010).

Tectonic processes lead to changes in the geography of the earth and to the evolution of depositional environments through time (Chamberlin, 1909; Scotese, 2004; Wilson, 1968). The geographic distribution of Phanerozoic black shales is largely independent of latitude but instead related to the distribution of continental masses (Irving et al., 1974; Klemme and Ulmishek, 1991). The distribution of continents (and ocean basins) controls the position of landmasses relative to climate belts, the opening and closure of gateways (i.e., basin connectivity), and hence ocean circulation. Ocean circulation affects seawater temperature, oxygenation, and nutrient content. The preservation of organic matter on the seafloor of marine basins appears to be aided by a latitudinal position of continents that obstructs meridional ocean circulation. This position, typical of the Mesozoic, inhibits the formation and spread of cold, oxygenated, high-latitude deep water which promotes the destruction of organic matter in deeper water. Global climate also exerts an important control in this regard, because bottom water cannot be colder than the coldest surface water; bottom water during greenhouse climates will therefore contain lower initial oxygen content than present-day bottom water (Berger, 1974).

It has long been recognized that tectonic processes have the potential to affect global climate (Chamberlin, 1897), over both extremely short (e.g., Storey et al., 2007) and long (e.g., Raymo and Ruddiman, 1992) timescales. Climate, which is also forced by subtle cyclic variations in the earth's axis and orbit (de Boer and Smith, 1994), plays a fundamental role in the evolution of sedimentary environments on earth. Seafloor spreading and mountain building, both driven by tectonic processes, control earth's climate over long timescales. These two processes lead to changes in CO_2 input by volcanism and dissociation of subducted limestones, and to changes in CO_2 removal by weathering of silicates and organic matter burial (Berner, 1991). Volcanism related to plate tectonic processes can also drive rapid climate change both directly due to faster seafloor spreading rates (Berner et al., 1983), increasing length of oceanic ridges, and extrusion of large igneous provinces (e.g., Kerr, 1998; Sinton and Duncan, 1997), and indirectly due to greenhouse gas generation as a consequence of increased seawater temperatures (e.g., Dickens et al., 1995; Hesselbo et al., 2000) and contact metamorphism (e.g., McElwain et al., 2005; Storey et al., 2007).

Relative sea level, which depends both on global and local tectonics and on climate, controls the size and distribution of paralic and shallow marine environments, where most ancient black shales were deposited (Arthur and Sageman, 2005; Hedges and Keil, 1995; Laws et al., 2000; Walsh, 1991; Wignall, 1991). It is therefore unsurprising that intervals of relatively widespread black shale deposition should coincide with eustatic highstands. High sea levels favor the deposition of black shales (Duval et al., 1998) by expanding sunlit shallow marine environments where primary productivity is high and export paths short. Moreover, during transgressions and early highstands, coarser grained siliciclastic material is trapped in nearshore environments, such as, estuaries, reducing excessive dilution of organic matter on the shelf. The composition, including organic matter type and content, and texture of marine sediments are thus a function of depositional environment and of allogenic forcing mechanisms acting on them.

2.4 GEOGRAPHIC DISTRIBUTION OF ORGANIC MATTER-RICH SHALES

2.4.1 Background

In the early scientific literature, the main debate concerning the origin of black shales focused on whether they had been deposited in shallow or deep water (Cluff, 1981). Hard (1931), for example, interpreted the Devonian black shales of New York as having been deposited in shallow water under toxic and saline conditions, whereas Clarke (1904)

---

[5] Oil-prone kerogens have higher capacities for hydrocarbon generation per unit organic carbon than gas-prone kerogens. Although gas-prone source rocks generate large amounts of gas at high maturity, late stage gas generation and cracking of residual oil/bitumen in oil-prone source rocks can account for more gas generation than gas-prone source rocks (Dembicki, 2013).

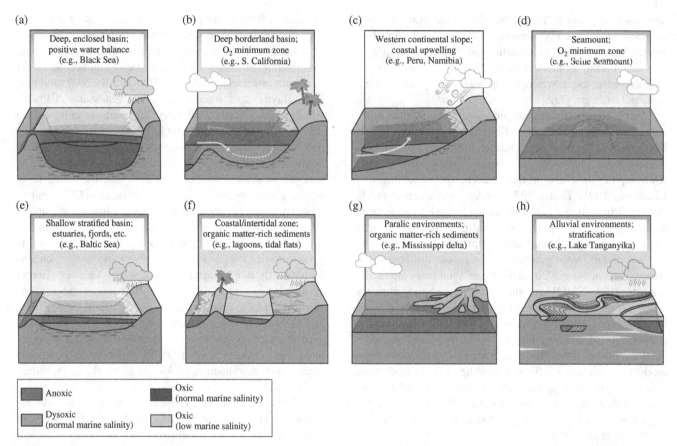

FIGURE 2.3 Summary of the environments of accumulation of organic matter showing an idealized basin physiography, water mass distribution and properties, and prevalent climate (rain cloud indicates positive water balance, cloud streamers indicate offshore winds). This figure is an adaptation and expansion of figure 1 in Arthur and Sageman (1994, p. 507).

favored a deepwater environment similar to the bottom of the modern Black Sea. Pettijohn (1975, p. 284) summarized this debate as follows: "[t]he origin of black shales has been much debated. *Certainly they were deposited under anaerobic conditions.* [emphasis added] How such conditions were achieved is less certain. [...] Some writers contend that black shales were deep-marine (geosynclinal) sediments; others have postulated comparatively shallow waters, either lagoonal or marine."

It is obvious from Pettijohn's remark that reducing environments, whether shallow or deep, were considered the key control behind the deposition of black, organic matter-rich mud. The idea that black shales are the product of sedimentation in reducing environments was reinforced early on by the study of black muds in the Black Sea (Pompeckj, 1901; Schuchert, 1915) and in Norwegian fjords (Strøm, 1939). The Black Sea (Fig. 2.3a) has been used extensively as a model for the deposition of ancient epicontinental and open ocean black shales. However, the presence of a halocline in the Holocene Black Sea and the fact that it is significantly deeper (ca. 2000m) than ancient epicontinental seas (ca. 100m), and therefore characterized by a different depth-to-width ratio, preclude its use as an analog for shelf or open ocean sedimentation in the geologic record. Tyson (2005, p. 29) summed up this idea by stating that the "Black Sea is [...] a freak of paleogeography and has very specific circumstances that are unlikely to be common in the geological record." Although some authors argued for an open marine origin for black shales, the view that black shales were deposited in restricted basins conducive to strongly reducing conditions, and for which the Black Sea may be a good analog, was largely prevalent. Twenhofel (1939), for example, argued for an open marine origin for ancient black shales in general, but favored the Black Sea as an analog for the Paleozoic shales of northwest Europe and the Appalachian basins of North America. Fleming and Revelle (1939), who discussed the role of oceanographic processes on dissolved oxygen distribution in the water column, also gave the Black Sea and the borderland basins of Southern California as examples of modern environments of black mud deposition (Fig. 2.3a and b), and thus emphasized the role of sill depth in controlling the rate of renewal of bottom water and oxygen replenishment.

While early authors placed great emphasis on the role of anoxia, some noted the importance of primary productivity in generating organic matter-rich sediments. Goldman

(1924), for example, suggested that the rate of supply of organic matter was important in oxygenated settings. Trask (1932) analyzed a very large amount of samples and concluded that upwelling zones were favorable settings for the deposition of organic matter-rich sediments. Brongersma-Sanders (1971) also discussed in detail the effects of upwelling on sediment composition. Importantly, Brongersma-Sanders noted that upwelling is a countercurrent system that creates a nutrient trap. This trap leads to high fertility of a basin or coast and, where the subsurface water ascends toward the photic zone, to high productivity. Parrish (1987) predicted the geographic distribution of ancient upwelling zones and compared that distribution with the distribution of organic matter-rich rocks. She concluded that as many as half the world's black shales may have been deposited in upwelling zones.

In addition to the debate concerning the physiography of ancient environments of black mud accumulation, which is probably one of the longest running controversies in geology, another intense debate arose, this time concerned whether unusually high primary productivity in the photic zone or unusual chemical conditions in the water column, namely, anoxia, provide the first-order control on the accumulation of organic matter-rich sediments in the ocean (Demaison, 1991; Pedersen and Calvert, 1990). As a result of this debate, models of black shale deposition are traditionally divided into two end-member types: one of enhanced supply and the other of enhanced preservation of organic matter. More recently, however, some authors recognized the interdependent roles of primary productivity, microbial metabolism, and sedimentation rates (e.g., Bohacs et al., 2000, 2005; Sageman et al., 2003; Tyson, 2005).

Although Van Waterschoot van der Gracht (1931) proposed a link between changes in ocean circulation on a global scale and the deposition of black shales (cf. Chamberlin, 1906), most early authors thought of black shales as the product of local processes. Indeed, it was not until Cretaceous black shales were recovered in a number of Deep Sea Drilling Project (DSDP) sites in the 1970s that it became widely recognized that basin physiography was not a sufficient explanation for some ancient black shale successions. The discovery of widespread organic matter-rich horizons in the deep sea represented a breakup with the notion that ancient black shales were the product of local conditions in marginal, restricted basins. Bernoulli (1972) recognized the similarities between Tethyan Cretaceous sediments exposed on land in the Mediterranean region and sediments recovered by drilling in the North Atlantic, and suggested that black shale horizons now exposed on land and horizons recovered by drilling are coeval. The discovery of Cretaceous black shales of the same age in the Pacific greatly extended the geographic range of those horizons, and led to the suggestion that the deposition of black shales during the Early Cretaceous might have been a worldwide oceanographic phenomenon (Jackson and Schlanger, 1976, p. 925). As a result, Schlanger and Jenkyns (1976) proposed that the occurrence of black shale horizons globally was due to the expansion of the oxygen minimum layer in the ocean as a consequence of the Late Cretaceous transgression and a reduced supply of oxygen to deep water due to an equable climate (Fig. 2.4), the so-called oceanic anoxic events (OAEs). The term OAE is unfortunate because it implies ocean-wide anoxia, although it has been noted that this is not the spirit of the term (Arthur et al., 1990). Indeed, the original concept (Fig. 2.4) included several environments of black shale deposition. At first, the idea that the deep sea could become anoxic was rejected by geochemists

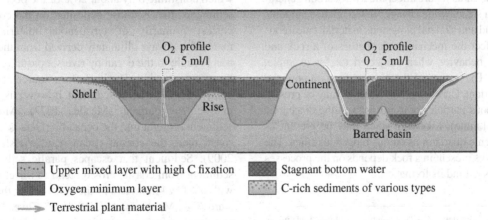

FIGURE 2.4 Ocean stratification during an oceanic anoxic event as proposed by Schlanger and Jenkyns in 1976 (their figure 2). The oxygen minimum layer is expanded and intensified. The shoaling of the upper boundary of the oxygen minimum layer translates into a geographic expansion of shallow seafloor impinged on by the oxygen minimum. The sinking of its lower boundary results in the seafloor at the top and flanks of oceanic rises being impinged by the oxygen minimum layer. The concept included a barred basin setting in which abundant terrestrial plant debris accumulated "in their early opening stages by rivers and turbidity currents."

(Broecker, 1969). However, the presence of "widespread" organic matter-rich horizons in the deep sea is sometimes invoked as evidence for episodes of ocean-wide stagnation and/or anoxia, and much modeling effort has been put into creating increasingly complex models that attempt to recreate global conditions that explain all occurrences of organic matter-rich sediments in ancient oceans. While certain stratigraphic intervals are characterized by frequent and/or widespread black shale horizons, the correlation of individual layers is almost always questionable and the petrologic characteristics of the black shales varied. This suggests that multiple processes are behind the deposition of not exactly coeval organic matter-rich sediments during OAEs (Hay, 1988; Trabucho-Alexandre et al., 2011); black shales are the product of both local and global conditions (cf. Trabucho-Alexandre, 2011). The danger in creating "fully detailed models of complex systems is ending up with two things you don't understand—the system you started with, and your model of it" (Paola and Leeder, 2011).

2.4.2 Controls on the Geographic Distribution of Black Shales

The petrologic characteristics and widespread geographic distribution of black shales suggest that processes rather than environments control their accumulation.[6] For this reason, a discussion concerning their geographic distribution should focus on the processes that result in the deposition of organic matter-rich mud in each environment. In particular, we are interested in linking the petrologic characteristics of shales, that is, their varying composition and texture, with depositional process and environment. This is important because regional changes in shale composition and texture due to paleoöceanographic and paleogeographic controls affect, among other things, the economic potential of a rock. Changes in organic matter content *and* in its nature affect the hydrocarbon generation potential of a shale, while changes in the relative amounts of siliciclastic mud and skeletal biogenous material, calcareous and siliceous, affect the mechanical properties of a rock and consequently its behavior when subjected to, for example, hydraulic stress. The nature of the phases that compose the rock is also important. Carbonate, for example, may be present as porous biogenous particles or as porosity-filling diagenetic crystals (e.g., dolomite). Likewise, silica may be present as detrital or authigenic quartz or as porous biogenous silica. The nature of the phases present in a rock depends on the processes and environments behind its formation.

In continental environments, the composition of shales is largely dominated by terrigenous material. In terms of composition and texture, shales deposited in continental environments are typically shales *sensu stricto*, that is, laminated and/or fissile organic matter-rich *siliciclastic* mudstones, and their organic matter is terrigenous. In the marine environment, however, sediments only rarely come from a single source (Fig. 2.1). Most marine sediments are a mixture of biogenous and terrigenous particles of various grain sizes, with an additional hydrogenous (authigenic) and/or (very minor) cosmogenous component. Biogenous debris, which may be either calcareous or siliceous, may form a significant proportion, if not the majority, of the inorganic fraction of a shale. The composition of marine shales follows a general pattern that is related to basin physiography, namely, water depth, and ocean circulation (Fig. 2.5).

The relationship between depositional setting and the texture of shales is more complicated. For example, although we often think that sediment grain size is a function of distance to shore, this is typically not the case, and there are many examples of fine-grained shores and relatively coarse-grained deepwater deposits (e.g., Rine and Ginsburg, 1985; Stow, 1985b). In neritic environments below effective wave base, which are most affected by variations in terrigenous input and in relative sea level, the variation in seafloor texture is the most unpredictable, because a significant fraction of the seafloor is covered in either relict or palimpsest sediment (Emery, 1968a; Shepard, 1932). The texture of shales is largely independent of environment and therefore more difficult to predict in terms of its geographic distribution than composition.

The sediments of continental margins are different both in quantity and quality from those on deeper seafloor. Almost 90% of the total volume of all marine sediment is associated with continental margins, that is, shelves, slopes, and rises, which constitute only about 20% of the ocean's area. Neritic sediments, that is, those deposited on the continental shelf, consist primarily of terrigenous material. Terrigenous material is always ultimately derived from the continent, and it is brought to the ocean by rivers, coastal erosion, and, to a lesser extent, wind. The immediate source of the terrigenous component of a marine sediment, however, is often within the marine environment (Meade, 1972). Most terrigenous sediment brought into the ocean by rivers is deposited where rivers meet the coastal ocean (e.g., Walsh and Nittrouer, 2009). Sediment that escapes paralic sediment traps and sediment from coastal erosion tend to travel along the shore within a few kilometers of the coast rather than moving seaward (e.g., Manheim et al., 1970; McCave, 1972). Sediment resuspended from the shelf bottom and sediment transported laterally from offshore constitute the main source of suspended matter on the shelf away from the mouths of large rivers. Many ancient black shales were deposited on broad continental shelves at times when sea levels were much

---

[6] An environment has been defined as "the complex of physical, chemical, and biological conditions under which a sediment accumulates," (Krumbein and Sloss, 1963, p. 234) and as "a spatial unit in which external physical, chemical, and biological conditions and influences affecting the development of a sediment are sufficiently constant to form a characteristic deposit" (Shepard and Moore, 1955, p. 1488).

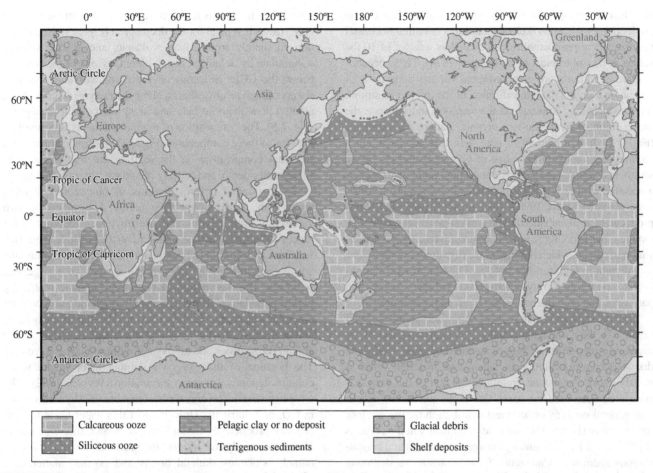

FIGURE 2.5 The general pattern of sediment cover of the seafloor. This pattern has been known more or less in its present state since the work of Murray and Renard (1891) following the voyage of H.M.S. *Challenger*. The main sediment facies are pelagic clay and calcareous ooze. Pelagic clay, also known as red clay, is typical for the deep seafloor, whereas the calcareous facies outlines oceanic rises and platforms. Biosiliceous deposits, characteristic for the seafloor beneath areas of high fertility, are superimposed on the previously mentioned topographically controlled dichotomy. The sediment that results in pelagic clay deposits, which is mostly eolian and cosmogenous dust associated with hydrogenous minerals, is present everywhere in the ocean, but it is masked whenever another component is present, because its mass accumulation rates are so low. Terrigenous sediment is present along the continental margins. In front of large rivers and submarine canyons, terrigenous sediment can penetrate the ocean basin to a considerable distance from the shelf. Even though this map shows sharp boundaries between well-defined facies, the seafloor is actually covered by a complex pattern of overlapping mixtures of the various components of fine-grained marine sediments (Fig. 2.1).

higher than today (Fig. 2.2). Terrigenous sediment is efficiently trapped in nearshore environments during highstands, and thus large areas of those shelves must have been starved of new terrigenous sediment input; their terrigenous mud was probably largely derived from longshore and relict offshore sources. Nevertheless, unlike black shales recovered in the deep sea by ocean drilling, which are never more than tens of centimeters in thickness, epicontinental black shales are usually meters to tens of meters thick, for example, the Devonian Chattanooga Shale (ca. 9 m in central Tennessee) and the Pliensbachian–Toarcian shales of northwest Europe (>100 m in northeast England). The thickness of these successions does not necessarily equate to high terrigenous input, because the successions often represent very long periods of time represented in the rock record by gaps and by mud beds that were deposited relatively quickly (e.g., Baird, 1976; Macquaker and Howell, 1999; Schieber, 1994, 2003; Trabucho-Alexandre, 2014). The gaps reflect periods during which no sediment accumulated, either because there was no sediment to accumulate, or because sediment was being removed from one area of the shelf to be deposited elsewhere on the shelf or in deeper water.

In addition to terrigenous material, neritic sediments almost always contain biogenous material (Fig. 2.1), which is a product of high fertility of surface water along continental margins. This is due to a high position or breakdown of the thermocline, which normally functions as a barrier to nutrient transport between deep and shallow, sunlit water. Upward mixing of nutrient-rich water containing dissolved silica leads to high productivity in the photic zone and to a high proportion of silica-secreting plankton in the upwelling biota. The production of siliceous skeletal material is maximal in coastal regions, where productivities can exceed by a factor of 10 the values in the subtropical gyres (Berger, 1974). These oceanographic conditions are also favorable for the deposition of organic matter and the intensification of the oxygen minimum layer, which may impinge on the seafloor and enhance the preservation of hydrogen-rich organic matter-rich sediments (Fig. 2.4). It is important to note here that anoxia is not a requirement for the preservation of organic matter in marine sediments; rather, under such conditions, more hydrogen is associated with carbon in the organic matter (Demaison, 1991; Pedersen and Calvert, 1990), which means that the shales thus produced have an enhanced hydrocarbon-generating potential. On the other hand, these conditions are adverse for the preservation of calcite, which minimizes the importance of carbonate as a diluent in high fertility settings. Dilution in such settings depends on the proximity of terrigenous sediment sources and pathways, and on biogenous silica input and dissolution.

Oceanic sediments, that is, those deposited beyond the shelf break, on the continental margin generally consist of an increasing proportion of biogenous material away from land due to decreasing dilution by siliciclastic material, which is mostly trapped nearshore. Indeed, about half of the deep seafloor area is covered by oozes, that is, by planktonic debris. The general outlines of sediment distribution on the seafloor are relatively simple. The main facies boundary in the ocean is the CCD, that is, the boundary between calcareous and noncalcareous sediments. Calcareous facies characterize shallower oceanic environments and submarine highs, whereas the "red clay" facies is typical for the deep ocean (Fig. 2.5). Red clays, derived mainly from eolian, volcanic, and cosmic sources accumulate by default in distal, barren regions of the seafloor below the CCD. Siliceous ooze accumulates under surface waters of high fertility, that is, along the margins of continents, along a periequatorial belt, and along the polar front regions (Fig. 2.5). The composition of marine sediments is normally a mixture of these components (Fig. 2.1). The main controls on sediment composition on the seafloor are thus distance to shore, water depth, and fertility of surface water (Fig. 2.6).

As mentioned before, the flux of biogenous sediment through the water column is mainly determined by two variables: productivity and destruction. Together with processes that redistribute sediment on the seafloor, these two variables also control the nature and distribution of sediments on the seafloor away from point sources of terrigenous sediment. The destruction of planktonic organic matter by bacterial oxidation during its transit through the water column dominates at depths of 300–1500 m. The supply of organic matter depresses oxygen content due to decay in deep water and on the seafloor and an oxygen minimum layer develops. The position of the oxygen minimum layer in the water column depends on ocean circulation (Wyrtki, 1962). The oxygen minimum layer is further characterized by a maximum in CO_2 and nutrients. Upwelling of this water leads to high primary productivity in surface waters and, therefore, to an abundant supply of biogenous sediment to the seafloor, namely, siliceous skeletal debris and organic matter. The oxygen minimum below upwelling zones is also more intense. If the rates of organic matter supply are sufficiently high and oxygen levels sufficiently low, the preservation of organic matter in the sediment is favored. Upwelling occurs at divergent oceanic fronts, over submarine topographic highs, and adjacent to continental margins, particularly on the western sides of continental masses. The distribution of ancient open marine black shales has been shown to correspond closely to the distribution of predicted upwelling zones (Parrish, 1982, 1987; Parrish and Curtis, 1982), and the deposition of black shales on the pericontinental shelves of the North Atlantic during OAEs has also been shown to be related to upwelling of nutrient-rich water (e.g., Trabucho-Alexandre et al., 2010).

Nutrient-rich surface waters have sufficient silica available to support the production of siliceous skeletal material. Siliceous ooze is an often forgotten component of marine fine-grained sedimentary rocks, and biogenous silica is typically not differentiated but grouped with clastic silica (e.g., quartz and feldspar) in ternary diagrams reflecting shale composition (e.g., Boak, 2012; Gamero-Diaz et al., 2013; Passey et al., 2010). Because silica-secreting plankton tends to proliferate in settings conducive to high organic productivity, organic matter-rich sediments are often siliceous (Hay, 1988). Many Mesozoic black shales that

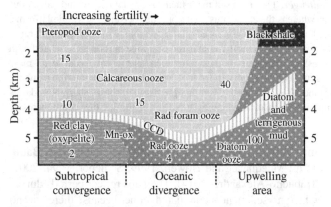

FIGURE 2.6 Distribution of major facies in a depth–fertility frame based on sediment patterns in the Eastern Central Pacific (Berger, 1974). Numbers are typical sedimentation rates in m Ma^{-1}. The CCD is the main facies boundary in deep-sea deposits. The composition of oozes varies according to water depth, fertility, and latitude, and depends on whether surface ocean currents have a tropical or polar origin. Black shales, or sapropelites, according to other authors, occur in high fertility settings on the relatively shallow seafloor of continental margins (shelf and upper slope). Other black shale occurrences are related to oceanic topographic highs (viz. seamounts, rises, etc.) and to sediment gravity flows transporting fine-grained material toward deeper water.

occur around the former Tethys and North Atlantic Oceans and on Indo-Pacific seamounts that occupied a periequatorial position at the time of deposition are actually black cherts. Black, organic matter-rich cherts are not always laminated.

Most of the skeletal debris produced by planktonic organisms never reaches the seafloor, and most of the debris that does reach the seafloor is nonetheless dissolved. This is particularly the case for siliceous skeletal material, because seawater is undersaturated with respect to biogenous silica at all water depths. Silica corrosion is greatest in surface waters due to elevated temperature, whereas carbonate dissolution is greatest at depth (Berger, 1974). There is a negative correlation between silica and carbonate distribution patterns on the seafloor (Fig. 2.5), which has been attributed to opposing chemical requirements for preservation (Correns, 1939): increasing productivity leads to decreasing preservation of calcite and to increasing accumulation of silica.

Seawater is also undersaturated with respect to all forms of calcium carbonate. At a critical level of undersaturation, dissolution rates of calcium carbonate increase rapidly and, below the CCD, which is the level at which the rate of supply of carbonate is balanced by its rate of dissolution, calcium carbonate does not accumulate on the seafloor. High fertility along the equator in the Pacific leads to a depression of the CCD by some 500m (Seibold and Berger, 1996). Increased fertility leads to an increased supply of calcareous skeletal debris to the seafloor in excess of the increased supply of organic matter, because while the calcareous debris transits to the seafloor, organic matter tends to be destroyed on its way down. Along continental margins, however, high productivity raises the CCD (Seibold and Berger, 1996). In fertile areas along continental margins, the high supply of organic matter to the relatively shallow seafloor leads to increased benthic activity and to the development of much CO_2 in sediment pore waters, which produces carbonic acid. For this reason, carbonate debris is dissolved even at depths of a few hundred meters on continental slopes, and pericontinental black shales tend to contain little carbonate.

In the ocean, most biogenous material is produced in the top layers of the water column and arrives at the seafloor as a rain of particles. These particles are mostly aggregates, and a large fraction of these consists of fecal pellets of various sizes and in various stages of disintegration. Aggregates sink faster through the water column than their constituent particles, which would take years to settle to the average depth of the seafloor. Because all biogenous particles are subject to dissolution and/or remineralization in the water column, if it were not for the aggregation mechanism, most would never reach the seafloor. Despite this mechanism, the proportion of primary production that leaves the photic zone (export production) is small. Along the continental margin, higher export factors and shorter distances to the seafloor enhance the burial of organic matter.

Sediment trap studies indicate that for every 100 gC that is produced in the sunlit layer of the ocean, about 30 gC reach the

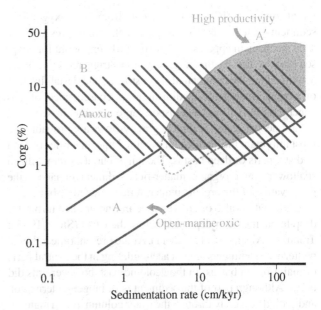

FIGURE 2.7 Correlation between marine organic carbon and sedimentation rates. The three fields, A, A', and B are based on data derived from Neogene and Quaternary sediments deposited in open ocean environments (A), upwelling zones (A'), and anoxic environments (Black Sea and Mediterranean sapropels and modern Black Sea sediments). The stippled area in field A' indicates coastal upwelling and the open area at the lower end of the field indicates equatorial upwelling. Based on figures 2 and 3 in Stein (1986).

seafloor on the shelf and upper slope, but only 1 gC reaches the deep seafloor (Berger et al., 1989). Most of the sinking material is oxidized and remineralized. Deep-sea sediments have low concentrations (C_{org} <0.25%) of organic matter. Below the equator, the concentrations increase slightly, because upwelling along the equator results in a greater supply of organic matter to the seafloor. The highest concentrations of organic matter, however, are linked to coastal upwelling along continental margins (Fig. 2.6). The large supply of organic matter along the continental margin, as well as the physiography of the basin, generates an oxygen minimum zone as a consequence of high oxygen demand and, in the case of restricted basins, low oxygen replenishment (Fig. 2.3), and this may result in anoxic conditions on the seafloor. Under such conditions, the preservation of organic matter is enhanced because anaerobic bacteria are less efficient in destroying organic matter.

The relationship between organic matter content and sedimentation rates in oxic and anoxic Neogene marine sediments gives insight into the processes that control the organic carbon content of marine sediments (Fig. 2.7). There is a lack of correlation between sedimentation rate and C_{org} in anoxic conditions, and a good correlation in oxic conditions (Stein, 1990). In the case of anoxic sediments, if there is sufficient input of organic matter, the organic matter is preserved even when sedimentation rates are low. In the case of oxic sediments, relatively rapid burial is required to preserve a significant portion of the organic matter delivered to the seafloor. Rapid burial removes the organic matter from the oxygen-rich sediment–water interface, thereby enhancing preservation. This mechanism appears to be particularly important for deep-sea black shales. Sediment gravity flows triggered by tectonic activity along the continental shelf, storms, and destabilization of organic matter-rich sediment as a consequence of gas generation in pore water transport large amounts of sediment toward deeper water. Sediment gravity flows are probably the most important mechanism for moving large quantities of mud to distal parts of deep basins. The quick burial of remobilized shallower water organic matter-rich sediment enhances the preservation of the organic matter in fine-grained turbidites.

Carbonate is also often preserved in fine-grained turbidites despite having been deposited below the CCD (Stow, 1985a; Trabucho-Alexandre et al., 2011). At the present time, organic matter-rich sediments are not accumulating in the central parts of major ocean basins. In the geologic past, however, this did occur. Although part of the sediment is of biogenic derivation and settled vertically through the water column, much material is redeposited (e.g., Degens et al., 1986; Stow et al., 2001; Trabucho-Alexandre et al., 2011; van Andel et al., 1977).

2.5 ORGANIC MATTER-RICH SHALE DEPOSITIONAL ENVIRONMENTS

Mud is everywhere, and life ubiquitous. Therefore, it is not surprising that black shales may be deposited in a wide range of sedimentary environments from the bottom of lakes to the abyssal plains of the ocean. The interpretation of ancient environments of black shale deposition has been influenced by studies of modern environments where organic matter-rich sediments are currently accumulating (Fig. 2.3 and Table 2.1). However, most ancient black shales appear to have been deposited in shallow marine epicontinental environments for which we have no modern analogs (cf. Arthur and Sageman, 1994).

2.5.1 Continental Depositional Environments

Organic matter-rich rocks deposited in continental environments account for more than 20% of current worldwide hydrocarbon production (Bohacs et al., 2000; Potter et al., 2005). Following the colonization of land by plants in the Devonian, organic matter becomes an important component in continental sediments. Mud in fluvial sequences mainly accumulates by vertical accretion as overbank deposits, which are produced when mud is deposited during floods in ephemeral ponds marginal to the main channel, and in oxbow lakes, which are persistent lakes formed by abandonment of meander loops.

Mud deposited in lakes often has organic matter contents that are significantly above the average for sediments in general. This organic matter comprises reworked terrestrial vegetation, from riparian environments, marginal macrophyte swamps, and phytoplankton (Talbot and Allen, 1996). Large lakes can contain a range of depositional environments including deltaic, coastal, and deepwater environments. Lakes are extremely sensitive to changes in climate and consequent changes in accommodation space (Bohacs, 1998). The ratio of accommodation space creation, that is, basin subsidence, to sediment/water input, which is controlled by climate, is the fundamental control on the stratigraphy of lakes (Carroll and Bohacs, 1999). Accommodation space determines water depth, which is a factor behind oxygen deficiency, depositional environments, and facies; climate determines the biota, temperature, and salinity of a lake (Potter et al., 2005).

The processes that favor enrichment in organic matter of lacustrine sediments depend on several factors that are ultimately linked to the type of lake in which the sediments are deposited (cf. Bohacs et al., 2000). A key factor is the availability of nutrients, which support primary production. Nutrients are brought in by land surface flow from the lake catchment area and by eolian transport. In many lakes, seasonal overturn recycles nutrients into surface waters. Permanently stratified lakes require an external nutrient source to support primary productivity. In alkaline lakes, such as lakes in tropical Africa, productivity is enhanced due to the abundance of carbonate ions that are available for incorporation by primary producers in addition to atmospheric CO_2 (Kelts, 1988).

The oxygenation of lake waters occurs primarily via exchange with the atmosphere, although some oxygen is a byproduct of photosynthesis. When a lake is thermally or chemically stratified, oxygen in bottom waters cannot be replenished (Fig. 2.3h). Oxygen is depleted by oxidation of sinking organic matter and the waters become anoxic, thereby favoring the preservation of organic matter. The extent and duration of bottom water anoxia in a lake depend on the frequency and intensity of mixing. In highly productive lakes, neither stratification nor permanently anoxic conditions are needed for the preservation of organic matter (Talbot, 1988). Bohacs et al. (2000) and Carroll and Bohacs (1999, 2001) discuss the processes and environments of organic matter-rich sedimentation in different types of lakes in great detail.

Lake deposits occur in several settings, but are most common in rift, intramontane, and foreland basins. Lake deposits of rift basins with rapid subsidence are more likely to be thick and well preserved (Potter et al., 2005). The lower Permian Whitehill Formation, South Africa, contains lacustrine black shales that are thought to originate from the accumulation of freshwater algae under anoxic, fresh-to-brackish-water conditions in a protorift basin in southwestern Gondwana (Faure and Cole, 1999). Lake Tanganyika in the East African Rift is a good example of a modern lacustrine environment of black shale deposition (Demaison and Moore, 1980). It is a large, deep lake with sedimentary environments ranging from deltas and narrow carbonate platforms in shallow water to deepwater fans. The equatorial location of the lake and its great depth result in high surface productivity

TABLE 2.1 Characteristics of shales and of major shale depositional environments (after Bohacs, 1998, table 1 and Potter et al., 2005, Table 5.1)

	Fluvial/ floodplain	Lake			Coastal	Deltaic	Shelf	Continental slope/ basin
		Underfilled	Balanced	Overfilled				
Lithology	Siliciclastic	Chemical (evaporites) biogenous	Biogenous chemical	Siliciclastic coal	Coal siliciclastic	Siliciclastic	Siliciclastic biogenous	Biogenous (siliciclastic)
TOC (%)	2 (1.8–50)	1.5 (0.2–15)	15 (2–30)	7 (0.5–45)	9.2 (4–17)	6 (3–34)	4.6 (1.1–20)	7 (1–27)
Type of organic matter	III	I	I	I/III	III, III-II	III, III-I	I/II	II ± sulfur
	HI = 150 (230–445)	HI = 400 (10–600)	HI = 900 (600–1100)	HI = 600 (50–700)	HI = 188 (35–599)	HI = 280 (170–520)	HI = 530 (165–825)	HI = 500 (150–800)
Typical thickness (m)	10–70	1–5	5–15	10–30	5–50	20–200	3–40	10–130
Major controls	Ratio of load to discharge; gradient	Supply + H$_2$O << accommodation. Very climate sensitive	Supply + H$_2$O = accommodation	Supply + H$_2$O >> accommodation; minor role for climate	Ratio of inshore wave/tidal power to supply		Water depth, bottom energy, and supply	Supply, slope stability, and bottom currents
Current systems	Overbank flows/shifting channels/ floods; suspension in strongly turbulent, unidirectional flows	Mostly wind-driven, but also turbidity currents (deep lakes), stratification, and overturning	Stable shoreline	Oscillating, but prograding shoreline	Longshore currents, tides, and storms; mud deposition behind barriers except for fluid mud along open coastlines	Diverse depending on type but includes overbank flows, shifting channels, longshore currents, tides, and storms	Distal riverine plumes, oceanic currents, storms, and tides; surface, midwater flows, and bottom currents	Resedimentation and contour currents on slope and proximal basin; some midwater suspensions
Organic matter input	Land plants, algae	Algae, bacteria	Algae (±land plants)	Land plants, algae			Marine algae (±some plants)	Marine algae, bacterial mats(?)
Fossils	Plant debris, spores, pollen, and rare vertebrates and invertebrates	Sparse, restricted fauna (high salinity or deep water)	Great diversity of pelagic and bottom fauna; water depth and climate permitting	Modest diversity; plant debris and rare vertebrates	As in alluvial, but also some brackish marine invertebrates		Open marine benthic and pelagic; some fine plant debris; spore abundance provides distance to shoreline	Open marine pelagics and limited benthics
Lateral continuity	Limited by channel cutouts and valley width	Carbonates, evaporites, and black shales most widespread	Carbonates and black shales most widespread	Sheets with channel cutouts on prograding shorelines	Limited except for some large lagoons and for open coastlines	Similar to alluvial except for widespread delta front deposits and some bay fills	Widespread lobes and sheets at highstands	Restricted on slope, but widespread in basin

and in the stratification of the water column. The waters of the lake are therefore increasingly dysoxic below 100 m, and the sediments have up to 12% organic matter and hydrogen index values up to 600 (Potter et al., 2005).

2.5.2 Paralic Depositional Environments

Paralic environments are transitional environments on the landward side of a coastline that are shaped by a complex interaction of processes; they form an intricate mosaic of closely associated facies between land and ocean. The morphology of a coast, which controls what sedimentary environments are present, is a function of wave energy, tidal power/range, and relative sea level change (Boyd et al., 1992; Dalrymple et al., 1992). The quantity and type of sediment supplied from land, alongshore, and offshore are also important.

Much, if not most, terrigenous mud is trapped in paralic environments; deltas, estuaries, lagoons, tidal flats, and so on, form the depocenters of many ancient shale successions. Grabau (1913, pp. 483–484) thought that ancient marine black shales (his sapropelargillytes) with a wide geographic distribution, namely the early Toarcian black shales of northwest Europe, had been deposited in coastal lagoons and/or on extensive mudflats on marginal epicontinental seas exposed to some extent at low tide (cf. French et al., 2014). The abundance of mud in a paralic environment depends on the relative magnitude of fine-grained fluvial sediment input versus sediment reworking by nearshore currents. The distribution of organic matter-rich sediments in paralic environments is typically patchy, and the organic matter content is variable and often dominated by terrestrial components. The preservation of organic matter depends on local water column oxygen deficiency and on relatively rapid burial. Excessive dilution by siliciclastic materials works against organic enrichment in paralic environments.

Mud-rich deltas form where the work carried out by waves, tides, and associated nearshore current systems is insufficient to winnow mud. Although tide-dominated deltas can also trap some of the mud brought by the rivers associated with them, mud-rich deltas are typically river-dominated. Mud in a delta is deposited in its subaerial environments, namely, lakes, swamps, and abandoned distributaries, and as interdistributary bay fills. Mud in tide-dominated deltas and in estuaries is trapped in broad mudflats that grade into marshes or evaporative flats depending on regional climate. In the upper delta plain, interdistributary areas may contain coals or carbonaceous shales deposited in swamp and marsh environments. For example, in the Mahakam Delta, Indonesia, plant debris from palms and mangroves, supported by the tropical climate, accumulates alongside mud in tidally influenced interdistributary areas (Reading and Collinson, 1996).

Mud that escapes these sediment traps is transported offshore and deposited on the subaqueous prodelta, which is the main site of mud accumulation in deltaic systems. When lower density river water flows out over saltwater (hypopycnal flow), suspended mud flocculates and is deposited on the subaqueous prodelta. Flocculation and biogenic pelletization result in mud aggregates that settle rapidly and proximally. During periods of high discharge, sediment-laden water may be denser than saltwater and the river becomes hyperpycnal (Bhattacharya, 2010). Prodelta hyperpycnites, the deposits of these bottom-hugging flows, may represent more than half of shelf mud (Bhattacharya, 2009).

Mud deposited in the prodelta is prone to resuspension during storms, but also by fair-weather waves and tides. This resuspended mud may migrate along the shelf forming dilute, hyperpycnal geostrophic fluid mud belts (Bhattacharya, 2010) and wave-enhanced sediment gravity flows (Macquaker et al., 2010b). In saltwater basins, rapid sedimentation of flocculated mud and methane formation in organic matter-rich sediments can result in high pore pressure and reduced mud density, which favor slope failures and resedimentation processes. Such failures occur at slopes as shallow as 0.2° (Potter et al., 2005).

Mud is deposited behind barriers in wave-dominated estuaries and lagoons, and on the bottom of fjords and other shallow-silled basins (Fig. 2.3e and f). Mud also accumulates in coastal swamps and mudflats along the sides of estuaries. Estuaries act as mud traps because filter-feeders remove suspended organic and inorganic particles greater than $3\mu m$ from the water column, and convert them into dense, mucus-bound fecal pellets and pseudofeces (Eisma, 1986; Newell, 1988; Pryor, 1975).

Climate is an important control on sedimentation in shallow, barred basins (Fig. 2.3e and f). In humid and temperate climates, mud will be dark colored and rich in organic matter, including plant debris washed in by rivers. In arid climates, mud will be lighter in color, organic matter lean, and alternate with evaporite beds.

Where nearshore wave power and associated nearshore current systems are high, mud is transported away from the river mouth along the shelf. Downdrift many modern large rivers, where the supply of mud is sufficient to dampen wave power and tidal currents, mud is deposited on open coastlines. The Chenier Plain of southwestern Louisiana is a sediment wedge formed by the westward moving mud stream of the Mississippi and Atchafalaya deltas (Gould and McFarlan, 1959). Variations in mud supply related to distributary channel activity (avulsions) cause shifts in the shoreline: the shoreline migrates seaward when supply is abundant and landward when supply is reduced. The longest nearshore Holocene mud belt is located along the open, high-energy northeastern coast of South America. The Amazon brings more than 10^9 tons of sediment to the Atlantic per year (Milliman and Meade, 1983). Most of this sediment is mud and a large fraction moves along the South American coast (Meade, 1994), part in suspension and part as large migrating mud banks (Rine and Ginsburg, 1985). Despite several modern examples (e.g., Anthony, 2008; Frey et al., 1989; Nair, 1976; Rine and Ginsburg, 1985), ancient equivalents of muddy open coastlines are not well documented in the literature. Walker (1971) studied Devonian marine mudstones in Pennsylvania that pass upward into mudstones with rootlets and mud cracks without a sand body at the paleoshore.

2.5.3 Shallow Marine Depositional Environments

Mud-dominated facies are the most abundant of all ancient shallow marine deposits (Johnson and Baldwin, 1996). On the shelf, siliciclastic sediment is supplied from adjacent land and, away from the mouths of large rivers, by reworking of seafloor sediment. Skeletal debris is an additional sediment source on shallow marine environments, but it only becomes dominant when siliciclastic sediment supply is low and biogenous sediment supply high. In the geologic record, shallow marine mudstones covered large epicontinental areas in response to sea-level rise. Examples of this type of setting include Paleozoic sequences of Africa, Europe, and North America, the upper Jurassic Kimmeridge Clay, and the Mesozoic of the Western Interior Seaway of North America (Johnson and Baldwin, 1996, and references therein).

Shallow marine, neritic, or shelf environments are those in which the seafloor is within the photic zone and periodically reworked by storms, that is, shallower than ca. 200 m. Most Neogene continental shelf sediments are relict in composition but modern in texture (Emery, 1968b; Milliman et al., 1972); the sediments were brought to the shelf during lowered sea levels associated with glacial episodes but have been reworked by present-day currents. The seafloor of the pericontinental shelf dips seaward from the shoreface to the shelf break at low angles of 0.1–1°, and the width of the shelf varies from tens to hundreds of kilometers. Because of their gentle slope, shelves are greatly influenced by changes in sea level.

A large proportion of preserved ancient marine sediments, including many black shales, was deposited in epicontinental marine environments. Epicontinental black shales are the typical subjects of shallow versus deepwater origin debates. At present, most shallow marine environments consist of relatively narrow pericontinental shelves, and there is a strong relation between water depth and distance to land. Shallower deposits thus show stronger terrigenous signatures than deeper deposits. During times of higher sea level in the geologic past, however, wide epicontinental shelves were common and covered vast areas of the continental crust. In epicontinental shelves, the relation between water depth and distance to land breaks down, and shallow water deposits remote from river input of terrigenous material and freshwater may have sedimentological, geochemical, and paleobiological signatures more characteristic of "deep" water (Hallam, 1967).

The organic matter content of sediments on the shelf seafloor is typically higher than in the deep sea (about three times higher, at present). Indeed, organic matter-rich deep-sea sediments are usually redeposited organic matter-rich shelf sediments (e.g., Dean et al., 1984; Degens et al., 1986). The water column of the shelf receives nutrients from runoff and coastal upwelling and thus is thus very fertile. Because water depths are relatively shallow, the export path for organic matter is much shorter and organic particles are less likely to be extensively oxidized en route to the seafloor. Sedimentation rates on the shelf are high in comparison to those of the deep ocean, which aids in the preservation of organic matter in relatively well-oxygenated environments, but retrogradation of clastic systems and trapping of terrigenous sediment in nearshore environments during transgressions and early relative sea level highstands minimize excessive dilution of organic matter settling to the seafloor of the shelf.

Shelf sediments below areas with strong upwelling are typically rich in organic matter and opal. For example, sediments on the SW African shelf, particularly off Walvis Bay, Namibia, have up to 20% organic carbon and up to 70% opal from the frustules of diatoms (Seibold and Berger, 1996). Fish debris and other vertebrate phosphate remains are also abundant. Upwelling is due to a combination of the cold, coastal Benguela Current with persistent offshore winds blowing in a northwest direction (Fig. 2.3c). Oxygen-poor, nutrient-rich water ascends from a depth of about 200 m and mixes in the photic zone with oxygenated water causing high productivity along a narrow coastal strip (Demaison and Moore, 1980). A zone of oxygen depletion, parallel and close to the coastline, is created on the shelf by the high oxygen demand due to the decomposition of large quantities of planktonic organic matter resulting from the Benguela Current upwelling.

However, not all areas of upwelling and high productivity in the ocean are associated with the intensification of oxygen minima and with the deposition and preservation of organic matter on the seafloor. This is the case in areas where oxygen supply exceeds the biochemical oxygen demand, for example, offshore Antarctica, offshore southeastern Brazil, and offshore Japan and the Kuril Islands (Summerhayes et al., 1976).

Shelf bathymetry can favor the preservation of organic matter-rich sediments. Bottom waters can become isolated from the well-mixed surface layer in bathymetric lows and may become oxygen-deficient through aerobic oxidation of organic matter. Moreover, topographic lows are traps for low-density organic matter. During winter, however, effective wave base is deeper and the water column is better mixed than in summer. Indeed, oxygen-deficiency in Phanerozoic shallow epi- and pericontinental shelves must have been seasonal.

Organic matter-rich shales may also form in epicontinental carbonate platforms. There are no modern examples, but these carbonate platforms were widespread in the geologic past. Some of these platforms may represent the interior of carbonate shelves or ramps, while others were truly epeiric platforms covering areas in the order of millions of square kilometers (Wright and Burchette, 1996). Intraplatform basins in such settings are rimmed by low gradient, ramp-like margins, and water depths in these basins were shallow (<150m). During transgressive and/or highstand phases, intraplatform basins became stratified and cyclic suboxic or anoxic sediments developed in the basin center. Organic matter-rich sediments may form in these settings, and a number of examples of black shales deposited in intraplatform basins are known, particularly in the Jurassic and Cretaceous of the Middle East (Burchette, 1993; Droste, 1990).

2.5.4 Deep Marine Depositional Environments

Deep marine environments extend from the shelf break to the abyssal seafloor. The deep marine continental margin, that is, the slope and rise, consists of thick accumulations of terrigenous sediment mixed with marine biogenous material; beyond the continental margin, the deep ocean is characterized by extensive facies belts dominated by biogenous sediment (Fig. 2.5), and little terrigenous sediment reaches the deep sea under modern conditions (e.g., Meade, 1994, his figure 2.8).

In terms of depositional processes, there are three different facies in the deep sea: pelagites, turbidites, and contourites (Stow, 1985a); the processes that are responsible for the enrichment in organic matter and its preservation in deep-sea sediments vary according to facies. Black pelagites form under regions of high surface productivity, where the rate of supply of organic matter exceeds aerobic oxidation in the water column. Rapid downslope resedimentation of organic matter-rich shelf sediments favors the preservation of organic matter in black turbidites on the deep ocean floor (e.g., Dean et al., 1984). High sedimentation rates associated with sediment gravity flows lead to a rapid burial of organic matter and therefore to its removal from the upper oxygenated part of the sediment column where degradation of organic matter is particularly aggressive (Stow et al., 2001). Low oxygen levels in pore water and low predator pressure in deep marine environments result in the absence of deep burrowing, characteristic of the *Nereites* ichnofacies, which also favors the preservation of organic matter in black turbidites.

The consumption of oxygen by biochemical processes in a layer of relatively small replenishment of oxygen by advective movement results in an oxygen minimum layer (Wyrtki, 1962). Where this layer comes into contact with the seafloor, the sediments are typically organic matter-rich, because the preservation of organic matter is favored (Fig. 2.4).

The occurrence of low oxygen water masses in the modern ocean appears to be much less widespread than during the Mesozoic. There are probably many reasons behind this, but two are likely to be the most important: seawater temperature and paleogeography. The modern Atlantic Ocean, for example, is a corridor for meridional circulation of cold, oxygen-rich water formed in the polar regions; this thermohaline circulation counters the expansion and intensification of the oxygen minimum layer. In comparison, the northwest Indian Ocean contains very little oxygen at depths between 200 and 1200m (Southam et al., 1982). Where oxygen concentrations in the oxygen minimum layer fall below $0.5\,ml\,l^{-1}$ and the layer impinges on the continental slope, sediments on the seafloor are typically rich in organic matter ($2\% \leq C_{org} \leq 20\%$).

The California continental borderland consists of a series of basins of varied sizes, separated by submarine ridges, sills, and islands (Gorsline, 1978). High nutrient levels in surface waters due to the combined effect of climate and oceanography support high, but variable, primary productivity. An oxygen minimum zone, which partly results from the oxidation of sinking pelagic organic matter, impinges upon this structurally complex continental margin. The basins are silled below a depth of about 500m and contain predominantly dysoxic water due to the interplay of oxygen-deficient deepwater flow and bottom topography (Savrda et al., 1984). The preservation of significant thicknesses of organic matter-rich sediments in these basins is favored by the impingement of their seafloor by the oxygen minimum layer (Fig. 2.3b). The Miocene Monterey Formation of southwestern California comprises a large volume of organic matter-rich siliceous and phosphatic (hemi)pelagic sediments. The sediments were deposited under similar conditions in relatively quiet deep water in a fault-bounded complex of borderland basins separated by islands and banks (Pisciotto and Garrison, 1981). Basins adjacent to shore received abundant terrigenous sediment, whereas basins farther offshore are sediment starved.

Seamounts and other oceanic rises affect ocean circulation patterns and result in local high fertility of seawater and consequent high productivity (e.g., Boehlert and Genin, 1987). The preservation of organic matter is generally favored on the seafloor of seamounts because the length of the export path is reduced (Fig. 2.3d). The preservation of organic matter is also favored where the oxygen minimum layer intersects the flanks of seamounts and other submarine topographic highs (Fig. 2.4). Cretaceous black shales in the Pacific were deposited on seamounts, oceanic plateaus, and other submarine topographic highs as a consequence of local high productivity. Other black shales were deposited as a result of high productivity associated with the passage of the submarine highs beneath the equatorial belt of high fertility (Waples, 1983).

2.6 CONCLUSION

Black shales are mixtures of terrigenous, biogenous, and hydrogenous sediment in which organic matter constitutes at least 0.5% of the material. The composition, texture, structures, and fossil content of a shale depend on depositional environment, that is, on a complex interplay of physical, chemical, and biological variables and processes that control the production, erosion, transport, deposition, and diagenesis of mud. Diagenetic processes act on the sediment and result in changes to its composition and/or texture. Patterns of vertical and lateral variability, which can be observed on a variety of scales in shale successions, are an expression of the dynamic nature of the processes behind their development, and preserve a record of the evolving depositional environments in which they formed.

Although shales are the most ubiquitous component of the stratigraphic record, the distribution of black shales in the Phanerozoic is predominantly limited to six stratigraphic intervals.

Black shales may be deposited in a wide range of sedimentary environments from the bottom of lakes to the abyssal plains of the ocean; however, most ancient black shales appear to have been deposited in shallow marine epicontinental environments for which we have no modern analogs.

ACKNOWLEDGMENTS

Nikki N. Bos is thanked for his help in producing the figures. Howard A. Armstrong is thanked for his comments on an earlier version of this chapter, which have helped improve it.

REFERENCES

Allen JRL. *Principles of Physical Sedimentology*. London: George Allen & Unwin Ltd; 1985.

Anthony EJ, editor. High mud-supply shores. In: *Shore Processes and Their Palaeoenvironmental Applications*. Amsterdam: Elsevier; 2008. p 131–158.

Aplin AC, Macquaker JHS. Mudstone diversity: origin and implications for source, seal, and reservoir properties in petroleum systems. AAPG Bull 2011;95:2031–2059.

Arthur MA. North Atlantic Cretaceous black shales: the record at site 398 and a brief comparison with other occurrences. In: Sibuet J-C, Ryan WBF, Arthur MA, Barnes RO, Habib D, Iaccarino S, Johnson D, Lopatin B, Maldonado A, Moore DG, Morgan GE, Réhault J-P, Sigal J, Williams CA, editors. *Initial Reports of the Deep Sea Drilling Project*. Washington, DC: U.S. Government Printing Office; 1979. p 47:719–751.

Arthur MA, Sageman BB. Marine black shales: depositional mechanisms and environments of ancient deposits. Ann Rev Earth Planet Sci 1994;22:499–551.

Arthur MA, Sageman BB. Sea-level control on source-rock development: perspectives from the Holocene Black Sea, the Mid-Cretaceous western interior basin of North America, and the Late Devonian Appalachian basin. In: Harris NB, editor. *The Deposition of Organic-Carbon-Rich Sediments: Models, Mechanisms, and Consequences*. Tulsa: SEPM; 2005. p 82:35–59.

Arthur MA, Jenkyns HC, Brumsack H-J, Schlanger SO. Stratigraphy, geochemistry, and paleoceanography of organic carbon-rich cretaceous sequences. In: Ginsburg RN, Beaudoin B, editors. *Cretaceous Resources, Events and Rhythms. Background and Plans for Research*. Dordrecht: Kluwer Academic Publisher; 1990. 304. p 75–119.

Baird GC. *Coral encrusted concretions: a key to recognition of a "shale on shale" erosion surface*. Volume 9, Lethaia; 1976. p 293–302.

Berger WH. Deep-sea sedimentation. In: Burk CA, Drake CL, editors. *The Geology of Continental Margins*. Berlin: Springer-Verlag; 1974. p 213–241.

Berger WH, Smetacek VS, Wefer G. *Productivity of the Ocean: Present and Past*. New York: Wiley-Interscience; 1989.

Berner RA. Burial of organic carbon and pyrite sulfur in the modern ocean: its geochemical and environmental significance. Am J Sci 1982;282:451–473.

Berner RA. A model for atmospheric CO_2 over phanerozoic time. Am J Sci 1991;291:339–376.

Berner RA. GEOCARBSULF: a combined model for phanerozoic atmospheric O_2 and CO_2. Geochimica et Cosmochimica Acta 2006;70:5653–5664.

Berner RA, Lasaga AC, Garrels RM. The carbonate-silicate geochemical cycle and its effect on atmospheric carbon dioxide over the past 100 million years. Am J Sci 1983;283:641–683.

Bernoulli D. North Atlantic and Mediterranean Mesozoic facies: a comparison. In: Hollister CD, Ewing JI, Habib D, Hathaway JC, Lancelot Y, Luterbacher H, Paulus FJ, Poag CW, Wilcoxon JA, Worstell P, editors. *Initial Reports of the Deep Sea Drilling Project*. Volume 11, Washington, DC: US Government Printing Office; 1972. p 801–822.

Bhattacharya JP. Hyperpycnal rivers and prodeltaic shelves in the Cretaceous seaway of North America. J Sediment Res 2009;79:184–209.

Bhattacharya JP. Deltas. In: James NP, Dalrymple RW, editors. *Facies Models 4*. St. John's, Newfoundland: Geological Association of Canada; 2010. p 233–264.

Boak J. Common wording vs. historical terminology. AAPG Explorer 2012;33:42–43.

Boehlert GW, Genin A. A review of the effects of seamounts on biological processes. In: Keating BH, Fryer P, Batiza R, Boehlert GW, editors. *Seamounts, Islands, and Atolls*. Washington, DC: American Geophysical Union; 1987. Geophysical Monograph Series 43. p 319–334.

Bohacs KM. Contrasting expressions of depositional sequences in mudrocks from marine to nonmarine environs. In: Schieber J, Zimmerle W, Sethi PS, editors. *Shales and Mudstones I: Basin Studies, Sedimentology and Paleontology*. Volume 1, Stuttgart: Schweizerbart; 1998. p 33–78.

Bohacs KM, Carroll AR, Neal JE, Mankiewicz PJ. Lake-basin type, source potential, and hydrocarbon character: an integrated sequence-stratigraphic–geochemical framework. In: Gierlowski-Kordesch EH, Kelts KR, editors. *Lake Basins Through Space and Time*. Tulsa: AAPG; 2000. p 3–34.

Bohacs KM, Grabowski Jr GJ, Carroll AR, Mankiewicz PJ, Miskell-Gerhardt KJ, Schwalbach JR, Wegner MB, Simo JA. Production, destruction, and dilution—the many paths to source-rock development. In: Harris NB, editor. *The Deposition of Organic-Carbon-Rich Sediments: Models, Mechanisms, and Consequences*. Tulsa: SEPM; 2005. p 61–101.

Bois C, Bouche P, Pelet R. Global geologic history and distribution of hydrocarbon reserves. AAPG Bull 1982;66:1248–1270.

Boyd R, Dalrymple R, Zaitlin BA. Classification of clastic coastal depositional environments. Sediment Geol 1992;80:39–150.

Bradley DC. Passive margins through earth history. Earth Sci Rev 2008;91:1–26.

Bralower TJ, Thierstein HR. Low productivity and slow deep-water circulation in mid-Cretaceous oceans. Geology 1984;12:614–618.

Brewer PG, Spencer DW. Distribution of some trace elements in black sea and their flux between dissolved and particulate phases. In: Degens ET, Ross DA, editors. *The Black Sea—Geology, Chemistry, and Biology*. Tulsa: AAPG; 1974. AAPG Memoir 20. p 137–143.

Broecker WS. Why the deep sea remains aerobic. Geol Soc Am Abstract Prog 1969;7:20–21.

Brongersma-Sanders M. Origin of major cyclicity of evaporites and bituminous rocks: an actualistic model. Marine Geol 1971;11:123–144.

Burchette TP. Mishrif formation (Cenomanian-Turonian), Southern Arabian Gulf: carbonate platform growth along a Cratonic basin margin. In: Simo JAT, Scott RW, Masse JP, editors. *Cretaceous Carbonate Platforms*. Tulsa: AAPG; 1993. AAPG Memoir 56. p 185–199.

Campbell CV. Lamina, laminaset, bed and bedset. Sedimentology 1967;8:7–26.

Carroll AR, Bohacs KM. Stratigraphic classification of ancient lakes: balancing tectonic and climatic controls. Geology 1999;27:99–102.

Carroll AR, Bohacs KM. Lake-type controls on petroleum source rock potential in nonmarine basins. AAPG Bull 2001;85:1033–1053.

Chamberlin TC. A group of hypotheses bearing on climatic changes. J Geol 1897;5:653–683.

Chamberlin TC. On a possible reversal of deep-sea circulation and its influence on geologic climates. J Geol 1906;14:363–373.

Chamberlin TC. Diastrophism as the ultimate basis of correlation. J Geol 1909;17:685–693.

Clarke JM. *Naples Fauna in Western New York*. New York: New York State Museum; 1904. New York State Museum Memoir 6. p 199–454.

Cluff RM. Mudrock fabrics and their significance–reply. J Sediment Petrol 1981;51:1029–1031.

Condie KC. Supercontinents and superplume events: distinguishing signals in the geologic record. Phys Earth Planet Inter 2004;146:319–332.

Correns CW. Pelagic sediments of the North Atlantic Ocean. In: Trask PD, editor. *Recent Marine Sediments*. Tulsa: AAPG; 1939. p 373–395.

Dalrymple RW, Zaitlin BA, Boyd R. Estuarine facies models: conceptual basis and stratigraphic implications. J Sediment Petrol 1992;62:1130–1146.

Dean WE, Arthur MA, Stow DAV. *Origin and geochemistry of Cretaceous deep-sea black shales and multicolored claystones, with emphasis on deep sea drilling project site 530, Southern Angola Basin*. In: Hay WW, Sibuet J-C, Barron EJ, Boyce RE, Brassell SC, Dean WE, Huc A-Y, Keating BH, McNulty CL, Meyers PA, Nohara M, Schallreuter RE, Steinmetz JC, Stow D, Stradner H, editors. Washington, DC: U.S. Government Printing Office; 1984. 75. p 819–844.

de Boer PL. Intertidal sediments: composition and structure. In: Eisma D, editor. *Intertidal Deposits. River Mouths, Tidal Flats, and Coastal Lagoons*. Boca Raton, FL: CRC Press; 1998. p 345–K.

de Boer PL, Smith DG. Orbital forcing and cyclic sequences. In: de Boer PL, Smith DG, editors. *Orbital Forcing and Cyclic Sequences*. Oxford: Blackwell Scientific Publications; 1994. p 1–14.

Degens ET, Ross DA, editors. The black sea—geology, chemistry, and biology. *AAPG Memoir 20*. Tulsa: AAPG; 1974.

Degens ET, Emeis KC, Mycke B, Wiesner MG. Turbidites, the principal mechanism yielding black shales in the early deep Atlantic Ocean. In: Summerhayes CP, Shackleton NJ, editors. *North Atlantic Palaeoceanography*. London: Geological Society; 1986. p 361–376.

Demaison G. Anoxia vs. productivity: what controls the formation of organic-carbon-rich sediments and sedimentary rocks?: discussion. AAPG Bull 1991;75:499.

Demaison GJ, Moore GT. Anoxic environments and oil source bed genesis. AAPG Bull 1980;64:1179–1209.

Dembicki H. Shale gas geochemistry mythbusting. *Search Discover* 2013:80294.

Dickens GR, O'Neil JR, Rea DK, Owen RM. Dissociation of oceanic methane hydrate as a cause of the carbon isotope excursion at the end of the Paleocene. Paleoceanography 1995;10:965–971.

Dow WG. Petroleum source beds on continental slopes and rises. AAPG Bull 1978;62:1584–1606.

Droste H. Depositional cycles and source rock development in an epeiric intra-platform basin: the Hanifa formation of the Arabian Peninsula. Sediment Geol 1990;69:281–296.

Duval BC, Cramez C, Vail PR. Stratigraphic cycles and major marine source rocks. In: de Graciansky P-C, Hardenbol J, Jacquin T, Vail PR, editors. *Mesozoic and Cenozoic Sequence Stratigraphy of European Basins*. Tulsa: SEPM; 1998. p 43–51.

Eisma D. Flocculation and de-flocculation of suspended matter in estuaries. Neth J Sea Res 1986;20:183–199.

Emerson SR, Archer D. Calcium carbonate preservation in the ocean. Philos Trans Royal Soc Lond A Math Phys Sci 1990;331:29–40.

Emerson S, Hedges JI. Processes controlling the organic carbon content of open ocean sediments. Paleoceanography 1988;3:621–634.

Emery KO. Relict sediments on continental shelves of the world. AAPG Bull 1968a;52:445–464.

Emery KO. Positions of empty pelecypod valves on the continental shelf. J Sediment Petrol 1968b;38:1264–1269.

Eppley RW, Peterson BJ. Particulate organic matter flux and planktonic new production in the deep ocean. Nature 1979; 282:677–680.

Erbacher J, Huber BT, Norris RD, Markey M. Increased thermohaline stratification as a possible cause for an ocean anoxic event in the Cretaceous period. Nature 2001;409:325–327.

Faure K, Cole D. Geochemical evidence for lacustrine microbial blooms in the vast Permian Main Karoo, Paraná, Falkland Islands and Huab basins of southwestern Gondwana. Palaeogeogr Palaeoclimatol Palaeoecol 1999;152:189–213.

Fleming RH, Revelle R. Physical processes in the ocean. In: Trask PD, editor. *Recent Marine Sediments*. Tulsa: AAPG; 1939. p 48–141.

Frakes LA, Francis JE, Syktus JI. *Climate Modes of the Phanerozoic*. Cambridge: Cambridge University Press; 1992.

French KL, Sepúlveda J, Trabucho-Alexandre J, Gröcke DR, Summons RE. Organic geochemistry of the early Toarcian oceanic anoxic event in Hawsker Bottoms, Yorkshire, England. Earth Planet Sci Lett 2014;390:116–127.

Frey RW, Howard JD, Han SJ, Park BK. Sediments and sedimentary sequences on a modern macrotidal flat, Inchon, Korea. J Sediment Petrol 1989;59:28–44.

Gamero-Diaz H, Miller C, Lewis R. sCore: a mineralogy based classification scheme for organic mudstones. SPE Int 2013;166284.

Garrels RM, Mackenzie FT. Sedimentary rock types: relative proportions as a function of geological time. Science 1969;163:570–571.

Ghadeer SG, Macquaker JHS. The role of event beds in the preservation of organic carbon in fine-grained sediments: analyses of the sedimentological processes operating during deposition of the Whitby Mudstone Formation (Toarcian, Lower Jurassic) preserved in northeast England. Marine Petrol Geol 2012;35:309–320.

Goldman MI. "Black shale" formation in and about Chesapeake bay. AAPG Bull 1924;8:195–201.

Gorsline DS. Anatomy of margin basins. J Sediment Petrol 1978;48:1055–1068.

Gould HR, McFarlan E Jr. Geologic history of the Chenier Plain, Southwestern Louisiana. Trans Gulf Coast Assoc Geological Soc 1959;9:261–270.

Grabau AW. *Principles of Stratigraphy*. New York: A.G. Seiler and Company; 1913.

Gradstein F, Ogg J, Smith A. *A Geologic Time Scale 2004*. Cambridge: Cambridge University Press; 2004.

Hallam A. Editorial. Mar Geol 1967;5:329–332.

Haq BU, Al-Qahtani AM. Phanerozoic cycles of sea-level change on the Arabian Platform. GeoArabia 2005;10:127–160.

Haq BU, Hardenbol J, Vail PR. Chronology of fluctuating sea levels since the triassic. Science 1987;235:1156–1167.

Hard EW. Black shale deposition in Central New York. AAPG Bull 1931;15:165–181.

Hay WW. Paleoceanography: a review for the GSA Centennial. Geol Soc Am Bull 1988;100:1934–1956.

Hay WW, Floegel S. New thoughts about the Cretaceous climate and oceans. Earth Sci Rev 2012;115:262–272.

Hay WW, Sibuet J-C, Barron EJ, Boyce RE, Brassell SC, Dean WE, Huc A-Y, et al. Introduction and explanatory notes, deep sea drilling project leg 75. In: Hay WW, Sibuet J-C, Barron EJ, Boyce RE, Brassell SC, Dean WE, Huc A-Y, Keating BH, McNulty CL, Meyers PA, Nohara M, Schallreuter RE, Steinmetz JC, Stow D, Stradner H, editors. *Initial Reports of the Deep Sea Drilling Project*. Volume 75, Washington, DC: U.S. Government Printing Office; 1984. p 3–25.

Hedberg HD. Gravitational compaction of clays and shales. Am J Sci 1936;31:241–287.

Hedges JI, Keil RG. Sedimentary organic matter preservation: an assessment and speculative synthesis. Mar Chem 1995; 49:81–115.

Hedges JI, Keil RG, Benner R. What happens to terrestrial organic matter in the ocean? Org Geochem 1997;27:195–212.

Hesselbo SP, Gröcke DR, Jenkyns HC, Bjerrum CJ, Farrimond P, Bell HSM, Green OR. Massive dissociation of gas hydrate during a Jurassic oceanic anoxic event. Nature 2000;406:392–395.

Hotinski RM, Bice KL, Kump LR, Najjar RG, Arthur MA. Ocean stagnation and end-Permian anoxia. Geology 2001;29:7–10.

Ingram RL. Fissility of mudrocks. Geol Soc Am Bull 1953;64:869–878.

Irving E, North FK, Couillard R. Oil, climate, and tectonics. Can J Earth Sci 1974;11:1–17.

Jackson ED, Schlanger SO. Regional syntheses, Line Islands Chain, Tuamotu Island Chain, and Manihiki Plateau, Central Pacific Ocean. In: Schlanger SO, Jackson ED, Boyce RE, Cook HE, Jenkyns HC, Johnson DA, Kaneps AG, Kelts KR, Martini E, McNulty CL, Winterer EL, editors. *Initial Reports of the Deep Sea Drilling Project*. Volume 33, Washington, DC: U.S. Government Printing Office; 1976. p 915–927.

Johnson HD, Baldwin CT. Shallow clastic seas. In: Reading HG, editor. *Sedimentary Environments: Processes, Facies and Stratigraphy*. Oxford: Blackwell Publishing; 1996. p 232–280.

Kelts K. Environments of deposition of lacustrine petroleum source rocks: an introduction. In: Fleet AJ, Kelts K, Talbot MR, editors. *Lacustrine Petroleum Source Rocks*. London: Geological Society; 1988. p 3–26.

Kerr AC. Oceanic plateau formation: a cause of mass extinction and black shale deposition around the Cenomanian-Turonian boundary? J Geol Soc 1998;155:619–626.

Kidder DL, Worsley TR. Phanerozoic Large Igneous Provinces (LIPs), HEATT (Haline Euxinic Acidic Thermal Transgression) episodes, and mass extinctions. Palaeogeogr Palaeoclimatol Palaeoecol 2010;295:162–191.

Klemme HD, Ulmishek GF. Effective petroleum source rocks of the world: stratigraphic distribution and controlling depositional factors. AAPG Bull 1991;75:1809–1851.

Krumbein WC, Sloss LL. *Stratigraphy and Sedimentation*. 2nd ed. San Francisco, CA: Freeman; 1963.

Lallier-Vergès E, Hayes JM, Boussafir M, Zaback DA, Tribovillard NP, Connan J, Bertrand P. Productivity-induced sulphur enrichment of hydrocarbon-rich sediments from the Kimmeridge Clay Formation. Chem Geol 1997;134:277–288.

Larson RL. Geological consequences of superplumes. Geology 1991;19:963–966.

Laws EA, Falkowski PG, Smith WO Jr, Ducklow H, McCarthy JJ. Temperature effects on export production in the open ocean. Glob Biogeochem Cycles 2000;14:1231–1246.

Lewan MD. Laboratory classification of very fine grained sedimentary rocks. Geology 1978;6:745–748.

Lobza V, Schieber J. Biogenic sedimentary structures produced by worms in soupy, soft muds; observations from the Chattanooga Shale (Upper Devonian) and experiments. J Sediment Res 1999;69:1041–1049.

MacKay DA, Dalrymple RW. Dynamic mud deposition in a tidal environment: the record of fluid-mud deposition in the Cretaceous Bluesky Formation, Alberta, Canada. J Sediment Res 2011;81:901–920.

Macquaker JHS, Howell JK. Small-scale (<5.0 m) vertical heterogeneity in mudstones: implications for high-resolution stratigraphy in siliciclastic mudstone successions. J Geol Soc 1999;156:105–112.

Macquaker JHS, Taylor KG, Gawthorpe RL. High-resolution facies analyses of mudstones: implications for paleoenvironmental and sequence stratigraphic interpretations of offshore ancient mud-dominated successions. J Sedim Res 2007;77: 324–339.

Macquaker JHS, Keller MA, Davies SJ. Algal blooms and "marine snow": mechanisms that enhance preservation of organic carbon in ancient fine-grained sediments. J Sediment Res 2010a; 80:934–942.

Macquaker JHS, Bentley SJ, Bohacs KM. Wave-enhanced sediment-gravity flows and mud dispersal across continental shelves: reappraising sediment transport processes operating in ancient mudstone successions. Geology 2010b;38:947–950.

Mancini EA, Tew BH, Mink RM. Petroleum source rock potential of Mesozoic condensed section deposits of Southwest Alabama. In: Katz BJ, Pratt LM, editors. *Source Rocks in a Sequence Stratigraphic Framework*. Volume 37, Tulsa: AAPG Studies in Geology; 1993. p 147–162.

Manheim FT, Meade RH, Bond GC. Suspended matter in surface waters of the Atlantic Continental Margin from Cape Cod to the Florida Keys. Science 1970;167:371–376.

Mantz PA. Bedforms produced by fine, cohesionless, granular and flakey sediments under subcritical water flows. Sedimentology 1978;25:83–103.

McAnally W, Friedrichs C, Hamilton D, Hayter E, Shrestha P, Rodriguez H, Sheremet A, Teeter A. Management of fluid mud in estuaries, bays, and lakes. i: present state of understanding on character and behavior. J Hydraul Eng 2007;133:9–22.

McArthur JM. Strontium isotope stratigraphy. In: Ratcliffe KT, Zaitlin BA, editors. *Application of Modern Stratigraphic Techniques: Theory and Case Histories*. Tulsa: SEPM; 2010. p 129–142.

McCave IN. Transport and escape of fine-grained sediment from shelf areas. In: Swift DJP, Duane DB, editors. *Shelf Sediment Transport: Process and Pattern*. Stroudsburg: Dowden, Hutchinson & Ross, Inc.; 1972. p 225–248.

McElwain JC, Wade-Murphy J, Hesselbo SP. Changes in carbon dioxide during an oceanic anoxic event linked to intrusion into Gondwana coals. Nature 2005;435:479–482.

McKee ED, Weir GW. Terminology for stratification and cross-stratification in sedimentary rocks. Geol Soc Am Bull 1953; 64:381–390.

Meade RH. Sources and sinks of suspended matter on continental shelves. In: Swift DJP, Duane DB, Pilkey OH, editors. *Shelf Sediment Transport: Process and Pattern*. Stroudsburg: Dowden, Hutchinson & Ross, Inc.; 1972. p 249–262.

Meade RH. Suspended sediments of the Modern Amazon and Orinoco Rivers. Quat Int 1994;21:29–39.

Meyer KM, Kump LR. Oceanic euxinia in earth history: causes and consequences. Ann Rev Earth Planet Sci 2008;36:251–288.

Migniot C. Étude Des Propriétés Physiques De Différents Sédiments Très Fins Et De Leur Comportement Sous Des Actions Hydrodynamiques. La Houille Blanche 1968;7:591–620.

Miller KG, Kominz MA, Browning JV, Wright JD, Mountain GS, Katz ME, Sugarman PJ, Cramer BS, Christie-Blick N, Pekar SF. The phanerozoic record of global sea-level change. Science 2005;310:1293–1298.

Milliman JD, Meade RH. World-wide delivery of river sediment to the oceans. J Geol 1983;91:1–21.

Milliman JD, Pilkey OH, Ross DA. Sediments of the continental margin off the Eastern United States. Geol Soc Am Bull 1972;83:1315–1334.

Morse JW, Mackenzie FT. *Geochemistry of Sedimentary Carbonates. Developments in Sedimentology 48*. New York: Elsevier; 1990.

Müller PJ, Suess E. Productivity, sedimentation rate, and sedimentary organic matter in the oceans—I. Organic carbon preservation. Deep-Sea Res 1979;26A:1347–1362.

Murray J, Renard AF. *Report on deep-sea deposits based on the specimens collected during the voyage of H.M.S. Challenger in the years 1872 to 1876*. London: Her Majesty's Stationery Office; 1891.

Nair RR. Unique mud banks, Kerala, Southwest India. AAPG Bull 1976;60:616–621.

Newell RIE. Ecological changes in Chesapeake Bay: are they the result of overharvesting the American Oyster, *Crassostrea virginica?*". In: *Understanding the Estuary: Advances in Chesapeake Bay Research*. Baltimore: Chesapeake Research Consortium; 1988. p 536–546.

North FK. Episodes of source-sediment deposition. J Petrol Geol 1979;2:199–218.

O'Brien NR. Shale lamination and sedimentary processes. In: Kemp AES, editor. *Palaeoclimatology and Palaeoceanography from Laminated Sediments*. London: Geological Society; 1996. p 23–36.

Östlund HG. Expedition "Odysseus 65": radiocarbon age of black sea deep water. In: Degens ET, Ross DA, editors. *The Black Sea—Geology, Chemistry, and Biology*. Tulsa: AAPG; 1974. AAPG Memoir 20. p 127–132.

Paola C, Leeder M. Simplicity versus complexity. Nature 2011;469:38–39.

Parrish JT. Upwelling and petroleum source beds, with reference to Paleozoic. AAPG Bull 1982;66:750–774.

Parrish JT. Palaeo-upwelling and the distribution of organic-rich rocks. In: Brooks J, Fleet AJ, editors. *Marine Petroleum Source Rocks*. London: Geological Society; 1987. p 199–205.

Parrish JT, Curtis RL. Atmospheric circulation, upwelling, and organic-rich rocks in the Mesozoic and Cenozoic eras. Palaeogeo Palaeoclim Palaeoecol 1982;40:31–66.

Passey QR, Bohacs KM, Esch WL, Klimentidis R, Sinha S. From oil-prone source rock to gas-producing shale reservoir—geologic and petrophysical characterization of unconventional shale-gas reservoirs. SPE Int 2010;131350.

Pedersen TF, Calvert SE. Anoxia vs. productivity: what controls the formation of organic-carbon-rich sediments and sedimentary rocks? AAPG Bull 1990;74:454–466.

Pettijohn FJ. *Sedimentary Rocks*. 3rd ed. New York: Harper & Row Publishers; 1975.

Pike J, Kemp AES. Silt aggregates in laminated marine sediment produced by agglutinated foraminifera. J Sediment Res 1996;66:625–631.

Piper DZ, Calvert SE. A marine biogeochemical perspective on black shale deposition. Earth-Sci Rev 2009;95:63–96.

Pisciotto KA, Garrison RE. Lithofacies and depositional environments of the Monterey Formation, California. In: Garrison RE, Douglas RG, editors. *The Monterey Formation and Related Siliceous Rocks of California*. Tulsa: SEPM; 1981. p 97–122.

Plint AG, Macquaker JHS, Varban BL. Bedload transport of mud across a wide, storm-influenced ramp: Cenomanian–Turonian Kaskapau Formation, Western Canada Foreland Basin. J Sediment Res 2012;82:801–822.

Pompeckj JF. Die Juraablagerungen Zwischen Regensburg Und Regenstauf. Geognostische Jahrbuch 1901;14:139–220.

Potter PE, Maynard JB, Depetris PJ. *Mud and Mudstones*. Berlin: Springer; 2005.

Prokoph A, Shields GA, Veizer J. Compilation and time-series analysis of a marine carbonate $\delta_{18}O$, $\delta_{13}C$, $_{87}Sr/_{86}Sr$ and $\delta_{34}S$ database through Earth history. Earth Sci Rev 2008;87:113–133.

Pryor WA. Biogenic sedimentation and alteration of argillaceous sediments in shallow marine environments. Geol Soc Am Bull 1975;86:1244–1254.

Raup DM, Sepkoski JJ. Periodic extinction of families and genera. Science 1986;231:833–836.

Raymo ME, Ruddiman WF. Tectonic forcing of late Cenozoic climate. Nature 1992;359:117–122.

Reading HG, Collinson JD. Clastic coasts. In: Reading HG, editor. *Sedimentary Environments: Processes, Facies and Stratigraphy*. Oxford: Blackwell Publishing; 1996. p 154–231.

Richter R. Die Großrippeln unter Gezeitenströmungen im Wattenmeer und die Rippeln im Pirnaer Turon. Senckenbergiana 1926;8:297–305.

Ridgwell A. A Mid Mesozoic Revolution in the regulation of ocean chemistry. Mar Geol 2005;217:339–357.

Rine JM, Ginsburg RN. Depositional facies of a mud shoreface in Suriname, South America—a mud analogue to sandy, shallow-marine deposits. J Sediment Petrol 1985;55:633–652.

Royer DL. CO_2-forced climate thresholds during the Phanerozoic. Geochim Cosmochim Acta 2006;70:5665–5675.

Sageman BB, Murphy AE, Werne JP, Straeten CAV, Hollander DJ, Lyons TW. A tale of shales: the relative roles of production, decomposition, and dilution in the accumulation of organic-rich strata, Middle–Upper Devonian, Appalachian Basin. Chem Geol 2003;195:229–273.

Savrda CE, Bottjer DJ, Gorsline DS. Development of a comprehensive oxygen-deficient marine biofacies model: evidence from Santa Monica, San Pedro, and Santa Barbara Basins, California Continental Borderland. AAPG Bull 1984;68:1179–1192.

Schieber J. Evidence for high-energy events and shallow-water deposition in the Chattanooga Shale, Devonian, central Tennessee, USA. Sediment Geol 1994;93:193–208.

Schieber J. Distribution and deposition of mudstone facies in the Upper Devonian Sonyea Group of New York. J Sediment Res 1999;69:909–925.

Schieber J. Simple gifts and buried treasures—implications of finding bioturbation and erosion surfaces in black shales. Sediment Rec 2003;1:4–8.

Schieber J. Discovery of agglutinated benthic foraminifera in Devonian black shales and their relevance for the redox state of ancient seas. Palaeogeo Palaeoclim Palaeoecol 2009;271:292–300.

Schieber J. Reverse engineering mother nature—shale sedimentology from an experimental perspective. Sediment Geol 2011;238:1–22.

Schieber J, Riciputi L. Pyrite ooids in Devonian black shales record intermittent sea-level drop and shallow-water conditions. Geology 2004;32:305–308.

Schieber J, Southard J, Thaisen K. Accretion of mudstone beds from migrating floccule ripples. Science 2007;318:1760–1763.

Schieber J, Southard JB, Schimmelmann A. Lenticular shale fabrics resulting from intermittent erosion of water-rich muds—interpreting the rock record in the light of recent flume experiments. J Sediment Res 2010;80:119–128.

Schlanger SO, Jenkyns HC. Cretaceous oceanic anoxic events: causes and consequences. Geol Mijnbouw 1976;55:179–184.

Schuchert C. The conditions of black shale deposition as illustrated by the Kupferschiefer and Lias of Germany. Proc Am Philos Soc 1915;54:259–269.

Scotese CR. A continental drift flipbook. J Geol 2004;112:729–741.

Seibold E, Berger WH. *The Sea Floor: An Introduction to Marine Geology*. 3rd ed. Berlin: Springer-Verlag; 1996.

Shepard FP. Sediments of the continental shelves. Geol Soc Am Bull 1932;43:1017–1039.

Shepard FP, Moore DG. Central Texas Coast sedimentation: characteristics of sedimentary environment, recent history, and diagenesis. AAPG Bull 1955;39:1463–1593.

Sheridan RE. Pulsation tectonics as the control of long-term stratigraphic cycles. Paleoceanography 1987;2:97–118.

Sinninghe Damsté JS, Rijpstra WIC, Kock-van Dalen AC, De Leeuw JW, Schenck PA. Quenching of labile functionalised lipids by inorganic sulphur species: evidence for the formation of sedimentary organic sulphur compounds at the early stages of diagenesis. Geochim Cosmochim Acta 1989;53:1343–1355.

Sinton CW, Duncan RA. Potential links between ocean plateau volcanism and global ocean anoxia at the Cenomanian-Turonian Boundary. Eco Geol 1997;92:836–842.

Sorby HC. On the application of quantitative methods to the study of the structure and history of rocks. Quart J Geol Soc 1908; 64:171–233.

Southam JR, Peterson WH, Brass GW. dynamics of anoxia. Palaeogeogr Palaeoclim Palaeoecol 1982;40:183–198.

Southard JB. Representation of bed configurations in depth-velocity-size diagrams. J Sediment Petrol 1971;41:903–915.

Stanley SM. *Earth System History*. New York: W.H. Freeman & Company, 1999.

Stein R. Organic carbon and sedimentation rate—further evidence for anoxic deep-water conditions in the Cenomanian/Turonian Atlantic Ocean. Mar Geol 1986;72:199–209.

Stein R. Organic carbon content/sedimentation rate relationship and its paleoenvironmental significance for marine sediments. Geo-Marine Lett 1990;10:37–44.

Storey M, Duncan RA, Swisher CC. Paleocene-Eocene thermal maximum and the opening of the Northeast Atlantic. Science 2007;316:587–589.

Stow DAV. Fine-grained sediments in deep water: an overview of processes and facies models. Geo-Mar Lett 1985a;5:17–23.

Stow DAV. Deep-sea clastics: where are we and where are we going?". In: Brenchley PJ, Williams BPJ, editors. *Sedimentology: Recent Developments and Applied Aspects*. London: Geological Society; 1985b. p 67–93.

Stow DAV, Reading HG, Collinson JD. Deep seas. In: Reading HG, editor. *Sedimentary Environments: Processes, Facies and Stratigraphy*. Oxford: Blackwell Publishing; 1996. p 395–453.

Stow DAV, Huc A-Y, Bertrand P. Depositional processes of black shales in deep water. Mar Petrol Geol 2001;18:491–498.

Strøm KM. Land-locked waters and the deposition of black muds. In: Trask PD, editor. *Recent Marine Sediments*. Tulsa: AAPG; 1939. p 356–372.

Suess E. Particulate organic carbon flux in the oceans—surface productivity and oxygen utilization. Nature 1980;288:260–263.

Summerhayes CP, de Melo U, Barretto HT. The influence of upwelling on suspended matter and shelf sediments off Southeastern Brazil. J Sediment Petrol 1976;46:819–828.

Talbot MR. The origins of lacustrine oil source rocks: evidence from the lakes of tropical Africa. In: Fleet AJ, Kelts K, Talbot MR, editors. *Lacustrine Petroleum Source Rocks*. London: Geological Society; 1988. p 29–43.

Talbot MR, Allen PA. Lakes. In: Reading HG, editor. *Sedimentary Environments: Processes, Facies and Stratigraphy*. Oxford: Blackwell Publishing; 1996. p 83–124.

Terwindt JHJ, Breusers HNC. Experiments on the origin of flaser, lenticular and sand-clay alternating bedding. Sedimentology 1972;19:85–98.

Tissot B. Effects on prolific petroleum source rocks and major coal deposits caused by sea-level changes. Nature 1979;277:463–465.

Topper RPM, Trabucho-Alexandre J, Tuenter E, Meijer PT. A regional ocean circulation model for the mid-Cretaceous North Atlantic Basin: implications for black shale formation. Clim Past 2011;7:277–297.

Tourtelot HA. Origin and use of the word "shale". Am J Sci 1960;258-A:335–343.

Trabucho-Alexandre J. *Mesozoic Sedimentation in the North Atlantic and Western Tethys; Global Forcing Mechanisms and Local Sedimentary Processes*. Utrecht: Universiteit Utrecht; 2011.

Trabucho-Alexandre J. More gaps than shale: erosion of mud and its effect on preserved geochemical and palaeobiological signals. In: Smith DG, Bailey RJ, Burgess PM, Fraser AJ, editors. *Strata and Time: Probing the Gaps in Our Understanding*, vol. 404. London: Geological Society; 2014. DOI:10.1144/SP404.10.

Trabucho-Alexandre J, Tuenter E, Henstra GA, van der Zwan KJ, van de Wal RSW, Dijkstra HA, de Boer PL. The mid-Cretaceous North Atlantic nutrient trap: black shales and OAEs. Paleoceanography 2010;25:PA4201.

Trabucho-Alexandre J, van Gilst RI, Rodríguez-López JP, de Boer PL. The sedimentary expression of oceanic anoxic event 1b in the North Atlantic. Sedimentology 2011;58:1217–1246.

Trabucho-Alexandre J, Dirkx R, Veld H, Klaver G, de Boer PL. Toarcian Black Shales in the Dutch Central Graben: record of energetic, variable depositional conditions during an oceanic anoxic event. J Sediment Res 2012a;82:104–120.

Trabucho-Alexandre J, Hay WW, de Boer PL. Phanerozoic environments of black shale deposition and the Wilson Cycle. Solid Earth 2012b;3:29–42.

Trask PD. *Origin and Environment of Source Sediments of Petroleum*. Houston, TX: Gulf Publishing Company; 1932.

Trusheim F. Rippeln in Schlick. Nat Museum 1929;59:72–79.

Twenhofel WH. Environments of origin of black shales. AAPG Bull 1939;23:1178–1198.

Tyson RV. Sedimentation rate, dilution, preservation and total organic carbon: some results of a modelling study. Org Geochem 2001;32:333–339.

Tyson RV. The "productivity versus preservation" controversy: cause, flaws, and resolution. In: Harris NB, editor. *The Deposition of Organic-Carbon-Rich Sediments: Models, Mechanisms, and Consequences*. Tulsa: SEPM; 2005. p 17–33.

van Andel TH, Thiede J, Sclater JG, Hay WW. Depositional history of the South Atlantic Ocean during the last 125 million years. J Geology 1977;85:651–698.

van Straaten LMJU. Longitudinal ripple marks in mud and sand. J Sediment Petrol 1951;21:47–54.

Van Waterschoot van der Gracht WAJM. Permo-carboniferous orogeny in South-Central United States. AAPG Bull 1931; 15:991–1057.

Walker RG. Nondeltaic depositional environments in the Catskill clastic wedge (Upper Devonian) of central Pennsylvania. Geol Soc Am Bull 1971;82:1305–1326.

Walsh JJ. Importance of continental margins in the marine biogeochemical cycling of carbon and nitrogen. Nature 1991;350:53–55.

Walsh JP, Nittrouer CA. Understanding fine-grained river-sediment dispersal on continental margins. Marine Geol 2009;263: 34–45.

Waples DW. Reappraisal of anoxia and organic richness, with emphasis on Cretaceous of North Atlantic. AAPG Bull 1983; 67:963–978.

Weaver CE. *Clays, Muds, and Shales. Developments in Sedimentology 44*. Amsterdam: Elsevier; 1989.

Wells JT, Coleman JM. Longshore transport of mud by waves: northeastern coast of South America. Geol Mijnbouw 1978;57:353–359.

Wignall PB. Model for transgressive black shales? Geology 1991;19:167–170.

Wilson JT. Static or mobile earth: the current scientific revolution. Proc Am Philos Soc 1968;112:309–320.

Wilson PA, Norris RD. Warm tropical ocean surface and global anoxia during the mid-Cretaceous period. Nature 2001;412:425–429.

Wright VP, Burchette TP. Shallow-water carbonate environments. In: Reading HG, editor. *Sedimentary Environments: Processes, Facies and Stratigraphy*. Oxford: Blackwell Publishing; 1996. p 325–394.

Wyrtki K. The oxygen minima in relation to ocean circulation. Deep Sea Res 1962;9:11–23.

3

GEOCHEMICAL ASSESSMENT OF UNCONVENTIONAL SHALE GAS RESOURCE SYSTEMS

DANIEL M. JARVIE

Worldwide Geochemistry, LLC, Humble, TX, USA

3.1 INTRODUCTION

The global search for shale gas resource systems is based on the astounding success in North America as up to 750 trillion cubic feet (tcf) is estimated to be technically recoverable according to the US Energy Information Administration (EIA, 2011). The current surplus of gas due primarily to shale gas production has kept natural gas prices modest, aiding American industries and consumers as well as aiding the US quest for lowered dependence on overseas sources of energy. In addition, the United States reported the lowest level of carbon dioxide (CO_2) emissions in 20 years (EIA, 2012).

Shale gas resource systems are typically characterized by organic-rich, gas-window mature, very low porosity, and ultralow permeability mudstones. Typical storage capacities range from about 4–14% porosity and is a combination of porosity created by organic matter decomposition (organoporosity) as well as classical storage (e.g., matrix porosity). It is also possible that some storage is in and among atomic pores within the organic matter itself (Locke and Winans, 2013; Orendt et al., 2011) due to sorption of generated petroleum. Additional storage is often found along expulsion conduits and matrix pores. Fractures are present but are commonly calcite-filled and do not contribute significant storage volume.

Nanodarcy permeability is predominant in these plays with values ranging from tens to hundreds of nanodarcies. Such low permeability restricts the flow of petroleum particularly compared to conventional reservoir rocks that have five to nine orders magnitude higher permeabilities. In this chapter, conventional reservoir systems are considered to be equal to or greater than 0.10 mD (Williams, 2012), whereas unconventional systems are considered to be less than 0.10 mD. Shale gas systems often have permeabilities less than 100 nD such as found in the Barnett Shale of the Fort Worth Basin, Texas (Loucks et al., 2009; Reed and Loucks, 2007).

Mudstone and shale nomenclature are used interchangeably in this text and are not necessarily referring to principal mineralogical components or even particle size. Most of the ongoing plays are mudstones based on particle size, but may have quite variable mineralogical contents.

What shale resource systems lack in porosity and permeability, they counter with massive areal and volumetric extents. One of their defining characteristics of an unconventional resource is being a continuous accumulation over extensive distances that generally do not have obvious structural components as targets (Zou, 2012). They may extend over thousands of square miles with thicknesses ranging from tens to hundreds of feet. This volume of continuous mudstone can store trillions of cubic feet of gas, despite their overall low porosity when reservoir PVT properties are considered (Jarvie et al., 2007). However, due to this storage and ultra-low permeability, it is necessary to shatter the rock matrix, enabling gas flow to reach the well bore. This is achieved by high-energy stimulation, that is, pumping high rates of slick water (freshwater with surfactants) with high amounts of proppant to open and prop open fractures as well as diverting energy, optimally creating a dendritic fracture network with extensive reservoir volume contact. Nonetheless, flow from shale gas reservoirs is typically very low when compared to conventional gas reservoir rocks. In order to be able to develop these systems commercially,

Fundamentals of Gas Shale Reservoirs, First Edition. Edited by Reza Rezaee.
© 2015 John Wiley & Sons, Inc. Published 2015 by John Wiley & Sons, Inc.

it is absolutely necessary to stimulate or fracture a sufficient volume of rock to release large volumes of gas. In general, recoveries from these systems range from 10 to about 20% of the gas in place (GIP), although higher numbers are often cited.

Gas flow requires not only large amounts of GIP and sufficient storage but also a brittle rock fabric as well as extensive knowledge of rock mechanics and stress fields in the prospect area to enable stimulation of sufficient volume of rock. Even with high-energy stimulation, the typical shale gas well will ultimately produce only a few billion cubic feet (bcf) of gas, and as such, requires drilling hundreds to thousands of wells to recover large amounts of gas, for example, over 15,000 wells have been drilled into Barnett Shale gas reservoirs to date yielding about 15 tcf of gas. Prospective shale gas areas may be located beneath commercial and residential areas limiting drilling access. Where such plays are located in areas where drilling costs are extremely high, the economics of drilling a gas well(s) to produce a few bcf of gas may not be commercially viable at a given price level. Low natural gas prices (ca. $3.69/mcf on average in 2013) in North America have currently resulted in diminished drilling of shale gas wells except in high-return areas such as the Marcellus Shale of the Appalachian Basin, Northeastern United States.

The search for shale gas outside of North America has identified promising shale gas systems, but to date has only resulted in marginal commercial success. However, the earliest wells are typically the most expensive and least productive, but as companies learn what is necessary to obtain high gas flow from a given system, commercial development often follows.

In addition, there has been much resistance to drilling such wells that require high-energy stimulation that has often cited as concern for the environment, which is a concern for all including the companies hoping to develop such a system. Commercial success requires environmental success too, otherwise development would be halted. Operations in North America have been safe with over 60,000 shale resource wells drilled and stimulated. While groundwater contamination has occurred by ongoing geological processes and are often reported in shallow water wells prior to drilling shale gas wells, these have often been cited as being derived from the high-energy stimulation efforts. Shale gas reservoirs are often at depths from 6000 to 15,000 ft or more, which is one to three miles below the nearest freshwater aquifers. Adequate and effective regulations on cementing and testing of the upper portion of the well bore where it passes through freshwater aquifers should be the focus of tighter regulation. For example, the largest and most tragic oil field operation in recent time has been the massive blow out in the Gulf of Mexico, where the casing cement job was the cause of various failures and also likely improper well testing (National Commission on the BP Deepwater Horizon Oil Spill and Offshore Drilling, 2011). Regulations need to be enforced, but are often based on outdated information. As such, they should be continually evaluated for their efficacy with regard to present-day drilling operations and updated as necessary.

As shale gas development requires hundreds to thousands of wells, this also puts a strain on infrastructure such as roads and water use when wells are stimulated with freshwater. Noise pollution is likely a large concern albeit over a short period of time. Both can be dealt with by appropriate actions of governments and drilling operations.

Seismic activity or small-scale earthquakes have also been associated with shale gas development, although it is actually saltwater disposal wells that have caused such seismicity. When disposal wells are drilled near preexisting faults, low-magnitude earthquakes can occur. One of the most widely reported seismic events occurred at the Dallas/Fort Worth (DFW) Airport in 2008–2009, having a maximum moment magnitude scale (M_w or M) of 3.3 and was likely due to injection of saltwater into a disposal well (Frolich and Potter, 2013). However, these authors also reported that more than a decade before this event (1997), a M3.4 natural earthquake occurred within 100 km (ca. 62 miles) of DFW Airport. Frolich and Potter (2013) proposed five hypotheses for the origin and nature of small intraplate earthquakes as might be originated from saltwater injection wells. They cite that the consensus of earthquake seismologists agrees that small earthquakes (ca. M2.0) are ubiquitous and not indicative of the occurrence or onset of larger earthquakes. Often increased seismicity is related to increased recording stations being deployed, so many more low-level quakes are recorded. They also cite that all 50 states in the United States experience earthquakes, but are small and go undetected. In Texas, most M3.5 earthquakes go unrecorded unless near areas of dense population where seismographs are located (Frolich and Potter, 2013). Avoidance of injection well-induced earthquakes can be accomplished by understanding regional stress patterns and not locating such wells in areas with preexisting faults (Frolich and Potter, 2013).

The societal and political power of developing shale gas resources in a friendly and environmentally sound manner is a tremendous asset in terms of supplying energy needs and also reducing carbon dioxide levels. Natural gas is the most environment-friendly carbon-based energy source as combustion emits lower carbon dioxide and other oxide emissions. The major source of carbon dioxide emissions in the United States is electricity generation, and reduction of CO_2 emissions is due in part to increased implementation of gas-fired power plants. In 2012, the United States had its lowest carbon dioxide emissions in over 20 years (EIA, 2012). Natural gas power is certainly not a green solution as there remains a carbon footprint; however, it does give us additional time to develop green energy resources. Shale gas is a tremendous resource for our ongoing energy needs resulting in lower emissions and lower costs for both consumers and industrial concerns.

Finally, it should be noted that additional societal benefit comes from the royalties paid to mineral right owners, whether landowners or governments. In North Texas, independent school districts received approximately $51 million, municipalities $86 million, and the University of Texas $5 million in royalty payments, bonuses, and tax revenues from Barnett Shale operations in 2010 alone (The Perryman Group, 2011). An independent petroleum company paid $271 million in royalties in 2010 to individuals, cities, counties, school districts, and other public and private entities; this same firm has paid over $1.1 billion since beginning operations in the Barnett Shale (The Perryman Group, 2011).

3.2 OBJECTIVE AND BACKGROUND

The key objective of this chapter is to describe shale gas resource systems in terms of their organic geochemical characteristics. Shale gas resource systems are petroleum-source rocks, that is, organic matter from which petroleum is generated, that also act as a reservoir rock. Organic matter is typically only a minor constituent of the total rock matrix that consists primarily of clays, silicates, and carbonates in varying proportions. It will often only be about 5% by mass or about 10% of the volume of a typical marine petroleum-source rock. It is this organic matter that yields petroleum, that is, gas and oil, upon increasing temperature exposure due to various sources of heat such as burial and heat flux from the mantle, hydrothermal heating, and heating by radioactive decay. Insoluble organic matter or kerogen is formed at temperature below 90°C, and petroleum generation occurs over a temperature range of about 90–150°C with peak generation occurring at about 120°C for marine carbonates and about 135°C for marine shales using BP's organofacies "A" and "B" kinetic data (Pepper and Corvi, 1995) at 2°C/ma.

While kerogen cracking is thought of as being the principal source of hydrocarbons, they are in fact primarily derived from decomposition of soluble bitumen (Behar and Jarvie, 2013; Behar et al., 2008a), the primary product of kerogen cracking. In laboratory experiments, kerogen yields less than 35% hydrocarbons, whereas bitumen cracking yields the remainder of generated hydrocarbons. Hydrocarbons are strictly carbon- and hydrogen-bearing molecules, whereas nonhydrocarbons, such as resins and asphaltenes, are the primary constituents of bitumen. To distinguish insoluble kerogen cracking from petroleum cracking, the terms primary and secondary cracking are used, respectively.

Kerogen is derived from deposited, preserved, and insolubilized biomass, which can be deposited in lacustrine, marine, fluvial–deltaic, or terrestrial settings. However, the predominant setting for highly productive shale gas systems to date has been in marine depositional settings. With such systems typically having original hydrogen indices (HI) from 350 to 700 mg/g TOC, these source rocks are classified as Type II kerogen based on chemical and visual analyses that reflect the depositional settings (Jones, 1984; Tissot and Welte, 1978). These kerogens are organic-rich, typically greater than 2.00 wt% TOC with sufficient hydrogen content to form abundant amounts of hydrocarbons. In order to explain the high content of gas in commercial shale gas systems, there must be sufficient organic matter to yield petroleum upon cracking, but also to account for retention of secondary products, for example, bitumen. Retention is a function of various factors such as permeability and pressure, but also the affinity of the polar constituents of bitumen and petroleum, that is, resins and asphaltenes, as well as hydrocarbons to be adsorbed and imbibed in the organic matter matrix. The molecular size and structural complexity of organic matter also act as a molecular sieve retaining a range of petroleum compounds. This combined retention by adsorption and absorption may be best referred to as sorption (Levine, 2013).

The most labile portion of kerogen cracks in the oil window as do most of the asphaltenes and a substantial portion of the resins. Saturated hydrocarbons begin to crack in the late oil window depending on molecular size and branching. This is most obvious in condensates where the hydrocarbon composition is highly concentrated in the C_{20-} carbon range. Refractory kerogen, C_{20-} hydrocarbons, hydrocarbon gases, and alkylated aromatics crack in the gas window. This portion of cracking occurs above about 150°C (Claypool and Mancini, 1989). This not only results in enhanced volumes of gas but also removes the more polar and viscous components of petroleum that tend to restrict petroleum flow from such tight rocks with sorptive kerogen and bitumen.

The extent of kerogen conversion is indicated by thermal maturity, although this must be in the context of understanding the kinetics of decomposition of organic matter. Oil window maturity indicates that the primary product is oil, but gas is also generated albeit in substantially lesser quantities depending on the source rock type with Type III source rocks generating more gas relative to oil (Tissot and Welte, 1978). Thus, gas kicks during drilling can be found in the oil window despite lower volumes of generated gas, and as such, do not necessarily indicate a shale gas play. It is in the gas window where shale gas resource systems become commercially productive. This is due to the volume of gas being highest due to both kerogen and retained petroleum cracking. In addition, this process also removes the larger, more polar molecules, thereby enhancing flow out of such low permeability rock.

3.3 KEROGEN QUANTITY AND QUALITY

A key component in the volume of petroleum that can be generated is the relative quantity and quality of organic matter preserved from the deposited biomass. Organic richness is

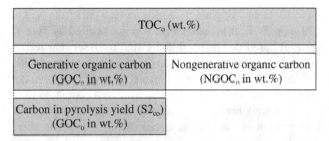

FIGURE 3.1 A diagrammatic model of TOC showing a hydrogen-rich portion as generative organic carbon (GOC) and a hydrogen-poor portion as nongenerative organic carbon (NGOC). The percentage of GOC is determined from the original hydrogen content. Subscript "o" indicates original and "pd" indicates present-day.

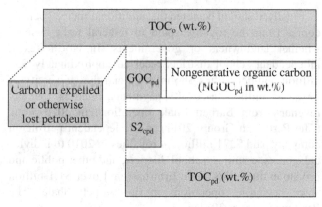

FIGURE 3.2 In this diagrammatic model of TOC, there is a net loss of organic carbon as a result of expulsion by the amount of carbon in expelled petroleum. There is additional organic carbon loss in the sample retrieval, handling, storage, and processing steps. Subscript "o" indicates original and "pd" indicates present-day, and "cpd" means corrected present-day.

quantified by the amount of organic carbon, which excludes any inorganic carbon in carbonates. The quantity of organic carbon is measured and reported as total organic carbon (TOC) and is reported in weight percent of organic carbon in the rock matrix. The volume of the rock occupied by TOC is approximately double the weight percent due to its low density relative the inorganic rock matrix.

A simple diagrammatic model of immature (and oil-free) TOC referred as original TOC (TOC_o) shows a somewhat arbitrary differentiation between hydrogen-rich organic carbon and hydrogen-poor organic carbon (Fig. 3.1). The hydrogen-rich organic carbon is referred to as the generative organic carbon (GOC) as it is this portion of TOC that generates petroleum with increasing thermal exposure (maturation). The remaining portion of the TOC_o is referred to as nongenerative organic carbon (NGOC), which has no commercial petroleum generation potential. Cooles et al. (1986) referred to these constituents as labile and inert, which is less descriptive and somewhat inaccurate as the so-called inert organic carbon functions as an adsorbent and appears to play a catalytic role in directing hydrocarbon product distributions (Alexander et al., 2009; Fuhrmann et al., 2003).

While the quantity of organic carbon is important to measure a source rock's capacity to generate petroleum, hydrogen content must be assessed to determine the amount of petroleum that can be generated from a given amount of TOC. Hydrogen is always the limiting factor in gas generation, for example, while alkanes take at least two hydrogens per carbon, methane requires four. A qualitative measure of the hydrogen content of a source rock is determined by pyrolysis of the source rock. This is part of the data measured from the Rock-Eval®[1] or HAWK[SM2] pyrolysis instruments as the hydrogen index (HI). Such pyrolysis results in cracking of the GOC portion of the TOC, providing an indirect indication of its hydrogen content by calibration with hydrocarbon or other rock standards. The pyrolysis yield, referred to in instrument terminology as S2 (originally, signal 2 from the instrument), is the remaining or present-day petroleum generation potential. Except for immature source rocks, this value is the present-day or remaining petroleum generation potential ($S2_{pd}$). It is reported in mg petroleum potential per gram of rock (mg/g rock). Prior to generation of petroleum, this pyrolysis yield reflects the original GOC (GOC_o) portion of TOC and is an indication of the original petroleum generation potential ($S2_o$) (see Fig. 3.1). The $S2_o$ can be converted to organic carbon by multiplying by 0.085 (assuming 85% carbon in petroleum), which is then the carbon in GOC_o. The original HI (HI_o) value is $S2_o/TOC_o \times 100$ (mg/g TOC).

This original petroleum potential ($S2_o$) can also be converted to barrels of oil equivalent per acre-foot by simple conversion. A pyrolysis generation potential of 1 mg petroleum/g rock would be equivalent to a generation potential of about 22 boe/acre-foot assuming a rock density of 2.7 g/cc and a petroleum density of 0.85 g/cc. If the thickness of the source rock is known, this can be converted to boe/section remembering that it is an indication of the total petroleum generation potential and does not account for incomplete maturation of kerogen. If $S2_o$ is known, the actual amount of petroleum generated at a given thermal maturity can be determined by subtracting $S2_{pd}$ from $S2_o$. The extent of kerogen conversion or transformation ratio (TR) can be calculated as, e.g., $1176\,(HI_o - HI_{pd})/(HI_o(1176 - HI_{pd}))$ or a simple version as $(HI_o - HI_{pd})/HI_o$.

When petroleum is expelled, there is a net loss of organic carbon due to the carbon in the expelled petroleum (Fig. 3.2). There is also carbon loss from cuttings, SWC, or core due

---

[1] Rock-Eval® is a registered trademark of Institut Francais du Petrole. All rights reserved.
[2] HAWK[SM] is a sales mark of Wildcat Technologies. All rights reserved.

to volatilization and evaporation of hydrocarbons during drilling operations, sample retrieval, handling, storage, and sample preparation prior to analysis.

3.4 SAMPLE TYPE AND QUALITY

Oftentimes, geochemical reconnaissance work is completed on archived cuttings that have been stored for years to decades. Core is sometimes available, but in the past cores were seldom taken in shales. Comparison of archived cuttings to fresh cuttings and core often show lower values in basic geochemical measurements such as TOC, Rock-Eval, and vitrinite reflectance for the older samples (Jarvie et al., 2007; Steward, 2007). This was noted in Barnett Shale wells drilled by Mitchell Energy & Development Corp. (MEDC) when wells were offset and cored (Steward, 2007).

In 2011, an independent oil company, Gunn Oil Co., drilled a well in Fisher County, Texas, that offset a well drilled in the 1980s, both of which penetrated Pennsylvanian period shales. A comparison of fresh core and cuttings to the archived cuttings shows significant changes in geochemical results (Fig. 3.3). In Figure 3.3, inorganic carbonate carbon (CC) is higher in the older samples suggesting oxidation. TOC, free-oil yield (S1), and pyrolysis yield (S2) are all significantly higher in the fresh samples compared to the 20-year-old cuttings. Note that oil yields (S1) may be lower depending on pyrolysis instrumentation, for example, Rock-Eval 6 instruments have 25–35% lower free-oil yields as shown by the Norwegian Petroleum Directorate (NIGOGA, 2000). The organic carbon dioxide (S3) released from kerogen is about the same in all samples. While it might be inferred that there was a difference in organofacies between the wells, the projected HI values from a linear fit of pyrolysis yields (S2) to TOC was 430 and 423 mg/g TOC, respectively, suggesting no organofacies differences. As suggested earlier, the source of difference may be the result of oxidation of the old samples as the carbonate carbon yields increased by a factor of 3. Although the flame ionization detector (FID) does not respond to carbon dioxide, it is likely that the increase in nonhydrocarbon gases such as carbon dioxide released into the S2 effluent during pyrolysis dilutes, and thereby reduces the FID response in the old cuttings. In other work, highly weathered outcrop samples may cause the FID flame to be extinguished.

Other data published by the Arkansas Geological Survey shows that cuttings data on the Fayetteville Shale often reflect a mixture of overlying organic lean rock (cavings) with Fayetteville Shale. Values typical of the Fayetteville Shale are often reached about 30–50 ft into the shale (Li et al., 2010). The opposite trend carries through the base of the Fayetteville Shale into underlying organic lean intervals that gradually decrease after about 30–50 ft out of the shale. These are not errors in formation tops but rather cavings of overlying sediments thereby affecting the geochemical results.

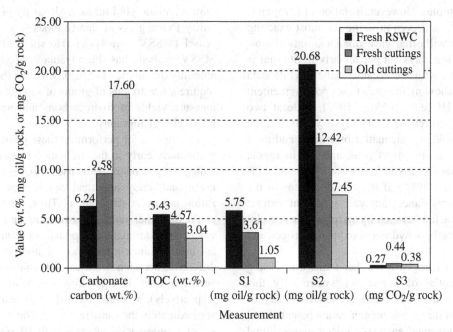

FIGURE 3.3 When fresh rotary sidewall core (RSWC) and cuttings are compared in source rock intervals, RSWC TOC values are higher likely due to dilution of the cuttings as part of the drilling process, generally over 10 ft intervals, and variability in TOC over the cuttings interval. When 20-year old cuttings are compared to fresh cuttings and fresh RSWC TOC values, both in the corresponding source rock unit, the TOC is higher in fresh samples.

3.5 KEROGEN TYPE AND COMPOSITIONAL YIELDS

Typical shale gas resource systems have HI_o values of about 350–700 mg/g TOC typical of Type II kerogen (Espitalie et al., 1977; Jones, 1984). At such values, only 30–60% of their original TOC can be converted to petroleum with the difference being the percentage of $NGOC_o$ (Jarvie, 2012a).

A detailed visual and chemical classification scheme modified from Jones (1984) and Hunt (1995) provides various characteristics of different kerogen types and the primary products through the oil window (Table 3.1). All kerogen types except Type IV can yield commercial amounts of petroleum that will ultimately crack to gas. Type IV organic matter will only yield minor amounts of dry gas but could act as a carbon catalyst.

Prior to the development of pyrolysis instruments for source rock analysis in the 1970s (Barker, 1974; Espitalie et al., 1977), atomic hydrogen and oxygen to carbon ratios (H/C and O/C) were commonly used for chemical classification of kerogen type following the conventions used in the coal industry. Coal petrologists describe coals as Type I, II, or III depending on organoclasts identified in kerogen. These organoclasts had specific ranges of H/C ratios as determined by elemental analysis. Atomic H/C ratio is more precise than HI in assessing hydrogen content due to working with isolated kerogen, thereby eliminating any clay adsorption effects as well as any oil and bitumen in the pyrolysis peak as is common with pyrolysis of organic-rich–source rock samples. However, isolation of kerogen is not a simple process and even utilizing the most exacting isolation procedures will often leave small amounts of inorganic matter including clays but particularly pyrite that is intimately associated with the organic matter. While data has been published showing the sometimes poor agreement between H/C and HI (e.g., Baskin, 1997), at least two things can affect measurement of pyrolysis yields: (1) the presence of extractable organic matter or organic additives that carryover into the pyrolysis peak and (2) inorganic adsorption effects particularly in leaner source rocks (<2% TOC) (Espitalie et al., 1984). If there is double as to the relative hydrogen abundance, analyze a solvent extract rock sample or extracted kerogen by pyrolysis or run elemental analysis for carbon, hydrogen, oxygen, nitrogen and sulfur (CHONS).

Visual analysis of kerogen is another tool for assessing the petroleum potential of a resource system. By this measurement, the population of hydrogen-rich macerals provides an indication of the petroleum generation potential and can be used with elemental analysis to determine original H/C values (Baskin, 1997). These can be converted to HI values using the formula of Orr (1981), that is, $HI = (694 (H/C - 0.29) - (800\ O/C))$.

Primary organic macerals found in kerogen include gas-prone vitrinite, oil-prone exinite and alginite, as well as hydrogen-poor inertinite. These organoclasts as identified by Alpern (1980) can be used to evaluate their combined oil and gas potential (Fig. 3.4). Also using the approach of Baskin (1997), the original H/C ratio can be estimated. Marine-source rocks will often have a mixture of these organoclasts, and this can alter the distribution of products expected at a given thermal maturity.

While kerogen type using hydrogen indices is generally acceptable particularly with larger data sets on extracted rock samples, in new plays both elemental and visual kerogen analyses should be completed in addition to TOC and pyrolysis analyses on whole and extracted rock samples. As the kerogen isolation step has been utilized for elemental analysis of CHONS, it should then also be used for visual kerogen analysis and vitrinite reflectance as visual inspection of the maceral colors will aid the determination of the autochthonous vitrinite population.

While thermal maturity is often used to provide an indication of the expected products based on the level of conversion of kerogen or these individual organoclasts, Waples and Marzi (1998) demonstrated vitrinite reflectance is not a universal indication of the level of conversion of a given kerogen. If feasible, it would be preferable to complete compositional analysis via laboratory maturation techniques to determine what products are present at a given level of transformation (conversion) of kerogen. Such maturation techniques in closed systems yield products akin to those generated by geological processes. Closed system techniques include gold tubes, hydrous pyrolysis cells, or high-purity Pyrex glass or quartz tubes (e.g., microscale-sealed vessel (MSSV) pyrolysis) (Horsfield et al., 1989, 2015). MSSV analysis has the advantage of working with small amounts of sample (1–5 mg), whereas the other techniques require a few to tens of grams of sample and can be used to measure yields of hydrocarbons and nonhydrocarbons by liquid chromatography.

One reason for performing these maturation experiments particularly early in the evaluation of new plays is because bulk pyrolysis instruments use a FID. While a FID is a very useful and handy analytical tool, it responds to carbon ionization not hydrogen content. Thus, the hydrogen potential could be either under or overestimated in some systems, where either aromatic compounds or saturated hydrocarbons are in abundance relative to one another. For example, the difference between two 6-carbon atom compounds, benzene versus hexane, is eight hydrogen atoms (C_6H_6 vs. C_6H_{14}, respectively), whereas the FID will respond to each with approximately the same response. This was demonstrated in select source rocks where a high HI source rock (487 mg petroleum potential/g TOC) yielded primarily gas (Alum Shale), and a moderate HI (310 mg petroleum potential/g TOC) yielded primarily oil (Nigerian Type II/III source

TABLE 3.1 Kerogen-type descriptions[a]

Organic matter classification	Kerogen description	Coal petrography	Maceral description	Kerogen-type assignment	H/C	O/C	HI	OI	Depositional environment	Internal structures	Principal products
Sapropelic	Algal amorphous	Alginite		Types I, II	>1.4	~0.1	>700	10–40	Anoxic (saline); usu. lacustrine; rare marine	Finely laminated	Oil, gas
Sapropelic	Amorphous algal	Liptinite (exinite)	Amorphous	Types I, II	1.2–1.4	~0.1	350–700	20–50	Anoxic usu. marine	Laminated; well bedded	Oil, gas
Sapropelic	Hebaceous		Sporinite Cutinite Resinite	Type II	1.0–1.2	~0.2	200–350	40–80	Variable; often deltaic	Poorly bedded	Gas, oil Gas, oil Oil, gas
Humic	Woody	Vitrinite	Telinite Collinite	Type III	0.7–1.0	~0.4	50–200	50–150	Mildly oxic; shelf/slope; coals	Poorly bedded; bioturbated	Gas
Humic	Coaly	Inertinite	Fusinite Micronite Sclerotinite	Types III, IV	0.4–0.7	~0.3	<50	20–200	Highly oxic	Massive; bioturbated	Dry gas

[a] Adapted and modified from Jones (1984) and Hunt (1995).

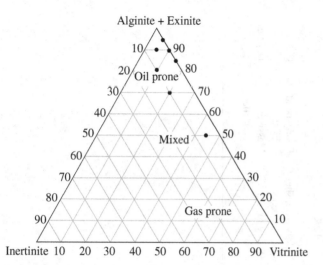

FIGURE 3.4 A ternary plot of recorded percentages of primary organoclasts from a visual kerogen assessment are plotted, and the relative abundance of oil prone versus gas prone organic matter is visualized.

rocks) (Jarvie, 2000, Compositional Kinetics of Select Petroleum Source Rocks, unpublished data).

3.6 THERMAL MATURITY

Knowing the TOC and relative hydrogen contents from H/C, HI, or organoclasts is key to evaluating the overall petroleum generation potential. With this information, an indication of the maximum temperature exposure of a sediment is also needed, that is, measurement of thermal maturity. Thermal maturity is used to describe the organic matter decomposition as being in the oil or gas window. While the oil window is often thought to be exclusive of gas, some gas is generated in all source rocks. However, the oil window has predominantly liquid petroleum products with varying amounts of gas relative to oil depending on kerogen type, whereas the gas window has only remnants of liquid petroleum and much more gas depending on how mature the system is.

Vitrinite reflectance is an indication of the maximum paleo-temperature exposure to which a sediment has been exposed. This can be quite different from bottom-hole temperature (BHT) values as sediments are often buried deep and subsequently uplifted to lower temperatures. While temperature is the key factor in petroleum generation and related increase in reflectivity of vitrinite particles, heating rate also plays a role and can result in variations of upward of 10°C in peak generation rates.

Vitrinite is an organoclast derived from catagenetically altered woody plant tissues, that is, fossilized wood particles. When such particles are identified as autochthonous in an isolated kerogen or whole rock mount, a light beam is shined onto the polished surface and the amount of light reflected recorded by a photomultiplier. If vitrinite is present, a histogram of the readings is recorded with the mean value and standard deviation of the indigenous population recorded. Thus, between 1 and 50 readings are completed depending on the presence of vitrinite particles with the microscopist determining the indigenous population of readings. The mean value is then often mapped by geologists in construction of thermal maturity maps.

As expected, such particles are very common in Type III organic matter deposited in fluvial-deltaic or terrestrial settings. For example, coals, coal lenses, coaly shales, or near-shore organic lean samples are generally the best samples for determination of vitrinite reflectivity. In Type II organic matter that has been derived from organic matter deposition in deep marine settings, vitrinite particles are far less common. If present, it must be ascertained whether these particles are autochthonous or allochthonous. Type II kerogens, from which all the highly commercial North American shale gas has been derived, have minimal amounts of indigenous vitrinite resulting in difficulties in measurement and consequently, interpretation of thermal maturity.

The optimum methodology for performing vitrinite reflectance is to complete a well profile including samples from above the zone of interest at 500 ft intervals. Type III shales or coaly intervals are very helpful for constructing such a profile. While samples from coal lenses are ideal for a high concentration of indigenous vitrinite, it is also possible that these may cave into lower sections, thereby yielding lower reflectivity values indicative of the overlying coal lens(es). Samples of the shale of interest should also be measured and in particular the color and any fluorescence of organic particles noted. Lighter yellow and browns indicate a less mature indigenous population whereas darker colors help the microscopist focus on the more mature population. Often the organic-rich shale of interest will have values different from the profile either due to the presence of bitumen or pyrobitumen, which have different reflectivities from the morphologically similar vitrinite particles (Landis and Castaño, 1995).

An example maturation profile for a Barnett Shale well is illustrated in Figure 3.5. Fitting all the measurements including those in the Barnett Shale suggests a good correlation (dotted line) but with the Barnett Shale in the late oil window. When the fit is projected through the Barnett Shale using only the mean vitrinite reflectance measurements from the overlying sediments, a different interpretation is reached, that is, the Barnett Shale is in the early gas window at 1.10–1.25%Roe. To resolve this discrepancy, the chemical data such as HI_{pd} values help determine the correct reading. In this well because the HI values are less than 100 mg/g TOC with over 4% TOC, it is interpreted that the Barnett Shale is in the gas window matching the uphole profile (dashed line). The Barnett Shale at this early gas window thermal maturity still contains small amounts of bitumen and pyrobitumen.

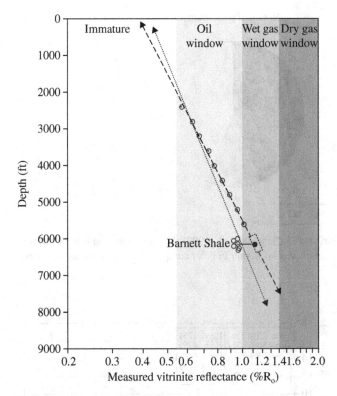

FIGURE 3.5 A vitrinite reflectance maturation profile above and through the Barnett Shale provides a relatively good fit at an R^2 of 0.93; however, there is noticeable reduction in the Barnett Shale itself. When a fit is projected using only samples above the Barnett Shale, a much better fit is obtained ($R^2 = 0.98$) and the Barnett Shale is projected to be in the early gas window. This projection is supported by the present-day hydrogen indices of the Barnett Shale and thereafter by production.

As polishing is a key to achieving reliable reflectivity measurements and the fact that neither bitumen nor pyrobitumen are amenable to polishing, they reflect less light, thereby accounting at least in part for the lower reflectivity. In fact, this well has now produced over 2 bcf of gas with fair amounts of natural gas liquids (NGLs).

It is recommended that not only vitrinite measurements be completed for thermal maturity determinations but also chemical data from TOC, pyrolysis, and any elemental analysis be acquired. Pyrolysis provides a chemical measure of thermal maturity, which is the temperature at maximum evolution of the pyrolysis (S2) peak, named T_{max}. T_{max} is dependent on kerogen type but also the shape of the pyrolysis peak. At late gas window maturity, there is no distinguishable pyrolysis peak that yields accurate T_{max} readings. However, at such high maturities, knowing the kerogen type aids this chemical assessment as the HI or H/C values will decrease with increasing thermal maturity. If an original hydrogen index is known or estimated, the extent of kerogen conversion provides a means of estimating thermal maturity as well using the kerogen transformation ratio. TR is most commonly used as a kerogen conversion ratio, but there are multiple definitions for TR resulting in confusion on the formulation (Espitalie et al., 1984; Jarvie, 2012a; Pelet, 1985; Tissot and Welte, 1978). Comparable values for the oil and gas windows are shown in Table 3.2 for a low sulfur Type II kerogen.

TABLE 3.2 Thermal maturity ranges or values for a low sulfur Type II kerogen

Zone	$\%R_o$	T_{max} (°C)
Black oil	<0.95	<450
Volatile oil	0.95–1.20	450–464
Gas condensate	1.20–1.29	465–470
Wet gas	1.30–1.59	>470
Dry gas	>1.60	

As condensates are often present albeit in minute amounts in shale gas systems, they are amenable to light hydrocarbon analysis but not biomarkers that have been cracked in the early gas window. The light hydrocarbons can be used to characterize the source organofacies as well as thermal maturity using various parameters. In addition, they can be used to predict intrinsic GOR values (Jarvie, 2001; Mango and Jarvie, 2001) or as improved to include higher molecular weight kinetically controlled distributions of iso-alkanes (Jarvie, 2000, Compositional Kinetics of Select Petroleum Source Rocks, unpublished data).

An additional means to assess thermal maturity from light hydrocarbons condensates is by diamondoid analysis (Dahl et al., 2013). They may also be used to assess the extent of oil cracking. Diamondoids are very stable and become predominant at high thermal maturity much as do aromatic hydrocarbons, where they can be used as indicators of thermal maturity with proper calibration.

As shown in Jarvie et al. (2007), it is recommended that various maturity parameters be risked in combination to ascertain the interpreted thermal maturity and relationship to products generated at given maturity values.

3.7 ORGANOPOROSITY DEVELOPMENT

As organic matter is converted to petroleum and carbonaceous char, pores are created that are filled with petroleum (Jarvie, 2006; Loucks et al., 2009; Reed and Loucks, 2007). As kerogen is converted to petroleum, there is a reduction in its mass and volume. However, this reduction is countered by swelling of the organic matter matrix due to the presence of solvents, that is, petroleum (Ertas et al., 2006). Even chemically isolated kerogen often contains free hydrocarbons that are evident in GC fingerprints of kerogen extracts or in pyrolysis gas chromatography. This is also confirmed, for example, by fingerprinting of early mature Bakken

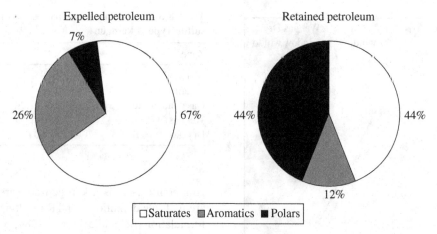

FIGURE 3.6 Data from laboratory maturation of the Toarcian Shale, Paris Basin (Sandvik et al., 1992) showing expulsion fractionation as the expelled oil is higher in saturated and aromatic hydrocarbons, and the retained oil is higher in polar compounds (resins and asphaltenes).

Shale samples in the Parshall and Sanish fields of the Williston Basin with no observable organoporosity. The pyrolysate of Bakken Shale has a visible shoulder indicative of petroleum being "carried-over" into the pyrolysis or kerogen peak. Barker (1974) referred to this peak as a "micro-reservoir" and this appears to be correct due to simple retention of more polar constituents of petroleum within the kerogen matrix. However, upon treatment with organic solvents, this shoulder is removed and the pyrolysis yield reduced by about 10% in most cases. Completion of a GC fingerprint of this material shows the presence of C_5–C_{40} paraffins and GOR averaging 400 standard cubic feet of gas per stock tank barrels of oil (scf/stb) (Jarvie et al., 2011). The larger, more viscous and polar compounds are the predominantly retained petroleum as shown experimentally by Sandvik et al. (1992), but their experiments also showed substantial percentages of saturated and aromatic hydrocarbons as well as polar constituents in the retained oil (Fig. 3.6). Once the kerogen and retained petroleum are cracked to gas, the pores within the organic matter are very obvious under SEM (Loucks et al., 2009; Reed and Loucks, 2007). The presence of pores is necessary to explain the storage of this portion of the unexpelled petroleum even though it is not usually observable in the oil window under SEM imaging of argon ion-milled sections.

Organoporosity development is hypothesized to be a function of TOC and the extent of kerogen conversion (Jarvie, 2012a; Jarvie et al., 2007). As both TOC and maturity increase, the potential for organoporosity increases due to conversion of organic matter to petroleum. The amount of calculated organoporosity increases depending on the overall TOC, the amount of TOC as GOC, and thermal maturity (Fig. 3.7). This calculation assumes the same HI_o for all TOC values (500 mg/g) and densities of 1.0 g/cc for GOC and 1.4 g/cc for NGOC, which is based on density data from

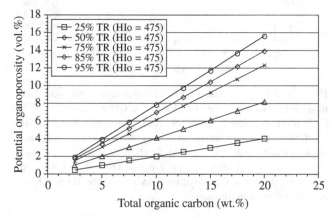

FIGURE 3.7 Computing the potential for organoporosity development takes into account original TOC, original hydrogen index, and the extent of kerogen transformation. This shows the potential for organoporosity development at various levels of transformation (maturity) for different amounts of TOC with an original HI of 475 mg/g (40% convertibility).

Okiongbo et al. (2005) and Vandenbroucke and Largeau (2007). Thus, simple arithmetic and fingerprinting of stored petroleum suggest there is organoporosity in the oil window although it has not been identified as such.

In some shale gas resource plays, the organoporosity storage appears to be the predominant means of storage rather than matrix storage, such as in the Barnett Shale. Barnett Shale. Although this is not necessarily a detriment to development of shale gas, it does hinder shale oil development as the petroleum is closely associated with the kerogen, which retains petroleum more tenaciously than inorganic matrix pores of low adsorptive affinity (Schettler and Parmely, 1991). This is one reason why higher maturity shales that have undergone 80% petroleum cracking are better shale gas producers, whereas the better shale oil systems are those where oil has been expelled into a nearby nonsource lithofacies, that is, a hybrid system (Jarvie, 2012b).

3.8 GAS CONTENTS

Gas contents are derived from measuring gas as it desorbs from canistered core with a projection of lost gas contents. Gas contents are a key component in the determination of sufficient volumes of gas for commercial development. However, these are necessarily completed postdrill as canistered core samples are required for accurate gas desorption and adsorption experiments. An alternate predrill approach has been utilized in various North American shale gas plays with good success by computing the total generation potential of the formation based on restored generation potentials, the percentage of primary oil versus primary gas in the original generation potential, estimated expulsion efficiency, and finally retained petroleum gas cracking yields (Jarvie et al., 2007). This stochastic model is then converted to mcf/acre-foot and multiplied by formation thickness less expulsion losses to obtain gas content per section, that is, bcf/square mile.

However, core desorption experiments are necessary to confirm the commerciality of these predrill calculations. One foot sections of core are placed in gas desorption canisters as quickly as possible upon retrieval from the coring tool. Gas is then desorbed through time ultimately yielding the desorbed gas content. A key estimation from this technique is the loss of gas content, that is, gas that escaped prior to enclosure of the core section in a canister. As this technique evolved from the coal-bed methane (CBM) industry, a US Bureau of Mines (USBM) technique has been utilized (Diamond and Levine, 1981). This technique extrapolates a linear fit of the early desorbing gases to time zero to compute the amount of missing gas. However, in most highly productive shale gas systems, more than 50% of the gas is in the form of free gas, that is, it is not adsorbed in the shale gas reservoir. As these gases escape exponentially, the USBM technique will often underestimate the amount of lost gas especially from high pressure, higher porosity (7–14%) systems such as the Haynesville Shale in East Texas–Northern Louisiana.

Data from a pressure core drilled in the normally pressured New Albany Shale, Illinois Basin, Kentucky, USA, demonstrates the exponential gas loss (A. Young, 2001, Carbon-number fractionations between sources and associated oils, unpublished manuscript) (Fig. 3.8). The logarithmic fit does not include the initial value (negative in this construct), but it is simply a projection from the positive data points. The USBM methodology would be a straight-line projection from early points in the curve, obviously yielding a lower estimated lost gas content. The lost gas content can

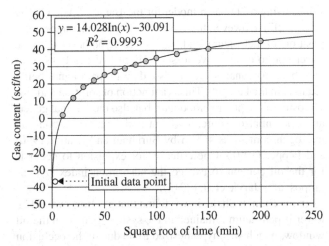

FIGURE 3.8 Gas desorption data from a New Albany Shale that was pressure cored in the 1970s. The logarithmic fit of data includes only the positive gas desorption yields, but the original gas content value is shown relative to the fit using the natural log equation of best fit to the data points and extrapolation to time=0.

be a substantial portion of higher porosity and overpressured systems such as the Haynesville Shale.

3.9 EXPULSION–RETENTION OF PETROLEUM

Expulsion from source rocks and migration accounts for the petroleum found in conventional reservoirs. During expulsion, there is a fractionation of petroleum as a result of various geochemical functions such as the fugacity of gases relative to liquids, molecular size, and polarity of molecules in low permeability mudstones. Conventional reservoir rocks containing unaltered petroleum always have lower amounts of resins and asphaltenes and higher amounts of saturates and aromatics than their corresponding source rocks. Expulsion fractionation is also shown in the experimental data of Sandvik et al. (1992) by the variable percentages of saturates, aromatics, and polars (resins and asphaltenes) (see Fig. 3.6). It is this retained petroleum in shale reservoirs that is higher in the polar resins and asphaltenes and residual kerogen that crack and accounts for the yield of gas at elevated thermal maturities.

Expulsion thresholds have been difficult to estimate but it appears that more oil is retained than often estimated. This was suggested in the case of the Barnett Shale where it is estimated that only about 55% of petroleum was expelled (Jarvie et al., 2007). Sandvik et al. (1992) estimated that approximately 10 g of petroleum per 100 g of total organic matter would be retained in the source rock, whereas Pepper (1991) estimated up to 200 mg/g TOC, although values were estimated to be much lower for gaseous hydrocarbons. A value of 200 mg oil/g TOC was used to match the gas-in-place values using a 2D basin model for the Barnett Shale (Jarvie et al., 2007). However, this is variable depending on the composition of petroleum and values are hypothesized to be lower due to evaporative losses of light hydrocarbons prior to laboratory analysis and also due to petroleum being retained in the kerogen. This is a function of sorption and low viscosity of the polar compounds, but also the original generation potential of source rocks in both organic carbon and hydrogen content as shown by Burnham and Braun (1990) and Pepper (1991). One criterion for expulsion to occur is for the oil content to exceed the adsorption index (AI) as proposed by Jarvie et al. (2013).

In the Barnett Shale, it was found that gas flows would be lower if petroleum remained in the system, that is, in the oil window, which was hypothesized to be due to the occlusion of nanodarcy permeability by more viscous and polar petroleum constituents (Jarvie et al., 2007). Thus, it is important to have these constituents cracked as completely as possible to limit their occluding capacity and to obtain increased pressure for optimum gas flow. This is applicable to a lesser extent in shale oil resource system plays.

3.10 SECONDARY (PETROLEUM) CRACKING

The decomposition of organic matter into petroleum is dependent primarily on organic matter composition and structure as well as the temperature regime in which it has resided. It is well known that kerogen decomposes into petroleum under increasing temperature by cracking of the weakest bonds first; as the temperature continues to increase under deeper burial, the more refractory (difficult to break) bonds begin to crack until the final products are methane and carbonaceous char.

What is commonly misunderstood is the difference between petroleum versus oil cracking. Oil cracking is often defined as alkane (paraffin) cracking (e.g., Behar and Vandenbroucke, 1996; Burnham et al., 1997; Fabuss et al., 1964; Ford, 1986; Tsuzuki et al., 1999), which is quite different from petroleum cracking. As defined earlier, petroleum is any of the secondary products resulting from kerogen decomposition, which include bitumen, oil, and gas. Bitumen cracking begins nearly contemporaneously with its generation from kerogen (Behar et al., 2008a, b). Similarly, it has been shown experimentally that asphaltenes decompose over the temperature range of kerogen by asphaltene pyrolysis (di Primio et al., 2000). This is substantiated by the empirical observation of decreasing asphaltene content in more mature source rock extracts. The same is true of resins shown experimentally (Behar et al., 2008a, 2010) and again empirically in fractionation yields from oils of various maturities. At the end of the oil window, cracking has largely shifted from cracking of these polar compounds to the cracking of alkanes and ultimately to gas cracking and demethylation of aromatics leading to a graphitic-like structure of metamorphosed kerogen and carbonaceous char.

3.11 UPPER MATURITY LIMIT FOR SHALE GAS

While thermal maturity above the latest oil window is indicative of increasing amounts of gas, there does appear to be an upper limit for increasing gas content. From drilling results in areas above about 3.5%R_o, no commercial production has been achieved, although an independent gas company's well, the Hallwood Petroleum Right Angle Minerals #1, flowed about 600 mcf of gas per day at about 3.7%R_o from a Fayetteville Shale well in the eastern portion of the Arkoma Basin, Arkansas. While drilling tests in high maturity areas have not been extensive, no commercial production has occurred in the Fayetteville and Marcellus shales at such high maturities. The reasons for this are unknown, but it is certainly not the stability of methane alone under related temperature regimes at such levels of thermal maturity. However, alteration of methane may be transpiring in the presence of hot mineral matrix, associated gases, and water. For example, it was postulated by Barker and Takach (1992) that methane oxidation by water could be a possible mechanism for loss of methane and increased yield of carbon dioxide as well as hydrogen. In their discussion, this was preceded by the formation of methane as a result of reaction of graphitic carbon with water. An additional alternative explanation could be alteration of the rock fabric and destruction of pores at such levels of maturity. However, in the experimental work of Barker and Takach (1992), loss of methane occurred under laboratory conditions where the initial concentration of methane was not dependent on *in situ* storage, that is, gas and sandstone were combined in a pressurized cylinder and heated. In their experiment at ca. 3.5%R_o, methane volume was reduced 50% over the starting volume (Fig. 3.9). In this particular experiment, only methane was present so no additional gas was generated from refractory kerogen or other organic matter as would occur in a source rock. These results suggest that higher yields of methane occur at equivalent vitrinite reflectance values under 3.5%R_o. While the empirical and experimental evidence of Barker and Takach (1992) are not proof of methane destruction at high thermal maturity and temperatures, when combined with the poor drilling results, certainly suggests greater risk for commercial amounts of gas in such high maturity areas.

On the other hand, it has been shown experimentally that methane gas generation at elevated thermal maturities does occur (e.g., Behar and Jarvie, 2013; Behar et al., 2008a; Erdmann and Horsfield, 2006; Mahlstedt and Horsfield, 2012). Their separate experimental results show that kerogen can generate about 30% of its total potential for gas at

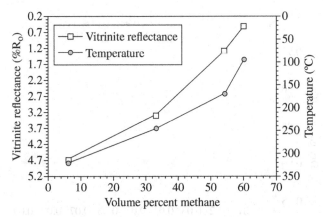

FIGURE 3.9 Data from Barker and Takach (1992) laboratory experiments on methane loss in a conventional reservoir sandstone. At 3.5%R_o, there is approximately a 50% reduction in the volume of methane.

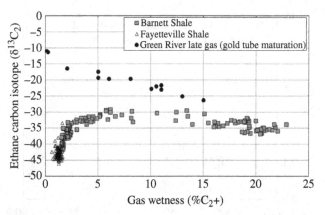

FIGURE 3.10 Using data and graphic routine from Zumberge et al. (2012), ethane rollover or isotopic reversal is shown to occur at about 5% gas wetness ratio (GWR). Laboratory maturation data in closed system using gold tubes on the Green River Oil Shale does not show rollover as it follows the normal thermal decomposition trend to very heavy values. This indicates that the mechanism is different when rollover occurs from the classical thermogenic decomposition mechanism.

maturities above 2.00%R_o. This is largely from gas cracking and dealkylation of aromatics as shown by Fusetti et al. (2010a, b).

3.12 GAS COMPOSITION AND CARBON ISOTOPES

Gas composition and carbon isotopic measurements on methane, ethane, and propane provide an indication of gas type and thermal maturity. Gas type in shale gas reservoirs may be oil-associated gas, nonassociated gas, or biogenic gas. Biogenic gas is typically limited to shallow shale reservoirs usually fractured and in contact with freshwater plus temperatures less than 80°C. The dry gas ratio (DGR) ($C_1/(C_1–C_4)$) provides a simple ratio indicative of thermal maturity in most shale gas systems as do carbon isotopes of methane, ethane, and propane. Gas becomes drier with increasing thermal maturity in a closed system such as a shale resource play. There is a two-step process in gas generation in the oil and gas windows with an initial phase dominated by C_2–C_5 gases and a later phase dominated by cracking of these wet gases to methane and pyrobitumen (Tian et al., 2008). Because almost all shale gas systems usually only contain autochthonous gas, there is typically no concern of mixing of gas from a different source rock. Nonetheless, DGR and methane isotopes should be used with caution for mixed kerogen types.

Formation of pyrobitumen at gas window maturities appears to play a role in lowering energy required to break carbon–carbon bonds and enhancing methane formation (Pan et al., 2012). This role may be related to bond angle strain that increases the likelihood of carbon–carbon bond breakage (Sheiko et al., 2006). Product rearrangements due to adsorption on carbon have also been shown by Fuhrmann et al. (2003) and Alexander et al. (2009).

Zumberge et al. (2012) demonstrated ethane isotopic reversal, that is, ethane becoming lighter isotopically, using a plot of $\delta^{13}C_2$ versus a wet gas ratio (WGR). In their data, ethane isotopic reversal is obvious below ca. 5% WGR or 95% DGR at about 1.50%R_o. Laboratory data acquired from closed system pyrolysis of Green River Oil Shale does not show the ethane carbon isotopic reversal noted by Zumberge et al. (2012) (Jarvie and Behar, 2010, Carbon isotopes of late gas generation, unpublished data), suggesting that this reversal reaction is not governed by classical cracking schemes (Fig. 3.10).

When DGR is compared to the ratio of *iso-* and *normal-*butane (*i*-C_4 to *n*-C_4), there is a distinct change in the highly correlative trend that occurs above ca. 95% DGR (Fig. 3.11). The *i*-C_4/*n*-C_4 ratio increases from low values of about 0.5 to a high of about 2.0, but subsequently reverses to a low of about 0.2 at very high DGR values. Hill et al. (2007) suggested that this reversal in *i*-C_4/*n*-C_4 ratio was due to the onset of oil cracking.

As butane cracking is a potential precursor to ethane, using a ratio of *i*-C_4 to C_2 relative to DGR shows a transition zone where the DGR increases rapidly (Fig. 3.12). The DGR increases from about 83 to 92% that is indicative of transition from the earliest gas window to the dry gas window. Higher BTU gas will be found at the lower end of the transition zone and lower BTU will be found at the upper end of the transition zone. This is also shown by a cross plot of *i*-C_4/*n*-C_4 to C_2/C_3, where there is a trend of increasing *i*-C_4/*n*-C_4 that reverses with the C_2/C_3 ratio increasing rapidly to values of 25–40 (Fig. 3.13). This graphic shows a maturation trend going from black and volatile oil to condensate and lower BTU gas, ultimately transitioning to dry gas.

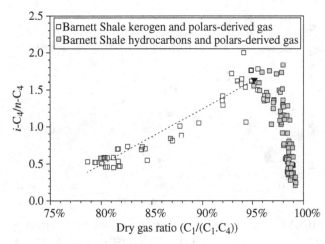

FIGURE 3.11 The dry gas ratio (or as shown by Zumberge et al. (2012) the wet gas ratio), there is an offset in the i-C_4/n-C_4 ratio moving from labile kerogen and polar cracking to refractory secondary cracking at about 95% DGR. Data from Zumberge et al. (2012).

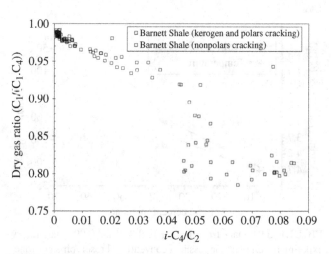

FIGURE 3.12 As i-C_4 cracking is one likely source of methane and propane and n-C_4 a source of ethane, the ratio of i-C_4/C_2 shows a distinct transition zone when compared to the DGR. Increased C_1 or DGR is noted above ca. 83–92% DGR. Data from Zumberge et al. (2012).

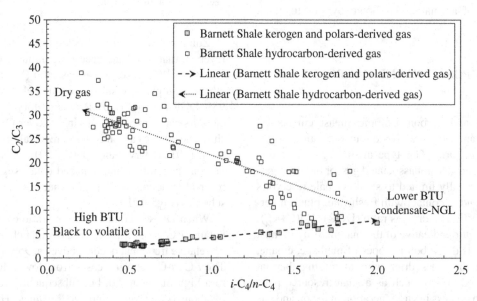

FIGURE 3.13 When the i-C_4/n-C_4 ratio is compared to the C_2/C_3 ratio, the reversal in i-C_4/n-C_4 occurs at about a C_2/C_3 of 10 where C_2 begins to dominate the ratio of C_2/C_3 indicative of n-C_4 cracking predominating over i-C_4 cracking. Data from Zumberge et al. (2012).

Characterization of primary and secondary cracking was undertaken by Lorant et al. (1998) using closed system, gold tube pyrolysis results. They showed intervals of kerogen and petroleum cracking based on ratios of ethane and propane, both molar yields and carbon isotopic values. As cracking of kerogen and petroleum proceeds, the fundamental geochemical difference is the increase in bond energies that result in cracking reactions requiring more energy (specifically increased temperature) to occur. Using a graphical scheme derived from Prinzhofer and Huc (1995) and Lorant et al. (1998) illustrates that, cracking proceeds from kerogen and generated bitumen cracking to and through paraffin cracking. This leads ultimately to light paraffin and gas cracking as well as demethylation of aromatics at late wet gas to dry gas window thermal maturities (Fig. 3.14). The latter process is one of the potential mechanisms for the formation of isotopically lighter ethane as demethylation would yield carbon isotopic signatures indicative of the original biomass and could readily recombine to form ethane. The important point of this scheme for explorationists is the changing range of petroleum products being generated from black to volatile oil to condensate-wet gas and finally dry gas. The graphical representation also illustrates the difference between ethane and propane carbon isotopic

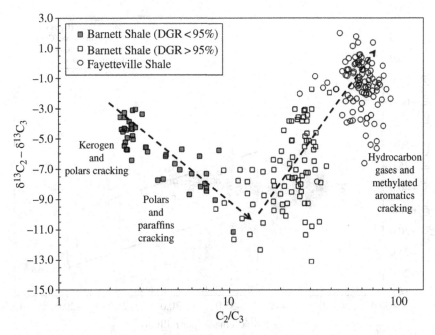

FIGURE 3.14 Utilizing the graphical scheme of Lorant et al. (1998), the change in C_2/C_3 and the decrease in the isotopic difference between $\delta^{13}C_2$ and $\delta^{13}C_3$ ratio is suggested to relate to gas cracking with other cracking processes preceding this event. Data from Zumberge et al. (2012).

values that increase during oil window cracking, but this difference is reduced at high maturity (>1.5%R_o).

A common tool used for thermal maturity assessments of gases or oil-associated gases is a comparison of methane and propane isotopic values to those of ethane as shown by Ellis et al. (2003) and Schoell et al. (2001) (Fig. 3.15). Typically, carbon isotopes become heavier (less negative) with a linear increase in thermal maturity and provide another means of assessing thermal maturity, when calibrated to shale gas in a given prospect area. Thus, data at increasing thermal maturity would plot along the trend lines toward −15‰. However, as identified by Zumberge et al. (2012), this is not true for either ethane or propane carbon isotopes above about 1.5%R_o. In the rollover zone, the graphical representation shows the gases becoming less mature, which is misleading given the fact that this occurs with increasing maturity, which is also noted by the data plotting nearest 1.5%R_o in this data set.

In addition, while many graphical representations of carbon isotopic data provide a simple linear relationship between ethane versus methane or propane, the relationship is anything but linear particularly when considering that both kerogen and petroleum are cracking (Tang et al., 2000). When ethane and propane rollover occurs, the direction of the isotopic shift reverses causing very mature samples to appear less mature or even immature as shown in output from GOR-Isotopes[SM3] (GeolsoChem) (Fig. 3.16). This is a result of isotopic fractionation but also combination kerogen types, that is, mixed Type II/Type III kerogens and secondary cracking of petroleum.

Reasons for the increasingly light values of ethane have been hypothesized to occur as a result of oil cracking (Xia et al., 2013) or via water reforming reactions (Zumberge et al., 2012). Hao and Zou (2013) have provided a detailed assessment of various scenarios and have tentatively concluded that various compositional and isotopic changes are due to a variety of factors. These include occurrence in a closed system with poor expulsion efficiencies and reactions occurring among a number of different constituents ranging from kerogen to various petroleum fractions. Such evidence was also presented by Tilley et al. (2011). This is consistent with the primary storage mechanism (organoporosity) that aids retention of petroleum by adsorption on organic carbon. Isotopic reversal of ethane has also been shown to occur under hydrous pyrolysis conditions (Gao et al., 2014).

3.13 ADDITIONAL GEOCHEMICAL ANALYSES FOR SHALE GAS RESOURCE SYSTEM EVALUATION

For initial wells drilled for shale gas prospects, science wells are an essential ingredient leading to a better understanding of new plays. One thing that has been learned with all the shale gas plays in North America is that no two behave in exactly the same way. Usually in a new area, a vertical well is drilled and cored through the shale of interest. Segments of the core are preserved in gas desorption canisters where the gas contents can be determined. These data

---

[3]GOR-Isotopes[SM] is a sales mark of GeolsoChem. All rights reserved.

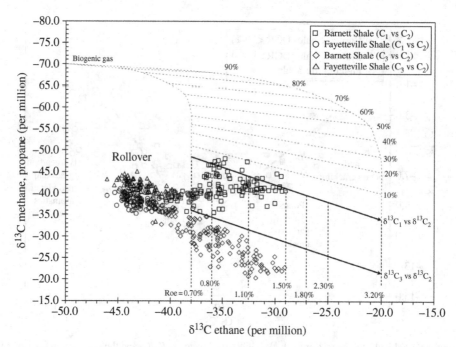

FIGURE 3.15 A commonly used linear relationship for thermal maturity utilizes $\delta^{13}C_3$ vs. $\delta^{13}C_2$ and $\delta^{13}C_1$ vs. $\delta^{13}C_2$ carbon isotopic values (Ellis et al., 2003; Schoell et al., 2001). However, during rollover this plot yields lower thermal maturity than actual due to the reversal of ethane carbon isotopes. Data from Zumberge et al. (2012). It is also necessary to calibrate the maturity lines to specific source rocks.

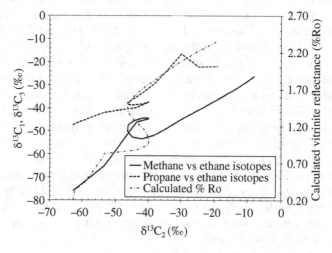

FIGURE 3.16 Tang et al. (2000) provide the basis for a nonlinear interpretation of thermal maturity from methane, ethane, and propane isotopes for a Type II kerogen, but including the effect of petroleum cracking.

in combination with core analyses provide the basis for evaluating the magnitude of the shale gas resource. Additional analyses include mineralogical and rock mechanics as well as geochemistry.

A basic analytical program for geochemistry includes taking mud gas samples for gas composition and carbon isotopes. Rock samples are analyzed for carbonate carbon, TOC, pyrolysis, vitrinite reflectance, visual kerogen analyses, and pyrolysis/gas chromatography. For marginally mature shale resource systems, it is essential to determine the quantity and quality of any remaining petroleum in the shale by solvent extraction, fingerprinting, liquid chromatography, and biomarkers, the latter more for maturity assessment.

Analytical techniques that have not been widely used include combined white light reflectance and fluorescence characteristics of the macerals that can be used to better identify normal vitrinite (vitrinite and inertinite reflectance and fluorescence (VIRF); Newman, 1997; Newman et al., 2000). Fluorescence alteration of multiple macerals (FAMM) is another maturity technique (Wilkins et al., 1995, 1998).

Pyrolysis gas chromatography (Lis et al., 2008) and GC/MS/MS (D. Rocher, 2012, QQQ thermal maturity, unpublished data) have also been used for maturity assessments.

Advanced well logging techniques include new geochemical tools such as extended hydrocarbon analysis, HAWK well-site pyrolysis, and rock-oil fingerprinting. The DQ1000™ (Fluid Inclusion Technologies, Broken Arrow, OK, USA) is used to assess product composition, water saturation, and barriers or seals. It measures petroleum compounds up to ten carbon atoms including saturated and aromatic hydrocarbons. HAWK^{SM} (Wildcat Technologies, Humble, TX, USA) is an automated pyrolysis instrument designed for well-site use to measure oil and kerogen contents and thermal maturity as well as discriminating GOC and NGOC components of organic carbon. Its output includes mass and volume of TOC and yields in barrels of oil equivalent. It is the only pyrolysis instrument that has multipoint calibration insuring good results over a broad range of pyrolysis yields and forces recalibration when standard values are not obtained. GC-Tracer™ (Weatherford International, Houston, TX, USA) provides continuous surface gas analysis during drilling using a membrane extraction to obtain optimum results. PEERI (Covina, CA, USA) has developed a laser-based analytical system for carbon isotopic measurement on methane, ethane, and propane at the well site. Geoservices has developed the FLAIR system for quantitative gas concentrations of C_1–C_5 gases and methane carbon isotope measurements in the field (Niemann and Breviere, 2010).

3.14 OIL AND CONDENSATE WITH SHALE GAS

The presence of black and even volatile oil tends to diminish gas production from shales. First, oil is certainly more viscous and has larger molecules some of which are polar meaning enhanced adsorption in nanodarcy pore throats. Second, if oil remains in the system, the maximum amount of gas has not been generated as the predominant source of gas is not kerogen conversion but conversion or cracking of secondary products, that is, bitumen and oil. Condensates do not significantly alter gas production, but such liquids are generally produced in very low amounts. Condensates are dominantly nonpolar, C_{20-} saturated and aromatic hydrocarbons. The polar materials, resins and asphaltenes, have been largely eliminated by thermogenic cracking processes. Of course, the economics of oil versus gas prices has pushed the development of shale resource systems toward oil in North America.

In shale gas systems, GOR increases as a function of thermal maturity. Oil cracking data from Claypool and Mancini (1989) show a constant gas-to-oil ratio (GOR) from 0.86 to 1.15%R_o of 1000 scf/stb. However, GOR values do vary considerably from 0.60 to 1.20%R_o with low values early in the oil window at about 100 scf/stb and values at 1.20% in the 3500 GOR range depending on kerogen type or mixing of kerogen types; this assumes no alteration processes or other gas input either biogenic or thermogenic.

There is a small drop in GOR in the early–middle oil window. This is the transition phase of gas being derived directly from kerogen to that being derived from secondary products. Such an effect was noted on pyrolysis of coal with and without bitumen, where the presence of bitumen resulted in a decrease in gas yields (Tiem et al., 2008). Ultimately, GOR increases exponentially above approximately 1.40%R_o; this is the zone of C_{20-} alkane cracking, gas cracking, and demethylation of aromatics. This is obvious from gas chromatographic fingerprints of condensates, where above this maturity the ratio of compounds less than 20 carbon atoms (C_{20-}) accounts for all of the C_{6+} oil content.

Condensate production occurs in small amounts with many of the highly mature shale gas systems. For example, the Mitchell Energy T.P. Sims #2 Barnett Shale gas well has a measured and calculated thermal maturity of about 1.60%R_o, but has produced about 7000 barrels of light condensate with a gas-to-condensate ratio (GCR) of ca. 250,000:1. Similarly, some high maturity (2.0%R_o) Haynesville Shale wells yield small amounts of condensate.

Prediction of GOR or GCR becomes an important consideration when the economics of oil are more favorable than natural gas. A reference table modified from Whitson and Brulé (2000) shows a range of GOR values with various other chemical properties (Table 3.3). GOR increases exponentially and rapid increase occurs above about 3000 scf/stb. At very high GOR values, there are still minor amounts of wet gases and condensable hydrocarbons present. The yield is the amount of liquids available given the GOR value.

TABLE 3.3 Approximate physicochemical properties of various petroleum products with projected GOR and yields

Product	C_1 (%)	C_7+ (%)	°API	GOR (scf/stb)	Yield (bbls/mmcf)
Black oil	<60	>35	<39	<1,500	
Volatile oil	61–80	15–34	40–49	1,500–3,499	667–285
Gas condensate	80–84	19–10	50–54	3,500–4,999	286–200
Wet gas	85–90	10–0.10	55–66	5,000–99,999	199–10
Dry gas	>90	<0.10	>60	>100,000	<10

These properties require calibration to specific source rocks or by organofacies within a given source rock.

3.15 MAJOR SHALE GAS RESOURCE SYSTEMS

The major commercial shale gas resource systems all currently reside in North America. These systems span a variety of geological ages and basin types with variable lithofacies. One dominant characteristic has been the occurrence of such systems in Type II oil prone–source rocks. These source rocks all have had high petroleum generation, of which the retained portion was ultimately cracked to gas and carbonaceous char at post-oil window thermal maturities. The predominant shale gas systems are listed in Table 3.4. Recoverable gas estimates are from EIA (2011).

There are some distinct differences among these systems, but all are dominated by brittle rocks having low porosity and permeability. TOC values vary due not only to thermal maturity but also due to varying organic productivity and preservation of organic matter. The Montney Shale has distinctly lower TOC values but is considered as a hybrid system with modest source potential but more conventional, but tight reservoir lithofacies. The other silica-rich systems have biogenic sources of silica such as radiolarians and sponge spicules. When present-day TOC values are compared to porosity, there is a positive correlation except in the Montney Shale that shows negative correlation.

For comparison purposes only, relative production decline curves were constructed for each play listed in Table 3.4 (Fig. 3.17). These decline curves are purely for a relative comparison and represent at best the P50 values for gas yields. Often "monster" (highly productive) wells reported in press releases are the best wells (P10), and certainly not representative of all wells in a basin let alone averages. With the price of natural gas being low in the United States, current wells typically only reflect the most economic wells that are high flow rate with good return on investment. It is important to note also that there is tremendous variability in the results from given wells in the same basin or even in the same area.

The most obvious issue noted in Figure 3.17 is the high decline rates for all systems, although the Montney Shale system is somewhat lower being more of a hybrid play with

TABLE 3.4 Illustrative production decline curves with input parameters on selected shale gas resource systems in North America

System	Period	Basin	System type	Est. recoverable gas (tcf)
Marcellus Shale	U. Devonian	Appalachian	Tight mudstone	410
Muskwa Shale	U. Devonian	Horn River, NW Alberta, Cordova	Tight mudstone	145
Haynesville Shale	U. Jurassic	East Texas-North Louisiana Salt	Tight mudstone	75
Montney Shale	Triassic	Western Canada Sedimentary	Hybrid mudstone	43
Barnett Shale	L. Mississippian	Fort Worth	Tight mudstone	43
Fayetteville Shale	L. Mississippian	Arkoma (Oklahoma)	Tight mudstone	32

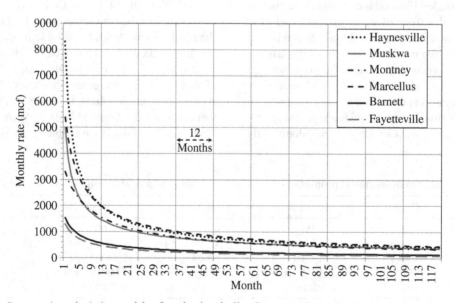

FIGURE 3.17 Construction of relative models of production decline for comparison of major North American shale gas systems.

TABLE 3.5 Average (P50) production yields of selected shale gas resource systems in North America based on relative production decline curves[a]

System	First month decline (%)	First year decline (%)	Exponent (m)	IP first 30 days (mcf)	Cumulative 30 years (bcf)
Marcellus	15	64	1.40	5400	5.955
Haynesville	27	77	1.40	8300	5.787
Muskwa	23	71	1.60	4789	4.789
Montney	10	54	1.25	3300	4.439
Barnett	15	64	1.40	1500	1.696
Fayetteville	15	65	1.30	1300	1.296

[a] See input listed in Table 3.4 and Fig. 3.17.

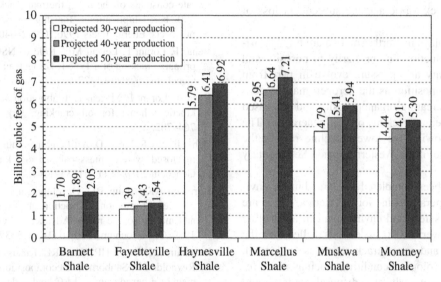

FIGURE 3.18 Based on the relative production decline models in Figure 3.17, gas yields are computed for 30, 40, and 50 years for each shale gas system.

more matrix than organic porosity (Jarvie, 2012a). It is also unusual in that it has the lowest overall TOC values averaging only ca. 1.50%, although some thin intervals are up to 4%.

First-year decline rates range from 54 to 77% using strong hyperbolic decline models with the specified exponent (m) (Table 3.5) (Fig. 3.18). Initial production (IP) yields are highly variable with the highest values inferred to be due to both higher storage capacity and pressures. It is estimated that there is 80% free gas in the Haynesville, whereas the Barnett Shale has about 55%, the remainder in both cases being adsorbed gas. The Haynesville Shale averages about 9% porosity with strong overpressure of about 0.80 psi/ft, whereas the Barnett Shale averages about 4.5% porosity with slight overpressure at about 0.52 psi/ft.

3.16 CONCLUSIONS

The best shale gas systems in North America are all characterized by having greater than 1.00% present-day TOC values with original values almost always exceeding 2.00%. This is due in part because only a portion of the TOC can be converted to petroleum, the generative organic carbon. In marine shales, GOC amounts to about 35–60% of the original TOC, and is a function of HI_o. It is this portion of the kerogen that generates petroleum consisting of both hydrocarbons and nonhydrocarbons, the latter of which predominate at low to moderate thermal maturity. While kerogen accounts for petroleum formation, it is the cracking of retained petroleum that accounts for the bulk of gas generation.

Assessment of thermal maturity is a key aspect of identifying targets for shale gas production. Typical methods include T_{max} and vitrinite reflectance measurements, but now commonly include gas composition and carbon isotopic assessments. Thermal maturity assessments from these data involves a number of caveats and all data must be utilized to interpret what product will be found in a given well or region of a shale resource play.

Sample type and age (storage time) affect organic geochemical results. Cuttings yield lower TOC, oil, and pyrolysis yields than core chips or sidewall core (SWC) samples.

This appears to be due to mixing and dilution by organic lean sediments overlying organic-rich shales. The best shales often have an organic-lean carbonate overlying them, which functions as a seal and frac barrier. TOC, oil, and pyrolysis yields, and T_{max} values are shown to be reduced with long-term storage (20 years or more), whereas carbonate contents increase. This suggests weathering-induced oxidation effects.

Based on both experimental and empirical results, excess thermal maturity appears to reduce the amount of gas found in shale reservoirs. This is hypothesized to be due to oxidation of methane, but may also be due to destruction of the porosity in the rock fabric and the subsequent loss of retained gas.

The top resource plays in North America are highly variable in terms of production due primarily to porosity and pressure. Both porosity and pressure evolve in part from organic matter decomposition as the kerogen mass is converted to petroleum creating organoporosity and increased pressure from petroleum cracking or gas generation. The shale gas system showing the lowest decline rate is the hybrid Montney Shale of the Western Canada Sedimentary Basin.

Shale drilling has been completed safely and under environmentally sound operations in North America. There are always concerns for safety and environment in all drilling operations, but with over 50,000 shale wells drilled over the past 10 years a safe and sound track record is evident. In cases where known geological conditions such as near preexisting intraplate plates, saltwater disposal wells should either not be drilled or handled differently. Regulations for cementing of casing protecting freshwater aquifers should always be updated for any changes in known operational or geological conditions just as safety measures are taken in any other industry.

Overall, shale gas resource systems provide a remarkable energy resource that is the best carbon-based resource that we have in terms of carbon dioxide and other emissions as well as being quite inexpensive over the course of the last 100 years. It represents a solution to reduced energy dependence, if not near complete independence for North America. Hopefully, this resource can be adequately and safely tapped elsewhere to provide comparable energy benefit to global energy consumers.

REFERENCES

Alexander R, Dawson D, Pierce K, Murray A. Carbon catalyzed hydrogen exchange in petroleum source rocks. Organic Geochemistry 2009;40:951–955.

Alpern B. Petrographie du kerogene. In: Durand B, editor. *Kerogen—Insoluble Organic Matter from Sedimentary Rocks*. Paris: Editions Technip; 1980. p 339–383.

Barker C. Pyrolysis techniques for source rock evaluation. AAPG Bull 1974;58 (11):2349–2361.

Barker C, Takach NE. Prediction of natural gas composition in ultradeep sandstone reservoirs. AAPG Bull 1992;76:1859–1873.

Baskin DK. Atomic H/C ratio of kerogen as an estimate of thermal maturity and organic matter conversion. AAPG Bull 1997;81: 1437–1450.

Behar F, Jarvie D. Compositional modeling of gas generation from two shale gas resource systems: Barnett Shale (United States) and Posidonia Shale (Germany). AAPG Memoir 2013;103: 25–44.

Behar F, Vandenbroucke M. Experimental determination of the rate constants of the n-C_{25} thermal cracking at 120, 400, and 800 bar: implications for high-pressure/high-temperature prospects. Energy Fuels 1996;10:932–940.

Behar F, Lorant F, Lewan MD. Role of NSO compounds during primary cracking of a Type II kerogen and Type III lignite. Org Geochem 2008a;39:1–22.

Behar F, Lorant F, Mazeas L. Elaboration of a new compositional kinetic schema for oil cracking. Org Geochem 2008b;39: 764–782.

Behar F, Roy S, Jarvie D. Artificial maturation of a Type I kerogen in closed system: mass balance and kinetic modelling. Org Geochem 2010;41 (11):1235–1247.

Burnham AK, Braun RL. Development of a detailed model of petroleum formation, destruction, and expulsion from lacustrine and marine source rocks, Advances in Organic Geochemistry 1989. Org Geochem 1990;16 (1–3):27–39.

Burnham AK, Gregg HR, Ward RL, Knauss KG, Copenhaver SA, Reynolds JG, Sanborn R. Decomposition kinetics and mechanism of n-hexadecane-1,2-$^{13}C_2$ and dodec-1-ene-1,2-$^{13}C_2$ doped in petroleum and n-hexadecane. Geochim Cosmochim Acta 1997;61 (17):3725–3737.

Claypool GE, Mancini EA. Geochemical relationships of petroleum in Mesozoic reservoirs to carbonate source rocks of Jurassic Smackover formation, Southwestern Alabama. AAPG Bull 1989;73:904–924.

Cooles GP, MacKenzie AS, Quigley TM. Calculation of petroleum masses generated and expelled from source rocks. Advances in Organic Geochemistry 1985. Org Geochem 1986;10 (1–3):235–245.

Dahl JEP, Moldowan JM, Moldowan S. Determination of thermal maturity and extent of oil cracking in tight shales using extract biomarker and diamondoid concentrations and distributions. Unconventional Resources Technology Conf (URTeC), URTeC paper 1574980; 2013. 5 p.

di Primio R, Horsfield B, Guzman-Vega MA. Determining the temperature of petroleum formation from the kinetic properties of petroleum asphaltenes. Nature 2000;406:173–176.

Diamond WP, Levine JR. Direct from the Norwegian North Sea. Geochim Cosmochim Acta 1981;70:3943–3956.

Ellis L, Schoell M, Uchytil S, Brown A. Mud Gas Isotope Logging (MGIL): a new field technique for exploration and production. Oil Gas J 2003;101 (21):32–41.

Erdmann M, Horsfield B. Enhanced late gas generation potential of petroleum source rocks via recombination reactions: evidence from the Norwegian North Sea. Geochim Cosmochim Acta 2006; 70, 15:3943–3956.

Ertas D, Kelemen SR, Halsey TC. Petroleum expulsion Part 1. Theory of kerogen swelling in multicomponent solvents. Energy Fuels 2006;20:295–300.

Espitalie J, Madec M, Tissot B, Mennig JJ, Leplat P. Source rock characterization method for petroleum exploration: Proceedings of the 9th Offshore Technology Conference, Volume 3, Paper 2935; 1977. pp 439–444.

Espitalie J, Madec M, Tissot B. Geochemical logging. In: Voorhees KJ, editor. *Analytical Pyrolysis—Techniques and Applications*. Boston: Butterworth; 1984. p 276–304.

Fabuss BM, Smith JO, Satterfield CN. Thermal cracking of pure saturated hydrocarbons. In: McKetta JJ, editor. *Advances in Petroleum Chemistry and Refining*. New York: John Wiley & Sons, Inc.; 1964. p 157–201.

Ford TJ. Liquid-phase thermal decomposition of hexadecane: reaction mechanisms. Ind Eng Chem Fundam 1986;25:240–243.

Frolich C, Potter E. What further research could teach us about "Close Encounters of the Third Kind": intraplate earthquakes associated with fluid injection. In: Chatelier J-Y, Jarvie DM, editors. *AAPG Memoir 103, Critical Assessment of Shale Resource Plays*. Tulsa, OK: American Association of Petroleum Geologists; 2013. p 109–119.

Fuhrmann A, Thompson KFM, di Primio R, Dieckmann V. Insight into Petroleum Composition Based on Thermal and Catalytic Cracking. 21st IMOG, poster presentation; Krakow, Poland; 2003.

Fusetti L, Behar F, Bounaceur R, Marquaire PM, Grice K, Derene S. New insights into secondary gas generation from oil thermal cracking: methylated monoaromatics. A kinetic approach using 1,2,4-trimethylbenzene. Part I: a free-radical mechanism. Org Geochem 2010a;41:146–167.

Fusetti L, Behar F, Grice K, Derene S. New insights into secondary gas generation from oil thermal cracking: methylated monoaromatics. A kinetic approach using 1,2,4-trimethylbenzene. Part II: a lumped kinetic scheme. Org Geochem 2010b;41:168–176.

Gao L, Schimmelmann A, Tang Y, Mastalerz M. Isotope rollover in shale gas observed in laboratory pyrolysis experiments: insight to the role of water in thermogenesis of mature gas. Org Geochem 2014;74:59–65.

Hao F, Zou H. Cause of shale gas geochemical anomalies and mechanisms for gas enrichment and depletion in high-maturity shales. Mar Pet Geol 2013;44:1–12.

Hill RJ, Jarvie DM, Zumberge J, Henry M, Pollastro RM. Oil and gas geochemistry and petroleum systems of the Fort Worth Basin. AAPG Bull 2007;91 (4):445–473.

Horsfield B, Disko U, Leistner F. The micro-scale simulation of maturation: outline of a new technique and its potential applications. Geol Rundsch 1989;78:361–374.

Horsfield B, Leistner F, Hall K. Microscale sealed vessel pyrolysis, In: Grice K, editor. *Principles and Practice of Analytical Techniques in Geosciences*. Cambridge: Royal Society of Chemistry; 2015. 600p, pp. 209–250.

Hunt JM. *Petroleum Geochemistry and Geology*. 2nd ed. New York: Freeman WH & Co.; 1995. p 743.

Jarvie DM. Williston Basin petroleum systems: inferences from oil geochemistry and geology. Mount Geol 2001;38 (1):19–41.

Jarvie DM. Hydrocarbon generation and storage in the Barnett Shale, Ft. Worth Basin, SPE Houston Chapter meeting; May 2006; Houston, TX. Available at http://www.wwgeochem.com/resources/Jarvie+-+SPE+Houston+April+19+2006+presentation+-+submitted.pdf. Accessed November 12, 2013.

Jarvie DM. Shale resource systems for oil and gas: Part 1—shale gas resource systems, In: Breyer J, editor. *Shale Reservoirs—Giant Resources for the 21st century*. AAPG Memoir 2012a; 97:69–87.

Jarvie DM. Shale resource systems for oil and gas: Part 2—Shale oil resource systems, In: Breyer J, editor. *Shale Reservoirs—Giant Resources for the 21st century*. AAPG Memoir 2012b; 97:89–119.

Jarvie DM, Hill RJ, Ruble TE, Pollastro RM. Unconventional shale-gas systems: The Mississippian Barnett Shale of north-central Texas as one model for thermogenic shale-gas assessment. AAPG Bull 2007;91 (4):475–499.

Jarvie DM, Coskey RJ, Johnson MS, Leonard JE. The Geology and Geochemistry of the Parshall Field Area, Mountrail County, North Dakota. In: Robinson JW, LeFever JA, Gaswirth SB, editors. *Rocky Mountain Association of Geologists' (RMAG) The Bakken-Three Forks Petroleum System in the Williston Basin*. Denver: RMAG; 2011. p 229–281.

Jarvie DM, Jarvie B, Courson D, Garza T, Jarvie J, Rocher D. Geochemical tools for assessment of tight oil reservoirs, Geochemical tools for the assessment of shale oil plays, AAPG/SEG/SPE/SPWLA Hedberg Conference: Critical Assessment of Shale Resource Plays; Austin, TX; December 10–15, 2010; AAPG Memoir 103, Critical Assessment of Shale Resource Plays. Chatelier J-Y, Jarvie DM, editors, extended abstract; 2013; 2p.

Jones RW. Comparison of Carbonate and Shale Source Rocks. In: Palacas J, editor. *Petroleum Geochemistry and Source Rock Potential of Carbonate Rocks*. Volume 18, AAPG Studies in Geology; 1984. p 163–180.

Landis CR, Castaño JR. Maturation and bulk chemical properties of a suite of solid hydrocarbons. Org Geochem 1995;22: 137–149.

Levine JA. Physical sorption of gases: implications for resources, reserves, and production from sorbed gas reservoir systems, AAPG/SEG/SPE/SPWLA Hedberg Conference: Critical Assessment of Shale Resource Plays; Austin, TX; December 10–15, 2010; AAPG Memoir 103, Critical Assessment of Shale Resource Plays. Chatelier J-Y, Jarvie DM, editors, extended abstract; 2013. 4p.

Li P, Ratchford ME, Jarvie DM. Geochemistry and thermal maturity analysis of the Fayetteville Shale and Chattanooga Shale in the Western Arkoma Basin of Arkansas, Arkansas Geological Survey, Information Circular 40, DFF-OG-FS-EAB/ME 012. 2010.

Lis GP, Mastalerz M, Schimmelmann A. Increasing maturity of kerogen type II reflected by alkylbenzene distribution from pyrolysis-gas chromatography-mass spectrometry. Org Geochem 2008;39:440–449.

Locke DE, Winans RE. X-ray characterization of type I Green River Oil Shale Kerogen, presentation at Houston Organic

Geochemical Society meeting; August 22, 2013; Houston, TX; 2013.

Lorant F, Prinzhofer A, Behar F, Huc A-Y. Carbon isotopic and molecular constraints on the formation and the expulsion of thermogenic hydrocarbon gases. Chem Geol 1998;147:249–264.

Loucks RG, Reed RM, Ruppel SC, Jarvie DM. Morphology, Genesis, and Distribution of nanometer-scale pores in siliceous mudstones of the Mississippian Barnett Shale. J Sediment Res 2009;79:848–861.

Mahlstedt N, Horsfield B. Gas generation at high maturities (>Ro=2%) in gas shales. Search Discovery Article 2012; 40873:21.

Mango FD, Daniel JM. GOR from oil composition, 20th International Meeting on Organic Geochemistry, Nancy, France; September 10–14, 2001; Abstracts Vol 1; 2001. pp 406–407.

National Commission on the BP Deepwater Horizon Oil Spill and Offshore Drilling. *National Commission on the BP Deepwater Horizon Spill and Offshore Drilling*, UNT Digital Library; 2011. Available at http://digital.library.unt.edu/ark:/67531/metadc132999/. Accessed November 13, 2013.

Newman J. New approaches to detection and correction of suppressed vitrinite reflectance. Aust Petrol Prod Expl Assoc J 1997;37:524–535.

Newman J, Eckersley KM, Francis DA, Moore NA. Application of vitrinite-inertinite reflectance and fluorescence to maturity assessment in the East Coast and Canterbury Basins of New Zealand, 2000 New Zealand Petroleum Conference Proceedings; 2000. pp 314–333.

Niemann M, Breviere J. Continuous isotope logging in real time while drilling, AAPG Search and Discovery Article 90110 from AAPG Hedberg Conference. Application of Reservoir Fluid Geochemistry; June 8–11, 2010; Vail, CO; 2010. Available at http://www.searchanddiscovery.com/abstracts/pdf/2010/hedberg_vail/abstracts/ndx_niemann.pdf. Accessed October 10, 2012.

NIGOGA. *The Norwegian Industry Guide to Organic Geochemical Analyses*, 4.0 ed.; May 30, 2000; 102p. Available at http://www.npd.no/engelsk/nigoga/default.htm. Accessed November 11, 2013.

Okiongbo KR, Aplin AC, Larter SR. Changes in type II kerogen density as a function of maturity: evidence from the Kimmeridge Clay Formation. Energy Fuels 2005;19 (6):2495–2499.

Orendt AM, Birgenheier LP, Solum MS, Pugmire RJ, Facelli JC, Locke D, Chapman K, Seifert S, Winans R, Chupas P. Structural characterization of Green River Oil Shale core segments and the kerogen isolated from these segments, 31st Oil Shale Symposium; September 18, 2011. Available at http://www.yatedo.com/ajax/redirect?url=http%3A%2F%2Fwww.costar-mines.org%2Foss%2F31%2FF-pres-sm-sec%2F10-2_Orendt-Anita.pdf. Accessed August 18, 2013.

Orr WL. Comments on pyrolytic hydrocarbon yields in source-rock evaluation. In: Bjoroy M, editor. *Advances in Organic Geochemistry*. Chichester: John Wiley & Sons Ltd; 1983. p 775–787.

Pan C, Jiang L, Liu J, Zhang S, Zhu G. The effects of pyrobitumen on oil cracking in confined pyrolysis experiments. Org Geochem 2012;45:29–47.

Pelet R. Evaluation quantitative des produits formes lors de l'evolution geochimique de la matiere organique. Revue Inst Francais du Petrole 1985;40 (5):551–562.

Pepper AS. Estimating the petroleum expulsion behavior of source rocks: a novel quantitative approach. In: England WA, Fleet AJ, editors. *Petroleum Migration: The Geological Society Special Publication 59*. London: The Geological Society; 1991. pp 9–31.

Pepper AS, Corvi PJ. Simple models of petroleum formation. Part I: oil and gas generation. Mar Pet Geol 1995;12:291–320.

Prinzhofer AA, Huc AY. Genetic and post-genetic molecular and isotopic fractionations in natural gases. Chem Geol 1995;126: 281–290.

Reed R, Loucks R. Imaging nanoscale pores in the Mississippian Barnett Shale of the northern Fort Worth Basin, AAPG Annual Convention; Long Beach, CA; April 1–4, 2007. Available at http://www.searchanddiscovery.com/abstracts/html/2007/annual/abstracts/lbReed.htm. Accessed November 12, 2010.

Sandvik EI, Young WA, Curry DJ. Expulsion from hydrocarbon sources: the role of organic absorption: Advances in Org Geochem 1991. Org Geochem 1992;19 (1–3):77–87.

Schettler Jr. PD and Parmely CR. 1991, Contributions to total storage capacity in Devonian shales, SPE paper 23422, 9p.

Schoell M, Ellis L, Muehlenbachs K, Coleman DD, Underdown I. Gas isotope analyses while drilling (GIAWD): an emerging technology for exploration and production, AAPG National Convention; June 3–6, 2001; Sunver, CO, abstract book, p A180, CD.

Sheiko SS, Sun FC, Randall A, Shirvanyants D, Rubinstein M, H-I Lee, Matyjaszewski K. Adsorption-induced scission of carbon-carbon bonds. Nature 2006;440(9):191–194.

Steward DB. The Barnett Shale play: phoenix of the Fort Worth Basin, a history. In: Paniszczyn F, editor. *The Fort Worth Geological Society & The North Texas Geological Society*. Kansas City, MO: Covington Group; 2007. p 202.

Tang Y, Perry JK, Jenden PD, Schoell M. Mathematical modeling of stable carbon isotope ratios in natural gases. Geochim Cosmochim Acta 2000;64 (16):2673–2687.

The Perryman Group. The impact of the Barnett Shale on business activity in the surrounding region and Texas: an assessment of the first decade of extensive development, August 2011 report prepared for The Fort Worth Chamber of Commerce; 2011. 148p.

Tian H, Xiao X, Wilkins RWT, Tang Y. New insights into the volume and pressure changes during the thermal cracking of oil to gas in reservoirs: implications for the in-situ accumulation of gas cracked from oils. AAPG Bull 2008;92 (2):181–200.

Tiem VTA, Horsfield B, Sykes R. Influence of in-situ bitumen on the generation of gas and oil in New Zealand coals. Org Geochem 2008;39:1606–1619.

Tilley B, McLellan S, Hiebert S, Quartero B, Veilleux B, Muehlenbachs K. Gas isotope reversals in fractured gas reservoirs of the western Canadian foothills: mature shale gas in disguise. AAPG Bull 2011;95:1399–1422.

Tissot BP, Welte DH. *Petroleum Formation and Occurrence*. 2nd ed. New York: Springer Verlag; 1978. p 699.

Tsuzuki N, Takeda N, Suzuki M, Yokoi K. The kinetic modeling of oil cracking by hydrothermal pyrolysis experiments. Intl Jour Coal Geol 1999;39:227–250.

US Energy Information Agency (EIA). Review of energy resources: US shale gas and shale oil plays; July 2011. 105p.

US Energy Information Agency (EIA). Today in energy, US energy-related CO_2 emissions in early 2012 lowest since 1992; August 1, 2012.

Vandenbroucke M, Largeau C. Kerogen origin, evolution, structure. Org Geochem 2007;38:719–833.

Waples DW, Marzi RW. The universality of the relationship between vitrinite reflectance and transformation ratio. Org Geochem 1998;28 (6):383–388.

Whitson CH, Brulé MR. Phase behavior. Soc Petro Eng SPE Monograph 2000;20:240.

Wilkins RWT, Wilmshurst JR, Hladky G, Ellacott MV, Buckingham CP. Should fluorescence alteration replace vitrinite reflectance as a major tool for thermal maturity determination in oil exploration? Organic Geochemistry 1995;22:191–209.

Wilkins RWT, Buckingham CP, Sherwood N, Russell NJ, Faiz M, Kurusingal J. The current status of the FAMM thermal maturity technique for petroleum exploration in Australia. Aust Petrol Prod Expl Assoc J 1998;38:421–437.

Williams KE. The permeability of overpressure shale seals and of source rock reservoirs is the same, AAPG Search and Discovery Article 2012;40935:10. Available at http://www.searchanddiscovery.com/documents/2012/40935williams/ndx_williams.pdf. Accessed October 20, 2013.

Xia X, Chen J, Braun R, Tang Y. Isotopic reversals with respect to maturity trends due to mixing of primary and secondary products in source rocks. Chem Geol 2013;339:205–212.

Zou C, editor. Unconventional continuous petroleum accumulation. In: *Unconventional Petroleum Geology*. Beijing: Petroleum Industry Press; 2012. p 27–60.

Zumberge J, Ferworn K, Brown S. Isotopic reversal ("rollover") in shale gases from the Mississippian Barnett and Fayetteville formations. Mar Pet Geol 2012;31:43–52.

4

SEQUENCE STRATIGRAPHY OF UNCONVENTIONAL RESOURCE SHALES

ROGER M. SLATT

Institute of Reservoir Characterization, School of Geology and Geophysics, Sarkeys Energy Center, University of Oklahoma, Norman, OK, USA

SUMMARY

In this chapter, I have examined and interpreted the vertical stratigraphy, and sometimes lateral attributes, of the Barnett Shale, Woodford Shale, New Albany Shale, Marcellus Shale, Haynesville Shale, Eagle Ford Shale, LaLuna Shale, and Brown Shale. It is interpreted that water depths during deposition of these shales varied from nearshore to basin; therefore, depositional processes also varied, as evidenced by sedimentary features.

All of these shales exhibit stratigraphic zonation indicating at least two scales of predictable relative sea-level cyclicity. On this basis, a general sequence stratigraphic model is established that consists of a basal erosion surface of underlying strata (sequence boundary, SB), which can be combined with a younger transgressive surface of erosion (TSE), generally overlain by an organic-rich transgressive systems tract (TST) capped by a condensed section/maximum flooding surface (CS/mfs), which is overlain by downlapping highstand systems tract (HST) deposits. In the case of Paleozoic shales, presumed third-order cycles are superimposed on longer frequency second-order cycles; in Mesozoic, and probably Cenozoic, shales, biostratigraphically identifiable fourth-order cycles are generally superimposed on the third-order cycles.

A principal application of sequence stratigraphy of shales is the ability to regionally correlate and map sequence stratigraphic intervals in a systematic manner. This ability then allows for relating sequence stratigraphic characteristics to geomechanical or geochemical characteristics for sweet-spot identification.

4.1 INTRODUCTION

There are vast quantities of shales worldwide, which are now in various stages of exploration and development because of their potential as oil and gas resources. Shales have for many years been considered the principal hydrocarbon source rock as well as an effective seal rock. But it was only after horizontal drilling and hydraulic fracturing technologies were demonstrated to be effective at releasing hydrocarbons from these tough, impermeable rocks that they became recognized as reservoir rocks—or as they are often referred to—"unconventional resource shales."

There are many factors that affect the storage and flow capacity of these resource shales, including (i) mineralogic composition, which affects geomechanical properties and the "artificial fracability" of the shales; (ii) regional to local structure, which affects fracture patterns and distribution; (iii) orientation of the modern stress field, which affects placement and orientation of horizontal wells; (iv) organic matter type, distribution, and degree of maturity (burial history,) which affect the oil and gas content at a given location; and (v) origin and depositional history of the shales (summarized in Abouelresh and Slatt, 2012a, b; Slatt, 2011; Slatt et al., 2012).

4.2 GENERAL SEQUENCE STRATIGRAPHIC MODEL FOR UNCONVENTIONAL RESOURCE SHALES

It is this latter property of the origin and depositional history of the shales that is the primary consideration for this chapter, especially within the context of a standard and well-known

Fundamentals of Gas Shale Reservoirs, First Edition. Edited by Reza Rezaee.
© 2015 John Wiley & Sons, Inc. Published 2015 by John Wiley & Sons, Inc.

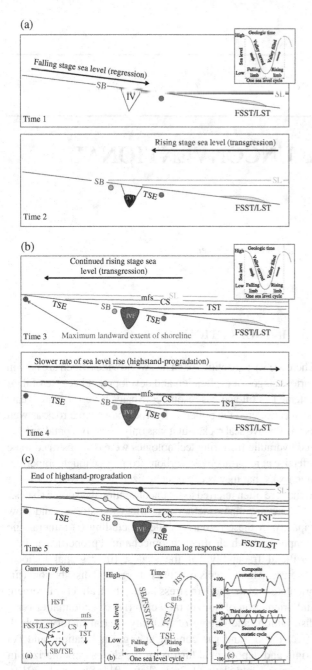

sequence stratigraphic model (Haq et al., 1984). In an earlier paper, Slatt and Rodriguez (2012) suggested a stratigraphic commonality among the resource shales and provided a general sequence stratigraphic model that is applicable to shales at a variety of chronostratigraphic scales. In this chapter, we have placed this model into a five-step time frame (Time 1 to Time 5, beginning with onset of sea-level drop (Time 1) (Fig. 4.1). During Time 1, the shoreline moves basinward with the falling stage of sea level, generating an erosion surface (sequence boundary, SB). Seaward of the shoreline, falling stage tract deposit (FSST) and lowstand systems tract deposit (LST) form. With the onset of transgression during Time 2, the shoreline advances landward and may generate a transgressive surface of erosion (TSE) (i.e., "ravinement surface"), which merges with the SB. Time 3 represents that time interval in which the shoreline transgresses to its most landward position; during the transgression, progressively finer grained sediments (both detrital and biogenic) will be deposited vertically at each point on the sea floor to give a transgressive systems tract (TST), capped by the most organic-rich interval, the condensed section (CS), with its top surface, the maximum flooding surface (mfs). At Time 4, the relative rate of sea-level rise and/or the supply of clastic sediment to the marine environment increases, giving rise to the progradational highstand systems tract (HST). Time 5 represents the end of the relative sea-level cycle. For a depositional cycle that forms *landward* of the maximum seaward extent of the shoreline (Fig. 4.1, Time 1), a resulting gamma ray log will look similar to that shown in Figure 4.1a. The sharp-based surface—which is quite common at the top of strata which immediately underlie many unconventional resource shales (discussed in the following)—represents the combined SB/TSE. For a depositional cycle that forms *seaward* of the maximum seaward extent of the shoreline, lowstand systems tract deposits will form the base of the unconventional resource shale, so the resulting gamma ray log will look similar to that shown in Figure 4.1a (dashed box at base of sequence). In that case, the base of the FSST/LST will sit on a correlative conformity.

FIGURE 4.1 Generalized sequence stratigraphic model of unconventional resource shale as shown in five time-steps (Time 1 to Time 5). SB, sequence boundary; FSST, falling stage systems tract; LST, lowstand systems tract; TSE, transgressive surface of erosion; TST, transgressive systems tract; CS, condensed section; mfs, maximum flooding surface; HST, highstand systems tract. The time steps (a–c) are described in the text. A conceptual gamma ray log is shown on (A) both for stratigraphic sequences that formed landward of the minimum position of the shoreline (TST sits directly on SB/TSE) and seaward of the minimum position of the shoreline (FSST/LST sits below the TST). (B) A relative sea-level curve illustrating the relative times within a sea-level cycle when each component is formed. (C) Second- and third-order cycles and a composite relative sea-level curve by superimposition of these two orders of cyclicity. After VanWagoner et al. (1990).

4.3 AGES OF SEA-LEVEL CYCLES

Sequence stratigraphy concepts indicate that relative sea-level (sea level due to a combination of eustacy, tectonics, and sediment supply) varies in a cyclical manner. Although exact age ranges of the cycles are not agreed by all, approximate durations, as summarized by Miall (1997) and noted in the preceding, are of second (10–25 Myr duration), third (1–5 Myr duration), and fourth order (100,000–500,000 yr duration). As presented in the following for the different resource shales, at least two of these scales can usually be identified due to superimposition of two orders of cyclicity (Fig. 4.1c). However, particularly with Paleozoic shales, the age range of fossils is often too broad to age-date stratigraphic surfaces or intervals that formed during relatively short-lived cycles, usually precluding the ability to definitively correlate strata within a high-frequency, time-stratigraphic framework. Paleozoic shales can generally be resolved at third-order cycles (~1–5 Myr duration) superimposed on a second-order cycle (~10–30 Myr duration) (Fig. 4.1c). Owing to greater biostratigraphic resolution, Mesozoic and Cenozoic shales can be resolved at fourth-order cycles (100,000–300,000 years duration) superimposed on a third-order cycle, as demonstrated in the following.

4.4 WATER DEPTH OF MUD TRANSPORT AND DEPOSITION

It has long been assumed that because of their fine-grained nature, precursor muds were deposited in quiet, "relatively deep" ocean waters. More recent studies have shown that siliciclastic muds can be deposited in tidal mud flats (Rine and Ginsburg, 1985) and shelf to upper slope water depths (Loucks and Ruppel, 2007) as well as in deep basins. In addition to "hemipelagic settling" of mud particles, hyperpycnal flows (Bhattacharya and MacEachern, 2009; Mulder and Chapron, 2011) and turbidites can also transport sediment from continental to shelf/slope/basin environments; storm and contour currents can rework mud deposited by these processes.

Many resource shales exhibit microsedimentary structures such as graded beds, cross-laminations, and cross-beds, indicating current transport along the sea floor (Fig. 4.2) (Abouelresh and Slatt, 2012a, b). Such transport requires particles larger than clay size, as these would tend to be buoyed upward by a turbulent current. Schieber et al. (2007; Schieber and Southard, 2009) have demonstrated that flocculation of clay-size particles occurs in the laboratory and in nature which provides floccules large enough to be transported by currents (i.e., hydraulically equivalent to silt and sand size particles). Floccules have been preserved in many Mesozoic and Paleozoic shales (O'Brien and Slatt, 1990; Slatt and O'Brien, 2011). Hyperpycnite muds have been documented for the Cretaceous Lewis Shale (Soynika and Slatt, 2008) and other rocks in the Cretaceous western interior seaway deposits (Bhattacharya and MacEachern, 2009), as well as in the modern Sea of Japan, where a transport distance of 700 km has been documented for a hyperpycnal flows (Nakajima, 2006).

It would seem that microfossils such as radiolarian and coccoliths might offer the best chance of deposition through the water column as individual particles (i.e., marine snow; Bennett et al., 1991) since they are not electrostatically charged, as are clay particles. However, even high concentrations of biogenic particles can move along the seabed and erode underlying mud (Abouelresh and Slatt, 2012a, b).

The presence of phosphate minerals in shales is often attributed to upwelling currents. The upwelling model proposes that cold, deep, oxygen-deficient, phosphate-rich water is drawn along the sea floor until it reaches the continental slope, where it rises to the shelf edge. Organisms thrive on the phosphate nutrients, and generate "algal blooms" (i.e., modern "red tides") and further deplete dissolved oxygen, creating eutrophication of the water mass and deposition of the phosphates. A second model postulates that phytoplankton productivity is increased due to seasonal nutrient input from continental weathering and runoff during times of broad shallow seas (Lash and Blood, 2011; Rimmer et al., 2004), particularly if the basin is silled (Molinares-Blanco, 2013).

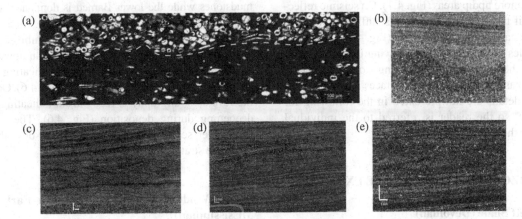

FIGURE 4.2 Thin section photographs of sedimentary structures in the Barnett Shale. (a) Irregular (erosional) bottom surface of a siliceous sponge spicule laminae. (b) Close-up view of scour surface at base of light-colored siltstone bed. (c) Ripple stratification, note the clay materials (black) delineating the ripple marks. (d) Low-angle cross lamination. (e) Hummocky lamination. Figures from Abouelresh and Slatt (2012a, b). Reprinted with permission of Central European Journal. Geosciences http://www.degruyter.com/

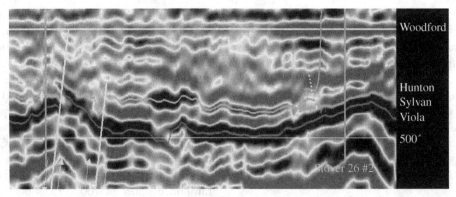

FIGURE 4.3 Woodford seismic interval showing the highstand downlap seismic pattern onto the Hunton unconformity. After May and Anderson (2011) who have provided permission to publish this figure.

4.5 CRITERIA TO IDENTIFY SEQUENCES AND SYSTEMS TRACTS

Because of the lack of high-frequency chronostratigraphy, other criteria must be used to recognize stratigraphic sequences superimposed at different scales. Such criteria include the gamma ray log responses described for Figure 4.1, stratigraphic changes in fauna, mineralogy and organic matter (TOC and geochemical biomarkers), and seismic responses such as downlap of HST onto an mfs (Fig. 4.3).

In this chapter, emphasis is placed upon the commonly observed upward increase in API units to the CS and mfs, followed by an upward decrease in API units of the HST (Fig. 4.1). This systematic pattern is the result of systematic variations in the aforementioned criteria. The TST/CS is generally enriched in clays and organic matter, and HSTs tend to be enriched in somewhat coarser-grained, detrital carbonate and/or quartz. Laterally, the TST is deposited either horizontally or with a slight seaward dip, while the younger HST deposits are more prone to downlap, especially in their more updip areas (Fig. 4.1). On seismic reflection records, if the vertical resolution is sufficient, then the HST downlap pattern can be identified (Fig. 4.3). Also, published examples presented in the following uniformly refer to the base of the resource shale as sitting atop an unconformity surface based upon erosional surfaces and biostratigraphy. These details are not presented in the following for each example, so the reader is referred to the individual references for their description.

4.6 PALEOZOIC RESOURCE SHALE EXAMPLES

4.6.1 Barnett Shale (Devonian)

The Barnett Shale is a prolific gas and oil producer in the Fort Worth Basin of Texas, thus it has been studied extensively (summarized in Abouelresh and Slatt, 2012a, b; Singh, 2008; Slatt et al., 2012). It is several hundred feet (m) thick. The Barnett overlies an erosional unconformity (SB) marking the top of the Ellenburger (and sometimes the Viola) Limestone (Fig. 4.4) (Baruch et al., 2012). It was deposited over a 22 Myr time interval based on conodont ages (Loucks and Ruppel, 2007; Pollastro, 2007), thus making it a second-order depositional sequence. A high gamma ray, organic-rich interval immediately overlies the unconformity, and above that interval, the strata generally provide a lower gamma ray log response (Fig. 4.4).

Fourteen to sixteen higher frequency cycles comprise the Barnett Shale, so by dividing the total number of high-frequency sequences by 22 Myr, these are interpreted as third-order sequences (app. 1.5 Myr average) (Abouelresh and Slatt, 2012a, b; Singh, 2008). Several sequences thin/pinchout toward the south, away from the northerly source area (Fig. 4.4). *At the second-order scale*, the Forestburg Limestone separates the Barnett into lower and upper intervals toward the northeast (Fig. 4.5). The Forestburg and higher frequency sequences downlap onto the basal organic-rich, high API gamma ray interval. The upper Barnett is dominated by calcareous mudstones while the lower Barnett is dominated by siliceous mudstones (Abouelresh and Slatt, 2012).

Third-order sequences generally exhibit a "cleaning-upward" gamma ray log response due to an upward increase in calcareous mudstones and fossils, indicating an upward shallowing of water during deposition (Fig. 4.6). Occasionally, the reverse stratigraphy is recognized, indicating an upward deepening during deposition (Fig. 4.6). The second- and third-order scales conform to the earlier-described general sequence stratigraphic model (Fig. 4.1).

4.6.2 Woodford Shale (Late Devonian–Early Mississippian)

The Woodford Shale is the main hydrocarbon source rock in the oil/gas-rich state of Oklahoma, USA. It occurs in the Anadarko, Arkoma, Ardmore, and Marietta basins and the

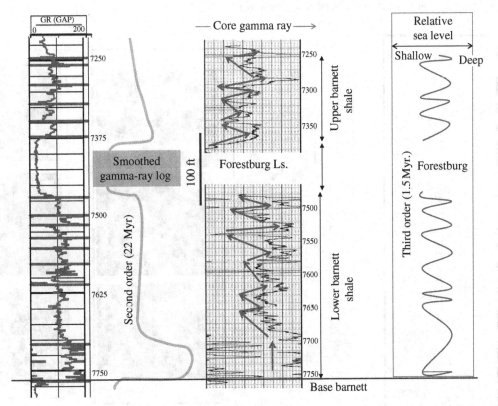

FIGURE 4.4 Well log and core gamma scan of Barnett Shale. The smoothed gamma ray curve is the second-order Barnett sequence. The arrows depict couplets of upward increasing-then decreasing gamma ray log, which represent third order sequences; these sequences are depicted by a relative sea level curve on the right. The Forestburg Limestone in this well separates the lower from the upper Barnett Shale. Modified from Slatt et al. (2012). Reprinted with permission of AAPG, whose permission is required for further use.

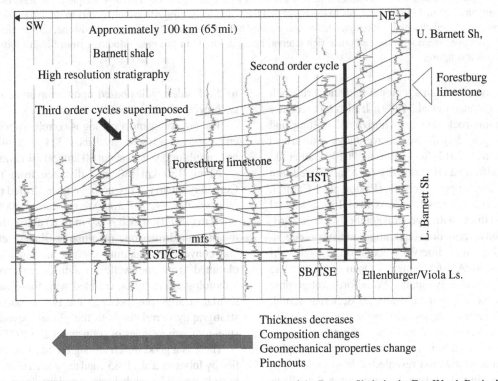

FIGURE 4.5 Approximate 100 km (65 mi.) long stratigraphic cross section of the Barnett Shale in the Fort Worth Basin, Texas, USA. The second-order sequence consists of a lower, high gamma ray API interval (TST/CS) overlying the top of the eroded Ellenburger/Viola Limestone and downlapping HST deposits onto the mfs. Third-order sequences are shown to be thinning and pinching out in the depositionally downdip (southwest) direction. Modified from Singh (2008) who provided permission to publish this figure.

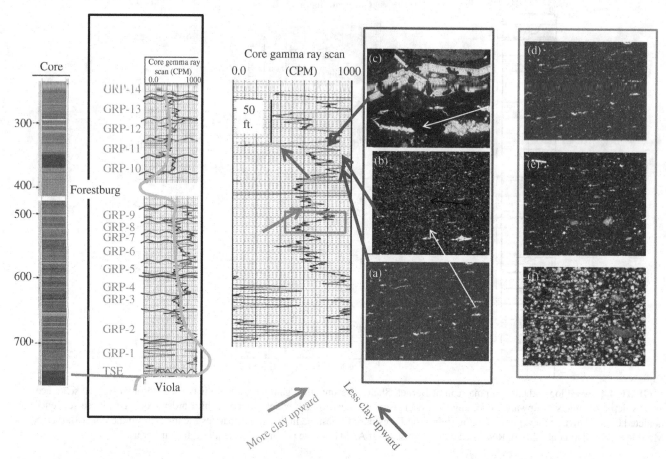

FIGURE 4.6 Core gamma scan and basic core description for the well shown in Figure 4.4. Two high-frequency sequences are highlighted, one that is increasing upward in API (arrow points to the right) and the other decreasing API (arrow points to the left). Thin sections of the first sequence show an upward change in lithologies from clay-organic rich mudstone to fossiliferous mudstone; thin sections of the second sequence show the opposite trend with the carbonate rich interval being dolomite in this case. Modified from Singh (2008) who provided permission to publish this figure.

Cherokee Platform, all within a regional fold and thrust belt tectonic setting (Johnson et al., 1989). It also has recently become a reservoir rock owing to horizontal drilling and artificial fracturing, leading to its extensive study (summarized in Molinares-Blanco, 2013; Serna-Bernal, 2013; Slatt et al., 2012). The Woodford has traditionally been subdivided into lower, middle, and upper intervals (Paxton and Cardott, 2008); total thickness varies from less than 100 to several hundred feet (m) thick, with a general southerly (basinward) thickening. Based on conodont biostratigraphy, it was deposited over a 29 Myr time interval (second-order cycle). At one location—the McAlester Cemetary Quarry—a complete 120 m (400 ft) thick, steeply dipping, Woodford stratigraphic section is exposed (Paxton and Cardott, 2008; Serna-Bernal, 2013). The Woodford overlies an erosional unconformity (SB) at the top of the Hunton Limestone. An outcrop gamma ray log, obtained with a Scintrex Scintillometer™, coupled with geologic characterization revealed at least eight high-frequency cycles (Fig. 4.7), superimposed upon the 120 m (400 ft) stratigraphic interval. These are presumed equivalent to third-order cycles based upon an average of 3.6 Myr for each cycle (29 Myr/8 = 3.6 Myr average).

The outcrop gamma ray log is correlative with another outcrop gamma ray log obtained about 32 km (20 mi.) away as well as to a subsurface well about 80 km (50 mi.) away (Fig. 4.8).

About 130 km (80 mi.) distance from the McAlester Cemetary Quarry, a research well was drilled behind another quarry—the Wyche Farm Quarry (Portas, 2009)—to a depth of 55 m (180 ft), which is 0.6 m (2 ft) above the contact with the Hunton Limestone (Molinares-Blanco, 2013). Spectral and conventional gamma ray log and an FMI™ log were obtained from the well, in addition to core description. Palynological analysis, coupled with these data, revealed a similar stratigraphy between the two areas (Fig. 4.9). The stratigraphy correlates with the global Devonian transgressive–regressive cycles of Johnson et al. (1985) (Fig. 4.9).

There is a good match among the third-order cycles identified by Johnson et al. (1985), and those seen in the two Woodford logs. In particular, global transgressions during upper Frasnian time (shown as LK and UK in Fig. 4.9) correlate well with the

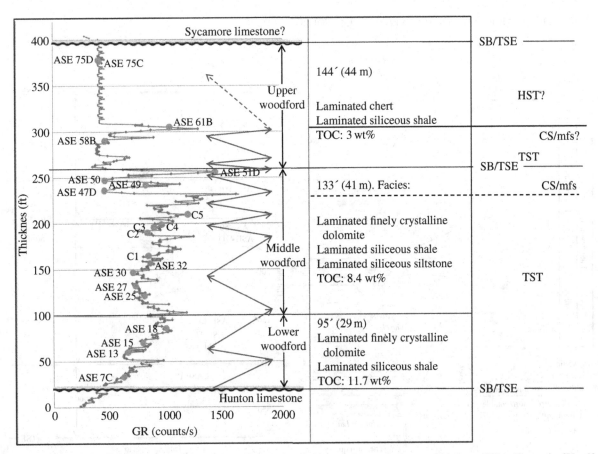

FIGURE 4.7 Outcrop gamma ray log of the Woodford Shale in the McAlester Cemetary Quarry, Oklahoma, USA. The entire Woodford is a second-order depositional sequence; and eight higher frequency sequences are shown by the red arrows. A SB/TSE corresponding to a global fluctuation in sea level occurs within the second-order sequence. Lettered numbers alongside the gamma ray log are sample locations. General features of the lower, middle, and upper Woodford are listed. Modified from Serna-Bernal (2013) who provided permission to publish this figure.

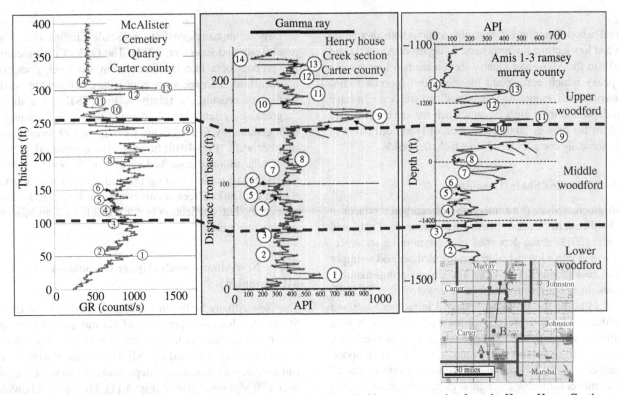

FIGURE 4.8 McAlester Cemetary Quarry outcrop gamma ray log compared with a gamma ray log from the Henry House Creek outcrop section and the Amis 1-3 Ramsey well gamma ray log. Modified from Paxton et al. (2007) and Serna-Bernal (2013), both of whom provided permission to publish this figure.

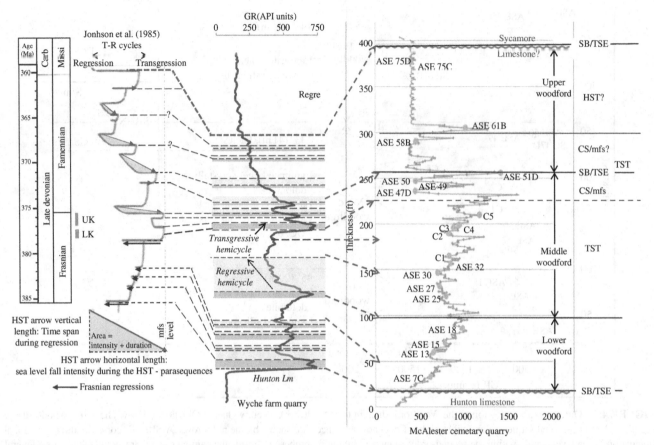

FIGURE 4.9 McAlester Cemetary Quarry outcrop gamma ray log compared with a gamma ray log from the Wyche Farm Quarry well. Both logs compare favorably with the Devonian relative sea-level curve of Johnson et al. (1985), including not only the larger transgressions (labeled UK and LK), but also smaller cycles. Modified from Serna-Bernal (2013) who provided permission to publish this figure.

interval in both Woodford stratigraphic sections with alternating high and low gamma ray responses. At depths of ASE47D and ASE50 in the McAlester Quarry section are two low gamma ray peaks which correspond to very clean, porous radiolarian-rich beds, suggesting that these two strata might have been deposited from algal blooms. Such marker horizons are excellent for long-distance correlation of Woodford strata in the tectonically complex terrain in which they reside.

4.6.3 Marcellus Shale (Devonian)

The Marcellus Shale is an emerging hydrocarbon producer in the northeastern United States. According to Lash and Engelder (2011) it was deposited as a second-order sequence. It is divided into a lower Union Springs member and an upper Oatka Creek member (Fig. 4.10). The Union Springs member overlies an unconformity surface, interpreted as a SB/TSE (Fig. 4.1) at the top of the Onondaga Limestone. Lash and Engelder (2011) interpret the Union Springs member as a third-order depositional sequence which is composed of a lower, upward-increasing API gamma ray TST and an upper, upward-decreasing API gamma ray "regressive systems tract," with an mfs constituting the highest gamma ray shale.

Deposition corresponds to a middle Eifelian rise in sea level (Lash and Engelder, 2011). The Oatka Creek member, the product of a late Eifelian rise in sea level, comprises another third-order sequence consisting of a basal "regressive surface of erosion," a relatively thin TST and a thicker "regressive systems tract." In some wells, high-frequency, high gamma ray API TSTs are more readily observable than in other wells (Fig. 4.10); however, the gamma ray patterns are similar to those described earlier for the other Paleozoic shales. Small-scale cleaning-upward gamma ray intervals within the Oatka Creek member may be fourth-order parasequences, but verification is not possible without biostratigraphic resolution.

4.6.4 New Albany Shale (Upper Devonian–Lower Mississippian)

The New Albany Shale, in the Illinois Basin of the United States, is a longtime producer of natural gas. Bohacs and Lazar (2010) have indicated that the New Albany Shale, which overlies the eroded top (SB/TSE) of the Mount Vernon Limestone, is a second-order depositional sequence deposited over a 20 Myr time interval (Fig. 4.11). They have subdivided

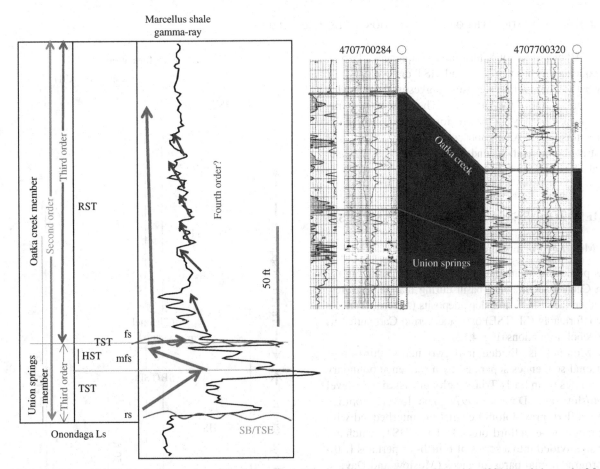

FIGURE 4.10 Marcellus Shale gamma ray logs. The log on the left is modified from Lash and Engelder (2011) and shows the Marcellus as a second-order sequence with two third-order sequences and a number of higher frequency sequences superimposed. The logs on the right show lateral variability in thickness and log response. Modified from Lewis et al. (2011) who provided permission to publish this figure.

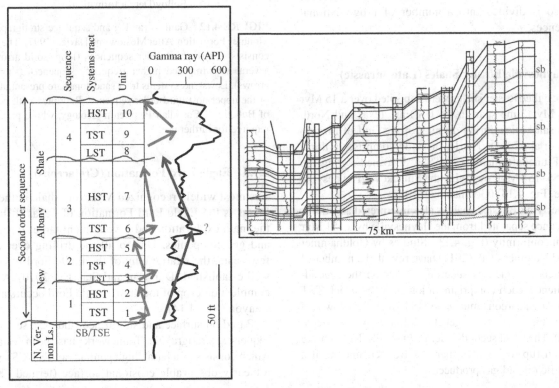

FIGURE 4.11 New Albany Shale gamma ray log showing it as a second-order depositional sequence with four third-order sequences and systems tracts superimposed. The well log cross section on the right is of the Marcellus Shale over a 75 km (47 mi.) interval showing a lower frequency downlap pattern onto a high TOC interval. After Bohacs and Lazar (2010). Reprinted with permission from the Houston Geological Society whose permission is required for further use.

the New Albany into four third-order sequences. Two of these sequences consist of TST and HST deposits and the other two also contain LST deposits, suggesting a somewhat more basinal extent of the latter two. The same gamma ray log patterns as observed in the previously mentioned shales occur for this shale. Also, Bohacs and Lazar (2010) have demonstrated an overall second-order downlap pattern onto the basal TST (Fig. 4.11).

4.7 MESOZOIC RESOURCE SHALE EXAMPLES

4.7.1 Montney Formation (Early Triassic)

The gas-producing Montney Formation was deposited in the western Canada sedimentary basin during marine transgression as shoreface, shelf, and slope deposits (Adams, 2009). It sits unconformably (SB/TSE) upon the Permo-Carboniferous Belloy/Debolt formations (Fig. 4.12).

The Montney is divided into two major third-order depositional sequences separated by a sequence boundary that correlates to an Early Triassic global eustatic sea-level fall (Moslow and Davies, 1997). The lower Montney consists of dark gray dolomitic siltstone interbedded with shales; it comprises a third-order TST and HST, which are further subdivided into a series of fourth- or perhaps fifth-order progradational parasequences (Moslow and Davies, 1997). The upper Montney consists of siltstone with interlaminated fine-grained sandstones; it consists of, from the base toward the top, third-order LST turbidites (which form reservoirs in part of the basin) and a TST/HST, which also is divided into a number of progradational parasequences.

4.7.2 Haynesville/Bossier Shales (Late Jurassic)

The Haynesville/Bossier shales were deposited over a 15 Myr (140–155 Myr) time span in the US East Texas and North Louisiana salt basins during opening of the Gulf of Mexico (Hammes et al., 2011). The Haynesville comprises a third-order TST unconformably (SB/TSE) deposited during the interval 155–151 Myr, overlying the Smackover/Haynesville Limestone (Fig. 4.13) (Goldhammer, 1998).

The overlying Bossier Shale is also interpreted as a third-order sequence separated from the Haynesville by a 151 Myr regional unconformity (Fig. 4.13). Studies by Goldhammer (1998) and Hammes et al. (2011) have revealed a number of high-frequency sequences superimposed upon the second-order sequence, each comprising a lower, organic-rich TST and an upper calcareous mudstone HST (including low API gamma ray/high-density diagenetic dolomite marker beds) (Fig. 4.13). The basal second-order Haynesville TST is more organic rich (up to 7 wt.%) than the Bossier, making it a more prolific oil and gas producer.

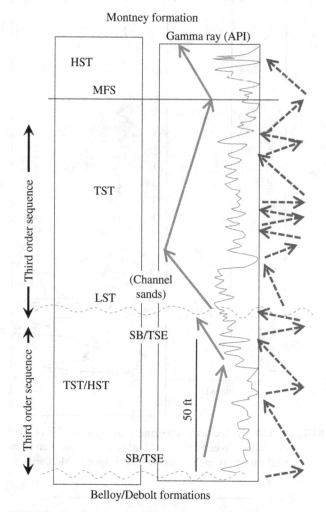

FIGURE 4.12 Gamma ray log and sequence stratigraphy of the Montney Formation. After Moslow and Davies (1997). The Montney consists of two third-order sequences (long, solid arrows) with several superimposed higher frequency sequences (short dashed arrows). Lowstand systems tract sandstones are present at the base of the uppermost third-order sequence. Reprinted with permission of Bulletin of Canadian Petroleum Geology, whose permission is required for further use.

4.7.3 Eagle Ford Formation (Cretaceous)

The most widely recognized Mesozoic shale in the United States is the Eagle Ford Formation of Texas. It has been the main exploration and development target for many oil and gas companies. Due to all the drilling over the past few years, the stratigraphy of the Eagle Ford has become well established (Donovan and Staerker, 2010). A superb, complete outcrop of the entire Eagle Ford occurs at Lozier Canyon (Fig. 4.14).

Based on surface and subsurface studies, a second-order sequence stratigraphic framework has been established which consists of a basal, high gamma ray TST/CS overlying a clearly observable erosional surface (termed "K63SB"

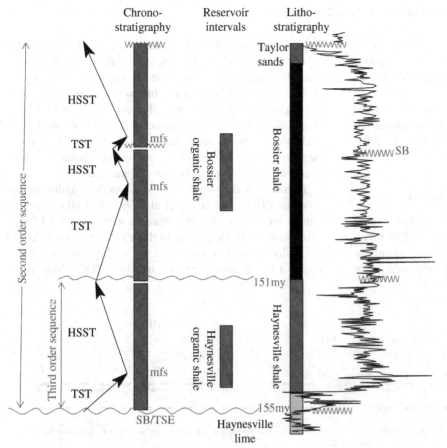

FIGURE 4.13 Sequence stratigraphy of the Haynesville and Bossier Shales (Vorce, 2011). The two third-order sequences together form a second-order sequence. Reprinted with permission of Houston Geological Society, whose permission is required for further use.

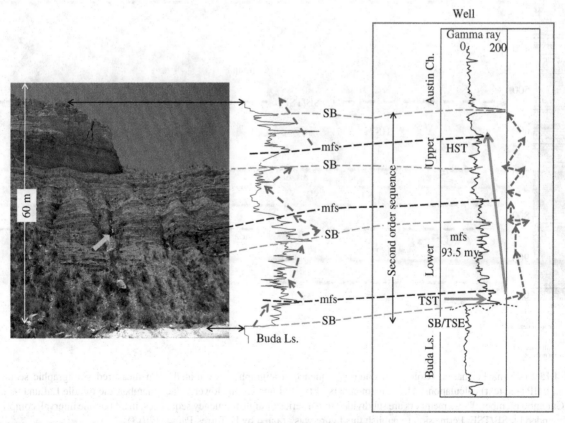

FIGURE 4.14 Left figure shows the Lazar Canyon outcrop of the Eagle Ford Shale. The gamma ray log in the middle is an outcrop gamma ray log obtained by scientists scaling the cliff face (see thick arrow); the right well log shows the second- (long solid arrows) and third-order (short dashed arrows) sequence stratigraphy of the Eagle Ford. Modified from Donovan and Staerker (2010). Reprinted with permission of the Gulf Coast Association of Geological Societies, whose permission is required for further use.

by Donovan and Staerker, 2010) of the underlying Buda Limestone. Above this interval, the gamma ray log response diminishes progressively upward, indicating an HST. Superimposed on this second-order sequence are a number of third-order sequences; three additional SBs and mfss have been recognized on the outcrop and extended to subsurface wells; more high-frequency sequences occur throughout the Eagle Ford at the outcrop and subsurface sites.

4.7.4 LaLuna Formation (Upper Cretaceous)

The LaLuna Formation is the main hydrocarbon source rock in the Middle Magdalena Valley basin of Colombia, South America. It is currently being investigated as unconventional resource shale. Limited outcrop and subsurface studies have subdivided the LaLuna into the lower Salada (black shales, black mudstones, black calcareous claystone, black limestone layers with internal pyritized concretions), middle Pujamana (claystone, mudstone, gray shale, and cherts), and upper Galembo (calcareous shales with limestone layers and nodules) members. There are no well logs available, but based upon geochemical analysis of outcrop samples, a preliminary sequence stratigraphic framework has been established (Fig. 4.15) (Torres-Parada, 2013).

Included in this framework is a residual hydrocarbon potential (RHP) plot (Fang et al., 1993) that allows stratigraphic intervals to be interpreted as varying upward from oxic to anoxic or vice versa. In this case, the entire LaLuna Formation records a lower interval of high and variable, overlain by an upper, interval of lower TOC. Both TOC

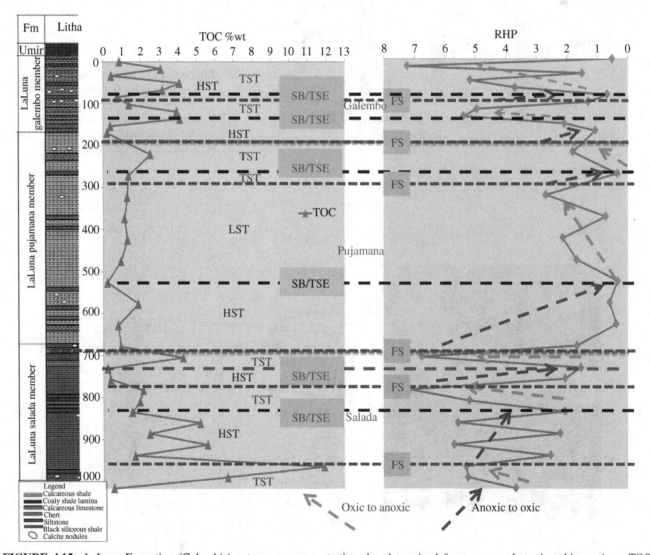

FIGURE 4.15 LaLuna Formation (Colombia) outcrop sequence stratigraphy, determined from measured stratigraphic sections, TOC content, and RHP (see text) calculations. The three members of the LaLuna are the lower Salada member, the middle LaLuna member, and the upper Galembo member. These members are subdivided into a series of high-frequency sequences, in all but one interval, comprising TST and HST bounded by SB/TSE. Permission to publish this figure was granted by E. Torres-Parada (2013).

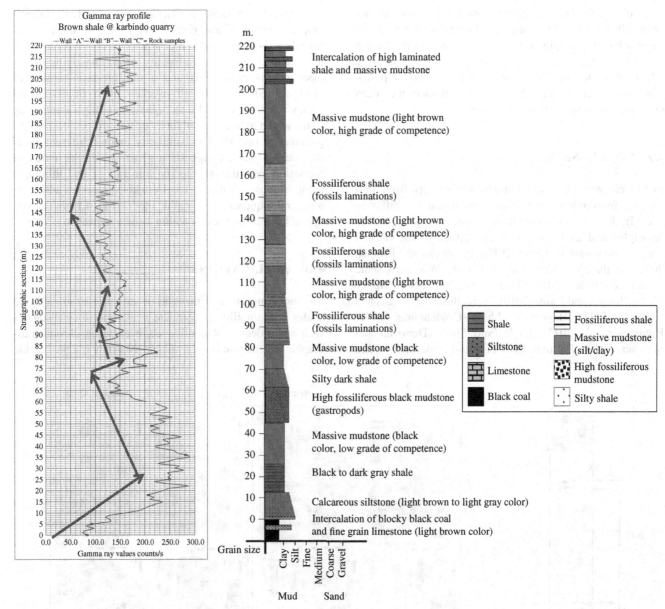

FIGURE 4.16 Measured stratigraphic section and outcrop gamma ray log of the Brown Shale in Sumatra, Indonesia. The three-colored intervals of the gamma ray log indicate three different measured sections combined into this composite log. Arrows denote sequences and systems tracts. Permission to publish this figure was provided by Brito (2014).

and RHP variations provide a subdivision into a number of higher frequency depositional sequences; one possible interpretation is provided in Figure 4.16.

4.8 CENOZOIC RESOURCE SHALE EXAMPLE

Examples of Cenozoic resource shales are not common, perhaps because many of them have not been sufficiently buried to be geochemically mature. One shale of current interest is the Eocene-Oligocene Brown Shale, which occurs in Sumatra, Indonesia; it is thought to be the major hydrocarbon source rock for many of the younger, stratigraphically shallower reservoirs there (Katz and Dawson, 1997). It has been interpreted as a lacustrine, syn-rift shale by Katz and Dawson (1997). There are very few wells that have penetrated the Brown Shale, and to our knowledge, there is only one lengthy outcrop within the active Karbindo coal mine. Along the vertical mine walls, an outcrop gamma ray log was obtained along with outcrop description; these measurements revealed increased gamma ray log response in the lower portion of the lower zone, followed upward by a variable gamma ray response in the upper interval of the lower zone (Fig. 4.16).

To date, a final sequence stratigraphic framework has not been fully developed for this shale, but it does have similar gamma-log stratigraphy at least at two scales. If the Brown Shale is entirely lacustrine in origin, it suggests that the effects of sea level cyclicity can extend well into the paleo landward direction either through connections to the ocean or by the lowering of fluvial base level.

4.9 CONCLUSIONS

In this chapter, we have compared stratigraphy from well logs and from outcrop/core of a variety of shales of differing age. Traditional and conventional sequence stratigraphic principles and analysis have been applied to the Barnett Shale (Abouelresh and Slatt, 2012a, b; Loucks and Ruppel, 2007; Singh, 2008; Slatt et al., 2012); the Woodford Shale (Molinares-Blanco, 2013; Serna-Bernal, 2013); New Albany Shale (Bohacs and Lazar, 2010), Marcellus Shale (Lash and Engelder, 2011), Haynesville Shale (Goldhammer, 1998; Hammes et al., 2011), Eagle Ford Shale (Donovan and Staerker, 2010); LaLuna Shale (Torres-Parada, 2013), and the Brown Shale (Brito, 2014). Comparison of typical stratigraphic sequences and gamma ray logs from these shales reveals a similarity in their characteristics—most notably a combined basal unconformity atop underlying strata (sequence boundary-SB) and a transgressive surface of erosion (TSE)—which are overlain by a fining-upward shaley interval capped by an organic-rich, high gamma ray shale, which in turn is overlain by an upward-decreasing API gamma ray pattern. This similarity among different-aged/environment shales suggests a generally similar mode of formation even though specific mineralogic and lithologic compositions may differ. The formative processes adhere to a general sequence stratigraphic model (Fig. 4.1) that can be applied at multiple scales within a shale sequence.

4.10 APPLICATIONS

The most important application of sequence stratigraphy to shales is the ability to correlate strata over long distances with some degree of confidence, even when chronostratigraphic information is not available. Not only is a predictable

FIGURE 4.17 Woodford gamma ray log showing subdivision of high-frequency sequences and systems tracts. GRP-4 and GRP-5 comprise a TST–HST (Brittle–Ductile) couplet. Properties of average P-wave acoustic impedance, average % TOC, average fracture gradient, and isopach maps show lateral variations in these properties across the study area. Such maps can lead to improved drilling decisions and locations. After Amorocho-Sanchez (2012) who provided permission to publish this figure.

vertical stratigraphy common and interpretable, but over long distances sequences and parasequences can be expected to downlap and terminate onto an organic-rich condensed section (Figs. 4.5 and 4.11 are examples).

Another important application is the ability to relate stratigraphic features to geomechanical, geochemical, and other features of shale strata, and through sequence stratigraphic correlations, to be able to map those properties. An example is shown in Figure 4.17 where P-wave impedance, TOC, fracture gradient, and isopach thickness of two Woodford Shale intervals (labeled GRP-4 and GRP-5) were mapped on the basis of logs from several wells (Amorocho-Sanchez, 2012). Taken together, these two intervals comprise a high-frequency sequence consisting of a lower, organic rich, high gamma ray CS/mfs interval (GRP-4)

overlain by an upper HST interval which exhibits a lower gamma ray response. In terms of geomechanical properties, Slatt and Abousleiman (2011) refer to these as "Brittle–Ductile couplets."

Although it may seem that drilling into a CS/mfs (ductile strata) might provide the opportunity to release more hydrocarbons than in more Brittle HST strata, these intervals tend to not fracture as well, and drilling can be more hazardous in the organic-rich zones (Figs. 4.18 and 4.19). By contrast, drilling into the Brittle zone of a Brittle–Ductile couplet has the opportunity to produce longer vertical fractures that will release more hydrocarbons (sweet spot). In addition, proppant will tend to keep fractures open in the Brittle zone, but not as readily in the ductile zone (Terracina et al., 2010).

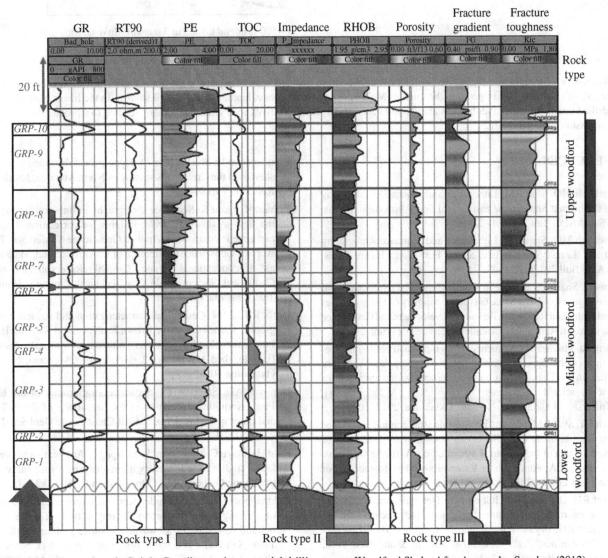

FIGURE 4.18 Properties of a Brittle–Ductile couplet, potential drilling target, Woodford Shale. After Amorocho-Sanchez (2012).

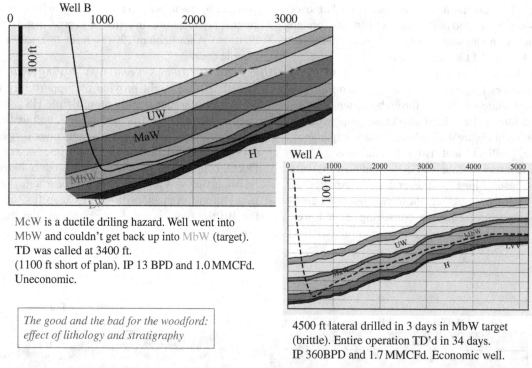

FIGURE 4.19 Stratigraphically targeted drilling within Woodford Brittle–Ductile couplets. In the upper figure, the well was drilled into the brittle (MbW) horizon, but dipped into the ductile (McW) drilling hazard zone, leading to early well shut-in. In the lower figure, the well was drilled into the ductile layer but was able to pull back into the brittle (MbW) layer and remain there, leading to a productive well. With the permission of Killian (2012).

REFERENCES

Abouelresh MO, Slatt RM. Lithofacies and sequence stratigraphy of the Barnett Shale in the east-central Fort Worth Basin, Texas, USA. AAPG Bull 2012a:1–22.

Abouelresh MO, Slatt RM. Lithofacies and sequence stratigraphy of the Barnett Shale in east-central Fort Worth Basin, Texas. AAPG Bull 2012b;96:1–22.

Adams C. Shale gas activity in British Columbia, exploration and development of BC's Shale Gas areas, Ministry of Energy, Mines and Petroleum Resources, Resource Development and Geoscience Branch. 3rd Annual Unconventional Gas Tech. Forum; 2009.

Amorocho-Sanchez JD. Sequence stratigraphy and seisic interpretation of the upper Devonian-lower Mississippian Woodford Shale in the Cherokee Platform: a characterization approach for unconventional resources [Unpublished M.Sc. thesis]. University of Oklahoma, Norman, Oklahoma; 2012. 109 p.

Baruch ET, Slatt RM, Marfurt KJ. Seismic stratigraphic analysis of the Barnett Shale and Ellenburger unconformity southwest of the core area of the Newark East field, Fort Worth Basin, Texas. In: Breyer JA, editor. Shale Reservoirs—Giant Resources for the 21st Century. 2012. AAPG Memoir 97; p 403–418.

Bennett RH, O'Brien NR, Hulbert MH. Determinants of clay and shale microfabric signatures: processes and mechanisms. In: Bennett RH, Bryant WR, Hulbert MH, editors. *Microstructure of Fine-Grained Sediments: From Mud to Shale*. New York: Springer-Verlag; 1991. p 5–32.

Bhattacharya JP, MacEachern JA. Hyperpycnal rivers and prodeltaic shelves in the Cretaceous seaway of North America. J Sediment Res 2009;79:184–209.

Bohacs K, Lazar R. Sequence stratigraphy in fine-grained rocks at the field to flow-unit scale: insights for correlation, mapping and genetic controls. In: Applied Geoscience Conference, 2010, Applied Geoscience Conference of US Gulf Region, Mudstones as Unconventional Shale Gas/Oil Reservoirs, Houston Geological Society Shale Gas Technical Program, Houston, Texas; February 8–9, 2010.

Brito, R.J., 2014, Geological characterization and sequence stratigraphic framework of the Brown Shale, central Sumatra Basin, Indonesia: Implications as an unconventional resource [Unpublished MS thesis]. University of Oklahoma, Norman, Oklahoma. 91p.

Donovan AD, Staerker TS. Sequence stratigraphy of the Eagle Ford (Boquillas) formation in the subsurface of South Texas and outcrops of West Texas. Gulf Coast Assoc Geol Soc Trans 2010;60:861–899.

Fang H, Jianyu C, Yongchuan S, Yaozong L. Application of organic facies studies to sedimentary basin analysis: a case study from the Yitong Graben, China. Org Geochem 1993;20:27–42.

Goldhammer RK. Second-order accommodation cycles and points of "stratigraphic turnaround": implications for carbonate buildup reservoirs in Mesozoic carbonate systems of the East Texas Salt Basin and south Texas. In: Demis WD, Nelis MK, editors. *West Texas Geological Society Annual Field Conference*

Guidebook. Volume 98–105, West Texas Geologic Society Publication, Midland, Texas; 1998. p 11–28.

Hammes U, Hamlin HS, Ewing TE. Geologic analysis of the upper Jurassic Haynesville shale in east Texas and west Louisiana. AAPG 2011;95:1643–1666.

Haq BU, Hardenbol J, Vail PR. Mesozoic and Cenozoic chronostratigraphy and cycles of sea-level change. In: Wilgus CK, Hastings BS, St. Kendall CGC, Posamentier HW, Ross CA, Van Wagoner JC, editors. Sea-Level changes—an integrated approach. 1984. SEPM Spec. Pub l. No. 42; p 71–108.

Johnson JG, Klapper G, Sandberg CA. Devonian eustatic fluctuations in Euramerica. Geol Soc Am Bull 1985;96:567–587.

Johnson KS, Amsden TW, Denison RE, Dutton SP, Goldstein AG, Jr. Rascoe B, Sutherland PK, Thompson DM. Geology of the southern Midcontinent. Oklahoma Geological Survey Special Publication 89-2; 1989. p 1–53.

Katz BJ, Dawson WC. Pematang-Sihapas! Petroleum system of central Sumatra. Proceeding of the Conference on Petroleum Systems of SE Asia and Australasia; Jakarta, Indonesia; 1997. p 685–698.

Killian BJ. Sequence stratigraphy of the Woodford Shale, Anadarko Basin, Oklahoma: implications on regional Woodford target correlation [M.S. thesis]. University of Oklahoma; 2012. 102 p.

Lash GG, and Blood R. Chemostratigraphic trends of the Middle Devonian Marcellus Shale, Appalachian Basin: Preliminary Observation: AAPG Search and Discovery Article #80198; 2011.

Lash GG, Engelder T. Thickness trends and sequence stratigraphy of the Middle Devonian Marcellus Formation, Appalachian basin: implications for Acadian foreland basin evolution. AAPG Bull 2011;95:61–103.

Lewis E, Behling M, Pool S, An overview of Marcellus and other Devonian Shale Production in West Virginia. AAPG Search and Discovery Article #10372; 2011.

Loucks RG, Ruppel SC. Mississippian Barnett Shale: lithofacies and depositional setting of a deep-water shale-gas succession in the Fort Worth Basin, Texas. AAPG Bull 2007;91:579–601.

May JA, Anderson DS. Mudrock reservoir deposition and stratigraphy: not homogenous, not boring. AAPG Search and Discovery #90122; 2011.

Miall AD. *The Geology of Stratigraphic Sequences*. Berlin: Springer-Verlag; 1997. p 433.

Molinares-Blanco C. Stratigraphy and palynomorphs composition of the Woodford Shale in the Wyche Farm Shale pit, Pontotoc County, Oklahoma [M.S. thesis]. Norman, Oklahoma: University of Oklahoma; 2013. 90 p.

Moslow TF, Davies GR. Turbidite reservoir facies in the lower Triassic Montney Formation, east-central Alberta. Bull Can Petrol Geol 1997;45:507–536.

Mulder T, Chapron E. Flood deposits in continental and marine environments: character and significance. In: Slatt R, Zavala C, editors. *Sediment transfer from shelf to deep water revisiting the delivery mechanisms*. Volume 61, Tulsa: AAPG Stud Geol; 2011. p 1–31.

Nakajima T. Hyperpycnites deposited 700 km away from river mouths in the Central Japan Sea. J Sediment Res 2006;76:60–73.

O'Brien NR, Slatt RM. *Argillaceous rock atlas*. New York: Springer-Verlag; 1990. p 141.

Paxton ST, Cardott BJ. Oklahoma gas shales field trip, October 21 & 23, 2008: Oklahoma Geological Survey Open File Report 2-2008; 2008. 110 p.

Paxton ST, Cruse A, Krystyniak A. Fingerprints of global sea level change revealed in hydrocarbon source rock? 2007, http http://www.searchanddiscovery.com/documents/2006/06095paxton/images/paxton.pdf (accessed February 5, 2013).

Pollastro RM. Geologic framework of the Mississippian Barnett Shale, Barnett–Paleozoic total petroleum system, Bend Arch-Fort Worth Basin, Texas. AAPG Bull 2007;91:405–436.

Portas R. Characterization and origin of fracture patterns in the Woodford Shale in southeastern Oklahoma for application to exploration and development [Master's thesis]. Norman, Oklahoma: University of Oklahoma; 2009. 113 p.

Rimmer SM, Thompson JA, Goodnight SA, Robl TL. Multiple controls on the preservation of organic matter in Devonian-Mississippian marine black shales geochemical and petrographic evidence. Palaeogeography, Palaeoclimatology, Palaeoecology 2004;215:125–154.

Rine JM, Ginsburg RN. Depositional facies of a mud shoreface in Suriname, South America. J Sediment Petrol 1985;55:633–652.

Schieber J, Southard JB. Bedload transport of mud by floccule ripples—direct observation of ripple migration processes and their implications. Geology 2009;37:483–486.

Schieber J, Southard J, Thaisen K. Accretion of mudstone beds from migrating floccule ripples. Science 2007;318:1760–1763.

Serna-Bernal A. Geological characterization of the Woodford Shale, McAlister Cemetary Quarry, Criner Hills, Oklahoma [Unpublished M.S. thesis]. University of Oklahoma, Norman, Oklahoma; 2013. 141 p.

Singh P. Lithofacies and sequence stratigraphic framework of the Barnett Shale, northeast Texas [Ph.D. dissertation]. The University of Oklahoma; 2008. 181 p.

Slatt RM. Important geological properties of unconventional resource shales. Cent Eur J Geosci 2011;3:435–448.

Slatt RM, Abousleiman Y, Merging sequence stratigraphy and geomechanics for unconventional gas shales. Leading Edge 2011;30(3):1–8. Special section: Shales.

Slatt RM, O'Brien NR. Pore types in the Barnett and Woodford gas shales: contribution to understanding gas storage and migration pathways in fine-grained rocks. AAPG Bull 2011;95:2017–2030.

Slatt RM, Rodriguez ND. Comparative sequence stratigraphy and organic geochemistry of gas shales: commonality or coincidence? J Nat Gas Sci Eng 2012;8:68–64.

Slatt RM, Philp P, O'Brien N, Abousleiman Y, Singh P, Eslinger EV, Perez R, Portas RM, Baruch ET, Marfurt KJ, Madrid-Arroyo S. Pore- to regional- scale, integrated characterization workflow for unconventional gas shales. In: Breyer JA, editor. Shale Reservoirs—Giant Resources for the 21st Century. 2012. AAPG Memoir 97; p 127–150.

Soynika OA, Slatt R. Identification and microstratigraphy of hyperpycnites and turbidites in Cretaceous Lewis Shale, Wyoming. Sedimentology 2008;55:1117–1133.

Terracina JM, Turner JM, Collins DH, Spillars SE, Proppant selection and its effect on the results of fracturing treatments performed in shale formations. SPE Annual Technical Conference and Exhibition, September 19–22, 2010, Florence, Italy: SPE 135502; 2010. 17 p.

Torres-Parada E. Unconventional gas shale assessment of LaLuna Formation in the central and south areas of the Middle Magdalena Valley basin, Colombia. [Unpublished M.S. thesis]. University of Oklahoma, Norman, Oklahoma; 2013.

Van Wagoner JC, Mitchum RM, Campion KM, Rahmanian VD. Siliciclastic sequence stratigraphy in well logs, cores and outcrops; concepts for high resolution correlation of time and Facies. AAPG Methods in Exploration Series, no.7; 1990. p 53.

Vorce CL. Depositional and sequence stratigraphic framework for the Haynesville and Bossier Shales. Houston Geol Soc Northsiders Luncheon Mtg.; 2011.

5

PORE GEOMETRY IN GAS SHALE RESERVOIRS

ADNAN AL HINAI AND REZA REZAEE
Department of Petroleum Engineering, Curtin University, Perth, WA, Australia

SUMMARY

Assessing shale formations is a major challenge in the oil and gas industry. The complexities are mainly due to the ultralow permeability, the presence of a high percentage of clay, and the heterogeneity of the formation. Knowledge and understanding of rock properties, including pore geometry, permeability, and fluid distribution are essential for determining shale's hydrocarbon storage and recovery. This chapter discusses the microstructural characterization of gas shale samples through mercury injection capillary pressure (MICP), low-field nuclear magnetic resonance (NMR), and nitrogen adsorption (N_2). High resolution focused ion beam–scanning electron microscopy (FIB–SEM) image analysis is used to further support the experimental pore structure interpretations at submicron level. The chapter focuses on three key areas: (i) comparisons of pore size distribution (PSD), (ii) recognizing the relationship between pore geometry and permeability, and (iii) effects of clay occurrence on fluid transport properties.

MICP and N_2 are destructive techniques used as PSD measurements. MICP is capable of characterizing the PSD in the range of mesopores (5 nm < pore diameter > 50 nm: intra- and inter-clays) to macropores (pore diameter > 50nm: inter-grains and discontinuities) while N_2 can be applied to pores less than 2 nm. NMR is a nondestructive technique that is performed under room conditions. It supposes that the sample is fully or partially water saturated.

In contrast with MICP PSD that provides only "connected" pore throats as tube shapes and no pore body sensu stricto, NMR PSD provides full experimental characterization of pore geometry, the size of the pore body behind the throats, and the isolated pores. The pore body to pore-throat ratio is a characteristic that controls fluid flow. The connectivity in the pore system can be represented by the pore body to pore-throat size ratio: the lower the ratio, the lower the connectivity; hence the lower will be the permeability/fluid flow. The results demonstrated a complex geometry of the pore network from clay-rich rocks.

The siliceous and organic-rich gas shales studied are marked by a strong component of clay minerals, mostly made of kaolinite and illite/smectite (I/S) mixed layers. Three types of shales can be classified according to their clay content: (i) low I/S but high kaolinite, (ii) high I/S but low kaolinite, and (iii) high I/S and high kaolinite. It is understood that I/S acts as a fluid trapping mineral by increasing the pore geometry complexity (surface to volume ratio increase) but generates low porosity made up of microporosity. Kaolinite acts as fluid storage by clogging pores and helps to keep high porosity made up of relatively larger pores.

The combination of MICP, N_2, and NMR forms an ideal approach to overcome each of their individual limits in terms of pore size resolution and the external influences (dehydration/hydration state or sample preparation).

5.1 INTRODUCTION

5.1.1 Gas Shales and Their Challenges

Gas shale systems comprise fine-grained sedimentary rocks that are mainly consolidated from clay-sized mineral grains. It is known as "mudstone" in the category of sedimentary

---

Fundamentals of Gas Shale Reservoirs, First Edition. Edited by Reza Rezaee.
© 2015 John Wiley & Sons, Inc. Published 2015 by John Wiley & Sons, Inc.

rocks. Fissile and laminated attributes (Zahid et al., 2007) distinguishes shale from other "mudstones." That is, the rock is made up of many thin parallel layers, and the rock readily splits into thin pieces along the layers.

The systems are typically organically rich; a higher total organic content (TOC) shale commonly has a higher adsorbed gas content (Boyer et al., 2006). Fracture stimulation is required for the systems to economically produce gas (King, 2010). Fractures are created easily in silica-rich and carbonate-rich shales when compared to clay-rich shales, and total porosities are larger in clay-rich shales than in silica-rich shales (Bustin et al., 2008; Ross and Marc Bustin, 2009).

One of the most important and difficult variables to determine is the *in situ* permeability (Shaw et al., 2006), which is controlled by the pore structure (Bustin et al., 2008). Rock typing in terms of the hydraulic process (Rushing et al., 2008) from porosity-permeability cross-plots is not practical in gas shale reservoirs because the dynamic range for porosity in shales is very narrow compared to the conventional reservoirs.

Undeniably, a fluid's efficiency in flowing through the pore system (hydraulic conductivity and permeability) will also depend on the fluid–solid interactions, the tortuosity of the pore network, intrinsic structures such as veins, faults, or bedding (i.e. heterogeneities), and the anisotropic aspects of these characteristics. Currently, the only way to extract gas from gas shale is through extensive hydraulic fracturing (Gale et al., 2007), and the gas recovery efficiency will depend on the flow and trap properties of the gas shale. It is therefore crucial to understand the pore structures of gas shale. As yet, there is no clear understanding of how these pore systems are connected.

5.1.2 Pore Size Classification

The word "size" is associated either with diameter, if a pore throat is considered as cylindrical, or with width, if a pore throat is characterized as a thin slot. Generally, characterization of the pore-throat size of a rock sample (Nelson, 2009) requires the choice of (i) a method of measurement, (ii) a model for converting the measurement to a dimension, and (iii) a parameter to represent the resulting pore size distribution. For instance, MICP uses the Washburn equation to determine a dimension associated with a specific saturation of the invading fluid or an inflection point on a graph of pressure versus the volume of the invading fluid.

A range of classifications are available in the literature to describe the pore system. In general, they can be categorized based on petrographic, depositional, and hydraulic rock types. Petrographic rock types are geologically classified using image acquisition techniques. The depositional types are explained by their core, as categorized by sedimentary structure, composition, and sequence stratigraphy that are determined by the depositional environment. Hydraulic rock

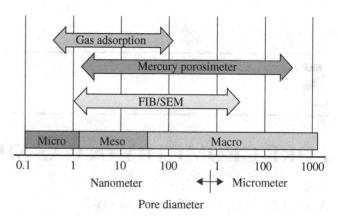

FIGURE 5.1 Pore size scale based on the methodology utilized to characterize the pore size distribution. The values are based on IUPAC Classification.

types are defined in terms of the physical rock property characteristics, such as flow and storage properties, which are controlled by the pore geometry (Rushing et al., 2008).

In this study, the pore size classification (Fig. 5.1) has been adopted from the International Union of Pure and Applied Chemistry (IUPAC), which was established by Rouquerol et al. (1994). The classification is based on three groups: micropores that include pores less than 2 nm diameter, mesopores that comprise pores with diameters between 2 and 50 nm, and macropores that include pores with diameters larger than 50 nm.

5.2 SAMPLES CHARACTERISTICS

5.2.1 Sample Collection

A total of 31 samples from three formations: named here as PCM, PKM, and CCM have been sampled in this study. The sample collection and sequence of laboratory experiments conducted are shown in Table 5.1.

5.2.2 Mineral Composition

Bulk X-ray diffraction (XRD) analysis was performed on 23 samples, using a Siemens D500 automated powder diffractometer to characterize their mineral composition and content. The XRD results show that all the shale collections are siliceous matrix dominated, with the highest quartz content in CCM and PCM formations compared to PKM, at 53.28, 36.75, and 19.4%, respectively (Fig. 5.2). CCM and PCM also record occurrence of K-feldspars while PKM is rich in pyrite. The remaining mineral contents are clay minerals; formations PCM and PKM are mostly composed of mixed I/S, with averages of 15.5 and 26.4%, respectively, while formation CCM exhibits a high presence of kaolinite of approximately 20%, and about 6% mixed I/S. Detailed XRD results are shown in Appendix 5.A.

TABLE 5.1 Laboratory methods applied

Formation	Sample number	Depth (m)	XRD	MICP	N$_2$	NMR	SEM
PCM	1	1618	x	x			
	2	1614	x	x			
	3	400.8		x			
	4	2650		x			
	5	3771	x	x			
	6	3792	x	x			
	7	2294		x			
	8	2780	x	x	x	x	x
	9	2782		x	x	x	x
	10	2790		x		x	
	11	2817	x		x		
	12	2825		x	x	x	
	13	2794	x	x	x	x	
	14	2806	x	x	x	x	x
	15	2813	x		x	x	
	16	2831	x		x		x
CCM	17	1947	x	x	x	x	
	18	1246	x	x	x	x	x
	19	1384	x	x	x		
	20	1152	x	x	x	x	x
	21	1160	x	x	x	x	
	22	1650	x		x		x
	23	1454	x	x	x	x	
	24	1410	x	x	x	x	
	25	1855	x	x	x	x	
	26	1436	x	x	x	x	
	27	1949	x	x	x	x	
PKM	28	3793		x			x
	29	3799		x			x
	30	3800	x	x			
	31	3793	x	x			

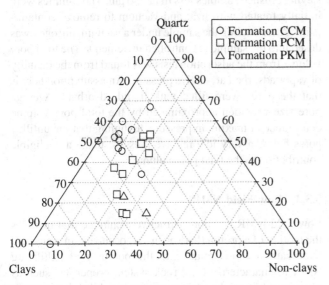

FIGURE 5.2 Ternary plot of the average weight percentage of mineral composition of the CCM, PCM, and PKM formations.

5.3 EXPERIMENTAL METHODOLOGY

5.3.1 Capillary Pressure Profile

Gas shale reservoirs play a major role in exploration and production because they are deemed to be both source rock and cap rock. They display good sealing characteristics due to their small pore throats, which are responsible for creating high capillary pressures (Al-Bazali et al., 2005). To understand capillary pressure behavior, mercury intrusion experiments are normally conducted.

The MICP technique is used to determine various quantifiable aspects of a porous medium such as pore diameter, total pore volume, surface area, and bulk and absolute densities (Burdine et al., 1950; Chen and Song, 2002; Kale et al., 2010a, b) as a function of pressure, correlated with permeability in some rocks (Dastidar et al., 2007; Ma et al., 1991; Owolabi and Watson, 1993; Swanson, 1981) and rock typing in shale by integrating geological cores (Kale et al., 2010a).

MICP was performed on 24 dry samples of an average weight of 8 g with a Micromeritics autopore IV porosimeter. MICP provides the porosity of the connected pores from the volume of mercury injected within the pore network under high pressure, and the capillary pressure curves from the injected volume of mercury under incremental increase of applied pressure. Pore-throat size distribution down to 3 nm in diameter (i.e., maximum of 60,000 psi) can be derived from the capillary pressure curves. The pore-throat radius can be found by Laplace-Washburn equation (Washburn, 1921):

$$R = \frac{2\sigma \cos\theta}{P_c} \quad (5.1)$$

where P_c is the entry pressure (psi), σ is the interfacial tension (dynes/cm), θ is the contact angle (degrees), and R is the pore-throat radius (um). The minimum capillary entry pressure is the capillary pressure at which the non-wetting phase starts to displace the wetting phase, confined in the largest pore throat within a water-wet formation. The capillary entry pressure can be major, particularly for shales with very small pore throats (permeability) (Al-Bazali et al., 2005). The entry pressure is inversely proportional to the size of the pore in which mercury will intrude (radius).

Figure 5.3 shows the capillary pressure curve during the injection process. At the lower injection pressure, the mercury starts to enter the large pores and then starts to plateau. The bend "apex" or the inflection point proposed by Swanson (1981) is where the pressure curve starts to have a steep slope toward the higher capillary pressure, illustrating the smaller pore throats, micropores, or nanopores when dealing with tight gas or gas shale rocks.

However, mercury intrusion experiments alone do not provide full experimental characterization of pore geometry (Chen and Song, 2002), because they operate by injecting pressure

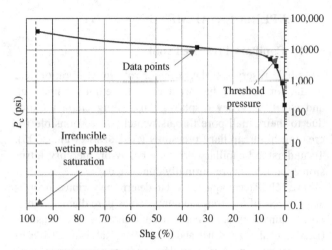

FIGURE 5.3 Typical capillary pressure profile.

incrementally into the porous media and recording the injected volume at each step. This type of pressure-controlled instrument measures the pore-throat size (pore entry radii) and does not detect the size of the pore body behind the throats (Burdine et al., 1950; Churcher et al., 1991; Heath et al., 2011).

5.3.2 Nitrogen Adsorption (N_2)

The low pressure adsorption measurement allows us to understand the PSD and study the parameters which control the adsorbed gas capacity, such as surface area and microporosity. Low pressure adsorption measurement has been used extensively in surface chemistry analysis for characterization of porous materials, and more recently has been adopted for characterization of the nanopores in the shale samples (Chalmers et al., 2012; Kuila and Prasad, 2011; Ross and Marc Bustin, 2009)

Low-pressure nitrogen adsorption (LPNA) measurement is used to quantify the amount of gas adsorbed at different relative pressures (P/P_0), where P is the gas vapor pressure in the system and P_0 is the saturation pressure of the adsorbent. The Brunauer–Emmett–Teller (BET) method is the most widely used procedure for determination of the surface area of porous samples (Brunauer et al., 1940). Equivalent surface area is calculated using the BET equation (Quantachrome Instruments, 2008):

$$\frac{1}{W((P/P_0)-1)} = \frac{1}{W_mC} + \frac{C-1}{W_mC}\left(\frac{P}{P_0}\right) \quad (5.2)$$

where W is the weight of gas adsorbed at a relative pressure (P/P_0) (P being the gas vapor pressure in the system and P_0 the saturation pressure of the adsorbent), W_m is the weight of monolayer nitrogen adsorbed to the sample. The C constant is related to the energy of adsorption and its value shows the magnitude of the adsorbent/adsorbate interactions.

The total pore volume is derived from the amount of vapor adsorbed at a relative pressure close to unity, by assuming that the pores are then filled with liquid adsorbate. The average pore size could be estimated from the total pore volume determined at maximum pressure, by assuming that the pores which would not be filled below a relative pressure of 1 have a negligible contribution to the total pore volume. For example, assuming cylindrical pore geometry, the average pore radius (r_p) can be expressed as:

$$r_p = \frac{2V_{ads}}{S} \quad (5.3)$$

where V_{ads} is the total amount of nitrogen adsorbed and S is the surface area (Quantachrome Instruments, 2008).

The distribution of pore volume with respect to pore size is called pore size distribution (PSD). Usually the BJH model (Barret et al., 1951) and DH model (Dollimore and Heal, 1964) are used for determining PSD using nitrogen adsorption for the shale layers. The actual pore size in both of these models is calculated using the thickness of the adsorbed layer and the Kelvin equation (Gregg and Sing, 1991):

$$\ln\left(\frac{P}{P_0}\right) = \frac{2\gamma V_m}{RTr_K}\cos\theta \quad (5.4)$$

where P is the gas vapor pressure, P_0 is the saturation pressure of the adsorbent, γ is the surface tension of nitrogen at its boiling point (77 K), θ is the contact angle between the adsorbate (liquid nitrogen) and the adsorbent, V_m is the molar volume of liquid nitrogen, R is the gas constant, T is the boiling point of nitrogen (77 K), and r_K is the Kelvin radius of the pore.

Micromeritics TriStar II 3020 was used to determine the quantity of nitrogen adsorbed. The samples were prepared by sieving crushed samples less than 250 μm. The samples were initially treated with heat and vacuum to remove contaminants. After cooling the sample under a vacuum, nitrogen was dosed into the sample at controlled increments. The total pore volume from the adsorption tests was found from the quantity of vapor adsorbed at relative pressure. An assumption here is that the pores were filled with liquid adsorbate. Average pore size was then approximated by the total pore volume determined at maximum pressure, assuming that the unfilled pores below a very low relative pressure make a negligible contribution to the total pore volume.

5.3.3 Low-Field NMR

Low-field NMR is a nondestructive technique that involves the motion of a proton (hydrogen 1H) occurring in water and hydrocarbon fluids relative to the porous rock. Utilizing NMR to characterize fluid rock system properties, such as porosity, pore size distribution, and permeability, has become popular in the industry (Bowers et al., 1993; Coates et al.,

1999; Gabriela and Lorne, 2000; Glorioso et al., 2003; Grunewald and Knight, 2011; Hidajat et al., 2003; Kenyon et al., 1995; Minh and Sundararaman, 2006).

NMR T_2 relaxation time was conducted on 16 partially saturated and brine-saturated core-plugs (3.8 cm diameter and 4–8 cm long) using a low-field Maran Ultra-Spectrometer 2 MHz from Oxford Ltd. Low-field NMR is a nondestructive technique that involves the motion of the proton (Hydrogen 1H) occurring in water and hydrocarbon fluids relative to the porous rock. The relaxation time T_2 was acquired during a Carr–Purcell–Meiboom–Gill (CPMG) spin-echo pulse sequence (see Dimri et al., 2012 for more details). The transverse relaxation time is mainly controlled by the pore geometry and diffusion transport as

$$\frac{1}{T_2} = \frac{1}{T_{2\text{Bulk}}} + \rho_2 \left(\frac{S}{V}\right) + \frac{D(\gamma GTE)^2}{12} \quad (5.5)$$

where ρ_2 is the surface relaxivity related to mineral interaction with fluid (in Pm/s), T_2 is the transverse NMR relaxation time, and $T_{2\text{bulk}}$ is the transverse relaxation time of the bulk water only (in s), defined as a constant at a specific temperature and constant water viscosity. S/V (in Pm^{-1}) is the ratio of pore surface to pore fluid volume and is defined as a pore geometry index. The last part of the equation represents the diffusion aspect of the spin echo with D for the molecular diffusion coefficient (in cm^2/s), γ being a constant of the gyromagnetic ratio of a proton in (in MHz/T), G being the field-strength gradient (in G/cm), and TE being the inter-echo spacing used in the CPMG sequence. Since no static magnetic gradient field was applied during the CPMG sequence, Equation 5.4 can be simplified to the second part of the equation as a function of the pore geometry and surface relaxivity. More details on the principles of NMR are described in the work of Coates et al. (1999).

The core plugs were first measured as received (i.e., partially saturated) before performing injection under a hydrostatic pressure of 3.5 MPa of brine (20 g/l KCl) over several days to resaturate the core plugs and repeat NMR acquisition. Prior to weighing, the excess fluid on the surface of the core plug was removed by rolling the sample on white printing paper twice along the landscape length. White paper was used as opposed to paper towel because the paper towel could draw out more of the fluids in the pores close to the surface of the plugs.

The plugs were then wrapped tightly with a transparent plastic wrap to keep the fluids intact with the plugs and to prevent the fluid spreading through the container.

5.3.4 Image Acquisition and Analysis

SEM and FIB–SEM were used to support the different types of porosities, recognized from the experimental techniques, and to visualize the distribution and type of clay minerals.

SEM imaging was conducted using a Zeiss Neon 40EsB and Philips XL40. The Zeiss Neon 40EsB is equipped with a field emission gun with a maximum extra-high tension (EHT) voltage of 30 kV. Individual samples were mounted upon pin-type mounts prior to coating with a thin layer of platinum, to ensure surface conductivity. Samples were introduced into the SEM for secondary electron imaging using an EHT of 5 kV. Mineralogy and pore size were visually identified with the resultant images.

The FIB instrument works in a similar way to SEM; instead of a beam electron, FIB uses a Ga$^+$ primary ion beam that hits the surface of the sample and sputters a small amount of materials that leaves the surface as either secondary ions (i^+ or i^-) or neutral atoms (n^0). The signal from the sputtered ions or secondary electrons is collected to form an image of the surface of the sample and gives information on the topography and material characteristics (Fibics, 2011). The system works repetitively; first the images are registered and are interpolated normally to the slice (direction) and the SEM beam creates a 2D image of the sample. The ion beam removes a thin layer of material on the surface of the sample, creating a new surface that is aligned with the previous slice, the SEM then generates an image again and the process is repeated (Butcher and Lemmen, 2011).

A piece of ±20 × 5 mm size from the sample 10 was embedded in resin and the surface was polished up to 1200 grit. The sample was placed on an aluminium stub using a silver dab and coated with silver and carbon to reduce electron charging and energy drift. The sample was placed on the dual beam stage at an angle of 52 degrees and a working distance of 4 mm, and the chamber was vacuumed (Fig. 5.4). Platinum (20 × 20 × 2.5 μm) was deposited on the region of interest using 30 kV and 0.28 nA energy beam (Fig. 5.5). A large trench was made around the platinum coat at various beam currents and voltages (Fig. 5.5). The large trench reduces

FIGURE 5.4 Illustration of the sample stage tilted at an angle of 52°.

94 PORE GEOMETRY IN GAS SHALE RESERVOIRS

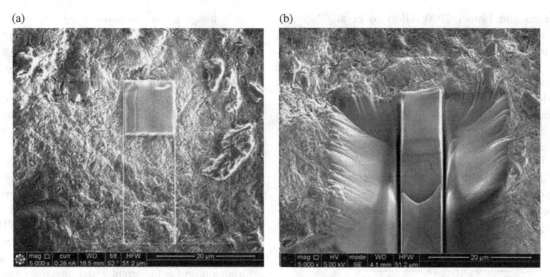

FIGURE 5.5 (a) Image from FIB–SEM showing the platinum coating (rectangular section) (b) Image showing the rough cut of the trench.

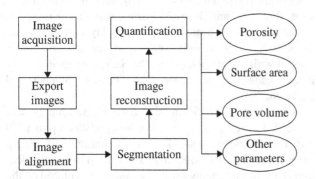

FIGURE 5.6 Flowchart of general image analysis procedures.

the debris around the surface to be imaged. The final cut was performed at 30 kV and 0.93 pA energy beam to provide a fine clean cut. The milling process for a 10 × 10 × 7 μm specimen size was carried out at an energy beam of 2 kV and 1.4 nA using back-scattered electrons. The FIB–SEM image acquisitions and pore size analysis focus on a small volume area of the sample that is not necessarily representative of the core plug results from laboratory methods.

The general steps involved in pore space image analysis from FIB–SEM are illustrated in Figure 5.6. Filtering (Talabi et al., 2008) is applied to the sample to improve the image quality and to reduce noise. Image cutting removes any surface patches along with the outer edges of the sample for the analysis that might have been mishandled. The objective of segmentation is to simplify and/or alter the representation of an image, to make it more meaningful and easier to analyse. The process involves converting a gray-scale image to a binary image composed of two types of pixels: black and white (Dougherty and Lotufo, 2003) by categorizing two populations based on the intensity (i.e., dividing the images into two phases, pores and solid phase). Segmentation is achieved by a thresholding process that utilizes two coefficients, T_0 and T_1 (Prodanovic et al., 2006), lower and higher attenuation, respectively. Also, T_0 and T_1 values would correspond to phase one or phase two. Once the image is segmented, image analysis can be done.

Porosity can be determined directly from the segmented image, counting the sum of the segmented voxels of the pore space divided by the total image volume (Al-Raoush and Willson, 2005):

$$\phi = \frac{V_{\text{segmented pores}}}{V_{\text{total image}}} \times 100\% \tag{5.6}$$

The pore surface area is found by counting the number of surface voxels between void and solid in each element. The slice of the CT image would be made up of voxels, that is, volume elements; hence, the volume can be determined by counting the void blocks. In other words, the number of voxels belonging to the body can be calculated as the sum of 2D areas multiplied by the Z spacing, in image analysis software (Boudier, 2014). The 3D sphericity of an object can be assumed being extension of 2D circularity, and can be determined from the ratio of volume over area. The sphericity, as well as the circularity, is maximal and equals 1 for a sphere:

$$S^3 = \frac{36 \cdot \pi \cdot V^2}{A^3} \tag{5.7}$$

where S is the sphericity, V is the volume, and A is the area.

In image analysis, shape factor—a dimensionless quantity—is determined to describe the shape of the element (independent of its size). The measure signifies the degree of deviation from an ideal shape, that is, for pore space, an ideal shape would be a circle, where the value of the shape factor would be 1. The shape factor is given by

$$G = \frac{VL}{A_s^2} \quad (5.8)$$

where A is the surface area of the pore or throat block, V is block volume, and L is the block length. This is equivalent to

$$G = \frac{A}{P^2} \quad (5.9)$$

where A is the cross-sectional area and P is perimeter.

5.4 ADVANTAGES AND DISADVANTAGES OF EXPERIMENTAL PSD METHODS

In some instances, it is difficult to obtain suitable core plugs for evaluating the petrophysical properties of reservoir rocks. In this situation, MICP and N_2 adsorption tests can be carried out. The advantages of the MICP technique are: it directly measures the pore volume through the mercury volume injected; it requires small rock cuttings or fragments; results are obtained relatively quickly with reasonable accuracy; and finally very high capillary pressure ranges can be achieved. The measurement does not require a completely symmetrical sample, but it is commonly limited to 1 cm^3.

MICP and N_2 are destructive techniques but quite often used as relevant PSD measurements for gas shales (Clarkson et al., 2013). MICP is capable of characterizing the PSD in the range of mesopores (5 nm < pore diameter > 50 nm: intra- and interclay) to macropores (pore diameter > 50 nm: intergrains and discontinuities) while N_2 relates to pores less than 2 nm. The pore diameter classification in this study is based on IUPAC classification (Rouquerol et al., 1994).

The N_2 method also requires crushing the sample into powder to fully wet the surface of the sample, which could affect the original pore structure of the solid matrix. Another inconvenience of the MICP and N_2 methods for evaluating the properties of shale is the occurrence of a double layer (or Stern layer). This layer is directly related to the surface clay-bound water that reduces porosity and pore radius.

NMR is a nondestructive technique and it supposes that the sample is fully or partially water saturated (i.e., in a preserved condition) for proper porosity assessment. In contrast to MICP PSD that provides only "connected" pore throats as tube shapes and no pore body sensu-stricto, NMR PSD provides full experimental characterization of pore geometry (Chen and Song, 2002), the size of the pore body behind the throats (Burdine et al., 1950; Churcher et al., 1991; Heath et al., 2011) and the isolated pores (e.g., if the sample kept its original fluid).

5.5 PERMEABILITY MEASUREMENT

Liquid or gas can be used to measure permeability on core plugs in laboratories. Usually, gas preferred as the sample preparation is relatively simpler and the measurement duration is also much shorter. However, at low average pressure gas slippage occurs, which causes Darcy Law to produce high permeability values and requires Klinkenberg corrections. The corrected permeability is then termed "liquid permeability" or "Klinkenberg permeability."

Both steady state and unsteady state methods are used for permeability measurement as outlined by Luffel (1993). Permeability measurements are done using resin disks, where a piece of the rock is embedded in resin (Egermann et al., 2004, 2006; Lenormand and Fonta, 2007; Lenormand et al., 2010). For rocks with permeabilities >1 mD, the initial coating is done with a high viscosity resin. This would prevent the resin from invading the pores. For lower permeabilities, a low viscosity resin allows the partial invasion over a small distance and a good sealing. After the sample is embedded in the resin, the sample is cut into slices, with thicknesses ranging from 1 to 5 mm and their surfaces are polished. Samples with predicted low permeabilities undergo the measurement using a modified steady-state method with gas flow rate measured at the outlet; this minimizes or eliminates any potential errors due to system leaks.

The resin disc is placed between two end pieces of the core holder. The tightness is ensured by applying a load using a hydraulic press. The entry can be connected to several vessels of different volumes. The outlet is either open to the atmosphere or closed with a small volume. Inlet and outlet pressures are also measured.

The gas permeability is derived from pressure and flow rate. The average pressure $<P> = (P_{in} + P_{out})/2$ used was in the range of 14.5–101.5 psi through 5 pressure periods. The average gas permeability $<K_g>$ at a single point steady-state measurement was found using the Jones–Owens technique (Jones and Owens, 1980). The measurements were conducted at a net confining pressure of 1015 psi. The microscopic flow can be described as "average" gas permeability:

$$K_g = K_l \left(1 + \frac{b}{P}\right) \quad (5.10)$$

where K_g is the average gas permeability, K_l is the Klinkenberg corrected permeability that is determined from the intercept, and b is the gas slippage factor that is computed from the slope of the gas permeability versus the reciprocal average pressure plot. The Klinkenberg plot is used to determine the liquid permeability and also take into account the gas slippage effects. Gas slippage factor is given by:

$$b = \frac{m}{K_l} \quad (5.11)$$

where m is the slope.

5.6 RESULTS

5.6.1 Pore Size Distribution from MICP Experiments

The capillary pressure curves of each tested sample summarize the relationship between the volume of mercury intrusion and the pressure applied on the sample at different stages of increasing pressure. Mercury saturation started at low pressure, entering the large pores, and progressively invaded smaller pores as the pressure increased. Most of the samples collected exhibited 100% mercury saturation, close to 60,000 psi (413 MPa) and close to the instrument's testing limits (i.e., 3 nm pore throat diameter). Four samples (S20, S21, S23 and S24) did not reach saturation beyond 60,000 psi. These samples also show a distinctive profile at low pressure. Indeed, they recorded a much lower entry pressure (<200 psi) starting to invade their pore structures, while the other samples needed an entry pressure that is an order higher (>1000 psi) (Figs. 5.7, 5.8, and 5.9).

PCM and PKM samples recorded a porosity range of 3–7.7%, with a questionable sample (Sample 4) showing up to 13.8% (Table 5.2). CCM samples have higher porosity values compared to PCM and PKM, ranging from 4.2 to 9%. The general peak pore throat radius size shows PKM at 3 nm, smaller than PCM which has values around 6 ± 2 nm (Figs. 5.10, 5.11, and 5.12). More specifically, pore throat distribution reveals a second minor population in PKM that have a pore throat size >1 μm that is easily invaded by mercury injection at low pressure. If all the PCM samples record similar porosities around 3.2%, differences appear in their pore throat sizes. Samples 1 and 2 have the smallest pore throat at 4.7 nm, with low permeability at around 144 nD, while the other samples are >6 nm.

Excluding samples 4, 20, 23, and 24, an average modal pore throat radius for the whole data set was found to be

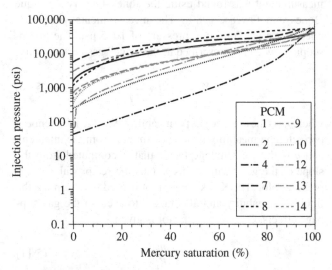

FIGURE 5.7 MICP results from PCM: mercury injection pressure as a function of mercury saturation.

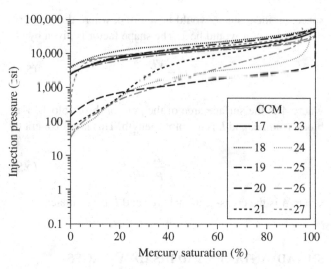

FIGURE 5.8 MICP results from CCM: mercury injection pressure as a function of mercury saturation.

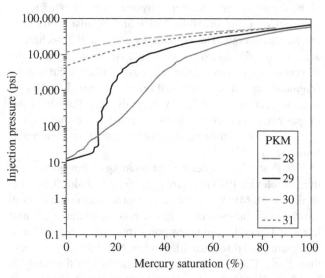

FIGURE 5.9 MICP results from PKM: mercury injection pressure as a function of mercury saturation.

around 5.2 ± 1.8 nm, with some differences between each formation, at 6 nm for PCM, >5 nm for CCM, and >3 nm for PKM (Figs. 5.10, 5.11, and 5.12). The four samples were excluded from the average calculation because they anonymously showed pore throat sizes >20 nm. For all the three formations, the mesopores ranged from 62 to 75% of the total volume and the remaining 25–38% belong to the macropore volumes (Table 5.2).

There is no relationship between MICP porosity and the modal pore throat radius. However, a general trend exerts an expected relationship between the entry pressure and the modal pore throat radius (Fig. 5.13). The previously excluded samples also exhibited a higher macroporosity contribution obtained from pore throat radius >50 nm, based on the

TABLE 5.2 MICP results and pore volume percentages based on IUPAC classification from the sample collection

Formation	Sample ID	Porosity (%)	Peak radius (um)	Median diameter by volume (um)	Threshold pressure (psi)	Percentage of pore volume	
						Meso pores	Macropores
PCM	1	6.2	0.007	0.01	1,394	93	7
	2	7.7	0.03	0.05	246	51	49
	4	13.8	0.2	0.32	45	21	79
	7	5.7	0.005	0.01	5,701	>95	0
	8	3.78	0.005	0.01	6,196	83	17
	9	3.05	0.005	0.02	5,783	85	15
	10	3.17	0.007	0.01	2,377	68	32
	12	3.03	0.006	0.08	680	45	55
	13	3.54	0.008	0.03	1,795	63	37
	14	3.57	0.007	0.03	1,701	64	36
CCM	17	7.22	0.005	0.02	6,115	86	14
	18	8.43	0.003	0.01	10,027	88	12
	19	6.52	0.004	0.01	7,347	86	14
	20	8.5	0.026	0.15	469	15	85
	21	9.02	0.008	0.03	1,259	59	41
	23	8.08	0.057	0.12	218	38	62
	24	8.14	0.021	0.11	1,019	34	66
	25	4.18	0.006	0.02	4,460	80	20
	26	7.41	0.006	0.01	6,725	91	9
	27	6.98	0.005	0.01	5,895	90	10
PKM	28	2.87	0.004	0.06	1,355	50	50
	29	6.79	0.003	0.04	5,261	52	48
	30	2.4	0.002	0.01	11,585	>95	0
	31	2.4	0.003	0.01	4,942	>95	0

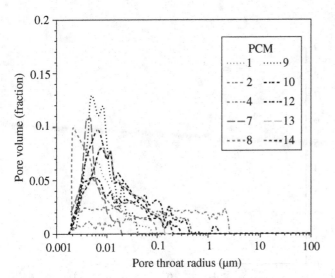

FIGURE 5.10 Converted previous capillary pressure curves into equivalent pore throat radius as a function of the pore volume fraction or porosity for PCM samples.

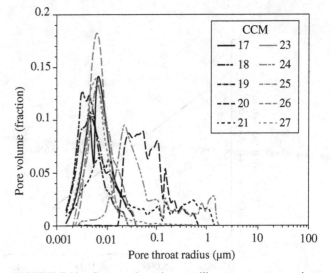

FIGURE 5.11 Converted previous capillary pressure curves into equivalent pore throat radius as a function of the pore volume fraction or porosity for CCM samples.

IUPAC classification. Note that microporosity (<2 nm pore size) is not achievable by MICP method.

5.6.2 Pore Size Distribution from Nitrogen Adsorption Experiments

A summary of the results collected from low pressure adsorption measurements and the IUPAC pore percentages are shown in Table 5.3.

The average pore radius for the whole data set was found to be around 7.5 ± 2.9 nm, with CCM showing larger pore sizes (9.2 ± 2.4 nm) compared to PCM (5.2 ± 1.5 nm).

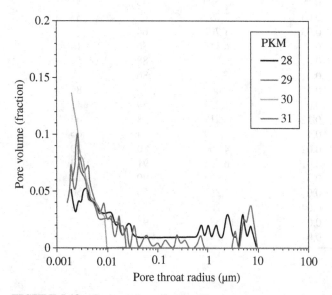

FIGURE 5.12 Converted previous capillary pressure curves into equivalent pore throat radius as a function of the pore volume fraction or porosity for PKM samples.

Pore volumes and surface areas obtained from N_2 tests show a relatively narrow range both for the PCM and CCM formations, with the exception of samples 17 and 27 from formation CCM. The average BET surface area is similar for the two tested formations (CCM and PCM) at around 5.5 ± 2 mm²/g. However, the total pore volume for CCM is higher than it is for PCM, with them measuring 2 ± 0.9 cm³/100 g and 1.39 ± 0.2 cm³/100 g, respectively. All of these samples are characterized by a very high contribution of the mesopore size, at 78.5 ± 7.3% of the total porosity (Figs. 5.14, 5.15, and 5.16).

There is an inverse relationship between pore size and BET surface area, and all samples show an increase in micropore volume with decreasing average pore diameter (Fig. 5.17).

5.6.3 NMR T_2 Relaxation Time

In NMR, porosity is directly related to the amount of hydrogen atom (protons) present in the sample. The hydrogen atoms are proportional to proton density and correspond to the initial amplitude of the spin. Under partially saturated conditions, the average porosity was found to be 4 ± 1.3%. The samples have systematic monomodal distribution, with a relaxation time (T_2) centred around 0.35 ± 0.03 ms (Figs. 5.18 and 5.19).

After saturation, the previous population showed a shift toward longer T_2, centred between 0.5 and 0.6 ms, with the exception of samples 24 and 26 which showed a shift up to 3 ms (Figs. 5.20 and 5.21). This first population is defined as the short relaxation time (T_{2S}). A second population defined by long relaxation time (T_{2L}) was also recorded for some of the samples: PCM samples recorded a T_{2L} at 24.5 ± 6.5 ms while CCM had a T_{2L} around 18.2 ± 9 ms (Figs. 5.20 and 5.21).

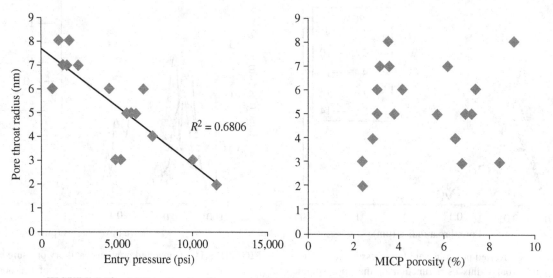

FIGURE 5.13 (a) Pore throat radius versus entry pressure, (b) pore throat radius versus porosity.

TABLE 5.3 Nitrogen adsorption results and pore volume percentages based on IUPAC classification from the gas shale collection

Formation	SAMPLE ID	BET surface area (m²/g)	Total pore Vol. (cm³/100g) at maximum pressure	Average pore radius (nm)	% DFT model (porosity)		
					Micropore Vol.	Meso pore Vol.	Macropore Vol.
PCM	8	5.4	1.54	5.7	5.4	85.7	8.9
	9	7.6	1.67	4.4	4.4	93.0	2.7
	11	2.3	0.99	8.5	1.1	84.0	15.0
	12	4.3	1.19	5.6	1.0	93.0	6.3
	13	4.9	1.28	5.2	5.6	72.0	22.5
	14	7.8	1.57	4.0	11.2	78.6	10.3
	15	6.0	1.28	4.3	10.1	75.8	14.0
	16	7.8	1.55	4.0	11.3	75.8	13.0
CCM	17	8.7	3.04	7.0	3.0	78.2	18.8
	18	3.4	1.83	10.8	1.9	63.7	34.4
	19	2.8	1.49	10.8	2.7	76.2	21.1
	20	2.8	1.42	10.5	2.6	77.2	20.2
	21	7.7	2.69	7.0	2.5	81.8	15.7
	22	3.4	1.39	8.2	3.0	77.9	19.1
	23	2.0	1.04	10.3	2.9	74.7	22.4
	24	2.1	1.36	13.1	0.0	66.4	33.6
	25	2.0	0.96	9.6	0.0	78.2	21.8
	26	6.3	3.09	9.9	0.0	80.0	20.0
	27	18.0	3.6	4.0	11.7	79.3	9.1
PKM	28.0	10.0	1.6	6.3	14.8	82.9	2.4
	29.0	5.4	1.1	8.4	9.3	87.0	3.7

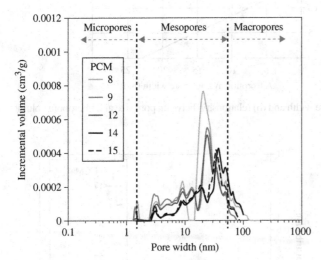

FIGURE 5.14 Pore size distribution from nitrogen adsorption tests with IUPAC boundaries of micro-, meso-, and macropores for PCM samples.

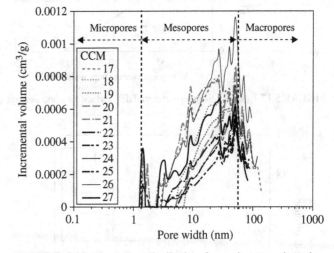

FIGURE 5.15 Pore size distribution from nitrogen adsorption tests with IUPAC boundaries of micro-, meso-, and macropores for CCM samples.

Saturated samples recorded total NMR porosity of 12.3 ± 3%. The average total NMR porosity contribution of the T_{2s} is 87 and 93% for PCM and CCM, respectively. The T_{2L} contributed 13 and 7% for PCM and CCM, respectively.

This secondary population T_{2L} is more likely the effect of the macropores filled by brine during the saturation process, with porosity reaching values of >10% when it was only 4 ± 1.5% under partial-saturation conditions (Table 5.4).

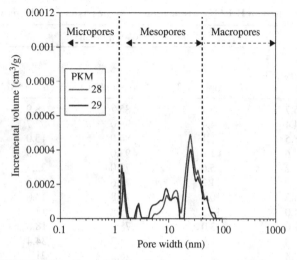

FIGURE 5.16 Pore size distribution from nitrogen adsorption tests with IUPAC boundaries of micro-, meso-, and macropores for PKM samples.

5.6.4 Scanning Electron Microscopy

The visual interpretation of mineralogy and porosity in PCM and CCM was conducted using two distinctive SEM imaging modes. SEM imaging, consisting of both SE and BSE images, reveals quartz and clay particles as the dominant components of the shale (Fig. 5.22). These results are in agreement with the XRD results. The images suggest that quartz is typically found as large discrete particles intermixed with clay. The clay particles are displayed as repeating layers of flat platelets, typical of illite and/or kaolinite. Secondary minerals, such as iron-rich materials, are visible within the images and are typically highlighted during BSE imaging.

5.6.5 Focused Ion Beam/Scanning Electron Microscopy

FIB–SEM image acquisition and analysis was conducted on sample 10 from the PCM formation to visually inspect the

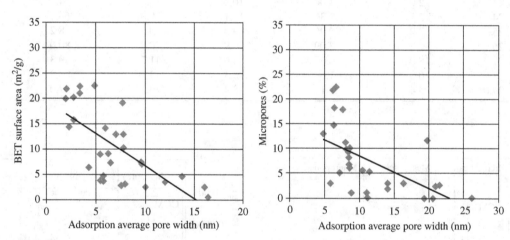

FIGURE 5.17 (a) Relationship between BET surface area and average pore width and (b) relationship between pore size and micropore volume.

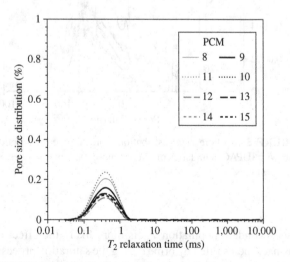

FIGURE 5.18 NMR T_2 relaxation time distribution of partially saturated PCM samples.

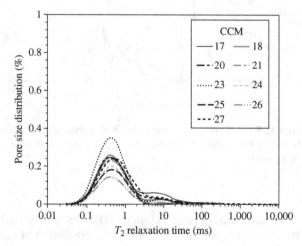

FIGURE 5.19 NMR T_2 relaxation time distribution of partially saturated CCM samples.

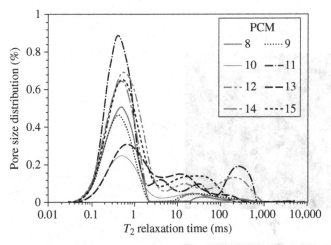

FIGURE 5.20 NMR T_2 relaxation time distribution of fully saturated PCM samples.

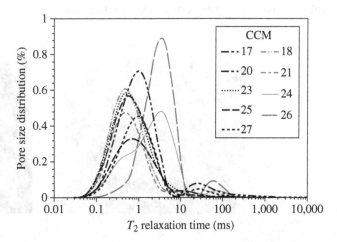

FIGURE 5.21 NMR T_2 relaxation time distribution of fully saturated CCM samples.

TABLE 5.4 NMR CPMG T_2 relaxation time results from the PCM and CCM samples collection

	Partially saturated		Brine saturated				
Samples	T_{2s} (ms)	Total porosity (%)	T_{2s} (ms)	Porosity contribution @ T_{2s} (%)	T_{2L} (ms)	Porosity Contribution @ T_{2L} (%)	Total porosity (%)
8	0.29	3.2	0.46	97	30.1	3	11.4
9	0.38	5.2	0.43	89	27.3	11	10.8
10	0.4	2.2	0.43	82	22.6	18	6.7
12	0.32	1.9	0.5	80	12	20	14.2
13	0.36	2.3	0.56	68	29.2	32	11.55
14	0.32	2.7	0.51	93	20.6	7	14
15	0.32	2.5	0.56	79	30.2	21	18.5
17	0.35	5.8	0.89	92	26.1	8	15.75
18	0.35	5.2	0.4	96	12	4	10.80
20	0.35	5.2	0.5	100	—	—	12.76
21	0.36	4.8	0.45	96	8.4	4	14.20
23	0.36	6.5	0.45	100	—	—	11.30
24	0.36	3.6	2.73	96	64.3	4	14.00
25	0.4	3.5	0.63	100	—	—	18.50
26	0.36	2.8	3.06	85	64.3	15	9.63
27	0.36	5.3	0.89	93	26.1	7	11.40

Two T_2 populations are recorded: a short T_{2s} and a long T_{2L}.

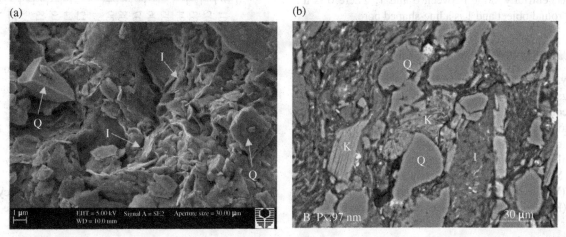

FIGURE 5.22 (a) SE image of sample 14 displaying dominant illite and quartz particles (b) BSE image of sample 20 displaying dominant kaolinite, Illite, and quartz particles.

102 PORE GEOMETRY IN GAS SHALE RESERVOIRS

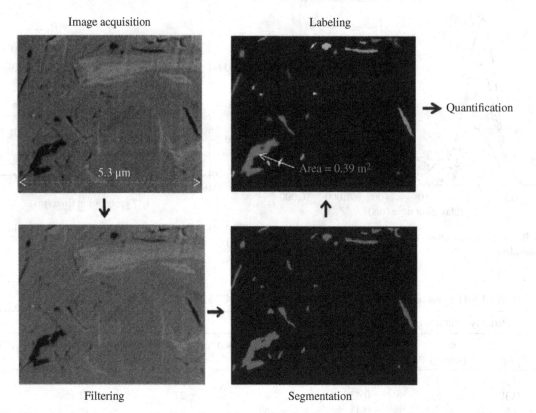

FIGURE 5.23 Illustration of the image analysis conducted for sample 10 following the general image analysis procedures (as shown in Fig. 5.6).

pore characteristics and to support experimental analysis. Figure 5.23 provides a brief illustration of the image analysis conducted for sample 10 following the general image analysis procedures (as shown in Fig. 5.6). The total porosity from sample 10 was found to be 3.56% and the majority of the pore sizes were in the range of 0.05 μm (Fig. 5.24), comparable to what was obtained from MICP porosity (3.17%). From the image example, it is obvious that the pores are not an ideally shaped circle. Hence, the average shape factor was found to be 0.35, where a circle is equal to 1. In addition, the average eccentricity was found to be 0.86, which describes how elongated the pores are. An object can have an eccentricity value between 0 and 1, where 0 is a perfectly round object and 1 is a line-shaped pore.

5.6.6 Capillary Pressure and Permeability

Generally, permeability is measured in laboratories using core plugs. In some cases, however, it is difficult to obtain suitable core plugs. In these instances, other approaches can be used to predict permeability. These are chiefly based on mathematical and theoretical models. Predicted MICP permeabilities are compared with those measured permeabilities. Models evaluated in this study include the Kozeny–Carman (Wyllie and Gregory, 1955) and Swanson (1981), Winland (Kolodzie, 1980), Jorgensen (1988), Pape et al. (1998),

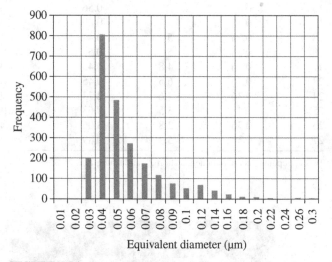

FIGURE 5.24 Pore size distribution of sample 10 from image analysis.

Rezaee et al. (2006), Katz and Thompson (1986), Pittman (1992), and Dastidar et al. (2007) methods.

A total of 10 samples from the PCM formation were used for permeability measurements (Fig. 5.25).

Generally, for gas shale formations, the accuracy of the MICP-based permeability methods is expected to be low. As

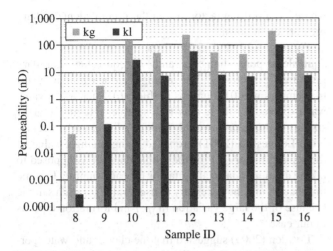

FIGURE 5.25 Measured permeability of selected PCM samples.

TABLE 5.5 Ranking of predicted MICP permeability

Rank	Method	Equation references	MSE	Std deviation	R^2	SUM
1	Rezaee R_{50}	Rezaee et al. (2006)	2	1	4	7
2	Pittman R_{25}	Pittman (1992)	3	4	1	8
3	Winland R_{35}	Kolodzie (1980)	5	3	2	10
4	Dastidar (OU Method)	Dastidar et al. (2007)	6	2	3	11
5	Pape	Pape et al. (1998)	7	6	5	18
6	Kozney-Carman	Wyllie and Gregory (1955)	1	5	8	14
7	Swanson	Swanson (1981)	4	8	7	19
8	Jorgensen	Jorgensen (1988)	8	7	6	21

a quantitative comparison, the authors rank the mean square error (MSE) and the standard deviation (σ) in ascending order, and the coefficient of determination (R^2) in descending order. The final ranking of the suitable model is done through a cumulative rank of each MSE, σ and R^2. Table 5.5 summarizes the ranking of each MICP permeability method.

5.7 DISCUSSION

5.7.1 Porosity and PSD Comparisons

The bulk porosity from MICP, N_2, and NMR presents some differences (Tables 5.2, 5.3, and 5.4). Indeed, the porosities from MICP and N_2 are mostly around 3–3.5%, with more variability recorded in the N_2 method. The porosities from NMR are much higher, between 6 and 15%, which corresponds to three times higher than for the MICP or N_2 methods. This scale of difference with NMR has also been described by other research that has analyzed mudstones (Hildenbrand and Urai, 2003). Possible explanations of these higher values of porosity from NMR come from the fact that

(i) NMR measures both connected and isolated pores. Further, MICP porosity measures connected pores and thus misses out on the pores located within the grains and the clay-bound water spaces.
(ii) Samples were dried out for MICP experiments, inducing potential clay shrinkages, while the NMR re-saturation with artificial brine was prone to generate clay swelling and cracks.

The MICP PSD did not seem to be able to record three types of pores as the N_2 did, with micro-, meso-, and macropore distribution (Tables 5.2 and 5.3). MICP advocates substantial pore volume percentages in the meso- and macropore range. Samples 7, 30, and 31 obtained from MICP analysis have the largest mesopore volume and the least macropore volume. The slight inconsistency between MICP and N_2 is because MICP only quantifies pore throat sizes and not the pore bodies, whereas N_2 quantifies both of them.

PSD analysis using NMR and MICP methods gives similar results, with pore distribution made of meso- and macropores. While N_2 was able to record the microporosity (pore size <2 nm), the pressure injection of mercury was not strong enough to override the strong capillary pressure of such small pore throat sizes. Considering that partial saturation of the core plugs kept the fluids inside the micro- and mesopores intact due to strong capillary pressure and clay bound water, the NMR was also not able to record the microporosity. The general contributions of MICP and NMR to analysis of meso- and macropores are similar. MICP showed 68% mesopores and 32% macropores, and NMR showed 63% mesopores and 37% macropores. A drawback of the NMR T_2 measurements is that some of the nanopore signals (i.e., microporosity) that are typical in shales cannot be detected by the low-field NMR, with T_2 <0.03 ms).

5.7.2 Interchanging MICP with NMR Data

5.7.2.1 Correlating NMR and MICP
The procedures we used to determine the NMR T_2 equivalent pore diameter are similar to those suggested by Lowden (2009) with slight modifications. The approach is based on a single phase, where the wetting phase is similar to the saturating phase. No static magnetic gradient field was applied during the CPMG sequence, and $1/T_2$ bulk is so small that Equation 5.5 can be reduced to the second part of the equation as a function of the pore geometry and surface relaxivity. Samples

with small pores will relax much faster compared to samples with larger pores. The T_2 distribution then defines the PSD:

$$\frac{1}{T_2} = \rho_2 \left(\frac{S}{V}\right) \quad (5.12)$$

1. Determine the T_2 equivalent pore diameter (size) using the following formula:

$$D = T_2 \cdot \rho_2 \quad (5.13)$$

where D is the MICP pore throat diameter (μm), ρ_2 is the surface relaxivity (commonly denoted as the scaling factor) (μm/ms), and T_2 is the transverse NMR relaxation time (ms). The scaling factor in this study was obtained from NMR and MICP laboratory measurements. It was found by taking the dominant pore diameter (MICP) and dominant T_2 relaxation time (NMR). The procedures are as the following:

2. Determine the weighted incremental pore volume from the T_2 distribution. This is found by multiplying the pore volume at each data point by the ratio of MICP/NMR bins (pressure points). This weighted incremental volume versus the T_2 equivalent pore diameter can be plotted along the same axis with pore diameter versus normalized incremental volume from the MICP data (Fig. 5.26).
3. Determine the cumulative pore volume from the NMR data and find the NMR capillary pressure derived from:

$$P_{NMR} = \frac{4 \cdot \sigma \cdot \cos\theta \cdot b}{T_2 \text{eqv} \times \text{pore diameter}} \quad (5.14)$$

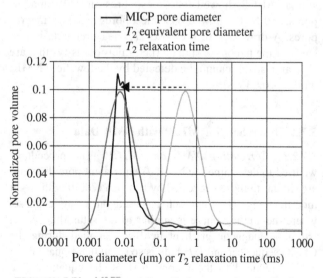

FIGURE 5.26 MICP and T_2 equivalent pore size distribution.

where b is a shifting factor to shift the T_2 equivalent capillary pressure up or down.

Figures 5.27 and 5.28 show the results of the T_2 equivalent pore diameter and T_2 equivalent capillary pressure, respectively. The average surface relaxivity for the samples was 0.02 μm/ms (Table 5.6). Such a low value was expected for these clay-rich samples. Having a larger surface–volume ratio would produce a lower surface relaxivity. Looyestijn (2001) proposed using a single value of the scaling factor (ρ_2) for the whole data set as opposed to taking an individual scaling factor for each sample. In some instances it may be necessary to use a value for each sample because of the large variation in (ρ_2) and formation type, as in our case.

Lowden (2009) suggested that the clay-bound water portion from the NMR T_2 distribution be removed when deriving the capillary pressure, because when computing the cumulative pore volume (NMR), porosity at the maximum diameter should equal zero, and porosity at the minimum diameter should equal the total NMR porosity.

However, the T_2 relaxation times (PSD) for shales are so low that it is difficult to distinguish or separate the clay-bound water from the capillary-bound water. Furthermore, these high clay content materials have kerogen, a material that contains hydrogen atoms that could possibly be picked up by the NMR signals. The approach described by Zhi-Qiang et al. (2005) is also not applicable for the shale samples in our study because their samples had dominant T_2 relaxation times above 10 m/s, whereas the dominant T_2 relaxation time in our study was about 1 m/s.

5.7.2.2 T_2 Relaxation Time from MICP Data
One could also extract the T_2 relaxation time from the MICP data. There are two approaches that would give the same results. The first is through the use of the Equation 5.13.

1. Multiply the T_2 relaxation time with surface relaxivity.
2. Plot T_2 versus weighted incremental pore volume (from NMR) and MICP Pc equivalent T_2 versus normalized intrusion volume (from MICP) on the same plot. The x-axis will be illustrated in terms of time (μm/ms).

The second approach is based on the use of the surface–volume ratio. This ratio can be determined from the MICP data and then substituted into Equation 5.12. The steps are as follows:

1. Determine the surface–volume ratio from the MICP data.
2. With the known surface relaxivity, determine the MICP-equivalent T_2 relaxation time with a scaling factor of $c = 3.6$.

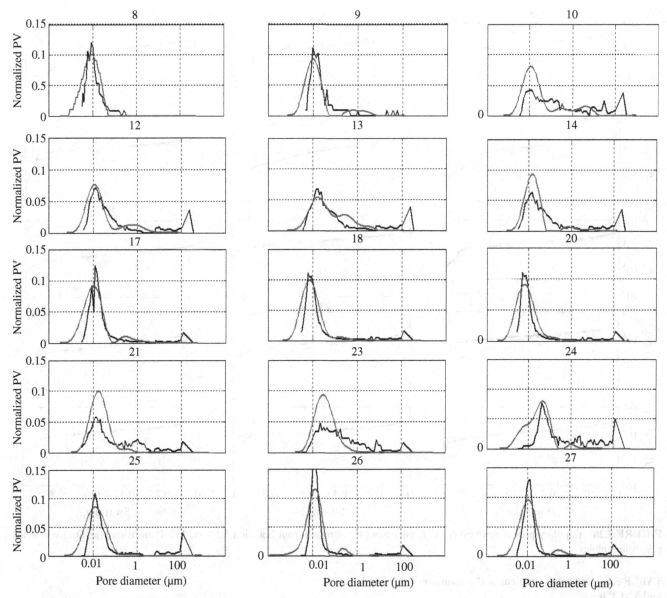

FIGURE 5.27 Pore diameter: NMR T_2-equivalent pore diameter extracted from Equation 5.13 and MICP pore diameter for selected gas shale samples.

$$\text{MICPequiv} \cdot T_2 = \left[\frac{1}{\rho_2 \cdot (S/V)} \right] \cdot c \quad (5.15)$$

3. Plot NMR T_2 versus the weighted incremental pore volume (from NMR) and the MICP Pc-equivalent T_2 versus the normalized intrusion volume (from MICP) on the same plot.

P_{MICPTh} is the threshold pressure from MICP, P_{NMRTh} is the threshold pressure from T_2 equivalent pressure, ρ_2 is the surface relaxivity, and b and c are shifting factors.

For the majority of the samples, threshold pressure from the NMR-derived capillary pressure was slightly higher compared to the MICP-derived result (Table 5.6). However, a good trend was observed between MICP and NMR (Fig. 5.29).

The advantages of predicting capillary pressure from NMR is that the capillary pressure readings can be continuously determined with depth (Volokitin et al., 2001), and permeability models requiring capillary pressure can be used to estimate permeability. The method is also nondestructive compared to MICP. On the other hand, deriving the T_2 relaxation time from MICP data brings certain advantages, particularly in cases where it is difficult to obtain suitable core plugs for evaluating petrophysical properties of reservoir rocks. The MICP technique measures the pore volume directly and only requires a small rock cutting or

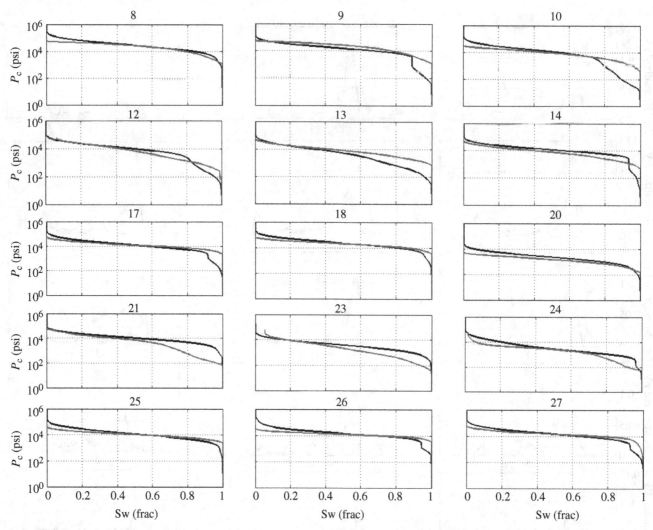

FIGURE 5.28 Capillary pressure curves: NMR T_2-equivalent Pc extracted from Equation 5.14 and MICP intrusion pressure for selected gas shale samples.

TABLE 5.6 Extracted supplemental parameters from NMR and MICP tests

Sample number	P_{MICPTh} (psi)	P_{NMRTh}	$\rho 2$ μm/ms	Shift factor (b)	Scaling factor (c)
8	6,169	5633	0.020	0.11	3.60
9	2,337	3390	0.021	0.11	3.50
10	680	1203	0.032	0.11	4.00
12	1,795	1382	0.023	0.11	3.70
13	1,795	430	0.029	0.19	3.90
14	1,701	2850	0.027	0.11	3.90
17	6,115	2384	0.011	0.11	3.90
18	10,027	5203	0.015	0.11	3.60
20	469	427	0.104	0.01	2.20
21	1,259	838	0.036	0.09	4.00
23	218	294	0.253	0.08	3.80
24	1,019	1279	0.015	0.09	4.00
25	4,460	2506	0.019	0.10	3.60
26	6,725	8242	0.004	0.11	3.80
27	5,895	2625	0.011	0.11	4.20

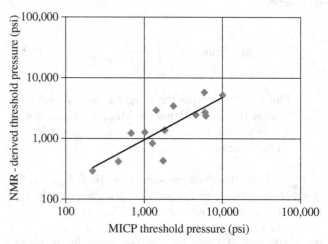

FIGURE 5.29 Threshold pressure from MICP and NMR.

fragments; results are obtained relatively quickly with reasonable accuracy, and very high capillary pressure ranges can be achieved.

5.7.3 Pore-Body to Pore-Throat Size Ratio: Pore Geometry Complexity

Pore-body to pore-throat ratio is an important characteristic that controls fluid flow. The connectivity in the pore system can be represented by the pore-body to pore-throat size ratio; the lower the ratio, the lower the connectivity, and so the lower the permeability/fluid flow will be. Determining the exact physical shape of the pores is both difficult and time-consuming, and it also requires demanding test equipment.

The pore body–pore throat size ratio was derived from the Coates equation (Coates et al., 1999):

$$k_{coates} = \left[\left(\frac{\phi}{C} \right)^2 \left(\frac{FFI}{BVI} \right)^2 \right] \quad (5.16)$$

with ϕ being total porosity (%), FFI being the free fluid index (or movable water), and BVI being the bound volume of irreducible water. C is a constant parameter usually used to "tune" the NMR log analysis from the Coates equation. However, behind this constant lies the concept of pore geometry defined as: pore-throat to pore-body size ratio. A "strong" geometry will be characterized by a low C, representing a very small pore throat compared to the pore body size, which will require a lot of pressure to overcome the strong induced capillary pressure and will increase the fluid trapping effect during/after flow experiments. The results will be a very low permeability when C is low, and vice versa. Typically, sandstones have $C = 10$, which can decrease when clay minerals occur. Clay-rich rocks should have a very low C and a strong complex pore geometry.

The Coates model requires the values of FFI and BVI from the T_2 distribution. Based on the literature, the classical T_2 cut-off is set at 33 ms for sandstone reservoirs (Coates et al., 1999). For these shale samples, we made the assumption that T_2 relaxation response from the partially saturated samples (samples received from the core storage) was only due to irreducible fluids (CBW and capillary-bound water), considering that all the mobile water was evaporated in the core storage condition. The difference in T_2 response between saturated and partially saturated brine is mostly due to mobile fluids. Figure 5.30 shows how to extract the cut-off, on a commutative NMR signal curve as a function of T_2 relaxation time. The NMR signals are common along a short T_2 range, where the same irreducible water signals are recorded from both saturation states of the sample, until a point of divergence where mainly the mobile water controls T_2 distributions. This point of divergence corresponds to T_2 cut-off (Fig. 5.30).

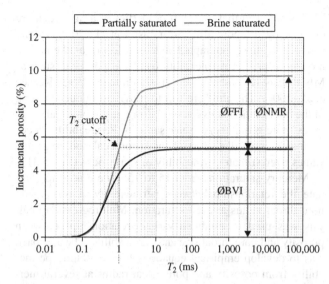

FIGURE 5.30 This figure shows how to extract T_2 cut off for a shale sample under different saturation status.

TABLE 5.7 Computed pore-body to pore-throat size ratio (C) on the Cmf samples from NMR dataset calibrated against gas permeability measurements

Sample ID	K_g (nD)	NMR porosity (%)	C (constant)
8	0.05	11.38	0.001
9	3.10	10.75	0.29
10	144	6.68	4.47
12	238	17.65	3.46
14	46.6	11.55	0.71
15	336	10.57	4.41

The best C parameter from the Coates equation can then be derived to match the computed permeability from NMR against the measured gas permeability (K_g). All the C constants are relatively low, at $<<10$ (Table 5.7), demonstrating a complex geometry of the pore network, as expected from clay-rich rocks. However, among the PCM sample collection, the range of gas permeability is directly a function of the pore geometry C, with the highest permeability exhibiting the highest C constants. This illustrates weaker geometry and/or higher pore connectivity that ease the fluid to flow through the pore network in some samples from the same formation.

5.7.4 Pore Throat Size and Permeability

Porosity and permeability relationships are qualitative in nature; particular rocks may exhibit high porosity, but ultralow permeability. Porosity and measured permeability of the samples in the study exhibit a weak correlation. This is not unexpected, given that the porosity symbolizes the pore volume and the permeability reflects the pore throat size in the system (Al Hinai et al., 2013).

It is believed that transport properties of a tight rock are dictated by the pore structure (Bustin et al., 2008). Many researchers have attempted and developed mathematical models to predict permeability based on pore size such as MICP tests (Dastidar et al., 2007; Pittman, 1992, Rezaee et al., 2006). The aim here was to provide an understanding of the interrelation between fluid flow and the physical properties (pore geometry) of shale samples. Permeability values estimated show that most of the theoretical and empirical values overestimate permeability of shale rocks (Fig. 5.31).

Mercury saturations from 15 to 75% were used to evaluate the relationship between permeability, porosity, and pore throat radius at each saturation. Multiregression analysis was used to establish various relationships between porosity and pore throat size and permeability. The approach was to develop empirical equations for calculating permeability from porosity and pore throat radius at several mercury saturation percentiles. The best correlation for the samples studied is:

$$\log k = 37.255 - 6.345 \log \varnothing + 15.227 \log R_{75}$$

where k is permeability in nanodarcy, porosity in percentage, and R_{75} is the pore throat size in microns when a sample is 75% saturated with mercury. Pore throat radius does not display exclusivity at some definite mercury saturation levels. Every rock may vary in R values depending on its pore structure and geometry. The authors stress that the comparisons between the measured and predicted permeability based on pore throat size made are indefinite, but they serve as an indication of the method that would function better in evaluating the permeability of a gas shale formation (Al Hinai et al., 2013).

5.7.5 Mineralogy

These siliceous and organically rich shales are marked by a strong component of clay minerals, with up to 56% kaolinite and I/S mixed layers. The I/S mixed layers are slightly swelling clays with structural attributes from both illite and smectite clay minerals.

For the samples studied, three types of shales can be classified according to their clay types: (i) low I/S but high kaolinite (CCM); (ii) high I/S but low kaolinite (PCM); and (iii) high I/S and high kaolinite (PCM and PKM). The images obtained from SEM show that kaolinite is dominant clay in CCM and part of PCM. The I/S clays dominant in PCM and PKM are mostly found around the pores and coating the grains or often plugging the pore throats. It is therefore expected to see I/S clay-bearing formations that are fluid-trapping mechanisms leading to degraded flow dynamics.

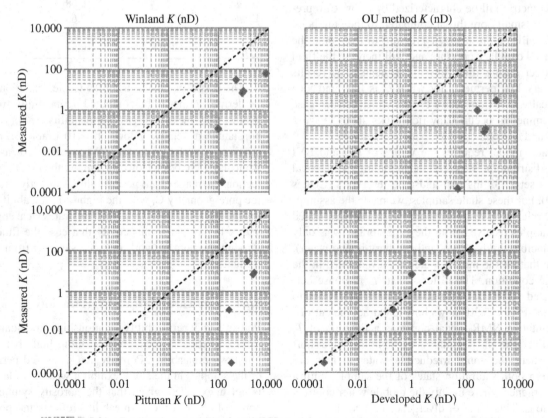

FIGURE 5.31 Predicted permeability from MICP models versus laboratory measured permeability.

5.7.5.1 MICP Porosity and the Presence of Clays

Two groups of samples were identified from capillary pressure profiles, showing distinct entry pressures at low pressure and pore saturation at the highest pressure of mercury injection. These two groups are distinctive in their content of I/S clays. The dominant group, mostly comprising PCM and PKM formations, with high I/S content, records high entry pressure due to low pore throat radius. The second group of samples (CCM) with low I/S (but high kaolinite) has lower entry pressure and never reaches full saturation, with up to 60,000 psi of pressure applied on the mercury. This illustrates that I/S clays are prone to degrade fluid flow efficiency much more than kaolinite-rich clay formations, as shown in the I/S that clogs part of the pore throats, leading to a "tight" pore structure (Fig. 5.32). This contrasts with the kaolinite-rich formations that record the lowest entry pore pressure and the highest porosities (Table 5.8).

5.7.5.2 Mixed Illite/Smectite Effects on Surface Area from N2 Experiment

The BET surface areas were found to be 2–8 m²/g for both PCM and CCM formations, with the exception of sample 27 (<10 m²/g). There was no direct relationship found between the BET surface area, the occurrence of I/S, and the total clay content. Based on the IUPAC pore classification, N_2 pore volumes showed a pore range from 60 to 90% mesopores, 6 to 34% macropores, and a small portion of 1 to 11% micropores (Fig. 5.33).

In addition, the N_2 adsorption and desorption curves for the samples tested indicate that the change in I/S-to-kaolinite ratio has an effect on the profile (i.e., shape of the curve) (Fig. 5.34). The samples with the least kaolinite content have a narrow separation, quasisuperimposed, between the adsorption and desorption curves at high relative pressure, while samples with higher kaolinite content showed high separation. In addition, the quantity of N_2 adsorbed increases as kaolinite increases. If we compare samples 21 and 25, we can see that sample 25 has 13.8% kaolinite, and sample 21 has a twofold higher concentration at 26.8%. The amount of N_2 adsorbed in mmol/g increased from 0.27 (sample 25) to 0.77 (sample 20).

This behavior is also confirmed in samples 8, 13, and 27, where the I/S-to-kaolinite ratio effect ranges from high to low. At a low ratio of I/S to kaolinite (i.e., high kaolinite content), the separation and quantity of N_2 adsorbed is large. In other words, if the desorption profile is similar to the adsorption profile, the N_2 is adsorbed at the same rate and amount as increasing pressure is released during the desorption process. When a separation appears between the N_2 profiles, typically more N_2 is released during desorption than during adsorption, and there is very little fluid-trapping effect. It is basically easy to release gas that is typically related to the amount of kaolinite and is stored in the pore network. A high amount of kaolinite will store a lot of gas during adsorption and will desorb much more quickly and at a high rate as soon as the pressure decreases.

5.7.5.3 Effects of Clay on T_2 Relaxation Time

CCM has a lower percentage of mixed I/S clays but a higher kaolinite percentage compared to the PCM formation (Appendix 5.A: XRD results). Furthermore, the presence of mixed I/S influences the NMR response (Fig. 5.33). As the clay component of mixed I/S increases, the T_2 relaxation time tends to decrease, with corresponding smaller pore sizes or a restricted environment (e.g., higher specific surface area and/or stronger grain surface relaxation effect on proton spin). This contrast with kaolinite shows little influence on

TABLE 5.8 MICP threshold pressure of the samples with clay content

ID	Porosity MICP (%)	Entry pressure (psi)	Quartz	Mixed illite/smectite (20%S)	Kaolinite
1	6.2	1,394	15.1	6.4	28.3
3	7.7	246	14.7	12	29.1
8	3.78	6,196	24.5	19.8	2.6
13	3.54	1,795	53	11.7	0.8
14	3.57	1,701	41.3	14.9	0.8
17	7.22	6,115	51.1	5.8	19.7
18	8.43	10,027	47.2	—	27.8
19	6.52	7,347	54.1	—	25.6
20	8.5	469	56.1	—	21.5
21	9.02	1,259	45.6	—	26.8
23	8.08	218	67.1	—	11.3
24	8.14	1,019	60.5	3	22.6
25	4.18	4,460	67.2	—	13.8
26	7.41	6,725	50.5	4.1	25.1
27	6.98	5,895	34.2	7.2	17.6
30	2.4	11,585	23.3	25.1	16.3
31	2.4	4,942	15.5	27.7	10.5

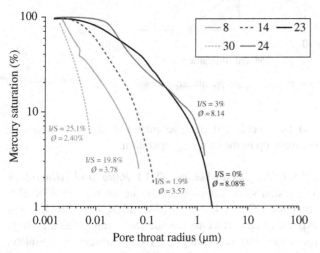

FIGURE 5.32 Effects of I/S on mercury saturation process.

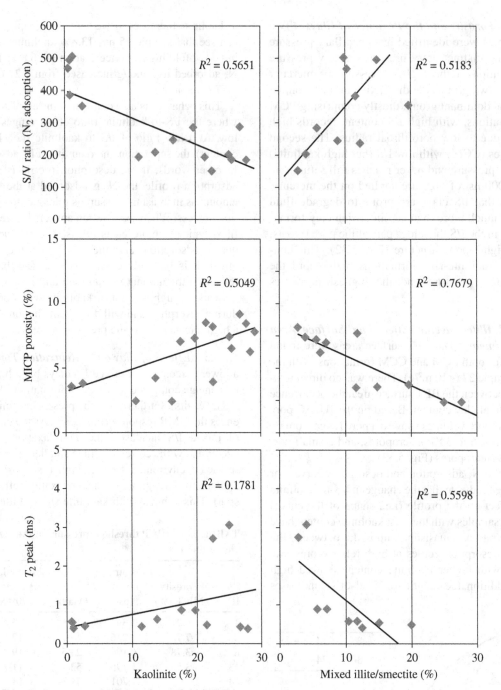

FIGURE 5.33 Influence of I/S and kaolinite on various parameters.

the T_2 values that remains constant, regardless of the amount of kaolinite.

The presence of the swelling clay (Smectite) could have been responsible for blocking the pore throats, or in a more general view, the pore connectivity that allows the fluids to access neighboring pores during the saturation process, leading to lower T_2 amplitude values. The long T_2 represents macropores that are potentially new cracks induced by the artificial brine reactivity during resaturation with the shales, and by mechanical damage inherited during the sample recovery up to the sample preparation.

5.7.5.4 Surface-to-Volume (S/V) Ratio and Mineralogy

The tested shale samples contain an average of 37% clay content for CCM, 42% for PCM, and 51% for PKM. It was expected that surface-to-volume (S/V) ratio would be high for these clay-rich samples. To illustrate this postulate, nitrogen adsorption measurements seem more appropriate to

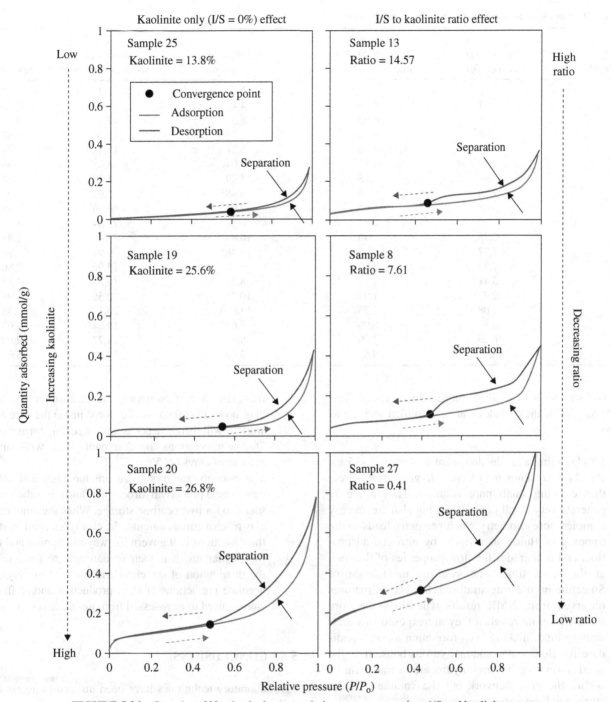

FIGURE 5.34 Quantity of N_2 adsorbed versus relative pressure at various I/S and kaolinite content.

use than MICP data, as MICP determines only the connected pores, missing the isolated pores. Using the BET method, the surface area was determined from the N_2 adsorbed volume at maximum relative pressure. N_2 results presented higher average pore throat radii for CCM (9.2 ± 2.4 nm) samples compared to PCM ones (5.2 ± 1.5 nm).

Similarly, the total pore volume is determined to be 2 ± 0.9 cm³/g and 1.4 ± 0.2 cm³/g for CCM and PCM, respectively (Table 5.9). Both the presence of kaolinite and mixed I/S show a reasonable trend with the S/V (Fig. 5.33). Kaolinite decreases with the increase in S/V, while mixed I/S increases with the increase in S/V. This demonstrates that a larger I/S content is found in shales that exhibit larger S/V ratio (Howard, 1991).

5.7.5.5 Clay Influence on Fluid Flow Properties

When gathering all the experimental results from MICP, NMR, and N_2 adsorption along with mineralogical information from

TABLE 5.9 Surface-to-volume ratio from N_2 experiments

Sample ID	BET surface area (m²/g)	Total pore Vol. (cm³/100 g) at maximum pressure	Average pore radius (nm)	Average pore width (4V/A) (nm)	S/V ratio
8	5.43	1.54	5.66	11.32	3.53
9	7.57	1.67	4.41	8.82	4.53
11	2.34	0.99	8.5	17	2.36
12	4.28	1.19	5.57	11.14	3.60
13	4.91	1.28	5.21	10.42	3.84
14	7.79	1.57	4.04	8.08	4.96
15	5.98	1.28	4.29	8.58	4.67
16	7.79	1.55	3.985	7.97	5.03
17	8.66	3.04	7.02	14.04	2.85
18	3.39	1.83	10.82	21.64	1.85
19	2.75	1.49	10.8	21.6	1.85
20	2.77	1.42	10.485	20.97	1.95
21	7.7	2.69	6.98	13.96	2.86
22	3.41	1.39	8.18	16.36	2.45
23	2.03	1.04	10.28	20.56	1.95
24	2.08	1.36	13.085	26.17	1.53
25	2	0.96	9.62	19.24	2.08
26	6.26	3.09	9.86	19.72	2.03
27	18.02	3.6	3.99	7.98	5.01

XRD and structures from SEM images, three general fluid flow behaviors of these shales can be identified and are as follows.

1. I/S clay mineral as the dominant clay phase (>15%):
   I/S clays are known to have a larger surface area, thus creating small pore volumes. They not only generate very small pore throats but also create very complex pore geometry. Such geometry leads to the trapping of fluids, as attested by nitrogen adsorption, and it degrades the flow properties of the rock at the same time. The swelling mechanism of Smectite in the interstratified I/S clay structures observed from NMR results will block the pore throats even more when they are exposed to water/drilling mud, making such formation a poor candidate for flow water and/or hydrocarbons. It is the ideal formation to trap hydrocarbon/water fluids within the pore network but the volume of fluid storage will be very low.

2. Kaolinite clay minerals as the dominant clay phase (>15%):
   The kaolinite-rich gas shales present the opposite behavior to the I/S-rich clay formations. The entry pore pressure is much smaller, which eases the flow dynamics, with bigger pores and a smooth pore geometry that avoid too much fluid trapping as illustrated by MICP and N_2 adsorption methods (Fig. 5.35). This low trapping effect and high storage capacity seen with nitrogen adsorption make such formations ideal to fluid flow. However, if no trapping mechanism comes into play, no hydrocarbon will be stored inside the pore network. It will therefore act as a basic sealing formation.

3. Quartz minerals as the dominant phase with small amount of clays (<15%):
   The porosity and pore size are too high and clays cannot hold properly hydrocarbon fluids. It is the worst scenario for hydrocarbon storage. When the amount of clay reach a critical amount, the clay types combined to their locations will govern the way of trapping and the flow dynamics. It is then fundamental to understand the distribution of the clays and the type of clays, if accurate predictions of fluid production and/or fluids storage need to be assessed from gas shale reservoirs.

5.8 CONCLUSIONS

Five laboratory techniques have been utilized to assess the full pore size structures of gas shale formations at the core sample scale: MICP, N_2 adsorption, low-field NMR, SEM, and FIB–SEM. The following conclusions can be reached about the pore structure assessment of these gas shale formations:

1. MICP is relatively fast, seems to be a reliable method to understand the pore throat size distribution down to 3 nm, and determine most of the porosity involved in the fluid transport, despite the dry state of gas shale samples.

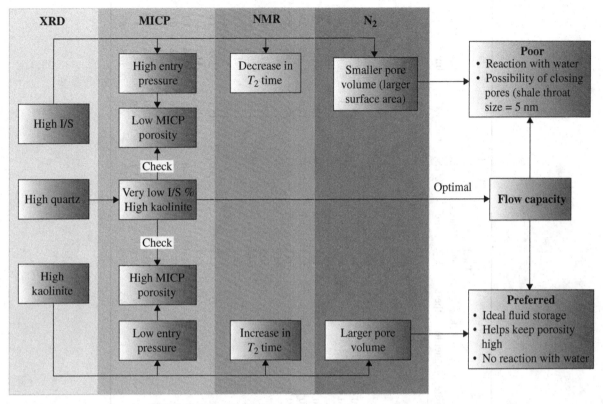

FIGURE 5.35 Flow chart summarizing flow capacity scenarios.

2. N_2 adsorption analysis can effectively reveal information about the PSD in the micropore range (<2 nm) that is not accessible by MICP and limited in low-field NMR resolution.
3. When comparing low pressure N_2 adsorption pore volumes with MICP, a discrepancy was found when classifying the pore size. The inconsistency comes from MICP that quantifies pore throat sizes and not the pore bodies, while N_2 pore volumes provide both pore size types.
4. It is suggested that NMR is an applicable nondestructive method to examine the water content and the pore body size distribution to further characterize the pore geometry of gas shale deposits. The porosity from NMR is always larger than MICP and N_2 porosity. NMR measures both connected and unconnected pores at all scales. However, as with MICP, the microporosity remains difficult to access with a 2 MHz NMR apparatus and the NMR logging tools.
5. Clay-rich rocks show complex pore geometry that directly controls the permeability within the same formation, while porosity and PSDs remain quasi-similar.
6. SEM images have supported the discerned trends in mineralogical samples obtained by XRD. The large dominating quartz particles are found throughout the samples of PCM and CCM. Typical clay platelets of Illite and I/S were found alongside the quartz particles. Additionally, secondary minerals were shown to be intermixed with the dominant quartz and clay particles. Macro- and mesoporosity was displayed in the images, with voids as large as a few micron (~3 μm) being viewed at particle boundaries.
7. MICP porosity involves only the connected porosity that is responsible for the fluid transport property. MICP porosity is proportional to kaolinite content and the connectivity of the system favors kaolinite.
8. The increased content of I/S restricts the flow of mercury into the sample, thus the high displacement pressure and the pore throat size distribution range decreases.
9. I/S traps fluid by increasing the pore geometry complexity (S/V increases) but generate low porosity, that is, microporosity.
10. The increase in I/S presence in the rock shows an increase in S/V ratio from N_2 adsorption tests, and the opposite behavior is found with an increase in kaolinite.

A combination of MICP, N_2, and NMR is an ideal approach to overcoming each of the methods' individual limitations on pore size resolution and external influences.

APPENDIX 5.A XRD Results

Sample ID	Quartz	K-Feldspar	Siderite	Pyrite	Natro-jarosite	Ankerite/FE-dolomite	Plagioclase	Calcite	Dolomite	Fluorapatite	Barite	Magnetite	Kerogen/others	Total non-clay	Smectite	Mixed illite/smectite (20%S)	Illite+MICA	Kaolinite	Chlorite	Total clay
1	15.1	6.3	1.4	1.3									17.9	42		6.4	13.3	28.3	10	58
2	14.7	4.7	0.5	0.9									22.6	43.4		12	8.2	29.1	7	56.3
5	34.1	1.3		1.6									8.3	55.1		27.6	3.7	4.3	9.3	44.9
6	37.3	2.7		1.2									8.1	56.3		32.5	3.2	2.1	5.7	43.5
8	24.5	4	0.8	3.7			8.3	1.5						44		19.8	24.9	2.6	7.1	56
11	49.2	4	1.4	1.7		1.2	5.3	0.5		0.2	0.4			66	1.7	12.4	14.3	0.9	4.8	34.1
13	53	3.3	1	2.1		0.2	5.7	4		0.4	0.4			69	1.7	11.7	13.9	0.8	3.1	31.1
14	41.3	3.6	1.5	3.1		0.7	4.6	3.3				0.9		59.1	1.6	14.9	18.9	0.8	5.2	40.7
15	53.8	4	1.2	1.7			8.1	0.9	0		0.7			72.1	0.9	10.4	12.4	0.5	2.9	27.9
16	44.5	3.6	2.3			1.8	7.6	0.6	0.4	0.2				66.8	1.7	9.7	16.0	1.2	4.3	33.1
17	51.1	5.3	2.4	0.9		1	7.8	1.2	1.1		0.5			58.8	1.9		11.8	19.7	3.9	41.2
18	47.2	6.9	0.5				10.8	2.1						56.5		5.8	14.6	27.8	1.1	43.5
19	54.1	2.5	0.4		1.9									60.1			13.5	25.6	0.8	39.9
20	56.1	6.5	0.6	0.6	2.5									66.1			11.5	21.5	0.9	33.9
21	45.6	6.5	0.8	0.3	2.9									57.1			14.7	26.8	1.4	42.9
22	52.5	3	0.6	0.2	3.9									57.5		9.0	15.1	16.6	1.8	42.5
23	67.1	9.1	0.5	0.2	1.2									78.7			9.7	11.3	0.3	21.3
24	60.5	1.3	0.3	1.5	1.8									64.8		3.0	8.5	22.6	1.1	35.2
25	67.2	3.6	0.5	0.6	1.2									74			11.6	13.8	0.6	26
26	50.5	2.4	0.4		2.1									54.1		4.1	13.7	25.1	3.0	45.9
27	34.2	3.2	20.5		0.8									58.7		7.2	12.0	17.6	4.5	41.3
30	23.3	0.8		10.4	0.8		3.6	0.5					8	46.6		25.1	4.9	16.3	7	53.3
31	15.5	0.8		3.7			2.9	21.9					6.5	51.3		27.7	4.5	10.5	6	48.7

REFERENCES

Al-Bazali TM, Jianguo Z, Martin EC, Mukul MS. Measurement of the sealing capacity of shale caprocks. Paper presented at SPE Annual Technical Conference and Exhibition; October 9–12, 2005; Dallas, TX. DOI: 10.2118/96100-MS.

Al Hinai A, Rezaee R, Saeedi A, Lenormand R. Permeability prediction from mercury injection capillary pressure: an example from the Perth Basin, Western Australia. APPEA J 2013;53:31–36.

Al-Raoush RI, Willson CS. Extraction of physically realistic pore network properties from three-dimensional synchrotron X-ray microtomography images of unconsolidated porous media systems. J Hydrol 2005;300:44–64.

Barret EP, Joyner LG, Halenda PP. "The determination of pore volume and area distribution in porous substances: computations from nitrogen isotherms. J Am Chem Soc 1951;73:373–380.

Boudier T. 3D Processing and Analysis with ImageJ. Paris: Université Pierre et Marie Curie. http://imagejdocu.tudor.lu/lib/ (accessed April 16, 2014).

Bowers MC, Ehrlich R, Howard T, Kenyon W. Determination of porosity types from NMR data and their relationship to porosity types derived from thin section. J Soc Pet Eng 1993;13:1–14.

Boyer C, Kieschnick J, Suarez-Rivera R, Lewis RE, Waters G. Producing gas from its source. Oilfield Rev 2006;18 (3):36–49.

Brunauer S, Deming LS, Deming WE, Teller E. On a theory of the van der Waals adsorption of gases. J Am Chem Soc 1940;62 (7):1723–1732.

Burdine NT, Gournay LS, Reichertz PP. Pore size distribution of petroleum reservoir rocks. J Pet Technol 1950;2 (7):195–204.

Bustin RM, Bustin AMM, Cui A, Ross D, Pathi VM. Impact of shale properties on pore structure and storage characteristics. Paper presented at SPE Shale Gas Production Conference; November 16–18, 2008; Fort Worth, TX. DOI: 10.2118/119892-MS.

Butcher AR, Lemmen HJ. Advanced SEM technology clarifies nanoscale properties of gas accumulations in shales. The American Oil and Gas Reporter. July2011.

Chalmers GR, Bustin RM, Power IM. Characterization of gas shale pore systems by porosimetry, pycnometry, surface area, and field emission scanning electron microscopy/transmission electron microscopy image analyses: examples from the Barnett, Woodford, Haynesville, Marcellus, and Doig units. AAPG Bull 2012;96:1099–1119.

Chen Q, Song Y-Q. What is the shape of pores in natural rocks? J Chem Phys 2002;116:8247–8250.

Churcher PL, French PR, Shaw JC, Schramm LL. Rock properties of berea sandstone, baker dolomite, and Indiana limestone. SPE International Symposium on Oilfield Chemistry; 1991; Anaheim, CA.

Clarkson CR, Solano N, Bustin RM, Bustin AMM, Chalmers GRL, He L, Melnichenko YB, Radlin AP, Blach TP. Pore structure characterization of north American shale gas reservoirs using USANS/SANS, gas adsorption, and mercury intrusion. Fuel 2013;103:606–616.

Coates GR, Xiao L, Prammer MG. NMR Logging Principles and Applications. Houston: Halliburton Energy Services; 1999.

Dastidar R, Sondergeld CH, Rai CS. An improved empirical permeability estimator from mercury injection for tight clastic rocks. PetroPhysics 2007;48 (3):186–190.

Dimri VP, Srivastava RP, Vedanti N, editors. Fractal Models in Exploration Geophysics Applications to Hydrocarbon Reservoirs. Amsterdam/Boston: Elsevier; 2012. Handbook of Geophysical Exploration: Seismic Exploration, Vol. 41; p 1–165.

Dollimore D, Heal GR. An improved method for the calculation of pore-size distribution from adsorption data. J Appl Chem 1964;14:109–114.

Dougherty ER, Lotufo RA. Hands-on Morphological Image Processing. Washington, DC: SPIE Press; 2003.

Egermann P, Doerler N, Fleury M, Behot J, Deflandre F, Lenormand R. Petrophysical measurements from drill cuttings an added value for the reservoir characterization process. Abu Dhabi International Conference and Exhibition; Abu Dhabi: Society of Petroleum Engineers; 2004.

Egermann P, Doerler N, Fleury M, Behot J, Deflandre F, Lenormand R. Petrophysical measurements from drill cuttings: an added value for the reservoir characterization process. SPE Reserv Eval Eng 2006;9:302–307.

Fibics. Introduction: Focused Ion Beam Systems. 2011. Available at http://www.fibics.com/fib/tutorials/introduction-focused-ion-beam-systems/4/. Accessed June 14, 2014.

Gabriela AM, Lorne AD. Petrophysical measurements on shales using NMR. SPE/AAPG Western Regional Meeting; 2000; Long Beach, CA.

Gale JFW, Reed RM, Holder J. Natural fractures in the barnett shale and their importance for hydraulic fracture treatments. AAPG Bull 2007;91 (4):603–622.

Glorioso JC, Aguirre O, Piotti G, Mengual J-F. Deriving capillary pressure and water saturation from NMR transversal relaxation times. SPE Latin American and Caribbean Petroleum Engineering Conference; Port-of-Spain, Trinidad and Tobago: Society of Petroleum Engineers; 2003.

Gregg SJ, Sing KSW. Adsorption, Surface Area and Porosity. London: Academic Press; 1991. p 303.

Grunewald E, Knight R. A laboratory study of NMR relaxation times in unconsolidated heterogeneous sediments. Geophysics 2011;76:G73–G83.

Heath JE, Dewers TA, McPherson BJOL, Petrusak R, Chidsey TC, Rinehart AJ, Mozley PS. Pore networks in continental and marine mudstones: characteristics and controls on sealing behavior. Geosphere 2011;7 (2):429–454.

Hidajat I, Singh M, Mohanty KK. NMR response of porous media by random walk algorithm: a parallel implementation. Chem Eng Commun 2003;190 (12):1661–1680.

Hildenbrand A, Urai JL. Investigation of the morphology of pore space in mudstones—first results. Mar Pet Geol 2003;20:1185–1200.

Howard JJ. Porosimetry measurements of shale fabric and its relationship to illite/smectite diagenesis. Clays Clay Mine 1991;39 (4):355–361.

Jones FO, Owens WW. A laboratory study of low-permeability gas sands. SPE J Pet Technol 1980;32:1631–1640.

Jorgensen DG. Estimating permeability in water-saturated formations. Log Anal 1988;296:9.

Kale S, Rai C, Sondergeld C. Rock typing in gas shales. Paper presented at SPE Annual Technical Conference and Exhibition; September 19–22, 2010a; Florence, Italy. DOI: 10.2118/134539-MS.

Kale S, Rai C, Sondergeld C. Petrophysical characterization of Barnett shale. Paper presented at SPE Unconventional Gas Conference; February 23–25, 2010b; Pittsburgh, Pennsylvania. DOI: 10.2118/131770-MS.

Katz AJ, Thompson AH. Quantitive prediction of permeability in porous rock. Phys Rev 1986;34:8179.

Kenyon WE, Takezaki H, Straley C, Sen PN, Herron M, Matteson A, Petricola MJ. A laboratory study of nuclear magnetic resonance relaxation and its relation to depositional texture and petrophysical properties—Carbonate Thamama Group, Mubarraz Field, Abu Dhabi. Middle East Oil Show; Bahrain; 1995; Copyright 1995; Society of Petroleum Engineers, Inc.

King, GE. Thirty years of gas shale fracturing: what have we learned? Paper presented at SPE Annual Technical Conference and Exhibition; September 19–22, 2010; Florence, Italy. DOI: 10.2118/133456-MS.

Kolodzie S. Analysis of pore throat size and use of the Waxman-Smits equation to determine OOIP in Spindle Field, Colorado. Society of Petroleum Engineers, 55th Annual Fall Technical Conference, Paper 9382; 1980.

Kuila U, Prasad M. Surface area and pore-size distribution in clays and shales. SPE Annual Technical Conference and Exhibition; Denver, CO: Society of Petroleum Engineers; 2011.

Lenormand R, Fonta O. Advances in measuring porosity and permeability from drill cuttings. SPE/EAGE Reservoir Characterization and Simulation Conference; Abu Dhabi: Society of Petroleum Engineers; 2007.

Lenormand R, Bauget F, Ringot G. Permeability measurement on small rock samples. International Symposium of the Society of Core Analysts; 2010; Halifax, Canada.

Looyestijn WJ. Distinguishing fluid properties and producibility from NMR Logs. Paper presented at the 6th Nordic Symposium on Petrophysics; 2001; Norway.

Lowden B. Some Simple Methods for Refining Permeability Estimates from NMR Logs and Generating Capillary Pressure Curves. ResLab-ART: Suffolk; 2009.

Luffel DL. Advances in Shale Core Analysis. Gas Research Institute: Houston; 1993.

Ma S, Jiang M-X, and Morrow NR. Correlation of capillary pressure relationships and calculations of permeability. Paper presented at SPE Annual Technical Conference and Exhibition; October 6–9, 1991; Dallas, TX. DOI: 10.2118/22685.MS.

Minh CC, Sundararaman P. NMR petrophysics in thin sand/shale laminations. SPE Annual Technical Conference and Exhibition; 2006; San Antonio, TX.

Nelson PH. Pore-throat sizes in sandstones, tight sandstones, and shales. AAPG Bull 2009;93 (3):329–340.

Owolabi OO, Watson RW. Estimating recovery efficiency and permeability from mercury capillary pressure measurements for sandstones. Paper presented at SPE Eastern Regional Meeting; November 2–4, 1993; Pittsburgh, Pennsylvania. DOI: 10.2118/26936-MS.

Pape H, Clauser C, Iffland J. Permeability prediction for reservoir sandstones and basement rocks based on fractal pore space geometry. Paper presented at 1998 SEG Annual Meeting; September 13–18, 1998; New Orleans, LA. DOI: 10.1190/1.1820060.

Pittman ED. Relationship of porosity and permeability to various parameters derived from mercury injection-capillary pressure curve for sandstone. AAPG Bull 1992;76:191.

Prodanovic M, Lindquist WB, Seright RS. Porous structure and fluid partitioning in polyethylene cores from 3D X-ray microtomographic imaging. J Colloid Interface Sci 2006;298 (1):282–297.

Quantachrome Instruments. Autosorb AS-1/ASWin Gas Sorption System Operation Manual. Quantachrome Instruments: Boynton Beach, FL; 2008.

Rezaee MR, Jafari A, Kazemzadeh E. Relationships between permeability, porosity and pore throat size in carbonate rocks using regression analysis and neural networks. J Geophys Eng 2006;3:370.

Ross DJK, Marc BR. The importance of shale composition and pore structure upon gas storage potential of shale gas reservoirs. Mar Pet Geol 2009;26 (6):916–927.

Rouquerol J, Avnir D, Fairbridge CW, Everett DH, Haynes JM, Pernicone N, Ramsay JDF, Sing KSW, Unger KK. Recommendations for the characterization of porous solids. Pure Appl Chem 1994;66 (8):1739–1758.

Rushing JA, Newsham KE, Blasingame TA. Rock typing-keys to understanding productivity in tight gas sands. Paper presented at SPE Unconventional Reservoirs Conference; February 10–12, 2008; Keystone, CO. DOI: 10.2118/114164-MS.

Shaw JC, Reynolds MM, Burke LH. Shale gas production potential and technical challenges in western Canada. Paper presented at Canadian International Petroleum Conference; June 13–15, 2006; Calgary, Alberta. DOI: 10.2118/2006-193.

Swanson BF. A simple correlation between permeabilities and mercury capillary pressures. SPE J Pet Technol 1981;33 (12):2498–2504.

Talabi OA, Alsayari S, Blunt MJ, Dong H, Zhao X. Predictive pore scale modeling: from 3D images to multiphase flow simulations. Paper presented at SPE Annual Technical Conference and Exhibition; September 21–24, 2008; Denver, CO. DOI: 10.2118/115535.MS.

Volokitin Y, Looyestijn WJ, Slijkerman WFJ, Hofman J. A practical approach to obtain primary drainage capillary pressure curves from NNR core and log data. Petrophysics—Houston 2001;42:334–343.

Washburn EW. Note on a method of determining the distribution of pore sizes in a porous material. Proc Natl Acad Sci USA 1921;7:115–116.

Wyllie MRJ, Gregory AR. Fluid flow through unconsolidated porous aggregates: effect of porosity and particle shape on Kozeny-Carman constants. Ind Eng Chem 1955;47 (7):1379–1388.

Zahid S, Bhatti A, Khan H, Ahmed T. Development of unconventional gas resources: stimulation perspective. Paper presented at Production and Operations Symposium; March31 to April 3, 2007; Oklahoma City, Oklahoma. DOI: 10.2118/107053-MS.

Zhi-Qiang M, Yu-Dan H, Xiao-Jun R. An improved method of using NMR T2 distribution to evaluate pore size distribution. Chin J Phys 2005;48 (2):412–418.

6

PETROPHYSICAL EVALUATION OF GAS SHALE RESERVOIRS

MEHDI LABANI AND REZA REZAEE

Department of Petroleum Engineering, Curtin University, Perth, WA, Australia

6.1 INTRODUCTION

For many years, shale formations were viewed as a hydrocarbon source rock or cap rock. Due to this traditional point of view, only geochemical analysis has been routinely performed on the shale layers. However, for sweet spot mapping of the gas shale layers, it is necessary to know about petrophysical and geomechanical properties as well as geochemical ones. The main focus of this chapter is the petrophysical evaluation methods of shale formations. In the first section, the key properties for evaluation of potential gas shale intervals are defined, and then the available techniques for measuring these parameters will be discussed. The chapter will be wrapped up with the common well log signatures of the gas shales and ways to interpret them for finding petrophysical properties of shale intervals.

6.2 KEY PROPERTIES FOR GAS SHALE EVALUATION

Shale is a fine-grained detrital sedimentary rock, formed by the consolidation of clay (less than 4 μm) and silt (between 4 and 62.5 μm) sized particles into rock layers of ultralow permeability. In general, shales are characterized by finely laminated rocks and/or fissility approximately parallel to the bedding (Serra, 1988). This definition gives the lowest opportunity for shale as a reservoir. However, the right combinations of geological, geochemical, petrophysical, and geomechanical properties would result in a productive gas shale interval. In the following sections, the key properties for a shale formation to be considered as a potential gas shale layer are discussed.

6.2.1 Pore System Characteristics

Pore system characteristics are very important for evaluation of gas shale reservoirs. Pores within the matrix of the gas shale reservoirs are smaller than pores in the conventional reservoirs (Nelson, 2009). The main types of pore spaces in productive gas shale systems are matrix porosity, either associated with mineral particles or organic matter, and fracture porosity, either natural or induced fractures (Wang and Reed, 2009). In gas shale systems, matrix pores along with natural networks of fractures provide the flow of gas from matrix to induced fractures during production. Generally, in describing the matrix pore size in shales the pores are all considered to fall within the nanopore range (Javadpour, 2009; Javadpour et al., 2007; Loucks et al., 2009). There is a pore classification system for materials that contain nanometer-scale porosity developed by Rouquerol et al. (1994). According to this pore classification, micropores are <2 nm diameter, mesopores are between 2 and 50 nm, and macropores are >50 nm. Mesopores and micropores are economically important to gas shale production because of their large contribution to porosity and storage sites for methane (Keller et al., 2011). Recently, Loucks et al. (2012) defined a new pore size classification for matrix-related mudrock pores; however, herein the Rouquerol's classification has been used.

To find out more about the pore system of organic-rich shale layers, it is necessary to know about the accessible pore volume or effective porosity, pore size distribution (PSD), pore shape, and specific surface area.

An inverse relationship exists between pore size and surface area (Beliveau, 1993). Recent study by Labani et al, (2013) on the potential gas shales from Western Australia (WA) has shown that the micropores have a higher contribution to surface area than mesopores, whereas macropores contribute the least. Organic matter (OM) characteristics (quantity, quality, and maturity of OM) and mineralogical composition control micro/mesopore volume. For the gas shales studied from the Perth Basin, WA, summation of the micro- and mesopore volumes correlate with total organic carbon (TOC) and thermal maturity indicator, that is, T_{max} (Fig. 6.1a). On the other hand, clay content controls the micro- and mesopore volumes because clay minerals are normally associated with high content of micropores (Ross and Bustin, 2007a) while finding a relationship between quartz content and micro/mesopore volume is difficult (Fig. 6.1b).

6.2.2 Organic Matter Characteristics

Organic matter (OM) quantity (TOC), organic matter maturity, and organic matter quality are the determining parameters for gas production from gas shales. Loucks et al. (2009), Ambrose et al. (2010), and Curtis et al. (2010) suggested and demonstrated that nanopores in the OM form the major connected or effective pore network in some gas shale systems. This mode of porosity possibly evolved with the thermal transformation of organic matter (Modica and Lapierre, 2012). Variability in the source of organic matter could greatly influence amount of porosity developed during thermal transformation. Loucks et al. (2012) proposed that type II kerogen may be more prone to the development of OM pores than type III kerogen due to the higher hydrogen content. On the other hand, TOC content can affect on the amount of adsorbed gas capacity; therefore, it could be a controlling factor for the total gas content as well (Jacobi et al., 2008).

Thermal maturity is an important parameter for commercial gas production if the shale has considerable organic content. The common viewpoint for mapping sweet spots is locating the shale layers with higher thermal maturity. It is believed that insufficient gas is generated at lower thermal maturity to fill the pore space, and besides that, oil in the nanopore system can block the movement of gas (Cluff et al., 2007).

6.2.3 Permeability

Permeability of gas shale which is extremely low, from subnanodarcys to microdarcys, is a function of mineralogy, sample type, porosity, confining pressure, and pore pressure (Bustin et al., 2008; Cui et al., 2009). During production, gas shale wells typically show a rapid initial decline (gas flows through fractures) followed by a slow, gradual decline (gas desorption from pore wall into microfracture) (Fig. 6.2) (Soeder, 1988). Therefore, two effective permeabilities exist for gas shale: matrix permeability and fracture permeability. Matrix permeability is expected to be a combination of diffusive flux in very small pores and advective flux in larger pores and fractures (Javadpour et al., 2007).

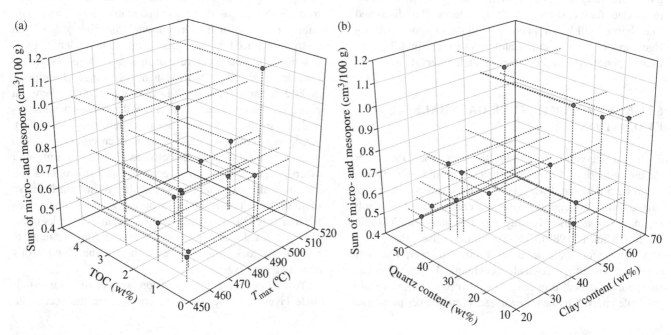

FIGURE 6.1 Three-dimensional scatter plot showing the relationship between sum of micro- and mesopore volume with (a) TOC and T_{max} and (b) quartz and clay content for the Perth samples.

6.2.4 Gas Storage Capacity

Gas in place is often the critical factor for evaluating the economics of a gas shale system. The gas storage mechanism in the shale layers is different from conventional reservoirs. In shales, natural gas can be stored in two main ways: free gas in pores and fractures and condensed gas in a form of adsorbed gas on the surface of organic material and clay minerals.

It is believed that much of the gas is stored in the adsorbed state for gas shale. For example, some studies suggest that 50% of the total gas storage in the Devonian shale exists as a condensed adsorbed phase (Lu et al., 1995). The relative importance of each mode of gas storage is determined by pore space characteristics, organic matter characteristics, mineralogical composition, and geological conditions (Allen et al., 2009; Ross, 2007).

Size and surface area of the pores have an effect on the mode of gas storage. Higher surface area means higher potential sites for methane adsorption; therefore, considering Figure 6.1, micropores and mesopores are the better sites for methane adsorption, while macropores are conducive to free gas storage because of the low effect on surface area and the larger pore volume.

The presence of organic matter in shales lowers density, increases micro/mesoporosity, provides the source of gas, alters the wettability, and facilitates adsorption (Zhang et al., 2012). A generally positive correlation of methane sorption capacity with TOC in shales has been observed in previous studies (Lu et al., 1995; Ross and Bustin, 2009). Zhang et al. (2012) showed that gas sorption capacities of kerogen decrease in the following order: type III > type II > type I (Fig. 6.3a). They attributed the differences in gas sorption capacities among different kerogen types to changes in chemical structures, and stated that aromatic-rich kerogens have a stronger affinity with methane than kerogens containing more aliphatic organic matter. They also showed that the shale samples with higher thermal maturity have a higher capacity for methane adsorption. As can be seen in Figure 6.3b, the Barnett shale samples with maturity of 0.58 and 0.81% R_o have similar gas sorption capacities, whereas a sample with 2.1% R_o has an obviously higher adsorbed gas capacity.

Mineralogical composition can greatly affect the adsorbed gas capacity. Based on a study by Ross and Bustin (2009), clay minerals, especially illite and montmorillonite, have a large adsorbed gas capacity due to the presence of greater micropore volume and surface area.

Finally, geological conditions like depth, reservoir temperature, reservoir pressure, and moisture content could have an effect on the amount of adsorbed, free, and dissolved gas. The adsorption capacity of the shales increases as the temperature decreases, which is expected since gas adsorption

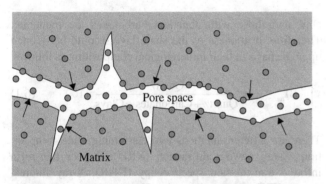

FIGURE 6.2 Nanoscale schematic of gas molecule locations in the gas shale reservoirs.

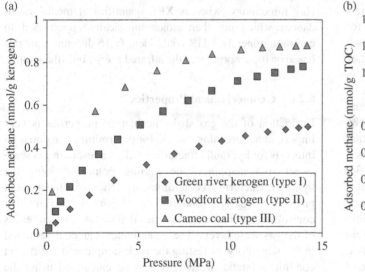
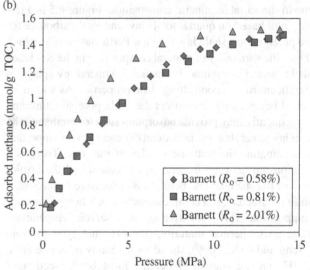

FIGURE 6.3 Effect of organic matter type (a) and maturity (b) on adsorbed gas capacity (data from Zhang et al., 2012).

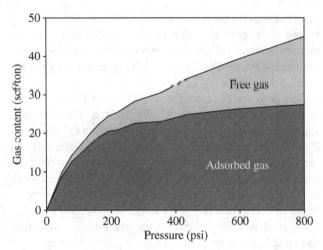

FIGURE 6.4 Variation of adsorbed gas versus free gas as a function of pressure for a potential gas shale sample from the Perth Basin, WA.

on solids is an exothermic process (Lu et al., 1995). Reservoir pressure, meanwhile, has an important impact on the adsorbed gas capacity. Adsorption is a very efficient mechanism for storing gas at low pressures, while at high pressures the role of free gas is highlighted (Alexander et al., 2011). It is due to the nature of gas adsorption on the shale layers which increases rapidly at relatively low pressures; thereafter, the adsorption capacity plateaus as the system reaches gas saturation (Fig. 6.4) (Ross and Bustin, 2007b). Moisture content competes with the methane molecules for adsorption sites (Bustin and Clarkson, 1998); therefore, the gas shale with higher moisture content should have lower gas adsorption capacity.

6.2.5 Shale Composition

Petrophysical evaluation of gas shale reservoirs is complex due to the variable mineral composition. Figure 6.5 is a ternary plot based on quartz, total clay, and total carbonate for the potential gas shale layers in the Perth Basin, WA, which shows the variability of mineral composition in the Kockatea Shale and Carynginia Formation. Mineralogy plays a significant role in controlling shale properties. As was mentioned before, clays can affect the shale pore structure and consequently may provide adsorption sites for methane. The nonclay minerals especially quartz content is very important for estimating the brittleness index of the rock. There is a relationship between mineralogical content and the brittleness of the shale layers. Brittleness, a measure of the rock's ability to fracture, is a complex function of lithology, mineral composition, TOC, effective stress, reservoir temperature, diagenesis, thermal maturity, porosity, and type of fluid (Wang and Gale, 2009). Based on the study by Jarvie et al. (2007) on the Barnett shale, the most brittle section of Barnett has abundant quartz, the least brittle has abundant

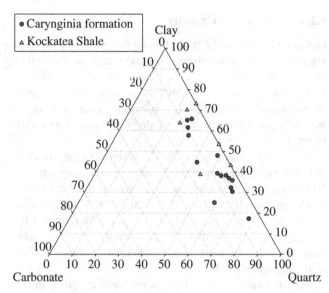

FIGURE 6.5 Variability of mineralogical composition in the potential gas shale layers of Perth Basin.

clay, and those with abundant carbonates are moderate. Therefore, brittleness of the shale layers could be defined upon the basis of their mineralogical composition as follows:

$$BI_{mineralogy} = \frac{Quartz}{Quartz + Carbonates + Clays} \times 100 \qquad (6.1)$$

There are different methods for determining shale composition: X-ray powder diffraction (XRD), Fourier transform infrared transmission spectroscopy (FTIR), X-ray fluorescence (XRF), energy-dispersive X-ray spectroscopy setting on the scanning electron microscope (EDS-SEM), and thin section analysis (TS). Among these techniques XRF, XRD, and FTIR are used more than others for determining shale composition. XRD could determine the bulk mineralogy and clay mineralogy, whereas XRF quantifies elemental abundances which are then stoichiometrically apportioned to common minerals. FTIR could identify 16 different minerals based on the absorption of the infrared energy onto the sample.

6.2.6 Geomechanical Properties

Evaluation of the gas shale mechanical properties is very important as screening criteria for determining the potential intervals for hydraulic fracturing and, as a result, in gas shale sweet spot mapping. The starting point for doing the hydraulic fracturing is determination of the rock's mechanical properties. Young's modulus and Poisson's ratio are two controlling mechanical properties that dictate the brittleness of the gas shale layers. These parameters can be determined in the laboratory by testing the rock sample under different conditions (static method) or can be calculated using the dipole sonic log data (refer to Section 6.2.5).

6.3 PETROPHYSICAL MEASUREMENTS OF GAS SHALE RESERVOIRS

Petrophysical measurement of gas shale formations is of critical importance for finding out about potential gas shale intervals for economic gas production. In conventional reservoir rocks, characterizing petrophysical parameters like porosity, fluid saturation, and permeability are very well documented, and API methodologies are widely adopted (API, 1998). However, there are not any well-established laboratory methods specific to gas shale reservoirs, and sometimes there is not any consistency among the results (that are) reported from different commercial laboratories (Sondergeld et al., 2010). Also, using (the) conventional methods for gas shale reservoirs has some limitations. For example, the Dean Stark method, which is a routine procedure for water saturation determination of conventional reservoirs, is unable to separate free water from bound water in gas shale; therefore, it is not possible to calculate effective saturation, effective porosity, and clay-bound water volume (Handwerger et al., 2011). In the following section, the available laboratory methods used specifically for petrophysical measurements of gas shale samples are explained.

6.3.1 Pore Structure Evaluation Techniques

In order to clarify the complex pore structure of the shales, researchers have utilized different fluid invasion techniques including low-pressure gas adsorption analysis using nitrogen and carbon dioxide, helium pycnometry, and mercury porosimetry. Effective porosity of the shale matrix is determined by mercury immersion of the sample (for bulk volume determination) coupled with helium pycnometry (for grain volume determination) (Chalmers et al., 2012). Measuring the shale matrix porosity is usually performed on the crushed samples, due to two reasons: (i) Luffel and Guidry (1992) proposed that low porosity of the shale samples might be the result of incomplete penetration of the pore network by helium under the helium porosimetry method. Crushing the shale samples increases the area accessible by gas and thereby increases the accuracy of the measurement, and (ii) the presence of microfractures due to the laminated structure of the organic-rich shales might affect the measurement of matrix porosity. Crushing eliminates both these microfractures and also core-induced artefacts.

Since the pore sizes of the gas shale vary between nanometer and micrometer scale, crushing the samples in millimeter size range does not affect the pore structure, and the crushed sample porosity can be considered as representative of the matrix porosity. Any contention arises from different protocols for crushing and sieving the shale samples. In different reviews (Chalmers et al., 2012; Karastathis, 2007; Luffel and Guidry, 1992; Ross and Bustin, 2009), there are different numbers for the size of the crushed samples.

The final size of the crushed samples should be higher than the size of the grains. According to the classification of sedimentary rocks based on the grain size, maximum grain size of the shales is about 62.5 μm in diameter; therefore, 250 μm would be a proper lower range measure for the crushed sample size. The upper range could be selected as 2 mm which is the maximum grain size of the sandstones.

Before starting the helium pycnometry, the crushed samples should be heated to remove the moisture content of the shale samples. The main concern during heating of the shale samples is preserving organic materials and the clay-bound water. Analysing the evaporated components of the shale samples shows that by heating up to around 120°C, only free water evaporates from the matrix of the shale samples (Easley et al., 2007; Handwerger et al., 2011).

Low pressure CO_2 and N_2 isotherms (<18.4 psia) can give useful information about pore volume, pore surface area, PSD, and pore shape of the shale samples (Quantachrome, 2008). Between 1 and 2 g of ground sample (<250 μm) is degased in the evacuated oven prior to analysis. In different reviews, there are different values for time and temperature that are required for degasing the samples (Clarkson et al., 2012a, b, 2013; Ross and Bustin, 2009). Low pressure adsorption of CO_2 is useful for characterizing microporosity, while nitrogen adsorption is useful for characterizing meso- and part of macroporosity. Low-pressure gas adsorption analysis cannot determine pores greater than 300 nm in diameter (Clarkson et al., 2011).

Mercury porosimetry provides PSD, total pore volume or porosity, the skeletal and apparent density, and specific surface area (Giesche, 2006). Similar to low-pressure adsorption measurement, the samples should be evacuated before the test to remove moisture and possible gas content. Mercury intrusion is useful for characterizing meso- and macroporosity. Therefore, combining mercury data with low-pressure gas adsorption data allows for the determination of full pore size spectrum of gas shale reservoirs. Figure 6.6 shows the representative PSD using mercury porosimetry and gas adsorption data for a gas shale sample in the Perth Basin, WA. As can be seen in this figure, combining these two techniques yields a full pore size spectrum of the shale samples from micro- to macropore. By comparing the PSD in the overlapped area (mesopore), it is clear that the position of the peaks does not match precisely. For all analyzed samples, mercury porosimetry suggests a lower mode pore diameter compared to that obtained from nitrogen adsorption. The possible explanations for this observed shift are (i) mercury intrusion measures the pore throats and not the actual pore size, and therefore the measured pore size using mercury would be smaller than that obtained from the nitrogen adsorption and (ii) in mercury porosimetry for accessing the smaller pore diameters, mercury injection pressure should increase. The experimental results show that the mercury injection pressure for accessing pore

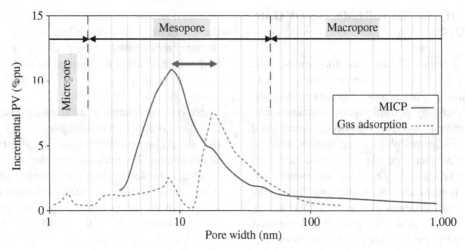

FIGURE 6.6 A comparison between pore size distribution defined by incremental pore volume using mercury porosimetry and gas adsorption data (the arrows show the difference between peak positions in the mesopore area) for a gas shale sample from the Perth Basin, WA.

diameters around 3 nm is about 60 kpsi. This high pressure, as suggested by Giesche (2006), could compress the sample and subsequently decrease the measured pore throat size especially at smaller sizes.

6.3.2 Fluid Saturation Measurement

There are two common methods for fluid extraction used by commercial labs in conventional reservoirs: Dean Stark and the retort method. Both methods are documented in the recommended practice 40 (API, 1998). These methods can be used for gas shale; however, they are performed on the crushed shale samples. The technique used by the Gas Research Institute (GRI) to measure water and hydrocarbon saturation is by performing Dean Stark method with toluene (Luffel and Guidry, 1992). There are some limitations with this technique; although crushing the shale samples facilitates removing the pore fluid with the use of toluene, it still presents some difficulties in the extraction of fluid from nanopores (Handwerger et al., 2011). Besides, this technique cannot differentiate between free water and bound water, and typically only a combination of these two volumes is reported (Sondergeld et al., 2010).

The retort distillation method is a fluid extraction technique that involves sequential heating of the sample under atmospheric pressure. During the heating process water and oil are vaporized, collected, and measured by condensing the fluids into a graduated cylinder. The commercial procedure available for performing the retort method on tight shales is as follows (Handwerger et al.; 2011):

- Measuring the bulk density by mercury immersion
- Crushing the samples and weighing
- Measuring the gas filled porosity using helium pycnometry
- Heating the samples continuously up to succession of three characteristic retort temperatures ($T_1 = 250°F$, $T_2 = 600°F$, and $T_3 = 1300°F$)
- Measuring the volumes of pore water and oil after having recovered fluid volumes from a retort experiment
- Measuring the gas volume by subtracting the recovered fluids volume (water + oil) from measured pore volume
- Calculating total pore volume by summation of extracted pore fluids and gas filled porosity
- Calculating appropriate saturation by dividing fluid volume by total pore volume

It is worth mentioning that any water released before T_1 is free water. A significant quantity of water that is released up to T_2 can be capillary-bound/clay-bound water and finally water extracted by T_3 is the indicator of structural water. The retort method also provides a measure of the free oil (measured at T_2) and the bound oil at T_3.

Apart from these two methodologies, it is possible to extract the water content of the gas shale samples using the thermogravimetric analyzer (TGA), based on the simple assumption that two fluid phases are present in the shale sample: gas and water. During thermogravimetric analysis, crushed samples are heated in an inert atmosphere and the mass loss due to pore fluid vaporization is recorded in real time. Then, water saturation can be measured by relating the mass loss in the sample to the water evaporation only (Handwerger et al., 2012). The temperature steps used for TGA are similar to those used in the retort method.

6.3.3 Permeability Measurement

Permeability is one of the most difficult properties to measure in gas shale reservoirs. There are several methods for permeability measurement/estimation for shale formations:

1. Unsteady-state techniques: Pulse decay (performed on the plug sample) and pressure decay (performed on the crushed sample)
2. Permeability determination from desorption test
3. Use of mercury injection capillary pressure (MICP) data
4. Use of nuclear magnetic resonance (NMR) data

Routine methods for measuring permeability in the laboratory are based on steady-state flow. However, if the permeability is low, it needs a long period of time for establishing steady-state flow; therefore, unsteady state flow is preferred for gas shale. A pulse decay method to measure low permeability has been introduced by Brace et al. (1968). This methodology uses a cylindrical sample under hydrostatic confining pressure, which is connected to two fluid reservoirs. At the start of the experiment, the fluid pressure in the upstream reservoir is suddenly increased. When fluid flows from the upstream reservoir, its pressure in that reservoir declines with time. Similarly, when fluid flows from the sample into the downstream reservoir, its pressure builds up with time (Jones, 1997). The sample permeability can be calculated from the observed pressure decay in the upstream or the pressure buildup in the downstream reservoir.

Another approach to measure matrix permeability is through pressure decay with helium using crushed shale samples. In this method, shale core samples are crushed and then a narrow sieve cut is used to obtain a relatively uniform particle size. Figure 6.7 shows a schematic (diagram) of the laboratory equipment used in this method. By expanding helium from reference cell to sample cell, pressure suddenly drops due to the dead space in the sample cell. After that, it decays with time to a lower pressure as helium moves into the matrix pores of the crushed samples (Luffel, 1993). This observed pressure decay can be used to determine gas permeability. The permeability values for both methods (pulse decay and pressure decay) can be determined using a simulation-based history matching or analytical solution (Cui et al., 2009; Darabi et al., 2012; Jones, 1997).

Permeability can be determined through the desorption test from fresh cores retrieved from the wells. The cores are put into specially designed canisters at reservoir temperature to desorb gas at ambient pressure. The cumulative volume of released gas is measured with time. Primarily, these data can be used to evaluate the gas content of the gas shale; however, the gas desorption rate from freshly cut cores can also be related to matrix permeability and diffusivity of the shale samples (Cui et al., 2009).

Permeability can also be estimated from MICP data. As was mentioned earlier, mercury intrusion measures the pore throats. Pore throats provide the path of fluid flow to the pore body; therefore, it could give an idea about permeability as well. The most common methodology for permeability estimation through mercury injection data is the Swanson (1981) method. According to this method, permeability is a function of capillary pressure and mercury saturation at the apex of a hyperbolic log–log mercury injection plot. It is worth mentioning that the Swanson method and all other methodologies for permeability estimation through using mercury injection are intended for sandstone (Pittman, 1992), tight gas sands (Rezaee et al., 2012), and carbonates (Rezaee et al., 2006) and therefore applying them to the shale samples needs careful attention.

Nuclear magnetic resonance measures pore body size in terms of transverse relaxation time or T_2, which is the required time for protons to return (back) to their original situation after being affected by an external magnetic field.

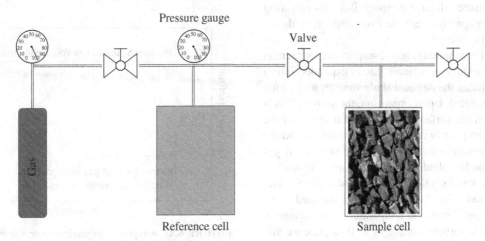

FIGURE 6.7 Schematic sketch of a pycnometer apparatus used for permeability measurement of the crushed shale samples.

NMR T_2 distribution allows differentiation between free fluid and bound fluid. The ability of NMR to distinguish between bound and free fluid increases the capability of NMR to estimate the formation permeability (Rezaee et al., 2012). Two common models for estimating permeability are the free-fluid (or Coates model) and mean T_2 (or SDR model) (Coates et al., 1999). $T_{2cutoff}$ (the) T_2 value which differentiates between free fluid and bound fluid, and T_{2gm}, the geometrical mean of T_2 distribution, are the effective parameters for the Coates model and mean T_2 model, respectively. Small pore sizes in the shales resulted in fast T_2 relaxation time and small $T_{2cutoff}$ and T_{2gm} compared to normal values for sandstone or carbonates. Coates et al. (1999) proposed a $T_{2cutoff}$ of 33 ms for sandstone and 92 ms for carbonates, while Sigal and Odusina (2011) showed that for the gas shale most of the NMR signals occur below 20 ms. There are many challenges with the interpretation of NMR signals for gas shale samples that make it difficult to differentiate between free fluid and bound fluid and, as a result, to estimate permeability. As Sigal and Odusina (2011) proposed, some of these challenges are as follows:

- Although T_2 mode is on the order of 1 ms for brine response and 10 ms for methane, water and gas NMR signals are very similar, and it is difficult to separate between gas and brine response.
- Resolving the effect of adsorbed gas in the determined porosity from NMR is an issue.
- The effect of temperature on methane relaxivity should be investigated.

6.3.4 Adsorbed Gas Measurement

Gas storage evaluation of the gas shale is performed through two different measurements. Measuring the free gas component is the same as with the conventional reservoirs and can be done using helium pycnometry or other conventional methods that measure effective porosity. But for measuring the adsorbed gas capacity, there are two common methods: volumetric and gravimetric.

The physics of a volumetric gas adsorption experiment is simple: a given amount of sorptive gas is expanded into a vessel which includes the degased shale sample, and which has first been evacuated. Upon expansion, the sorptive gas is partly adsorbed on the surface (external and internal) of the sorbent material, and partly remains as gas phase around the adsorbent. By measuring mass balance, the amount of gas being adsorbed can be calculated if the free space volume of the adsorbent is known (Keller and Staudt, 2005). The gravimetric method consists of exposing a degased shale sample to a pure gas at constant temperature. The change in the weight of the adsorbent sample as well as pressure and temperature is measured when equilibrium is reached (Keller and Staudt, 2005). This method allows for the direct measurement of the amount adsorbed, which is not the case for volumetric method.

The adsorption isotherm is the relationship between the amount of adsorbed gas and the gas pressure at constant temperature (Lu et al., 1995). Therefore, by repeating adsorbed gas measurement at different pressures, the Langmuir isotherm will be obtained (Fig. 6.8). According to the Langmuir equation, the adsorbed gas capacity (G_s) can be expressed as follows:

$$G_s = \frac{V_l P}{P + P_l} \qquad (6.2)$$

where V_l and P_l are the Langmuir volume and pressure respectively, and P is the reservoir pressure. Due to the exothermic nature of adsorption phenomena, it is necessary to do the adsorbed gas measurement at the reservoir temperature to simulate the reservoir conditions exactly. Zhang et al. (2012) and Ross and Bustin (2007a) derived some formulas for determining the effect of temperature on the adsorbed gas capacity. Zhang et al. (2012) proposed the following relationships between the Langmuir pressure and temperature based on the organic matter type:

$$\ln(P_l) = 5.89 - \frac{1241}{T} \text{ for type I kerogen} \qquad (6.3)$$

$$\ln(P_l) = 9.75 - \frac{2628}{T} \text{ for type II kerogen} \qquad (6.4)$$

$$\ln(P_l) = 11.06 - \frac{3366}{T} \text{ for type III kerogen} \qquad (6.5)$$

where P_l is the Langmuir pressure in MPa and T is temperature in Kelvin. However, as was mentioned by Zhang et al. (2012), the aforementioned correlations were obtained from

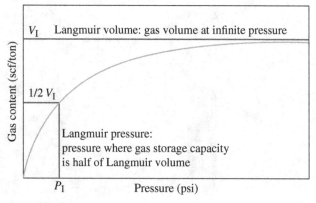

FIGURE 6.8 A typical Langmuir isotherm showing the quantity of adsorbed gas at a given pressure.

limited data points at low maturity conditions, and more data are required to find a general relationship for different types of kerogen.

6.4 WELL LOG ANALYSIS OF GAS SHALE RESERVOIRS

Well log data are valuable sources of information for reservoir characterization. Like the other parts of gas shale evaluation; well log analysis of these reservoirs is complex, and it needs unconventional as well as the routine conventional well logs. In this section, there is a brief explanation of the well log signatures and well log interpretation in the gas shale reservoirs.

6.4.1 Well Log Signatures of Gas Shale Formations

6.4.1.1 Resistivity Log The measurement of formation resistivity is of primary importance in well logging since it is a definitive method for identifying hydrocarbons and quantifying the water saturation. The resistivity of the rocks depends on the following:

- Fluid resistivity in the pore spaces
- Fluid saturation
- Rock lithology and the percentage of conductive minerals and the rock anisotropy
- Overburden pressure and pore pressure
- Temperature

Usually in the gas shale reservoirs, it is expected that the resistivity of the rock will increase due to the presence of hydrocarbon and organic matter. This assumption is correct only if the thermal maturation of the formation is high enough to result in hydrocarbon generation. Conversely, Anderson et al. (2008) showed that some gas producing shale layers can have high electric permittivities or lower resistivity. The cause of the high permittivity has been attributed to the presence of conductive minerals, such as pyrite or graphite, that build up as a result of kerogen transformation and exposure to elevated temperature and pressure. Interestingly, so far this observation has yielded mixed results: some gas shale samples have not shown such high permittivities while others have. This might be due to the different depositional environment of the gas shale layers or different thermal history of the formation.

Cation-exchange capacity (CEC) of the clay minerals is another property that has an effect on the resistivity of the shale layers. CEC value varies with the surface area of the clays. This means that the difference between the conductivity of clay species should be related to surface area (Rider, 1991). Smectite has a far greater specific surface area than the other clays and is therefore more conductive (Passey et al., 2010).

The effect of CEC on (the) shale conductivity depends on the salinity of the formation water. If the formation water salinity is greater than sea water salinity, the effect of excess conductivity due to clay minerals is small (Passey et al., 2010).

Anisotropy of the gas shale is an effective parameter in the interpretation of resistivity log and water saturation estimation in shale layers. Chemali et al. (1987) reported a disparity between laterolog and induction resistivity measurements in shales. Induction devices are sensitive only to the horizontal resistivity (R_h) of the formation, while laterolog measures a combination of both horizontal and vertical resistivity (R_v). Due to the vertical transverse isotropy (VTI), R_v is expected to be higher than R_h. Now the challenge is to find whether the true resistivity of gas shale is closer to R_h or R_v (Miller, 2010).

Taking these limitations into account, using resistivity log in gas shale layers needs closer attention, and it is not possible to predict a universal response for the resistivity log in the gas shale compared to conventional reservoirs.

6.4.1.2 Gamma Ray Log The gamma ray (GR) log is a type of tool that is used to measure the formation of natural GR radioactivity. There are basically two main types of GR tools:

- Natural GR tool (NGT): It measures the general GR emissions of all the radioactive elements (potassium, uranium, and thorium) together.
- Spectral GR tool (SGR): It differentiates GR emissions from (the) three main individual radioactive elements.

Amongst the sediments, shales have by far the strongest GR radiation. Due to this fact, the GR log is principally used to derive shale volume quantitatively.

The potassium content of the clay mineral species varies considerably. Illite contains the greatest amount of potassium, while kaolinite has very little or none (Dresser Atlas, 1983). The consequence of this is that clay mixtures with a high kaolinite or high smectite content will have lower potassium radioactivity than clays made up essentially of illite. However, since most clays are mixtures of several clay minerals, the differences discussed earlier are muted. The average shale has a potassium content of about 2–3.5% (Rider, 1991).

Uranium forms unstable soluble salts that are present in sea water and rivers. Uranium content has a positive relationship with the TOC deposited under marine conditions (Fertl and Reike, 1980). In lacustrine settings, due to the paucity of uranium there is not any relationship between uranium and TOC (Bohacs and Miskell-Gerhardt, 1998); therefore, in these cases, GR could be used as a clay volume indicator but not TOC content. It should be noted that the use of uranium is suitable for gas shale reservoirs that do not have uranium-enriched minerals like apatite (Kochenov and Baturin, 2002). In these reservoirs, elevated uranium could not be used to predict TOC.

Unlike (the) uranium, thorium is extremely stable and will rarely pass into solution; thus, its concentration can be directly attributed to the provenance (source area) of the accumulated sediment. The relative immobility of thorium, as a stable, conserved, trace element in the marine environment, compared to the transient mobility of uranium due to fluctuations in oxidation–reduction potential is a relationship that can be used to delineate the possible sequence stratigraphy in the target gas shale layer (Jacobi et al., 2008).

6.4.1.3 Neutron Log Neutron log is a porosity log (NPHI) that measures the amount of hydrogen in a formation. Like the other conventional well log data, neutron log interpretation in the gas shale layers is a complex task and needs many parameters to be considered:

- Hydrogen in the organic matter
- Hydrogen in the structure of clay minerals (hydroxyl groups)
- Hydrogen in water and hydrocarbons present in the formation

Figure 6.9 shows the responses of GR, NPHI, and deep resistivity in the Carynginia Shale which is a potential gas shale layer in the Perth Basin, WA. As can be seen in this figure, due to the clay effect, NPHI log response shows the higher value in the lower and upper Carynginia Shale, while in the middle section of the Carynginia, which is a sandy shale member, NPHI values decrease. It is also expected that NPHI log response will be reduced in the gas shale layers due to the lower hydrogen index (HI) of gas and organic matter compared to water, although quantifying the effects of reducing porosity due to lack of hydrogen in gas and organic matter is quite complex (Glorioso and Rattia, 2012). To some extent, this effect can be observed on the lower Carynginia Shale in Figure 6.9. Between 2435 and 2475 m the resistivity is higher and NPHI is lower compared to the lower part of this section (i.e., between 2475 and 2520 m), possibly due to the presence of gas.

Apart from the aforementioned parameters based on the studies by Zhao et al. (2007) and Labani and Rezaee (2012), neutron porosity decreases with increasing thermal maturity in the gas shale layers. The following explanations can justify this relationship:

- HI of generated hydrocarbons in the final stages of thermal maturity (i.e., gas window) is lower than oil window products; for example, HI of dry gas is less than that of/in wet gas.
- By increasing thermal maturity, smectite converts to illite, and HI of transformed illite is lower than that of/in smectite.
- Reduction of the water saturation at the high thermal maturity levels causes a relatively lower HI values for the shale layers.

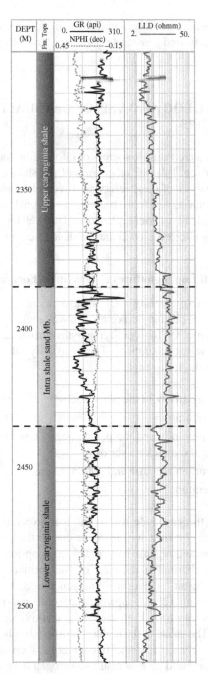

FIGURE 6.9 Typical well log response from a well in the Carynginia Shale, Perth Basin, WA.

6.4.1.4 Density Log The density log measures the formation bulk density. Density log has been used for source rock evaluation for a long time (Schmoker, 1979; Schmoker and Hester, 1983; Passey et al., 1990). The bulk density log data can be successfully employed in porosity modeling of the gas shale layers if the mineral composition (or matrix density) is properly determined using mineralogical tools (Vernik and Milovac, 2011; Alexander et al., 2011).

The density of the organic matter is low (typically 1.1–1.4 g/cm^3) compared to the matrix density (2.6–2.8 g/cm^3) of the shale layers. Due to this low density value, the presence of organic matter can decrease the measured bulk density of the formation. Moreover, high levels of gas content can reduce the bulk density of the gas shale layers. The presence of pyrite (FeS$_2$) and siderite (Fe$_2$CO$_3$) found in the organic-rich shale can elevate formation density. The most favorable environments for siderite formation are reducing freshwater systems (potential environment for kerogen type I), while pyrite commonly occurs in marine sediments (potential environment for kerogen type II) (Passey et al., 2010; Lim et al., 2004).

The density log can give a qualitative indication for estimating the thermal maturity of the gas shale layers as well (Labani and Rezaee, 2012). For example, in some wells of the Perth Basin, there is a decreasing trend for density log responses with increasing thermal maturity in the potential gas shale layers. Although this relationship is not so strong, it seems compatible with what occurs in the gas shale layers during thermal maturity evolution. By increasing thermal maturity in the organic-rich shale layers, the following changes may happen:

- Changes in the type of saturated fluid from brine to gas
- Changing of the heavier components of hydrocarbon into the lighter ones and finally dry gas
- Generation of porosity in the organic matter due to its thermal transformation (Loucks et al., 2009)
- Increase in pore pressure due to mineral transformation (smectite to illite) and hydrocarbon generation

All of these transformations can result in decreasing density of the formation with increasing thermal maturity. It is worth mentioning that sometimes the use of the density log and NPHI log as a thermal maturity indicator is not possible. For example, the presence of heavy minerals could increase the density and hide the decreasing effect of thermal maturity. Therefore, it could be said that conventional logs can only be used for thermal maturity estimation if the lithology of the formation does not vary significantly over the interval of interest.

6.4.1.5 Photoelectric Factor Log

Photoelectric factor (PEF) is a kind of density tool that measures the PEF absorption of a formation. The photoelectric absorption index is used principally for lithological determination. This log is mainly controlled by mean atomic number of the formation. It is slightly influenced by formation porosity; however, the effect is not enough to hinder correct matrix identification when dealing with simple lithologies (one-mineral matrix). PEF log is little affected by the fluid in the pores. Shales have photoelectric values somewhere between limestone and dolomite, but they should be clearly distinguished using (the) GR log. It is believed that organic-rich shales have low photoelectric values compared to normal shales due to the low PEF values of kerogen (Boyer et al., 2006), but there are many mineralogical complexities, and tracking PEF changes versus organic matter is not possible most of the time.

6.4.1.6 Sonic Log

The sonic log measures the speed of sound waves in rocks. Numerical studies suggest that *in situ* rock parameters such as mineral composition and TOC, as well as the interaction among them, can significantly influence the sound wave velocities of the organic-rich rocks. The presence of organic matter in gas shale rocks reduces both the density and the compressional and shear wave velocities, and hence the acoustic impedance, while increasing the velocity anisotropy (Zhu et al., 2011). Besides that, the presence of gas and high clay-bound water, which is common in shales, can decrease the sonic wave velocity.

The main application of acoustic measurements for gas shale evaluation is to provide (the) mechanical properties for gas shale reservoirs. Full waveform sonic log (shear and compressional) can be used for determining the Poisson's ratio, Young's modulus, shear modulus, bulk modulus, yield strength, and compressive strength, all of which are important for determining the brittle shale intervals (i.e., favorable intervals for hydraulic fracturing) (Grieser and Bray, 2007; Alexander et al., 2011). Cross-dipole shear sonic log can be used for determining velocity anisotropy of the gas shale formations. Velocity anisotropy is an important parameter that is of interest in geomechanical applications related to reservoir characterization. A high level of velocity anisotropy is primarily due to the lenticular distribution of kerogen and preferred orientation of clay mineral parallel to the bedding plane (Vernik and Milovac, 2011). Velocity anisotropy can give an idea of the formation permeability due to the higher crack density accompanied with the laminated organic matter.

6.4.1.7 Pulsed Neutron Mineralogy Log

Petrophysical evaluation of unconventional reservoirs mainly depends on determining the mineralogy of the shale layers. The pulsed neutron mineralogy tool is a kind of unconventional log for determining the mineralogy of the formation. This tool, accompanied with a natural GR spectroscopy tool, can determine the concentrations of elements available in the matrix of the gas shale layers including aluminium, carbon, calcium, iron, gadolinium, potassium, magnesium, sulfur, silicon, thorium, titanium, and uranium (Pemper et al., 2006). Each mineral in the matrix requires a specific amount of each element based on stoichiometry. Currently, the following minerals can be quantified by pulsed neutron mineralogy along with the spectral GR: illite, smectite, kaolinite, chlorite, glauconite, apatite, zeolites, halite, anhydrite, hematite, pyrite, siderite, dolomite, calcite, K-feldspar, plagioclase, quartz, and organic carbon (Franquet et al., 2012). This log can give an idea of the geomechanical properties of the shale formations, regarding the defined relationship between mineralogical content and brittleness of the shale layers (refer to Section 6.2.5).

6.4.1.8 Nuclear Magnetic Resonance Log Nuclear magnetic resonance (NMR) log provides useful information for petrophysical study of the hydrocarbon-bearing intervals. Free fluid porosity (effective porosity), rock permeability, and bound fluid volume could be obtained by processing and interpreting the NMR log data (Labani et al., 2010). Unlike neutron, density, and acoustic porosity data which are affected by all components of the reservoir rock, NMR has a signal that contains no contribution from the rock matrix and only responds to the hydrogen associated with pore-filling fluids; because of that, NMR porosity does not need to be calibrated with lithology or lithofacies changes (Coates et al., 1999). This is particularly beneficial in gas shale reservoirs where matrix calibration is difficult due to a high degree of heterogeneity. NMR properties of different reservoir fluids are quite different from each other. These differences make it possible to type hydrocarbons and, as a result, to determine the density of pore-filling fluid (Coates et al., 1999; Sigal and Odusina, 2011). The density of formation fluid can be used later for TOC and kerogen density estimation (refer to Section 6.2.5).

6.4.2 Well Log Interpretation of Gas Shale Formations

The first step for petrophysical evaluation of the gas shale reservoirs is to define a petrophysical model that serves as a basis for well log interpretation. Hitherto, different petrophysical models have been introduced for gas shale reservoirs (Passey et al., 1990, 2010; Ambrose et al., 2010; Ramirez et al., 2011). Like the conventional reservoirs, shales are assumed to consist of two main components: solid matrix and pore space. The organic matter is assumed to be part of the solid matrix. Figure 6.10 shows a simple petrophysical model for the gas shale reservoirs. This model is used for the following well log analysis to determine petrophysical parameters of gas shale reservoirs.

6.4.2.1 Determination of Total Porosity Typical gas shale reservoir porosities are low, often in the range of 3–10%. Porosity calculations using only conventional log measurements may have significant uncertainties due to the variable mineralogies, variable amounts of low-density organic material, and fluids present in these reservoirs (Franquet et al., 2012). Total porosity can be determined using NMR log data. Comparison of NMR total porosities with core porosities in several shale plays has shown good agreement (Jacobi et al., 2009).

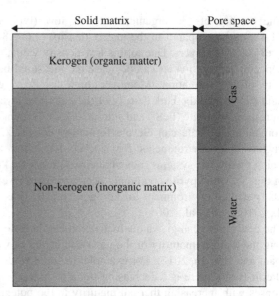

FIGURE 6.10 Simple petrophysical model showing the volumetric constituents of gas shale matrix and pore space.

The density log is commonly used to calculate the total porosity of a formation. Equation 6.6 states the bulk density of a clean rock.

$$\rho_b = \rho_{ma}(1-\phi) + \rho_f \phi \qquad (6.6)$$

Equation 6.6 can be written in the form of Equation 6.7 if a TOC component is added to it (Sondergeld et al., 2010):

$$\rho_b = \rho_{ma}\left(1-\phi-V_{TOC}\right) + \rho_f \phi + \rho_{TOC} V_{TOC} \qquad (6.7)$$

Since the TOC term is generally provided as a weight fraction (w_{TOC}) (e.g., Passey et al., method), the TOC volume fraction (V_{TOC}) has to be converted to a weight fraction:

$$V_{TOC} = \frac{w_{TOC}}{\rho_{TOC}} \rho_b \qquad (6.8)$$

Then Equation 6.8 can be written as follows:

$$\phi = \frac{(\rho_{ma} - \rho_b) + \rho_b\left(w_{TOC} - \rho_{ma}\dfrac{w_{TOC}}{\rho_{TOC}}\right)}{\rho_{ma} - \rho_f} \qquad (6.9)$$

where ϕ is the total density porosity, ρ_{ma} is the solid matrix density, ρ_b is the bulk density, w_{TOC} is TOC weight fraction, ρ_{TOC} is organic material or kerogen density, and ρ_f is the fluid density.

In a simple model, a fixed value can be used for fluid density. It can be assumed to be 0.5 g/cm³ for gas shale and 0.8 g/cm³ for oil shale. However, considering Figure 6.10, fluid density can be estimated using the following equation if water saturation, S_w, is known:

$$\rho_f = \rho_g(1-S_w) + \rho_w S_w \qquad (6.10)$$

Due to the high heterogeneity of the shale layers, it is not possible to determine a specific value for ρ_{ma}, although a default value of 2.65 g/cm³ can be considered since quartz and most clays have a density close to 2.65 g/cm³. But to get an accurate matrix density, it has to be measured using a mineralogy logging tool. Using the percentage of different minerals from the mineralogy logging tool, ρ_{ma} could be computed using the following formula:

$$\rho_{ma} = (1-K)\sum_{i=1}^{n}\left(Min_i \times \rho_i\right) + K\rho_k \quad (6.11)$$

where Min_i and ρ_i are the volume percentage and density of mineral i, respectively, and K and ρ_k are the volume percentage and density of kerogen, respectively.

Considering Figure 6.10, Formula (6.11) could be simplified into the following form:

$$\rho_{ma} = (1-K)\rho_{nk} + K\rho_k \quad (6.12)$$

where ρ_{nk} is the nonkerogen density.

Van Krevelen in 1961 showed that the density of vitrinite varies as a function of thermal maturity (Van Krevelen, 1961). Density may change from 1.27 g/cm³, for low maturity vitrinite, to 2.25 g/cm³ for pure graphite. This is due to loss of volatility in vitrinite. Ward (2010) reported that vitrinite reflectance can be used to estimate kerogen density. Equation 6.13 can be utilized to convert vitrinite reflectance (R_o) to kerogen density:

$$\rho_k = 0.342 R_o + 0.972 \quad (6.13)$$

Total shale porosity can also be calculated from the sonic log using a model similar to the one used for density log. Wyllie et al. (1956) proposed a linear time–average or weighted–average relationship between porosity and transit time for clean and consolidated formations with uniformly distributed small pores:

$$DT = DT_{ma}(1-\phi) + DT_f \phi \quad (6.14)$$

Equation 6.14 can be rewritten if we add a TOC component to it:

$$DT = DT_{ma}\left(1-\phi-V_{TOC}\right) + DT_f \phi + DT_{TOC} V_{TOC} \quad (6.15)$$

Since the TOC term is generally provided as weight fraction (w_{TOC}), it has to be converted to weight fraction (see Eq. 6.8). Then Equation 6.15 can be rearranged as follows:

$$\phi_{sonic} = \frac{(DT - DT_{ma}) + \left(\dfrac{w_{TOC}}{\rho_{TOC}} \times \rho_b\right) \times (DT_{ma} - DT_{TOC})}{DT_f - DT_{ma}}$$

$$(6.16)$$

where DT is rock transit time (us/ft), DT_{ma} = matrix transit time, DT_f = fluid transit time, and DT_{TOC} = kerogen transit time. Kerogen transit time for coal is reported to be approximately 120 (us/ft).

6.4.2.2 Determination of Water Saturation
Considering the complexities of shales, the Archie equation may seem too simple for estimating the water saturation of these kinds of reservoirs. However, the Archie equation has been accepted as an industrial standard for water saturation determination of the gas shale layers based on porosity and resistivity logs:

$$S_w = \sqrt[n]{\frac{aR_w}{\phi^m R_t}} \quad (6.17)$$

Determining some parameters of the Archie equation in gas shale is not as easy as in conventional reservoirs:

- Salinity of the formation water and thus pore water resistivity, R_w
- Archie parameters of the gas shale layers (a, m, and n).

In general, formation water salinity of the shale formations cannot be obtained directly since these layers do not produce formation water normally. Existing data indicates that great variability occurs over short vertical distances; therefore, compared to conventional reservoirs, it is difficult to determine a fixed value for R_w and, as a result, the validity of the Archie formula for estimating water saturation is questionable (Luffel and Guidry, 1992; Sondergeld et al., 2010). Besides that, the Archie model does not differentiate the electrical contribution of different types of water saturating the shale matrix and uses a single value for water resistivity. Obviously, this simplification can turn out to be erroneous when different electrical contribution exists from clay-bound water and free water. With conventional reservoirs, water resistivity value can be obtained in both porous and permeable reservoirs that have a bottom water leg. Glorioso and Rattia (2012) proposed that for gas shale reservoirs, water resistivity could be calculated over non-kerogen intervals (intervals with no kerogen content). Within these intervals, it can be assumed that water saturation would be high because there is not any organic matter for generating hydrocarbon; therefore, the lean shale intervals could be similar to the water saturated intervals in the conventional reservoirs. Cementation exponent (m), saturation exponent (n), and tortuosity factor (a) have been discussed in depth for the conventional reservoirs, but there are limited reviews for the gas shale. In conventional reservoirs, formation water provides paths for electric currents, while in shale formations, due to presence of large amount of interconnected clays accompanied by formation water, there are more paths for them. These extra paths increase the ease of electric current flow in shale (Yu and Aguilera, 2011). This phenomenon would be reflected by a reduction in formation factor and, as a result, in cementation exponent to a value smaller than 2 (Zhao et al., 2007; Ramirez et al., 2011).

In cases where within a shale formation, there are both lean shale intervals and organic-rich shales, the simplified Archie equation, $S_w = (R_o/R_t)^{1/n}$, can be used to quantify gas saturation. Within lean shale intervals where water saturation is basically high, rock resistivity (R_o) is low (similar to the wet zone in a conventional reservoir rock), whereas within TOC-rich, gas-mature shale intervals water saturation is low and thus rock resistivity (R_t) is high in comparison with the lean shale.

The use of this approach for the log data shown in Figure 6.9 has resulted in a very low gas saturation of about 10% for the upper lean part of the shale interval at a depth of about 2310 m and a gas saturation of more than 50% at a depth of about 2450 m for high TOC shale interval. The saturation exponent (n) was considered to be 1.7 based on Luffel and Guidry (1992), who report that a saturation exponent of 1.7 for shales provides a good match to core-derived water saturation.

6.4.2.3 Determination of TOC

There are two main methodologies for *in situ* TOC determination in the gas shale layers: the pulsed neutron mineralogy tool and the Passey ($\Delta \log R$) methodology.

Pulsed Neutron Mineralogy Tool The pulsed neutron mineralogy tool can determine the amount of carbon in the formation. The most important matrix minerals containing carbon are calcite, dolomite, and siderite. Therefore, excess carbon can then be interpreted as organic carbon, hydrocarbon, coal, or organic matter (Jacobi et al., 2009) using the following relationship:

$$C_{TOC} = C_{Measured} - C_{Calcite} - C_{Dolomite} - C_{Siderite} \quad (6.18)$$

The elemental ratio of silicon to carbon determines whether this excess carbon is coal or not. To determine whether the carbon is oil or organic matter, a cut-off value for uranium is used. If the uranium is above the minimum value, the excess carbon is assumed to be organic matter; otherwise, it should be hydrocarbon. The minimum uranium cut-off is from 4 to 7 ppm for most gas shale layers (Pemper et al., 2009). Measuring *in situ* carbon for TOC estimation using the pulsed neutron mineralogy tool is preferable compared to other techniques where TOC is determined from well log data.

Passey ($\Delta \log R$) Methodology This is a practical methodology first developed by Passey et al. (1990) for identifying and calculating TOC in organic-rich rocks using well logs. This method employs overlaying of a properly scaled porosity log (generally the sonic transit time curve) on a resistivity curve (preferably from a deep reading tool) and then calculating the separation between these two curves by defining a baseline:

$$\Delta \log R = \log_{10}\left(\frac{R}{R_{baseline}}\right) + 0.02 \times (\Delta t - \Delta t_{baseline}) \quad (6.19)$$

Baseline is determined when sonic and resistivity directly overlaid each other or they just tracked each other. According to the assumptions of this technique, this condition will exist at the organic-lean interval. The amount of TOC can then be determined from the following relationship by knowing the level of maturity (LOM):

$$TOC = \Delta \log R \times 10^{(2.297 - 0.1688 \times LOM)} \quad (6.20)$$

Although this methodology is used extensively for TOC determination in the shale layers, there are many uncertainties in its evaluation. This method requires similar clay minerals or similar conductive minerals (e.g., pyrite) in both organic-lean shale (baseline) and the organic-rich interval. Extensive vertical heterogeneity of the shale layers may result in very high uncertainty for the calculated TOC. Moreover, this method requires knowledge of the LOM for converting the apparent $\Delta \log R$ to a quantitative TOC. In exploration wells the LOM may not be known or may also change with depth (Pemper et al., 2009).

Furthermore, according to the $\Delta \log R$ technique, an increase in the resistivity and sonic transit time is also a function of hydrocarbon saturation. Passey et al. (1990) concluded that an increase in the amount of hydrocarbon at the higher thermal maturity level could be correlated to the present TOC content of the rock. However, this assumption seems not to be correct all of the time. Theoretically, the amount of hydrocarbon in the pores relies on both the maturity level and initially deposited TOC (iTOC), and not on the amount of TOC present in the rock. Analysis of the data reported by Modica and Lapierre (2012) confirms this idea. As can be seen in Figure 6.11, for the data points reported for the Mowery Shale in the Powder River Basin of Wyoming, thermal maturity and initial TOC have the higher effect on the generated hydrocarbons than the present TOC. Therefore, generated hydrocarbons and, as a result, separation between sonic and resistivity logs, could be correlated to iTOC and not present TOC. Although there is a relationship between iTOC and TOC, this relationship is not a global relationship and depends on the thermal maturity of the data points (Fig. 6.12), and therefore should be (separately) determined for different case studies.

6.4.2.4 Determination of Kerogen Density

Kerogen density can be determined from geochemical data but if geochemical data is not available it can be determined using the following log-based procedure which uses NMR and density logs accompanied by the pulsed neutron mineralogy data (Jacobi et al., 2008; Vernik and Milovac, 2011):

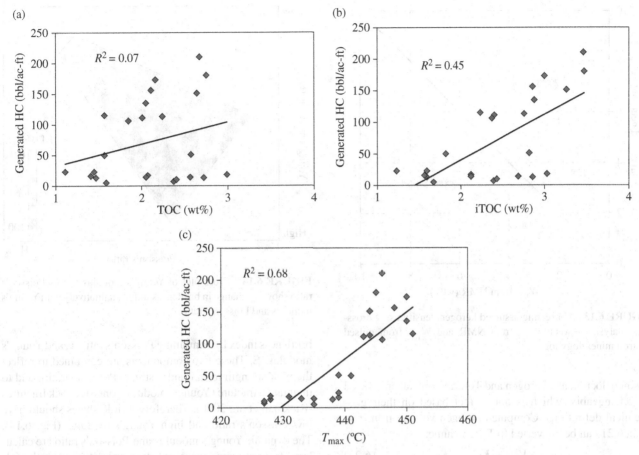

FIGURE 6.11 Cross-plot analysis between generated hydrocarbons with (a) TOC, (b) iTOC, and (c) thermal maturity indicator (T_{max}) (data from Modica and Lapierre, 2012).

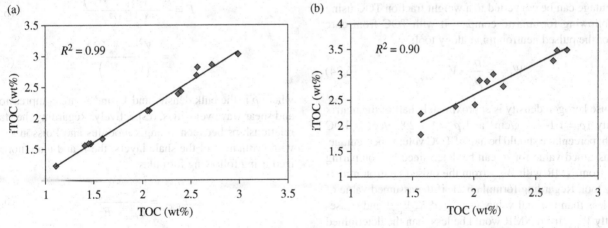

FIGURE 6.12 Cross-plot analysis between iTOC and TOC for (a) data points in oil window and (b) gas window.

Kerogen volume could be computed using the below formula by rearranging formula 6.12:

$$K = \frac{\rho_{nk} - \rho_m}{\rho_{nk} - \rho_k} \quad (6.21)$$

ρ_{nk} is determined using mineralogy logs, ρ_k is assumed as a fixed value between 1.1 and 1.4 g/cm^3, and ρ_m is computed from bulk density and total NMR porosity:

$$\rho_m = \frac{\rho_b - \rho_f \phi}{1 - \phi} \quad (6.22)$$

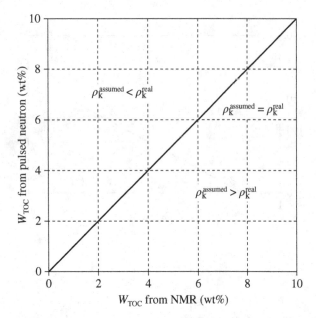

FIGURE 6.13 Correcting assumed kerogen density using cross-plot analysis between W_{TOC} from NMR and W_{TOC} from pulsed neutron mineralogy log.

In some documents, kerogen and TOC are (mistakenly) used interchangeably, which is not correct based on their geochemical definitions. Computed kerogen volume from formula 6.21 can be converted to TOC volume:

$$V_{TOC} = KC_k \quad (6.23)$$

C_k is a constant value between 0.7 and 0.85 and depends on the maturity level and type of kerogen. Finally, TOC volume percentage can be converted to a weight fraction TOC using the following formula to compare it with TOC from core data or the pulsed neutron mineralogy tool:

$$W_{TOC} = \frac{\rho_k}{\rho_m} \times V_{TOC} \quad (6.24)$$

Because kerogen density is approximately half of the matrix density ($\rho_k = 1.1-1.4$ g/cm³ and $\rho_m = 2.6-2.8$ g/cm³), TOC weight percentage should be half of TOC volume percentage. The assumed value for ρ_k can be determined by comparing W_{TOC} from NMR with W_{TOC} from the pulsed neutron mineralogy tool. Regarding formula 6.21, if the assumed value of ρ_k is less than the real value, estimated K, V_{TOC}, and consequently W_{TOC} from NMR would be less than the determined value from the mineralogy log and vice versa. If the assumed value of kerogen density is correct, then these two curves should provide similar results (Fig. 6.13).

6.4.2.5 Determination of Brittleness Index

Brittle shales are more likely to be naturally fractured and will also be more likely to respond well to hydraulic fracturing treatments. Rickman et al. (2008) and Grieser and Bray (2007) defined a

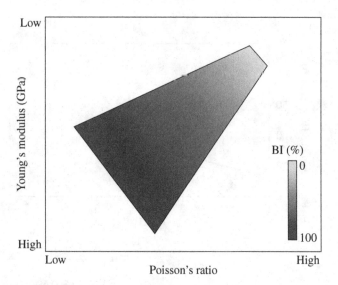

FIGURE 6.14 Cross plot of Young's modulus versus Poisson's ratio showing change in brittleness index qualitatively with Young's modulus and Poisson's ratio.

brittleness index by combining Poisson's ratio, ν, and Young's modulus, E. These two components are combined to reflect the rock strength to fail under stress (Poisson's ratio) and to maintain a fracture (Young's modulus) once the rock fractures (Rickman et al., 2008). Therefore, brittle shales should have low Poisson's ratio and high Young's modulus (Fig. 6.14). The dynamic Young's modulus and Poisson's ratio are calculated from compressional and shear velocities using the following formulas:

$$E = \frac{\rho V_s^2 \left(3V_p^2 - 4V_s^2\right)}{V_p^2 - V_s^2} \quad (6.25)$$

$$\nu = \frac{V_p^2 - 2V_s^2}{2\left(V_p^2 - V_s^2\right)} \quad (6.26)$$

where ρ is the bulk density and V_p and V_s are compressional and shear wave velocities, respectively. Regarding the stated relationships between Young's modulus and Poisson's ratio with brittleness of the shale layers, the E and ν are normalized using following formulas:

$$E_{brittle} = \frac{E - E_{min}}{E_{max} - E_{min}} \quad (6.27)$$

$$\nu_{brittle} = \frac{\nu - \nu_{max}}{\nu_{min} - \nu_{max}} \quad (6.28)$$

The brittleness index (BI) is then defined as average of $E_{brittle}$ and $\nu_{brittle}$:

$$BI_{sonic} = \frac{E_{brittle} + \nu_{brittle}}{2} \times 100 \quad (6.29)$$

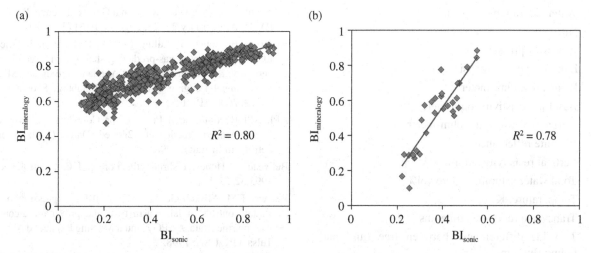

FIGURE 6.15 Cross-plot analysis between $BI_{mineralogy}$ and BI_{sonic} from two wells in Perth Basin, WA. $BI_{mineralogy}$ calculated using mineralogy log data for well (a) and XRD analysis data for well (b).

As is clear, this analysis requires full wave-form sonic data, including shear wave and compressional wave velocities. In some cases, shear velocity is not available in the data set; therefore, it should be estimated from the compressional velocity data. Using available empirical formulas for the shale layers like Castagna et al. (1985) formula, shear velocity can be estimated as follows:

$$V_s = 0.862 \times V_p - 1.172 \tag{6.30}$$

where V_p and V_s are compressional and shear velocities respectively, in kilometers per second.

Figure 6.15 shows the cross-plot analysis between BI_{sonic} and $BI_{mineralogy}$ using formula 6.1 for two wells in the Perth Basin, WA. As (it is) expected, although there is a relatively good correlation coefficient between brittleness indexes from different methodologies, for both cases $BI_{mineralogy}$ is higher than BI_{sonic}. And as mentioned earlier, brittleness is a complex function of different parameters, not only mineralogy; therefore, it seems that BI_{sonic} could give a better idea for determining prospect layers for doing hydraulic fracturing.

6.4.2.6 Determination of Velocity Anisotropy

Velocity anisotropy can take account of stratification, which is very important for evaluating potential of the shale layers for hydraulic fracturing. The relative anisotropy of the formation could be computed by measuring the difference between shear wave velocity in X and Y directions (Tang et al., 2001). The difference between V_{sx} and V_{sy} can give an idea about the relative anisotropy of the formation, which, for simplicity, could be called the anisotropy index. The anisotropy indices in X and Y directions are calculated by the following formulas, respectively:

$$\text{Anisotropy index}_x = \frac{|V_{sx} - V_{sy}|}{V_{sx}} \tag{6.31}$$

$$\text{Anisotropy index}_y = \frac{|V_{sx} - V_{sy}|}{V_{sy}} \tag{6.32}$$

Finally, anisotropy of the formation can be calculated by taking a simple average from the anisotropy indices in X and Y directions:

$$\text{Average anisotropy} = \frac{\text{Anisotropy index}_x + \text{Anisotropy index}_y}{2} \tag{6.33}$$

NOMENCLATURE

a	Archie equation constant
°C	Degree Celsius
C_k	Stoichiometric constant relating kerogen to TOC in Formula
CO_2	Carbon dioxide
DT	Sonic transit time, μs/ft
DT_f	Fluid transit Time, μs/ft
DT_{ma}	Matrix transit Time, μs/ft
DT_{TOC}	Kerogen transit time, μs/ft
E	Young's modulus, Pa
G_s	Adsorbed gas capacity, scf/ton
K	Volume percent of kerogen, vol%
m	Cementation exponent

n	Water saturation exponent
N_2	Nitrogen
P	Reservoir pressure, psi
P_l	Langmuir pressure, psi
R	Resistivity, ohm-meter
$R_{baseline}$	Baseline resistivity, ohm-meter
R_h	Horizontal resistivity, ohm-meter
R_o	Vitrinite reflectance
R_v	Vertical resistivity, ohm-meter
S_w	Total water saturation, pore vol%
T	Temperature, K
T_2	Transverse relaxation time, ms
$T_{2cutoff}$	T_2 value differentiates between free fluid and bound fluid, ms
T_{2gm}	Geometrical mean of T_2 distribution, ms
V_l	Langmuir volume, scf/ton
V_p	Compressional wave velocity, m/s
V_s	Shear wave velocity, m/s
V_{sx}	Shear velocity in X direction, m/s
V_{sy}	Shear velocity in Y direction, m/s
V_{TOC}	Volume percentage of TOC, vol%
W_{TOC}	Weight percentage of TOC, wt%
$\Delta \log R$	Separation between resistivity and sonic, log scale
Δt	Sonic transit time, microseconds/ft
$\Delta t_{baseline}$	Baseline sonic transit time, microseconds/ft
μm	Micrometer
ν	Poisson's ratio
ρ_i	Density of mineral i, g/cm^3
ρ_b	Bulk density, g/cm^3
ρ_f	Fluid density, g/cm^3
ρ_g	Gas density, g/cm^3
ρ_m	Solid matrix density, g/cm^3
ρ_w	Formation water density, g/cm^3
ρ_k	Kerogen density, g/cm^3
ρ_{nk}	Non-kerogen density, g/cm^3
ϕ	Porosity, decimal fraction

REFERENCES

Alexander T, Baihly J, Boyer C, Clark B, Waters,G, Jochen V, Calvez JL, Lewis R, Miller CK, Thaeler J, Toelle BE. Gas shale revolution. Oilfield Rev 2011;23:40–55.

Allen N, Aplin A, Thomas M. Introduction to gas shale storage [Internal presentation]. University of Calgary; 2009.

Ambrose RJ, Hartman RC, Campos MD, Akkutlu IY, Sondergeld C. New pore-scale considerations for gas shale in place calculations. SPE Unconventional Gas Conference; Pittsburgh, PA, USA, February 23–25, 2010. SPE131772; 2010.

Anderson B, Barber T, Lüling M, Sen P, Taherian R, Klein J. Identifying potential gas-producing shales from large dielectric permittivities measured by induction quadrature signals. SPWLA 49th Annual Logging Symposium; Edinburg, Scotland, May 25–28, 2008, SPWLA.

API. API Recommended Practice for Core-Analysis Procedure, Recommended Practice 40. 2nd ed. Dallas (TX): American Petroleum Institute; 1998.

Beliveau D. Honey, I shrunk the pores!. J Can Petrol Technol 1993;32:15–17.

Bohacs, KM, Miskell-Gerhardt K. Well-log expression of lake strata; controls of lake-basin type and provenance, contrasts with marine strata. AAPG Annual Meeting Expanded Abstracts, Tulsa, OK, USA; 1998.

Boyer C, Kieschnick J, Suarez-Rivera R, E, Lewis R, Waters G. Producing gas from its source. Oilfield Rev 2006;18:36–49.

Brace WF, Walsh JB, Frangos WT. Permeability of granite under high pressure. J Geophys Res 1968;73:2225–2236.

Bustin RM, Bustin AMM, Cui X, Ross D, Pathi VM. Impact of shale properties on pore structure and storage characteristics. SPE Gas Shale Production Conference; Fort Worth, TX, USA. SPE119892; November 16–18, 2008.

Bustin RM, Clarkson CR. Geological controls on coal-bed methane reservoir capacity and gas content. Int J Coal Geol 1998;38:3–26.

Castagna JP, Batzle ML, Eastwood RL. Relationships between compressional-wave and shear-wave velocities in clastic silicate rocks. Geophysics 1985;50:571–581.

Chalmers GR, Bustin RM, Power IM. Characterization of gas shale pore systems by porosimetry, pycnometry, surface area, and field emission scanning electron microscopy/transmission electron microscopy image analyses: Examples from the Barnett, Woodford, Haynesville, Marcellus, and Doig units. AAPG Bulletin 2012;96:1099–1119.

Chemali R, Su SM, Gianzero S. The effect of shale anisotropy on focused resistivity devices. SPWLA 28th Annual Logging Symposium. Society of Petrophysicists & Well Log Analysts, London, England; June 29–July 2, 1987.

Clarkson CR, Freeman M, He L, Agamalian M, Melnichenko YB, Mastalerz M, Bustin RM, Radliński AP, Blach TP. Characterization of tight gas reservoir pore structure using USANS/SANS and gas adsorption analysis. Fuel 2012a;95:371–385.

Clarkson CR, Jensen JL, Blasingame T. Reservoir engineering for unconventional reservoirs: What do we have to consider? North American Unconventional Gas Conference and Exhibition; The Woodlands, TX, USA. SPE145080; June 14–16, 2011.

Clarkson CR, Solano N, Bustin RM, Bustin AMM, Chalmers GRL, He L, Melnichenko YB, Radliński AP, Blach TP. Pore structure characterization of North American gas shale reservoirs using USANS/SANS, gas adsorption, and mercury intrusion. Fuel 2013;103:606–616.

Clarkson CR, Wood J, Burgis S, Aquino S, Freeman M. Nanopore-structure analysis and permeability predictions for a tight gas siltstone reservoir by use of low-pressure adsorption and mercury-intrusion techniques. SPE Reserv Eval Eng 2012b;15:648–661.

Cluff RM, Shanley KW, Miller MA. Three things we thought we understood about gas shale, but were afraid to ask. AAPG Annual Convention, Long Beach, CA, USA; April 1–4, 2007.

Coates GR, Xiao L, Prammer MG. *NMR Logging: Principles and Applications*. Houston (TX): Haliburton Energy Services Publication; 1999.

Cui X, Bustin AMM, Bustin RM. Measurements of gas permeability and diffusivity of tight reservoir rocks: Different approaches and their applications. Geofluids 2009;9: 208–223.

Curtis ME, Ambrose RJ, Sondergeld CH. Structural characterization of gas shale on the micro- and nano-scales. Canadian Unconventional Resources and International Petroleum Conference; Calgary, AB, Canada. SPE137693; October 19–21, 2010.

Darabi H, Ettehad A, Javadpour F, Sepehrnoori K. Gas flow in ultra-tight shale strata. J Fluid Mech 2012;710:641–658.

Dresser Atlas. *Log Interpretation Charts*. Houston (TX): Dresser Atlas Publication; 1983.

Easley TG, Sigal R, Rai C. Thermogravimetric analysis of Barnett shale samples. International Symposium of the Society of Core Analysts; Calgary, AB, Canada, September 10–12, 2007. Society of Core Analyst (SCA).

Fertl WH, Rieke HH. Gamma ray spectral evaluation techniques identify fractured shale reservoirs and source-rock characteristics. J Petrol Technol 1980;32:2053–2062.

Franquet JA, Bratovich MW, Glass RD. State-of-the-Art Openhole gas shale logging. SPE Saudi Arabia Section Technical Symposium and Exhibition; Al-Khobar, Saudi Arabia. SPE160862; 2012.

Giesche H. Mercury porosimetry: A general (practical) overview. Particle Syst Character 2006;23:9–19.

Glorioso JC, Rattia AJ. Unconventional reservoirs: Basic petrophysical concepts for gas shale. SPE/EAGE European Unconventional Resources Conference and Exhibition; Vienna, Austria. SPE153004; March 20–22, 2012. SPE/EAGE.

Grieser WV, Bray JM. Identification of production potential in unconventional reservoirs. Production and Operations Symposium; Oklahoma City, OK, USA. SPE106623; March 31–April 3, 2007. SPE.

Handwerger DA, Keller J, Vaughn K. Improved petrophysical core measurements on tight shale reservoirs using retort and crushed samples. SPE Annual Technical Conference and Exhibition; Denver, CO, USA. SPE147456; October 30–November 2, 2011. Society of Petroleum Engineers (SPE).

Handwerger DA, Willberg DM, Pagels M, Rowland B, Keller J. Reconciling retort versus dean stark measurements on tight shales. SPE Annual Technical Conference and Exhibition, San Antonio, TX, USA. SPE159976; October 8–10, 2012. SPE.

Jacobi DJ, Breig JJ, Le Compte B, Kopal M, Hursan G, Mendez FE, Bliven S, Longo J. Effective geochemical and geomechanical characterization of gas shale reservoirs from the wellbore environment: Caney and the Woodford Shale. SPE Annual Technical Conference and Exhibition; New Orleans, LA, USA. SPE124231; October 4–7, 2009. SPE.

Jacobi DJ, Gladkikh M, LeCompte B, Hursan G, Mendez F, Longo J, Ong S, Bratovich M, Patton GL, Shoemaker P. Integrated petrophysical evaluation of gas shale reservoirs. CIPC/SPE Gas Technology Symposium—Joint Conference, Calgary, AB, Canada. SPE114925; June 16–19, 2008. SPE.

Jarvie DM, Hill RJ, Ruble TE, Pollastro RM. Unconventional shale-gas systems: The Mississippian Barnett Shale of north-central Texas as one model for thermogenic shale-gas assessment. AAPG Bull 2007;91:475–499.

Javadpour F. Nanopores and apparent permeability of gas flow in mudrocks (Shales and Siltstone). J Can Petrol Technol 2009;48:16–21.

Javadpour F, Fisher D, Unsworth M. Nanoscale Gas Flow in Gas shale Sediments. J Can Petrol Technol 2007;46:55–61.

Jones SC. A Technique for faster pulse-decay permeability measurements in tight rocks. SPE Formation Eval 1997;12: 19–26.

Karastathis A. Petrophysical measurements on tight gas shale [MSc, thesis]. University of Oklahoma; 2007. 117 pp.

Keller JU, Staudt R. Gas adsorption equilibria: Experimental methods and adsorptive isotherms. New York: Springer; 2005. 422 pp.

Keller LM, Holzer L, Wepf R, Gaser P. 3D geometry and topology of pore pathways in Opalinus clay: Implications for mass transport. Appl Clay Sci 2011;52:85–95.

Kochenov AV, Baturin GN. The paragenesis of organic matter, phosphorus, and uranium in marine sediments. Lithol Mine Resour 2002;37:107–120.

Labani MM, Kadkhodaie-Ilkhchi A, Salahshoor K. Estimation of NMR log parameters from conventional well log data using a committee machine with intelligent systems: A case study from the Iranian part of the South Pars gas field, Persian Gulf Basin. J Petrol Sci Eng 2010;72:175–185.

Labani MM, Rezaee, R. Thermal maturity estimation of gas shale layers from conventional well log data: A case study from Kockatea Shale and Carynginia Formation of Perth Basin, Australia. SPE Asia Pacific Oil and Gas Conference and Exhibition, Perth, WA, Australia. SPE158864; October 22–24, 2012. SPE.

Labani MM, Rezaee R, Saeedi A, Al-Hinai A. Evaluation of pore size spectrum of gas shale reservoirs using low pressure nitrogen adsorption, gas expansion and mercury porosimetry: A case study from the Perth and Canning Basins, Western Australia. J Petrol Sci Eng 2013;112:7–16.

Lim DI, Jung HS, Yang SY, Yoo HS. Sequential growth of early diagenetic freshwater siderites in the Holocene coastal deposits, Korea. Sedimentary Geol 2004;169:107–120.

Loucks RG, Reed RM, Ruppel SC, Hammes U. Spectrum of pore types and networks in mudrocks and a descriptive classification for matrix-related mudrock pores. AAPG Bull 2012;96:1071–1098.

Loucks RG, Reed RM, Ruppel SC. Jarvie DM. Morphology, genesis, and distribution of nanometer-scale pores in siliceous mudstones of the Mississippian Barnett Shale. J Sedimentary Res 2009;79:848–861.

Lu XC, Li FC, Watson AT. Adsorption measurements in Devonian shales. Fuel 1995;74:599–603.

Luffel DL. Advances in shale core analysis. Gas Research Institute Report. GRI-93/0297. Houston (TX): Gas Research Institute; 1993. 138 pp.

Luffel DL, Guidry FK. New core analysis methods for measuring reservoir rock properties of Devonian Shale. SPE J Petrol Technol 1992;44:1184–1190.

Miller M. Gas shale evaluation techniques: things to think about. Internal Workshop on 'New perspectives on shales'. Oklahama University, USA; 2010.

Modica CJ, Lapierre SG. Estimation of kerogen porosity in source rocks as a function of thermal transformation: Example from the Mowry Shale in the Powder River Basin of Wyoming. AAPG Bull 2012;96:87–108.

Nelson PH. Pore-throat sizes in sandstones, tight sandstones, and shales. AAPG Bull 2009;93:329–340.

Passey QR, Bohacs K, Esch WL, Klimentidis R, Sinha S. From oil-prone source rock to gas-producing shale reservoir—geologic and petrophysical characterization of unconventional gas shale reservoirs. International Oil and Gas Conference and Exhibition in China; Beijing, China. SPE131350; June 8–10, 2010.

Passey QR, Creaney S, Kulla JB, Moretti FJ, Stroud JD. A practical model for organic richness from porosity and resistivity logs. AAPG Bull 1990;74:1777–1794.

Pemper RR, Han X, Mendez FE, Jacobi D, LeCompte B, Bratovich M, Feuerbacher G, Bruner M, Bliven S. The Direct measurement of carbon in wells containing oil and natural gas using a pulsed neutron mineralogy tool. SPE Annual Technical Conference and Exhibition, New Orleans, LA, USA. SPE124234; October 4–7, 2009.

Pemper RR, Sommer A, Guo P, Jacobi D, Longo J, Bliven S, Rodriguez E, Mendez F, Han, X. A new pulsed neutron sonde for derivation of formation lithology and mineralogy. SPE Annual Technical Conference and Exhibition, San Antonio, TX, USA. SPE102770; September 24–27, 2006. SPE.

Pittman ED. Relationship of porosity and permeability to various parameters derived from mercury injection-capillary pressure curves for sandstone. AAPG 1992;76 (2):191–198.

Quantachrome. *Autosorb As-1/Aswin Gas Sorption System Operation Manual*. Boynton Beach (FL): Quantachrome Instruments; 2008.

Ramirez TR, Klein JD, Bonnie R, Howard JJ. Comparative study of formation evaluation methods for unconventional gas shale reservoirs: Application to the Haynesville Shale (Texas). North American Unconventional Gas Conference and Exhibition; The Woodlands, TX, USA. SPE144062; June 14–16, 2011. SPE.

Rezaee R, Saeedi A, Clennell B. Tight gas sands permeability estimation from mercury injection capillary pressure and nuclear magnetic resonance data. J Petrol Sci Eng 2012; 88–89:92–99.

Rezaee MR, Jafari A, Kazemzadeh E. Relationships between permeability, porosity and pore throat size in carbonate rocks using regression analysis and neural networks. J Geophys Eng 2006;3:370–376.

Rickman R, Mullen MJ, Petre JE, Grieser WV, Kundert D. A practical use of shale petrophysics for stimulation design optimization: All shale plays are not clones of the Barnett Shale. SPE Annual Technical Conference and Exhibition; Denver, CO, USA. SPE115258; September 21–24, 2008. SPE.

Rider MH. *The Geological Interpretation of Well Logs*. 2nd ed. Scotland: Whittles Publishing; 1991.

Ross DJK. Investigation into the importance of geochemical and pore structure heterogeneities for gas shale reservoir evaluation [PhD Thesis]. University of British Columbia, BC, Canada; 2007. 373 pp.

Ross DJK, Bustin RM. Characterizing the gas shale resource potential of Devonian-Mississipian strata in the Western Canada sedimentary basin: Application of an integrated formation evaluation. AAPG Bull 2007a;92:87–125.

Ross DJK, Bustin RM. Impact of mass balance calculations on adsorption capacities in microporous gas shale reservoirs. Fuel 2007b;86:2696–2706.

Ross DJK, Bustin RM. The importance of shale composition and pore structure upon gas storage potential of gas shale reservoirs. Mar Petrol Geol 2009;26:916–927.

Rouquerol J, Avnir D, Fairbridge CW, Everett DH, Haynes JH, Pernicone N, Ramsay JDF, Sing KSW, Unger K. Recommendations for the characterization of porous solids. International Union of Pure and Applied Chemistry. Pure Appl Chem 1994;68:1739–1758.

Schmoker JW. Determination of organic content of Appalachian Devonian shales from formation-density logs. AAPG Bull 1979;63:1504–1509.

Schmoker JW, Hester TC. Organic carbon in Bakken Formation, United States portion of Williston Basin. AAPG Bull 1983;67:2165–2174.

Serra O. Clay, Silt, Sand, Shales: A Guide for Well-Log Interpretation of Siliciclastic Deposits. Schlumberger Publication; 1988. p 609.

Sigal R, Odusina E. Laboratory NMR measurements on methane saturated Barnett Shale samples. Petrophysics 2011;52: 32–49.

Soeder DJ. Porosity and permeability of Eastern Devonian gas shale. SPE Formation Eval 1988;3:116–124.

Sondergeld CH, Newsham KE, Comisky JT, Rice MC, Rai CS. Petrophysical considerations in evaluating and producing gas shale resources. SPE Unconventional Gas Conference; Pittsburgh, PA, USA. SPE131768; February 23–25, 2010. SPE.

Swanson BF. A simple correlation between permeabilities and mercury capillary pressures. J Petrol Technol 1981;33: 2498–2504.

Tang XM, Patterson D, Hinds M. Evaluating hydraulic fracturing in cased holes with cross-dipole acoustic technology. SPE Reserv Eval Eng 2001;4:281–288.

Van Krevelen DW. Coal: typology-chemistry-physics-constitution. Amsterdam: Elsevier Science; 1961. 514 p.

Vernik L, Milovac J. Rock physics of organic shales. Leading Edge 2011;30:318–323.

Ward J. Kerogen density in the Marcellus Shale. SPE 131767 presented at SPE Unconventional Gas Conference. Pittsburgh, PA, USA, February 23–25, 2010.

Wang FP, Gale JF. Screening criteria for shale-gas systems. Gulf Coast Assoc Geological Soc Trans 2009;59:779–793.

Wang FP, Reed RM. Pore networks and fluid flow in gas shale. SPE Annual Technical Conference and Exhibition; New Orleans, LA, USA. SPE124253; October 4–7, 2009. SPE.

Wyllie MRJ, Gregory AR, Gardner LW. Elastic wave velocities in heterogeneous and porous media. Geophysics 1956;21:41–70.

Yu G, Aguilera R. Use of Pickett plots for evaluation of gas shale formations. SPE Annual Technical Conference and Exhibition; Denver, CO, USA. SPE 146948; October 30–November 2, 2011. Society of Petroleum Engineers.

Zhang T, Ellis GS, Ruppel SC, Milliken K, Yang R. Effect of organic-matter type and thermal maturity on methane adsorption in shale-gas systems. Org Geochem 2012;47:120–131.

Zhao H, Givens NB, Curtis B. Thermal maturity of the Barnett Shale determined from well-log analysis. AAPG Bull 2007;91:535–549.

Zhu Y, Liu E, Martinez A, Payne MA, Harris CE. Understanding geophysical responses of shale-gas plays. Leading Edge 2011;30:332–338.

7

PORE PRESSURE PREDICTION FOR SHALE FORMATIONS USING WELL LOG DATA

ABUALKSIM AHMAD AND REZA REZAEE

Department of Petroleum Engineering, Curtin University, Perth, WA, Australia

7.1 INTRODUCTION

Pore pressure in any sedimentary formation is defined as the pressure of the fluid contained in the pore space of the rocks, and can be either normal or abnormal pressure. Abnormal pressure is subclassified into abnormal high pressure (overpressure) and subnormal pressure. Knowledge of pore pressure regimes in any sedimentary basins is an integral part of the formation evaluation process in gas shale formations (Gretener, 1979). Appropriate evaluation of pore pressure is crucial for drilling and completion planning (Tingay et al., 2003).

In this chapter, definitions of important pore pressure-related terms are presented first, and then overpressure-generating mechanisms are explained in detail followed by overpressure estimation methods. In addition, the relationships between pore pressure distribution and tectonic elements in sedimentary basins are presented herein. These relationships were observed in a recent study that was conducted on the potential gas shale formations in the Perth Basin, Western Australia. Finally, the origins of overpressure in these shale intervals are explained and some examples are presented.

7.1.1 Normal Pressure

The normal hydrostatic pressure at any depth is defined as being the pore pressure equivalent to the hydrostatic pressure due to an open column of pore fluids that reaches from the surface to the vertical depth of the formation. In normally pressured formations, pore fluids communicate efficiently with the surface during burial. Therefore, the pore fluids are squeezed out by the normal compaction, and as a result, a normal hydrostatic pressure regime is established. In normally pressured sediments, the vertical effective stress continues to increase as the depth increases. The normal hydrostatic pressure and gradient can be calculated by using Equations 7.1 and 7.2, respectively, and graphic illustration of the normal pore pressure regime is presented in Figure 7.1:

$$P = \rho_w \times g \times z \quad (7.1)$$

In terms of pressure gradient,

$$\frac{dP}{dz} = \rho_w \times g \quad (7.2)$$

where P is the pore fluid pressure, ρ_w is the pore water density, g is the gravitational acceleration, and z is the vertical depth of the formation. For freshwater with a density of 1 g/cm³, the hydrostatic pressure gradient is 0.433 psi/ft.

7.1.2 Overpressure

Overpressure is defined as a formation pressure that is greater than the normal hydrostatic pressure of a column of pore fluids that reaches from the surface to the vertical depth of the formation. Sediment compaction is mainly caused by an increase in overburden stress, and the theory of compaction was well described by Terzaghi et al. (1996). The authors established an equation of equilibrium (Eq. 7.3). It was

Fundamentals of Gas Shale Reservoirs, First Edition. Edited by Reza Rezaee.
© 2015 John Wiley & Sons, Inc. Published 2015 by John Wiley & Sons, Inc.

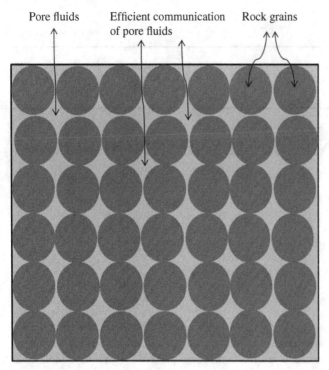

FIGURE 7.1 Example of a situation where pore fluids communicate efficiently and develop normal pore pressure regime in sedimentary basins.

discovered that the overburden stress S is supported by pore pressure P and the effective stress σ. An example of an overpressure situation is when the pore fluids are trapped within the pore space due to a lack of communications (low permeability barriers) between the sediments that are being compacted and the overlying sediments. This process is referred to as compaction disequilibrium or under-compaction.

$$S = \sigma + P \qquad (7.3)$$

It can be observed from Equation 7.3 that if overburden stress S at a certain point increases and water is allowed to escape, the effective stress σ increases and pore pressure P remains constant at hydrostatic pressure. However, if water is not allowed to escape as overburden stress S increases, both pore pressure P and the effective stress σ increase.

Overpressure is also generated by the increase in volume resulting from the expansion of the pore fluids such as hydrocarbon generation, heating, and expulsion/expansion of intergranular water during clay diagenesis. In fluid expansion processes, overpressure develops as the rock matrix restricts the escape of the pore fluids as the latter increase in volume.

A typical method of plotting pore pressure/stresses versus depth is illustrated in Figure 7.2. As illustrated in this real example, pore pressure P increases hydrostatically up to the top of the overpressure zone. Then the pore

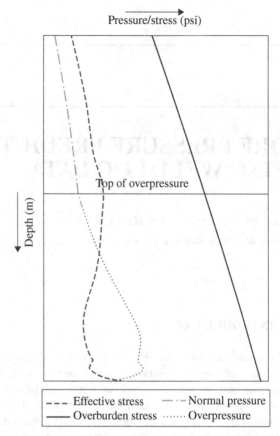

FIGURE 7.2 Real example of pore pressure and vertical stresses as functions of depth.

pressure increases abnormally near the base of the overpressure zone and returns to the normal trend below the overpressured section. The vertical effective stress σ may increase or decrease depending on the overpressure-generating mechanisms. In this particular example, the vertical effective stress decreases due to fluid expansion mechanisms which will be discussed in detail in Section 7.2.2.

7.2 OVERPRESSURE-GENERATING MECHANISMS

An accurate prediction of pore pressure involves proper understanding of overpressure-generating mechanisms as different origins of overpressure have different signatures on petrophysical properties of the formations. The theories of overpressure-generating mechanisms were well described by several authors such as Watts (1948), Draou and Osisanya (2000), Shunhua et al. (2006), and Butler (2011). The two main generating mechanisms of overpressure in sedimentary rocks can be classified as (1) loading mechanisms, for example, under-compaction (compaction disequilibrium) and lateral tectonics compression, and (2) unloading mechanisms (fluid expansion), for example, hydrocarbon generation, clay transformation, and aqua-thermal heating. Most researchers have observed that the under-compaction (compaction disequilibrium) accounts for the majority of overpressure situations that were encountered in sedimentary rocks. In under-compaction situations, overpressure is a result of the rapid loading of sediments with a lack of communication between the pore fluids and the overlaying sediments. Hence, pore fluids are trapped and become overpressured (Osborne and Swarbrick, 1997). The main overpressure-generating mechanisms are discussed in detail in Sections 7.2.1 and 7.2.2.

7.2.1 Loading Mechanisms

Loading mechanisms involve increases in compressive stresses. Loading mechanisms include under-compaction (compaction disequilibrium) where the sediments compact vertically and also include lateral loading (tectonic compression) where the sediments compact horizontally in tectonically active areas.

7.2.1.1 Under-Compaction (Compaction Disequilibrium)

In normal sedimentary environments, sediments compact and lose porosity as a result of an increase in the effective stress (grain to grain contact). Normal compaction creates efficient communication between the pore spaces and the water table; and hence, some of the pore fluids are squeezed out as a result of a normal increase in overburden pressure. Therefore, a normal pore pressure regime is established. This pressure trend can be defined by the hydrostatic pressure of the water that is contained in the pores (Draou and Osisanya, 2000). However, in many geological settings, compaction is hindered where many mechanical and geological variables that preclude the compaction process lead to pore fluids becoming overpressured. The ideal environment for overpressure generated by under-compaction is when the rate of sedimentation is faster than the rate at which the pore fluids are able to escape. Therefore, the pore fluids are trapped within the pore spaces and the porosity would be greater than it should be in normal compaction circumstances. As a result, the formation becomes overpressured due to the lack of conduits between the pore spaces and the overlaying formations (Eaton, 1975; Wallace, 1965). The main difference between overpressured formations caused by under-compaction and normally pressured ones is that, in overpressured formations, the pore fluids no longer have efficient communication with the water table.

7.2.1.2 Lateral Tectonic Loading

Lateral tectonic loading causes an increase in lateral stress as a result of compaction of the sediments horizontally in addition to the vertical compaction caused by an increase in overburden stress. The lateral stress associated with vertical stress causes overpressure if the pore fluids are not squeezed out by the compaction (Van Ruth et al., 2003). Another example of overpressure generated by tectonic compression is when a fault moves; the fault plane separates and lets the high pressure zones communicate with the surrounding lower pressure sand bodies (Fig. 7.3). However, when the fault closes, the charged sand releases its pressure into surrounding shales and develops overpressure (Osborne and Swarbrick, 1997). Unlike compaction disequilibrium, lateral tectonic loading can generate high magnitude of overpressure that may cause the vertical effective stress to decrease. This is attributed to the fact that in tectonically active areas, compaction is not controlled only by vertical effective stress (Bowers, 2002).

7.2.1.3 Wireline Logs' Response to Loading Mechanisms

The responses of wireline logs to overpressure generated by disequilibrium compaction are a constant transit time and a constant density (Ramdhan and Goulty, 2011). The effective stress

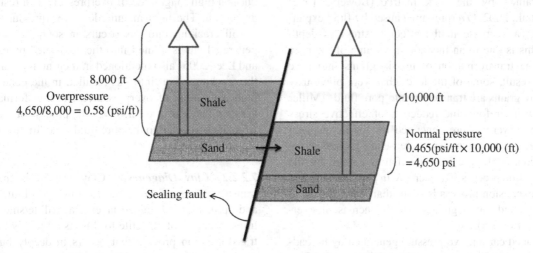

FIGURE 7.3 Graphic illustration of overpressure generated by lateral tectonic compression.

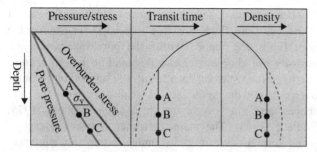

FIGURE 7.4 Graphic illustration of the response of wireline logs to overpressure generated by under-compaction.

across the charged interval increases or remains constant (Fig. 7.4). Moreover, the magnitude of pore pressure increase due to the under-compaction mechanism is less than or equal to the increase in overburden stress (Miller et al., 2002). In other words, the under-compaction mechanism cannot cause a decrease in effective stress. All responses for the aforementioned wireline logs are for the shale sequence, and it is critical to differentiate between shales and other formations prior to analyzing the logs response.

7.2.2 Unloading Mechanisms (Fluid Expansion)

Overpressure in sedimentary basins can be generated by unloading mechanisms. The process involves either expansion of the contained pore fluids or load transfer into pore fluids with minimal change in porosity at rates that do not allow the pore fluids to dissipate. Origins of fluid expansions that are mentioned in the literature include hydrocarbon generation, cracking of oil to gas, clay transformation, for example, smectite to illite, aqua-thermal heating, and cementation and mineral participation (Osborne and Swarbrick, 1997). Fluid expansion associated with the transformation of the load-bearing framework into pore fluids results in an increase in pore pressure when the expanded fluids have been constrained by the rock matrix (Bowers, 1995; Swarbrick et al., 2002. Overpressure caused by fluid expansion involves a decrease in the effective stress as depth increases. This is due to an increase in volume of the pore fluids and the transformation of matrix grains into pore fluids. As a result, some of the loads that were previously carried out by grains are transferred into pore fluids (Miller et al., 2002). Therefore, the reduction of effective stress resulting from overpressure generated by fluid expansion processes forces pore pressure to increase to a higher degree than the increase of pore pressure that is caused by the under-compaction process. The increase in pore pressure due to the fluid expansion process is faster than the decrease in effective stress and can be greater than the increase in overburden stress (Fig. 7.5).

As mentioned earlier, overpressure generated by unloading mechanisms involves the expansion of the pore fluids or

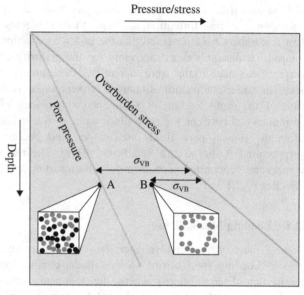

FIGURE 7.5 Graphic illustration of overpressure generation by unloading mechanisms, for example, the transformation of load-bearing grains or kerogen (black) into pore fluid (white).

the transformation of the load-bearing framework into pore fluids. The most significant unloading mechanisms presented herein include hydrocarbon generation, clay diagenesis, and aqua-thermal heating.

7.2.2.1 Hydrocarbon Generation
Hydrocarbon generation processes represent an effective mechanism to generate a large magnitude of overpressure. The processes include cracking from oil into gas and the transformation of kerogen into gas or oil. The volume of the expanded pore fluids during hydrocarbon generation depends on the type of the kerogen and the density of the hydrocarbon generated. As mentioned by Swarbrick et al. (2002), out of the many fluid expansion mechanisms, the cracking from oil into gas produces a high magnitude of overpressure as a result of fluid expansion. The high magnitude of overpressure generated by oil cracking into gas occurs in source rocks, and the generated gas is diluted into the connected pores. Hansom and Lee (2005) also mentioned in their numerical study that the cracking of oil into gas resulted in the generation of a high magnitude of overpressure. The transformation from kerogen into gas or oil involves the process of load transfer from the kerogen into the pore fluid in addition to the expansion of pore fluids.

7.2.2.2 Clay Diagenesis
Clay diagenesis includes the transformation of smectite to illite, kaolinite to illite, and illitization of mixed-layer clay (illite/smectite). The transformation of smectite to illite is a broadly known clay transformation process that occurs in deeply buried shale formations (Hower et al., 1976). In compacting shales under

diagenetic processes, smectite is stable and at least two water layers are preserved. There would be no loss of the interlayer water (dehydration) as the temperature of the interlayer water is below the threshold temperature of 71°C (Colten-Bradley, 1987). Within a temperature range of 71–81°C, clay becomes unstable and one of the interlayers water is released. For the other interlayer to be released, it requires a temperature range of 172–191°C (Boles and Franks, 1979; Hower et al., 1976). In other words, the conversion of smectite to illite eliminates a considerable amount of the smectite interlayer surface, which was hydrated when the clay was in the smectite phase. As a result, the volume of the shales' intergranular water increases and this increases the pore pressure thereafter. However, the increase in water volume resulting from clay transformation processes cannot generate a high magnitude of overpressure unless a perfect sealing exists (Osborne and Swarbrick, 1997). The related chemical reaction of this transformation produce major changes in the behaviors of subsurface rocks due to the release of a significant amount of water into the pore system (Draou and Osisanya, 2000). The chemical reaction of the transformation of smectite to illite is presented by Boles and Franks (1979) and stated in Equation 7.4.

$$\text{Smectite} + K^+ \rightarrow \text{Illite} + \text{Silica} + H_2O \quad (7.4)$$

All the processes of clay diagenesis are subject to temperature and create overpressure through the transfer of load-bearing into pore fluids and through the fluid expansion process, for example, release of water process.

7.2.2.3 Heating As depth increases, temperature increases and causes expansion of both the rock matrix and the pore fluids. According to Miller (1995), the increase in volume resulting from the rock expansion is one order less in magnitude than the increase in volume resulting from the expansion of pore fluids. Hence, the increase in volume resulting from rock expansion can be ignored. If pore fluids are heated while they are efficiently sealed, pore pressure could increase significantly. However, Luo and Vasseur (1992) concluded in their study that the expansion of pore fluids due to heating is not a significant contributor for generating a high magnitude of overpressure. The authors stated that in order to maintain overpressure generated by heating, the pore fluids must be sealed effectively. However, this condition cannot be met in real situations as there is no formation with zero permeability and when there is a leaking of the fluids in the system, this mechanism is neglected.

7.2.2.4 Wireline Logs' Response to Unloading Mechanisms The response of wireline logs to overpressure generated by unloading mechanisms is a decrease in effective stress, which produces a reversal in sonic transit time moving to a higher sonic transit times as depth increases. However, there

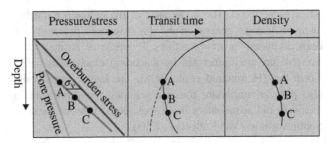

FIGURE 7.6 Schematic diagram of the responses of wireline logs to overpressure generated by unloading mechanisms.

would be no reversal in density log, and it often continues to increase but may reverse slightly at the bottom of the overpressured section. The responses of wireline logs to overpressure caused by unloading are presented in Figure 7.6.

The pore structure is classified into storage pores (pore spaces) and connecting pores (pore throats) (Bowers and Katsube, 2002). The effective porosity is the sum of all interconnected pores, whereas the total porosity is the sum of interconnected pores and the isolated pores. The storage pores affect the total porosity and the bulk density of a certain formation. These two petrophysical properties are attributed to the total volume of the net pore; thus, the storage pores are the major porosity contributor of shale. On the other hand, the connecting pores that control the flow within the pore system make very minor contributions to porosity. When overpressure in shale is generated by fluid expansion mechanisms, the response of the fluid expansion is basically an elastic opening (widening) of the connecting pores as a result of effective stress reduction (Bowers and Katsube, 2002; Cheng and Toksöz, 1979). This response is due to the fact that connecting pores have a low aspect ratio, and they are mechanically flexible and more harmonious than the storage pores. As a result, the porosity increases only by a very small amount (Hermanrud et al., 1998). In contrast, the aspect ratio of the storage pores is high and they are mechanically inflexible and scarcely affected by fluid expansion. Moreover, the bulk density is hardly influenced by fluid expansion responding to the low magnitude of porosity increase. Bowers and Katsube (2002) and Hermanrud et al. (1998) stated that the connecting pores have significant impacts on transport properties such as sonic velocity and electrical resistivity and thus affect sonic transit time and electrical resistivity logs. On the other hand, they have insignificant effects on density and neutron porosity logs.

7.2.3 World Examples of Overpressures

Overpressure exists in almost every geological environment of all ages. The event appears in all parts of the world. It is believed that the mechanism of under-compaction is the main cause of overpressure in young geological environment that experiences rapid sedimentation rates, for example, US Gulf Coast region

(Dickinson, 1951). The importance of the unloading mechanisms such as gas generation and clay diagenesis processes has been discussed by many authors. Example of basins where overpressure was generated by unloading mechanism is the North Sea (Hermanrud et al., 1998). The lateral stress could also play an important role for generating overpressure in relatively old sedimentary basin such as the Cooper Basin in South Australia (Van Ruth et al., 2003).

7.2.4 Overpressure Indicators from Drilling Data

The best technique to detect and assess overpressure is to study and combine all available pore pressure-relevant parameters. Depending only on one specific type of data can lead to erroneous interpretations. In conjunction with well log data analysis, the mud log data at the surface can also be used as an indicator for penetrating overpressured formations (Fertl and Timko, 1971). These measurements are discussed in detail in the following sections and include the drilling rate of penetration (ROP), gas show at the surface, kicks, mud weight, and the flow line temperature.

7.2.4.1 Drilling Rate of Penetration (ROP)
The evaluation of drilling performance parameters, specifically the ROP, is used to detect overpressure formations (Fertl and Timko, 1971). The ROP is inversely proportional to the differential pressure at the bottom hole between formation pressure and the hydrostatic pressure which results from the mud weight column. The advantage of using the ROP data over log data analysis for overpressure detection is the immediate availability during the drilling operations (Jorden and Shirley, 1966).

Field examples show the normal trend for the ROP reduces as depth increases with the other drilling parameters remaining constant. Under constant mud conditions while drilling normally pressured formations, the bottom hole differential pressure increases as a result of the increase in the formation pressure as depth increases: the effect is a reduction in the ROP. By contrast, when drilling overpressured sections under the same drilling conditions, the bottom hole differential pressure decreases and the ROP increases thereafter (Fertl and Timko, 1971). Therefore, under constant drilling conditions, it is believed that the ROP decreases while drilling normally pressured formations and deviates to a higher rate when encountering overpressured formations.

7.2.4.2 Gas Show
There are many possible sources of gas show in the mud returned to the surface. According to Fertl and Timko (1971), the origins of these gases could be (i) underbalanced conditions when drilling overpressure formations, (ii) gas released from the cutting, and (iii) gas-bearing rocks. The compounds of the returned gas can be studied, and certain components can be related to overpressure. Hence, a sudden increase of connection gas and gas in the mud may be an indication of overpressured formations.

7.2.4.3 Kicks
Kicks are also considered as overpressure indications when occurring in balanced drilling operations. The advanced drilling technology may require the drillers to adjust the mud weight to a narrow line between effective pressure control and the blowout, and this may cause sudden kick when penetrating overpressured formations. Thus, inappropriate pressure balance between mud weights and formation pore pressures may cause kicks (Fertl and Timko, 1971).

7.2.4.4 Mud Weight
Kicks may be best avoided by regular checks of the drilling mud properties, particularly for any drop in the mud weight, and this can be used as an indicator of gas cuttings and kicks.

7.2.4.5 Flow Line Temperature
Temperature measurement of the drilling flow line is a useful way of identifying changes in temperature while drilling petroleum wells. Several authors such as Lewis and Rose (1970) reviewed the heat conductivity and suggested that whenever encountering overpressured formations, formation temperatures increase. The observed changes in the flow line temperature is of the order by 2–10°F above the normal temperature when entering the overpressured environments. However, it is important to bear in mind that temperature increase can also be due to changes in lithology and the presence of salt dome (Fertl and Timko, 1971).

7.2.5 Identification of Shale Intervals

A typical definition of shale points to three main characteristics: (1) the clay content forms the load-bearing framework; (2) the pore size of shale, which is measured on the nanometer scale and permeability, which is measured by nanodarcy; and (3) the surface area of shale is large and adsorbs water easily. The basic data available for petrophysicists for shale identification are well log data. Shale pore pressure estimation from well logs is based on compaction theory, which requires establishing normal compaction trends (NCTs) within continuous shale and a uniform lithology section. In order to discriminate shale intervals from other lithology sections, a gamma-ray log that measures the rocks radioactivity is generally used to measure the clay contents (Fig. 7.7).

It was suggested by Fertl (1979) that the use of gamma ray for shale discrimination may cause errors as shale radioactivity varies significantly from one shale to another and not all shales are radioactive. Likewise, sandstone grains may have radioactive materials. Based on petrophysical properties of rocks, Katahara (2008) proposed a technique for distinguishing shale from sandstone, which is based on the difference between neutron porosity and density porosity. This approach is a more appropriate technique for measuring clay contents than using gamma ray, and it is a simplified model where quartz, clay minerals, and water are the only components of shale (Fig. 7.8). The shale falls within the triangle constrained by each end point of the three components. The distance from the water–quartz line indicates the clay contents, and clay content is proportional to the difference between neutron porosity and density porosity. However, the difference between neutron porosity and density porosity will remain commensurate with the clay content variation.

This can be better viewed on a neutron–density cross-plot (Fig. 7.9, left). The difference between neutron porosity and density porosity is a measure of clay content (Katahara, 2008). As can be seen in the figure, there is a break in the slope occurring at the threshold between sands and shales, and this can be used for shale discrimination. On a log plot (Fig. 7.9, right), the difference between the neutron and density porosities, which defines shale and sand intervals, is illustrated in the green and yellow windows, respectively.

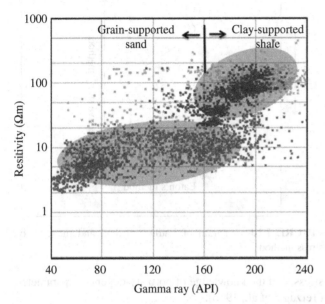

FIGURE 7.7 Shale discrimination based on the gamma ray log.

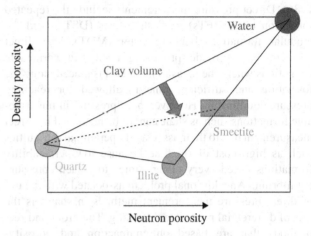

FIGURE 7.8 Graphic illustration of where the three components of shale stand on density porosity versus neutron density cross-plot.

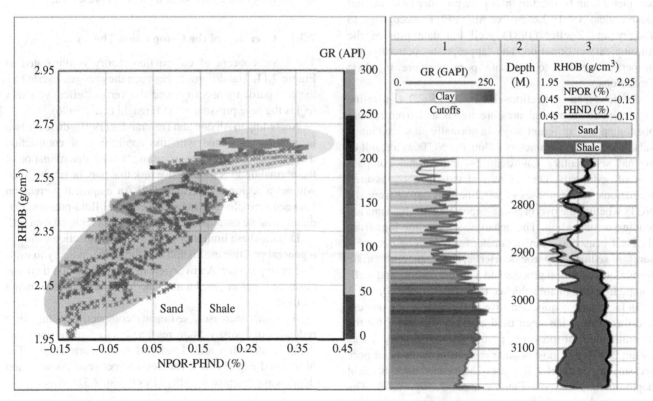

FIGURE 7.9 Shale identification based on the difference between neutron porosity and density porosity showing on cross-plot (left) and log-plot (right).

7.3 OVERPRESSURE ESTIMATION METHODS

Techniques used to detect abnormal pore pressure can be classified as (i) predictive and before drilling methods from offset wells and seismic data; (ii) during drilling from mud log data such as kicks, drilling ROP, and flow line temperature; and (iii) postdrilling methods from well logging data (Draou and Osisanya, 2000; Pikington, 1988).

In other words, formation pressure can be determined either by direct or indirect methods (Lesso and Burgess, 1986). Direct pressure measurements include the repeated formation tests (RFTs), drill stem tests (DSTs), and the modular formation dynamic tester (MDT). The direct measurements provide promising results in permeable formations where the measurement tool is placed along the formation and sufficient time is allowed for reaching pressure equilibrium. However, pore pressure in impermeable formations such as shale cannot be measured by direct measurement due to their associated operational difficulties such as high cost of rig time because low-permeability formations need very long time to reach pressure equilibrium. An additional problem associated with the use of direct pressure measurement methods in shales is the risk of differential pipe or tool sticking. Therefore, indirect methods that are based on compaction and porosity-dependent parameter concepts such as the applications of well logging and drilling data, where pressure-dependent parameters can be used to infer pore pressure (Alixant and Desbrandes, 1991; Lesage et al., 1991). According to Tanguy and Zoeller (1981), well log data provide the lithological information and appropriate petrophysical properties needed to estimate pore pressure in shale formations.

As described in Sections 7.2.1.3 and 7.2.2.4, wireline logs respond to normal pressure trends and overpressure phenomenon in different ways. In normally pressured intervals, wireline log parameters follow the NCTs as a result of normal sedimentary environments and normal compaction of sediments. On the other hand, in overpressured formations, the responses of wireline logs depart from the NCTs whether the overpressure-generating mechanisms are loading or unloading. The departure of wireline logs from the NCT is used as a key parameter for predicting overpressure in sedimentary rocks. Hence, both an appropriate formation evaluation process and a proper drilling and well-completion design are achieved (Tingay et al., 2003).

In fact, shale is quite sensitive to the compaction process and therefore, it has been used as a key parameter for the determination of pressure profile in sedimentary rocks (Muir, 2013). The most popular prediction methods for pore pressure are (i) the effective stress, also called the equivalent depth method and (ii) Eaton's method (Fig. 7.10). The fundamental concepts for estimating pore pressure in shale formations are the knowledge of overburden stress, effective stress, and the knowledge of porosity-dependent parameters (Terzaghi et al., 1996).

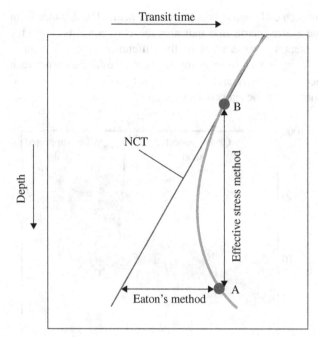

FIGURE 7.10 Diagram of Eaton's method and the effective stress method.

Prior to explaining the overpressure estimation methods, an overview of the compaction theory is presented.

7.3.1 Overview of the Compaction Theory

The basic concept of compaction theory is illustrated in Figure 7.11. The difference between the pressure exerted by the overburden stress (σ_{ob}) and the vertical effective stress (σ_v) is the pore pressure (P_p) (Terzaghi et al., 1996).

According to Alixant and Desbrandes (1991), there are two limitations associated with the application of compaction theory in determining pore pressure: "(1) the determination of the normal trend is a subjective task that may be troublesome without a regional experience. (2) An empirical correlation between petrophysical measurements and fluid-pressure gradients must be established on the basis of a regional data set."

Despite these limitations, the use of compaction theory is a general practice evaluation method in the industry to evaluate pore pressure. An overview of the compaction theory is described before explaining the overpressure estimating methods.

As depth increases, sediments compact, resulting in a reduction of porosity. Many researchers studied the porosity–depth relationship and developed many correlations. The most cited experimental relationship between porosity and depth was presented by Athy (1930) (Eq. 7.5):

$$\varnothing = \varnothing_0 e^{-bz} \quad (7.5)$$

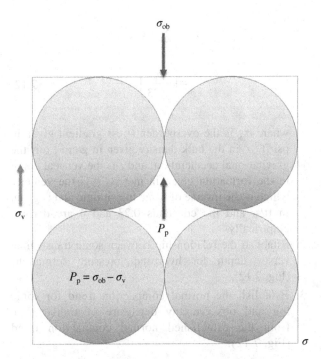

FIGURE 7.11 Basic concept of the compaction theory.

where $\varnothing_0$ is porosity at the surface, $\varnothing$ is the porosity at any specified depth z, e is the base for Napierian logarithms, and b is an empirical constant attained after fitting the exponential relationship of porosity versus depth.

Athy (1930) did not take into account the high porosity values in overpressured zones resulting from under-compaction. Therefore, Rubey and Hubbert (1959) expanded Athy's relationship to account for the vertical effective stress and remove the effects of high porosity points due to overpressure generated by disequilibrium, and they developed the following equation (Eq. 7.6)

$$\varnothing = \varnothing_0 e^{-\frac{C}{\rho_b - \rho_w}\sigma'_v} \qquad (7.6)$$

where C is an empirical constant, ρ_b is the bulk density of the formation, ρ_w is the density of the pore water, and σ'_v is the vertical effective stress.

It is obvious from Equations 7.5 and 7.6 that as sediments compact mechanically, the porosity decreases as a result of burial or vertical effective stress increase. Ramdhan and Goulty (2011) stated that mechanical compaction is basically a permanent plastic process, with a minor elastic component. The exponential decrease of porosity with depth (Eq. 7.5), or vertical effective stress (Eq. 7.6), indicates that the sediments become more resistant to mechanical compaction when porosity reduces and vertical effective stress increases.

Due to the rare use of the direct measurements for porosity in shale formations, the determination of the NCT is used to infer porosity and hence, detect any abnormality of pore pressure within the shales (Athy, 1930; Rubey and Hubbert, 1959). The NCTs obtained from Equations 7.5 and 7.6 are fitted on wireline log(s) data such as sonic transit time, electrical resistivity, and density. Among well log data, sonic transit time is the most commonly used as a porosity-dependent parameter. This is because it is usually available with good quality and less affected by the bad borehole conditions compared to other well log data.

For proper estimation of pore pressure in shales, it is important to establish a reliable NCT. According to Ramdhan and Goulty (2011), the three main techniques for defining NCTs from wireline logs are (1) direct plots of wireline log data versus depth, (2) plot of wireline log data versus effective stress, and (3) cross-plotting wireline log data.

7.3.2 Eaton's Method

The principle of Eaton's method is the comparison of the wireline log data and drilling data with the NCTs at the same depths. Eaton (1975) developed four equations for pore pressure estimation using well log and drilling data. Among pore pressure estimation methods that use log data, Eaton's method is the most widely used and has been in use in the industry for more than 25 years and found to be fairly reliable. The correlations can be used with different sources of data such as sonic, resistivity, conductivity, and corrected drilling exponent (Eqs. 7.7, 7.8, 7.9, and 7.10), respectively.

$$g_p = g_{ob} - (g_{ob} - g_n)\left(\frac{\Delta t_n}{\Delta t_o}\right)^x \qquad (7.7)$$

$$g_p = g_{ob} - (g_{ob} - g_n)\left(\frac{R_o}{R_n}\right)^x \qquad (7.8)$$

$$g_p = g_{ob} - (g_{ob} - g_n)\left(\frac{C_n}{C_o}\right)^x \qquad (7.9)$$

$$g_p = g_{ob} - (g_{ob} - g_n)\left(\frac{d_{co}}{d_{cn}}\right)^x \qquad (7.10)$$

where g is the pressure gradient, the subscript n denotes to the value of data parameters at normal compaction trend, and the subscript o denotes to the observed parameters, ob denotes to overburden, p denotes to pore pressure and the exponent x is Eaton's exponent, which can be adjusted based on regional experiences.

The departure of data from their normal compaction trends is used as a measure of pore pressure within the shale.

It should be noted that the aforementioned correlations were derived based on an empirical basis for Gulf of Mexico data taking into consideration the overpressure-generating mechanism in disequilibrium compaction. For this reason, it is also important to note that this method does not imply the particular overpressure-generating mechanisms, whether loading or unloading. Eaton (1975) also did not mention in his study how to determine the NCT; however, experience indicates that normal compaction trend curves can be established for sediments with normal pressure overlying the overpressured sections.

7.3.2.1 Hints for Using Eaton's Method

- Establish relationships of depth versus the porosity-dependent parameters, for example, logarithm of shale sonic transit time or shale resistivity for normally pressured formations.
- Establish the normal compaction trend in normally pressured clean shale.
- On sonic transit time versus depth plot, the observed relationship will be generally a linear relationship.
- On resistivity versus depth plot, the observed relationship will be a nonlinear relationship.
- The departure of data from normal compaction trends is used as a measure of pore pressure within the shale.

Example:

Using Eaton's method (sonic), estimate pore pressure gradient at depth of 3576 m TVD for Well #2.

Steps as shown in Figure 7.12:

1. Generate overburden gradient from the density log.

   The overburden stress is computed from the density log which measures the bulk density every 0.152 m. Thus, the calculation of the overburden stress is made at every depth step. The overburden stress is computed by using the Equation 7.11, and examples of these calculations are shown in Table 7.1.

   Note: If the density log is not available, generate the density log from sonic transit time by using Gardner's method (Gardner et al., 1974) (Eq. 7.12) or the other available methods.

$$\sigma_{ob} = \sum_{i=1}^{n} \rho_b g dz \qquad (7.11)$$

$$v = \left(\frac{\rho_b}{0.23}\right)^4 \qquad (7.12)$$

where σ_{ob} is the overburden stress gradient given in psi/ft, ρ_b in the bulk density given in g/cm^3, g is the gravitational acceleration, and z is the vertical depth to the formation given in meters, v is the velocity which is the inverse of sonic transit time and is given in ft/s, and the constants 0.23 and 4 are derived empirically.

2. Establish the relationship between sonic transit time versus depth for hydrostatic-pressure formations (Fig. 7.12).
3. Establish the normal compaction trend for sonic log and observe any departure of transit time from the established normal compaction trend (Fig. 7.12).

Estimate the pore pressure using the relevant Eaton's correlation (Eq. 7.7 and Fig. 7.13).

This real example was taken from Perth Basin in Western Australia. After testing different values for the x exponent in Eaton's correlation (Eq. 7.7), the study concluded that the best match between the estimated pore pressures and other relevant data could be found when using a value of 1.5 for the x exponent.

$$g_p = g_{ob} - (g_{ob} - g_n)\left(\frac{\Delta t_n}{\Delta t_o}\right)^{1.5}$$

$$g_p = 0.997 - (0.997 - 0.433) \times \left(\frac{65.74}{79.66}\right)^{1.5}$$

$$g_p = 0.574 \, \text{psi/ft}$$

p_p at 3576 m TVD

$$p_p = 0.574 \times 3576 \times 3.281 = 6733 \, \text{psi}$$

TABLE 7.1 Example of overburden stress calculations in Well #2

TVD (m)	RHOB (g/cm^3)	dz (m)	dσ_v (psi)	σ_{ob} (psi)	OB Grad (psi/ft)
2054.447				6237.591	0.926
2054.599	2.0734	0.1524	0.448	6238.039	0.926
2054.751	2.0953	0.1523	0.452	6238.491	0.926
2054.903	2.1167	0.1521	0.456	6238.947	0.926
2055.056	2.1376	0.1524	0.462	6239.409	0.926
...	...	...	...	...	...
3576.200	2.6463	0.1521	0.570	11692.695	0.997
3576.353	2.6463	0.1526	0.572	11693.267	0.997

OVERPRESSURE ESTIMATION METHODS 149

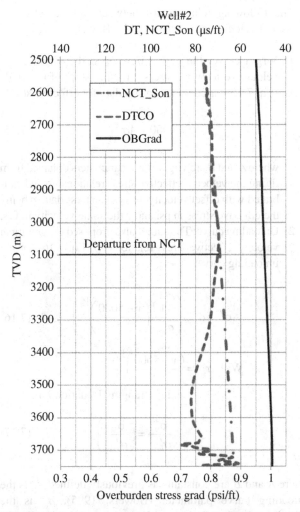

FIGURE 7.12 The relationship between sonic transit time, overburden stress gradient, and NCT versus depth for Well #2.

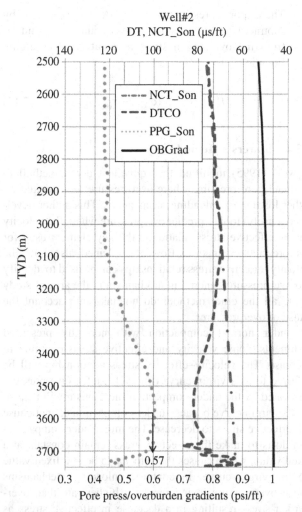

FIGURE 7.13 Estimated pore pressure profit using Eaton method (sonic) for Well #2.

7.3.3 Effective Stress Method

This method is also called equivalent depth method, and it further stresses the importance of the NCTs in shales. The principle of the effective stress method is that the overpressured shale has the same effective stress value with normally pressured shale that has the same porosity (Mouchet and Mitchell, 1989). Hints for using this method are provided in Section 7.3.3.1, and Figure 7.10 illustrates the application for using the effective stress method. It is important to note that the method is only applicable when the overpressure-generating mechanism is the disequilibrium compaction and the lithology at the two points of interest is the same. If overpressure is generated by unloading mechanisms, then the effective stress will be overestimated, and the pore pressure will be underestimated. Thus, equivalent depth method will fail due to deviation of the porosity–effective stress relationship from the normal trends.

7.3.3.1 Hints for Effective Stress Method
Referring to Figure 7.10, to apply the effective stress method, the following steps are followed:
- The sonic transit time at point A is the same as the sonic transit time at point B, which has a normal pressure gradient.
- The effective stress at point B is computed by subtracting the normal pressure at point B from the overburden stress value at the same point.
- Implementing the principle of the effective stress method, the effective stress at point B is as same as the effective stress at point A.

$$\sigma_{v(A)} = \sigma_{v(B)}$$

since $\sigma_v = \sigma_{ob} - P_p$ then

$$(\sigma_{ob} - P_p)_A = (\sigma_{ob} - P_n)_B \qquad (7.13)$$

- Then, pore pressure at point A is computed by subtracting the effective stress value at point B from the overburden stress at point A. Rearranging Equation 7.13,

$$P_{p(A)} = P_{n(B)} + (\sigma_{ob(A)} - \sigma_{ob(B)}) \quad (7.14)$$

7.3.4 Bowers's Method

Bowers (1995) modified the equivalent depth method to estimate pore pressure where overpressure is generated by either loading or unloading mechanisms. This author developed a useful tool to predict overpressure where the velocity versus effective stress relation is the key element used for overpressure estimation. Bowers explained also how the velocity–effective stress relationships can be used to identify the overpressure-generating mechanism in the area of study while all the other methods do not take into account the cause of overpressures.

Under normal compaction with normally pressured sediments, sonic velocity and effective stress continue to increase. The velocity–effective stress relationship will be referred to as a virgin curve (Fig. 7.14, left). Overpressure generated by the under-compaction mechanism will be also on the virgin curve because under-compaction cannot cause the effective stress to decrease. The most under-compaction can do is to make the effective stress remain constant at a fixed value, which causes the velocity to be at a fixed value on the virgin curve. In contrast, unloading mechanisms cause overpressure to increase at a higher rate than overburden stress resulting in a decrease in effective stress as depth increases, as well as a velocity reversal. As a result, the data inside the velocity reversal follow a different path called the unloading curve, whereas the data outside the velocity reversal stay on the virgin curve (Fig. 7.14, right). In case of any subsequent increase in effective stress, the velocity will track the unloading curve returning to the virgin curve (Bowers, 1995).

The following relationships between vertical effective stress and velocity were developed by Bowers:

1. Virgin curve: the author developed the virgin curve velocity–effective stress relationship for shale (Eq. 7.15) based on the in situ data for effective stress,

$$v = 5000 + A\sigma^B \quad (7.15)$$

where A and B are virgin curve parameters obtained from fitting the velocity–effective stress relationship and calibrated with offset velocity–effective-stress data, σ being the effective stress in psi and v the sonic velocity in ft/s.

2. Unloading curve: The author also proposed the empirical velocity–effective stress relationship (Eq. 7.16) for the unloading curve:

$$\sigma'_v = \sigma'_{max} \left(\frac{1}{\sigma'_{max}} \left(\frac{v - 5000}{A} \right)^{1/B} \right)^U \quad (7.16)$$

With $\sigma_{vc} = \left(\dfrac{v - 5000}{A} \right)^{1/B}$

and rearranging Equation 7.16,

$$\frac{\sigma'_v}{\sigma'_{max}} = \left(\frac{\sigma_{vc}}{\sigma'_{max}} \right)^U \quad (7.17)$$

where A and B are constants as previously defined, U is the unloading curve parameter (Bowers, 1995), σ_{vc} is the effective stress at which the velocity intersects with the virgin curve given in psi, and σ'_{max} is the maximum vertical effective stress given in psi. In the absence of major lithology changes, σ'_{max} is usually taken to be equal to the effective stress at which the velocity reverses.

Pore pressure caused by unloading can be then computed by deducting the vertical effective stress from the overburden stress.

7.3.4.1 Hints for Bower's Method

- The virgin curve parameters A and B are determined by fitting velocity versus effective stress data from the normal pressure section above the overpressured zone (Bowers, 1995). A regional normal pressure gradient is to be used.
- The normal trend line is determined from the virgin curve relationship.
- Eaton's method is used to compute the effective stresses along the normal trend.
- Pore pressure inside velocity reversal is calculated from the unloading curve relation, with U as the known parameter.

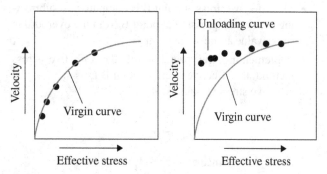

FIGURE 7.14 Velocity–effective stress relationship and shale behavior: the virgin curve (left) and the unloading curve (right).

7.4 THE ROLE OF TECTONIC ACTIVITIES ON PORE PRESSURE IN SHALES

This section stresses the importance of tectonic activities on pore pressure distribution in shale formations. A pore pressure study conducted on the Perth Basin has linked the variations in pore pressure gradients in the same intervals to the variations in tectonic intensity that took place within the same regions. Normal trends in pore pressure gradients were observed in regions that were situated in severely uplifted sections within tectonically active areas. On the other hand, pore pressure gradients increased in the same intervals where there was less intense tectonic activity. Regions with a lower intensity of tectonic activity showed an increase in pore pressure gradients when moving away from the center of the uplifting. This study demonstrated that there was a combination of mechanisms contributing to overpressure development driven by the complicated basin geology. Overpressure-generating mechanisms for the shale formations studied were attributed to fluid expansion and later tectonic loading. Both mechanisms have contributed to a different extent to overpressure development.

This section starts with a geological summary and an overview of the stress field direction in the area studied. In addition, observations of pore pressure profiles are presented both in tectonically active regions and reasonably stable areas. Furthermore, a thorough approach is presented to identify with certainty the causes underlying the overpressure development in this part of the Perth Basin, and that includes, firstly, the signatures of well log data being analyzed and initial thoughts being established as to what the potential overpressure-generating mechanisms are. Secondly, additional methodologies used to identify the cause of overpressure are discussed, and these involve analysis of sonic–density cross-plots, analysis of X-ray diffraction (XRD) and analysis of the natural gamma ray spectrometry (NGS) logs.

7.4.1 Geology of the Study Area

The Perth Basin is a north-northwest trending extension located in the southwest of Western Australia. The basin covers more than $100,000 km^2$ along the western coast of Australia and extends from Geraldton to Augusta. The basin contains sedimentary sequence that varies from Silurian to Pleistocene (Mory et al., 2005). The eastern boundary of the basin is defined by the Darling Fault, while the western end margin extends offshore to the edge of the continental crust in water depths of up to $4500 m$ (Iasky and Mory, 1994). Northampton block forms the northern boundary of the basin and the southern boundaries extend to the edge of the continental shelf. The focus of this study is on two different areas: (1) Dandaragan Trough and its adjacent terrace of similar characterization and (2) Beagle Ridge and its adjacent terraces.

The Dandaragan Trough is a major depocenter in the north Perth Basin that covers more than $5000 km^2$. It is virtually unfaulted syncline and has the thickest sediment accumulation of approximately $12,000 m$ (Crostella, 1995). The Mountain Bridge Fault bounds the western margin of the northern Dandaragan Trough. This fault showed a progressive down-dip decrease of dip angle from considerably high angles at the top ranges of 65° to 20° at the base of the sedimentary cover (Fig. 7.16, right). The mid-southern section of the trough is marked by Coomallo Fault that declines in dip from 62° at the surface to 42° at a depth of $10 km$ (Song and Cawood, 2000). The northern boundary of the Dandaragan Trough is marked by the Allanooka Transfer Fault, which separates the trough from the Allanooka High to the north. The eastern boundary of the Dandaragan Trough is bounded by Urella Fault and Darling Fault system, and the Cervantes Transfer Fault constitutes the southern boundary of the Dandaragan Trough and separates it from the Beermullah Trough (Fig. 7.15).

The structural history of the Perth Basin is dominated by two major phases of extension. An Early Permian phase created half-grabens that hinge around the Northampton High. A period of uplift and erosion in the Late Permian terminated this period of basin development. The second, and more extensive, phase of oblique extension occurred in the Late Jurassic to the Early Cretaceous period during the separation of the Australian Plate from Greater India and Africa. This phase caused extensive basin inversion and uplifting as well as the development of transfer faults which influenced the geometry and divided the basin into compartmentalized regions characterized as subbasins, ridges, and troughs of similar structure that reflect the present form of geological structure of the Perth Basin (Song and Cawood, 2000). The center of the uplift is near the coastal town of Jurien within Beagle Ridge where up to $8 km$ of section has been removed. Extensive faulting systems were identified within the Kockatea Shale in the Beagle Ridge and the adjacent Cadda Terrace. In addition, the severe erosion and uplifting that took place in these areas have removed considerable parts of the Kockatea Shale in some localities. The tectonic divisions and structural framework as well as maximum stress direction in the Perth Basin are illustrated in Figure 7.15.

The lower Triassic Kockatea Shale is one of the potential shale gas formations in the Perth Basin. The unit is made up of dark shale or siltstone with minor thin sandstone and limestone beds (Crostella and Backhouse, 2000). Well log data were used to evaluate pore pressure regimes in the Kockatea Shale. The log data that were used include sonic velocity, neutron porosity, density, and gamma ray. The boreholes which intersect the candidate formation were divided into groups according to their geographic and geological locations. Mud log data, for example, mud weight, well flows and kicks while drilling, and ROP, were also reviewed and correlated to the log data analysis. All data were combined and analyzed and the results revealed the

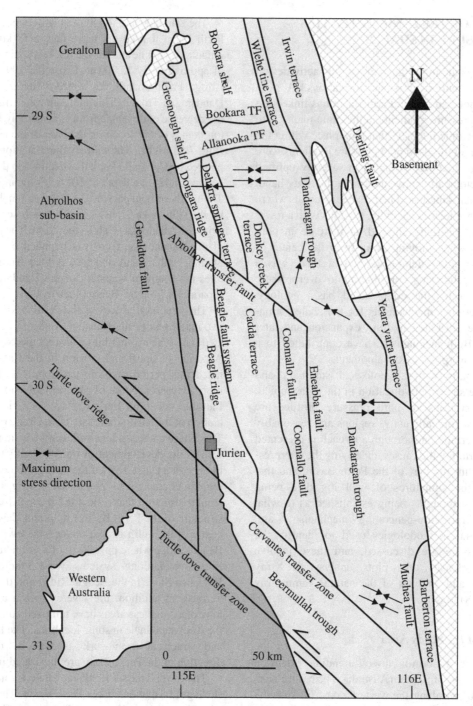

FIGURE 7.15 Schematic illustration of the regional geology and the principal stress field direction of Perth Basin. (Adapted from Mory and Lasky, 1996)

pore pressure profiles in the Kockatea Shale. The Kockatea Shale was buried to a greater deeper depth in some localities of the Perth Basin and has been severely uplifted in some other regions of the basin. Figure 7.17 shows a simplified contour map to the top of Kockatea Shale in the northern Perth Basin. Data from 35 boreholes in the Perth Basin have been reviewed and analyzed, and representative results are presented in Sections 7.4.3–7.4.5.

7.4.2 Stress Field in the Perth Basin

Various data obtained from different sources such as Van Ruth et al. (2003) suggest that the Perth Basin is broadly in a compressive stress regime where thrust faulting dominates. Recent tectonic stress data from the Perth Basin indicates that the stress regime in the Perth Basin is a transitional reverse fault to strike-slip fault stress where $S_{hmax} > S_v \approx S_{hmin}$

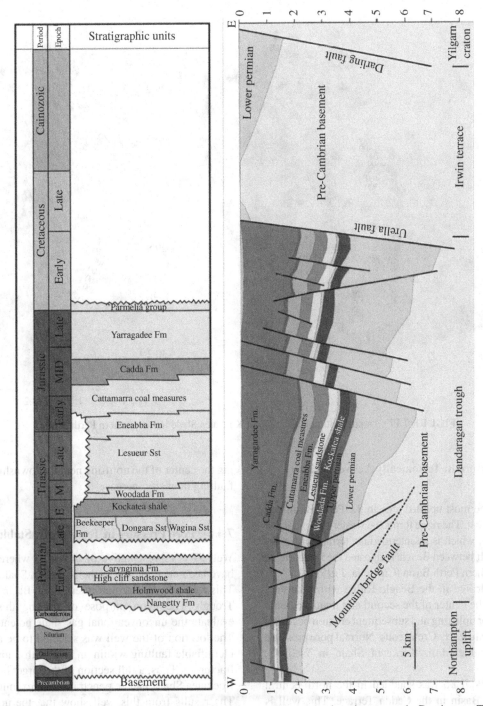

FIGURE 7.16 Northern onshore Perth Basin stratigraphy (left) and east–west structural section across the Dandaragan Trough and adjacent terraces (right). (Adapted from Mory and Lasky, 1996)

(King et al., 2008, Van Ruth et al., 2003). Regardless of the exact state of stress, most data suggest that the trajectory of the principal stress is in a horizontal plane and the main direction of the principal stress axis (S_{hmax}) is largely oriented EW (Fig. 7.15). The orientation of principal stress is perpendicular to the main north–south and northwest–southeast faults trends and this trend is consistent throughout the basin.

Unlike many other basins around the world, stress orientations in Perth Basin are not parallel to the direction of regional structure trends (Hillis and Reynolds, 2000). Stress orientation in this part of the Australian continent is likely to be induced by the forces exerted on the Indo-Australian plate boundary and do not appear to be influenced by regional structure trends.

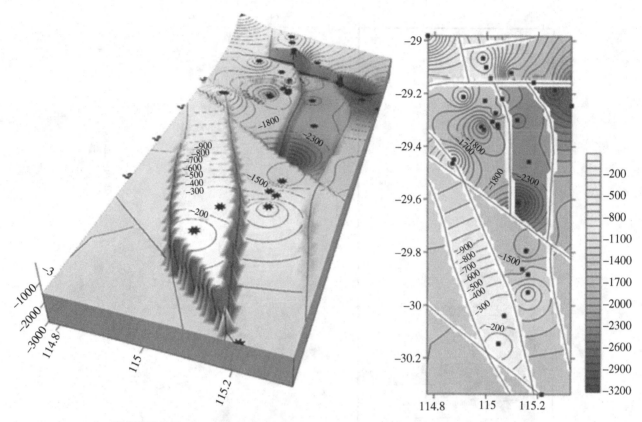

FIGURE 7.17 Depth contours of the top of Kockatea Shale in the northern Perth Basin.

7.4.3 Pore Pressure in Tectonically Active Regions (Uplifted Areas)

Beagle Ridge is the most uplifted area in the Perth Basin and has shallow basement. The ridge forms the western boundary of the Cadda Terrace, which is an intermediate terrace of north–south trending shelf between Beagle Ridge and the Dandaragan Trough in the northern Perth Basin (Crostella, 1995).

Well #5 was drilled in the Beagle Ridge within an area that is located at the center of the second and major tectonic phase where major uplifting and subsequent erosion occurred in the late Jurassic to Early Cretaceous. Normal pore pressure profile was encountered in Kockatea Shale in Well #5, (Fig. 7.18).

Furthermore, Well #6 was spudded in the north-central part of the Perth Basin in the Cadda Terrace. This well is located in a narrow block between the Lesueur and Cadda faults. The Kockatea Shale in this well was expected to be 900 m, but the well went through a normal fault and about 450 m is missing (Fig. 7.19). In addition, a continuous dipmeter set of data clearly indicates a faulted zone in the Kockatea Shale, and the core cut of the top of the Kockatea indicates a highly tectonized zone. Results obtained from this well indicate normal pore pressure regimes across the whole interval of Kockatea Shale. No signs of overpressure were observed in mud logging data. Note that, by approaching the center of the uplifting near the township of Jurien, the faulting intensity increases.

7.4.4 Pore Pressure in Tectonically Stable Regions

Well #1 is representative of the wells where overpressures have been encountered in the Kockatea Shale (Perth Basin). This well was drilled in northern Perth Basin in Dandaragan Trough. The main purpose of drilling this well was to evaluate the unconventional gas shale potential formations. The location of the well was selected to be away from any identifiable faulting within an area with a low incidence of fracturing. Thus, a full section of the target interval, namely, Kockatea Shale, can be penetrated and examined effectively. The results from this well show that the upper section of Kockatea Shale is normally pressured while the lower section of the same interval is overpressured. The pore pressure gradient at the upper section of Kockatea Shale is 0.44 psi/ft up to a depth of 2425 m TVD. Then the pore pressure gradient gradually increases with depth and reaches 0.56 psi/ft at depth of 2630 m TVD (Fig. 7.20). Mud log data show qualitative overpressure indications, for example, a continuous increase in the ROP and mud weight, over the lower section of Kockatea Shale. It should be noted there were no changes in other drilling parameters such as weight

THE ROLE OF TECTONIC ACTIVITIES ON PORE PRESSURE IN SHALES 155

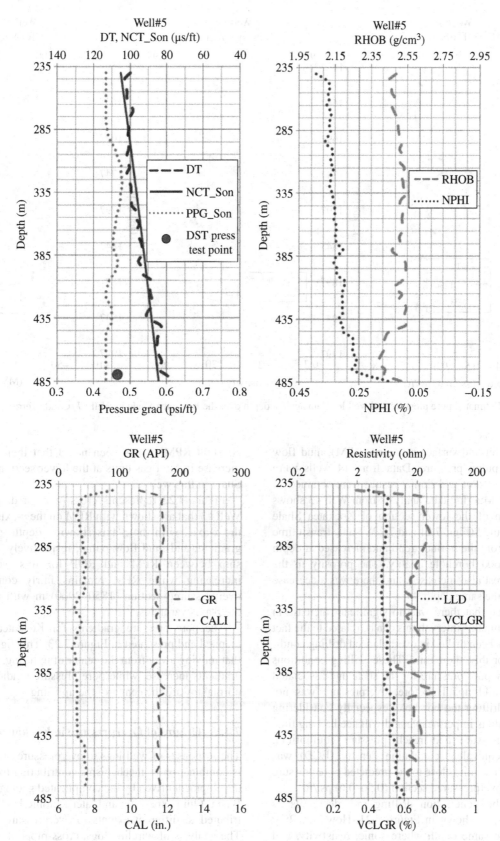

FIGURE 7.18 Estimated pore pressure and well log data against depth over the Kockatea Shale in Well #5 (Beagle Ridge—Perth Basin).

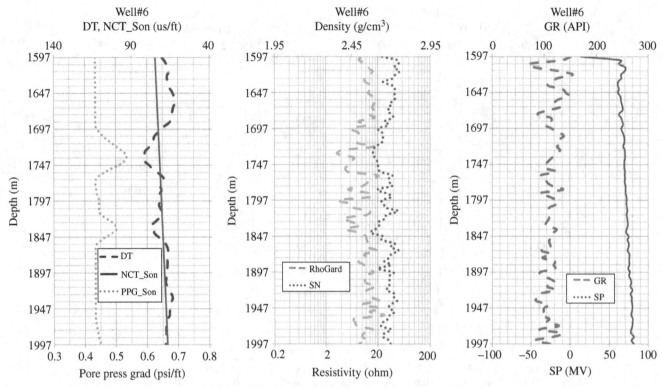

FIGURE 7.19 Estimated pore pressure and well log data against depth over the Kockatea Shale in Well #7 (Cadda Terrace—Perth Basin).

on bit (WOB), bit and surface rotations (RPM), mud flow rate, and mud pump pressure. Data from 14 wells were studied within this area and similar results were obtained.

Well log and Mud log data analysis from Well #3 shows every indication of the lower section of Kockatea Shale being overpressured (Figs. 7.21 and 7.22). Sonic transit time (DT) diverted from the normal trend and increased at depth 1870 m with good borehole quality. The porosity in the interval of interest also increased, and there was a decrease in resistivity at the same depth.

It can be seen that there was no significant increase in mud weight within this interval; this can be related to the fact that the well has been drilled in overbalanced drilling conditions in a reasonably short time. These drilling conditions would not allow pore pressure for the shale and associated thin beds to build up. Therefore, overpressure was not noticed while drilling the borehole, especially considering the fact that shale is impermeable and will need a long time to build up pressure and reach pressure equilibrium. However, the equivalent circulation density (ECD) was 0.52 psi/ft which is quite close to the predicted pore pressure gradient in the overpressured section (0.56–0.6) psi/ft.

Generally, the penetration drilling rate decreases as depth increases, as shown in Figure 7.21. However, ROP increased at the same depth where sonic, resistivity, and porosity logs deviated. This is an additional indicator for overpressure, supported by the data taken from the drilling report, which showed no changes in rotary speed and constant RPM. It has been noted that there was also an increase in total gas units at the lower section of Kockatea Shale in this well.

Figure 7.22 represents cross-plots of data taken from Well #3 that are shown with RPM on the y-axis and ROP on the x-axis, with the z-axis showing depth, pore pressure gradient, and mud flow pump, respectively. The relationship between RPM and ROP for this well has ROP increasing while RPM remains fairly constant in the overpressured section (1850–2000) m with no change in the z-axis variables.

The pore pressure gradients for Kockatea Shale were mapped and presented in Figure 7.23. This figure illustrates that moving away from the center of uplifting, pore pressure gradients increase, while pore pressure gradients approach normal toward the center of the uplifting.

7.4.5 Origins of Overpressure in Kockatea Shale

The responses of well logs to overpressure reveal that there is combination of mechanisms contributing to overpressure development driven by the complicated geology of the Perth Basin. Fluid expansion and later tectonic loading have contributed to different extents to overpressure development. The analysis of wireline logs, cross-plots of well log data, and the study of compositional variations imply that the fluid expansion mechanism plays an important role in the buildup of overpressure. The later tectonic loading is also believed to

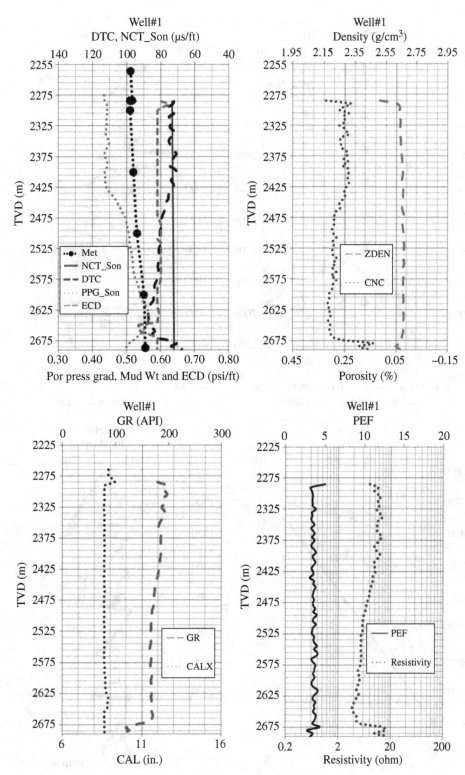

FIGURE 7.20 Estimated pore pressure, mud weight, and equivalent circulation density gradients as well as log data against depth over the Kockatea Shale in Well #1.

make a significant contribution to overpressure generation. This belief is reached from analyzing the complicated geology of the study area in conjunction with the trajectory of the principal stress (S_{hmax}) that acts in a horizontal plane and is largely oriented EW. The analysis was combined with analysis of well log data responses and with the high magnitude of overpressure that caused a reversal in the vertical effective stress (Fig. 7.24).

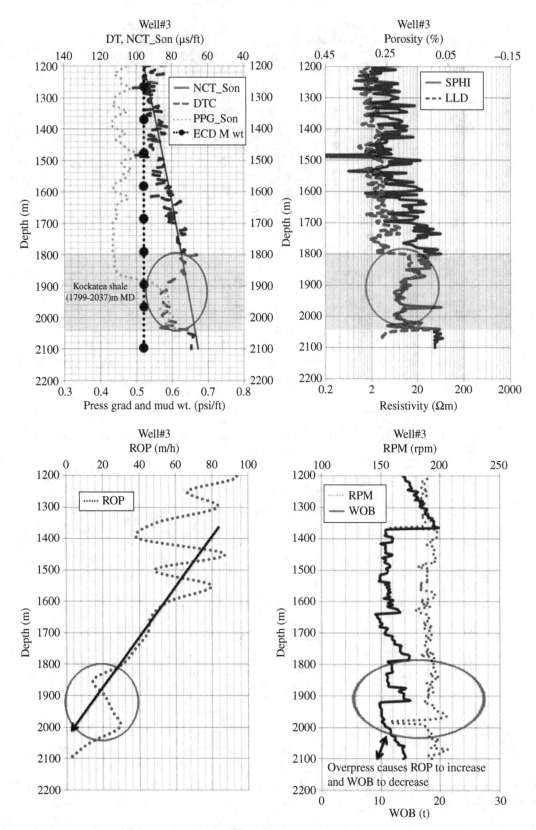

FIGURE 7.21 Estimated pore pressure, equivalent circulation density gradients, and well log data against depth over the Kockatea Shale in Well #3.

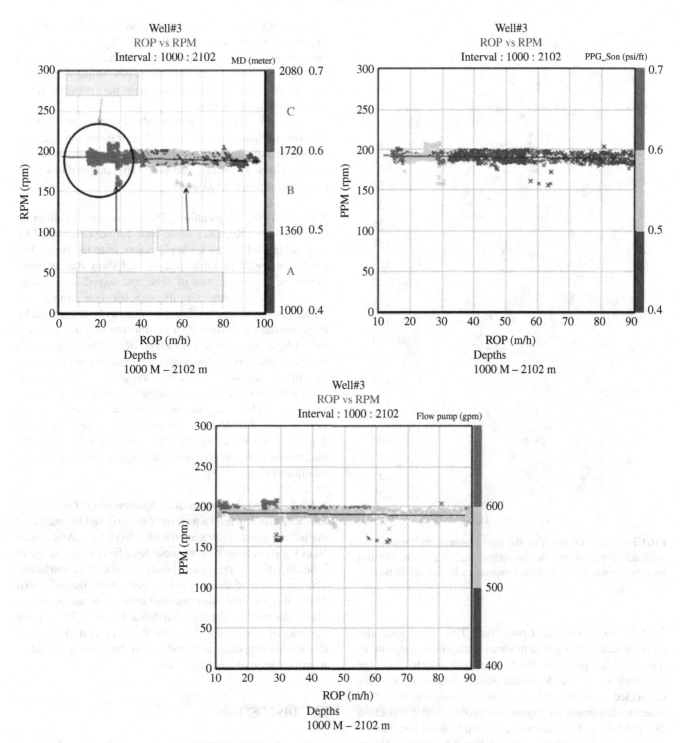

FIGURE 7.22 Cross-plots of mud logging data from Well #3.

7.4.5.1 Wireline Logs' Responses
The wireline logs available for analysis include density log, neutron porosity log, neutron density log, sonic transit time log, resistivity log, and gamma ray log. The identification of shale intervals was completed by the cross-plotting of density log versus the difference between neutron porosity and density porosity using a technique developed by Katahara (2008). In this study, the data showed a clear difference in the slope, and a threshold value of 0.15 was chosen for the difference between neutron porosity and density porosity ($\Phi_N - \Phi_D > 0.15$) to ensure that the overpressure analysis was conducted in shale intervals (Fig. 7.9, left).

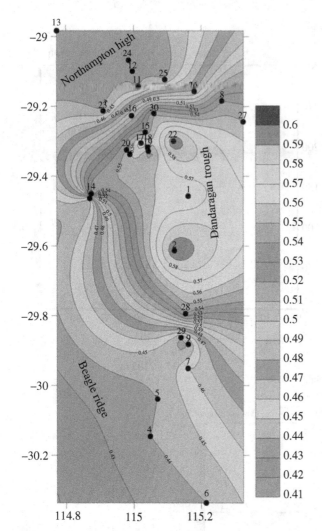

FIGURE 7.23 Contours of the pore pressure gradients of the Kockatea Shale in the Beagle Ridge, Cadda Terrace, Beharra Springs Terrace, and part of the Dandaragan Trough (Perth Basin).

7.4.5.2 Sonic–Density Cross-Plot

This study used the conventional well log data to identify the mineralogy and to quantify the possible fluid expansion mechanism for overpressure in the Kockatea Shale. Based on a study conducted by Dutta (2002) which discussed the impact of smectite diagenesis on compaction profiles and compaction disequilibrium, the study proposed two limiting compaction profiles; one for smectite and one for illite. Their smectite and illite curves were used in this study for a purpose other than the one which they were originally intended for.

Specifically, the two profiles were used herein to identify the causes of overpressure in the Kockatea Shale in the Perth Basin. To demonstrate this, a cross-plot of sonic transit time versus density in the shale intervals was analyzed to see where the overpressured data points fall. Figure 7.25 shows that the points with higher difference between neutron porosity and density porosity fall farther from the smectite-rich trend. It has been noticed that as depth increases, the neutron porosity–density porosity difference increases, and the ratio of smectite to illite decreases (illite percentage increases), and there are increases in the pore pressure gradients (unloading increase).

Further investigations were needed to ensure whether the patterns in Figure 7.25 are due to load transfer through clay transformation (unloading) rather than just variations in clay compositions. In order to do this, clay compositions were examined through XRD study.

7.4.5.3 X-Ray Diffraction

The proportions of different minerals in the Kockatea Shale were further determined by examining XRD on rock samples taken from the wells. The results presented in Tables 7.2 and 7.3 show that as depth increases, the proportion of illite and mixed-layer (illite-smectite) clays increases in the samples with minor or no presence of smectite clay. These results can be interpreted by the consistent transformation of smectite clay to illite and mixed-layer (illite–smectite) clays. Remarkably, as depth increases, the clay compositions are consistent with the process of clay transformations from smectite to either illite or to mixed-layer clay (illite–smectite), other than a random variation of clay composition within the formations under examination. In addition, the temperatures and the burial depths for the Kockatea Shale in these two wells are in the range at which the processes of the clay transformations mentioned occur.

7.4.5.4 Natural Gamma Ray Spectrometry Logs

Clay minerals identification was further investigated by analyzing the natural gamma ray spectrometry NGS logs. Analysis of NGS log plots from Well #1 indicates a higher percentage of illite clay in the overpressured section of the Kockatea Shale. The cross-plot of thorium versus potassium illustrates that illite is the principal clay contained in the examined Kockatea Shale interval. It can be seen from Figure 7.26 that the majority of data points plot in the illite segment of the chart that further supports the hypothesis of the possible complete transformation of smectite to illite.

7.5 DISCUSSION

The reasons why pore pressure should be studied for the potential gas shale potential intervals and the relevance of pore pressure to production potentials are discussed in Section 7.5.1. The overpressure detection and estimation method in shale intervals are discussed in Section 7.5.2. In addition, the relationship between pore pressure in Kockatea Shale and the tectonic compression phases that took place sequentially in the Perth Basin is discussed in Section 7.5.3. Furthermore, the origins of overpressure in the Kockatea

DISCUSSION 161

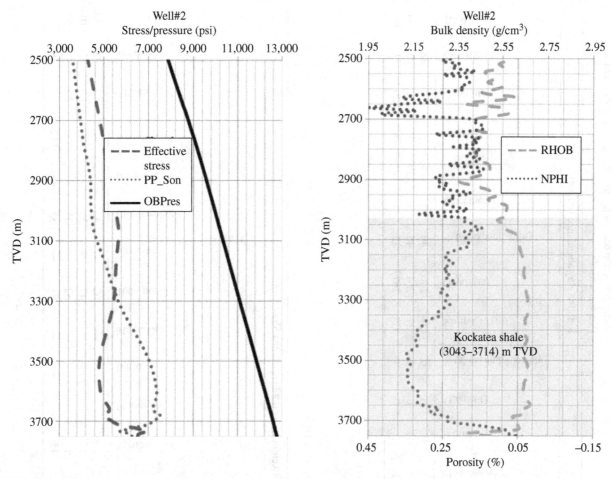

FIGURE 7.24 Well log responses to overpressure across the Kockatea Shale (Perth Basin).

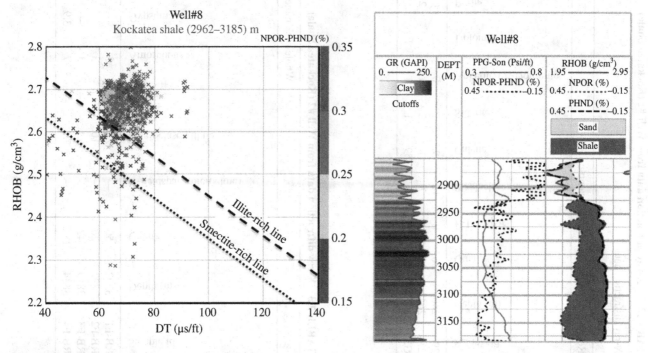

FIGURE 7.25 Sonic–density cross-plots for Kockatea Shale for Well #8 (left), (Neutron porosity–density porosity) is shown in color. Well logs plot and pore pressure profile versus depth for Well #8 (right).

TABLE 7.2 X-ray diffraction data from Well #1 (Kockatea Shale)

X-ray diffraction data

Relative clay abundance in bulk sample

Sample id	Depth (m)	Quartz	K-Feldspar	Plagioclase (albite)	Calcite	Siderite	Ankerite/FE-Dolomite	Dolomite	Pyrite	Fluorapatite	Barite	Magnetite	Total expandable clay
29	2400	36	8	12	9	1	0	2	0	0	0	0	0
27	2645	22	2	7	1	2	1	0	2	0	1	0	2
G3	2677	8	2	2	59	1	1	1	2	1	1	0	4

X-ray diffraction data—clay mineralogy (weight%) well #1 (Kockatea Shale)

Relative clay abundance (<4 μm)

Sample id	Depth (m)	% I/S expandability	Smectite	Illite–smectite	Illite	Kaolinite	Chlorite	Total clay
29	2400	40	1	19	19	38	23	100
27	2645	45	1	29	16	33	21	100
G3	2677	45	5	33	31	20	11	100

TABLE 7.3 X-ray diffraction data from Well #2 (Kockatea Shale)

Whole rock mineralogy (weight%) well #2 (Kockatea Shale)

Sample id	Depth (m)	Quartz	K-Feldspar (microcline)	Plagioclase (albite)	Calcite	Amorphous	Gypsum	Dolomite (calcite magnesium)	Pyrite	Fluorapatite	Barite	Magnetite	Total nonclay	Smectite	Illite/smectite	Illite	Kaolinite	Chlorite	Total clay	Grand total
RB-F1	3792	50	—	7.3	—	3	—	—	1.3	—	—	—	62	—	32	—	—	6.1	38	100
RB-F2	3798	20	—	10	9.7	3	—	—	4.3	—	—	—	47	—	35	—	16	1.6	53	100
RB-F3	3819	42	—	6.1	—	3	—	—	0.7	—	—	—	52	—	41	—	—	7.1	48	100
RB-F4	3834	18.2	—	7.1	—	7	3.9	—	15	—	—	—	51	—	35	—	11.8	2.8	49	100

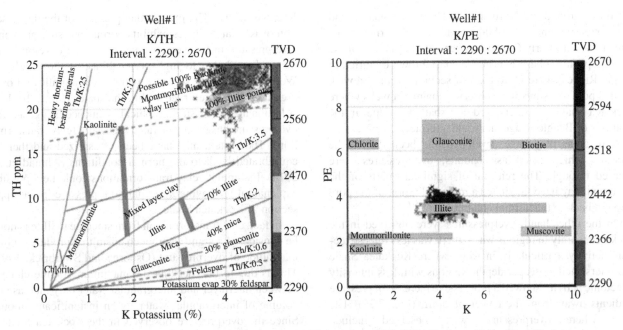

FIGURE 7.26 Natural gamma ray spectrometry (NGS) logs from Well #1 for Kockatea Shale (Perth Basin).

Shale are discussed in Section 7.5.4. An additional point for discussion is a verification of the estimated pore pressure in the absence of measured pore pressure such as RDT, MDT, and DST tests, which is discussed in Section 7.5.5.

7.5.1 Significance of Pore Pressure Study

Pore pressure evaluation in sedimentary rocks is not only significant for safe drilling operation and appropriate completion planning, but it is also important for formation evaluation analysis for the gas shale intervals. If pore pressure evaluation study is coupled with a study of other shale parameters such as the total organic contents (TOCs) and the thermal maturity of shales, this may be used for the identification of shale gas sweet spots. For an efficient formation evaluation process, it is important to know the pore pressure regime, which can be normal, subnormal, or abnormal high pressure (overpressure). The overpressure phenomenon is frequently detected in low permeability zones such as shale formations as shale tends to retain overpressure more when compared to other types of rocks. Furthermore, overpressure could be attributed to an indicator of commercial gas accumulation and therefore, give greater significance to the formation that is being evaluated.

7.5.2 Overpressure Detection and Estimation

In total, data from 35 boreholes in the Perth Basin were evaluated. The normal compaction trend method (NCT) was implemented to predict pore pressure in the Perth Basin in Western Australia. Well log and mud log data were the base of this study. Pore pressure was inferred indirectly using Eaton's equation from sonic logs utilizing the Interactive Petrophysics software. Overpressure was observed through the diversion of the porosity-dependent parameters from their normal trends. These data parameters include sonic transit time, resistivity, porosity, and density. Sonic transit time was preferred for pore pressure estimation of shales mainly due to it being less affected by bad borehole conditions than any other logs such as resistivity and density logs.

7.5.3 Pore Pressure and Compressional Tectonics

Investigations into the well log data from the Perth Basin have characterized the lower section of Kockatea Shale in some parts of the Perth Basin as being overpressured. On the other hand, similar sections in similar formations have been found to be normally pressured in other parts of the basin. It is believed that the severe uplifting and the subsequent erosion caused by compressional tectonics had a great impact on the pore pressure distribution in the Kockatea Shale. Normal pore pressure regimes were observed in the Kockatea Shale in areas that were rigorously uplifted and eroded. Uplifting and subsequent erosion due to tectonics compression removed significant portions of the Kockatea Shale in some regions and induced fractures within the Kockatea Shale in some other areas. This took place in localities that are in the immediate vicinity of the centers of the two major uplifting and extension phases. The first phase of extension is an Early Permian phase created half-grabens that hinge around the Northampton High. The second and more extensive phase of oblique extension occurred in the Late Jurassic to Early Cretaceous periods. The center of the second uplifting phase is near the coastal town of Jurien in Beagle Ridge, where up to 8 km of sections were removed. Normal pore pressures were observed mainly in wells where Kockatea Shale was intersected at a shallower depth in the localities of Beagle Ridge and Cadda Terrace. Overpressure in these areas may have developed and been due to the fractures which acted as conduits, overpressures were released through. The removal of significant parts of the formation may have facilitated a re-equilibration of the pore pressure back to the normal condition.

On the other hand, overpressures were observed in the Kockatea mainly in regions where there was less intense tectonic activity, particularly in wells where Kockatea Shale was intersected at deeper depth. Regions with less intensity in tectonic activity showed a progressive increase in pressure gradients away from the center of uplift (Fig. 7.23). The areas where overpressures were observed include Dandaragan Trough and adjacent terraces that have similar structures. The phenomenon of overpressure has been observed by the diversion of the effective stress-dependent parameters from their normal trends. The top of the overpressure zones are the depths where the diversion occurred and overpressures were confirmed by cross-checking the available data such as drilling reports, mud log data.

7.5.4 Overpressure-Generating Mechanisms

It is noted that there were combination mechanisms that contributed to overpressure development, driven by the complicated basin geology. Fluid expansion and later tectonic loading have contributed to different extents to overpressure development.

Among fluid expansion mechanisms, clay diagenesis is the mechanism most likely to have contributed to overpressure development in the lower section of the Kockatea Shale. This deduction was reached from the results that were obtained from the analysis of wireline log responses, sonic-density cross-plots, XRD, and NGS logs. The shale in Figure 7.25 is of a depth and temperature where it is likely for the smectite to be mostly transformed to illite. This figure shows points with a higher difference between neutron porosity and density porosity falling farther from the smectite-rich trend. It has been noticed that as depth increases, the difference between neutron porosity and density porosity increases while the ratio of smectite to illite decreases, and therefore, the pore pressure gradients increase (unloading increases). It should be noted that while there are a few data points where the difference of neutron porosity–density porosity approaches the smectite rich trend (the ratio of smectite to illite increases), these data points are not considered as being representative of the whole shale interval other than in the inter-bedded sandstone sections within the Kockatea Shale. The present-day pressure of the upper section of Kockatea Shales exhibits normal pressure gradients. Overpressure in this section may have been developed and then released due to high permeability of the overlaying Woodada sandstone sections, which have not allowed overpressures to be preserved within the shales (Fig. 7.16, left). Additionally, this stratigraphical sequence suggests that overpressure has been generated internally (clay transformation). The system may have been pressurized and then re-equilibrated back to the normal conditions (Ahmad et al., 2014). Incomplete clay transformations could be a possible explanation for having a normal pore pressure in the upper section of Kockatea Shale.

In fact, clay transformation from smectite to illite is not to be considered a significant mechanism that produces a high magnitude of overpressure (Osborne and Swarbrick, 1997). This is due to the fact that the maximum volume changes between interlayer and intergranular water only increase the volume of intergranular water by an insignificant amount. Since the overpressure observed in the Kockatea Shale is noticeably in high magnitude as noticed from the reversals of the vertical effective stress, the clay transformation mechanism could be combining with other overpressure-generating mechanisms.

The most likely mechanism to be associated with the clay transformation is the lateral tectonics compression. The main reasons for this deduction are (i) the forces induced by the principal stress (S_{hmax}) which act in a horizontal plane EW perpendicular to the main north–south and northwest–southeast faults trends (Fig. 7.15) and (ii) the positions and the trends of the main faults and the main transfer faults providing efficient lateral seals for the overpressure developed in the Kockatea Shale. The lateral tectonics explains the high magnitude of overpressure observed as the diagenesis effects cannot produce high overpressure. Lateral tectonics compression would have caused the vertical effective stress to be reversed and density log to change slightly or remain at a reasonably constant value because compaction is not reversible. Additionally, lateral tectonics causes an increase in neutron porosity (Fig. 7.24). More investigations are needed to ascertain whether clay transformation or lateral compression was the primary mechanism for generating overpressure.

7.5.5 Overpressure Results Verifications

Mud logging data including mud weight, drilling ROP, kicks, and so on were used to validate the results and replace the absence of direct pressure measurement (e.g., RFT, DST). In locations where there were diversions of the porosity-dependent parameters from their normal trends, mud logging data in these sections showed every indication of Kockatea Shale being overpressured. Overpressure-related parameters obtained from mud logging data include a continuous increase in each of the ROP, the mud weight, and the gas shows within these intervals. It is noted there were also no changes in the other drilling parameters such as WOB, bit and surface rotations (RPM), mud flow rate, and mud pump pressure.

7.6 CONCLUSIONS

- Normal pore pressure profiles were observed in the Kockatea Shale in areas that have been rigorously uplifted and where there was severe uplifting and erosion. Uplifting and erosion due to tectonics compression removed significant portions of the Kockatea Shale in some areas and induced fractures in some other areas. The removal of significant parts of the formation and the induced fractures have acted as communication channels and facilitated a re-equilibration of the pore pressure back to the normal state or condition.
- Overpressures were observed in the Kockatea mainly in areas where there was less intensity in tectonic activity and where Kockatea Shales intersected at a greater depth. Regions with less intensity of tectonic activity showed an increase in pressure gradients away from the center of uplift. The depth to the top of the overpressure zone is linearly related to the depth to the top of the Kockatea Shale.
- The occurrence of overpressures in Kockatea Shale that were buried to deeper depths and the analysis of data suggest that overpressures could be generated by a number of overpressure-generating mechanisms.
- The overpressures in the lower section of Kockatea Shale were developed internally due to clay transformation processes mainly by complete transformation of smectite clay to illite and mixed-layer clay (smectite/illite). This conclusion was reached by analyzing well logs, clay compositional variations, and the stratigraphical sequence. On the other hand, the upper section of Kockatea Shale showed a normal pressure profile as a result of either incomplete clay transformations or overpressure that initially developed and then re-equilibrated back to the normal conditions through overlaying high permeability Woodada sandstone. Woodada formation would not allow overpressure to develop, and it maintained normal pressure in the upper section of Kockatea Shale. This stratigraphical sequence also suggests that overpressure has been generated internally.
- The principal direction of the stress and the complicated structure of the northern Perth Basin indicate that the lateral tectonics compression mechanism is the other mechanism that is associated with the clay transformation mechanism. The main reasons for this claim are (i) the forces induced by the principal stress (S_{hmax}), which act in a horizontal plane EW perpendicular to the main north–south and northwest–southeast faults trends, (ii) the positions and the trends of the main faults and the main transfer faults providing efficient lateral seals for the overpressure developed in the Kockatea Shale, and (iii) the observed high magnitude of overpressure, suggesting that lateral tectonics compression has contributed significantly to overpressure development as the diagenesis effects cannot produce such high overpressure.
- It is difficult to ascertain whether either the lateral tectonic compression or the clay transformation is the major overpressure-generating mechanism in the lower section of Kockatea Shale. However, both mechanisms contributed to a different extent to overpressure development.
- The severe tectonic activity accompanying the final breakup of the continents and seafloor spreading are responsible for most of the major structural features of the Perth Basin and responsible also for the distribution of pore pressure in the basin.

NOMENCLATURE

Abbreviation	Log name (unit)
DTC	Compressional wave transit time log (µs/ft)
NCT_Son	Normal compaction trend obtained from sonic log (µs/ft)
GR	Gamma ray (API)
VCLGR	Volume of shale from gamma ray log (%)
RHOB	Bulk density log (g/cm^3)
RhoGard	Density calculated from Gardner's method (g/cm^3)
PPG_Son	Pore pressure gradient estimated from sonic log (psi/ft)
PP_Son	Pore pressure estimated from sonic log (psi)
OBPres	Overburden pressure/stress (psi)
DST	Drill stem pressure test in gradient (psi/ft)
ROP	Drilling rate of penetration (m/h)
ZDEN	Bulk density log (g/cm^3)
g_p	Estimated pore pressure gradient (psi/ft)
g_{ob}	Overburden pressure gradient (psi/ft)
g_n	Normal pore pressure gradient (psi/ft)
DT	Sonic transit time (µs/ft)
Δt_n	Normal sonic transit time (µs/ft)
Δt_o	Observed sonic transit time (µs/ft)
TVD	True vertical depth (m)
R_n	Normal resistivity (ohm)
LLD	Resistivity (ohm)
SN	Resistivity (ohm)
R_o	Observed resistivity (ohm)
PHND	Porosity obtained from enhanced density (%)
NPOR	Neutron porosity (%)
NPHI	Neutron porosity (%)
CNC	Borehole size corrected compensated neutron porosity (%)

σ_{ob}	Overburden stress gradient (psi/ft)
PEF	Photoelectric absorption factor log (unitless)
ρ_b	Bulk density (g/cm^3)
v	Sonic velocity (ft/s)
CAL and CALX	Caliper logs (inch)
Mwt	Mud weight (psi/ft)
ECD	Equivalent circulation density (psi/ft)

REFERENCES

Ahmad A, Rezaee R, Rasouli V. Significance of compressional tectonic on pore pressure distribution in Perth Basin. J Unconvent Oil Gas Resour 2014;7 (2014):55–61.

Alixant J-L, Desbrandes R. Explicit pore-pressure evaluation: concept and application. SPE Drill Eng 1991;6 (3):182–188. DOI: 10.2118/19336-pa.

Athy LF. Density, porosity, and compaction of sedimentary rocks. AAPG Bull 1930;14 (1):1–24.

Boles JR, Franks SG. Clay diagenesis in Wilcox sandstones of Southwest Texas: implications of smectite diagenesis on sandstone cementation. J Sediment Res 1979;49 (1):55–70.

Bowers G. Pore pressure estimation from velocity data: accounting for overpressure mechanisms besides undercompaction. SPE Drill Complet 1995;10 (2):89–95.

Bowers G. Detecting high overpressure. Lead Edge 2002;21 (2):174–177.

Bowers G, Katsube TJ. The role of shale pore structure on the sensitivity of wire-line logs to overpressure. Mem Am Assoc Petrol Geol 2002;76:43–60.

Butler R. 2011. Pore pressure analysis. Formation Evaluation Society of Australia. Available at http://www.fesaus.org/webcast/2009/06/RexButler/NonMember/. Accessed July 20, 2011.

Cheng CH, Toksöz MN. Inversion of seismic velocities for the pore aspect ratio spectrum of a rock. J Geophys Res Solid Earth (1978–2012) 1979;84(B13):7533–7543.

Colten-Bradley VA. Role of pressure in smectite dehydration—effects on geopressure and smectite-to-illite transformation. AAPG Bull 1987;71 (11):1414–1427.

Crostella A. An evaluation of the hydrocarbon potential of the onshore northern Perth Basin. Western Australia: Geological Survey of Western Australia, Report 1995;43:67.

Crostella A, Backhouse J. Geology and petroleum exploration of the central and southern Perth Basin. Western Australia: Geological Survey of Western Australia; 2000.

Dickinson G. Geological aspects of abnormal reservoir pressures in the Gulf Coast Region of Louisiana, USA 3rd World Petroleum Congress; May 28–June 6; the Hague, the Netherlands. 1951.

Draou A, Osisanya SO. New methods for estimating of formation pressures and fracture gradients from well logs. SPE Annual Technical Conference and Exhibition; January 1, 2000; Dallas, TX; 2000. Available at http://www.onepetro.org/mslib/app/Preview.do?paperNumber=00063263&societyCode=SPE. Accessed January 13, 2014.

Dutta NC. Deepwater geohazard prediction using prestack inversion of large offset P-wave data and rock model. Lead Edge 2002;21 (2):193–198.

Eaton BA. The equation for geopressure prediction from well logs. Fall Meeting of the Society of Petroleum Engineers of AIME; January 1, 1975; Dallas, TX; 1975. Available at http://www.onepetro.org/mslib/app/Preview.do?paperNumber=00005544&societyCode=SPE. Accessed January 13, 2014.

Fertl WH. Gamma ray spectral data assists in complex formation evaluation. Log Anal 1979;20 (5):3–37.

Fertl WH, Timko D. Parameters for identification of overpressure formations. Drilling and Rock Mechanics Conference; January 5–6; Austin, TX; 1971.

Gardner GHF, Gardner LW, Gregory AR. Formation velocity and density-the diagnostic basics for stratigraphic traps. Geophysics 1974;39 (6):770–780.

Gretener PE. *Pore Pressure: Fundamentals, General, Ramifications, and Implications for Structural Geology (Revised)*. AAPG: Oklohoma; 1979.

Hansom J, Lee M-K. Effects of hydrocarbon generation, basal heat flow and sediment compaction on overpressure development: a numerical study. Petrol Geosci 2005;11 (4):353–360.

Hermanrud C, Wensaas L, Teige GMG, Nordgard BHM, Hansen S, Vik E. Shale porosities from well logs on haltenbanken (offshore mid-Norway) show no influence of overpressuring. In: Law BE, Ulmishek GF, Slavin VI, editors. *Abnormal Pressures in Hydrocarbon Environments*. AAPG Memoir 70: Oklohoma; 1998. p 65–85.

Hillis RR, Reynolds SD. The Australian stress map. J Geol Soc 2000;157 (5):915–921.

Hower J, Eslinger EV, Hower ME, Perry EA. Mechanism of burial metamorphism of argillaceous sediment: 1. Mineralogical and chemical evidence. Geol Soc Am Bull 1976;87 (5):725–737.

Iasky RP, Mory AJ. Structural and tectonic framework of the onshore northern Perth Basin. Explor Geophys 1994;24 (4):585–592.

Jorden JR, Shirley OJ. Application of drilling performance data to overpressure detection. J Petrol Technol 1966;18 (11):1387–1394.

Katahara K. What is shale to a petrophysicist? Lead Edge 2008;27 (6):738–741.

King RC, Hillis RR, Reynolds SD. In situ stresses and natural fractures in the northern Perth Basin, Australia. Aust J Earth Sci 2008;55 (5):685–701.

Lesage M, Hall P, Pearson JRA, Thiercelin MJ. Pore-pressure and fracture-gradient predictions. SPE J Petrol Technol 1991;43(6). DOI: 10.2118/21607-pa.

Lesso Jr WG, Burgess TM. 1986. Pore pressure and porosity from MWD measurements. SPE/IADC Drilling Conference; January 1, 1986; Dallas, TX. Available at http://www.onepetro.org/mslib/app/Preview.do?paperNumber=00014801&societyCode=SPE. Accessed January 13, 2014.

Lewis CR, Rose SC. A theory relating high temperatures and overpressures. J Petrol Technol 1970;22 (1):11–16.

Luo X, Vasseur G. Contributions of compaction and aquathermal pressuring to geopressure and the influence of environmental conditions (1). AAPG Bull 1992;76 (10):1550–1559.

Miller TW. New insights on natural hydraulic fractures induced by abnormally high pore pressures. AAPG Bull 1995;79 (7):1005–1018.

Miller TW, Luk CH, Olgaard DL. The interrelationships between overpressure mechanisms and in situ stress. Mem Am Assoc Petrol Geol 2002;76:13–20.

Mory AJ, Iasky RP. 1996. Stratigraphy and structure of the onshore northern Perth Basin. Western Australia: Western Australia Geological Survey, Report 46.

Mory AJ, Haig DW, McLoughlin S, Hocking RM. Geology of the Northern Perth Basin, Western Australia—a field guide. Geological Survey of Western Australia Record 2005;2005:9.

Mouchet J-P, Mitchell AF. Abnormal Pressures While Drilling: Origins, Prediction, Detection, Evaluation. Volume 2, Paris: Editions Technip; 1989.

Muir W. 2013. Gas shale in Australia, Drivers and Roadblocks. Available at http://www.pnronline.com.au/article.php/99/1234. Accessed June 24, 2013.

Osborne MJ, Swarbrick RE. Mechanisms for generating overpressure in sedimentary basins: a reevaluation. AAPG Bull 1997;81 (6):1023–1041.

Pikington PE. Uses of pressure and temperature data in exploration and now developments in overpressure detection. SPE J Petrol Technol 1988;40(5). DOI: 10.2118/17101-pa.

Ramdhan A, Goulty NR. Overpressure and mudrock compaction in the Lower Kutai Basin, Indonesia: a radical reappraisal. AAPG Bull 2011;95 (10):1725–1744.

Rubey WW, Hubbert M. Role of fluid pressure in mechanics of overthrust faulting II. Overthrust belt in geosynclinal area of western Wyoming in light of fluid-pressure hypothesis. Geol Soc Am Bull 1959;70 (2):167–206.

Shunhua CAO, Yuhong X, Chang L, Gongrui Y, Ping YI, Jun CAI, Liqiang XU. Multistage approach on pore pressure prediction—a case study in South China Sea. International Oil & Gas Conference and Exhibition in China; January 1, 2006; Beijing, China; 2006. Available at http://www.onepetro.org/mslib/app/Preview.do?paperNumber=SPE-103856-MS&societyCode=SPE. Accessed January 13, 2015.

Song T, Cawood PA. Structural styles in the Perth Basin associated with the Mesozoic break-up of greater India and Australia. Tectonophysics 2000;317 (1–2):55–72. DOI: 10.1016/s0040-1951(99)00273-5.

Swarbrick RE, Osborne MJ, Yardley GS. Comparison of overpressure magnitude resulting from the main generating mechanisms. In: Huffman A, Bowers G, editors. Pressure Regimes in Sedimentary Basins and Their Prediction. AAPG Memoir 70: Oklohoma; 2002. p 1–12.

Tanguy DR, Zoeller WA. 1981. Applications of measurements while drilling. SPE Annual Technical Conference and Exhibition; January 1, 1981; San Antonio, TX; 1981. Available at http://www.onepetro.org/mslib/app/Preview.do?paperNumber=00010324&societyCode=SPE. Accessed January 13, 2015.

Terzaghi K, Peck RB, Mesri G. Soil Mechanics in Engineering Practice. New York: Wiley-Interscience; 1996.

Tingay MRP, Hillis RR, Morley CK, Swarbrick RE, Okpere EC. Variation in vertical stress in the Baram Basin, Brunei: tectonic and geomechanical implications. Mar Petrol Geol 2003;20(10):1201–1212. DOI: http://dx.doi.org/10.1016/j.marpetgeo.2003.10.003.

Van Ruth P, Hillis R, Tingate P, Swarbrick R. The origin of overpressure in "Old" sedimentary basins: an example from the Cooper Basin, Australia. Geofluids 2003;3 (2):125–131. DOI: 10.1046/j.1468-8123.2003.00055.x.

Wallace WE. 1965. Abnormal subsurface pressures measured from conductivity or resistivity logs. The Log Analyst V(4):1–13. Available at http://www.onepetro.org/mslib/app/Preview.do?paperNumber=SPWLA-1965-vVn4a2&societyCode=SPWLA. Accessed January 13, 2015.

Watts EV. 1948. Some aspects of high pressures in the D-7 zone of the Ventura. Available at http://www.onepetro.org/mslib/app/Preview.do?paperNumber=SPE-948191-G&societyCode=SPE. Accessed January 13, 2015.

8

GEOMECHANICS OF GAS SHALES

VAMEGH RASOULI
Department of Petroleum Engineering, University of North Dakota, Grand Forks, ND, USA

8.1 INTRODUCTION

Gas shale reservoirs have recently been the subject of many studies in various countries due to energy shortage and energy prices (Hartwig et al., 2009; Jiang et al., 2010; Jingzhou et al., 2011). Gas shale reservoirs are the second largest unconventional energy resource after heavy oil. The term "play" is used by the energy industry for the gas shales to indicate a specific area targeted for exploration as they believe that there is an economic quantity of natural gas located there. According to the United States Geological Survey (USGS) estimation, gas shale can hold up to 460 Tcf of gas (Ross and Bustin, 2008). However, production from gas shale reservoir as a source rock is very recent in Australia, where most gas shale reservoirs are located in Cooper Basin in South Australia, and the Perth and Canning basins in Western Australia. There are at least two main reasons why these types of reservoirs are more attractive compared to other energy resources: (i) its low levels of carbon dioxide (CO_2) emission and (ii) the very low level of sulfur dioxide contents. As a matter of fact, development of shale gas has the advantage of being a great resource, with long life and production cycle (Daixu et al., 2011; Xinjing et al., 2007). It is also a very good solution to cope with climate change and to increase the economic growth as well as energy security (Jiang et al., 2010; Ma et al., 2011). Among the different parameters indicating whether shales have potential to be a gas resource, organic matter abundance, type and thermal maturity, porosity–permeability relationships, pore size distribution, and brittleness and its relationship to mineralogy and rock fabric are some of the important ones (Josh et al., 2012).

Recent success in exploration and production from unconventional gas shale reservoirs can be attributed to scientific study, engineering innovation, advancements in technology risk taking, and so on. Although the concepts and technologies related to these reservoirs are now mature, and there is considerable experience given from similar basins worldwide, basin conditions are unrepeatable, so it will be very hard to apply a single model of productive gas shale to any other reservoirs even in the same field. Thus, every shale reservoir has its own separate exploration development and strategy. Here, development of gas shale depends mainly on geological, geochemical, and engineering studies (Montgomery et al., 2005). In fact, geological and geochemical assessments are some basic tasks for exploration and development of gas shale reservoirs. This is because geological analysis can identify and characterize the portions of shale, while geochemical data is used to explain shale potential as well as observing patterns of productivity. Engineering studies in these cases, on the other hand, help us to determine the geomechanical parameters of gas shale reservoirs for developing hydraulic fracture schemes to increase the productivity. However, substantial differences exist between gas shale reservoirs and conventional reservoirs in their geology, mechanical aspects, and mechanism of fracture initiation, which have not yet been well understood, leading to the lack of quantitative technical basis for design and implementation of hydraulic fracturing in this type of reservoir (Gregory et al., 2011). The aim of this chapter is to present different petrophysical, mechanical, and seismological studies carried out on gas shale reservoirs. This is expected to provide a better understanding of the impact of

---

Fundamentals of Gas Shale Reservoirs, First Edition. Edited by Reza Rezaee.
© 2015 John Wiley & Sons, Inc. Published 2015 by John Wiley & Sons, Inc.

170 GEOMECHANICS OF GAS SHALES

these parameters on various applications during different phases of a field life including exploration, drilling, and completion.

8.2 MECHANICAL PROPERTIES OF GAS SHALE RESERVOIRS

The characteristics of shales have been mainly studied on many occasions as a seal or overpressure region (e.g., Dewhurst and Hennig, 2003; Dewhurst et al., 1998, 1999; Yang and Aplin, 2007) or wellbore instability problematic area (e.g., Detournay et al., 2006; Horsrud, 2001; Sarout and Detournay, 2011). Many studies have also been conducted on preventing pore pressure build up around the wellbore caused by the shale–drilling fluids interaction (Bol and Woodland, 1992; Ewy and Stankovich, 2000; Schlemmer et al., 2002; Tare et al., 2000; Van Oort et al., 1995; Yu et al., 2001). However, there is still limited information and knowledge regarding the geomechanical parameters of gas shale reservoirs.

8.2.1 Gas Shale Reservoir Properties under Triaxial Loading

Hydraulic fracturing is a stimulation technique used in many situations to enhance productivity. Therefore, geomechanical properties of gas shale reservoirs need to be studied in much further detail since they play a vital role in gaining a better understanding of fracture initiation and propagation, as well as fracture reopening in this type of reservoir (e.g., Britt and Schoeffler, 2009). Brittleness of shale reservoir is a very important aspect in a hydraulic fracturing operation since those shales with stiffness less than a certain value cannot be considered for hydraulic fracturing (Britt and Schoeffler, 2009). Thus, a measurement of the strength and stiffness of such shales is very important since it can determine the brittleness of these shales and help us to initiate the fractures and keep them open. In addition, the state of *in situ* stresses needs to be defined. This includes the magnitude and orientation of principal stresses in the field and stress regime. This is because having knowledge about the orientation of principal stresses with respect to plane of foliation in shale is very important due to the anisotropic behavior induced by a weak plane of bedding. Geomechanical parameters that are important for assessment of shale behavior consist of elastic parameters of Young's modulus and Poisson's ratio and strength properties including friction coefficient, cohesive strength, and unconfined compressive strength (UCS). Triaxial compression and ultrasonic velocity tests are performed on shale plug samples with diameters of 1.5 or 2.0 in. a length to diameter ratio of 3.8–5 cm order to estimate the mechanical properties of shales. The triaxial compression test, which is a destructive lab experiment, is

FIGURE 8.1 A conventional triaxial stress frame.

used to determine static Young's modulus and Poisson's ratio, whereas ultrasonic velocity tests are used to determine the equivalent dynamic properties ratio. Figure 8.1 shows a typical triaxial testing set up where the sample is placed into a hook cell and subjected to a constant confining pressure.

Then the axial stress is increased until the rock reaches the failure point. Usually, 4–5 tests at different confining pressures are performed on ideally identical samples to draw the Mohr circles and then determine the failure envelope and extract mechanical properties of the sample. As usually there is limited access to shale samples, multistage tests are performed on one sample only to estimate mechanical properties (ISRM, 1978). In this type of test, the sample, which is subjected to a certain confining pressure, is axially loaded until deviation is observed in the stress–strain curve. At this point the loading is stopped, the axial load is released and the larger confining pressure is applied to the sample. The experiment is now repeated for the second confining pressure. A similar procedure is repeated for 4–5 stages and the sample is taken to the failure point at the final stage of loading. There has been a large debate about the advantages and disadvantages of a multistage versus single-stage test; however, multistage tests are perhaps the only available option when there is no access to sufficient and identical samples. A view of a typical stress–strain curves corresponding to a multistage triaxial test is shown in Figure 8.2.

In doing a triaxial test on shale samples, preservation of the cores after it is retrieved is a big challenge. Loss of pore

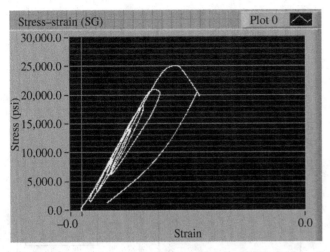 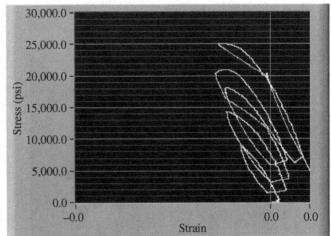

FIGURE 8.2 Axial stress–axial strain (left) and axial stress–radial strain (right) curves corresponding to a typical multistage triaxial test.

water from low porosity shales with appreciable amounts of clay generally results in strengthening of the material. This will introduce a significant increase in the associated strength and stiffness parameters of the sample (Ghorbani et al., 2009). In addition, drying of shales can induce high capillary pressures, causing softer specimens to be destroyed (Horsrud et al., 1998b). Thus, for partially saturated shales such as gas shales, it is suggested that materials be cling-filmed, wrapped in tin foil, and then waxed, either as whole cores or as core plugs and tested in as short a time period as possible as wax is permeable to air/water on longer time frames. If the samples are preserved for lab tests, static and dynamic tests can be conducted. Typically, shales dominated by mixed layer of illite–smectite with Young's Modulus of 1–3 GPa and a UCS of 8 MPa cannot be considered for hydraulic fracturing. So here triaxial tests can play an import role in describing a shale formation from the hydraulic fracturing point of view. According to Josh et al. (2012), shales with higher cohesive strength and UCS containing silts and approximately 30% clay mainly illite, which is the most thermally stable clay mineral, are often seen in gas shales reservoirs. The silty nature of gas shales and clay composition has an impact on the mechanical properties of such shales. However, those shales with clay content of up to 60% and smectites are usually the weakest of the common clay minerals, resulting in the low Young's modulus and weak strength parameters. Rickman et al. (2008) suggest that low modulus and high Poisson's ratio shales are generally too ductile to be prospective. They have found that low Young's modulus gas shales are those with relatively high clay content. They categorized nonprospective shales as a group having very high clay content and generally exhibiting visible laminations to the naked eye, whereas prospective shales in many cases are isotropic since their elastic properties in different directions are mostly the same. Britt and Schoeffler (2009) indicated that a significant increase in displacement during the triaxial test in shale core samples is due mainly to the ductility of the formation at high closure stresses and not the failure of the core sample. Therefore, if the sample is isotropic and laminations are not visible, core plug orientation may not be important in terms of its effect on mechanical response of the shale. This means that either vertical plugs from the whole core or horizontal plugs from sidewall cores can be used for determining the rock mechanical properties of the samples.

8.2.2 True-Triaxial Tests

In a conventional triaxial test, as explained in the previous section, the sample is subjected to an isotropic confining pressure during the experiment. It should be noted, however, that in real field condition the rocks are under three different independent stresses known as principal stresses. These are usually the vertical stress, due to the weight of overburden rocks, and two horizontal stresses perpendicular to each other, one being maximum and the other one minimum. Simulating the rock failure using the conventional method where a cylindrical sample is used for the experiment means that the effect of the intermediate stress is ignored in the rock's failure response. This could have a significant effect on rock behavior, in particular shales which are made of VTI material where the effect of having three independent stresses on the sample at different orientations could be more pronounced in terms of impact on its failure response. In order to test a sample under three independent stresses, that is, true-triaxial stress conditions, a cube shape sample is needed. Figure 8.3 shows a view of a true-triaxial stress cell (TTSC) which has been designed for estimation of rock strength parameters and various petroleum-related applications including hydraulic fracturing and sanding analysis. Samples of up to 300 mm can be tested using this cell and the magnitude of external stresses that can be applied to a 100 mm sample is approximately 15 MPa. A pore pressure of 21 MPa can be applied. Figure 8.3 (right) shows the view

FIGURE 8.3 View of the true-triaxial stress cell used for advance rock mechanics experiments (top) and a 100 mm sample placed in the cell for hydraulic fracturing test (bottom).

of the cell and a 100 mm sample placed in the cell before closing the top lid to apply vertical stress. In this figure (right), the horizontal rams through which the two independent horizontal stresses are applied to the sample are shown. The drilled hole in the sample center is connected to the outside of the cell using a pipe which is the path for injecting the fracturing fluid, when performing a hydraulic fracturing experiment.

Figure 8.4 shows the plot of the maximum differential stress ($\sigma_1-\sigma_3$) versus strain for a sample tested under true-triaxial stress conditions (Minaeian et al., 2013). In this figure, ε_1, ε_2, and ε_3 are the strains along three principal stresses and ε_v is the volumetric strain. Through this type of

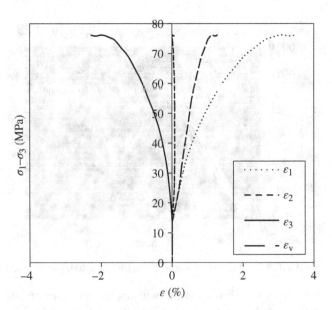

FIGURE 8.4 A typical differential stress versus strain curve for a sample tested under true-triaxial stress condition (Modified from Minaeian et al., 2013).

experiment on shale samples, it is possible to obtain VTI mechanical parameters (i.e., Poisson's ratio and Young's modulus) along the lamination planes and perpendicular to that. These parameters are the input to well design and planning when drilling in gas shales.

8.2.3 Gas Shale Reservoir Properties under Ultrasonic Tests

Velocity-based modeling is quite a well-known method used in geophysical and seismic exploration in order to characterize different properties of subsurface layers. Such applications in gas shale reservoirs are very new; therefore limited knowledge is available on how gas shales may respond to geophysical attributes. The lack of adequate knowledge about mechanical properties of gas shales, as explained before, is partly due to the difficulties associated with sample preservation for lab experiments. In addition to this, evaluation of gas shale reservoirs is further complicated due to the anisotropic nature of shales.

With the exception of a few attempts, most of the lab test reports are on samples tested under conditions different from their original *in situ* conditions. Sarout and Guéguen (2008) measured P- and S-wave velocities and estimated the dynamic and static properties of shale under ultrasonic tests. However, they could not control the pore pressure during deformation so the results presented by them were only approximate. Josh et al. (2012) in their ultrasonic measurements under isotropic stress conditions found that samples showed variations in velocity, and the elastic coefficients (Cij) as well as the P-wave and S-wave anisotropy parameters.

According to them weak shale has moderate porosity (~20%) and low velocities which increase by 10–15% with increasing stress. They indicated that the dynamic stiffness is three to five times the static stiffness for 100% water saturated shales. They also showed that the velocity of low porosity shale is almost two times higher than that of the weak shale which increases by approximately 5–8% with increasing isotropic stress. Kuila et al. (2011) indicated that a hard shale shows significant anisotropy in wave propagation but this anisotropy for the P-wave is much higher than that of the S-wave. They attributed this to low clay content and the laminated nature of the shale. In addition, it is well known that water or gas saturation has a significant effect on the P-wave velocity in a porous media. This impact will be increased as the degree of saturation increases. However, this straightforward concept may not be valid for gas shale reservoirs since there are publications reporting different effects of gas saturation in shale (Hsu and Nelson, 1993; Lashkaripour and Passaris, 1993; Rozkho, 2010). The possible reason for this might be the kerogen type, mineralogy, heterogeneity of the shale as well as its pore structure. Hence, it would not be rational to draw any conclusions about the mechanical parameters of shale based on the behavior of the P-wave velocity. Ghorbani et al. (2009) pointed out the significant increase in P-wave and S-wave velocity of partially saturated shale samples. Naturally, the S-wave velocity will not be affected by fluids, but because the rigidity of the shale increases as the fluid saturation decreases, even S-wave velocities will change. Nagra (2002), who did studies on shale gas reservoirs, concluded that as the water content decreases, different shales show different behavior. For instance, with low water content, shale seems to be a ductile–brittle material with some strain hardening. At very low water content, opalinus clay will be entirely brittle material with strength reduction toward its residual strength. In terms of stiffness, Young's modulus significantly increases as the saturation decreases. Considering the situation described earlier, care must be taken when assessing gas shale reservoirs containing different amounts of clays. This is mainly because of the negative impact of clays on the saturation, strength, and P-wave and S-wave velocity. It is still possible to relate the variation in P- and S-wave velocity to stiffness and strength of gas shale reservoir rocks, but care must still be taken (e.g., Gray, 2010).

8.2.4 Nanoindentation Tests on Gas Shale Plays

Shale is widely recognized as one of the most heterogeneous and complicated materials. Due to the variation of its composition, prediction of its elastic properties is a very hard task to accomplish. Having information about the mechanical properties of shale is critical as it can be helpful for drilling stability design and seismic interpretation (Ahmadov, 2011). Besides, hydraulic fracture will be significantly improved by acquiring enough information about the mechanical properties of shale. However, due to mechanical and chemical instability of shales, retrieving larger core samples for conventional mechanical testing will be barely possible. Hence, a new approach in the determination of the mechanical parameters of shale is really necessary in order to obtain the required information from small-scale sample of shale.

Nanoindentation testing (NIT) has been developed in the past few decades. The testing method uses a high-resolution electronic instrument to measure applied loads and value of displacement of an indenter (Hay and Pharr, 2000). The aim of NITs is to calculate Young's modulus and hardness of the sample from load-displacement recording. A small size sample (~mm) and a very small load are used in this type of test, so indention area will be as small as a few nanometers. Depth of penetration as well as area of indentation is also recorded using known geometry of an indenter tip. This technique is widely used to determine mechanical properties of metals, ceramics, polymers, etc. In conventional uniaxial or triaxial tests performed to obtain Young's modulus and mechanical properties of different samples, a relatively large sample (~cm) and high load is required to get accurate results. However, in NITs, a very small sample and very low force (~mN) are used to calculate mechanical properties of the samples. Thus, this technique can be a very good choice for determination of/determining the mechanical properties of shale samples since retrieving large samples from this type of formation is barely possible.

Abousleiman et al. (2007) used NITs on 10 Woodford outcrop shale samples to calculate elastic properties. X-ray diffraction (XRD) mineralogy data of these samples showed quartz content varying from 27 to 53 vol%, clay content ranging from 18 to 43 vol%, kerogen content varying from 11 to 18 wt%, and porosity ranging from 0.16 to 0.19 fraction volume. Indentations were done both parallel and perpendicular to the bedding plane. They exhibited low range of Young's moduli because the sample might be weakened due to chemical weathering at shallow depth. Zeszotarski et al. (2004) used nanoindentation technique to measure mechanical properties of kerogen. Elastic properties were measured both parallel and perpendicular to the bedding plane.

Samples used in their study showed total organic carbonate (TOC) content of 22 wt% and maximum temperature of 421°C. They indicated that Young's modulus of shale has an inverse relationship with TOC content. They measured hardness and indentation modulus on three orthogonal faces of shale sample cube and reported no evidence of anisotropy in the mechanical properties of kerogen. Ahmadov et al. (2009) performed nanoindentation measurements on a sample from organic-rich formation from 3800 m depth showing kerogen content in the range of 7–21 vol%. All imaging and measurements were done orthogonal to the horizontal bedding plane. They calculated Young's modulus of organics in the range of 10–15 GPa. Kumar (2012) conducted nanoindentation measurements on a total number of 144 organic-rich shale samples. His measurements were performed on different types of shales. He concluded that:

- Woodford samples show Young's modulus of 23–80 GPa. Samples with lower E values (23–30 GPa) showed either high concentration of TOC or high clay content; whereas samples with higher E values (60–80 GPa) showed relatively low TOC, porosity, and clay content. Hardness of this formation varies from 0.54 to 7.2 GPa and shows higher hardness values (6–7 GPa) due to high quartz content and low TOC and porosity values.
- Barnett samples gives Young's modulus of 39–78 GPa while higher E values were found to be either relatively low in TOC and porosity, or high in carbonate content.
- Haynesville samples have a Young's modulus of 31–79 GPa where samples with higher E were found to be either high in carbonate or low in TOC. Hardness for Haynesville samples with average value of 1.1 ± 0.6 GPa is attributed to the high carbonate content of samples.
- Eagle Ford samples were found to have a Young's modulus of 31–57.5 GPa. Samples with higher carbonate content exhibit higher Young's moduli. Hardness of these samples obtained was between 0.45 and 1.5 GPa.
- Ordovician shale with high average carbonate content of 73 ± 4 wt% showed Young's modulus of 49–57 GPa. They showed a hardness range from 1 to 1.3 GPa.

Kumar (2012) also indicated that for all shale plays, samples with high TOC and high porosity exhibited low Young's modulus, whereas samples with low TOC, low porosity, and high carbonate content showed high Young's modulus values. Hardness on the other hand shows negative correlation with porosity and clay content as well as poor correlation with TOC.

8.2.5 Scratch Tests

The uniaxial compressive strength (UCS) test is the most conventional method in the lab to measure the strength of rocks. UCS plays a key role in the design of underground structures, as well as ensuring the stability of the drilled wells in civil, mining, and petroleum engineering. The standard procedure used to determine the UCS has already been documented by ASTM (2010) and ISRM (Ulusay and Hudson, 2007). Many publications report on the application, advantages, and disadvantages of the UCS test (Bieniawski, 1968; Broch and Franklin, 1972; Hawkes and Mellor, 1970; Hudson et al., 1972; Jaeger et al., 1976; Wawersik and Fairhurst, 1970). It is well known that the UCS test suffers from several drawbacks as it requires cores of intact rock and considerable time for sample preparation, for example, polishing the sample ends and precise length to diameter ratio. In addition, the water content and the irregularity of the ends of the samples can also cause errors in the measurements results (Bieniawski, 1968; Dey and Halleck, 1981; Farmer, 1992; Hoek and Brown, 1980; Hudson et al., 1972). Furthermore, the UCS test can be only used for testing homogeneous and intact rock and not for heterogeneous, damaged, layered, or fractured rocks. This is due to the existence of the weakest plane, that is, joint or a pre-existing crack, in the core sample which determines the failure of the rocks. Point load, indentation, or Schmidt hammer tests are alternative tests which are used to obtain an estimation of rock strength (Bieniawski, 1974; Broch and Franklin, 1972; Chau and Wong, 1996; Rusnak and Mark, 1999; Szwedzicki, 1998). In fact, point load and indentation tests are very useful as they can be used to assess rock strength with very small size samples, while the Schmidt hammer test even allows testing the strength of the outcrops. Yet, all of these indirect methods suffer from many drawbacks. For instance, both the point load and the Schmidt tests can be remarkably affected by the elastic properties, the sample size, and the water content of the samples (Aydin and Basu, 2005; Thuro et al., 2001; Tsiambaos and Sabatakakis, 2004; Tsur-Lavie and Denekamp, 1982).

The nanoscratch test is one of the most recent methods used for determination of UCS. This test requires a small-scale sample for measurement and does not suffer from any of the disadvantages mentioned earlier for the direct and indirect method of UCS determination (Richard et al., 2012). In fact, it has been indicated and proved that the UCS of rocks can be reliably assessed from nanoscratch tests performed with a sharp cutter, and at a shallow depth of cut to prevent any significant chipping of the rock.

Scratch tests have been the subject of various studies discussing the effect of rock characteristics on drilling performance (Glowka, 1989; Nishimatsu, 1972; Deliac, 1986; Duc, 1974). During the scratch test, depth of cut and cutter velocity remains constant, while magnitude and orientation of the force acting on the cutter are measured (Detournay and Defourny, 1992; Duc, 1974; Fairhurst and Lacabanne, 1957). Two cutting mechanisms, ductile and brittle, usually occur in this type of test and depend mostly on the depth of cut (Chaput, 1992; Huang and Detournay, 2008; Huang et al., 2012; Richard, 1999; Richard et al., 1998). In fact, at shallow depth of cut, ductile regime will be the dominate mechanism while at larger depth of cut, brittle failure occurs. In the brittle regime fracture toughness controls the cutting force, whereas the UCS controls the cutting force in the ductile regime (Richard et al., 1998). Therefore, the scratch test should be performed under ductile regime. Considering the intrinsic specific energy associated with the cutting process, the inclination of the force acting on the cutting face, and the friction coefficient mobilized across the wear flat and the nominal wear flat area, UCS measurement can be done through the use of the scratch tests.

Several studies have been done to reveal the application and reliability of the scratch tests in different fields of study fields. Ulm and James (2011) showed that such test can be used in oil well cements cured at high temperatures and pressures. They found that increase of strength and toughness of different oil well cement baseline formulations can be related to the water-to-binder ratio for a series of cementations materials. Richard et al. (2012) performed nanoscratch tests under kinematic condition and concluded with a summary of extensive experimental testing on more than several hundred rocks that this test can provide very reliable results, while it also has many advantages over the conventional method of UCS measurements. They showed that the scratch tests can even be used to determine the UCS of different shale samples. Thus, considering the overall limitations in reserving shale core samples from unconventional oil and gas reservoirs, scratch tests can be a very good choice in such undesirable situations.

8.3 ANISOTROPY

Seismic acquisition is frequently used in petroleum, mining, and civil engineering in order to estimate various geomechanical and geotechnical parameters. Both surface and borehole seismic acquisitions are used for this purpose depending on the required applications.

The continuity logging method of measurements between boreholes has been used for detecting and characterizing the waves and exploring their potential applications in different geological environments, and is currently being used for reservoir continuity in hydrocarbon fields. The concepts and applications of this method are well documented by Liu et al. (1992), Dresen and Ruter (1994) for coal-seams, and by Krohn (1992), Turpening et al. (1992), Parra et al. (1996) for oil/gas reservoirs. The technique is now known as crosshole continuity logging or crosshole seismic logging. Previous applications of crosshole surveys include examples from the Conoco borehole test facility, Oklahoma (Lines et al., 1992, 1995; Liu et al., 1991) and the Gypsy test site, Oklahoma (Parra et al., 1996; Turpening et al., 1992). Synthetic examples were given by Zhong and Worthington (1994) and Parra (1996). Lou and Crampin (1991, 1992) have provided a theoretical basis for channel wave propagation in anisotropic media. To obtain information about the anisotropy of rocks located between the wells, three-component geophones are used. In this type of acquisition, numerous sources are used to propagate shear and compressional waves in media between the wells and several three-component receivers cover the given interval. The offset, or interval distance, between the source and the receiver boreholes depends on the type of measurements. The raypaths are almost horizontal in crosshole seismic acquisitions (Cole, 1997; Hardage, 1992).

Implementing this type of acquisitions makes it possible to pick the first arrival time of P- and S-waves velocity in vertical and horizontal directions, so it will be possible to obtain the anisotropy and Thomsen parameters in different media.

Surface seismic measurements like reflection and refraction acquisitions can be designed to use three-component geophones for first arrival picking of P- and S-wave. Compressional and shear wave velocity models used in processing surface seismic data in many cases do not consider the anisotropic behavior of subsurface formations. Although migration is used to locate events in their exact locations, this method does not directly account for the characteristics of anisotropic formations like shale. As determined by Thomsen (1986), anisotropy for nearly vertical wave propagation is mostly governed by parameter δ, which is a complex combination of elastic parameters (Thomsen, 1986), and appears to be sensitive to the conformity of the contact regions between clay particles, as well as to the extent of disorder in their orientation (Sayers, 2005). However, the importance of parameter ε increases with increasing horizontal component of the propagation path. Event location accuracy in surface seismic reflection is known to be fairly robust using the regularly assumed isotropic velocity model (Tsvankin, 1996), but this can be further improved in some instances by determining ε and δ to account for velocity anisotropy (Eisner et al., 2011). After performing initial processing, the velocity model is used through a/the stacking and migration process, to locate the subsurface events in their exact locations. Since, in most cases, the velocity is assumed to be isotropic, the model has to be modified to account for the anisotropic nature of the earth (Maxwell et al., 2010).

8.3.1 Anisotropy in Gas Shale Reservoirs

As stated by Thomsen (1986), the anisotropy observed in shales is caused by a combination of the preferred orientation of clay platelets, anisotropic and other isotropic minerals, and the preferred orientation of fissures. However, despite the fact that the key minerals in shales are highly anisotropic, the overall anisotropy of the formation is weak (Maxwell et al., 2010), and can be characterized as VTI with a vertical axis of symmetry (Sayers, 2005). To determine the anisotropy parameters of gas shale reservoirs, VTI parameters in vertical direction (TIV) need to be determined. VTI parameters are five independent elastic constants that can be determined as follows:

$$C_{11} = \rho V_{\text{ph}}^2 \tag{8.1}$$

$$C_{33} = \rho V_{\text{pv}}^2 \tag{8.2}$$

$$C_{66} = \rho V_{\text{sh}}^2 \tag{8.3}$$

$$C_{44} = \rho V_{sv}^2 \qquad (8.4)$$

$$C_{12} = C_{11} - C_{66} \qquad (8.5)$$

$$C_{55} = C_{44} \qquad (8.6)$$

where ρ is the density of rocks, V_{ph} and V_{pv} are the horizontal and vertical P-wave velocities and V_{sh} and V_{sv} are horizontal and vertical S-wave velocities. The estimation of the fifth independent elastic constant C_{13} requires various measurements of P- and S-wave velocity at an off-axis angle. Thus, this parameter can be estimated by the method presented by Helbig (1994).

It is convenient to describe anisotropy in terms of the anisotropy parameters of Thomsen (1986). Thomsen introduced a more effective and scientific measure of anisotropy in 1986. He introduced three constants ε, γ, and δ as effective parameters for measuring anisotropy. According to Thomsen, the δ parameter is the most critical measure of anisotropy, and it does not involve the horizontal velocity in its definition. Therefore measuring δ is very important for processes like depth imaging. Thomsen's parameters are presented as follows:

$$\varepsilon = \frac{C_{11} - C_{33}}{2C_{33}} \qquad (8.7)$$

$$\gamma = \frac{C_{66} - C_{44}}{2C_{44}} \qquad (8.8)$$

$$\delta = \frac{(C_{13} + C_{44})^2 - (C_{33} - C_{44})^2}{2C_{33}(C_{33} - C_{44})}. \qquad (8.9)$$

While Thomsen's parameter ε takes only positive values, δ is observed to have both positive and negative values, a phenomenon that is not well understood. Parameter δ appears to be sensitive to the conformity of the contact regions between clay particles as well as to the extent of disorder in their orientation (Sayers, 2005). Anisotropy for wave propagation in a vertical direction is mostly governed by this parameter. Downhole applications, utilizing dominantly horizontal propagation, on the other hand, are more affected by parameter ε, which does not seem to be correlated to δ (Thomsen, 1986). However, as the signal travels to surface stations with larger offsets, the propagation path involves an increasing horizontal component, therefore increasing the influence of ε.

If these anisotropy parameters can be determined through seismic measurements by first arrival time picking of P- and S-wave in horizontal and vertical directions, VTI parameters of shales can be determined. Table 8.1 presents the results of several attempts made in the past to apply this type of analysis to study the anisotropy of gas shale reservoirs.

As location accuracy depends on the quality of the stack of all seismic traces used for processing, utilization of an anisotropic velocity model should result in a better move-out of the traces, which should subsequently result in a positive effect on both the amplitude of a seismic event as well as the signal-to-noise ratio.

TABLE 8.1 Anisotropic parameters obtained for various types of shales in different studies

Types of shales	ε	γ	δ
Jurassic shale (Hornby, 1995)	0.24	0.47	0.11
Kimmeridge shale (Hornby, 1995)	0.38	0.58	0.20
Cretaceous shale (Jones and Wang, 1981)	0.28	0.39	0.06
Maikop clay (Slater, 1997)	0.25	0.73	0.1
Pirre shale (White et al., 1983)	0.2	0.18	0.3
Mesaverde clayshale (Roberson and Corrigan, 1983)	0.33	0.73	0.58
Wills Point shale (Roberson and Corrigan, 1983)	0.22	0.32	0.28

8.4 WELLBORE INSTABILITY IN GAS SHALE RESERVOIRS

The strength properties of bedded rocks have been known for many years. In fact, the results of a number of studies demonstrated how most of the sedimentary rocks, such as shale, display a strong anisotropy in their strength behavior resulting in instability in the weaker strength direction (Colak and Unlu, 2004; Donath, 1964; Horino and Ellickson, 1970; Kwasniewski, 1993; McLamore and Gray, 1967; Ramamurthy, 1993). Therefore, compressive strength and deformability of these rocks need to be estimated as important design parameters. In fact, deformation and strength behavior of these materials are strongly dependent on loading orientation with respect to their bedding planes. As was mentioned in previous section, this type of behavior is referred to as VTI behavior (Hudson and Harrison, 1997), observed in many types of formations including shale rocks, in which five independent stiffness parameters need to be determined. Jaeger (1960) gave a thorough analysis of various loading scenarios that explain bedding failure. Considerable attention has been paid to the strength differences, failure modes, and failure criteria of anisotropic rocks (Bagheripour et al., 2011; Gatelier et al., 2002; Jyh Jong et al., 1997; Nasseri et al., 2003; Nova, 1980; Saroglou and Tsiambaos, 2008; Tavallali and Vervoort, 2010; Tien et al., 2006). A considerable amount of research work has have also been carried out to measure the strength anisotropy of various rock types, for example, Donath (1964), Chenevert and Gatline (1965), McLamore and Gray (1967), and Hoek (1968) on shale. As a result of these studies, it was noted that the maximum strength observed was at angles $\beta = 0°$ or $\beta = 90°$, whereas the minimum strength was found to be happening at an approximate angle of $\beta = 30°$. Here, β is the angle between plane of weakness (i.e., foliation or bedding planes) and the direction of the maximum load applied to the sample.

The applications of plane of weakness in oil and gas drilling were introduced by Aadnoy (1988). In modeling highly inclined boreholes, he investigated the effects of wellbore inclination, anisotropic elastic rock properties, anisotropic stresses, and anisotropic rock strength. As a result of this study, it was seen that if the borehole wall is in the same plane as the borehole axis and the normal axis to the bedding plane, one Mohr–Coulomb envelope applies for all borehole angles. This is the least serious and therefore the preferred case. On the other hand, if the least in situ stress is normal on the plane of the borehole axis and the axis is normal on the bedding plane, the directional-shear-strength properties come into play. Now, the borehole has a potential collapse problem in the inclination range $15° < \gamma < 35°$. It should be noted that this applies only to sedimentary rocks with a plane of weakness.

Because of the specific geomechanical properties of shale (high pore pressure, alignment of phyllosilicates due to overburden diagenesis), slip surfaces may exhibit significantly more potential to fail as compared to stronger rock units, such as limestone and sandstone. For this reason, shale instability is an important design factor in drilling practices.

8.4.1 Structurally Controlled Instability

Structurally controlled instability due to slippage of plane of weakness is likely to happen when drilling in shales, similar to other layered formations. In drilling into shale formations, borehole orientation, with respect to the direction of in situ stresses, in addition to the magnitude of the in situ stresses and the location of failure on the borehole wall with respect to the bedding plane orientation should be determined. According to Al-Ajmi and Zimmerman (2009), failure criterion does not significantly influence the optimal drilling trajectory. This conclusion has also been reported in a number of publications (Chen et al., 1996; Djurhuus and Aadnoy, 2003; Kårstad and Aadnoy, 2005; Moos et al., 1998; Zhou et al., 1996). Many other studies have also attempted to assess those wells drilled through laminated (anisotropic) rocks (Aadony, 1988; Fjaer et al., 1992; Niandou et al., 1997; Ong and Roegiers, 1993; Singh et al., 1989; Yang and Gray, 1970). Chenevert and Gatlin (1965) studied the mechanical anisotropy of laminated sedimentary rocks and determined that formation compressive strength and stiffness can vary significantly as the angle between the direction of the axial load and bedding planes varies. Aadony (1988) developed a simulator to study the wellbore instability of highly inclined borehole in VTI rock formations. Ong and Roegiers (1993) studied the influence of anisotropic stress on borehole stability using an anisotropic strength criterion for assessing compressive failure. They indicated that wellbore stability was significantly influenced by the mechanical anisotropy of the rock. Gazaniol et al. (1994) found that rock strength anisotropy has a great influence on wellbore instability and that a wellbore could fail along the bedding planes if the trajectory of the well is not properly selected. Last and McLean (1995) discussed that drilling perpendicular to bedding planes was beneficial as it improved wellbore stability in Cusiana Field, where they performed their studies. They pointed out that wellbore stability was affected by the relative angle between the wellbore and the bedding planes. Similar conclusions were reached by Skelton et al. (1995) where they showed that the tangent section inclination and azimuth of the wellbore should be perpendicular to the bedding dip and strike of the formations, respectively, in order to avoid formation layers from sliding along their bedding planes. Niandou et al. (1997) observed that during lab testing of the Tournemire shale, failure was caused by extension and sliding of bedding planes or shear band development within the shale matrix. Okland and Cook (1998) noted that for the Draupne Formation, the "angle of attack" between the wellbore and the bedding planes should always exceed 20° so as to improve wellbore stability. Willson et al. (1999) noted that bedding plane slippage could result from unfavorable interaction between *in situ* stresses, well trajectory, and bedding planes. They also pointed out that the reduced strength (friction or cohesion) acting on the bedding plane could also result in greater and sometimes catastrophic instability. Russel et al. (2003) found that it is important to determine the relative angle between the well trajectory and the rock structure because this angle dictates the stability of the formation when drilling close to the bedding dips or at unfavorable angles to fracture planes.

According to Aadnoy et al. (2009), orientation of plane of weakness with certain wellbore plane inclinations can cause the borehole to become unstable. To show this, let us assume that the *in situ* principal stresses (i.e., σ_v, σ_H, and σ_h) are associated with the coordinate system (x', y', z'), as shown in Figure 8.5. The z'-axis is parallel to σ_1, x'-axis is parallel to σ_2, and y'-axis is parallel to σ_3. These virgin formation stresses should be transformed to another coordinate system (x, y, z), to conveniently determine the stress distribution around a borehole. Figure 8.5 shows the (x, y, z) coordinate system, where the z-axis is parallel to the borehole axis, the x-axis is parallel to the lowermost radial direction of the borehole, and the y-axis is horizontal (after Al-Ajmi and Zimmerman, 2009). This transformation can be obtained by a rotation of α around the z'-axis, and then a rotation of angle

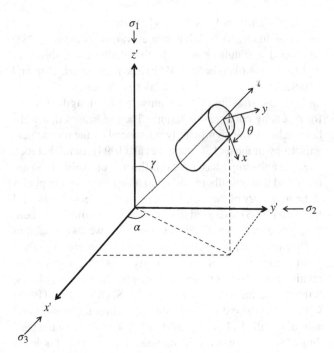

FIGURE 8.5 Stress transformation system for a deviated well (After Al-Ajmi and Zimmerman 2009).

γ around the y'-axis. Using the stress transformation equation, the virgin formation stresses expressed in the (x, y, z) coordinate system become:

$$\sigma_x = (\sigma_H \cos^2 \alpha + \sigma_h \sin^2 \alpha) \cos^2 \beta + \sigma_v \sin^2 \beta \quad (8.10)$$

$$\sigma_y = (\sigma_H \sin^2 \alpha + \sigma_h \cos^2 \alpha) \quad (8.11)$$

$$\sigma_z = (\sigma_H \cos^2 \alpha + \sigma_h \sin^2 \alpha) \sin^2 \beta + \sigma_v \cos^2 \beta \quad (8.12)$$

where α and γ, respectively, are the projected orientation of the wellbore with respect to the x-axis and the deviation of the wellbore from the vertical axis. According to Aadnoy et al. (2009), directions that expose the plane of weakness and cause the wellbore to be instable are given by:

$$\sigma_H (\mathrm{Sin}^2 \alpha + \mathrm{Cos}^2 \alpha \mathrm{Cos}^2 \gamma) + \sigma_h (\mathrm{Cos}^2 \alpha - \mathrm{Sin}^2 \alpha \mathrm{Cos}^2 \gamma) \pi \;\; \sigma_v \mathrm{Sin}^2 \gamma \quad (8.13)$$

Equation 8.3 can only be used if the bedding inclination range determined by the compressive strength data is between 10° and 30°. If this is the case and the inclination of the well is between 10° and 30°, in the normal, strike slip, and reverse faulting regime, well with azimuth of 0°–40° will fail due to bedding plane exposal. Thus, in the entire drilling operation aiming at drilling a deviated well, the attack angle (orientation of wellbore with respect to the bedding plane, e.g., attack angle of 90° is when the wellbore is perpendicular to the bedding plane) needs to be defined.

If there is a strong stress contrast between σ_x and σ_y, the above analysis typically holds true. However, for a small stress contrast and within the sensitive borehole/bedding orientation, other failures may occur, which depend on the degree of the weakness of the planes. Thus, determination of a safe mud window and a safe direction of drilling need to be carried out, taking into account the above equations when directional drilling is planned in formations containing weakness plane. It should be remembered that in anisotropic stress field, wells may be stable for some azimuths, but fail under another drilling direction; so critical parameters like planes of weakness, normal stress, and the angle between the borehole and bedding plane should be calculated. Besides, it should be noticed that for many bedded rocks, the critical angle between borehole and bedding is 10–30°. For zero or 90° the wells are more stable.

8.4.2 Instability due to Directional Dependency of Geomechanical Parameters

The use of mechanical earth models (MEMs) has been well established in the oil and gas industry. It is the modeling of the earth's mechanical properties coupled with the regional in situ earth effective stress (Barton et al., 1998). When applied correctly, the model can be used to understand how the earth will react when subjected to a drilling scenario (Van Oort et al., 2001). It is well known that there are correlations between rock's physical properties obtained from petrophysical logs and its elastic and mechanical properties. The MEM uses this basis and extracts rock elastic and mechanical properties as well as the state of stresses from data obtained in one or a number of wells in a field (Rasouli et al., 2011). The results are presented as continuous logs and the output of the model is used for different field studies including safe mud weight window (MWW) determination during drilling, hydraulic fracturing studies and sanding analysis, reservoir depletion and injection, etc. Figure 8.6 presents the workflow used in building a MEM. Figure 8.7 shows an example of the constructed MEM in a shale gas well. Good understanding of the failure mechanism of the formations is essential in order to construct a reliable MEM. This becomes further complicated in VTI formations like shale as mechanical properties are direction dependent; therefore, more parameters need to be considered in failure analysis.

8.4.2.1 Estimation of Stiffness (Elastic) Properties in VTI Formation
The sonic data provides the necessary input for calculations of isotropic and anisotropic parameters. Isotropic dynamic stiffness parameters of rocks are calculated as a function of compressional slowness, fast shear slowness, and density (ρ). Elastic properties including Young's modulus (E), Poisson's ratio (v), and shear and bulk

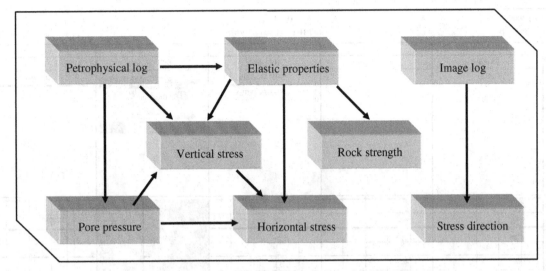

FIGURE 8.6 The flowchart for constructing a mechanical earth model.

moduli (G, K) can be estimated using the following equations (Fjaer et al., 1992):

$$E_{Dyn} = \rho V_s^2 \frac{3(V_p/V_s)^2 - 4}{(V_p/V_s)^2 - 1} \quad (8.14)$$

$$\nu_{Dyn} = \frac{1}{2} \frac{(V_p - V_s)^2 - 2}{2(V_p - V_s)^2 - 1} \quad (8.15)$$

$$G_{Dyn} = \rho V_s^2 \quad (8.16)$$

$$K_{Dyn} = \rho \left(V_p^2 - \frac{4}{3} V_s^2\right) \quad (8.17)$$

The subscript Dyn used in above equations indicates that these are dynamic properties as obtained from log data.

Anisotropic dynamic properties of rocks, on the other hand, are calculated based on the assumption of a VTI medium with the axis of symmetry oriented parallel to the wellbore. For a VTI medium, the elastic stiffness matrix describing the rock properties is represented as follows:

$$[C_{ij}] = \begin{bmatrix} C_{11} & C_{12} & C_{13} & 0 & 0 & 0 \\ C_{12} & C_{11} & C_{13} & 0 & 0 & 0 \\ C_{13} & C_{13} & C_{33} & 0 & 0 & 0 \\ 0 & 0 & 0 & C_{44} & 0 & 0 \\ 0 & 0 & 0 & 0 & C_{44} & 0 \\ 0 & 0 & 0 & 0 & 0 & (C_{11} - C_{12})/2 \end{bmatrix}. \quad (8.18)$$

As can be seen in Equation 8.18 and mentioned in Section 3.1, to completely characterize a VTI medium, five independent stiffness tensors C_{ij} need to be determined (Nye, 1985). In vertical wells with flat bedding planes, C_{33} represents the vertical propagation of P-wave, C_{44} is the vertically polarized shear wave and C_{66} is the horizontally polarized shear wave (Norris and Sinha, 1993). After calculation of these five independent parameters, anisotropic rock strength properties can be derived as a function of directional Young's modulus and Poisson's ratio. In fact, based on C_{ij}, one vertical (E_{33}) and one horizontal (E_{11}) Young's modulus together with two Poisson's ratio (ν_{31}, vertical, ν_{12}, horizontal) are calculated for each depth point based on Equations 8.19–8.22. Bulk modulus and shear modulus can then be derived using the conventional expressions.

$$E_{33} = C_{33} - \frac{2C_{13}^2}{C_{11} + C_{12}} \quad (8.19)$$

$$E_{11} = C_{11} + \frac{C_{13}^2(C_{12} - C_{11}) + C_{12}\left(-C_{33}C_{12} + C_{13}^2\right)}{C_{11}C_{33} - C_{13}^2} \quad (8.20)$$

$$\nu_{31} = \nu_{32} = \frac{C_{13}}{C_{11} + C_{12}} \quad (8.21)$$

$$\nu_{12} = \frac{C_{33}C_{12} - C_{13}^2}{C_{33}C_{11} + C_{13}^2} \quad (8.22)$$

Elastic parameters obtained with assumption of VTI differ from those estimated by the assumption of isotropy, so the results of these two assumptions result in two different MEMs.

8.4.2.2 UCS in VTI Formation

For isotropic materials, various correlations have been proposed based on studies in different fields where the UCS of rocks can be derived as a function of other properties such as elastic parameters (Chang et al., 2006). One can use these correlations to obtain a continuous log of the UCS of formations. The log produced

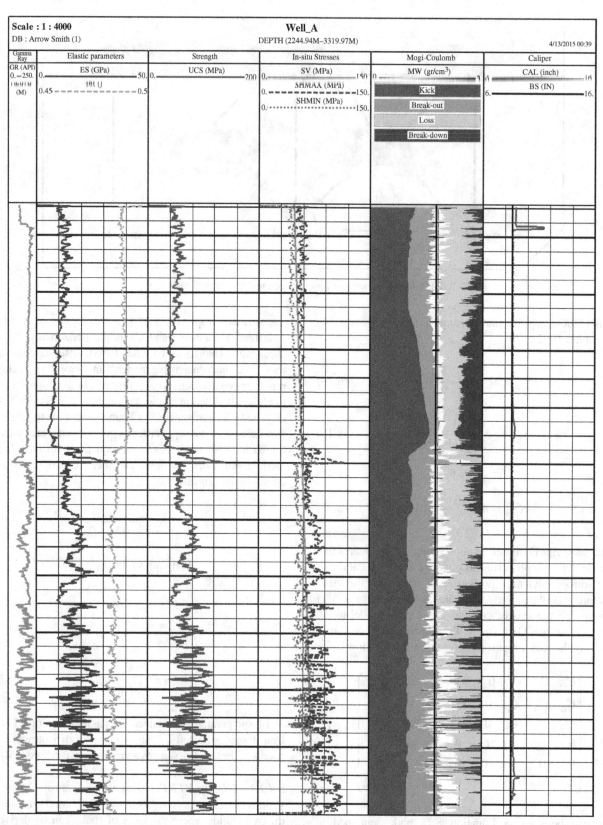

FIGURE 8.7 An example of the MEM for a gas shale well. 1st track from left: gamma ray. 2nd track: Young's modulus (solid dark curve) and Poisson's ratio (dashed light curve). 3rd track: UCS. 4th track: vertical stress (solid curve), minimum horizontal stress (SHMIN) (pointed curve), and maximum horizontal stress (SHMAX) (dashed curve). 5th track: mud weight windows including kick and breakouts in the left and mud loss and induced fracture to the right. The used mud weight to drill this section is shown in black. Last track shows the caliper.

can then be calibrated against core test data if any is available. However, for VTI material, the following equation presented by Wilson et al. (2007) should be used to calculate the UCS of rocks relative to the bedding planes.

$$\text{UCS}_\vartheta = \text{UCS}_{max}(\cos\vartheta + k_1 \sin\vartheta)(1 - \sin\vartheta\cos\vartheta) \\ \left[1 - 2\sin\vartheta\cos\vartheta\left(1 - \frac{4k_2}{\sqrt{2}(1+k_1)}\right)\right] \quad (8.23)$$

Where UCS_ϑ is the UCS at ϑ with consideration of bedding effect and ϑ is the angle between the stress concentration to bedding (e.g., $\vartheta = 0$ represents loading perpendicular to bedding). UCS_{max} is the maximum strength at any orientation and k_1 and k_2 are defined as:

$$k_1 = \frac{q_\text{II}}{q_\perp} \quad (8.24)$$

$$k_2 = \frac{\text{UCS}_{min}}{\text{UCS}_{max}} \quad (8.25)$$

where q_II is the strength when the bedding is parallel to the sample axis, $q_\perp$ is the strength when bedding is perpendicular to the sample axis, and UCS_{min} is the minimum strength at any orientation.

8.4.2.3 Estimation of In Situ Stresses in VTI Formation

Two equations are often used to estimate far-field (?) effective horizontal stress magnitudes using effective overburden stress term (this stress is easily obtained by integrating formation bulk density from surface to depth). Traditional equations used to determine horizontal in situ stress magnitude assume an isotropic poroelastic medium (Fjaer et al., 1992):

$$\sigma_h = \frac{v}{(1-v)}(\sigma_v - \alpha P_p) + \alpha P_p + \frac{E_{sta}}{(1-v^2)}(\varepsilon_x + v\varepsilon_y) \quad (8.26)$$

$$\sigma_H = \frac{v}{(1-v)} \cdot (\sigma_v - \alpha \cdot P_p) + \alpha \cdot P_p + \frac{E_{sta}}{(1-v^2)} \cdot (\varepsilon_y + v \cdot \varepsilon_x)$$
$$(8.27)$$

However, considering VTI behavior of reservoir rocks, magnitude of in situ stress should be obtained using the following equations (Higgins et al., 2008):

$$\sigma_h = \frac{E_{11}}{E_{33}} \frac{v_{31}}{(1-v_{12})}(\sigma_v - \alpha P_p) + \alpha P_p + \frac{E_{11}}{(1-v_{12}^2)}\varepsilon_h + \frac{E_{11}v_{12}}{(1-v_{12}^2)}\varepsilon_H$$
$$(8.28)$$

$$\sigma_H = \frac{E_{11}}{E_{33}} \frac{v_{31}}{(1-v_{12})}(\sigma_v - \alpha P_p) + \alpha P_p + \frac{E_{11}}{(1-v_{12}^2)}\varepsilon_H + \frac{E_{11}v_{12}}{(1-v_{12}^2)}\varepsilon_h$$
$$(8.29)$$

The minimum horizontal stress obtained from above formulae can be calibrated against direct measurements of extended leak-off test (XLOT), a standard leak-off test (LOT), or a mini-frac test (Yamamoto, 2003; Zoback et al., 2003).

8.4.2.4 Mud Weight Window in VTI Condition

Calculating all of the parameters mentioned earlier including, elastic, strength, and in situ stress parameters under VTI condition, mud weight can be estimated by different failure criteria. In drilling engineering practice, a linear poroelasticity stress model in conjunction with a rock strength criterion is commonly used to determine the optimum mud pressure required to stabilize the wellbore. During drilling, borehole collapse and drilling-induced fractures are the two main wellbore instability problems that often lead to the need for fishing, stuck pipe, reaming operations, sidetracking, and loss of circulation. These problems can often be eliminated by selecting a suitable mud weight. This is typically carried out using a constitutive model to estimate the stresses around the wellbore coupled with a failure criterion to predict the ultimate strength of reservoirs rocks. Therefore, the main aspect of wellbore stability analysis is the selection of an appropriate rock strength criterion. Numerous simple, and now common, triaxial criteria have been proposed during the last few decades in which intermediate and minor principal stresses are equal ($\sigma_1 > \sigma_2 = \sigma_3$) (Bieniawski, 1974; Fairhust, 1964; Franklin, 1971; Hobbs 1964; Hoek and Brown, 1980; Johnston, 1985; Mohr, 1900; Murrel, 1965; Ramamurthy et al., 1985; Sheorey et al., 1989; Yudhbir et al., 1983). The triaxial criteria show close agreement with results of triaxial tests and are frequently used in stability analyses of rock structures. However, they ignore the influence of intermediate principal stress on ultimate strength of rocks causing unrealistic prediction of stability for structures. For instance, Mohr–Coulomb strength criterion is the most commonly used triaxial criterion used for the determination of rocks strength. This criterion suffers from two major limitations: (i) it ignores the nonlinearity of strength behavior and (ii) the effect of intermediate principal stress is not considered in its conventional form. Thus, the criterion overestimates the minimum mud pressure because it neglects the effect of the intermediate principal stress (McLean and Addis, 1990). Vernik and Zoback (1992) found that Mohr–Coulomb criterion is not able to provide realistic results to relate the borehole breakout dimension to the in situ stresses in crystalline rocks. Thus, they recommended the use of a strength criterion to consider the effect of the intermediate principal stress. Zhou (1994) developed a numerical model to determine the borehole breakout dimensions based on various rock failure criteria. He found that the Mohr–Coulomb criterion tends to predict larger breakouts than are predicted by criteria that incorporate the effect of σ_2. Song and Haimson (1997) concluded that the Mohr–Coulomb criterion did a poor job in the prediction of breakout dimensions, while the criteria that consider the strengthening effect of intermediate principal stress were in much better agreement with the experimental observations. Ewy (1999) concluded that the Mohr–Coulomb criterion is too conservative in the prediction of minimum mud pressure required to stabilize the wellbore.

Hoek–Brown triaxial failure criterion is another well-known criterion successfully applied to a wide range of rock for almost 30 years (Cai, 2010; Carter et al., 1991; Douglas, 2002). Zhang and Radha (2010) used the three-dimensional (3D) Hoek–Brown strength criterion developed by Zhang and Zhu (2007) for wellbore stability analysis. Their study showed that the minimum mud pressures obtained were in better agreement with observed incidents than those obtained by the Mohr–Coulomb criterion. Despite successful applications of the Hoek–Brown criterion in a number of cases, it was indicated that the intermediate principal stress needs to be included in the wellbore stability analysis (Al-Ajmi and Zimmerman, 2006). For instance, Single et al. (1998) pointed out the effect of σ_2 in underground excavations applications and suggested a modification to the Hoek–Brown criteria which is frequently used. Fjaer and Ruistuen (2002) developed a numerical model to simulate rock failure tests for a granular material. Their simulations showed that σ_2 has an influence on rock strength that is in rough agreement with several previously published sets of experimental data.

In order to account for the effect of the intermediate principle stress in rock failure response, many true-triaxial or polyaxial failure criteria such as those by Drucker and Prager (1952), Mogi (1967, 1971), Lade and Duncan (1975), Zhou (1994), Benz et al. (2008), and You (2009) have been developed. The results obtained from these criteria have shown that the value of σ_2 has a considerable effect on rock strength (Mogi, 2007). However, most of these criteria are mathematically subject to some limitations and yield physically unreasonable solutions. For instance, the Mogi criterion (Mogi, 1971) yields two values of σ_1 at failure for the same value of σ_2 when it is used to predict the strength of some types of rock (Colmenares and Zoback, 2002; You, 2009). Wiebols and Cook (1968) derived a failure criterion by calculating the shear strain energy associated with microcracks in the material. This model predicts a strengthening effect of the intermediate stress, but it requires the knowledge of the coefficient of sliding friction between crack surfaces—a parameter that cannot be determined experimentally. Furthermore, numerical methods are required for implementation of this criterion. Desai and Salami (1987) introduced a 3D failure criterion that requires more than six input parameters, and Michelis (1987) proposed another criterion in which four constants are involved (Hudson and Harrison, 1997; Pan and Hudson, 1988). In general, 3D failure criteria that contain numerous parameters or require numerical evaluation are difficult to apply in practice, particularly for wellbore stability problems. Recently, Al-Ajmi and Zimmerman (2005) introduced a 3D failure criterion called Mogi–Coulomb criterion. This failure criterion is a linear failure envelope in the Mogi domain containing two parameters that can be directly and simply related to the two Coulomb strength parameters, the cohesion and the friction angle. The Mogi–Coulomb criterion neither ignores the strengthening effect of intermediate stress, nor does it predict a strength as unrealistically high as does the other criteria. Figure 8.8 compares the results of three failure criteria namely Mohr–Coulomb, Hoek–Brown as well as Mogi–Coulomb criteria in prediction of mud weight window according to the VTI parameters obtained before.

In this Figure 8.8, the first track is the depth and gamma ray log. In the second track the MWW is shown. The red profile to the left shows the mud weight corresponding to kick. The brown profile is the mud weight below which breakouts or shear failure will occur. On the other side, if the used mud weight exceeds the blue or green profiles, the model predicts mud loss and induced fracture in the formation, respectively. Therefore, the white area in this track represents the safe MWW for drilling. As is seen from this figure, this window is changing as a function of depth and it is likely that it disappears at some depths meaning that there is practically no safe window to drill. In this case, the driller should take actions such as excessive hole cleaning when drilling in this zone. From this figure, the important conclusion is prediction of different safe MWW when applying different failure criteria. In fact, a model which provides the most comparable prediction with reality is the most reliable model. The observation regarding wellbore instability or failure during drilling is captured using caliper logs or image logs such as formation micro imager (FMI). The caliper log (HCAL) corresponding to this well is shown in the last track of Figure 8.8. As is seen in this track, a 6-inch bit was used to drill this section of the well. Any change in caliper data from this size is an indication of wellbore ovalisation or breakouts. From the caliper logs shown in this figure, severe breakouts are observed within the intervals of 4306–4314 m and 4322–4358 m, as well as from 4400 to 4421 m. As is seen from this figure, Mohr–Coulomb criterion overestimates the rock strength and results in larger values for the lower bound of the safe mud weight windows compared to the other two failure criteria. This could be linked to the fact that in this criterion, the effect of the intermediate stress is ignored. Although Hoek–Brown and Mogi–Coulomb criteria predict the breakouts observed from caliper data more realistically, and the latter criterion appears to give a better match with the observed failures from calipers. This can be related to the fact that the Mogi–Coulomb criterion considers the effect of the intermediate stress in failure analysis which is happening in reality. It can be concluded that Mogi–Coulomb criterion is a better failure criterion to determine the safe MWW. This can be related to the fact that the Mogi–Coulomb criterion considers the effect of the intermediate stress in failure analysis which is happening in reality.

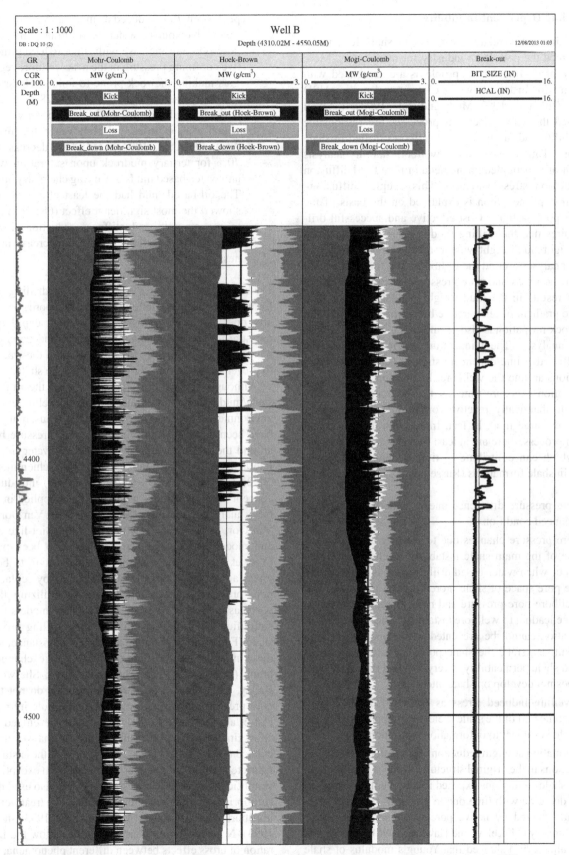

FIGURE 8.8 Determination of safe MWW for Well B using three different failure criteria and assuming VTI properties.

8.4.3 Time-Dependent Instability

Wellbore instability-related problems in shale formations have plagued the petroleum industry for many years. More than 90% of drilling-related problems are associated with shale formation instability which costs up to 1 million USD annually (Bol et al., 1992; Mody and Hale, 1993). It is well established that time-dependent processes are responsible for instability and failures in the wellbores drilled in shale formations. Time-dependency in wellbore stability analysis is a result of the coupled phenomena of pore fluid diffusion and formation stress variation. This coupled diffusion–deformation phenomenon is explained on the basis of the theory of poroelasticity. Cost-effective and successful drilling requires that the drilling fluid pressure be maintained within a tight mud weight window dictated by the stress and pressure analyses around the wellbore. The time-dependent nature of the stress and pore pressure variation around the wellbore results in the mud weight window varying with time. The gradient of temperature between the drilling mud and the rock formation is also an important issue in wellbore stability analyses. The temperature gradient will significantly affect the time-dependent stresses and pore pressure distributions around the wellbore. In addition, mud salinity and formation exposure time need to be considered while drilling in chemically reactive formations such as shale, using water-based mud. In fact, In addition to the thermal diffusion process, there are at least four other major mechanisms which can contribute to time-dependent wellbore stability in shale formations (Russell et al., 2008):

1. Pore pressure difference due to underbalanced/overbalanced conditions.
2. Pore pressure changes due to osmotic effect. This is one of the main shale instability mechanisms which occur when water-based drilling fluid is injected into the pore space of shale. Pore pressure raises the near wellbore pore pressure and reduces the true overbalance leading to wellbore instability. The pressure penetration cannot be prevented with standard filtration additives, since the shale pores are extremely small and shale permeability is very low and thus filter cake does not develop on shale intervals.
3. Swelling induced stress as the ions in the solvent become part of the shale skeleton component when the shale is subject to deformation restriction.
4. Formation strength reduction as a result of entering of the ions in the original structure of shales. The strength of shale formations exposed in a borehole is expected to decrease with time due to physical–chemical alteration caused by native pore water and mud filtrate chemistry (McLellan and Hawkes, 1995). Remvik and Skalle (1993) showed that Young's modulus of shale from the North Sea is reduced by 20–60% and the peak strength is reduced as much as 50% due to reaction with deionized water. Mud filtrate leak-off causes pore pressure increase with time, reducing the effective confining pressure in the near wellbore region and thus making the rock prone to failure. Horsrud et al. (1998a) presented data that show a 28% reduction in peak strength for triaxially loaded silty claystone which was exposed to fresh water for five days. Amanullah et al. (1994) noted a UCS decrease of up to 70% for tertiary mudrock upon saturation with oil- and water-based muds of varying chemistry and water. The oil-based mud had the least effect while water showed the most significant effect (68%). In addition to the strength and stiffness reduction, increasing rock–fluid interactions leads to a decrease in brittleness of mudrock.

Oil-based mud mitigates the problem of drilling in shale formations since penetration into the shale pore space generally does not occur. This is due to high capillary entry pressure for the nonaqueous fluid phase and good osmotic membrane which enables the salt content of the water phase to prevent osmotic transfer of water into the shale.

Although nonaqueous fluids minimize the unfavorable shale/mud interactions, thus improving wellbore stability, environmental concerns restrict their use. Thus, many studies have been conducted on preventing pore pressure build up around the wellbore caused by the shale–water mud-based interaction (Ewy and Stankovich, 2000; Schlemmer et al., 2002; Tare et al., 2000). Chenevret (1969) introduced the concept of water activity which has been applied in the stability analysis extensively (Sherwood, 1993; Van Oort, 1997; Van Oort et al., 1996; Yew et al., 1989). Biot-like analysis and model-based water activity have also been proposed (Yew et al., 1989). Assuming water advection to be negligible, a simplified model was developed by replacing the pore pressure with chemical potential utilizing the Biot poroelastic model (Yu et al., 2002). These models were too simplified to be used to simulate the swelling problem of shale (Frydman and Fontoura, 2001). For instance, some of these models consider shale as a perfect ion exchange membrane (Bol et al., 1992; Sherwood, 1993; Sherwood and Bailey, 1994; Yew et al., 1989) and others do not take the pore pressure advection around the bore hole into account (Yu et al., 1989). Although these newly developed models allow time-dependent pressure and stress changes to be calculated (Ghasemi and Diek, 2001), most of these studies are restricted within the poroelastic domain with exceptions and simplifications. Several investigators have also used the nonequilibrium thermodynamic approach in the treatment of the transport process in shales (Mody and Hale, 1993; Sherwood, 1993). Nonequilibrium thermodynamics allow the incorporation of cross effects between different phenomena, such as flux of a solution with different ionic species caused by the hydraulic gradients and/or chemical potential gradient of that species, as well as thermal and electrical gradients. Despite the attempts made to propose a generalized model for taking into account the effect of all the mechanisms involved in the instability of shale, there is not yet such a method available to be applied to drilling in shale formations. Therefore, the topic discussed here remains open to question and is likely to be the subject of further scientific research in the future.

ACKNOWLEDGMENT

The author would like to thank Dr Raoof Gholami for his effort in preparation of this chapter.

REFERENCES

Aadnoy BS. Modeling of the stability of highly inclined boreholes in anisotropic rock formations. SPE Drilling Engineering 1988. pp. 259–268.

Aadnoy BS, Hareland G, Kustamsi A, de Freitas T, Hayes J. Borehole failure related to bedding plane. In: The 43rd US Rock Mechanics Symposium and 4th U.S.-Canada Rock Mechanics Symposium and 4th US-Canada Rock Mechanics Symposium, Ashville, NC, USA, ARMA, 09-106; 28 June–1 July 2009.

Abousleiman Y, Minh T, Hoang S, Bobko C, Ortega A, Ulm F. Geomechanics field and laboratory characterization of the Woodford shale: the next gas play. In: Paper SPE 110120 presented at the SPE Annual Technical Conference Exhibition; November 11–14, 2007; Anaheim; 2007. 14pp.

Ahmadov R. Microtextural elastic and transport properties of source rocks [PhD Thesis]. Stanford: Stanford University; 2011.

Ahmadov R, Vanorio T, Mavko G. Confocal laser scanning and atomic force microscopy in estimation of elastic properties of the organic rich Bazhenov Formation. The Leading Edge 2009;28 (1):18–23.

Al-Ajmi AM, Zimmerman RW. Relationship between the parameters of the Mogi and Coulomb failure criterion. Int J Rock Mech Mining Sci 2005;42 (3):431–439.

Al-Ajmi AM, Zimmerman RW. Stability analysis of vertical boreholes using the Mogi–Coulomb failure criterion. Int J Rock Mech Mining Sci 2006;43:1200–1211.

Al-Ajmi AM, Zimmerman RM. A new well path optimization model for increased mechanical borehole stability. J Petrol Sci Eng 2009;69:53–62.

Amanullah M, Marsden JR, Shaw HF. Effects of rock-fluid interactions on the petrofabric and stress-strain behaviour of mudrocks. In: Paper 28030. Rock Mechanics in Petroleum Engineering, 29–31 August, Delft, the Netherlands; 1994.

ASTM, ASTM D7012-10 Standard Test Method for Compressive Strength and Elastic Moduli of Intact Rock Core Specimens under Varying States of Stress and Temperatures. ASTM; 2010.

Aydin A, Basu A. The Schmidt hammer in rock material characterization. Eng Geol 2005;81 (1):1–14.

Bagheripour MH, Rahgozar R, Pashnesaz H, Malekinejad M. A complement to Hoek–Brown failure criterion for strength prediction in anisotropic rock. Geomech Eng 2011;3:61–81.

Barton CA, Catillo DA, Moss DB, Peska P, Zoback MD. Charactering the full stress tensor based on observation of drilling induced wellbore failures vertical and inclined borehole leading to improved wellbore stability and permeability prediction. APPEA J 1998;466–488.

Benz T, Schwab R, Kauther RA, Vermeer PAA. Hoek–Brown criterion with intrinsic material strength factorization. Int J Rock Mech Mining Sci 2008;45 (2):210–22.

Bieniawski ZT. The effect of specimen size on compressive strength of coal. Int J Rock Mech Mining Sci 1968;5 (4):325–335.

Bieniawski ZT. Estimating the strength of rock materials. J South Afr Inst Min Metall 1974;74:312–20.

Bol GM, Woodland DC. Borehole stability in shales. In: SPE 24975. Presented at European Petrol Conference; November 16–18, 1992. Cannes; 1992.

Bol GM, Wong SW, Davidson, CJ, Woodland DC. Borehole stability in Shale, SPE, 24975; 1992.

Britt LK, Schoeffler J. The geomechanics of a shale play: What makes a shale prospective. In: SPE Eastern Regional Meeting, 23–25 September, Charleston, WV, USA, SPE 125525; 2009. 9 pp.

Broch E, Franklin JA. Point-load strength test. Int J Rock Mech Mining Sci 1972;9 (6):669–697.

Cai M. Practical estimates of tensile strength and Hoek–Brown strength parameter mi of brittle rocks. Rock Mech Rock Eng 2010;43:167–184.

Carter BJ, Scott Duncan EJ, Lajtai EZ. Fitting strength criteria to intact rock. Geotech Geol Eng 1991;9:73–81.

Chang CH, Zoback MD, Khaksar A. Empirical relations between rock strength and physical properties in sedimentary rocks. J Petrol Sci Eng 2006;51:223–237.

Chaput EJ. Observations and analysis of hard rocks cutting failure mechanisms using PDC cutters. Technical report, London: Imperial College; 1992.

Chau KT, Wong RHC. Uniaxial compressive strength and point load strength of rocks. Int J Rock Mech Mining Sci Geomech Abs 1996;33 (2):183–188.

Chen X, Tan CP, Haberfield CM. Wellbore stability analysis guidelines for practical well design. In: Proceedings of the SPE Asia Pacific Oil Gas Conference, October 28–31, 1996; Adelaide. Paper SPE 36972; 1996.

Chenevert ME. Shale alternation by water adsorption. SPE 2401. Presented at fourth conference on drilling and rock mechanics, 1970. . J Petrol Technol 1969;22(9):141–148.

Chenevert ME, Gatline C. Mechanical anisotropies of laminated sedimentary rocks. Soc Petrol Eng J 1965;5 (1):67–77.

Colak K, Unlu T. Effect of transverse anisotropy on the Hoek–Brown strength parameter 'mi' for intact rocks. Int J Rock Mech Mining Sci 2004;41:1045–1052.

Cole JH. The orbital vibrator, a new tool for characterizing interwell reservoir space. Leading Edge 1997;11:281–283.

Colmenares LB, Zoback MD. A statistical evaluation of intact rock failure criteria constrained by polyaxial test data for five different rocks. Int J Rock Mech Mining Sci 2002;39 (6):695–729.

Daixu T, Jinhai Z, Wang H. Technology analysis and enlightenment of drilling engineering applied in the development of Barnett shale gas in America. Sino-Global Energy 2011;16 (4):47–52.

Deliac EP. Optimisation des machines d'Abbatage à Pics [PhD Thesis]. Paris: Universite Pierre et Marie Curie; 1986.

Desai CS, Salami MR. A constitutive model and associated testing for soft rock. Int J Rock Mech Mining Sci 1987;24 (5):299–307.

Detournay E, Defourny P. A phenomenological model for the drilling action of drag bits. Int J Rock Mech Mining Sci Geomech Abs 1992;29 (1):13–23.

Detournay E, Sarout J, Tan C, Caurel J. Chemoporoelastic parameter identification of a reactive shale. In: Huyghe JM, Raats PAC, Cowin SC, editors. Symposium on Mechanics of Physicochemical and Electrochemical Interactions in Porous Media. Berlin: Springer; 2006. International Union of Theoretical and Applied Mechanics.

Dewhurst DN, Hennig A. Geomechanical properties related to top seal leakage in the Carnarvon Basin, Northwest Shelf. Aust Petrol Geosci 2003;9:255–263.

Dewhurst DN, Aplin AC, Sarda JP, Yang Y. Compaction-driven evolution of porosity and permeability in natural mudstones: An experimental study. J Geophys Res Solid Earth 1998;103 (B1):651–661.

Dewhurst DN, Aplin AC, Sarda JP. Influence of clay fraction on pore-scale properties and hydraulic conductivity of experimentally compacted mudstones. J Geophys Res Solid Earth 1999;104 (B12):29261–29274.

Dey T, Halleck P. Some aspects of size-effect in rock failure. Geophys Res Lett 1981;8 (7):691–694.

Djurhuus J, Aadnoy BS. In situ stress state from inversion of fracturing data from oil wells and borehole image logs. J Petrol Sci Eng 2003;38:121–130.

Donath FA. Strength variation, deformational behavior of anisotropic rocks. In: Judid WR, editor. *State of the Earth in the Earth's Crust*. New York: Elsevier; 1964. p 281–298.

Douglas KJ. The shear strength of rock masses [PhD Dissertation]. Sydney: The University of New South Wales; 2002. 496 pp.

Dresen L, Rüter H. Seismic Coal Exploration: Part B: In-seam Seismics. New York: Pergamon; 1994.

Drucker D, Prager W. Soil mechanics and plastic analysis or limit design. Appl Mathemat 1952;10:157–65.

Duc MNM. Contribution à l' étude de la taille des roches [PhD Thesis]. Paris: Universite de Paris VI; 1974.

Eisner L Zhang Y, Duncan PM, Mueller MC, Thornton MP, Gei D. Effective VTI anisotropy for consistent monitoring of microseismic events. The Leading Edge 2011;30(July):772–776.

Ewy R. Wellbore-stability predictions by use of a modified Lade criterion. SPE Drill Completion 1999;14:85–91.

Ewy RT, Stankovich RJ. *Pore pressure change due to shale–fluid interaction: measurements under simulated wellbore conditions*. In: Proceedings of Pacific rocks, Fourth North American rock mechanics symposium; July 31–August 3, 2000, Seattle/Rotterdam: Balkema; 2000. p 147–54.

Fairhurst C. On the validity of the Brazilian test for brittle materials. Int J Rock Mech Mining Sci 1964;1:515–46.

Fairhurst C, Lacabanne WD. Hard rock drilling techniques. Mine Quarry Eng 1957;23:157–161.

Farmer IW. *Main Lecture: Rock Testing — Deficiencies and Selection.* In: Rakowski Z, editor. Proceedings of the International Conference on Geomechanics 1992. Balkema, Hradec, Ostrava, Czecho-Slovakia. pp 3–7.

Fjaer E, Ruistuen H. Impact of the intermediate principal stress on the strength of heterogeneous rock. J Geophys Res Solid Earth 2002;107 (B2):1–10.

Fjaer E, Holt RM, Horsrud P, Raaen AM, Risnes R. Chapter 3: Geological aspects of petroleum related rock mechanics. Dev Petrol Sci 1992;33:89–108.

Franklin JA. Triaxial strength of rock material. Rock Mech 1971;3:86–98.

Frydman M, da Fontoura SAB. *Modeling aspects of wellbore stability in shales*. In: SPE 69529, SPE Latin American and Caribbean Petroleum Engineering Conference, March 25–28, 2001. Buenos Aires; 2001.

Gatelier N, Pellet F, Loret B. Mechanical damage of an anisotropic porous rock in cyclic triaxial tests. Int J Rock Mech Mining Sci 2002;39:335–354.

Gazaniol D, Forsans T, Boisson MJF, Piau J-M. *Wellbore failure mechanisms in shales: prediction and prevention*. In: Paper SPE 28851. Presented at the European Petroleum Conference; October 25–27, 1994. London; 1994.

Ghassemi A, Diek A. Effects of ion transfer on stress and pore pressure distributions around a borehole in shale. In: 38th NARM Symposium; Washington, DC: DC Rocks; 2001. pp 85–91.

Ghorbani A, Zamora M, Cosenza P. Effects of desiccation on the elastic wave velocities of clay-rocks. Int J Rock Mech Mining Sci 2009;46:1267–1272.

Glowka DA. Use of single-cutter data in the analysis of PDC bit designs: Part 1—development of a PDC cutting force model. J Petrol Technol 1989:41(8):797–799 (844–849).

Gray D. Principal stress estimation in shale plays using 3D seismic. In: Extended abstract for the GeoCanada 2010 conference "Working with the Earth" March 17, 2010, Montreal, Canada; 2010. 4 pp.

Gregory KB, Vidic RD, Dzombak DA. Water management challenges associated with the production of shale gas by hydraulic fracturing. Elements 2011;7 (3):181–186.

Hardage BA. *Crosswell Seismology and Reverse VSP*. London: Geophysical Press; 1992.

Hartwig A, Könitzer S, Schulz HM, Horsfield B. Investigations of the shale gas potential in NE Germany [abstract]. In: EGU General Assembly. (Held 19–24 April, 2009 in Vienna, Austria). Geophysical Research Abstracts 2009;11: EGU2009e8849.

Hawkes I, Mellor M. Uniaxial testing in rock mechanics laboratories. Eng Geol 1970;4 (3):177–285.

Hay JL, Pharr GM. *Instrumented Indentation Testing. ASM Handbook*. ASM International; 2000. p 232–243.

Helbig K. *Foundation of Elastic Anisotropy for Exploration Seismics*. Amsterdam: Pergamon; 1994.

Higgins S, Goodwin SS, Donal A. Anisotropic stress models improve completion design in the Baxter shale. In: SPE Annual Technical Conference and Exhibition, 21–24 September, Denver, CO, USA, SPE 115736; 2008.

Hobbs DW. The strength and stress–strain characteristics of coal in triaxial compression. J Geol 1964;72:214–231.

Hoek E. Brittle failure of rock. In: Stagg KG, Zienkiewicz OC, editors. *Rock Mechanics in Engineering Practice*. London: Willey; 1968.

Hoek E, Brown ET. *Underground Excavations in Rock*. London: Institute of Mining and Metallurgy; 1980.

Horino FG, Ellickson ML,*A method of estimating strength of rock containing planes of weakness*. Report of investigation 7449. Washington, DC: U.S. Bureau of Mines; 1970.

Hornby BE. The elastic properties of shales [PhD Thesis]. Cambridge University, London; 1995.

Horsrud P. Estimating mechanical properties of shale from empirical correlations. SPE Drill Completion 2001;16:68–73.

Horsrud P, Bostrom B, Sonstebo EF, Holt RM. Interaction between shale and water-based drilling fluids. In: SPE 48986. Presented at SPE annual technical conference and exhibition; September 27–30, 1998; New Orleans; 1998a.

Horsrud P, Sønstebø EF, Bøe R. Mechanical and petrophysical properties of North Sea shales. Int J Rock Mech Mining Sci 1998b;35:1009–1020.

Hsu SC, Nelson PP. *Characterisation of Cretaceous clay shales in North America*. In: Anagnostopoulos et al., editors. Proc. Int. Symp. on Geotech. Eng. of Hard Soils-Soft Rocks. Rotterdam: Balkema; 1993. pp 139–146.

Huang H, Detournay E. Intrinsic length scales in tool-rock interaction. ASCE Int J Geomech 2008;8 (1):39–44.

Huang H, Lecampion B, Detournay E. Discrete element modeling of tool-rock interaction I: Rock cutting. Int J Numer Anal Methods Geomech 2012 10.1002/nag.2113.

Hudson JA, Harrison JP. *Engineering Rock Mechanics: An Introduction to the Principles*. Oxford: Pergamon; 1997.

Hudson J, Crouch SL, Fairhurst C. Soft, stiff and servo-controlled testing machines: a review with reference to rock failure. Eng Geol 1972;6 (3):155–189.

ISRM. International Society for Rock Mechanics, Commission on Standardisation of Laboratory and Field Tests. Suggested methods for the quantitative description of discontinuities in rock masses. Int J Rock Mech Mining Sci Geomech Abs 1978;15:319–368.

Jaeger J. Shear failure of anisotropic rocks. Geol Mag 1960;97:65–72.

Jaeger J, Cook NG, Zimmerman R. *Fundamentals of Rock Mechanics*. London: Chapman & Hall; 1976.

Jiang B, Sun ZQ, Liu MQ. China's energy development strategy under the low carbon economy. Energy 2010;35:4257e64.

Jingzhou Zh, Chaoqiang F, Jie Zh. Evaluation of China shale gas from the exploration and development of North America shale gas. J Xi'an Shiyou Univ Natural Sci Edn 2011;26(2):1–7, 110.

Johnston JW. Strength of intact geomechanical materials. J Geotech Eng 1985;111:730–749.

Jones LE, Wang HF. Ultrasonic velocities in Cretaceous shales from the Williston Basin. Geophysics 1981;46:288–297.

Josh M, Esteban L, Delle Piane C, Sarout J, Dewhurst DN, Clennell MB. Laboratory characterization of shale properties. J Petrol Sci Eng 2012;88–89:107–124.

Jyh Jong L, Yang M-T, Hsieh H-Y. Direct tensile behavior of a transversely isotropic rock. Int J Rock Mech Mining Sci 1997;34:837–849.

Kårstad E, Aadnoy BS. Optimization of borehole stability using 3D stress optimization. In: Paper SPE 97149. Proceedings of the 80th SPE annual technical conference and exhibition; October 9–12, 2005; Dallas; 2005.

Krohn CE. Crosswell continuity logging for oil and gas field application. Leading Edge 1992;11–7:39–45.

Kuila U, Dewhurst DN, Siggins AF, Raven MD. Stress anisotropy and velocity anisotropy in low porosity shale. Tectonophysics 2011;503:34–44. DOI: 10.1016/j.tecto.2010.09.023.

Kumar V. Geomechanical characterization of Shale using Nano-Indentation [MSc thesis]. Oklahoma: University of Oklahoma; 2012.

Kwasniewski M. Mechanical behavior of anisotropic rocks. In: Hudson JA, editor. *Comprehensive Rock Engineering*. Volume 1, Oxford: Pergamon; 1993. p 285–312.

Lade P, Duncan J. Elasto-plastic stress–strain theory for cohesionless soil. J Geotech Eng Div ASCE 1975;101:1037–53.

Lashkaripour GR, Passaris EKS. Correlations between index parameters and mechanical properties of shales. In: 8th international congress on Rock Mechanics, 25–29 September, Tokyo, Japan, Vol 1; 1993. pp 257–261.

Last NC, McLean. Assessing the impact of Trajectory on Wells Drilled in an Overthrust Region. In: Paper SPE 30465. Presented at the SPE annual technical Conference and Exhibition; October 22–25, 1995; Dallas; 1995.

Lines L, Kelly KR, Queen JH. Channel waves in cross-borehole data. Geophysics 1992;57:334–342.

Lines L, Tan H, Queen JH, Rizer WD, Buller P, Cox D, Sinton J, Ballard J, Kokkoros G, Track A, Guerendel P, Harris J. Integrated reservoir characterization: beyond tomography. Geophysics 1995;60:354–364.

Liu E, Crampin S, Queen JH. Fracture detection using crosshole surveys and reverse vertical seismic profiling at the Conoco Borehole Test Facility, Oklahoma. Geophys J Int 1991;107:449–463.

Liu E, Crampin S, Roth B. Modeling channel waves with synthetic seismograms in an in-seam seismic survey. Geophys Prospect 1992;40:513–540.

Lou M, Crampin S. Dispersion of guided waves in thin anisotropic waveguides. Geophys J Int 1991;107:545–555.

Lou M, Crampin S. Guided-wave propagation between boreholes. Leading Edge 1992;11 (7):34–37.

Ma LW, Liu P, Fu F, Li Z, Ni WD. Integrated energy strategy for the sustainable development of China. Energy 2011;36:1143e54.

Maxwell SC, Bennett L, Jones M. Walsh J. Anisotropic velocity modeling for microseismic processing: Part 1—Impact of velocity model uncertainty. In: SEG Denver 2010 Annual Meeting, 17–22 October 17–22, Denver, Colorado.

McLamore R, Gray KE. The mechanical behavior of anisotropic sedimentary rocks. Transition Am Soc Mech Eng Service B 1967;89:62–67.

McLean M, Addis M. Wellbore stability: The effect of strength criteria on mud weight recommendations. In: Proceedings of the 65th annual technical conference and exhibition SPE. New Orleans. Paper SPE 20405; 1990.

McLellan PJ, Hawkes CD. Rock Mechanical and Transport Properties of Blackstone Formation Shales of the Rocky Mountain Foothills, Alberta, Poster Session, CSPG-CWLS Symposium—Exploration, Exploitation and Evaluation, Calgary; 1995.

Michelis P. True triaxial yielding and hardening of rock. J Geotech Eng Div (ASCE) 1987;113 (6):616–635.

Minaeian V, Rasouli V, Dewhurst D. 75th EAGE conference & exhibition incorporating SPE EUROPEC 2013; June 10–13, 2013. London; 2013.

Mody FK, Hale AH. *A borehole model to couple the mechanics and the chemistry of drilling fluid shale interactions.* Paper 25728. Proceedings of the SPE/IADC Drilling Conference. Richardson: Society of Petroleum Engineers; 1993.

Mogi K. Effect of the intermediate principal stress on rock failure. J Geophys Res 1967;72:5117–5131.

Mogi K. Fracture and flow of rocks under high triaxial compression. J Geophys Res 1971;76:1255–1269.

Mogi K. *Experimental Rock Mechanics*. London: Taylor & Francis; 2007.

Mohr O. Welcheumstä€nde bedingendieelastizita€ tsgrenze unddenbruch eines materials? VDI-Zeitschrift 1900;44:1524.

Montgomery Scott L, Jarvie DM, Bowker KA, Pallastro RM. Mississippian Barnett Shale, Fort Worth basin, north-central Texas: gas-shale play with multitrillion cubic foot potential. AAPG Bull 2005;89 (2):155e75.

Moos D, Peska P, Zoback MD. Predicting the stability of horizontal wells and multi-laterals—the role of in situ stress and rock properties. In: Proc. 73rd SPE Int. Conf. Horizontal Well Tech.; November 1–4, 1998; Calgary, AB, Canada. Paper SPE 50386; 1998.

Murrel SAF. The effect of triaxial stress systems on the strength of rock at atmospheric temperature. Int J Rock Mech Mining Sci 1965;3:11–43.

Nagra, Projekt Opalinuston: Synthese der geowissenschaftlichen Untersuchungsergebnisse—Entsor-gungsnachweis für abgebrannte Brennelemente, verglaste hochaktive sowie langle-bige mittelaktive Abfälle. Technical Report Number 2–3. Berlin: Nagra Technical; 2002. 560 pp.

Nasseri MHB, Rao KS, Ramamurthy T. Anisotropic strength and deformational behavior of Himalayan schists. Int J Rock Mech Mining Sci 2003;40:3–23.

Niandou H, Shao JF, Henry JP, Fourmaintraux D. Laboratory Investigation of the Behaviour of Tournemire Shale. Int J Rock Mech Mining Sci 1997;34 (1):3–16.

Nishimatsu Y. The mechanics of rock cutting. Int J Rock Mech Mining Sci 1972;9 (2):261–270.

Norrris A, Sinha B. Weak Elastic Anisotropy and tube wave. Geophysics 1993;58 (8):1091–1098.

Nova R. Failure of transversely isotropic rocks in triaxial compression. Int J Rock Mech Mining Sci Geomech Abs 1980;17: 325–332.

Nye JF. *Physical Properties of Crystals*, England: Oxford University Press; 1985.

Okland D, Cook JM. Bending related borehole instability in high angle wells. Proceeding of SPE/ISRM EuRock 1998, Norway; 1998. pp. 413–421.

Ong S, Roegiers JC. Influence of anisotropies in borehole stability. Int J Rock Mech Mining Sci 1993;30 (7):1069–1075.

Pan XD, Hudson JA. A simplified three dimensional Hoek–Brown yield criterion. In: Romana M, editor. *Rock Mechanics and Power, Plants*. Rotterdam: Balkema; 1988. p 95–103.

Parra JO. Guided seismic waves in layered poro viscoelastic media for continuity logging applications: model studies. Geophys Prospect 1996;44:403–425.

Parra JO, Zook BJ, Collier HA. Interwell seismic logging for continuity at the Gypsy test site, Oklahoma. J Appl Geophys 1996;35:45–62.

Ramamurthy T. Strength, modulus responses of anisotropic rocks. In: Hudson JA, editor. *Compressive Rock Engineering*. Volume 1, Oxford: Pergamon; 1993. p 313–329.

Ramamurthy T, Rao GV, Rao Ks. A strength criterion for rocks. In: Proceedings of the Indian Geotechnical Conference. Roorkee, vol. 1; 1985.pp. 59–64.

Rasouli V, Zacharia J, Elike M. The influence of perturbed stresses near faults on drilling strategy: a case study in Blacktip field, North Australia. J Petrol Sci Eng 2011;76:37–50.

Remvik F, Skalle P. Shale-fluid interaction under simulated downhole conditions, and its effects on borehole stability. Int J Rock Mech Mining Sci Geomech Abs 1993;30 (7):1115–1118.

Richard T. Determination of rock strength from cutting tests [MSc Thesis]. Minneapolis: University of Minnesota; 1999.

Richard T, Detournay E, Drescher A, Nicodeme P, Fourmaintraux D. *The scratch test as a means to measure strength of sedimentary rocks*. In: SPE/ISRM Eurock 98; Trondheim; Society of Petroleum Engineers. Number SPE 47196; 1998. pp 1–8.

Richard T, Dagrain F, Poyol E, Detournay E. Rock strength determination from scratch tests. Eng Geol 2012;147–148:91–100.

Rickman R, Mullen M, Petre E, Griese B, Kundert D. A practical use of shale petrophysics for stimulation design optimisation: All shale plays are not clones of the Barnett Shale. In: SPE Annual Technical Conference and Exhibition, 21–24 September, Denver, CO, USA, SPE115258. 2008

Roberson JD, Corrigan D. Radiation patterns of a shear-wave vibrator in near-surface shale. Geophysics 1983;48:19–26.

Ross DJK, Bustin RM. Characterizing the shale gas resource potential of Devonian–Mississippian strata in the Western Canada sedimentary basin: application of an integrated formation evaluation. AAPG Bull 2008;92:87–125.

Rozkho A. Unconfined compressive strength of partially saturated shales: The fracture mechanics approach. In: Extended Abstract for the 2nd EAGE Conference on Shale—Resource and Challenge, April 26, 2010, California; 2010. 3 pp.

Rusnak J, Mark C. Using the point load test to determine the uniaxial compressive strength of coal measure rock. In: Proceedings of the 19th International Conference on Ground Control in Mining, Queensland, Australia, April 10, 1999. pp 362–371

Russell KA, Ayan C, Hart NJ, Rodriguez JM, Scholey H, Sugden C, Davidson JK. Predicting and preventing wellbore instability using the latest drilling and logging technologies: Tullich field development, North Sea. In: Paper SPE 84269. Presented at the 2003 SPE Annual Technical Conference and Exhibition; October 5–8, 2003; Denver; 2003.

Russell T, Ewy E, Morton K. Wellbore stability performance of water based mud additives, In: SPE Annual technical conference and Exhibition, 116139, Colorado, USA; September 2008.

Saroglou H, Tsiambaos G. A modified Hoek–Brown failure criterion for anisotropic intact rock. Int J Rock Mech Mining Sci 2008;45:223–234.

Sarout J, Detournay E. Chemoporoelastic analysis and experimental validation of the pore pressure transmission tests for reactive shales. Int J Rock Mech Mining Sci 2011;48 (5):759–772.

Sarout J, Guéguen Y. Anisotropy of elastic wave velocities in deformed shales—part I: experimental results. Geophysics 2008;73 (5):75–89.

Sayers CM. Seismic anisotropy of shake. Geophys Prospect 2005;53:667 676.

Schlemmer R, Friedheim JE, Growcock FB, Bloys JB, Headlley JA, Polnaszek SC. Membrane efficiency in shale—an empirical evaluation of drilling fluid chemistries and implications for fluid design. In: SPE 74557. Presented at IADC/SPE drilling conference; February 26–28, 2002; Dallas; 2002.

Sheorey PR, Biswas AK, Choubey VD. An empirical failure criterion for rocks and jointed rock masses. Eng Geol 1989;26:141–159.

Sherwood JD. Biot poroelasticity of a chemically active shale. Proc Rock Soc Lond 1993;440:365–377.

Sherwood JD, Bailey L. Swelling of shale around a cylindrical wellbore In: Proceeding Rock Society of London 1994; A 444: 161–184.

Singh J, Ramamurthy T, Rao GV. Strength anisotropies in rocks. Indian Geotech 1989;19:147–166.

Single B, Goel RK, Mehrotra VK, Garg SK, Allu MR. Effect of intermediate principal stress on strength of anisotropic rock mass. Tunneling Underground Space Technol 1998;13 (1):71–79.

Skelton J, Hogg TW, Cross R, Verheggen L. Case history of directional drilling in the cusiana field in Colombia. In: Paper IADC/SPE 29380. Presented at the 1995 IADC/SPE Drilling Conference; February 28–March 2, 1995; Amsterdam, The Netherlands; 1995.

Slater CP. Estimation and modeling of anisotropy in vertical and walkaway seismic profiles at two North Caucasus oil field [PhD Thesis]. Edinburgh University; 1997.

Song I, Haimson BC. Polyaxial strength criteria and their use in estimating in situ stress magnitudes from borehole breakout dimensions. Int J Rock Mech Mining Sci 1997;34 (3–4):498.

Szwedzicki T. Indentation hardness testing of rock. Int J Rock Mech Mining Sci 1998;35 (6):825–829.

Tare UA, Mese AI, Mody FK. Interpretation and application of acoustic and transient pressure response to enhance shale instability predictions. In: SPE 63052. Presented at SPE annual technical conference and exhibition; October 1–4, 2000; Dallas, TX, USA; 2000.

Tavallali A, Vervoort A. Effect of layer orientation on the failure of layered sandstone under Brazilian test conditions. Int J Rock Mech Mining Sci 2010;47:313–322.

Thomsen L. Weak elastic anisotropy. Geophysics 1998;51: 1954–1966.

Thuro K, Plinninger RJ, Zah S, Schutz S. Scale effects in rock strength properties. Part 1: Unconfined compressive test and Brazilian test. In: Sarkka P, Eloranta P, editors. *ISRM Regional Symposium Eurock 2001; Espoo*. Finland/Balkema: ; 2001. p 169–174.

Tien YM, Kuo MC, Juang CH. An experimental investigation of the failure mechanism of simulated transversely isotropic rocks. Int J Rock Mech Mining Sci 2006;43:1163–1181.

Tsiambaos G, Sabatakakis N. Considerations on strength of intact sedimentary rocks. Eng Geol 2004;72 (3–4):261–273.

Tsur-Lavie Y, Denekamp SA. Comparison of size effect for different types of strength tests. Rock Mech 1982;15 (4):243–254.

Tsvankin I. P-wave signatures and notation for transversely isotropic media: An overview. Geophysics 1996;61:467–483.

Turpening WR, Chon YT, Peper REF, Szerbiak R, Schultz T, Thielmier G, Ballard R. Detection of bed continuity using crosswell data: Gypsy pilot site study. In: Proceedings of the Formation Evaluation and Reservoir Geology. SPE Annual Technical Conference and Exhibition, 4–7 October, Washington, DC, SPE 24710; 1992. pp 503–512.

Ulm F, James S. The scratch test for strength and fracture toughness determination of oil well cements cured at high temperature and pressure. Cement Concrete Res 2011;41:942–946.

Ulusay R, Hudson JA. *The Complete ISRM Suggested Methods for Rock Characterization, Testing and Monitoring: 1974–2006*. London: International Society for Rock Mechanics; 2007.

Van Oort E. Physical-chemical stabilization of shales. In: Paper SPE37263. SPE International Symposium on Oilfield Chemistry; February 18–21, 1997; Houston, TX, USA; 1997.

Van Oort E, Hale AH, Mody FK. Manipulation of coupled osmotic flows for stabilization of shales exposed to water-based drilling fluids. In: SPE 30499. Presented at SPE annual technical conference exhibition; October 22–25, 1995; Dallas, TX, USA; 1995.

Van Oort E, Hale AH, Mody FK, Roy R. Transport in shales and design of improved water-based shale drilling fluids. SPE Drill Completion 1996;11 (3):137–146.

Van Oort E, Nicholson J, D'Agostino J. Integrated borehole instability studies: Key to drilling at the technical limit and Trouble Cost reduction. In: Presented at the SPE/IADC Drilling Conference, 27 February–1 March, Amsterdam, the Netherlands; SPE/IADC 67763; 2001.

Vernik L, Zoback MD. Estimation of maximum horizontal principal stress magnitude from stress-induced well bore breakouts in the Cajon Pass scientific research borehole. J Geophys Res 1992;97 (B4):5109–5119.

Wawersik WR, Fairhurst C. A study of brittle rock fracture in laboratory compression experiments. Int J Rock Mech Mining Sci Geomech Abs 1970;7 (5):561–564.

White JE, Martineau-Nicoletis L, Monash C. Measured anisotropy in Pierre Shale. Geophys Prospect 1983;31:709–725.

Wiebols G, Cook N. An energy criterion for the strength of rock in polyaxial compression. Int J Rock Mech Mining Sci 1968;5: 529–549.

Willson SM, Last NC, Zoback MD, Moos D. Drilling in South America: stability approach for complex geologic conditions. In: Paper SPE 53940. Presented at the 1999 SPE Latin American & Caribbean Petroleum Engineering Conference; April 21–23, 1999; Caracas, Venezuela; 1999.

Willson SM, Edwards ST, Crook A, Bere A, Moos D, Peska P. Assuring stability in extended reach wells-analyses, practice and mitigations. In: SPE/IADC Drilling Conference, 20–22 February, Amsterdam, the Netherlands, SPE/IADC 105405; 2007.

Xinjing L, Suyun H, Keming C. Suggestions from the development of fractured shale gas in North America. Petrol Exploration Dev 2007;34 (4):392–400.

Yamamoto M. *Implementation of the extended leak-off test in deep wells in Japan*. In: Sugawra K, editor. Proceedings of the Third International Symposium on Rock Stress. Rotterdam: Balkema; 2003. pp 225e229.

Yang Y, Aplin AC. Permeability and petrophysical properties of 30 natural mudstones. J Geophys Res 2007;112:B03206. DOI: 1029/2005JB004243.

Yang JH, Gray KE. Behavior of anisotropic rocks under combined stresses. In: Paper SPE 3098. In: Presented at the 45th annual fall meeting of the society of petroleum engineers of AIME; October 4–7, 1970; Houston, TX; 1970.

Yew CH, Chenevert ME, Wang CL, Osisanya SO. Wellbore stress distribution produced by moisture adsorption. In: SPE 19536, TX, USA, December, 1990; 1989. 8 pp.

You MQ. True-triaxial strength criteria for rock. Int J Rock Mech Mining Sci 2009;46:115–127.

Yu M, Chen G, Chenevert ME, Sharma MM. Chemical and thermal effects on wellbore stability of shale formations. In: SPE 71366; September 30–October 3, 2001; New Orleans, LA; 2001.

Yudhbir RK, Lemanza W, Prinzl F. An empirical failure criterion for rock masses. In: Proceedings of the 5th international congress on rock mechanics, Balkema, Rotterdam, Vol. 1, 1983. pp. B1–8.

Zeszotarski JC, Chromik RR, Vinci RP, Messmer MC, Michels R, Larsen JW. Imaging and mechanical property measurements of kerogen via nano-indentation. Geochim Acta 2004;68 (20):4113–4119.

Zhang L, Radha KC. Stability analysis of vertical boreholes using a three-dimensional Hoek–Brown strength criterion. In: Proceedings of the GeoFlor-ida, January 21, 2010, Florida.

Zhang L, Zhu H. Three-dimensional Hoek–Brown strength criterion for rocks. J Geotech Geoenviron Eng ASCE 2007;133 (9):1128–1135.

Zhong L, Worthington MH. Modeling crosshole channel waves. J Seism Explor 1994;3:21–35.

Zhou S. A program to model the initial shape and extent of borehole breakout. Comput Geosci 1994;20:1143–1160.

Zhou SH, Hillis RR, Sandiford M. On the mechanical stability of inclined wellbores. SPE Drill Completion 1996;11:67–73.

Zoback MD, Barton CA, Brudy M, Castillo DA, Finkbeiner T, Grollimund BR, Moos DB, Peska P, Ward CD, Wiprut DJ. Determination of stress orientation and magnitude in deep wells. Int J Rock Mech Mining Sci 2003;40:1049e1076.

9

ROCK PHYSICS ANALYSIS OF SHALE RESERVOIRS

Marina Pervikhina[1], Boris Gurevich[1,2], Dave N. Dewhurst[1], Pavel Golodoniuc[2,3] and Maxim Lebedev[2]

[1] *CSIRO Energy Flagship, Perth, WA, Australia*
[2] *Department of Exploration Geophysics, Curtin University, Perth, WA, Australia*
[3] *CSIRO Mineral Resources Flagship, Perth, WA, Australia*

9.1 INTRODUCTION

Oil, gas condensate, and dry gas from organic-rich shales (ORSs) are transforming the energy outlook of the world economy. The US Energy Information Administration assessment from 2013 (EIA, 2013) estimates the technically recoverable resources of shale gas at 186,000 MTOE distributed between 41 countries with China having 15% of the total amount, Argentina 11%, Algeria 10%, United States 9%, Canada 8%, Mexico 7%, Australia 6%, South Africa 5%, and Russia 4%. The volumes of extractable shale oil accounted for 87 billion tons (Aguilera and Ragetzki, 2013; DERA, 2012). New technological advancements have made previously irrecoverable resources extractable. This fact in addition to more informed estimations of ORS resources may further drastically increase the assessments of recoverable shale oil and gas. A striking example is the US Bakken Shale formation: between 2008 and 2013, the estimate of shale oil in this formation increased more than fivefold to 2.5 billion tons (IEA, 2013).

Organic-rich shales were investigated for decades as source rocks. The focus of this research was to improve understanding of the processes that lead to chemical transformations of solid and immobile organic matter to mobile hydrocarbons (HCs). Another practically important goal was to get an insight into the microstructural changes that result from these maturation processes and allow versatile HC migration from the low-permeability source rocks into high-quality reservoirs. A number of studies investigated the microstructure of organic-rich source shales using high-resolution petrographic techniques to identify the microstructural changes caused by HC maturation processes. These studies aimed to observe possible paths of the HC transportation from the source to reservoir rocks. Some of these studies reported the existence of subvertical microcracks caused by the organic matter decomposition and related pore pressure generation (e.g., Meissner, 1978; Momper, 1978). Other researchers detected development of subhorizontal cracks parallel to bedding (e.g., Lewan, 1987; Price et al., 1984).

Such microstructural and petrographic studies of ORSs were complemented with laboratory measurements of acoustic wave velocities at ultrasonic frequencies (Vernik, 1993, 1994; Vernik and Landis, 1996; Vernik and Liu, 1997; Vernik and Nur, 1992). These earliest rock physics experiments on ORSs encountered a number of difficulties such as insufficient core material or its poor quality. Additional difficulties were caused by the necessity to establish completely new practices of sample storage and experimental procedures. These new practices of sample preservation are necessary as dewatering of clay minerals in clay-bearing cores can alter elastic properties of shales substantially. Experimental procedures need to be able to evaluate the extremely anisotropic elastic properties of organic-rich shales. Polar anisotropy (also called vertical transverse isotropy, VTI), which is intrinsic for shales without vertical fractures, requires measurements of ultrasonic velocities on at least three different samples cut normal, parallel, and at

Fundamentals of Gas Shale Reservoirs, First Edition. Edited by Reza Rezaee.
© 2015 John Wiley & Sons, Inc. Published 2015 by John Wiley & Sons, Inc.

45° to the bedding plane (e.g., Vernik and Nur, 1992) or on one sample in at least three directions (e.g., Dewhurst and Siggins, 2006; Wang, 2002).

These studies of source rocks established a foundation for the recent studies of ORSs as reservoir rocks but certainly did not answer all the questions that these challenging unconventional reservoirs raise. The ultimate ambitious goal of rock physics is to predict physical properties of overburden and reservoir rocks from their seismic response with at least a few well points (Avseth et al., 2005). ORSs play all roles in the unconventional reservoirs, sometimes serving simultaneously as reservoir, seal, and source rocks, and must be comprehensively investigated for key properties such as VTI anisotropy, velocity–porosity and porosity–permeability relations, fracture-induced azimuthal anisotropy, and so on. Establishing correlations of seismic velocities, VTI, HTI and orthorhombic anisotropy, attenuation and other seismic attributes with total organic carbon (TOC) content, organic matter maturity, hydrocarbon saturation, and permeability is practically important and controlled laboratory rock physics experiments are indispensable here.

This chapter reviews major developments in rock physics of ORSs and indicates the main outstanding questions. First, we bring together available published laboratory measurements on shales including those shales whose TOC content is unknown. Second, we review experimental and theoretical studies of anisotropic elastic properties of ORSs in connection with their TOC fraction, partial saturation, and maturity. We do not specifically look at the shale microstructure and how it relates with the maturation state of the shale as this topic is covered in another chapter of this book. However, we pay special attention to the effects of microstructure and maturity on elastic parameters of ORSs. Then application of the findings of rock physics modeling for predicting the seismic response of ORSs will be assessed from recent seismic surveys. Finally, an attempt to estimate orientation of vertical fracture sets permeating Bakken Shale from amplitude versus offset and azimuth (AVOAz) data will be discussed.

9.2 LABORATORY MEASUREMENTS ON SHALES: AVAILABLE DATASETS

Controlled laboratory measurements on ORS samples are crucial as, initially they link velocities in shales with their TOC content and maturity and, secondly, they are the best (if not the only) way to verify theoretical predictions. Despite their ubiquity and two decades of research, shales remain the least experimentally studied sedimentary rocks. To the best of our knowledge, the extensive research of Vernik and coauthors (Vernik, 1993; Vernik and Liu, 1997; Vernik and Nur, 1992) comprises most of the published laboratory ultrasonic measurements on organic-rich shales that are complemented with TOC content and maturity indicator measurements such as hydrogen index (HI) and vitrinite reflectance (R_0). Bocangel et al. (2013) reported dynamic elastic moduli, TOC content, and maturity of the Wolfcamp Shale from Midland Basin. Patrusheva et al. (2014) studied static and dynamic moduli of the Mancos Shale with known TOC content. A number of authors (e.g., Dewhurst et al., 2008; Hornby, 1998; Johnston and Christensen, 1995; Wang, 2002) studied shales as seals with no relation to the organic content. These ultrasonic measurements are performed on dry and saturated shales, drained or with controlled pore pressure. Some of these shales still comprise significant content of organic matter but without sufficient information on its fraction, texture, or maturity. Comparison of rock physics attributes of the organic-rich and organic-lean seal shales is interesting as these shales exhibit similar rock physics properties, namely, they are highly anisotropic, almost impermeable, and might be a source of abnormal pore pressure.

Here we bring together some published data on ultrasonic experiments on shales. We complement the ORS database with ultrasonic velocities obtained at different saturation conditions as they can shed some light on the effects of saturation on elastic properties of shales as well as on the effects of variations in inorganic matrix mineralogy on elastic properties of ORSs. Table 9.1 contains published information on saturation state, TOC content, and maturity indicators of shales used in this study.

9.3 ORGANIC MATTER EFFECTS ON ELASTIC PROPERTIES

The ORS dataset covers a broad range of TOC fractions (from 0 to 20.1%) and maturity levels (HI = 1 ÷ 692 and R_0 = 0.38 ÷ 3.5%). Elastic moduli broadly decrease with the increase of TOC content (Fig. 9.1). The fact that ORSs are strongly anisotropic and the Thomsen's anisotropy parameters broadly increase with the increase of kerogen volumetric fraction (Fig. 9.1) was also pointed out by Vernik and coauthors (Vernik, 1994; Vernik and Landis, 1996; Vernik and Liu, 1997; Vernik and Nur, 1992). This strong anisotropy of ORSs was explained with the fact that the kerogen forms lenticular beds in inorganic matrix and might be or not be load bearing depending on its fraction and maturation degree.

A number of theoretical approaches for quantitative modeling of elastic properties of shales have been developed (e.g., Carcione et al., 2011; Sayers, 2013a; Vernik and Nur, 1992). These models tackle the problem of how the TOC content affects elastic properties of ORSs. The answer to this question is not trivial. Hashin and Shtrikman upper and lower bounds (Hashin and Shtrikman, 1963) give a range of elastic moduli of mixture if the microstructure of the constituents is not known. But this range for ORS is quite broad as

TABLE 9.1 Shales used in this study

Shale and Basin (number of samples)	Saturation	TOC, wt%	Maturity Indicators		References
			HI	R_0, %	
Bakken Formation, Mississippian-Devonian, Williston Basin (15)	Preserved from major desiccation	5.9 ÷ 20.1	97 ÷ 584	0.61 ÷ 1.27	Vernik and Landis (1996); Vernik and Liu (1997)
Bazhenov Formation, Jurassic, Western Siberia (8)	Preserved from major desiccation	2.2 ÷ 7.4	200 ÷ 434	0.6 ÷ 0.78	Vernik and Landis (1996); Vernik and Liu (1997)
Monterey Formation, Miocene, Santa Barbara, and Santa Maria Basins (18)	Preserved from major desiccation	0.2 ÷ 18.2	316 ÷ 692	0.38 ÷ 0.5	Vernik and Landis (1996); Vernik and Liu (1997)
Niobrara Formation, Cretaceous, San Juan Basin (10)	Preserved from major desiccation	0.5 ÷ 2.1	7 ÷ 124	0.81 ÷ 1.46	Vernik and Landis (1996); Vernik and Liu (1997)
Kimmeridge Shale, North Sea (5)	Preserved from major desiccation	2.0 ÷ 8.6	65 ÷ 499	0.42 ÷ 1.25	Vernik and Landis (1996); Vernik and Liu (1997)
Japan Tertiary (1)	Preserved from major desiccation	2.1	438		Vernik and Landis (1996); Vernik and Liu (1997)
Lockatong Formation, Triassic, Newark Basin (7)	Preserved from major desiccation	0 ÷ 2.8	1 ÷ 4	2.58 ÷ 3.15	Vernik and Landis (1996); Vernik and Liu (1997)
Woodford Formation, Mississippian–Devonian, Anadarko Basin (6)	Preserved from major desiccation	2.3 ÷ 9.5	40 ÷ 634	0.46 ÷ 1.47	Vernik and Landis (1996); Vernik and Liu (1997)
Wolfcamp Formation, Permian, Midland Basin (9)		1.2 ÷ 3.6	179		Bocangel et al. (2013)
Jurassic Shale (1)	Saturated, drained				Hornby (1998)
Kimmeridge Shale (1)	Saturated, drained				Hornby (1998)
Chattanooga, Devonian–Mississippian, eastern Tennessee (2)	Dry				Johnston and Christensen (1995)
New Albany, Devonian–Mississippian, Illinois Basin (4)	Dry				Johnston and Christensen (1995)
Lower Antrim, Devonian–Mississippian, Michigan Basin (1)	Dry				Johnston and Christensen (1995)
Mannville (4)	Dry				Hemsing (2007)
Officer Basin (3)	Saturated, undrained	0.0			Dewhurst et al. (2008)
Bass Basin (2)	Saturated, undrained	<1.0			Dewhurst et al. (2008)
Africa Shale (5)	Saturated, undrained				Wang (2002)
North Sea Shale (3)	Saturated, undrained				Wang (2002)
Gulf Coast Shale (1)	Saturated, undrained				Wang (2002)
Hard Shale (5)	Saturated, undrained				Wang (2002)
Siliceous Shale (2)	Saturated, undrained				Wang (2002)
Mancos Shale, Cretaceous (4)	Dry	1.0			Patrusheva et al. (2014)

kerogen and shale elastic moduli exhibit significantly strong contrast. Consequently, to answer the question of the kerogen effects on elastic properties of ORSs, we have to assume some microstructure of organic and inorganic constituents. As the microstructure of the organic and inorganic constituents of ORSs depends on many factors such as depositional environment, compaction diagenesis, inorganic matrix mineralogy, and organic matter maturity, the microstructure of the constituents will probably change from shale to shale and the models should be chosen with respect to this microstructure.

Taking into account a strong alignment of illite and kerogen observed on SEM images, Vernik and Nur (1992) suggested using Backus averaging for a 1D-layered medium (Backus, 1962) to model elastic moduli of ORSs as follows:

$$c_{11}^* = \langle c_{11} - c_{13}^2 c_{33}^{-1} \rangle + \langle c_{33}^{-1} \rangle^{-1} \langle c_{33}^{-1} c_{13} \rangle^2 ; c_{33}^* = \langle c_{33}^{-1} \rangle^{-1} ;$$
$$c_{44}^* = \langle c_{44}^{-1} \rangle^{-1} ; c_{66}^* = \langle c_{66} \rangle ; c_{13}^* = \langle c_{33}^{-1} \rangle^{-1} \langle c_{33}^{-1} c_{13} \rangle \quad (9.1)$$

Here, the brackets indicate an average value of parameter, weighted by volume fractions of the constituents. Elastic

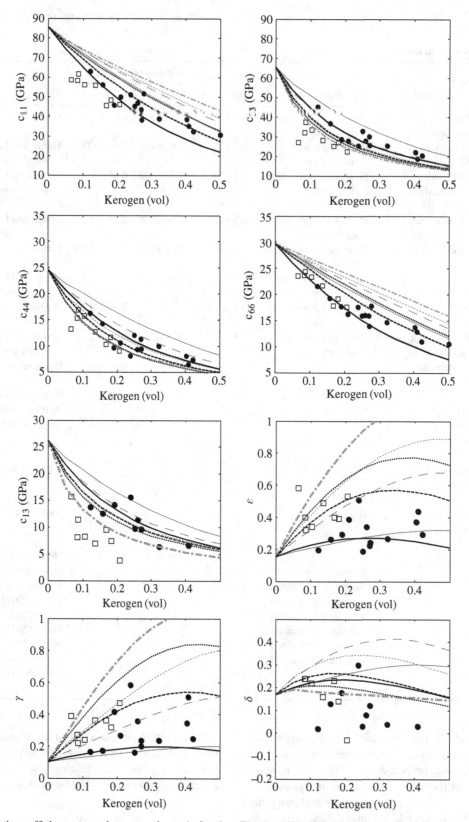

FIGURE 9.1 Elastic coefficients versus kerogen volumetric fraction. Elastic stiffness coefficients computed from the ultrasonic measurements reported by Vernik and Nur (1992) and Vernik and Liu (1997) in Bakken and Bazhenov shales are shown by solid circles and squares, respectively. Dashed-and-dotted line shows the theoretical prediction for a shale/kerogen-layered medium (Backus, 1962). Thick black lines show theoretical predictions for ellipsoidal inclusions of shale in kerogen matrix and thin lines show the results for ellipsoidal inclusions of kerogen in shale matrix. The solid, dashed, and dotted lines correspond to aspect ratio of inclusions of 0.3, 0.2 and 0.1. Anisotropic elastic coefficients of shale are assumed to be equal to c_{11} = 85.6 GPa, c_{33} = 65.5 GPa, c_{44} = 24.6 GPa, c_{66} = 29.7 GPa (Vernik and Landis, 1996), and c_{13} = 21.1 GPa (Sayers, 2013a). Isotropic elastic moduli of kerogen are chosen to be equal to K = 4.3 GPa and μ = 2.8 GPa (the mean between the upper and lower elastic moduli of kerogen reported by Yan and Han, 2013).

moduli of kerogen were chosen from the best fit with the experimental data as $c_{11} = c_{33} = 9.8\,\text{GPa}$, $c_{13} = 3.4\,\text{GPa}$, and $c_{44} = c_{66} = 3.2\,\text{GPa}$. Elastic moduli of shale without organic matter were chosen to be $c_{11} = 85.6\,\text{GPa}$, $c_{33} = 65.5\,\text{GPa}$, $c_{44} = 24.6\,\text{GPa}$, and $c_{66} = 29.7\,\text{GPa}$. The layered medium model describes the decrease in elastic moduli normal to bedding but shows much gentler decrease with increase of kerogen than the moduli calculated from experimentally measured velocities as can be seen in Figure 9.1. Vernik and Nur (1992) suggested that discontinuities in illite, distributed in the form of thin (<0.5 mm) lenticular beds rather than solid layers might be responsible for this discrepancy. To obtain a better fit, they modified the Backus model by replacing illite moduli in the Backus equations for the moduli parallel to bedding with effective moduli that incorporate effects of both kerogen and illite and can be used as a parameter to fit the experimental data.

Carcione et al. (2011) suggested an alternative model that is based on solid–solid anisotropic Gassmann's substitution suggested by Ciz and Shapiro (2007) and accounts for anisotropy of the shale framework by generalization of the empirical equation of (Krief et al., 1990)

$$c_{ij}^{m} = c_{ij}^{s}(1-\phi)^{A_{ij}/(1-\phi)} \qquad (9.2)$$

where indices m and s denote elastic coefficients of shale matrix and shale with zero porosity, respectively. By visual best fit, the fitting coefficients A_{ij} were assumed to be equal to 1.5 and 4 for the case when $i = j = 1$ and the cases when $i = j = 4$ or 6 and $i = 1, j = 3$, respectively.

The advantage of the method suggested by Carcione et al. (2011) is that the solid Gassmann substitution is independent of the microstructures. An obvious disadvantage is that the generalization of the Krief equations is purely empirical and requires two fitting parameters to obtain a good fit. Also, unlike Gassmann's original fluid substitution, solid substitution is an approximation, whose accuracy depends on the microstructure (Makarynska et al., 2010; Saxena et al., 2013).

Sayers (2013a) suggested using another effective medium method of modeling ORS elastic properties that does not require any fitting parameters. He used an anisotropic effective field theory developed by Sevostianov et al. (2005). The effective field theory accounts for the effect of ellipsoidal heterogeneities and allows one-particle solution for a transversely isotropic medium. Sayers (2013a) used the same elastic anisotropic coefficients of the shale and kerogen isotropic moduli as used in the original paper of Vernik and Nur (1992). Two cases have been considered. In the first case, the shale forms a continuous matrix, which hosts kerogen ellipsoidal inclusions with aspect ratios of 0.1, 0.2, and 0.5. In the second case, the kerogen forms the matrix, which hosts illite inclusions of different aspect ratios. Sayers demonstrated that the modeling results of the second case scenario are in a better agreement with the elastic moduli calculated from laboratory measurements of Bakken ORSs (Vernik and Nur, 1992).

Another reasonable possibility is to follow Sayers's (2013a) approach, but instead of the effective field theory, apply the anisotropic differential effective medium approach (Nishizawa, 1982). While moduli of shale can be chosen as in Vernik and Nur (1992), the fitting of the kerogen moduli can be avoided by using measured moduli recently reported by Yan and Han (2013). Isotropic elastic moduli of kerogen can be chosen to be $K = 4.3$ GPa and $\mu = 2.8$ GPa, which are the mean values between the upper and lower bounds of kerogen moduli (Yan and Han, 2013). The results for the two host/inclusion scenarios of shale/kerogen and kerogen/shale are shown in Figure 9.1 by red and black lines, respectively. The solid, dashed, and dotted lines correspond to aspect ratio of inclusions of 0.5, 0.2, and 0.1. Green dashed-and-dotted line shows the theoretical prediction for a shale/kerogen layered medium (Backus, 1962). One can see that the modeling results for the case of kerogen(host)/shale(inclusions) are in good agreement with the experimental data for Bakken Shale shown by solid circles. However, the experimentally measured elastic coefficients of Bazhenov Shale (Vernik and Landis, 1996) (open squares) exhibit a different trend with the decrease of kerogen volumetric fraction. The trend for the Bazhenov Shale cannot be modeled with the same coefficients of shale and kerogen as those used for the Bakken Shale. It is worth noting that Vernik and Landis (1996) also could not model the measured elastic moduli of the two different shales using the same moduli of inorganic shale and kerogen. They suggested using different moduli for the inorganic shale constituent.

Reliable estimation of elastic moduli of inorganic shale from its mineralogy and porosity is a long-standing problem. A detailed review of the works on this subject can be found in Pervukhina et al. (2011). Carcione et al. (2011) suggested inverting the moduli of the Bakken Shale with the known kerogen content (Vernik and Nur, 1992) for the moduli of its inorganic matrix. Carcione et al. (2011) reported that the inversion resulted in physically nonplausible results for 4 out of 11 samples. On top of that, the results exhibit strong scatter, namely, the samples from the adjacent depth 3271–3272 m show changes up to 15, 64, 35, 85, and 15% in the inverted c_{11}, c_{33}, c_{13}, c_{44}, and c_{66} moduli, respectively. Further refinements in kerogen modulus measurements and imaging of ORS microstructure might help choose the best way to obtain elastic moduli of inorganic shale constituent and rectification of the theoretical velocity–TOC relations.

9.4 PARTIAL SATURATION EFFECTS

Another important and not well-understood problem is how partial saturation of ORS affects their elastic properties. To model partial saturation, Carcione et al. (2011) assumed that ORSs can be modeled as a shale matrix with the pore space filled with kerogen, gas, and oil mixture. They used

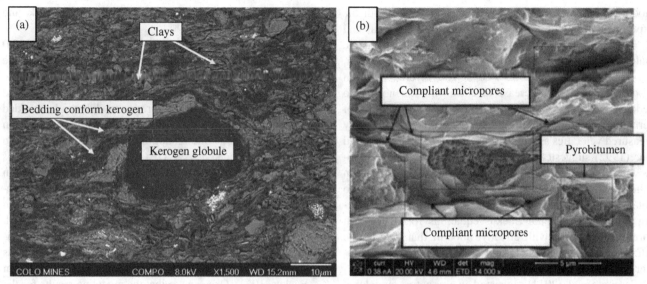

FIGURE 9.2 SEM images of immature and mature organic-rich shale: (a) FESEM image of Bakken Shale excavated from 7216 ft shows laminated texture of kerogen. (From Zargari et al., 2013.) (b) High-resolution backscatter SEM image of Haynesville shale from 12,781.2 ft. The amorphous organic matter, pyrobitumen, forms isolated inclusion in inorganic matrix. (From Lucier et al., 2011.)

the empirical model of Brie et al. (1995) to overcome the low-frequency limitation of the Gassmann equation, and to obtain effective fluid bulk moduli of oil and gas mixtures at all frequencies. With porosity of 40% filled with 25% of kerogen and 15% of an oil/gas mixture, saturation was shown to have a minor effect on elastic stiffness coefficients. The increase of oil saturation results in such small changes in elastic coefficients that acoustic velocities generally decrease due to the density increase. At a given porosity, the frequency effect on acoustic velocities was reported to be also small and the dispersion in velocities between the velocities at 1 MHz (unrelaxed state) and 25 Hz (relaxed state) to be below 2% for velocities normal to the bedding and 0.5% for those parallel to the bedding plane.

As laboratory measurements of ORS elastic properties with controlled water saturation are not available (to the best of our knowledge), these theoretical predictions have not been verified yet. However, water saturation can be recovered from wireline log data and such data complemented with sonic log measurements can be used for estimation of partially saturation effects on elastic properties of ORSs. Lucier et al. (2011) analyzed sonic log data from a vertical well drilled in NW Louisiana through the late Jurassic Bossier and Haynesville Shale, which are thermally overmature shales characterized with vitrinite reflectance R_0 of 2.0–2.8%. They found that the effect of gas saturation on V_p and V_s is significantly higher than the effect of TOC. The high-resolution backscatter SEM image of Haynesville Shale (Fig. 9.2b) shows that the texture of this shale is characterized by pyrobitumen inclusions embedded in a clay matrix and interconnected with a network of compliant cracks oriented along clay platelets that appear to be generated during hydrocarbon generation.

This microstructure is essentially different from the immature shale microstructure (an example of which can be seen in Fig. 9.2a). The effect of kerogen volumetric fraction on elastic properties of such gas shale must be negligible compared to the kerogen effect in immature shales. In immature shales, the effect of water saturation on P-wave velocity is very strong with a dramatic reduction in V_p with the increase of gas saturation by a few percent. The saturation effects can be clearly seen on V_p–V_s plot (Fig. 9.3a). Lucier et al. (2011) demonstrated that the deviation of Bossier and Haynesville Shales from the Castagna's trend for mudrocks (Castagna et al., 1993) can be explained by partial saturation.

This conclusion was obtained from well log data. To verify this conclusion on laboratory measurements, in Figure 9.3b we plot the compressional and shear velocities normal to bedding measured on dry, partially and fully saturated shales shown in Table 9.1. If TOC content is not known, the data are shown with open squares for dry samples with open triangles for partially saturated samples and with solid black triangles for fully saturated samples with drained or undrained experimental conditions. The experimental points from Vernik's dataset are color-coded with respect to TOC wt%. Unfortunately, quantitative information on saturation of these shales is unknown, and these shales are assumed to be partially saturated. The saturated shales generally follow Castagna's trend for mudrocks while the dry and partially saturated shales tend to exhibit 1000–1500 ft/s higher shear velocities for the same compressional velocity.

Dewhurst et al. (2012, 2013) showed that there was significant impact of partial saturation on dynamic elastic properties in clay-bearing silty shales (Fig. 9.4). These shales were from the Officer Basin in Western Australia

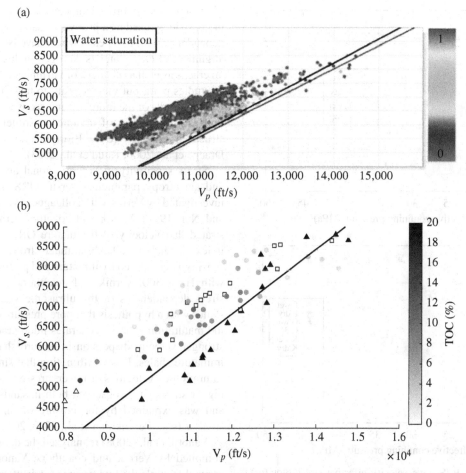

FIGURE 9.3 V_p–V_s plots of partially saturated shales: (a) Sonic log data from a Bossier/Haynesville well (from Lucier et al., 2011) and (b) laboratory measurements on dry (open squares), partially (open triangles and solid circles color coded with respect to TOC content), and fully saturated (solid triangles) shales shown in Table 9.1. The Castagna mudrock line (black) is shown in for reference in both plots.

and had clay content of ~30% (illite) with the rest comprising dolomite, K-feldspar, and quartz. Porosity was ~5%. Essentially, these were hard, laminated Proterozoic shales comprising mainly rigid grains and have similarities in composition to other gas shales (e.g., Tutuncu, 2012) except for the lack of organic matter. Modal pore sizes were in the region of 10 nm. These shales were tested in the fully preserved state (i.e., S_w = 100%) and a partially saturated state (S_w ~ 40%). The authors noted that P-wave velocity decreased with decreasing water saturation (i.e., increasing gas (air) saturation) but S-wave velocity increased by about 20%. These changes led to increases in Young's and shear moduli and a decrease in bulk modulus with increasing gas saturation. Large shear moduli changes have also been noted by Ghorbani et al. (2009) in more clay-rich shales and may be due to large effective stress increases associated with capillary suction in rocks with such small pore sizes (L. Laloui and A Ferrari, Personal Communication, 2013). It is currently unknown whether methane–organic matter interactions would result in different rock physics responses as compared to the water–air capillary system in these tests.

The laboratory measurements on shales whose water saturation ranges from 0 to 1 demonstrate significant effect of saturation on V_p that follow the same trend as the one obtained for sonic velocities of mature shales. Given that both sonic log and ultrasonic velocity show significant effects of partial saturation, theoretical models might need to be modified.

9.5 MATURITY EFFECTS

The thermal maturity determines the producible hydrocarbon type and, along with mineralogy of nonorganic constituents, is believed to control their brittleness. ORS thermal maturity estimation from surface seismic data is practically important for more robust reservoir characterization and reliable "sweet spot" detection. Evaluation of the ORS maturity from indirect seismic measurements might be achieved by using correlations between elastic properties, kerogen content, microstructure, and maturity obtained from controlled laboratory measurements. Several studies have been undertaken to identify key parameters that can be used to

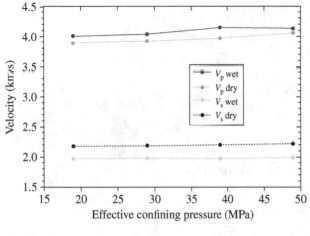

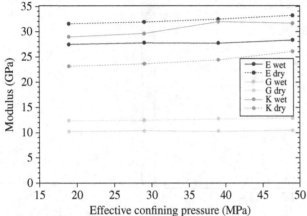

FIGURE 9.4 (top) P-/S-wave velocity effective stress plot for fully saturated (solid) and partially water saturated (dashed) silty shales. P-wave velocity decreases as water saturation decreases (gas saturation increases). S-wave velocity increases by ~20% under the same conditions. Effective stress has little effect; (bottom) Dynamic elastic moduli also vary little with effective stress, but with decreasing water saturation, Young's (E) and shear (G) moduli increase, while the bulk modulus (K) decreases. (From Dewhurst et al., 2012, 2013.)

understand relations between ORS maturity on their elastic properties and to eliminate the scatter caused by differences in their composition and microstructure.

The maturation process results in microstructural changes with the organic phase transformation from interconnected lenticular beds in immature shales into isolated inclusions embedded within the inorganic matrix. The TOC decreases with the maturity increase. Elastic moduli of ORSs increase with maturity due to (i) the decrease of TOC content, (ii) microstructural changes during maturation processes that lead to isolation of soft organic inclusions in a stiffer inorganic matrix, and (iii) changes in chemical and mechanical properties of inorganic constituents caused by exposure to high temperature and high pressure conditions, for instance, such as cementation or smectite–illite transformation.

Thomsen's anisotropy parameters ε and γ versus vitrinite index (widely used as a proxy for maturity) exhibit complex dependency with two peaks which have been explained with alignment of clay platelets, kerogen inclusions, subhorizontal cracks as well as changes of vitrinite alignment. A number of studies point out the necessity to take into account compaction trends of the inorganic matrix as well as the changes of elastic properties of organic constituent and the microstructural changes (e.g., Bjørlykke and Aagaard, 1992; Draege et al., 2006; Ruud et al., 2003).

Stress sensitivity of compressional and shear velocities and anisotropy parameters versus ORS maturity was also investigated by Vernik with colleagues (Vernik, 1994; Vernik and Nur, 1992). Vernik and coauthors graphically demonstrated that velocity of the mature ORSs with a hydrogen index (HI) below 200 exhibit much stronger nonlinear stress sensitivity compared to immature and partially mature shales with HI > 300. Vernik and coauthors explained stronger stress dependencies of the ultrasonic velocities in mature shales with a hypothesis that pore pressure developed due to the maturation process generates microcracks that result in stronger velocity dispersion in mature shales compared to immature shales. It is worth noting that similar exponential saturation of velocities at lower stresses to a linear trend at higher stresses was observed both in sandstones and shales and was explained by the closure of microcracks at low stresses (e.g., Pervukhina et al., 2010, 2011; Shapiro, 2003).

Vanorio et al. (2008) reanalyzed the data from the dataset compiled by Vernik and coauthors. Vanorio et al. (2008) argue that while the hydrogen index is often used as a proxy for kerogen maturity, this index actually mostly expresses hydrocarbon-generative potential and the quality of the rocks (Peters et al., 2004). In order to express kerogen maturity, Vanorio et al. (2008) used vitrinite reflectance. They demonstrated an increase in stress sensitivity of elastic properties with the increase of R_0 from 0.6 to 1.4% and a complex dependency of Thomsen's ε parameter against R_0 (Fig. 9.5).

Thomsen's ε parameter plotted against vitrinite reflectance shows two peaks, namely, one at R_0 ~0.6 and another at ~1.3 as can be seen in Figure 9.6, modified from Vanorio et al. (2008). The Thomsen's anisotropy parameters ε and γ are color-coded with respect to TOC content and shown with small, medium, and large circles representing the shales from 0 to 1500, 1501 to 3000, and 3001–4500 m depth intervals. Vanorio et al. (2008) explained the first peak (called "peak maturity") by the alignment of kerogen and clay particles, and the presence of subhorizontal microcracks developed as a result of kerogen maturation. For the first time, Vanorio et al. (2008) also identified the second peak (called "post-mature" peak) of Thomsen's ε parameter. They explained this second anisotropy peak in post-mature ORSs with the increase of the fraction of macerals of the vitrinite family, which due to aromatization processes, show higher density and more planar and aligned structure of the

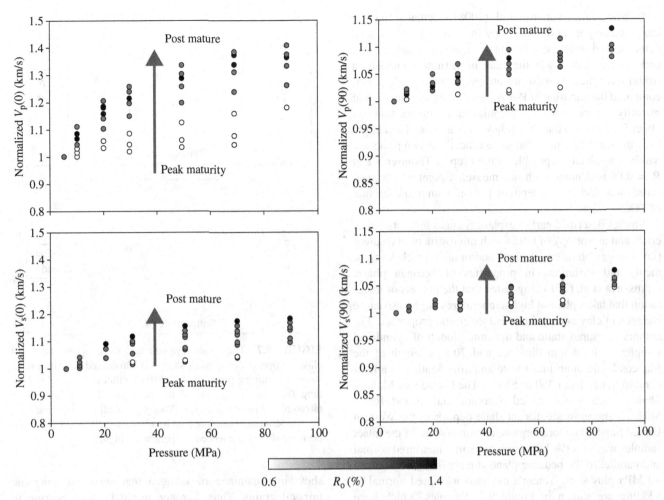

FIGURE 9.5 Normalized bedding normal (left) and parallel (right) P- and S-wave velocity as a function of confining pressure. Data are color-coded in terms of vitrinite reflectance index. (From Vanorio et al., 2008.)

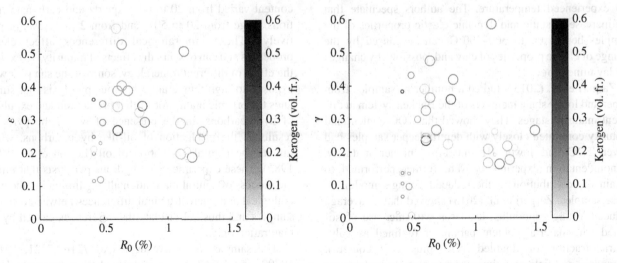

FIGURE 9.6 Thomsen's anisotropy parameters versus vitrinite reflectance. The data color coded with respect to kerogen volumetric fraction. The small, medium, and large circles correspond to depth ranges of 0–1500, 1501–3000, and 3001–4500 m, respectively. (Modified from Vanorio et al., 2008.)

carbon-rich rings. Vanorio et al. (2008) speculated that the intrinsic component of anisotropy in ORSs might be linked to the aligned structure of vitrinite and, thus, the early mature and mature shales with the peak in vitrinite composition would have the strongest anisotropy. Vanorio et al. (2008) compared the variation of P-wave anisotropy as a function of maturity and the changes in organic matter composition that affect R_0 and found that they follow the same trend and might have the same origin. At the same time, these two peaks are hardly statistically separable as the drop of Thomsen's ε at $R_0 = 0.88$ is denoted with one measured point which measured on a marl from a depth of 1121 m with a low fraction of TOC = 1.2%.

Studies described earlier explain changes in elastic properties and anisotropy of ORSs with microstructural changes (i.e., kerogen phase connectivity and/or microcrack development) and variations in properties of kerogen phase. Patrusheva et al. (2014) suggested that the process of maturation that takes place at high temperatures might also cause changes in clay composition and its elastic properties. The authors measured static and dynamic moduli of cylindrical samples with 38 mm diameter and 70 mm length of the Mancos Shale in an intact state and after heating to a temperature range from 360 to 550°C. The Cretaceous Mancos Shale, which is also called Niobrara, Gallup and Tocito Shale, is the most significant shale deposit in the Western United States. The total organic carbon content of the intact samples was ~1 wt%. Ultrasonic velocities measured normal and parallel to the bedding plane at a confining pressure of 40 MPa plus static Young's modulus measured normal to bedding are shown in Figure 9.7. Samples, which have experienced temperatures of 360–400°C, show higher velocities compared to the intact sample. However, the velocities measured on the samples that experienced higher temperatures of 400–550°C decrease with the increase of the experienced temperature. The authors speculate that the increase in static and dynamic elastic properties of the samples heated up to 360–400°C can be caused by the change of elastic properties of clay and, possibly, by changes in clay mineralogy.

Zargari et al. (2013) studied a number of samples from upper and lower shale members of the Bakken system at different maturity stages. They showed that TOC content and maturity correlated closely with depth; deeper samples had lower TOC and lower HI indicating higher maturity. Nanoindentation experiments were further performed to obtain the distribution of the reduced Young's moduli of these samples. Zargari et al. (2013) showed that an average reduced Young's modulus decreases with increase of so-called soft material content parameter (defined as volumetric fraction of doubled TOC plus clay content). Comparison of field emission scanning electron microscope (FESEM) images of mature and immature samples shows essential microstructural differences. Immature samples

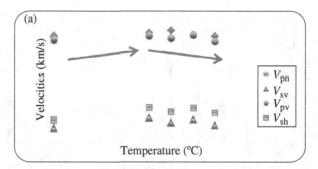

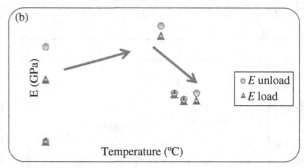

FIGURE 9.7 The results dynamic and static measurements of elastic properties of Mancos shale: (a) Ultrasound velocities measured at confining pressure of 40 MPa parallel and normal to bedding; (b) static Young's moduli measured parallel to bedding. Both ultrasonic velocities and static Young's moduli first increase with the temperature experienced up to approximately 360°C and then decrease if the experienced temperature is higher.

show microlaminae of kerogen that surround inorganic mineral grains. Thus, kerogen might be load bearing in immature samples. In more mature samples, kerogen seems to be isolated between inorganic grains (Fig. 9.2).

The investigated naturally matured samples of the Bakken Shale showed huge differences in mineralogy, namely, clay content varied from 20 to 60%, quartz and carbonate fractions range from 20 to 50% and from 2 to 15.5%, respectively). These mineralogical differences affect elastic properties and can obscure the effects of maturity. To exclude the effect of different mineralogy, some of the samples were exposed to high temperature hydrous pyrolysis that stimulates the organic matter for further production and expulsion of hydrocarbons. In the presence of water, the pyrolysis results in the production of oil-like hydrocarbons, which imitate the natural generation of oil (Lewan et al., 1979, 1985). These experiments on hydrous pyrolysis that mimic the process of natural maturation allow eliminating the mineralogical and microstructural differences between different samples and thus help isolate the differences caused by the maturation only.

The samples from five depths of 7216, 7221, 10479, 10792, and 10368 ft were subjected to hydrous pyrolysis. Then sample characterization and nanoindentation measurements were repeated and FESEM images reobtained.

Hydrous pyrolysis results in generation of hydrocarbons and evaporation of volatile hydrocarbons caused by high retort temperatures. Films of bitumen, bitumen globes, and bitumen flow channels were observed in the FESEM images of the samples after pyrolysis. These changes led to a decrease in average TOC from 12.7 ± 4.3 to 9.7 ± 1.6 and average HI from 400 ± 231 to 51.2 ± 16.8. The median Young's moduli also reduced from 25.2 ± 6.3 to 15.7 ± 5.3. To explain an apparent discrepancy with the well-known fact that more mature shales usually have higher elastic moduli, Zargari et al. (2013) speculated that bitumen, which had lower compared to kerogen elastic moduli and which would eventually migrate away in the process of natural maturation, was partially trapped in the kerogen and caused the reduction in the mean elastic moduli after pyrolysis.

We would like to say again that the effects of organic matter maturity on anisotropic elastic properties are complex and include microstructural transformations, decrease of TOC content, and changes in elastic moduli of organic and inorganic components. The lenticular distribution of kerogen in immature ORS is believed to be the reason for extraordinarily high Thomsen's ε and γ in ORSs. The general tendency of anisotropy to decrease with increasing vitrinite reflectance (Fig. 9.6) to some extent can be explained with isolation of the remaining kerogen in the pores of the inorganic matrix. Vanorio et al. (2008) identified a second maximum of Thomsen's ε at R_0 ~1.3 and explained it with the increase in elastic anisotropy of organic matter with maturity. The other two effects associated with the maturity are changes in the elastic properties of the inorganic matrix and the increase in stress sensitivity resulting from cracks induced during the maturation process. The inorganic clay matrix tends to become more anisotropic with increasing depth. Microcracks that had developed as a result of the maturation process are preferentially oriented along the bedding plane according to a number of microstructural studies of dry pyrolysis experiments (e.g., Kobchenko et al., 2011; Panahi et al., 2013; Yurikov et al., 2013). The presence of such subhorizontal cracks must also promote elastic anisotropy with maturity. As in reality elastic anisotropy decreases with the increase of maturity, the effects of stiffening of the inorganic clay matrix and the subhorizontal cracks caused by maturation must have minor effects on elastic anisotropy compared to the kerogen-laminated structure. To the best of our knowledge, no attempt to account for all these processes has been published.

9.6 SEISMIC RESPONSE OF ORSs

Carcione was the first to model the effect of a thin source-rock layer on AVO responses (Carcione, 2001). He concluded that the TOC content of a source rock layer can be determined with an AVO analysis. Carcione assumed that a transversely isotropic ORS layer of variable thickness is deposited between isotropic chalk (top) and sandstone (bottom). This structure models a typical structure in the North Sea, where a source rock with a thickness comparable to the seismic wavelength (<200 m) underlies a high velocity chalk. The shale, chalk, and sandstone were modeled as viscoelastic materials. The ORS was assumed to be a composite rock with horizontal layers of illite, kerogen, water, and oil, where the amount of oil depends on maturity of the rock. The modeled AVO response shows a strong decrease in the PP-reflection coefficient with an increase in the incidence angle. For a given TOC content and incident angle, the reflection coefficient dependency on ORS layer thickness exhibits an oscillatory character and the period of oscillations depends on the frequency of the seismic signal. Finally, for a given layer thickness and incidence angle, the reflection coefficient depends on the TOC content in a nonmonotonic way, having a minimum at a low volumetric kerogen content and above this value exhibiting a monotonic increase.

The ability to predict the TOC content from surface seismic data would significantly reduce the risk of developing ORS reservoirs. The first feasibility study into estimation of TOC content from real surface seismic data was undertaken by Løseth et al. (2011). They analyzed seismic data acquired on the Late Jurassic source rock formations from the North Sea (Draupne), Norwegian Sea (Spekk Formation), and Barents Sea (Hekkingen Formation) and could clearly identify the ORSs. This study also sheds some light on practical details of the possibility to convert the acoustic impedance (AI) into a TOC content map. This possibility exists as both laboratory measurements and log data analysis exhibit a strong nonlinear reduction in AI with an increase in TOC content and, hence, the AI of a source rock with TOC > 3–4 wt% might be half that of nonsource rocks. In ORSs thicker than the tuning thickness, these low AI values together with large difference of elastic properties in directions normal and parallel to bedding result in an AVO class 4 seismic response at the top of the source rocks. This AVO class 4 response is characterized by a high-amplitude negative reflection at the top of the reservoir and significant dimming with offset. The AVO class 4 responses are typical for source rocks and are not very common for other lithologies except coal, which has the same type of response. A high-amplitude positive reflection that also dims with offset (the AVO class 1 response) was observed at the base of the source rock reservoir. As the kerogen fraction varies across ORS formations, its vertical profiles might affect seismic response at the top and bottom boundaries. Løseth et al. (2011) found that the upward-increasing TOC profiles, which are typical for Jurassic Spekk and Draupne formations, result in a stronger top reflection compared to the basal reflection. For a downward-increasing TOC profile that was observed in the Hekkingen Formation, the top reflection is weaker than the basal one. These characteristics

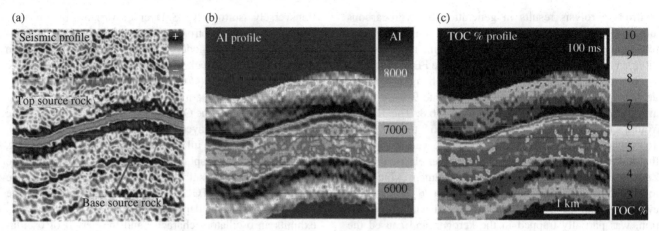

FIGURE 9.8 (a) Seismic section, (b) acoustic impedance inverted seismic section, (c) inverted seismic section where source rock interval is converted to TOC percent. (From Løseth et al., 2011.)

are specific for each formation and should be studied and calibrated for reliable and accurate conversions of AI into TOC content maps (Fig. 9.8).

The seismic response of the organic matter fraction might be different depending on the microstructure of the organic phase and its connectivity. Sayers (2013b) used the moduli of ORSs with (i) kerogen inclusions in shale matrix and (ii) shale inclusions in organic matter for inclusions with different aspect ratios calculated in Sayers (2013a) to investigate the effect of organic phase microstructure on the AVO response of ORSs. PP reflection coefficients for different angles of incidence for different microstructures and TOC fractions were calculated using the AVO theory of Schoenberg and Protazio (1992). Sayers showed that the decrease in the aspect ratio of inclusions results in an increase of PP reflection coefficient, that is, decreasing aspect ratio acts in the same way as increasing TOC content. Moreover, the connectivity of organic matter is also important as shale inclusions in an interconnected kerogen matrix results in a much stronger reflection coefficient compared to kerogen inclusions in an interconnected shale matrix. This work clearly showed that the AVO response is a function of both the composition and microstructure of organic-rich shales.

Estimation of the TOC content and/or the preferred orientation of fractures directly from acoustic impedance inversion is very challenging. Theoretical studies predict that the relationship between the acoustic impedance and the TOC content is nonunique as the texture and connectivity of the organic and inorganic components can strongly affect the seismic velocities at the same TOC fraction. This nonuniqueness along with possible variations of the TOC profile throughout an ORS layer would require calibration with laboratory measurements for reliable inversion results.

Notorious for their low permeability, ORS reservoirs are often permeated with subvertical fracture sets that permit the flow and facilitate hydrocarbon extraction. Such subvertical fracture sets with a preferred orientation might result in an azimuthal anisotropy and variations of the P–P reflection amplitude with azimuth. The detection of such subvertical fractures and insight into their preferable orientations is a practically important challenge for exploration and development of these unconventional reservoirs. Recently, Far and Hardage (2013) demonstrated the potential to invert P-wave seismic reflection data with variable offset and azimuth into the fracture parameters. The fractures cause an excess in shale compliance that was parameterized via second- and fourth-rank effective fracture compliant tensors using the model suggested by Sayers and Kachanov (1995). Far and Hardage (2013) assumed a general monoclinic symmetry of the subsurface, which accounts both for ubiquitous polar anisotropy and for azimuthal anisotropy caused by multiple fracture sets. Far and Hardage (2013) assumed that the fractures are characterized with rotationally invariant shear compliance and inverted synthetic data from wide azimuth (WA) and narrow azimuth (NA) arrays for the components of the tensors of effective fracture compliances. The eight parameters responsible for the excess compliances caused with the vertical fracture sets can be reliably estimated with the signal-to-noise ratio of 2, if the background medium properties are known (e.g., from the wells drilled through the formation). Relative errors in the estimation of the components of the second-rank and fourth-rank-effective fracture compliant tensors are different for the WA and NA synthetic data. WA arrays allow reliable estimation of all the second rank tensor components while, from an NA array, the same components can be obtained with much larger uncertainty. To apply this method to real fractured rocks, Far et al. (2013) used multicomponent P-wave AVOAz seismic data acquired over the Marcellus Shale in Pennsylvania, USA. AVOAz data from the interface between the Stafford Limestone and top Marcellus and the interface between the base Marcellus and Onondaga Limestone were inverted for fracture orientations and showed consistent results with FMI log observations. However, the strength of azimuthal anisotropy is not always a robust indicator of reservoir quality because the relationship between hydraulic and elastic properties of fractures is not unique. In particular, to the first order, the effect of fractures on elastic anisotropy is independent of fracture size, while their hydraulic conductivity is strongly scale dependent (Williams and Jenner, 2002). Also, reservoir quality is primarily controlled by fracture connectivity, which is not directly related to elastic anisotropy and might not have a direct expression in seismic data at all (Goodway et al., 2010; Parney et al., 2004).

9.7 CONCLUSIONS

We have given a snapshot of the state of the art in the rock physics of organic-rich shales. The TOC content, gas saturation, maturity, and subvertical fracture sets are the key parameters that affect elastic properties of ORSs. Scarce published laboratory experiments show a decrease in ORS velocities with the increase of kerogen volumetric fraction. This decrease can be modeled using anisotropic differential effective medium methods or by anisotropic effective field theory if the elastic properties of the inorganic matrix are known. A strong effect of partial gas saturation on compressional velocity was recently shown on log data from Bossier and Haynesville shales. The results of laboratory experiments on dry, partially saturated (preserved from major desiccation), and saturated (with controlled pore pressure) shale samples show the same trends as the log data. Experimental studies show that elastic anisotropy broadly decreases with increasing ORS thermal maturity. Maturation processes reduce TOC content in organic-rich shales, change microstructure and elastic properties of organic and inorganic constituents, and can introduce bedding parallel microcracks. To the best of our knowledge, no theoretical model takes into account all these superposing and competing effects for quantitative modeling of maturity effects.

Under favorable conditions, hydrocarbon source rocks can be identified on seismic data. Dramatic contrasts in their acoustic impedances compared to organic-lean rocks result in the strong negative reflection on the top of the ORS layer that dims with offset (Class 4 AVO response). Moreover, AI maps can potentially be used to build TOC content maps if the other parameters that affect ORS velocity such as microstructure and the elastic properties of organic and inorganic components can be assumed to be constant throughout the area of interest. Finally, subvertical fracture sets affect azimuthal velocities and turn elastic anisotropy symmetry from VTI to monoclinic in a general case. Effects of these fractures can be taken into account with the linear slip theory and P-wave AVOAz data from the top or bottom interfaces can be inverted to predict fracture set orientations.

REFERENCES

Aguilera RF, Ragetzki M. Shale gas and oil: fundamentally changing global energy markets. Oil Gas J 2013;111 (12):54–61.

Avseth P, Mukerji T, Mavko G, editors. Quantitative Seismic Interpretation: Applying Rock Physics Tools To Reduce Interpretation Risk. Cambridge (MA): Cambridge University Press; 2005.

Backus GE. Long-wave elastic anisotropy produced by horizontal layering. J Geophys Res 1962;67 (11):4427–4440.

Bjørlykke K, Aagaard P. Clay minerals in North Sea sandstones. In: Houseknecht DWP, Pittman ED, editors. Origin, Diagenesis, and Petrophysics of Clay Minerals in Sandstones. Tulsa (OK): SEPM Special Publication; 1992. p 65–80.

Bocangel W, Sondergeld C, Rai C. Acoustic Mapping and characterization of organic Matter in Shales. SPE, SPE: SPE 166331; 2013.

Brie A, Pampuri F, Marsala AF, Meazza O. Shear sonic interpretation in gas-bearing sands: SPE Annual Technical conference; 1995;30595:701–710.

Carcione JM. AVO effects of a hydrocarbon source-rock layer. Geophysics 2001;66 (2):419–427.

Carcione JM, Helle HB, Avseth P. Source-rock seismic-velocity models: Gassmann versus Backus. Geophysics 2011;76 (5):N37–N45.

Castagna JP, Batzle MI, Kan TK. Rock physics – The link between rock properties and AVO response in Offset-Dependent Reflectivity. In: Castagna JP, Backus M, editors. Theory and Practice of AVO Analysis. Tulsa (OK): Society of Exploration Geophysicists;.1993. pp 135–171.

Ciz R, Shapiro SA. Generalization of Gassmann equations for porous media saturated with a solid material. Geophysics 2007;72 (6):A75–A79.

DERA. Energy Study: Reserves, Resources and Availability of Energy Resources 2012. Deutsche Rohstoffagentur: Hannover; 2012.

Dewhurst DN, Siggins AF. Impact of fabric, microcracks and stress field on shale anisotropy. Geophys J Int 2006;165:135–148.

Dewhurst DN, Siggins AF, Kuila U, Clennell MB, Raven MD, Nordgård-Bolås HM. Rock Physics, Geomechanics and Rock Properties in Shales: Where Are The Links?, paper presented at Proceedings of the 1st SHIRMS conference; 2008.

Dewhurst DN, Maney B, Clennell B, Delle Piane C, Madonna C, Saenger E, Tisato N. Impact of saturation change on shale properties. Extended abstract for 3rd EAGE Shale Conference, Barcelona; 2012. 4p.

Dewhurst D, Bunger A, Josh M, Sarout J, Delle Piane C, Esteban L, Clennell B. Mechanics, Physics, Chemistry and Shale Rock Properties. Paper 13-0151, Proceedings of the 47th American Rock Mechanics Association Annual Conference, San Francisco; 2013. 11pp.

Draege A, Jakobsen M, Johansen TA. Rock physics modelling of shale diagenesis. Petrol Geosci 2006;12 (1):49–57.

EIA. Annual energy outlook. Washington DC: Energy Information Administration; 2013.

Far ME, Hardage B. Inversion of elastic properties of fractured rocks from AVOAz data Marcellus Shale example. Houston: SEG; 2013. p 3133–3137.

Far ME, Sayers CM, Thomsen L, Han DH, Castagna JP. Seismic characterization of naturally fractured reservoirs using amplitude versus offset and azimuth analysis. Geophys Prospect 2013;61 (2):427–447.

Ghorbani A, Zamora M, Cosenza P. Effects of desiccation on the elastic wave velocities of clay-rocks. Int J Rock Mech Min Sci 2009;46:1267–1272.

Goodway B, Perez M, Varsek J, Abaco C. Seismic petrophysics and isotropic-anisotropic AVO methods for unconventional gas exploration. The Leading Edge 2010;29:1500–1508.

Hashin Z, Shtrikman S. A variational approach to the theory of the elastic behaviour of multiphase materials. J Mech Phys Solids 1963;11 (2):127–140.

Hemsing D. Laboratory determination of seismic anisotropy in sedimentary rock from the Western Canadian Sedimentary Basin, University of Alberta. M.S. thesis; 2007.

Hornby BE. Experimental laboratory determination of the dynamic elastic properties of wet, drained shales. J Geophys Res 1998;103:29945–29964.

IEA. Resources to Reserves. Paris: International Energy Agency; 2013.

Johnston JE, Christensen NI. Seismic anisotropy of shales. J Geophys Res 1995;100 (B4):5991–6003.

Kobchenko M, Panahi H, Renard F, Dysthe DK, Malthe-Sorenssen A, Mazzini A, Scheibert J, Jamtveit B, Meakin P. 4D imaging of fracturing in organic-rich shales during heating. J Geophys Res 2011;116:B12201.

Krief M, Garat J, Stellingwerff J, Ventre J. A petrophysical interpretation using the velocities of P and S waves (full-waveform sonic). The Log Analyst 1990;31(Nov):355–369.

Lewan MD. Evaluation of Petroleum Generation by Hydrous Pyrolysis Experimentation. Philos Trans A Math Phys Eng Sci 1985;315 (1531):123–134.

Lewan MD. Petrographic study of primary petroleum migration in the woodford Shale and related rock units. In: Doligez B, editor. Migration of Hydrocarbons in Sedimentary Basins. Paris: Editions Technip; 1987. p 113–190.

Lewan MD, Winters JC, Mcdonald JH. Generation of oil-like pyrolyzates from organic-rich shales. Science 1979;203 (4383):897–899.

Løseth H, Wensaas L, Gading M, Duffaut K, Springer M. Can hydrocarbon source rocks be identified on seismic data? Geology 2011;39 (12):1167–1170.

Lucier AMH, Hofmann R, Bryndzia LT. Evaluation of variable gas saturation on acoustic log data from the Haynesville Shale gas play. NW Louisiana: The Leading Edge; 2011 March. pp 300–311.

Makarynska D, Gurevich B, Behura J, Batzle M. Fluid substitution in rocks saturated with viscoelastic fluids. Geophysics 2010;75 (2):E115–E122. DOI: 10.1190/1.3360313.

Meissner FF. Petroleum geology of the Bakken Formation, Williston Basin, North Dakota and Montana in The economic geology of the Williston Basin: Montana Geological Society. Williston Basin Symposium; 1978.

Momper JA. (1978). Oil migration limitations suggested by geological and geochemical considerations. In Physical Constraints on Petroleum Migration, AAPG, 1, B1-B60.

Nishizawa O. Seismic velocity anisotropy in a medium containing oriented cracks – transversely isotropic case. J Phys Earth 1982;30:331–347.

Panahi H, Kobchenko M, Renard F, Mazzini A, Scheibert J, Dysthe D, Jamtveit B, Malthe-Sorenssen A, Meakin P. A 4D synchrotron X-ray-tomography study of the formation of hydrocarbon-migration pathways in heated organic-rich shale. Spe J 2013;18 (2):366–377.

Parney B, Williams M, Jenner E. Azimuthal NMO as an indicator of natural fracturing Canadian Society of Exploration Geophysicists, Expanded Abstracts; 2004. pp. S035–S035.

Patrusheva N, Lebedev M, Pervukhina M, Dautriat J, Dewhurst DN. Changes in microstructure and elastic properties of Mancos shale after pyrolysis, in 76th EAGE Conference & Exhibition edited, EAGE, Amsterdam; 2014.

Pervukhina M, Gurevich B, Dewhurst DN, Siggins AF. Experimental verification of the physical nature of velocity-stress relationship for isotropic porous rocks. Geophys J Int 2010;181:1473–1479.

Pervukhina M, Gurevich B, Golodoniuc P, Dewhurst DN. Parameterization of elastic stress sensitivity in shales. Geophysics 2011;76 (3):B1–B9.

Peters KE, Walters CC, Moldowan JM. The Biomarker Guide. Volume 2, Cambridge: Cambridge University Press; 2004. Biomarkers and Isotopes in Petroleum Systems and Earth History.

Price LC, Ging T, Daws T, Love A, Pawlewicz M, Anders D. Organic metamorphism in the Mississippian-Devonian Bakken Shale, N. Dakota portion of the Williston Basin. In: Woodward F, Meissner F, Clayton LJ, editors. Hydrocarbon Source Rocks of the Greater Rocky Mountain Region. Denver: Rock Mountain association of Geologists; 1984. p 83–133.

Ruud BO, Jakobsen M, Johansen TA. Seismic properties of shales during compaction. In SEG Technical Program Expanded Abstracts, Dallas, TX; 2003. pp. 1294–1297. doi: 10.1190/1.1817522

Saxena N, Mavko G, Mukerji T. Change in effective bulk modulus upon fluid or solid substitution. Geophysics 2013;78: L45–L56.

Sayers CM. The effect of kerogen on the elastic anisotropy of organic-rich shales. Geophysics 2013a;78 (2):D65–D74.

Sayers CM. The Effect of Kerogen on the AVO Response of Organic-Rich Shales SEG. Houston (TX): SEG; 2013b. p 2762–2766.

Sayers C, Kachanov M. Microcrack-induced elastic wave anisotropy of brittle rock. J Geophys Res 1995;100:4149–4156.

Schoenberg M, Protazio J. Zoeppritz rationalized and generalized to anisotropy. J Seism Explor 1992;1:125–144.

Sevostianov I, Yilmaz N, Kushch V, Levin V. Effective elastic properties of matrix composites with transversely-isotropic phases. Int J Solids Struct 2005;42:455–476.

Shapiro SA. Elastic piezosensitivity of porous and fractured rocks. Geophysics 2003;68 (2):482–486.

Tutuncu AN. The Role of Mechanical, Acoustic and Permeability Anisotropies on Reservoir Characterization and Field Development for Two North American Fractured Unconventional Shale Reservoirs, ARMA 12-664, Proceedings of 46th US Rock Mechanics/Geomechanics Symposium, Chicago; 2012.

Vanorio T, Mukerji T, Mavko G. Emerging methodologies to characterize the rock physics properties of organic-rich shales. The Leading Edge 2008 June;780–787.

Vernik L. Microcrack-induced versus intrinsic elastic-anisotropy in mature HC-source shales. Geophysics 1993;58 (11): 1703–1706.

Vernik L. Hydrocarbon-generation – induced microcracking of source rocks. Geophysics 1994;59 (4):555–563.

Vernik L, Landis C. Elastic anisotropy of source rocks: implications for hydrocarbon generation and primary migration. Am Assoc Pet Geol Bull 1996;80 (4):531–544.

Vernik L, Liu XZ. Velocity anisotropy in shales: a petrophysical study. Geophysics 1997;62 (2):521–532.

Vernik L, Nur A. Ultrasonic velocity and anisotropy of hydrocarbon source rocks. Geophysics 1992;57 (5):727–735.

Wang Z. Seismic anisotropy in sedimentary rocks, part 2: laboratory data. Geophysics 2002;67 (5):1423–1440.

Williams M, Jenner E. Interpreting seismic data in the presence of azimuthal anisotropy; or azimuthal anisotropy in the presence of the seismic interpretation. The Leading Edge 2002 Aug; 21(8):771–774

Yan FH, Han D-H. Measurement of Elastic Properties of Kerogen SEG. Houston (TX): SEG; 2013. pp 2778–2782.

Yurikov A, Pervukhina M, Lebedev M, Shulakova V, Uvarova Y, Gurevich B. Fracturing of organic-rich shale during heating, in The 11th International SEGJ Symposium, edited, Yokohama; 2013.

Zargari S, Prasad M, Mba KC, Mattson ED. Organic maturity, elastic properties and textural characteristics of self resourcing reservoirs. Geophysics 2013;78 (4):D223–D235.

10

PASSIVE SEISMIC METHODS FOR UNCONVENTIONAL RESOURCE DEVELOPMENT

ALFRED LACAZETTE[1], CHARLES SICKING[2], RIGOBERT TIBI[1] AND ASHLEY FISH-YANER[1]

[1] Global Geophysical Services, Inc., Denver, CO, USA
[2] Global Geophysical Services, Inc., Dallas, TX, USA

10.1 INTRODUCTION

The purpose of this chapter is to provide a concise introduction and overview of passive seismic methods and practical applications of such methods for unconventional oil and gas work. We do not intend to provide a careful mathematical treatment of the subject or a comprehensive, thoroughly referenced review of the field and its history. Rather, the purpose is to explain briefly and concisely how these methods work, what their applications are, how to understand the methods in use by vendors, and how to interpret the results of passive seismic monitoring of hydraulic fracture treatments. This chapter also provides the most basic information on geomechanics, earth stress, and natural fractures required to understand passive seismic methods and interpret the results.

The current boom in unconventional resource development is almost entirely dependent on hydraulic fracture stimulation of tight petroleum and natural gas reservoirs. Successful fracture treatments require sound designs based on knowledge of the reservoir geomechanics and preferably with knowledge of frac behavior in the field being treated. In some cases, fracture treatments must be controlled in real time to ensure that the frac does not contact water-bearing strata or otherwise propagate in undesirable ways. Well spacing in a field is a function of the horizontal distance from a well that is effectively stimulated by a fracture treatment. This distance cannot be determined using only pumping data (e.g., amounts of fluid and proppant). Halving or doubling well spacing roughly doubles or halves (respectively) development costs of a field. The extent and manner of interaction of a fracture treatment with preexisting natural fractures provides an important control on frac effectiveness, location, and well spacing. Again, the nature and location of such interactions cannot be determined by pumping data and simulation alone. They must be observed. In summary, effective frac design and field development planning require knowledge of the actual behavior of hydraulic fractures in oil and gas reservoirs deep in the subsurface. Interactive control of hydraulic fractures can only be accomplished with real-time three-dimensional (3D) mapping.

As of this writing, passive seismic monitoring is the best method for characterizing the size, shape, and location of hydraulic fractures in oil and gas reservoirs and for understanding the interactions of hydraulic fractures with preexisting natural fracture systems. Frac engineers, development geologists, and reservoir engineers are the primary users of passive products.

Some basic definitions are required before proceeding further:

Active seismic methods—Seismic methods that record reflected or transmitted seismic waves produced by artificial explosions or vibrators on the earth's surface or in wells. The 2D and 3D reflection seismic surveys familiar to anyone in the upstream oil and gas industry are examples of active seismic products.

Passive seismic methods—Seismic methods that rely on energy emitted by rock movements in the subsurface. The ultimate source of the energy can be entirely natural (e.g., tectonic movements or earth tides), artificial (e.g., hydraulic fracture treatments), or both.

Fundamentals of Gas Shale Reservoirs, First Edition. Edited by Reza Rezaee.
© 2015 John Wiley & Sons, Inc. Published 2015 by John Wiley & Sons, Inc.

Earthquake—An event that generates seismic waves due to a sudden release of strain energy.

Microearthquake (MEQ)—An earthquake with a magnitude less than or equal to zero.

Passive seismic methods are often taken as synonymous with microseismic methods. However, the term *microseismic* implies methods relying only on small-magnitude earthquakes. Such studies typically use classical earthquake seismology methods to locate earthquakes by observing the differences in first arrival times of P (compressional)- and S (shear)-waves at multiple receivers.

We prefer to avoid the term *microseismic* to describe the subject of this chapter because it is restrictive and misleading. Recent developments in both the science and technology of hydraulic fracture monitoring show the importance of seismic energy sources other than microearthquake (MEQs). Such sources include long-duration, low-frequency emissions without distinct first arrivals, and P-waves produced by fluid pressure oscillations (water hammering). Also, not all monitoring technologies rely on classical methods of earthquake identification, specifically:

- Recent recognition of long-period, long-duration (LPLD) seismic activity and other types of activity shows that MEQs represent only a small fraction of the total seismic energy produced during hydraulic fracture treatments (Das and Zoback, 2013a, b). Also, we cannot rule out the occurrence of other as-yet-uncategorized types of seismic activity distinct from both MEQ and LPLD during hydraulic fracture treatments.
- Passive seismic monitoring technologies relying on seismic emission tomography (SET) do not identify MEQs via classical seismological methods that rely on the difference in P- and S-wave arrival times. Also, SET identifies sources of seismic energy other than MEQs. SET-based methods are becoming ever more widely used. As a result, these methods image energy from more types of activity than just MEQs, even over very short time intervals.

In conclusion, the term *passive seismic* is sufficiently general to cover all types of seismic activity and the results of all monitoring and analysis methods in current use.

The geothermal energy industry applied passive seismic methods in the 1970s and 1980s to monitor both hydraulic fracturing and natural fracture networks stimulated by injection and production. The oil and gas industry now uses passive seismic for these same tasks. The first application of passive seismic in the oil and gas industry was the Rangely experiment in 1976, which demonstrated induced seismicity in a petroleum field that resulted from injection (Rayleigh et al., 1976). The first successful, modern, downhole passive seismic monitoring of a fracture treatment in an unconventional reservoir occurred in the Barnett Shale in 2000 (Maxwell et al., 2002). To date, downhole monitoring has relied on traditional earthquake seismology methods to locate MEQs. The first application of SET to surface or shallow buried array data for hydraulic fracture treatment monitoring took place in June 2004 (Duncan, 2005; Lakings et al., 2006). The first attempts at extraction of fracture images from SET data followed rapidly (Geiser et al., 2006). Since then, the field of passive seismic monitoring has grown tremendously. Downhole methods have grown more sophisticated with new analytical methods and especially with the use of multiple downhole arrays for more precise imaging. Surface monitoring using SET is now offered by many service providers, and MEQ location and focal mechanism determination using SET are well-established methods (e.g., Duncan and Eisner, 2010; this reference provides a historical review of surface-based methods). Direct fracture imaging methods also have become more fully developed (e.g., Geiser et al., 2012) and have been extended to monitoring ambient seismic activity (natural continuous background activity) during 3D reflection seismic surveying (Lacazette et al., 2013). Methods for using passive seismic data for frac production and reservoir simulation have advanced tremendously in the past few years. Advancements in modeling approaches are greatly increasing the utility and hence value of passive seismic data.

Other applications of passive seismic methods deserve brief mention but are not the subject of this chapter and will not be considered further. Passive seismic methods are widely used in the mining and geotechnical industries to monitor the stability of excavations such as mines, waste disposal sites, and large structures such as dams and their foundations. Monitoring the integrity of water dams and levees is an especially important application of microseismic monitoring. A high-pressure influx of fluids into a levee or dam causes a significant increase in pressure in and along the structure. When the pressure reaches a failure threshold, preexisting fracture networks reactivate. The failure of the rocks or reactivation of fracture networks acts as a seismic source. Microseismic monitoring arrays then detect the seismic signatures of the dam or levee failure. The same applies to monitoring volcanos—a large influx of fluids (steam/lava/water) causes a significant pressure buildup in a volcano. Passive seismic monitoring is an important component of volcano warning systems. Passive methods are also used to monitor induced seismicity from oil and gas extraction and seismicity related to water disposal and other activities (e.g., Suckale, 2010).

10.2 GEOMECHANICS AND NATURAL FRACTURE BASICS FOR APPLICATION TO HYDRAULIC FRACTURING

10.2.1 Basics of Earth Stress and Strain

An elementary understanding of stress and strain is required to discuss seismic phenomena produced by hydraulic fracturing and to develop a basic understanding of the natural rock fractures systems with which hydraulic fractures interact. One application of passive seismic data is determining the stress state within an oil or gas reservoir. Stress controls both induced fracture propagation and reactivation of natural fractures. The discussion given here is general and elementary. Students needing a deeper understanding of geomechanics are referred to Engelder (1993), Jaeger et al. (2007), and Zoback (2010). Readers who already have a sound understanding of geomechanics and brittle structural geology should at least skim this section to ensure that they understand the terminology used in subsequent sections.

Unconventional oil and gas development concerns both the reservoir geology which developed over geologic time and engineering activities that are taking place today. Consequently, it is essential to distinguish *paleostress* from *neostress*. *Paleostress* refers to one or more ancient stress states that prevailed in the reservoir over geologic time. *Neostress* is the present-day stresses in the reservoir (also termed the *in-situ stress*).

Stress is defined as the force per unit area across a planar element. For example, if we push or pull on a rod parallel to the rod's length, then the stress in the rod is given by the force applied to the end of the rod divided by the cross-sectional area of the rod. In this case, we are defining our planar reference element as perpendicular to the rod's length. In a fluid at rest the stress (pressure) across any element is the same regardless of the element's orientation. In fluid that is being deformed viscous stresses resist the shape change. We will consider only the stresses in solids at rest. In solids, like rock, the stress on a planar element depends on its orientation. There are two kinds of stress: *normal* and *shear*. Normal stress is the stress directed perpendicular to the reference element. Shear stress is the stress directed parallel to the element. Consider three mutually perpendicular planar elements centered on a point. In a solid, we can rotate these elements in 3D until we find an orientation where the shear stresses on each element are zero. The three planes in these special orientations are termed the *principal planes* and the normal stresses on these planes (i.e., the stress acting perpendicular to the planes) are termed the *principal stresses*. Earth stress measurements show that in the earth's deep subsurface, such as in oil and gas reservoirs, all three stresses are always compressive. True tension is only observed near the earth's surface, around subterranean openings like mines and caves, and perhaps in rare special cases such as breccia blocks in fault zones. For this reason, we will follow the common geomechanical convention that gives compressive stress a positive sign. We will refer to the principal stresses as *S*max, *S*int, and *S*min for the maximum, intermediate, and minimum principal stresses.

Another common stress terminology refers to the stress axes that are mutually orthogonal and parallel and perpendicular to earth's surface. The use of this nomenclature usually assumes that the earth's surface is a principal plane. In this case, the stresses are termed SHmax, Shmin, and Sv for the maximum horizontal stress, the minimum horizontal stress, and the vertical stress, respectively. Note that Sv is computed easily from log data because it is equal to the weight of the overburden. Sv is computed at any depth by determining the average rock density from the depth of interest to the surface (e.g., from a density log run to the surface), then multiplying by the true vertical depth and acceleration of gravity.

If we choose a plane that is not a principal plane, that is, a plane at some angle to the principal planes, then we can determine the normal and shear stresses on the element by making vector sums of the principal stresses parallel and perpendicular to the element.

Effective stress is the stress in the solid skeleton of the rock minus the stress due to fluid pressure. The contribution of pore pressure to stress in the solid skeleton of a rock is proportional to *Biot's constant*, α. Biot's constant (also called the *poroelastic constant*) is always between zero and one and is always <1 for any consolidated material. Biot's constant can vary from 0.98 for unlithified clay to <0.1 or nearly zero for lithified rocks with little porosity. Since pore pressure acts outward in all directions, the effective stress in a given direction (S_E) in a porous rock with pore fluid is given by

$$S_E = S_T - \alpha P_p,$$

where S_T is the remote tectonic stress and P_p is the pore pressure in the rock. The walls of a hydraulic fracture move under the influence of the outward-acting frac fluid pressure and the inward-acting effective stress. Leakoff from a hydraulic fracture into the matrix therefore changes the effective stresses in the rock volume around the fracture. Depending on the problem at hand, we can ignore pore pressure and consider effective stress in a fracture-centered reference frame so that the effective stress on the fracture wall is the fluid pressure acting outward minus the stress acting inward. Engelder and Lacazette (1990) first discussed natural extensional rock fracture involving pore pressure and fluid-filled fractures that is equally applicable to extensional artificial hydraulic fracturing and extensional natural fracturing (see Engelder,1993 and Zoback, 2010 for detailed discussion of artificial hydraulic fracturing).

Strain is defined as the ratio of the change in length, area, or volume of an element divided by the initial length, area, or volume (respectively). Consequently, strain is dimensionless because the units cancel. Consider the example of a rod in tension or compression. The linear strain of the rod is the change in length of the rod divided by the initial length of the rod. The sign of strain depends on whether strain is computed with respect to the initial or deformed state of the object. Contractional (shortening) strain is commonly taken as positive in geomechanics, and extensional strain as negative. Hydraulic fracturing produces both recoverable (elastic) and permanent strain in the affected rock volume. Seismic activity, including MEQs, is a result of those strains. Although they are often treated separately, stress and strain are inextricably linked as a single phenomenon. Stress always causes strain and strain always causes a change in stress.

We will use the term *stress state* to refer to the orientations and *relative* magnitudes of the three principal stresses in the earth. Stress states in which the earth's surface is a principal plane, that is, when two of the principal stresses are parallel to the earth's surface and one is perpendicular, are referred to as *Andersonian* stress states (Fig. 10.1) after the structural geologist E.M. Anderson who first described them in 1905 (Anderson, 1905, 1951). The Andersonian stress states are as follows:

- Extensional (also termed normal) faulting—where Smin and Sint are horizontal and Smax is vertical. In terms of the horizontal and vertical stress nomenclature SHmax = Sint, Shmin = Smin, and Sv = Smax.
- Wrench (also termed strike-slip) faulting where Smax and Smin are horizontal and Sint is vertical. In terms of the horizontal and vertical stress nomenclature SHmax = Smax, Shmin = Smin, and Sv = Sint.
- Thrust (also termed contractional or reverse) faulting where Smax and Sint are horizontal and Smin is vertical. In terms of the horizontal and vertical stress nomenclature SHmax = Smax, Shmin = Sint, and Sv = Smin.

Andersonian stress states are commonly found in oil and gas reservoirs and prevail on average over large regions of the earth. However, borehole stress studies showed that non-Andersonian stress states are common on smaller scales within oil and gas reservoirs.

Extensional fractures of any kind including artificial hydraulic fractures propagate perpendicular to Smin. More precisely, the propagating tip of such a fracture is always perpendicular to the local Smin at the crack tip. Fractures will hence turn if the orientation of Smin varies. Many workers think of extensional hydraulic fractures as propagating parallel to Smax. Since Smax and Sint are both perpendicular to Smin which is of course true, but this type of thinking can lead to errors if the principal stresses are confused with SHmax and Shmin. The importance of correct stress nomenclature becomes clear when we consider hydraulic fracture propagation in Andersonian stress states.

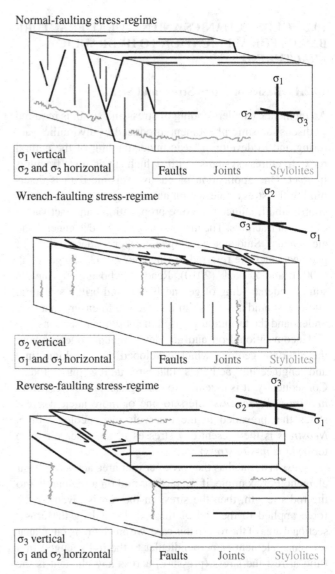

FIGURE 10.1 The Andersonian stress regimes and the general orientation of natural fractures that form in these regimes. Note that artificial hydraulic fractures will propagate in the same orientation as joints.

If we assume that we are hydraulically fracturing an idealized homogeneous medium with no discontinuities such as natural fractures, then in extensional and wrench faulting stress states hydraulic fractures propagate vertically. If a horizontal well is drilled parallel to Smin in these stress states, then extensional hydraulic fractures will propagate perpendicular to the wellbore which maximizes the stimulated reservoir volume (SRV), that is the reservoir volume affected by the frac. If a well is drilled perpendicular to Smin, then each frac stage produces a fracture that runs parallel to the wellbore resulting in generation of a single fracture and poor production.

In thrust faulting stress states Smin is vertical, so that Sint and Smax are Shmin and SHmax, respectively. Confusing the horizontal stresses with the principal stresses leads to the erroneous conclusion that hydraulic fractures will run vertically and parallel to SHmax. In a thrust faulting stress state, extensional hydraulic fractures propagate horizontally and the frac pressure is equal to the vertical stress Sv.

Although thrust faulting stress states are the least common worldwide, some potentially very large unconventional gas plays, such as some basins in China, are in active fold-thrust belts. This may appear to bode ill for unconventional production in these basins because idealized frac models predict that only a single horizontal fracture will be driven from a horizontal well. Very ductile rocks in a thrust faulting stress state may indeed require frac pressures in excess of Sv. However, current work in the microseismic community shows that hydraulic fracturing primarily affects the preexisting natural fracture system rather than generating idealized bi-wing, extensional hydraulic fractures. As discussed in subsequent sections, wells in thrust faulting stress states can be effectively stimulated at frac pressures below Sv if the frac exploits the natural fracture system. Also, wrench-faulting and normal-faulting stress states occur locally in active fold-thrust belts. These anomalous stress states result from geometric effects related to the complexities of folding and faulting.

The upper, brittle region of the earth's crust is a self-organized critical system (e.g., Leary, 1997) in a state of frictional equilibrium because of pervasive fracturing (Zoback, 2010). Frictional models provide the most accurate predictions of measured stress profiles in the crust (Zoback, 2010). Consequently, this equilibrium is very easy to disturb. Hydraulic fracturing disturbs the equilibrium, thereby stimulating microseismic activity.

Mechanical stratigraphy refers to the differing mechanical properties, stress states, and natural fracturing of different lithogical layers (e.g., Laubach et al., 2009). Stress states can vary strongly between different lithologies. When fracing a ductile layer (typically a stratum with high organic and clay content), frac pressures approximately equal to Sv can result from deformation of the layer over geologic time with attendant equalization of the stresses. The stress magnitudes and even stress orientations in a reservoir can vary strongly from lithology to lithology even when different lithologies are thinly interbedded (e.g., Evans et al., 1989). Such stress-state variations are very important both for interpreting passive seismic results and for frac design.

The World Stress Map (www.World-Stress-Map.org) is a worldwide map of publically available stress data.[1] The stress states are presented in terms of the Andersonian stress states because over large areas stresses are generally Andersonian. Inspecting the map shows that stress states can change rapidly across basins—from extensional to thrust, wrench and back again and can rotate across a basin. Similarly, the stress state can change substantially with location in an oil or gas field or within the volume of rock affected by a single frac. Because the earth's surface is a free surface, stresses are reliably Andersonian close to the surface but deep enough to be removed from near-surface topographic effects. Over large areas stresses tend to be Andersonian on average. However, borehole stress studies show that non-Andersonian stress states are common on small scales within oil and gas reservoirs.

10.2.2 Natural Fracture Basics and Interaction with Hydraulic Fractures

This section discusses basic natural fracture types, their orientations relative to each other, and their relevance to interpreting hydraulic fracture data. Lacazette (2000, 2009) provides a complete review of fracture nomenclature (also see the technical area of www.NaturalFractures.com). Textbooks on fractured reservoir evaluation are provided by Nelson (2001) and Narr et al. (2006).

10.2.2.1 Natural Fracture Types Natural rock fractures fall into three basic categories: joints, faults, and contractional fractures.

- *Joints* are extensional fractures that form perpendicular to Smin when the walls of the fracture move perpendicularly outward from the plane of the joint and the propagation direction (Engelder, 1987; Pollard and Aydin, 1988).
- *Faults* result from shear movements parallel to the fracture plane and at angles that range from parallel to perpendicular to the fracture propagation direction. Faults form approximately parallel to Sint and tend to initiate as conjugate pairs at ±30° to Smax and parallel to Sint.
- *Contractional fractures* form perpendicular to Smax by volume loss across a plane. The volume loss can result from crushing (*deformation bands*), grain rearrangement (*compaction bands*), or by chemical dissolution (*stylolites*). Although some workers object strenuously to terming stylolites as a type of fracture, it is well accepted in both the geological and engineering literature that chemical corrosion is an important fracture mechanism, and Fletcher and Pollard (1981) show that simply reversing the sign of extensional fracture mechanics equations correctly describes stylolite formation. For these reasons, stylolites must be classed as a type of fracture (a stress corrosion anticrack).

Note that extensional and shear fracturing can operate simultaneously as can contraction and shear. Figure 10.1 shows the average orientations of the natural fracture types that form in each of the Andersonian stress regimes. Different fracture types may transmit fluid differently depending on their types and properties.

---

[1] The world stress map is important for many activities ranging from oil and gas exploration to earthquake forecasting. The project benefits greatly from stress data contributed by the oil and gas industry. Please encourage your company or organization to contribute any stress data that you collect to the World Stress Map project.

Contained joints are joints that are restricted to individual mechanical layers. For example, brittle rocks such as limestones, cherts, or sandstones interbedded with shales may be jointed, while the softer shale interbeds are not.

Joints are an especially important feature of many unconventional reservoirs because they are generated by the high fluid pressures that develop during maturation of self-sealing reservoirs. These relationships are especially well-documented in the Appalachian Basin of the United States (Engelder and Lacazette, 1990; Engelder and Whitaker, 2006; Engelder et al., 2009; Lacazette and Engelder, 1992; McConaughy and Engelder, 1999). Because unconventional reservoirs are organic rich, self-sealing, and mature, they commonly have well-developed joints developed by poroelastic natural hydraulic fracturing.

Natural fractures are grouped into *sets* of fractures with a common type, orientation, and other characteristics. Thus we may speak of joint sets, conjugate fault sets, and so on. The complete natural fracture network in a reservoir (i.e., all the fracture sets) is referred to as the *fracture system*.

10.2.2.2 More about Natural Fracture Orientations

Natural fractures form in response to stresses that prevail at the time that they form. If the fractures are formed very recently in geologic time or if they are forming now in the reservoir, then the natural fracture orientations may be geometrically related to the neostress. However, in most reservoirs all, or at least some, of the natural fractures formed many millions of years earlier during ancient tectonic events so that the natural fracture orientations reflect the paleostress orientations. Such fractures generally do not have a regular geometric relationship to the neostress orientations and magnitudes, but may coincidentally have such a relationship. The natural fracture system in the Marcellus Shale Fm. in the United States is a famous example of such a coincidence (Engelder et al., 2009). An additional complexity is that ancient fractures are often reoriented and/or reactivated by tectonic events (such as folding and faulting) that postdate their original formation. In summary, it is not valid to draw conclusions about natural fracture orientations based only on the neostress state, without independent knowledge of the time and conditions of formation of the natural fractures.

Note that the planarity of a natural fracture, and hence the orientation variability of a set, is strongly dependent on the conditions of fracture formation. For example, joints that develop under conditions where Smin is much smaller than the other two stresses are very planar, whereas joints that form when the stresses are more equal are irregularly shaped and tend to hook into each other during propagation. The common approximation, as shown in Figure 10.1, that faults form as planar features that slip perpendicular to Sint is useful, but is not strictly correct. In reality, new faults form in a triaxial stress state so that new faults form as four orthorhombic sets none of which are oriented exactly as shown in Figure 10.1, that is, they do not slip perfectly perpendicular to Sint (e.g., Reches, 1983; Reches and Dieterich, 1983). The planarity of newly formed faults is also subject to control by stress ratios as is the case with joints.

10.2.2.3 Natural Fracture Reactivation during Hydraulic Fracture Treatments

The issue of natural fracture reactivation during hydraulic fracture treatments is important for many reasons including frac design, reservoir simulation, and interpreting passive seismic data. Current research is showing that natural fracture reactivation is commonly the most important mechanism of hydraulic fracture stimulation. Natural earthquakes and most MEQs produced by hydraulic fracturing result from slip on preexisting natural fractures (e.g., Doe et al., 2013; Eisner et al., 2010b; Moos et al., 2011; Williams-Stroud et al., 2012). As we will see in the discussions on MEQ focal mechanisms and stress inversion (Sections 10.3.2 and 10.6.2.2), this has bearing on the interpretation of reservoir stress.

Reactivation of existing fractures requires much less energy than generating new fracture surface area. Consequently, reactivation of existing natural fractures is common, especially farther from the wellbore where fluid pressures are lower. Natural fractures are reactivated during a hydraulic fracture treatment by two mechanisms: hydrojacking and hydroshearing, both of which generate seismic energy. Hydrojacking occurs when the fracture is forced open by the frac fluid. Hydroshearing occurs when the frac fluid pressure reduces the normal stress and hence the friction on natural fractures allowing them to slide under the influence of the reservoir stress. The stresses that produce sliding are generally a combination of the neostresses and stresses induced by the fracture treatment itself. Hydrojacking and hydroshearing generally occur together. Das and Zoback (2013b) present compelling passive seismic and other evidence for reactivation of even poorly oriented fractures during hydraulic fracture treatments.

The common and useful assumption that reactivated fractures slip in a plane perpendicular to Sint is not strictly true. Like newly formed faults, preexisting fractures slip under the influences of triaxial stresses so that fault movement may occur in a plane that is not perpendicular to Sint.

Figure 10.2 shows the simulated effect of natural fracture reactivation during a hydraulic fracture treatment. The images are map views of the simulated natural fracture population in the Marcellus Shale. Also shown are the simulated hydraulic fractures that took proppant, colored by stage. The simulation indicated that little new rock was broken by the frac. Instead, natural fracture reactivation was the primary mechanism of stimulation. The fracture population is a stochastic simulation that was calibrated with a passive seismic image of the fracture population in the reservoir. This imaging process is discussed in Section 10.5.5. (See Lacazette et al. [2014] for details of this study.) Look at

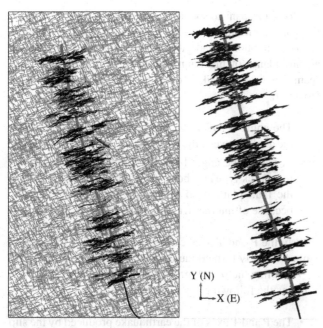

FIGURE 10.2 LEFT—Map view of a discrete fracture network (DFN) model of the Marcellus Shale formation showing the traces of the natural fracture network on a horizontal plane at the depth of the wellbore (gray) with simulated hydraulic fractures in 3D (black). Natural fracture reactivation was the primary mechanism of reservoir stimulation. RIGHT—The wellbore with only the hydraulic fractures from the diagram at left. See Lacazette et al. (2014) for more discussion.

Figure 10.2 and imagine that MEQs are generated only on the fractures that took proppant. Clearly, the microseisms will define a cloud, not a sharp, well-defined fracture plane. Also we should expect fluid leakoff into the fracture network and concomitant microseismic activity resulting from decreased friction on the natural fractures. Indeed, passive seismic data showed that this fracture treatment produced seismic activity that extended far beyond the propped fractures. Slip on fractures causes mismatch of asperities and increases the hydraulic conductivity of a fracture. Consequently, the microseismic activity in the natural fracture system should provide pathways for production and frac hits on adjacent wells. Indeed, frac hits on adjacent wells did occur during the simulated fracture treatment and simulation indicates productive contribution from the fracture network.

10.3 SEISMIC PHENOMENA

10.3.1 MEQs and Their Magnitudes

Most earthquakes result from the sudden release of stored elastic strain energy along fractures. The fracture may be formed by the slip event that produces the earthquake or may represent a preexisting feature.

An MEQ is defined in this article as an earthquake with magnitude less than or equal to zero. The *magnitude* that represents the measure of the size of an earthquake is expressed as the logarithmic function of the amplitudes of the generated seismic wave. As consequence of the logarithmic relationship, an increase of one unit in magnitude corresponds to 10-fold increase in seismic wave amplitudes. There exist different magnitude scales, including, among others, *local magnitude*, M_L, *body wave magnitude*, m_b, *surface wave magnitude*, M_S, and *moment magnitude*, M_w. The different magnitude scales are calculated from different seismic wave types observed at different frequencies. For that reason, the magnitude scales generally yield different values for the same earthquake. Region-specific or global conversion relations are available, which allow magnitudes to be converted from one scale to another.

A seismic source (e.g., a tectonic earthquake or nuclear explosion) is best described mathematically by its moment tensor. A seismic moment tensor can be expressed as a superposition of an isotropic component, a double-couple (DC) component, and a compensated linear vector dipole (CLVD) component. The isotropic part of the seismic moment tensor results from volume change in the source region. For example, nuclear explosions and MEQs caused by sudden extension of opening-mode fractures during fracture treatments are isotropic sources that involve a volume increase in the source region. A shear slip along a planar surface, the model for most earthquakes, is adequately described by the DC component. Processes known to produce CLVD components in seismic moment tensors include the opening and closing of tensile cracks due to fluid movements (e.g., Frohlich, 1994) and the superposition of two of more DC events that occur along faults with different orientations (Frohlich, 1994). Non-double-couple earthquake theory and observations are reviewed by Julian et al. (1998) and Miller et al. (1998).

Tectonic tremor (Shelly et al., 2007) consists of swarms of low-frequency earthquakes thought to be analogous to LPLD activity observed during hydraulic fracture treatments (Das and Zoback, 2013a, b).

10.3.2 Earthquake Focal Mechanisms

10.3.2.1 Focal Mechanism Basics Focal mechanism solutions of MEQs are a standard deliverable for most passive seismic studies of hydraulic fracture treatments. Earthquake focal mechanisms are used to:

- understand the stress that caused an MEQ,
- infer the overall stress state in the volume of rock that produced MEQs having focal mechanism solutions,
- constrain or identify the orientation of fractures that generated the MEQ.

A *focal mechanism solution*, or simply *focal mechanism*, is a description of the orientation (strike and dip) of the fault plane that slipped during an earthquake and the slip direction and slip sense of the fault and of the strain produced by the earthquake. The term *fault plane solution* is a synonym. The *moment tensor* of an earthquake is a complete mathematical description of the earthquake movement. The *moment tensor* is defined as "A mathematical representation of the movement on a fault during an earthquake, comprised of nine generalized couples, or nine sets of two vectors. The tensor depends on the source strength and fault orientation" (U.S. Geological Survey; http://earthquake.usgs.gov/learn/glossary/). The focal mechanism solution is a subset of the moment tensor if it represents only one component of the full moment tensor. Often, the focal mechanism solution considers only the simple DC shear component of the earthquake source; the moment tensor includes volumetric and other components as well. The details of earthquake mechanics are beyond the scope of this chapter. Thorough discussions of moment tensors are provided by Jost and Herrmann (1989) and Stein and Wysession (2002). Full moment tensors are not a common deliverable for hydraulic fracture monitoring studies, although very detailed studies may provide full moment tensors for some MEQs.

A DC focal mechanism solution is described completely by either:

- The 3D orientation of one of the two possible fault planes that produced the earthquake and the slip direction of the hangingwall of the fault. (The *hangingwall* is the fault block above the fault plane.) Only one fault plane needs to be described because the other plane is perpendicular to it and has the conjugate slip direction and slip sense.
- Two of the three principal strain axes, which are mutually orthogonal. These are P (contractional), N (neutral, also called B), and T (extensional). Generally, the P and T axes are provided.

Because the seismic radiation pattern of a slip event is symmetrical, a focal mechanism solution yields two possible fault plane solutions that cannot be distinguished from each other, as shown and described in Figure 10.3. Note that the pattern of polarities of first arrivals on any plane or line, such as the ground surface or a wellbore, does not reveal which nodal plane is the fault plane and which is the auxiliary plane (Fig. 10.4). Beach-ball diagrams of the focal mechanisms for each basic fault type are shown in Figure 10.5.

10.3.2.2 Relationship between Stress and Focal Mechanism Solutions
Focal mechanisms constrain but do not uniquely determine the principal stress orientations. The P, N, and T axes of a focal mechanism solution are often treated as approximations of the orientations of Smax, Sint, and Smin (respectively). If this assumption is made, then the nodal planes of the solution are the planes of maximum shear stress. These planes may seem to appear to be likely orientations for new faults to form. However,

- The angle between the two nodal planes is 90° so that the angle between the nodal planes and the P axis is 45°.
- In nature the angle between conjugate fault planes in most rock types is about 60°, so that the angle between the fault planes and Smax is about 30°. This occurs because of internal friction in the rock.

Clearly the P and T axes cannot be equivalent to Smax and Smin for newly formed faults.

We define the *transport plane* of a fault as the plane that contains the following:

- The P and T axes of the earthquake produced by the slip event,
- The slip line of the fault,
- Smax and Smin (this assumes a uniaxial stress state).

These relationships are shown in Figure 10.6. The drawings show the maximum stress axis within the transport plane of a slipping fault under uniaxial stress conditions. The angle between Smax and the fault plane is approximately equal to or sometimes less than 30° for newly formed faults. For older, reactivated fractures the angle is dependent on the

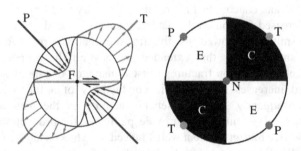

FIGURE 10.3 TOP—Schematic diagram showing the direction of initial movement for particles around the focus (F) of an earthquake on an E-W dextral wrench fault (gray), viewed from above. The particle motion is symmetrical, so based only on the particle motion either the blue line or the orthogonal black line could be the fault. (*Lacazette after U.S.G.S.*). The P and T axes are shown. The neutral particle motion axis is perpendicular to the P and T axes so it plots at the center of the diagram and is not shown. Some workers associate the P- and T-axes with the principal stresses Smax and Smin, although this a poor assumption. BOTTOM—A beach-ball plot showing the zones of compressional (C) and extensional (E) first motion in the seismic waves radiating outwards. The diagram also shows where the P, T, and N axes plot.

SEISMIC PHENOMENA 215

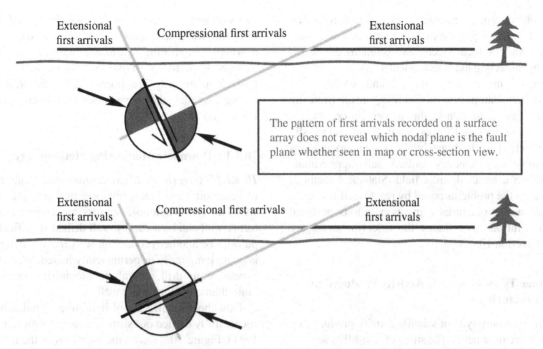

FIGURE 10.4 Vertical section showing the pattern of first arrivals from microearthquake on the surface. Either of the two nodal planes shown could represent the true fault plane. The pattern of first arrivals in the two cases is indistinguishable from first arrivals on surface, near surface, or downhole receivers.

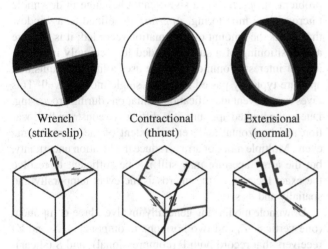

FIGURE 10.5 Beach-ball plots for the fault types expected in each of the Andersonian stress regimes. The two possible nodal planes for each radiation pattern are shown in gray and dark gray on the block diagrams and beach-ball plots. Note that the beach-ball plots only show orientation, not location.

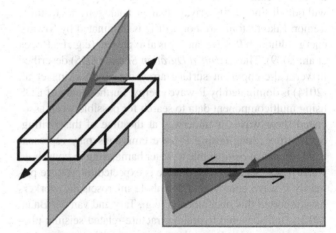

FIGURE 10.6 Left—Diagram of a slipping fault showing the direction of movement, the transport plane, (gray) and Smax (heavy gray arrow). Right—Detail of the transport plane shown in the figure at left. Black half-arrows show the slip-sense of the fault. Heavy gray arrows show Smax, which is contained in the transport plane. The angle between the fault plane and Smax is shown in dark gray.

effective stress (the difference between the normal stress on the fault and the fluid pressure within the fault) and can be anywhere in the range >0°, <90° but is usually <45°.

Focal mechanism solutions always place the P and T axes in the transport plane at 45° to the two nodal planes, one of which is the fault plane. Because the two nodal planes are orthogonal to each other, we know that the true orientation of Smax is likely on one side or the other of the P axis, but without independent knowledge, such as which is the true fault plane, we cannot determine where it is. We do know that the orientations of Smax and Smin lie in the transport plane. Consequently, only the intermediate stress axis is uniquely constrained by the focal mechanism solution. The intermediate stress axis is therefore the same as N, the neutral particle-motion axis. For this reason, mixed populations of focal mechanism solutions are sorted most easily by inspecting the N-axis orientations.

As discussed in Sections 10.2.2.2 and 10.2.2.3, the common and useful simplification that MEQs result from slip in a uniaxial stress field, that is, this movement occurs in a plane that contains the P and T axes, Smax, Smin, and the slip line, is not strictly true. Fortunately, the average orientations of orthorhombic sets of newly formed faults approximate faults formed in a uniaxial stress field. Statistical methods also can address the problems posed by reactivated fractures. These issues and procedures using populations of focal mechanism solutions to estimate the reservoir stress are discussed in Section 10.6.2.

10.3.3 Other Types of Seismic Activity Produced by Hydraulic Fracturing

MEQs are not the only type of seismic activity produced by hydraulic fracture treatments. The study of non-MEQ seismic phenomena is a new and rapidly developing field of study. Das and Zoback (2013a, b) have demonstrated that LPLD activity produced by fracture treatments can represent 40 times more energy than the MEQs produced by the hydraulic fracture treatment. LPLD is a long-lasting phenomenon without distinct first arrivals that is analogous to tectonic tremor. Like tectonic tremor, LPLD is dominated by S-wave energy although P-wave energy is also present (e.g., La Rocca et al., 2009). The *Extended Duration Signal* (*EDS*) described in vertical-component surface array data by Sicking et al. (2014) is dominated by P-wave energy. Further study of EDS using multicomponent data to search for possible weak associated shear waves is underway at the time of this writing. Such strong, long-lasting P-wave emissions may represent fluid resonance comparable to water hammering in household water pipes. Such fluid resonance is expected to produce primarily P-wave energy, and downhole microseismic workers have reported this phenomenon (e.g., Tary and van der Baan, 2012). Undiscovered hydraulic fracture-related seismic phenomena may exist. Regardless, it is rapidly becoming clear that MEQs represent only part of the seismic signal produced by fracture treatments of unconventional reservoirs, and in some reservoirs represent only a small part of the signal.

10.4 MICROSEISMIC DOWNHOLE MONITORING

Other than well production, the monitoring of microseismicity induced during reservoir simulation is the best means of assessing the stimulation effectiveness of unconventional reservoirs. For many years, downhole monitoring was the only method to monitor such MEQs and is still the most commonly used method. Surface and near-surface monitoring methods were developed in the early 2000s and are coming into ever-wider use. This section focuses on microseismic downhole monitoring. Surface monitoring is discussed in Section 10.5. Many of the principles used in downhole passive monitoring are borrowed from earthquake seismology. As such, they have been tested extensively and used for several decades.

10.4.1 Downhole Monitoring Methodology

10.4.1.1 Overview Microseismic downhole monitoring of reservoir stimulation involves one or more monitoring wells in which borehole sensors are operated. The monitoring well can be a nearby well drilled specifically for that purpose or another production well from which production is being temporally or permanently halted. Wells being very expensive to drill, downhole monitoring rarely involves more than one monitoring well.

Downhole monitoring of hydraulic stimulation has been successfully carried out since the early 1980s (e.g., Pearson, 1981). Figure 10.7 shows the locations of the treatment and monitoring wells of a typical downhole microseismic monitoring setup for a hydraulic fracturing project.

The use of a single well is a source of uncertainty because each earthquake is observed from only one direction. This problem of *aperture* is a significant challenge in downhole microseismic monitoring. *Aperture* is defined as a window that limits the amount of information recorded; it is the size and positioning of a survey needed to accurately image an area of interest. Common practice uses a limited acquisition aperture typically placed in only a single monitor well. This gives rise to event identification limitations during processing. Due to the limited aperture, it is difficult to constrain the location and horizontal and vertical extent of the microseismic event. Multiple monitor arrays reduce the location uncertainty, but the array aperture may still not be sufficient to resolve the exact location of the microseismic event temporally and spatially and its size.

Downhole monitoring generally involves three-component (one vertical, Z; and two horizontal components, X and Y) receivers that record both P (compressional)- and S (shear)-waves. The receivers are positioned and clamped to the casing along the wellbore. The length of the receiver array is the total aperture available for event location. Figure 10.8a shows typical waveforms for a microseismic event recorded using a 16-level downhole array of geophones. Consistent with the notion that the event represents predominantly a shear slip, the shear waves have much larger amplitudes than the compressional waves shown in Figure 10.8a. Since the receiver orientations are unknown, their orientations are determined before the beginning of the actual monitoring. This is achieved using data from perforation shots or string shots of known locations. Examples of perforation shot seismograms are displayed in Figure 10.8b. The calculated

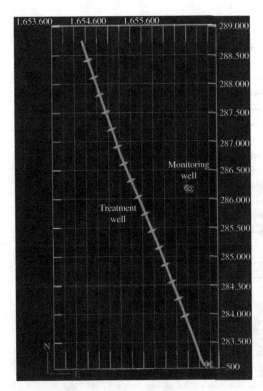
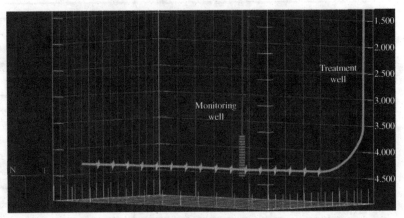

FIGURE 10.7 Map view (left) and profile view (right) showing the treatment and monitoring wells for a hydraulic fracturing project. The treatment and monitoring wellbores are shown in light gray and gray, respectively. Perforations are represented by stars along the lateral of the treatment well. Each disc along the monitoring well represents a three-component recording sensor.

sensor orientations allow the recorded data to be rotated from Z, X, Y coordinate system into vertical (Z), north–south (N), and east–west (E) components or, once the event location is estimated, into vertical (Z), radial (R), and transverse (T) components for further analyses. Sometimes, the data are rotated from the ZNE to the LQT ray coordinate system in which the L axis points in the direction of P-wave; the Q axis is in the ray plane, but perpendicular to L; and the T axis is perpendicular to both L and Q axes. The purpose of the rotation is to maximize the P-wave energy onto the L component, and the S-wave energy onto the Q and T components for adequate analyses.

10.4.1.2 Velocity Model Downhole monitoring requires velocity models for both P-wave and S-wave propagation because the time lag between the wave types provides a measure of distance. (S-waves are slower than P-waves.) Typically, perforation shots or string shots are used for velocity calibration after the initial velocity model has been developed from an existing sonic log or surface seismic data. Velocity calibration consists of adjusting the initial velocity model so that the calculated locations of perforation shots or string shots match the actual locations to an acceptable accuracy. Depending on the data from which the initial velocity is obtained, the various factors that make velocity calibration a requirement for an adequate microseismic monitoring campaign include (i) the fact that the sonic log only samples the region in the vicinity of the wellbore that may not be representative of the structure of the formation away from the wellbore, along the travel path of waves from the microseismic sources to the receivers; (ii) the effect of velocity anisotropy, particularly in reservoirs involving shale formations, which is known to be highly anisotropic (e.g., Sondergeld and Rai, 2011); and (3) the difference in frequency contents between microseismic waves and acoustic signals from the sonic source or the surface seismic data. A multistage stimulation in a highly heterogeneous environment sometimes requires a stage-by-stage calibration of the velocity model.

10.4.1.3 Locating MEQs The locations of recorded MEQs are typically estimated through an inversion or grid search approach involving P- and S-wave arrival times. When monitoring from a single well, the location process requires the determination of the direction of P- and/or S-wave particle motions (polarization angles). In this method, the difference between the P- and S-arrival times constrain the radial distance of the hypocenter, while the polarization angles provide the event backazimuth (Fig. 10.9). The polarization angles are obtained by analyzing the 3D particles motions of P- and/or S-waves. Figure 10.10 shows the locations of detected microseismic events for an example of downhole monitoring of a multistage

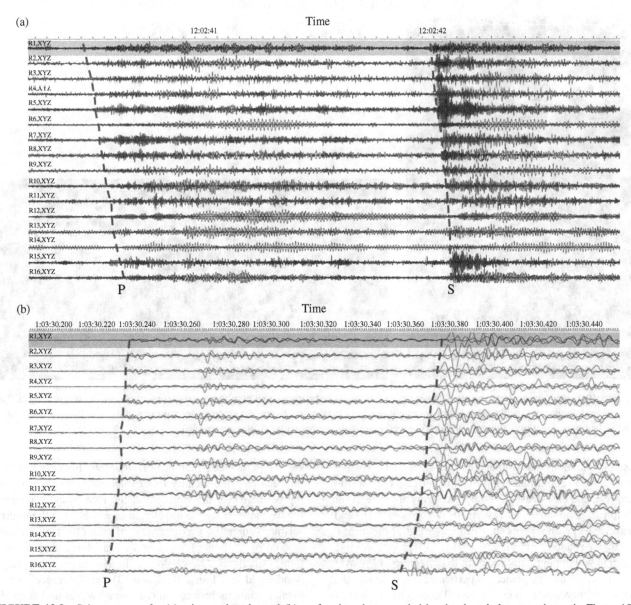

FIGURE 10.8 Seismograms of a (a) microearthquake and (b) perforation shot recorded by the downhole array shown in Figure 10.7. For each of the 16 receivers (R1–R16), waveforms for the three components (X, Y, and Z) are overlaid on each other. P and S wave arrivals are indicated with the dashed lines.

hydraulic fracturing campaign using P- and S-wave arrival times and back-azimuths obtained from polarization analyses.

Another approach to locating MEQs uses reverse time migration (Fish, 2012). Reverse time imaging is a less utilized technique for locating microseismic events, but is a more robust technique in the presence of noise. To overcome the difficulties associated with picking P- and S-wave arrivals, automatic techniques based on reverse time imaging eliminate the need for arrival identification. Reverse time imaging is capable, in principle, of focusing microseismic energy at its source position and at its trigger time, even when data are corrupted by high levels of noise (Artman et al., 2010; Xuan and Sava, 2010). The method relies on a grid search using semblance or some other measure of activity. In this respect, it resembles SET processing (Section 10.5.2).

The wavefield is inspected with the two-way wave equation around a known source location in space and time. Locations where wavefields focus are identified as event hypocenters. In a constant homogenous medium, we expect the wavefield to propagate evenly in all directions with respect to time, creating a spherical wavefront. In 2Ds (X,Y; X,Z; or Y,Z), the wavefront propagates circularly in time. If we eliminate a spatial variable by inspecting the wavefield propagation at a fixed point in space (e.g., X = Xsource or Z = Zsource), the wavefield propagates away from this source evenly in space. When we reverse the time axis of the

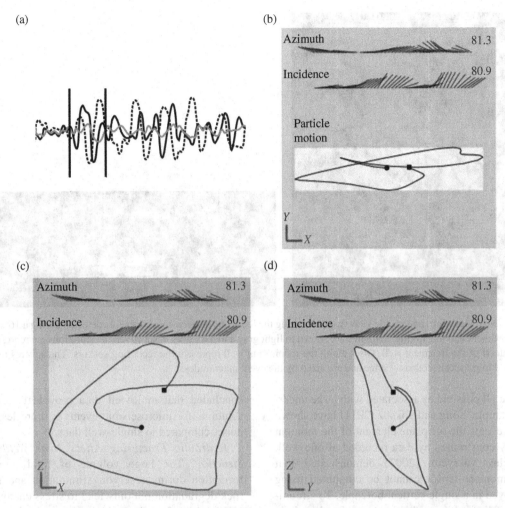

FIGURE 10.9 (a) X, Y, and Z component P waveforms for receiver R4 shown in Figure 10.8a. The two vertical bars delimit the data window used for the computation of the particle motion shown in (b, c, and d). (b) P wave particle motion in X–Y plane. The numbers indicate the estimated values of the ray incidence angle and event back-azimuth at the point of maximum linearity. (c) P wave particle motion in X–Z plane. (d) P wave particle motion in Y–Z plane.

wavefield so that it propagates beyond the source onset time, the wavefield defocuses and, again, propagates away from the source in X and Z. We refer to forward and reverse time as positive and negative time, respectively. In both negative and positive time, the wavefield appears as two cones in, for example, X, Z, t space. The cones point in opposite directions on the *t* axis. The apex of both cones located at the source origin in the X–t, Y–t, or Z–t domains. In a heterogeneous model or with real field data, the cone pattern is not perfect but is still recognizable. At the correct source onset time, in both homogenous and heterogeneous models, we expect to see a perfect focus in the X–Y, X–Z, or Y–Z domains. If we capture wavefield information from all directions, and assume that the velocity model and timing information is correct, the wavefield focuses to a point in X, Y, Z, and *t* at the correct location of the source in space and time. Note that any deviation from the true location of the source results from imperfect focusing of the wavefield due to source frequencies, aperture affects, incorrect timing, and/or velocity errors.

MEQs can also be located using triangulation methods, which are more practical with multiple arrays. Using P- and S-wave velocities and arrival times only, one may calculate the distance a wave has traveled to reach each receiver in the monitor array. Using these distances, spheres around each receiver with a radius of the distance to the earthquake from each receiver intersect at the event location. In theory, the method of triangulation should provide a unique solution, although in practice this method often yields multiple solutions in part because of the narrow aperture of the arrays.

10.4.1.4 Focal Mechanism Solutions Accurate determination of MEQ focal mechanisms is difficult with downhole data because of the poor azimuthal coverage of the source region due to limited number of monitoring wells. Consequently, the determination of some of the MEQ focal

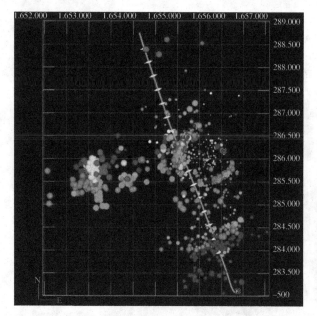
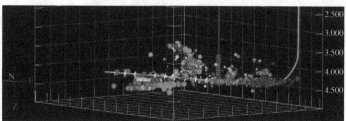

FIGURE 10.10 Map view (left) and profile view (right) showing the locations of microseismic events detected during a 16-stage fracturing campaign. The treatment and monitoring wellbores are shown in light gray and dark gray, respectively. Perforations are represented by white stars along the lateral of the treatment well. Discs along the monitoring well represent the recording sensors. The spheres representing the events are depicted in grayscale-coded by stage and are sized by the event magnitudes.

parameters, if at all possible, is associated with large uncertainties. For example, Song and Toksöz (2011) have shown that for distant events the off-plane element of the moment tensor is poorly constrained by data recorded at one well. Similarly, Václav Vavryčuk (2007) demonstrated that the complete moment tensor cannot be computed using receivers deployed in a single vertical borehole. To circumvent the issue of limited azimuthal coverage for downhole data recorded during a hydraulic stimulation in East Texas, Rutledge and colleagues assumed that a common focal mechanism occurs over the entire length of the activated fracture system; they computed a composite focal mechanism using polarity and amplitude information of all the microseismic events from different back-azimuths that were induced during that stimulation (Rutledge and Phillips, 2003; Rutledge et al., 2004). This approach works when the events do indeed share the same focal mechanism. However, for reservoirs with complex networks of failure planes having variable orientations, the approach may lead to a misrepresentation of the actual stress field.

Multiwell Downhole Monitoring. Downhole monitoring with more than one monitoring well has been shown to provide more accurate locations and focal parameters of the recorded microearthquakes, resulting primarily from the improved azimuthal coverage of the focal sphere compared to single-well monitoring. For example, Seibel and coworkers show decreased scatter for located hypocenters when going from a single to two and finally three downhole monitoring arrays (Seibel et al., 2010). Similarly, based on analyses of synthetic data, Du and Warpinski (2011) concluded that multiwell data provide better fault plane solutions for microseismic events and are less affected by noise, compared to single-well data.

Hydraulic Fracturing Alters Near Wellbore Velocity Structure. The large volume of fluid pumped into a formation during reservoir stimulation and the resulting rock degradation not only lead to stress changes that cause MEQs, but alter the velocity structure in the reservoir, especially in the near-wellbore region. Recent examples include a travel-time tomography study by Karimi et al. (2013) that shows decreased P- and S-wave velocities in a reservoir following stimulation; and a recent tomographic imaging study of a stimulated tight-sand reservoir in China that revealed a nearly 15% shear wave velocity decrease in the imaged area. The authors of this latter study interpret the low-velocity anomaly to result from pore pressure increase and/or rock degradation during the stimulation process (Huang et al., 2014).

10.4.2 Advantages and Disadvantages of Downhole Monitoring

In general, the main advantages of downhole monitoring over surface monitoring (discussed in the following section) are the proximity of the recording receivers to the signal source region, their depth well below from cultural and other surface noise, and the ease of performing real-time processing relative to surface data. Proximity to the source means that waves suffer less attenuation and scattering along their paths to downhole receivers than to

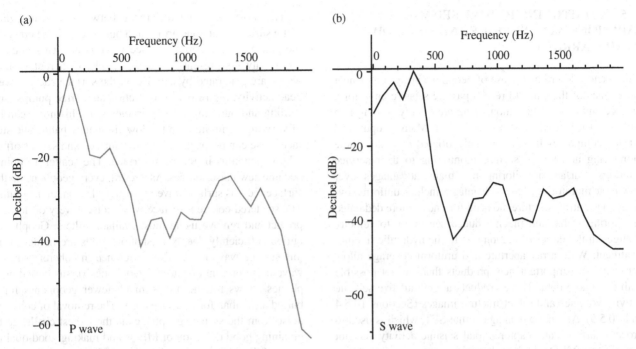

FIGURE 10.11 Frequency spectrum of (a) P and (b) S wave recorded by a downhole array.

surface receivers. This proximity allows even weaker events to be detected and preserves high frequencies (Figs. 10.8a and 10.11). Figure 10.11 shows that downhole data contain substantial energy at frequencies well above about 500 Hz for both P- and S-waves. By contrast, the frequency band hardly reaches 180 Hz for surface recordings of microseismic events occurring 3000 m below the surface. Fracturing jobs require pumps and other machinery which generates substantial noise. This noise propagates mainly as surface waves but attenuates quickly with depth and hence does not reach the receiver array in a downhole setting.

Many workers consider that downhole monitoring provides more accurate event depths than surface monitoring. Depth values are most accurate if the monitoring array straddles the target horizon. Positioning the array too high or low degrades the quality of both vertical and lateral locations. However, event depths can change radically with changes in processing parameters as can X, Y locations. Also, the ray path from an event to a downhole receiver is largely through the formation that is being treated. The frac treatment can substantially change the velocity of the target formation during treatment leading to substantial inaccuracies. Depending on the service provider and the location method used, downhole surveys can yield multiple solutions for individual hypocenters requiring the operator to choose a particular solution. Again, this problem results from limited array aperture. MEQs located from the surface using SET (Sections 10.5.2 and 10.5.3) are not subject to such ambiguity and the ray path from MEQ to the receivers is largely outside of the target horizon and hence is little affected by velocity changes in the target formation caused by the fracture treatment.

As mentioned earlier, the cost of downhole monitoring generally limits the method to one monitoring well. Because of this array limitation, we encounter wavefield attenuation effect in the recorded events. Wavefield attenuation often (and incorrectly) leads to a biased observation of the level of microseismic activities, with the region distant from the monitoring well incorrectly perceived as being less active. This results from higher detection threshold for the far region because of the relative remoteness of the receivers. Also, the inaccuracy of event location increases with distance from the monitoring well because of the small array aperture.

A major source of location uncertainty is that calculated particle motions can be very uncertain when using polarization angles for the determination of event back-azimuths with a single monitoring well (Fig. 10.9). Polarization angle uncertainties are generally due to near-receiver heterogeneities (Eisner et al., 2010a; Oye and Ellsworth, 2005) or noise (Oye and Ellsworth, 2005).

Downhole monitoring is especially practical for real-time monitoring because the small amount of raw data allows efficient real-time processing at the wellsite. However, final results are not delivered in real time and can differ substantially from the real-time results delivered onsite. As discussed in the next section, surface and shallow-buried array studies generate very large amounts of data that require extensive processing on powerful computers. Although such surface data can be processed in real time, additional costs and equipment are involved.

10.5 MONITORING PASSIVE SEISMIC EMISSIONS WITH SURFACE AND SHALLOW BURIED ARRAYS

This section describes the use of geophone arrays on or near the surface of the earth to record passive seismic emissions. Surface and near-surface monitoring are rapidly coming into wide use for passive seismic monitoring. Many geophysical service companies that traditionally offered only downhole monitoring are adding surface monitoring to their service offerings. Surface monitoring has many advantages over downhole monitoring. These advantages include uniform coverage, wide array aperture, no need for one or more dedicated monitoring wells, and much reduced sensitivity to velocity changes in the target formation due to the hydraulic fracture treatment. Wide array aperture and uniform coverage allow generation of important new products that are not possible with downhole data. These products are cumulative seismic activity volumes and direct fracture images (Sections 10.5.4 and 10.5.5). Another advantage is that SET, which is used to process surface data, captures total seismic activity and not just discrete MEQs. On the downside, surface and near-surface arrays suffer from reduced sensitivity resulting in higher detection threshold, limited frequency response, and lower signal-to-noise (S/N) ratio compared to downhole arrays. An additional problem is that landowner permits must be obtained over large areas to deploy a surface array, whereas downhole monitoring does not require such permitting.

10.5.1 Recording

Geophones may be installed on the surface or buried in shallow wells. Surface geophones are either planted directly in the ground surface or (preferably) are planted in the ground after scraping away 20–30 cm of soil and covering with the displaced soil. The soil slightly below the ground surface is firmer and provides better geophone coupling. Alternating layers of soil and coils of geophone cable while burying the phones, damps wind, rain, and other vibrations traveling down the cable to the geophones. However, burying the geophones adds substantially to deployment costs and hence is often not performed.

The geophones may either be single component (vertical) or three components (vertical and orthogonal horizontal components). The wells for shallow buried arrays are typically 20–100 m (65–330 ft) deep. The well depth depends on near-surface conditions and the depth to which surface waves can penetrate. One or more three-component sondes are cemented permanently into the well. Sondes that have only the vertical component are sometimes used. Hammer shots or small impulse-generating devices are used to orient multicomponent geophones and sondes deployed in shallow wells by striking the surface at eight azimuths around the device. Both surface and buried array receivers record continuously for many days.

The most important difference between recording data using surface geophone arrays and buried geophone arrays is the presence of coherent surface wave noise. This noise is almost always present on surface grids. Coherent surface waves are generated by activity on the surface such as well head activity, highways, construction sites, frac pumps, and drilling and pumping activity in the area. This noise can be of very high amplitude in the low-frequency band. The surface noise can propagate along the surface and scatter off of inhomogeneities in the near surface. The scattering points become new noise sources. As a result, every geophone on the surface records surface wave noise arrivals from all azimuths. The scattered coherent surface wave noise is very difficult to predict and remove using most available filters. Geophones buried sufficiently deeply in shallow wells are hidden from this surface wave noise. This difference in coherent surface wave noise present on surface geophones versus buried geophones allows for the use of many fewer geophones for a buried array than for a surface array. The removal of coherent noise from the surface geophones is the largest challenge to obtaining good detection of MEQs and making good-quality maps of the fracture networks using surface arrays.

While the quality of the surface geophone data suffers from the surface wave noise, the buried phone suffers from lower signal level and ghosting. The geophone on the surface has twice the signal amplitude for upcoming waves compared to the buried geophones because the free surface is essentially a perfect reflector and causes the amplitude of the up-going signal at the free surface to be multiplied by 2. The signal that hits the free surface is propagated back down with the same amplitude as the up-going signal. *Ghosting* occurs when the buried geophone records the up-going signal and then the down-going signal after the reflection off of the free surface. The delay between the up-going signal and the down-going signal (the "ghost" signal) at the buried geophone can be on the order of 10 ms for a very high velocity near surface and shallow geophone and be on the order of 100 ms for a very low velocity in the near surface and a deep geophone.

Consider these examples:

- In the Eagle Ford of South Texas the near surface velocity may be on the order of 1800 m/s (6000 ft/s) and the buried phone may be 90 m (300 ft) deep. The delay between the up-going and the down-going signals will be 100 ms.
- The Permian Basin of Texas has a very high velocity near surface layer that can be as thick as 365 m (1200 ft). The surface wave noise can penetrate hundreds of meters in this kind of rock, so the buried phone may need to be 300 m (1000 ft) deep in order to avoid surface noise. For the case of the geophone buried 300 m and the near-surface velocity of 6100 m/s (20,000 ft/s), the delay between the up-going and down-going signal will be 100 ms.

- For many areas such as the Marcellus Shale Fm., there is substantial topography and greater variation in the near-surface velocity. Buried geophones in the Marcellus are typically on the order of 30 m (100 ft) deep. For a near-surface velocity of 3700 m/s (12,000 ft/s), the delay between the up-going and down-going signal is 17 ms. This delay is such that the ghosting on the buried geophone interferes with the waveform of the signal and can lead to detection and location accuracy problems for the buried geophones. The ghosting also causes notches in the amplitude spectra which impacts moment tensor inversions. To improve correction of such ghosting, shallow buried arrays often have multiple receivers at different levels in each hole.

In summary, surface array data must be filtered to eliminate the coherent noise that is inherent to surface array recordings, and buried array data must be deghosted to improve location accuracy, focal mechanism solutions, and moment tensor inversions.

MEQs recorded with downhole sensors typically have strong amplitude in the frequencies above 100 Hz and show very broad band amplitude spectra (Fig. 10.11). For surface recordings, the emissions from the rock movements at depth must travel through the entire section of the earth to reach the geophones at the surface. This propagation path and distance results in attenuation and scattering, and there is a loss of the higher frequencies. Because higher frequencies attenuate more rapidly, these frequencies are lost. The dominant pass band of the rock column is below 60 Hz. However, frequencies below 15–20 Hz are often so contaminated with surface noise that they must be filtered out. Considering both low-frequency surface wave noise and attenuation of higher frequencies during wave propagation, the signal available for analysis at the surface is typically in the 20–60 Hz band. As described in the following sections, a complete suite of microseismic products can be generated using only this frequency band, although only larger magnitude MEQs are detected.

10.5.2 Seismic Emission Tomography

This section discusses processing of surface and near-surface passive seismic data. SET is the fundamental underpinning of the entire processing chain. SET is used for the detection and location of MEQs, focal mechanism analysis, imaging cumulative seismic activity, and constructing fracture images from the cumulative activity volumes.

10.5.2.1 Overview SET is the method of choice for detecting and locating seismic emissions from the subsurface. Seismic emissions from rock movements in the subsurface arrive at sporadic times and from various locations. In order to detect these emissions and locate them accurately, the recorded traces must be analyzed for all time windows over the time interval of interest (e.g., over the pumping time of a frac stage). Each time window must be focused (flattened) for every depth voxel in the volume of interest. The workflow and algorithms used for SET are the same as is used for prestack depth migration (PSDM) in surface reflection seismic data. The primary difference between PSDM and SET is that PSDM uses two-way travel times for focusing while SET uses one-wave travel times.

SET is performed as follows. First, the study volume is divided into voxels, and ray tracing is used to compute the travel time from the center of every voxel to every receiver (Fig. 10.12). Typical voxel sizes are 8–15 m (25–50 ft) on edge. Second, the traces from all receivers are aligned in time (flattened) using the one-way travel-time computations. This process aligns the traces as if they had all been emitted from the voxel of interest. In the last step, a short time window (typically 100 to a few hundred milliseconds) is stepped over the data through the time interval of interest and a measure of seismic activity is computed within the window by comparing all of the flattened traces. At each time step, the window is moved only a portion of its length so that overlapping windows are computed ensuring that all activity is captured. Different workers may use any of various measures of seismic activity including stacking, semblance, coherence, and cross-correlation methods. This procedure is repeated for every voxel resulting in a multidimensional data volume whose dimensions are the X,Y,Z coordinates of the depth voxels, the time step, and the activity measure. Consequently, a depth volume of activity is generated at each time step. For a single frac stage that lasts for 2.5 hours, there will be more than one hundred thousand depth volumes, and millions for the entire well job.

An important feature of SET is that it images total seismic activity within each time step, and thereby captures much more energy than conventional seismological methods that locate discrete MEQs only. Imaging total trace energy captures both more MEQ energy and seismic energy from other phenomena that are ignored by conventional methods, such as LPLD activity (Das and Zoback, 2013a, b) and EDS (Sicking et al., 2014).

SET captures more MEQ energy because seismological methods can only be used on MEQs that are sufficiently large and clear to identify as distinct events and that have clear, identifiable P-wave and S-wave arrivals. Earthquake frequency versus size distributions are linear when plotted on log–log plots (i.e., log of earthquake frequency vs. magnitude; magnitude is a logarithmic measure of signal amplitude). The slope and intercept of the line, respectively, termed b-value and a-value, may vary with location and failure type, but the distribution is linear. Large, clear earthquakes are consequently less abundant than smaller, less distinct earthquakes that cannot be utilized for conventional seismology. These smaller earthquakes combined emit more

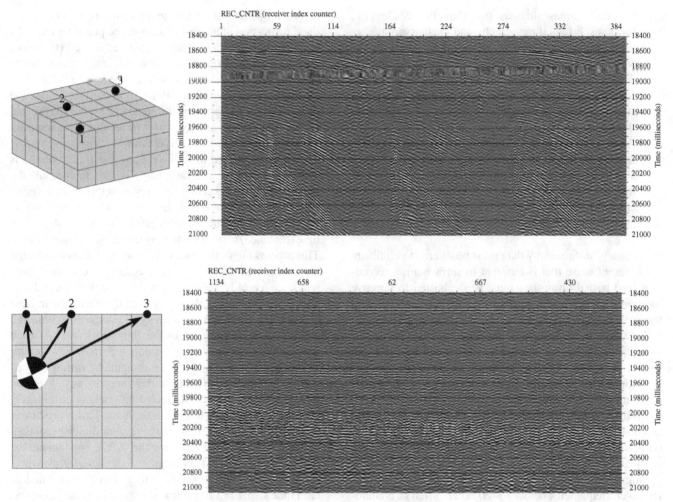

FIGURE 10.12 Focusing of the trace data using travel times from ray tracing. The microearthquake (MEQ) occurs in a specific voxel of the volume. The traces for the MEQ are flattened when the travel times for the voxel in which it occurred are used to time-shift the traces. After flattening, the traces are ready for the imaging step.

total energy than the relatively rare, large earthquakes. SET captures all seismic activity above the noise floor, which consists mainly of instrumental sensitivity and the ability of the rock column to transmit signal.

MEQs alone represent only a small fraction of the total seismic energy produced by hydraulic fracture treatments. SET also captures energy from other types of seismic phenomena. These types of phenomena include LPLD activity, long-duration P-wave emissions (EDS), and perhaps other, as-yet-undiscovered types of seismic emissions. LPLD and long-duration P-wave emissions do not have distinct first arrivals and may persist from many seconds to tens of minutes or even hours. For these reasons, such phenomena are ignored by conventional seismological methods. If we use fishing as an analogy, SET is like fishing with a net (you catch everything), whereas conventional earthquake seismology is like fishing with a lure (you only catch big fish of a few particular species). The downside is that trash (noise) may be caught in addition to fish. Even with effective removal of coherent noise at the trace-processing stage, random noise will produce some low-level false structure because random processes exhibit clustering and anticlustering. Although low-level signal may be part of this structure, the low end of the cumulative activity distribution must be clipped from stacked volumes to further minimize the effect of noise.

10.5.2.2 Parameters and Data Required for Set The quality of the results of the SET method depends on several factors:

1. Velocity model and statics
2. Design of geophone grid layout
3. Noise in the area of the grid, especially coherent surface wave noise
4. Lithology-dependent variation of the amount of signal generated in the rocks being fractured

10.5.2.3 Initial Velocity Model

The velocity model must be accurate in order to obtain correct location of MEQs, cumulative activity volumes, and fracture images. Often, a 1D P-wave velocity model is constructed from a nearby sonic log and the focusing is computed using a 1D velocity model. This works very well for the small area around the well being treated for focusing the emissions, especially if the strata are horizontal and relatively homogeneous. When there is a lateral velocity gradient in the velocity model, the location accuracy can be degraded. However, if the magnitude of the velocity gradient is known then beam steering methods can be used in SET to compute accurate locations. Figure 10.13 shows an example of a smooth lateral velocity gradient in the Eagle Ford and a complex 3D velocity model from a thrust-related fault bend-fold anticline (Lacazette et al., 2013). When the velocity has 3D complexity, the full-volume 3D interval velocity must be used for all aspects of focusing and imaging to obtain useful results. The complex 3D velocity model shown in Figure 10.13 was derived from PSDM of seismic reflection data in the Colombian Andes.

10.5.2.4 Velocity Calibration

The starting velocity model is estimated from available data, as described earlier. This starting model must be calibrated for microseismic imaging using a seismic source with a known location. The best calibration method is to record perforation shot waveforms and use the arrival times to estimate the velocity error. The depth error in the estimated perf shot location is a measure of the vertical velocity error and the error in the X, Y location is a measure of the lateral velocity gradient. By interactively focusing the perf shot and changing the velocity model, the focused location of the perf shot can be moved to match the known location of the perf shot in X, Y, and Z. When the focused location of the perf shot matches the known location of the perf shot, the velocity model is correct. Unlike perf shots that are located only on the wellbore, MEQs occur throughout the study volume and, therefore, provide an additional constraint on the velocity model. Although the absolute location of an MEQ is not known independently, the first arrivals from all MEQs should be flat after focusing regardless of where they are located in the volume of interest.

10.5.2.5 Statics

For surface recording arrays, statics are very important for achieving good-quality focusing. The velocity in the very near surface is never known accurately, so the flattening of the trace data has small errors regardless of the quality of the velocity model. There are various methods to get the residual shifts required to improve the flattening of the waveforms. These residual time shifts are called *statics* or *static corrections*. Residual or velocity statics accounts for near-surface velocity variations (i.e., weathering effects), while elevation statics account for topography. An example of a perf shot with and without the velocity and elevation statics applied is shown in Figure 10.14. Note the improvement in flattening after static correction. For this example, the residual statics were computed from the waveforms themselves. An alternative method is to use total receiver statics from surface reflection seismic data, but this method can only be used when the same array is used for both surface reflection and passive seismic recording. The static information computed from surface reflection data is not commonly available. Retrieving the refraction plus surface consistent residual statics from the surface reflection processing project will allow computation

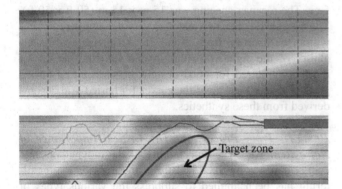

FIGURE 10.13 Top—Vertical section of a smooth lateral velocity gradient in the Eagle Ford Fm., Texas, U.S.A. Horizontal and vertical grid lines are 305 m (1000 ft) apart. Bottom—Vertical section showing part of a complex 3D velocity model from the Colombian Andes. The model was developed from pre-stack depth-migrated 3D reflection seismic. Line is 3.6 km (2.2 miles) long. Heavy grid lines are 1000 m (3281 ft) apart. When there is not a 3D interval velocity model available, the 1D velocity function from a sonic log can be combined with beam steering corrections to incorporate a lateral velocity gradient. There is a lateral gradient in the velocity in the Eagle Ford caused by the structural dip down to the basin.

FIGURE 10.14 Perf shot before (top) and after (bottom) the application of residual statics. Applying the residual statics has improved the alignment (flattening) of the traces. Horizontal grid lines represent 10 ms. Horizontal axis is traces in a spiral sort from the X,Y location of the perf shot.

of total receiver statics that will move the receivers to a constant flat datum. Using this total receiver static, a proper static for each receiver can be obtained that will allow improved flattening in microseismic focusing.

10.5.2.6 Grid Design The design of the grid for recording the passive data is important for detection and accurate location of both MEQ and fracture networks. Overall, the optimum grid design is as uniform a distribution of geophones as possible that covers the required aperture. The most uniform distribution is a face-centered hexagonal closest packed grid, which is only possible with a nodal recording system. For good-quality imaging, a radial aperture of 1.5 or even 1.6 times target depth is a good rule of thumb. In other words, the outer edge of the array should ideally be 1.5 times depth perpendicularly outward from the bottom edge of the subsurface image volume. Areas with fast velocity require larger aperture, while areas of slower velocity require smaller aperture. The optimum method for determining the required aperture is to use ray tracing and the velocity model for the area. Figure 10.15 shows the focusing and imaging for two different apertures. For the time window of the trace data, all voxels in the depth volume have been focused. For an aperture 1.6 times depth, the image of an MEQ is spherical. The vertical resolution is the same as the lateral resolution. For an aperture of 0.8 times depth, there is severe distortion and stretching in the depth dimension and the lateral resolution is much less.

Many surface grid designs use cable systems in the field and cannot distribute the receivers as uniformly as nodal recording systems. A common design is to use a star geometry with eight or more arms radiating outward from the well head. The star design is good for removing surface waves propagating out from the well head. However, normally there are many other surface wave noise sources so that the coherent surface wave noise cannot be removed adequately because of poor azimuth control. This grid design also suffers from poor sampling toward the ends of the arms. Other grid designs that use lines of receivers include parallel lines of receivers and orthogonal grids of receiver lines, both of which work quite well.

Buried grids normally will not have strong coherent surface wave noise because the geophones are below the depth of surface waves. This allows the density of the receivers in the grid to be much smaller than for surface array grids. Surface grids must have many more receivers than buried grids in order to predict and remove coherent surface wave noise. Prediction is accomplished by modeling noise as it propagates across the grid. The number of receivers required for surface grids is a factor of 10 higher than for buried grids.

Consider the results of a synthetic study carried out to obtain rule-of-thumb design criteria for buried grid arrays. A synthetic hypocenter was computed using a high-quality wave equation modeling code. Noise traces were generated by computing random number sequences, band passing them, and setting the root mean square (RMS) amplitude to different levels. The trace data generated for the synthetic had S/N levels that varied between 0.25 and 5.0. S/N is defined by the ratio of the amplitude of the peak of the signal to the RMS amplitude level of the noise. The synthetic was computed with a very high-density receiver grid on the surface. The receiver spacing was 15 m × 15 m (50 ft × 50 ft). The traces were then decimated to vary the number of traces within the aperture from 3 and 36. Analyzing the detectability of the signal provides insight into the density of receivers required for good designs of surface and buried grids. Figure 10.16 shows the curves for buried grid designs derived from these synthetics.

10.5.2.7 Trace Processing: The Critical Importance of Noise Removal The detectability of the radiated seismic energy depends on the S/N ratio of the trace data after filtering has been applied to suppress the various types of noise. For good detection, the signal level of the seismic energy arriving from the subsurface must be of similar amplitude to the noise in the traces after the coherent noise and other noise have been removed. Processing of the recorded seismic traces can improve the S/N in the trace data and is very important.

Under the same conditions, different rocks generate different levels of emissions when they crack or slip past each other. This is the signal that will be focused and analyzed. Some rock types generate very large amplitude signals, while others are very quiet. Similarly, some rock types generate

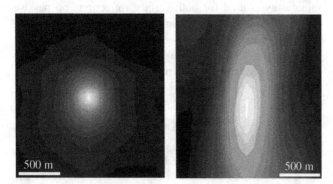

FIGURE 10.15 Vertical sections through a single depth volume (one time step) showing modeled focusing of a synthetic hypocenter for different apertures in the surface recording array. Vertical scale equals horizontal scale. The image at left was focused with an aperture of 1.625 times depth and is very good. The vertical resolution is the same as the lateral resolution. The figure at right shows the same synthetic hypocenter but with an aperture of 0.8125 times depth. The focusing in vertical direction is very poor. Overall, the vertical and lateral resolutions suffer substantially when the aperture is inadequate.

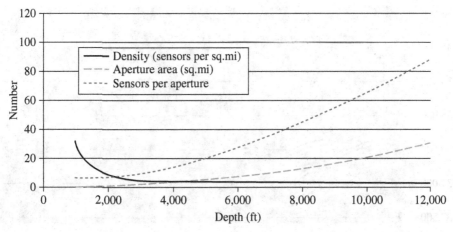

FIGURE 10.16 Graph showing the design criteria for a surface grid as a function of depth-to-target. The units of the vertical axis are those of the labeled curves. The density of the receivers goes up near the surface because the aperture shrinks, and there are a minimum number of receivers required within the aperture for good detectability. The aperture area is smallest near the surface and increases with depth. The sensors per aperture go down as the depth approaches the surface and then flattens because there is a minimum number of sensors required within the aperture.

primarily MEQs, while others generate different types of emissions such as LPLDs (Das and Zoback, 2013a, b) and EDS P-wave signals (Sicking et al., 2014). The type of emission is dependent not only on rock type and stress conditions but also on the orientations of failure planes relative to the stress field (Das and Zoback, 2013a, b).

The SET method is the primary method used for detecting and locating MEQs with surface and buried grid recordings. It is also the first step in the workflow for mapping cumulative seismic activity and direct imaging of fracture networks. Consequently, all SET-derived products are dependent on effective noise removal in the trace processing step, which is the first step of the processing sequence.

Figure 10.17 shows the raw trace data and the filtered trace data for the cables of a star array design. The raw trace data shows the surface wave noise propagating out from the well head as a linear move out from near to far on the cables. There is additional coherent surface wave noise propagating from outside of the array. This latter noise is not moving in line with the cables and shows move out from far offset to near offset on the second cable. This same noise on the first cable shows that the surface waves are propagating perpendicular to the cable. For good focusing, the surface wave noise must be removed. Both the surface noise from the well head and the surface noise from outside the grid have been removed.

Figure 10.18 shows large amplitude surface wave noise that is generated in between the arms of a star array. The noise hits the receiver lines broadside and has an apparent surface velocity higher than 10,000 ft/s. The imaging with the noise present versus with the noise removed is shown. If the recording grid were a pseudo-random design, this coherent noise would not impact the imaging to the same degree as it does for the star grid.

Removing the coherent surface wave noise is critically important for good imaging results from a surface recording. Other noise components of surface and buried grid data include excessive spikiness in the amplitude spectra and anomalous traces that are produced by noise sources very close to the individual geophone. The spectral spikiness is caused by the vertical propagation of the waves in a layered earth. The workflow for trace processing must kill the anomalous traces and remove the amplitude spectrum spikes from all traces. The anomalous high-amplitude traces and the amplitude spectra spikiness is also present in buried grid data. An example of trace processing and imaging improvement after processing is shown for data from a buried grid in Figure 10.19.

10.5.2.8 Focusing (Flattening)
All surface methods of passive seismic recording use SET to first focus the trace data for each of the voxels in the subsurface volume being searched. After the traces have been flattened, as described in Section 10.5.2.1 and shown in Figure 10.12, the detection and imaging processes can take different paths.

10.5.2.9 Imaging Methods
Stacking the Traces After focusing, the traces are stacked to produce a single trace. This stacked trace is then analyzed for windows in which the amplitude is higher than the background. This is frequently accomplished using the ratio of a short time window (short time aperture or STA) divided by a long time window (long time aperture or LTA) so that the method is referred to as STA/LTA. Flattening and then stacking the traces works well for seismic emissions having very similar wavelets on all of the unstacked traces. However, MEQs are typically slip-type, DC events so that the waveform varies in phase and duration across the recording array. Consequently, stacking the traces is not a good method for either MEQ detection or cumulative activity imaging.

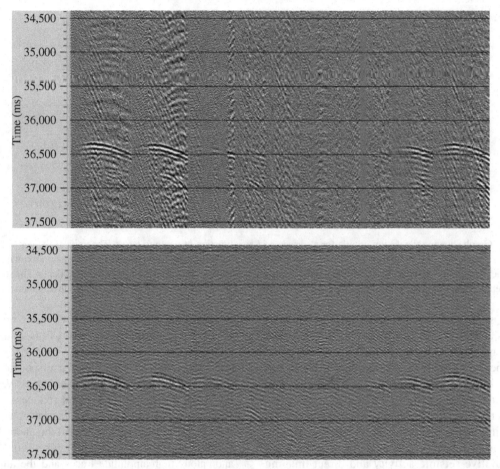

FIGURE 10.17 Data from a cabled array showing microseismic traces before (top) and after (bottom) noise filtering. The data contain a microearthquake. Noise generates artifacts in the coherence volumes which can distort the computed images. The linear noise from the well head is broader band and has a very slow velocity along the recording cables. The surface wave noise coming from outside of the array is moving along the cables with a higher velocity than the well head noise but is of lower frequency and narrower bandwidth.

As described in the following section, MEQs normally are analyzed using individual traces.

Cross Correlation with Nearby Traces Each trace can be cross correlated with the traces nearby and the cross correlations can be stacked. In this method, the phase changes as a function of offset and azimuth from the source location is removed by cross correlating each trace only with other traces that are within a specified distance. All of the traces that are cross correlated with each other will have the same wavelet except at the nodal points in the radiation pattern. Not considering background noise, the amplitudes of the traces at the nodal points will be close to zero amplitude and will not make a significant contribution. There will be as many cross-correlation functions as there are traces in the gather. The individual cross correlations can be stacked because they will have the phase changes removed. The stack of all of the correlation functions is used in the search for MEQs and can be used for fracture imaging. This method is more robust than many others but is more computationally intensive.

Cross Correlation with Source Function Each trace can be cross correlated with a synthetically derived source wavelet for that trace based on a modeled source slip function. In this approach, high-amplitude MEQs are located and analyzed for the source function. Using the radiation pattern of the MEQs from which the source functions are derived, source function wavelets are created for each trace in the gather. The cross correlation of the computed source wavelet for each trace with that trace produces a gather of cross-correlation functions that can be stacked. The stacked function is used in the search for time windows that have amplitudes above the background. This method produces detections for MEQs for each source type employed.

Coherence after Source Function Correction The coherence can be computed across all of the traces after adjustment of source waveforms on each trace based on a predetermined radiation pattern. These coherency results can be searched for amplitude highs in 5D space (X, Y, Z, t, A) that are closed in all dimensions. The locations and

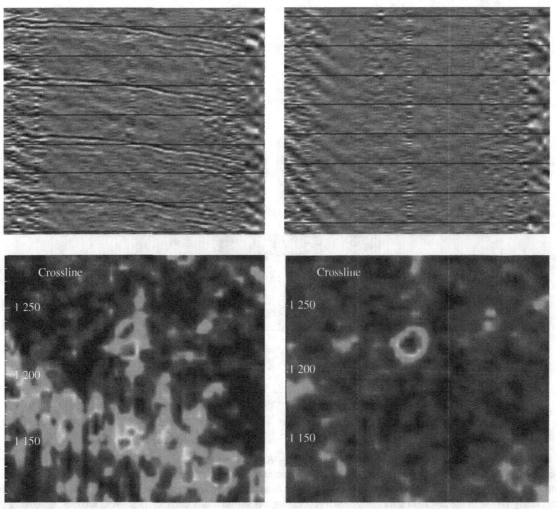

FIGURE 10.18 Top left—Traces showing coherent noise generated between two of the arms of a star grid array. Horizontal grid lines represent 10 ms intervals. The noise is very broad band and is hitting the cable from the side and at a very high apparent velocity. This noise is very difficult to remove using standard filters. Top right—The same traces with the noise removed. Bottom left—Depth slice of a depth volume showing semblance computed from the unfiltered traces. Scale in feet. Bottom right—The same depth slice after trace filtering. Note that a microearthquake hypocenter is imaged clearly in the filtered data.

times of these volumetric amplitude highs are the locations and times of MEQs.

Coherency and Semblance without Source Function Correction The coherency can be computed across all of the traces without adjustment for wavelet changes between the traces. This method does not correct for the wavelet changes with offset and azimuth and requires the presence of signal that is long duration and that is P-wave only. Documentation of this type of signal is shown by Sicking et al. (2014). This method is used to compute the time steps for cumulative activity imaging. The workflow for this method is to compute hundreds of thousands of semblance depth volumes, edit them both statistically and graphically, eliminate the volumes that contain larger amplitude slip dominated MEQs, and finally combine the remaining volumes into a final volume for use in computing the fracture networks.

10.5.3 MEQ Methods

10.5.3.1 MEQ Detection
To detect an MEQ, it must be located in X, Y, Z, and time. The accuracy of these measures depends on the grid design, aperture size, velocity model, and quality of calibration, as previously discussed. Various approaches are employed for detecting the occurrence of MEQs in surface microseismic data.

One method employs STA/LTA. For this method, the background amplitude in the trace data is computed by taking the RMS amplitude of a very long-time window, scanning the trace for the RMS amplitude over a short-time window that slides along the trace, and taking the ratio of the short

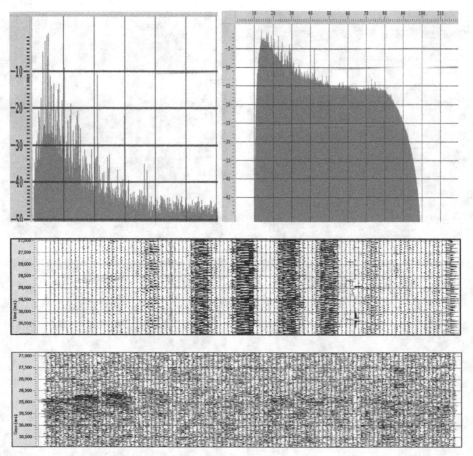

FIGURE 10.19 Example of traces from a shallow buried grid. Top—Amplitude spectra for a time window. Raw trace spectra shown at left, filtered spectra at right. Most microseismic traces show such pronounced spikes in the amplitude spectra. Vertical axis in dB, horizontal axis in Hz. Center and bottom—Traces before (center) and after (bottom) filtering. There are event arrivals in the center of the time window. This event is not clearly visible in the raw field trace data. Signal processing has substantially enhanced the event arrivals. Removing these spikes in the amplitude spectra and killing the excessively large amplitude traces has produced most of the improvement.

time sliding amplitude to the long-time window amplitude. This process is run on all traces of the surface microseismic grid. When this ratio is larger than a chosen value, a detection is registered for follow-up analysis. The detections for all of the traces are clustered and sorted to reduce the number of detections. For each MEQ detection, a more detailed focusing scan is employed to determine the location of the detection.

The second method of detecting MEQs is to inspect the individual depth volumes produced at each time step for high-activity voxels. The 5D volumes computed must be searched to obtain the optimal location for each of the detections. The same MEQ will generate detections in many of the X, Y, Z voxels. The multiple detections for each MEQ will be clustered and the detections must be analyzed to reduce the detections to the single MEQs. Typically, at least the initial identifications are performed by an autopicker because of the enormous amount of data that must be searched. Different service providers may or may not have a human quality control the final hypocenter picks and refine the spatial location and exact time of the pick.

10.5.3.2 Focal Mechanism and Moment Tensor Analysis
Most large MEQs are shear dominated. From these events, we can compute focal mechanisms and moment tensor information that constrains the stress field in the rocks. The data required for simple focal mechanism computations are the P-wave first motion polarity, azimuth, and take-off angle for each of the recording receivers. Adding amplitude information improves the quality of focal mechanism analysis and moment tensor estimates. Figure 10.20 shows the trace data, picks of first motion polarities, and the best-fit fault-plane solution for an MEQ.

10.5.4 Imaging Cumulative Seismic Activity

Cumulative activity imaging is an extension of standard SET processing. At the time of this writing, cumulative activity imaging is available from several different service

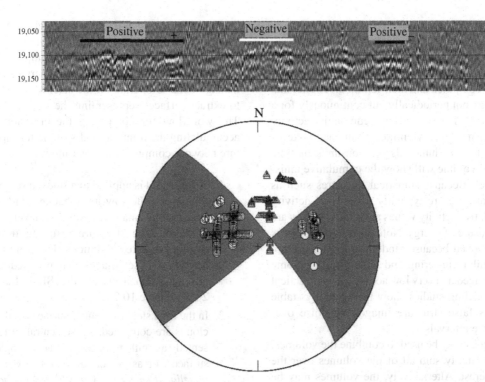

FIGURE 10.20 Focal mechanism example for a wrench-faulting microseismic event. Top—Traces sorted by azimuth, time in milliseconds. Note that the polarity of the first arrival changes with the azimuth. Bottom—Stereoplot (beach-ball plot) showing the polarity of the first arrivals plotted according to the azimuth of the observation and the ray take-off angle. Black and white quadrants are compressional and dilatational, respectively. The diagram shows the fault plane solution that best fits the polarity information.

providers although it is still not as widely used as MEQ hypocenter-based methods. Cumulative activity imaging stands in marked contrast to hypocenter-based methods. Rather than looking for relatively large but sparse events (MEQs), cumulative activity imaging methods accumulate total seismic trace energy over longer periods of time—minutes, hours, or even days. The resulting images define clouds of high activity (e.g., Duncan, 2005; Geiser et al., 2006; Lacazette et al., 2014). The clouds can be used as a measure of the SRV. Figure 10.21 is an example of a cumulative activity image. This method is tremendously more sensitive than MEQ-based methods because it both images total trace energy and accumulates that total activity over extended time periods (Lacazette et al., 2013, 2014; Sicking et al., 2014).

As discussed in Section 10.5.2.1, SET images total seismic activity in individual time steps and therefore captures more total energy than seismological methods that detect only MEQs using data from downhole arrays. Although SET can be used to detect MEQs by searching individual time steps in the (X, Y, Z, t, A) hypervolume, SET can also be used to image total microseismic activity over longer time periods. This is commonly done by summing the semblance volumes from each time step over a time interval of interest, such as the pumping time of a frac stage. This process enhances spatially stable signal and suppresses noise. Note that only random noise is suppressed. Coherent noise must be removed at the

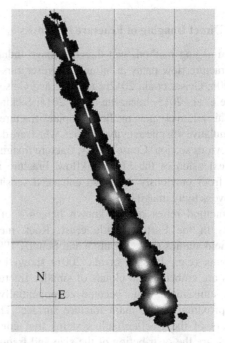

FIGURE 10.21 Map view of a depth slice of a cumulative activity volume from a fracture treatment in the Eagle Ford Fm. Slice ranges from above to slightly below the average depth of the wellbore. Color indicates the relative intensity of cumulative seismic activity (white = high, black = low). White rectangles along the wellbore denote perforated intervals. 1000 ft (305 m) grid for scale.

trace processing stage. Coherent noise remaining in the trace-processed signal is usually very apparent if not removed, indicating the need for more refined trace processing.

The key to this method is that a very large number of image volumes are computed and then summed. A voxel that emits seismic energy not periodically but continuously for a long period of time will have a very large cumulative activity value. The total seismic energy emitted from each voxel is accumulated over tens to hundreds of thousands of time windows. The final volume will show the cumulative emissions in each voxel. Because statistical measures such as semblance or covariance are typically used as an activity measure the cumulative activity values are relative—they are not an absolute measure of energy. Note that random noise is not completely removed because random processes are not uniform—they exhibit clustering and anticlustering. Some false structure is imaged at a very low activity level. Practical experience and modeling studies show that spatially stable signal overwhelms false structure imaged by noise over typical stacking time intervals.

Different strategies may be used to combine the volumes. The simplest is to simply sum all of the volumes over the time period of interest. Alternatively, the volumes may be edited statistically and then substacked into fewer volumes depending on product desired (Lacazette et al., 2014). The result is a single depth volume for the entire time period of interest, such as a frac stage, that shows the spatial distribution and relative intensity of activity.

10.5.5 Direct Imaging of Fracture Networks

Fracture imaging methods can directly image induced and natural fracture flow paths in oil and gas reservoirs (Geiser et al., 2006; Geiser et al., 2012; Lacazette and Geiser, 2013; Lacazette et al., 2013; Lacazette et al., 2014; Sicking et al., 2014). This relatively new method computes fracture images from cumulative seismic activity volumes, which are described in the previous section. Consequently, fracture imaging is yet another extension of the SET workflow. Fracture imaging benefits from previously described enhanced sensitivity of cumulative activity imaging.

The method relies on the known behavior of failure processes in the Earth's brittle crust. Rock mechanics theory, field studies (e.g., Vermilye and Scholz, 1998), and experiments (e.g., Janssen et al., 2001) show that large fractures are embedded in clouds of smaller fractures and seismic emissions that become exponentially more intense proximal to the main fracture surface. The three parameters of the fracture/fault systems that control the emissions are the distribution of the size and frequency of fractures, the fracture size distribution within the fracture/fault zone, and the fracture/fault kinematics. The model for fracture imaging predicts acoustic energy produced by the fracturing reaches a maximum in the vicinity of the main discontinuity and that the cloud of high activity is approximately centered on the main discontinuity.

Fracture imaging uses the above empirical observations on the relationship between seismic energy emission and the damage zone to operate on the stacked semblance volumes to extract surfaces representing the fracture/fault networks. The workflow for computing the emission volumes and accumulating them into a final volume for input to the fracture network computation is as follows:

1. A threshold is applied to a stacked semblance volume to remove the low-level background. Although the background may contain low-level signal, it also contains false structure resulting from noise. The result of this step is sinuous clouds of high cumulative activity. These images can be used to provide an alternative measure of the SRV (Lacazette et al., 2014; Figure 10.21).

2. In the final step, the central surfaces of the high-activity clouds are computed. These central surfaces are represented as either raster images, single-voxel thick surfaces, or as vector images. The vector images are *tessellated surfaces*. A *tessellated surface* is a continuous surface comprising triangles that share vertices and edges. The tessellated surfaces can be derived either from the raster images or directly from the stacked semblance volumes (Lacazette et al., 2014 Copeland and Lacazette, 2015)). Figure 10.22 shows the most energetic region of a fracture image that represents the likely extent of proppant distribution. Tessellated surfaces are commonly stored as ASCII data in the nonproprietary TSURF format, which is read and written by a wide variety of software packages.

More detailed description and discussion of nuances of the method are provided by Geiser et al. (2012), Sicking et al. (2012), Lacazette et al. (2013), and Lacazette et al. (2014).

10.5.6 Comparison of Downhole Hypocenters and Fracture Images

This section describes an example of hypocenters derived from conventional downhole monitoring and fracture maps determined by the methods described in Section 10.5.4. Figure 10.23 shows the near well zone for one frac stage. The downhole hypocenter detections are overlain on the fracture image mapped from the surface microseismic. There is excellent agreement between the fracture image and the downhole hypocenters. The azimuth of the fractures agrees with the SHmax and Shmin for this area. Figure 10.24 shows the fault tracks extracted from the surface reflection seismic data with overlays of the fracture network from surface microseismic for the entire well and all of the downhole hypocenters for the well. The large area fracture image maps

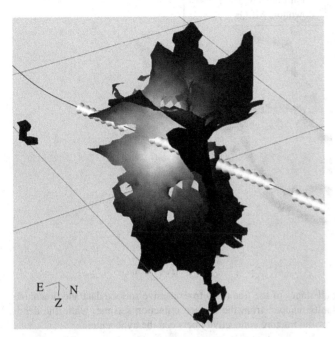

FIGURE 10.22 Oblique view of a tesselated surface representation of an image of a frac stage in the Eagle Ford Fm., Texas, USA with 1000 ft (305 m) grid for scale. The horizontal width of the surface measured perpendicular to the wellbore is approximately 108 m (355 ft). The image shows only the most energetic region of the fracture image connected to the wellbore. Color indicates the relative intensity of cumulative seismic activity (white = high, black = low). The black line represents the wellbore, the corrugated white cylinders represent perforated intervals (stages). Note that the geometric complexity of the frac increases and the cumulative activity decreases toward the outer edges of the frac where it quenches due to leakoff into the natural fracture system. Most of the frac emanates from one perforation cluster. See Lacazette et al. (2014) for additional discussion.

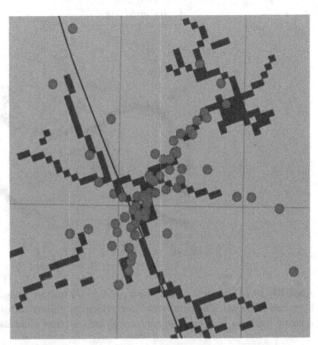

FIGURE 10.23 Map view of MEQ hypocenters (gray circles) and a single-voxel-thick depth slice of a fracture image (black) of a single frac stage in the Eagle Ford Fm., Texas, USA. The depth slice of the fracture image is at the level of the wellbore, the hypocenters are located throughout the imaged volume. The black line is the wellbore. N is up. The E–W dimension of the grid is 152 m (500 ft). The fracture image is from surface array data. The hypocenters are from downhole data. See Sicking et al. (2014) for detailed discussion.

are in good agreement with the fault tracks but show more details in the fracture network. The hypocenters generally line up along the fault track images but do not show as much detail in the fracture network as do the fracture images from the surface passive seismic data.

Comparing the fracture images to attributes computed from the surface reflection seismic shows that the active fracturing caused by the treatment occurs where the lithologies have less clay. Where there is more clay in the rocks, the fracturing was markedly less. Figure 10.25 shows the fracture images overlain on the acoustic impedance computed from the reflection seismic data. The gamma ray log is posted in the figure. Note that the sections of the log with high gamma ray values are where the fracturing is poorest, while fractures are better developed in regions of low gamma, which are more brittle. This is confirmed by comparing the impedance volume with fracture images, in which regions of higher impedance (higher brittleness) show better developed fractures. The top graphic shows a vertical cross section of the impedance volume with an overlay of the gamma ray log and the fracture images for each stage. The bottom graphic shows a depth slice of the impedance with overlay of the fracture images for each stage. There is good correlation of the fracture development with the impedance of the rocks and with the gamma ray log.

10.5.7 Summary

The attributes computed from surface passive seismic data includes MEQ hypocenters, focal mechanisms of especially strong, clear shear MEQs, solid volumes showing cumulative seismic activity, and direct images of fracture networks extracted from cumulative seismic activity volumes. The fracture networks computed from surface microseismic provide more details of the fractures than do simple hypocenter analysis. Unlike downhole methods, surface-based methods provide uniform coverage over arbitrarily large areas. However, surface-based methods do not have the broadband frequency content and sensitivity of downhole methods for MEQ detection and analysis.

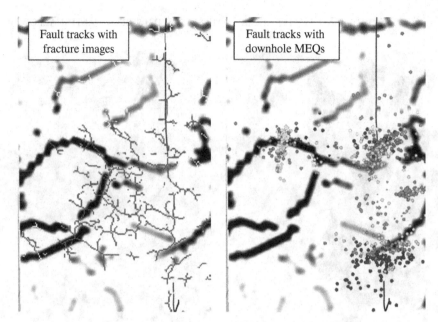

FIGURE 10.24 Fault track map with overlay of a fracture image for all stages of the frac (left) from passive surface data and downhole hypocenters (right). The hypocenters are primarily aligned with the faults mapped from the surface reflection seismic, while the depth imaging fractures show good alignment with the fault tracks and also show the fracture connectivity between the hypocenters.

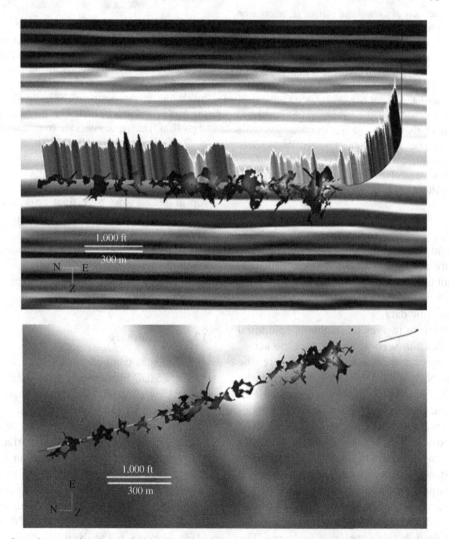

FIGURE 10.25 Top—Impedance section (1:1 vertical:horizontal scale) computed from surface seismic with overlay of gamma log and cumulative seismic emission volumes shows that fractures are much less developed in the high gamma log rocks and more developed in the low gamma log rocks. This indicates that fractures are better developed in the more brittle rocks. Image converted to grayscale from color scale. Bottom—Impedance volume sliced along the reservoir horizon shows that fractures are better developed in the higher impedance, more brittle rocks.

10.6 INTEGRATING, INTERPRETING, AND USING PASSIVE SEISMIC DATA

Passive seismic data is only useful if it adds value. Interpretation may be as simple as estimating well spacing by measuring the distance a cloud of MEQs extends from the wellbore. More sophisticated interpretation methods define the volume of rock affected by the fracture treatment and the volume of rock that is producing oil or gas (the two volumes are not necessarily the same). Stress orientation, stress compartmentalization, and stress changes induced by the fracture treatment can be determined. The ultimate application of passive seismic results is frac, well test, and reservoir simulations to develop optimum treatment and production methods and to more accurately forecast reserves and production. This section provides an overview of interpretation methods, applications, and examples of passive seismic interpretation.

10.6.1 General Considerations

10.6.1.1 Dry Seismicity Not all seismicity produced by a hydraulic fracture treatment results from the fracture fluid breaking new rock or infiltrating pre-existing natural fractures and causing slip. Tiltmeter surveys (e.g., Fisher and Warpinski, 2012) show that hydraulic fracture treatments produce measurable surface deformations even at depths exceeding 4.5 km (14,800 ft). This is due to inflation of the rock volume around the wellbore caused by injection of the frac fluid. This poroelastic strain wave propagates for thousands of feet horizontally as well as vertically. The wave generates seismicity by several mechanisms, primarily by increasing the shear stress state on pre-existing fractures and perhaps by increasing the fluid pressure in natural fractures by poroelastic mechanisms. (Pore-fluid pressure and stress in the solid skeleton of the rock are coupled via poroelasticity.) Also, fluid pressure can be transmitted through a connected natural fracture network far more rapidly than the frac fluid can physically move through the network (Lacazette and Geiser, 2013). In this case, there is a direct hydraulic connection to the wellbore even though frac fluid is not present. In some cases where chemical tracers have been used to track frac fluid, movement of tracer over distances of 1.5 km or more have confirmed that microseismic activity at long distances from the wellbore can represent frac fluid movement, not just transmission of a pressure wave (e.g., Geiser et al., 2012).

10.6.1.2 Seismic Activity Indicates Hydraulic Conductivity Seismic activity is an indicator (not proof) of hydraulically conductive fractures even if the microseismicity is dry. A key point about fluid flow in fractures is the dependence of fracture hydraulic conductivity on stress. A large body of published work accumulated since 1995 has shown that, in general, there is a positive correlation between the hydraulic conductivity of a fracture and the resolved shear stress on that fracture (e.g., Barton et al., 1995; Ferrill et al., 1999; Heffer, 2012; Heffer et al., 1995; Hennings et al., 2012; Morris et al., 1996; Sibson, 2000; Takatoshi and Kazuo, 2003; Tamagawa and Pollard, 2008). Drilling campaigns targeting highly stressed natural fractures can achieve spectacular results (e.g., Hennings et al., 2012). More highly stressed fractures also generate more seismic activity during fracture treatments, so that seismic activity is an indicator of the hydraulic conductivity of reactivated natural fractures whether or not they are hydraulically connected to the wellbore.

10.6.1.3 Bedding Parallel Features In most unconventional plays active today, bedding is either horizontal or close to horizontal. A thin horizontal cloud of MEQs or a horizontal fracture image is sometimes interpreted as a horizontal hydraulic fracture. This interpretation is untenable if the treating pressure is below Sv. If the treating pressure is below Sv, then the likely cause is either slip on a horizontal fault that was induced by the fracture treatment, or movement of fluid or at least fluid pressure through a highly permeable bed resulting in fracturing of the bed or reactivation of pre-existing fractures in the bed. In addition to matrix permeability, a common cause of strong bed-bounded permeability contrasts is contained jointing in which joints subperpendicular to bedding are well developed (Section 10.2.2.1). As discussed earlier, vertical bed-bounded joints are common in unconventional reservoirs. Contained joints develop preferentially in more brittle units, and brittle units are preferred fracing targets. Such features can produce very flat MEQ distributions, activity clouds, and fracture images if the bed that is carrying the fluid is thin relative to the resolution of the passive seismic data. This effect may be especially striking if the brittle bed is enclosed by rocks with high clay and organic matter content.

10.6.1.4 Microseismic Response to Hydraulic Fracturing in Contractional Faulting Stress States Contractional faulting stress states deserve special mention. Contractional faulting appears unfavorable for unconventional resource development because purely extensional hydraulic fractures run horizontally (Fig. 10.1). However, the authors of this chapter have performed several passive seismic studies of hydraulic fracture treatments in active fold-thrust belts. In these studies:

- Direct fracture imaging showed that virtually all imaged fractures were vertical even when MEQs showed slip on an active thrust fault near the treatment well during hydraulic fracturing.
- Wrench faulting focal mechanism solutions were common in the stimulated reservoir volume.
- The treatment pressures were well below Sv.

The simple explanation for these apparently incompatible observations is that frac fluid pressure primarily induced slip on pre-existing fractures by reducing friction. As discussed in Section 10.2.2.3, reactivating even poorly oriented fractures requires less energy than generating new fractures. Pre-existing fractures include vertical joints and wrench faults. Vertical joints are present in most strata, typically as orthogonal sets. Vertical wrench faults and joints commonly develop in active fold-thrust belts due to repeated local interchange of Smin and Sint during thrust-belt propagation. In summary, reservoirs in active contractional faulting tectonic regimes are full of vertical fractures that are easily reactivated when frac fluid pressure reduces friction on these fractures. Consequently, effective hydraulic fracture stimulation may be possible even in apparently unfavorable tectonic regimes. However, treatment pressures equal to Sv can only occur if horizontal fractures are being driven, which should be visible in passive seismic results. Horizontal fracturing is more common in soft rocks rich in clays and organic matter.

10.6.2 Interpreting Reservoir Stress from Focal Mechanisms

10.6.2.1 Introduction Reservoir stress, rock, properties, and natural fractures are the main influences on hydraulic fracture propagation. Pure hydraulic fractures run perpendicular to Smin and the stress state controls reactivation of natural fractures by the fracture treatment. Consequently, knowledge of reservoir stress is critical for frac design.

Borehole data are an excellent source of reservoir stress information and is especially valuable as a supplement to passive seismic data. Acoustic borehole images are the best and most reliable source of borehole stress data because they provide detailed, unambiguous images of breakout and image drilling-induced fractures. Commercial versions of these logs include UBI (Schlumberger), CBIL (Baker Atlas), and CAST (Halliburton). Crossed-dipole shear-wave logs can provide useful stress information, but are strongly influenced by borehole rugosity and rock features such as crossbedding and natural fractures. Oriented caliper data, including caliper data from electrical borehole imagers, provide only a crude constraint on borehole breakout and cannot reliably differentiate breakout from key seat and other types of borehole rugosity. Electrical borehole images are not a reliable breakout indicator in conductive rocks, especially shales. Induced fractures in oriented core can provide excellent stress information (Kulander et al., 1979, 1990). Various commercial mini-frac logging tools can provide quantitative stress estimates.

The downside of borehole stress data is that the borehole samples form only a small part of the reservoir. The stress state in a reservoir can vary on all scales from borehole- to field-scale. MEQ focal mechanism solutions provide stress data throughout a much larger part of the reservoir volume than borehole data. Consequently, passive seismic data can identify stress compartments where fracture treatments will behave differently. Focal mechanism solutions can identify stress changes caused by a fracture treatment during a fracture treatment (e.g., Neuhaus et al., 2012; Williams-Stroud et al., 2012a, b).

Focal mechanism solutions and the relationship between focal mechanism solutions and earth stress are described in Section 10.3.2. Again, a focal mechanism is a description of the strain produced by an MEQ; although the strain axes are often equated with the stress axes, this is not necessarily true. Consequently, a population of MEQ focal mechanisms can be used to provide an overall estimate of the strain in a volume of rock by simple averaging. Twiss and Unruh (1998) argue that such averages are good estimates of bulk strain, but provide poor stress estimates if substantial material rotation occurs. Such material rotation is not a consideration during hydraulic fracture treatments. On the time and volume scales of hydraulic fracture treatments, infinitesimal strain of the total rock volume is a good assumption.

In addition to simple averaging methods, a variety of methods used by structural geologists to evaluate populations of fault-slip data are equally applicable to evaluating populations of MEQ focal mechanisms produced by hydraulic fracture treatments. However, the outcrop or core data used by structural geologists is better constrained because the true fault orientation is known. These methods are reviewed by Allmendinger et al. (1989) and Marrett and Allmendinger (1990). Such methods are implemented in a variety of shareware and low-cost commercial software packages that can be downloaded for free or purchased online. Such packages include the popular Stereonet and FaultKin programs by R. Allmendinger of Cornell University. In this section, we will discuss simple, practical stress interpretation methods that users can apply easily with such software and provide some practical examples.

Interested readers should note that more sophisticated methods are available for estimating the stress axes from populations of earthquake focal mechanisms (e.g., Arnold and Townend, 2007; Gephart and Forsyth, 1984). These methods are applicable to microseismic data (e.g., Urbanic et al., 1993). The uncertainties in inverting focal mechanisms for stress are described by Abers and Gephart (2001).

10.6.2.2 Reservoir Stress
Averaging Focal Mechanisms and Fracture Complexity
Moment-tensor averaging methods are common in earthquake seismology (e.g., Jost and Herrmann, 1989) and, as previously discussed, are readily applied using various shareware or low-cost software packages intended for structural geologists. In the examples that we present here, we use the method of Tibi et al. (2013). The average

results of this method are identical to the standard averaging method for DC moment tensors, but provide additional information. The extra information is the fraction of the bulk strain that is simple shear and the fraction that is shape distortion without volume change. The simple shear fraction (like shearing a deck of cards) is termed the DC component (for double-couple). The shape distortion component is termed the CLVD component (for compensated linear vector dipole). This method is summarized in the following paragraphs, and applied to two case studies. Another way to think about the DC and CLVD components is as measures of complexity of movement planes and directions within the reservoir, and hence as a measure of fracture complexity induced by the fracture treatment.

The seismic consistency concept was introduced by Frohlich and Apperson (1992) as a measure of similarity of earthquakes within a given group. It is defined as the ratio of the scalar moment of the composite tensor, resulting from summation of the moment tensors of the events within the group, to the sum of scalar moments of tensors contributing to the composite tensor. The seismic consistency is 1.0 when the tensors are similar and 0.0 when they cancel one another. In general terms, as shown in Figure 10.26, a low value for the seismic consistency arises when a group consists of events of more than one type of focal mechanism. This characteristic makes the seismic consistency a viable parameter for assessing the stress field variability in a defined volume.

As illustrated in Figure 10.26, various studies have shown that the CLVD component of a seismic moment tensor can be caused by, among other mechanisms, the superposition of two or more shear events that occur along faults with different orientations (e.g., Frohlich, 1994; Tibi et al., 1999). Based on this notion, the complexity arising from changes in fracture orientations for an MEQ-generating fracture network can be assessed by evaluating the percentage of CLVD component in the composite moment tensor of the DC events induced in the reservoir.

Case Study 1. Figure 10.27 shows the locations along with the inferred fault plane solutions for 38 MEQs induced during a hydraulic fracture treatment. The DC focal mechanisms were determined using Snoke's (2003) code, by searching for the fault plane solutions that best fit the first motion polarities of P-waves (Tibi et al., 2013). The MEQs show three focal mechanism types. Most of the events

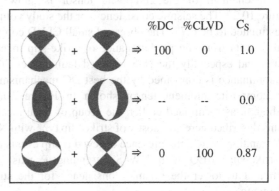

FIGURE 10.26 Examples of superposition of double-couple (DC) mechanisms (From Tibi et al., 2013). The two members of each combination have the same scalar seismic moment. The percentages of DC and compensated linear vector dipole (CLVD) for the resulting composite moment tensors are indicated, as well as the value of the seismic consistency (C_s).

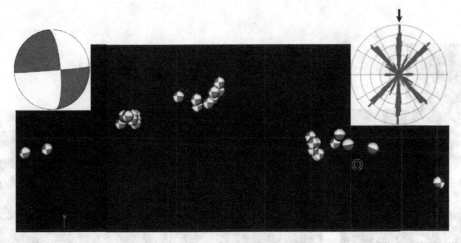

FIGURE 10.27 Map showing the locations and fault plane solutions of 38 microseismic events induced during a hydraulic fracturing treatment (From Tibi et al., 2013). White quadrants in the projections of the focal spheres are dilatational, while gray areas are compressional. Concentric circles indicate the well head of the vertical treatment well. Inset at the upper left corner of the map displays the best-fit DC mechanism of the composite moment tensor. Inset at the upper right corner shows the circular histogram of fracture segment orientations obtained from the reservoir-scale fracture images (Geiser et al., 2012). The arrow indicates the set of fracture segments oriented N–S, consistent with the strike of one of the nodal planes of the best DC mechanism.

represent strike-slip faulting along N–S or E–W striking nodal planes, with NE–SW trending P axes (Figure 10.13). The N–S striking nodal plane is consistent with the orientation of one of the three sets of fracture segments represented in the circular histogram shown in Figure 10.27, suggesting that this nodal plane represents the fault plane. The histogram was constructed from fracture images as described in Section 10.5.5. Four events are of dip-slip type (Figure 10.27). One event shows a strike-slip mechanism with NW–SE trending P axis (Figure 10.27). The moment tensors of the 38 individual events are summed to form the composite moment tensor of the group. The decomposition of the composite moment tensor yields a DC and a CLVD component of about 96% and 4%, respectively. The best DC mechanism of the composite tensor is shown in Figure 10.27. The seismic consistency for the study volume is estimated to be 0.77. The observed small CLVD component results from the predominance of strike-slip mechanisms and especially the N–S oriented fault planes. This predominance is confirmed by the best DC mechanism of the composite moment tensor shown in Figure 10.27. Although, as mentioned earlier, the group of focal mechanisms described here are mostly of strike-slip type with NE–SW trending P axes, the presence of few dip-slip events and a strike-slip MEQ with NW–SE trending P axis was sufficient to lower the seismic consistency for the study volume to 0.77.

Case Study 2. The locations of 339 energetic MEQs induced during another hydraulic fracturing campaign in a shale play and recorded using a surface array are displayed in Figure 10.28. The locations of the events shown in map view define predominantly NE–SW trends. Nevertheless, the inferred focal mechanisms appear to show significant variability. The summation and subsequent decomposition of the moment tensors of the events yield the best DC mechanism shown as inset in the upper right corner of Figure 10.28. The DC component of the composite moment tensor is only about 20%, whereas the CLVD component reaches 80%. The inferred seismic consistency of 0.25 is obviously very low. The high CLVD component is the consequence of the complexity of the fracture network as inferred from the changes in the orientation of the nodal planes. The large variability in terms of focal mechanism type, and hence stress field, results in the observed low seismic consistency.

The measure of fracture complexity and stress field variability discussed in this study may be used to infer implications for well placement and production as well as to develop strategies for field management.

An Example of Stress and Natural Fracture Compartmentalization Figures 10.29 and 10.30, respectively, show two sets of MEQ focal mechanisms and a summary map of microseismic results from a surface-based microseismic study in an active fold-thrust belt.

The two sets represent MEQs produced by fracture treatment of a single horizontal well. They were sorted easily using the N-axis orientations, which show less scatter than the P and T axes orientations (as is often the case, see Section 10.3.2.2). Note that the N axes are horizontal for the extensional focal mechanisms and vertical for the wrench-faulting focal mechanisms although the P and T axes have more scatter.

The focal mechanism solutions sort into two distinct sets: a set of extensional (normal) faulting focal mechanisms and a set of wrench faulting focal mechanisms. The DC/CLVD

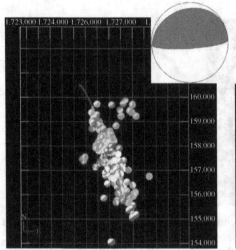
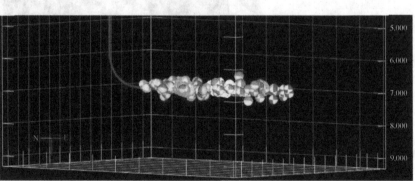

FIGURE 10.28 Map view (left) and profile view (right) showing the locations and fault plane solutions for 339 microearthquakes induced during a stimulation project. The solutions are depicted in grayscale coded by stage. Dark gray line represents the wellbore. Inset at the upper right corner of the map view shows the best DC mechanism of the composite moment tensor.

components of the two sets are 90%/10% for the extensional focal mechanisms and 99%/1% for the wrench faulting focal mechanisms. This suggests a higher level of complexity for the normal faulting mechanisms, although the significance of these differences is uncertain because the sample is small.

Nodal plane and fracture image data indicate that both sets of focal mechanisms represent reactivation of two sets of vertical orthogonal joints in the orientations of the wrench faulting nodal planes. No horizontal fracture images were observed, indicating that the subhorizontal nodal planes from the normal faulting solutions are auxiliary planes and not fault planes.

Plotting the focal mechanisms on a map with the fracture image and hypocenter data (Fig. 10.30) shows that the two sets of solutions occur in different parts of the reservoir. The extensional focal mechanism lies along a prominent fracture trend indicating that this trend is a fault. Note that the wellbore, which was drilled from south to north, was deflected when it encountered this zone. These observations suggest that the bit became trapped in a zone of fault damage that was weaker than the surrounding rocks. The region east of the well shows few hypocenters or fracture images east of the fault zone and contains a few normal faulting focal mechanisms. By contrast, the region west of the well contains abundant fracture images and wrench faulting focal mechanisms

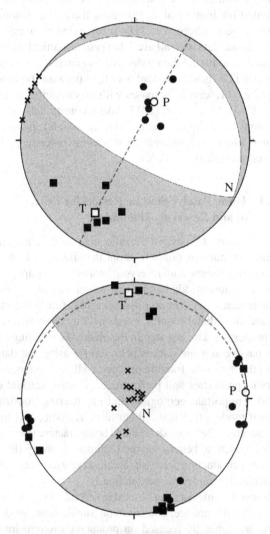

FIGURE 10.29 Stereonets plots showing two sets of focal mechanism solutions collected during a microseismic study. The two sets were separated using the N-axis orientations. P axes are shown as solid circles, T axes as solid squares, and N axes as Xs. The average P and T axes are shown as open circles and squares, respectively, and are named. The average N axis lies at the intersection of the nodal planes. The beach-ball plots for the average focal mechanisms are also shown. The average transport planes are shown as dashed great circles. Smax and Smin must lie along the transport plane. Top—Extensional (normal) faulting focal mechanism solutions—six solutions. Bottom—Wrench faulting focal mechanism solutions—13 solutions.

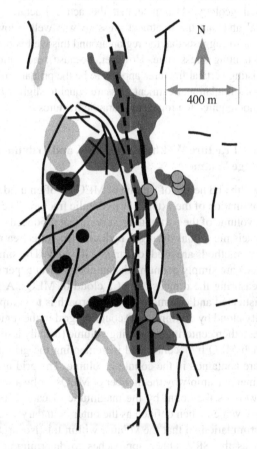

FIGURE 10.30 Schematic map showing key features of a microseismic data set in a fold-thrust belt. The thrusts (not shown) are moving northward (up in this map). Filled circles show the two sets of focal mechanism solutions from Figure 10.29: black circles—wrench faulting; gray circles—extensional faulting. Black lines are imaged fractures seen in a depth slice. Heavy N–S line is an especially strong structure that is interpreted as an extensional fault based on the focal mechanisms along it. Shaded areas indicate dense areas of microearthquake hypocenters. Dashed line is the wellbore, which was drilled from south to north.

scattered throughout it. The microseismic data therefore indicates two very different stress and natural fracture regimes. Because of these stress and natural fracture differences, an infill well drilled parallel to and west of the existing well would frac very differently than an infill well drilled to the east. Both wells would likely frac differently than the well in the diagram which was drilled down a fault zone. Also, the natural fracture permeability should be very different in both regions.

The likely geological explanation is that the region west of the wells is a complex fault stepover, while the region east of the well is a less damaged block in a normal faulting stress regime. Note that this study is from an *actively propagating* fold-thrust belt. This shows the variability in space and time of geological stress fields. Irregularities in faults and jostling of large fault blocks can produce complex and variable stress states very different from those indicated by the overall tectonic regime and regional geology. Also note that the active fractures were vertical and that the treatment pressure was well below Sv, which also suggests that the region around this well is not in a thrust-faulting stress state. However, because reactivation of preexisting vertical fractures appears to be the primary mechanism of stimulation, treatment pressure equal to slightly above Sv is not required to propagate hydraulic fractures.

10.6.3 Fracture Width, Height, SRV, and Tributary Drainage Volume

The width and height of a cloud of MEQs is often used as an approximation of the extent of a hydraulic fracture. The SRV is the volume of the part of the reservoir that produced strong microseismic signal during a hydraulic fracture treatment. Various methods are used to compute the SRV. The simplest methods are simply counting the number of events per stage or measuring the dimensions of a cloud of MEQs. A more sophisticated and commonly used approach is to compute a density cloud by defining a 3D counting grid in the region of interest, then centering a counting volume (usually a sphere) on each MEQ hypocenter and incrementing the grid nodes that are contained in the counting volume. The grid may be incremented simply by the number of MEQs, or by a scheme that weights the count by the magnitude of each MEQ. A density value is then defined as the outer boundary of effective stimulation so that the volume within this isosurface is taken as the SRV. Other approaches to determining SRV include fitting a shrink wrap (minimum estimate) or convex hull (maximum estimate) to a microseismic cloud after removing events thought to be outliers. A non-MEQ-derived measure of SRV is determined from cumulative seismic activity volumes by directly mapping the cumulative seismic activity at each voxel within a study volume, then defining a cutoff based on relative activity and requiring a direct connectivity to the wellbore (Section 10.5.4; Fig. 10.20; Lacazette et al., 2014).

It is important to remember that SRV is simply a volume defined by one or more measures of seismic activity—SRV is *not* a direct measure of the volume of the reservoir that is producing. Parts of the reservoir that exhibit microseismic activity may not contribute to production and the natural fracture system in the reservoir may contribute to production even if it did not produce seismic activity. For these reasons, Doe et al. (2013) introduced the term *Tributary Drainage Volume* (TDV). TDV is the volume of the reservoir that contributes production to the wellbore. Like SRV, TDV can be computed for individual stages or for the entire stimulated length of the wellbore. The TDV is estimated by integrating microseismic data with all available geomechanical and geological data to model the productive volume of the reservoir. Note that TDV is not constant over time because the volume of produced reservoir increases with production (Lacazette et al., 2014). Moos et al. (2011) take a conceptually similar approach to estimate frac effectiveness and productive volume albeit with alternative modeling procedures (and without using the term TDV).

10.6.4 Using Passive Seismic Results for Frac, Well-Test, and Reservoir Simulation

The full value of passive seismic data is obtained from learning as much as possible about the effects of hydraulic fracture treatments and reservoir properties. Simple measures of fracture width, height, and volume; identification of major features (e.g., faults) that take frac fluid; and stress state are immediately useful, but do not directly reveal reservoir properties. The last step in the chain of extracting value from passive seismic data is achieved by using the data to model the hydraulic fracture process, well performance, and reservoir properties and performance. Passive seismic data provides important geological and engineering constraints on such models that make them more realistic, and hence more accurate. Better models lead to better fracture treatments, better production, better estimated ultimate recovery (EURs), and better production forecasts and reserve estimates—all of which directly impact the bottom line.

Passive seismic data can be used in several ways to enhance frac, well performance, and reservoir simulations. Modeling efforts are generally focused on producing discrete fracture network (DFN) models of the hydraulic fractures, natural fractures stimulated by hydraulic fracturing, and the natural fracture network in the reservoir. DFN simulations informed by passive seismic results can provide better production forecasts than simple decline-curve analysis (e.g., Williams-Stroud et al., 2013). The most basic approach relies on a correspondence between clouds of MEQs observed during the fracture treatment and simulated clouds of MEQs from DFN hydraulic fracture simulations (e.g., see Dershowitz et al., [2010] for geomechanical details; Rogers et al., 2010). Additional detail is provided by using focal mechanism solutions to interpret the orientations of natural reservoir fractures and calibrate DFN models by using nodal planes as interpretations of reservoir fracture orientations (e.g., Eisner et al., 2010b; Williams-Stroud et al., 2012). Recently, direct fracture images have been used for DFN simulations both as planes that approximate the fracture images (Doe et al., 2013), by using tessellated fracture images to calibrate stochastic DFN models, and by directly incorporating the tessellated images as discrete features in DFN simulations (Lacazette et al., 2014). In contrast to DFN simulation, Moos et al. (2011) use a dual-porosity, dual-permeability simulator in conjunction with downhole hypocenter and natural fracture data collected in the wellbore to simulate stimulation and reservoir response in a shale reservoir.

10.7 CONCLUSIONS

Passive seismic data is currently the best method available for mapping the effects of hydraulic fracture treatments in unconventional oil and gas reservoirs. This contribution has reviewed how downhole and surface passive seismic studies differ, the major products of passive seismic studies, and how the products are generated from downhole and/or surface-derived microseismic data. Standard passive seismic products are MEQ locations, MEQ focal mechanism solutions, images of cumulative seismic activity, and fracture images derived from cumulative seismic activity volumes. At the time of this writing, the latter two products were available only from surface-based studies because uniform coverage is required to compute them. Useful information interpreted from these results include frac height, width, and volume; neostress orientation, stress compartmentalization, and stress changes caused by fracture treatments; how hydraulic fractures interact with the pre-existing natural fracture system; and large-scale features of the natural fracture flow system at reservoir scale. All of these products are unavailable from other methods. The greatest value of passive seismic data is achieved by using the results to build better frac, well-test, and reservoir simulators. Improved simulators lead to better, more cost-effective frac designs and development plans, better production, and better production forecasts and reserve estimates. Passive seismic monitoring is a rapidly advancing field. Engineering methodologies for applying the data are advancing equally rapidly. Passive seismic monitoring has been a growth industry for many years. Recent and continuing technological developments are likely to result in yet greater application of these methods in the future.

ACKNOWLEDGMENTS

This manuscript was improved greatly by the capable reviews of Amanda Klaus and Jan Vermilye.

REFERENCES

Abers GA, Gephart JW. Direct inversion of earthquake first motions for both the stress tensor and focal mechanisms and application to southern California. J Geophys Res Solid Earth 2001;106 (B11):26523–26540.

Allmendinger R, Gephart JW, Marrett A. Notes on fault-slip analysis. Geological Society of America Short Course "Quantitative Interpretation of Joints and Fault"; 1989. 59 p. (available online).

Anderson EM. Dynamics of faulting. Trans Edinburgh Geol Soc 1905;8:387–402.

Anderson EM. *Dynamics of Faulting and Dyke Formation with Application to Britain.* 2nd ed. Edinburgh: Oliver and Boyd; 1951. p 183.

Arnold R, Townend J. A Bayesian approach to estimating tectonic stress from seismological data. Geophys J Int 2007;170:1336–1356. DOI: 10.1111/j.1365-246X.2007.03485.x.

Artman B, Podladtchikov IB, Witten B. Source location using time-reverse imaging. Geophys Prospecting 2010;58:861–873.

Barton N, Stephansson O (eds). Rock Joints. Rotterdam: A. A. Balkema, 1990. p. 35–44.

Barton CA, Zoback MD, Moos D. Fluid flow along potentially active faults in crystalline rock. Geology 1995;23 (8):683–686.

Copeland D.M. and Lacazette, A., 2015 in press, Fracture surface extraction and stress field estimation from three-dimensional microseismic data. URTeC 2155064, 10 p.

Das I, Zoback MD. Long-period, long-duration seismic events during hydraulic stimulation of shale and tight-gas reservoirs—Part 1: Waveform characteristics. Geophysics 2013a;78(6 [Nov-Dec]): KS97–KS108.

Das I, Zoback MD. Long-period long-duration seismic events during hydraulic stimulation of shale and tight-gas reservoirs—Part 2: Location and mechanisms. Geophysics 2013b;78(No. 6 [Nov-Dec]:KS109–KS117.

Dershowitz WS, Cottrell MG, Lim D-H, Doe TW. A discrete fracture network approach for evaluation of hydraulic fracture stimulation of naturally fractured reservoirs. 44th U.S. Rock Mechanics Symposium and 5th U.S.-Canada Rock Mechanics Symposium, 27–30 June, Salt Lake City, Utah; 2010. 8 p.

Doe T, Lacazette A, Dershowitz W, Knitter C. Evaluating the effect of natural fractures on production from hydraulically fractured wells using Discrete Fracture Network models. Paper SPE 168823/URTeC 1581931, Proceedings, First Unconventional Resources Technology Conference. Society of Petroleum Engineers, 12–14 August, Denver, CO, USA; 2013.

Du J, Warpinski NR. Uncertainty in FPSs from moment-tensor inversion. Geophysics 2011;76:WC65–WC75. DOI: 10.1190/geo2011-0024.1.

Duncan PM. Is there a future for passive seismic? First Break 2005;23:111–115.

Duncan PM, Eisner L. Reservoir characterization using surface microseismic monitoring. Geophysics 2010;75(5 [September-October]):A139–A146.

Eisner L, Hulsey BJ, Duncan P, Jurick D, Werner H, Keller W. Comparison of surface and borehole locations of induced seismicity. Geophys Prospecting 2010a. DOI: 101111/j.1365-2478.2010.00867.x.

Eisner L, Williams-Stroud SW, Hill A, Duncan P, Thornton M. Beyond the dots in the box: Microseismicity-constrained fracture models for reservoir simulation. Lead Edge 2010b;29 (3):326–333.

Engelder T. Joints and shear fractures in rock. In: Atkinson BK, editor. *Fracture Mechanics of Rock*. London: Academic Press; 1987. p 27–69.

Engelder T. *Stress Regimes in the Lithosphere*. Princeton: Princeton University Press; 1993. p 457.

Engelder T, Lacazette A. *Natural hydraulic fracturing*. In: Barton N, Stephansson O, editors. Rock Joints: Proceedings of the international symposium on rock joints. Loen, Norway. Rotterdam/Brookfield, VT: A.A. Balkema. June 4–6, 1990.

Engelder T, Whitaker A. Early jointing in coal and black shale: Evidence for an Appalachian-wide stress field as a prelude to the Alleghanian orogeny. Geology 2006; 34(7 [July]):581–84.

Engelder T, Lash GG, Uzcátegui RS. Joint sets that enhance production from Middle and Upper Devonian gas shales of the Appalachian Basin. AAPG Bull 2009;93 (7):857–889.

Evans KF, Engelder T, Plumb RA. Appalachian stress study,1, A detailed description of in situ stress variations in Devonian shales of the Appalachian Plateau. J Geophys Res 1989;94 (B6):7129–7154.

Ferrill DA, Winterle J, Wittmeyer G, Sims D, Colton S, Argmstrong A, Morris AP. Stressed rock strains groundwater at Yucca Mountain, Nevada. GSA Today 1999;9:1–8.

Fish A. Microseismic velocity inversion and event location using reverse time migration [M.Sc. Thesis]. Colorado School of Mines; 2012. 85 p.

Fisher K, Warpinski N. Hydraulic fracture height growth: Real data. Society of Petroleum Engineering, SPE Production & Operations, USA, 145949; 2012. 19 p.

Fletcher RC, Pollard DD. Anticrack model for pressure solution surfaces. Geology 1981;9:419–424.

Frohlich C. Earthquakes with non-double-couple mechanisms. Science 1994;264:804–809.

Frohlich C, Apperson KD. Earthquake focal mechanisms, moment tensors, and the consistency of seismic activity near plate boundaries. Tectonics 1992;11:279–296.

Geiser PA, Vermilye J, Scammell R, Roecker S. Seismic used to directly map reservoir permeability fields. Oil Gas J 2006; published in two parts on 11 and 18 December.

Geiser P, Lacazette A, Vermilye J. Beyond 'dots in a box': an empirical view of reservoir permeability with tomographic fracture imaging. First Break 2012;30 (July):63–69.

Gephart JW, Forsyth DW. An improved method for determining the regional stress tensor using earthquake focal mechanism data: Application to the San Fernando earthquake sequence. J Geophys Res 1984;89 (B11):9305–9320.

Heffer KJ. Geomechanical mechanisms involving faults and fractures for observed correlations between fluctuations in flowrates at wells in North Sea oilfields. In: Spence GH, Redfern J, Aguilera R, Bevan TG, Cosgrove JW, Couples GD, Daniel J-M, editors. *Advances in the Study of Fractured Reservoirs*. Geological Society, vol 374. London: Special Publications, 2012. First published online August 28.

Heffer KJ, Fox RJ, McGill CA. Novel techniques show links between reservoir flow directionality, earth stress, fault structure and geomechanical changes in mature waterfloods. In: Society of Petroleum Engineers Annual Technical Conference & Exhibition, Dallas, TX. October 22–25, 1995. Society of Petroleum Engineers, USA, SPE 30711; 1995.

Hennings P, Allwardt P, Paul P, Zahm C, Reid Jr. R, Alley H, Kirschner R, Lee B, Hough E. Relationship between fractures, fault zones, stress, and reservoir productivity in the Suban gas field, Sumatra, Indonesia. Bulletin of the American Association of Petroleum Geologists 2012;96(4 [April]): 753–772.

Huang JW, Reyes-Montes JM, Zhao XP, Chu FD, Young RP. Quantifying reservoir stimulation using passive traveltime tomography. In: 76th EAGE Conference and Exhibition; 2014. DOI: 10.3997/2214-4609.20140722.

Jaeger JC, Cook GW, Zimmerman RW. *Fundamentals of Rock Mechanics*. 4th ed. Oxford: Blackwell; 2007. p 475.

Janssen C, Wagner FC, Zang A, Dresen G. Fracture process zone in granite: a microstructural analysis. Int J Earth Sci (Geol. Rundschau) 2001;90:46–59.

Jost ML, Herrmann RB. A student's guide to and review of moment tensors. Seismol Res Lett 1989;60(2 [April–June]):37–57.

Julian BR, Miller AD, Foulger GR. Non-double-couple earthquakes 1. Theory. Rev Geophys 1998;36 (4):525–549.

Karimi S, Baig A, Urbancic T. Passive imaging of hydraulic fracture stimulations. 2013 SEG Annual Meeting September 22–27, Houston, TX ; 2013. DOI: http://dx.doi.org/10.1190/segam2013-1342.1.

Kulander BR , Barton CC, Dean SL. The application of fractography to core and outcrop fracture investigations. Springfield: U.S. Department of Energy, Report METC/SP-79-3; 1979, 174 p.

Kulander BR, Dean SL, Ward BJ. Fractured core analysis: Interpretation, logging and use of natural and induced fractures in core. In: AAPG Methods in Exploration Series 8, American Association of Petroleum Geologists, Tulsa; 1990. 88p.

Lacazette, A. Natural fracture nomenclature, Disk 1, 13 pages. In: Thompson LB, editor. *Atlas of Borehole Images*. AAPG Datapages Discovery Series 4. Tulsa: American Association of Petroleum Geologists (2 compact disks); 2000. 13 p.

Lacazette A. Natural fracture nomenclature. In: Thompson LB, editor. *Atlas of Borehole Images*. 2nd ed. AAPG Datapages Discovery Series 4. Tulsa: American Association of Petroleum Geologists (compact disks); 2009.

Lacazette A, Engelder T. Fluid-driven cyclic propagation of a joint in the Ithaca Siltstone, Appalachian Basin, New York. In: Evans B, Wong T-F, editors. *Fault Mechanics and Transport Properties of Rocks; a festschrift in honor of W. F. Brace*. San Diego, CA: Academic Press; 1992. p 297–323.

Lacazette A, Geiser P. Comment on Davies et al., 2012—Hydraulic fractures: How far can they go? Mar Petrol Geol 2013;43: 516–518.

Lacazette A, Vermilye J, Fereja S, Sicking C. Ambient fracture imaging: a new passive seismic method. Unconventional Resources Technology Conference, Denver, CO, SPE 168849/URTeC 1582380; 2013. 10p.

Lacazette A, Dershowitz W, Vermilye J., Geomechanical and flow simulation of hydraulic fractures using high-resolution passive seismic images. SPE/AAPG/SEG Unconventional Resources Technology Conference, 25–27 August, Denver, CO, URTeC 1935902; 2014. 10 p.

Lakings JD, Duncan PM, Neale C, Theiner T. Surface based microseismic monitoring of a hydraulic fracture well stimulation in the Barnett Shale. In: SEG, 76th Annual International Meeting, Expanded Abstracts. 2006. p 605–6098.

La Rocca M, Creager KC, Galluzzo D, Malone S, Vidale JE, Sweet JR, Wech AG. Cascadia tremor located near plate interface constrained by S minus P wave times. Science 2009;323 (5914): 620–623.

Laubach SE, Olson JE, Gross MR. Mechanical and fracture stratigraphy. AAPG Bull 2009;93(11 [Nov]); 1423–1426.

Leary PC. Rock as a critical-point system and the inherent implausibility of reliable earthquake prediction. Geophys J Int 1997; 131:451e466.

Marrett R, Allmendinger RW. Kinematic analysis of fault-slip data. J Struct Geol 1990;12 (8):973–986.

Maxwell SC, Urbancic T, Steinsberger N, Zinno R. Microseismic imaging of fracture complexity in the Barnett Shale. In: Proceedings, Society of Petroleum Engineers Annual Technical Conference, Paper 77440. 2002.

McConaughy DT, Engelder T. Joint interaction with embedded concretions: Joint loading configurations inferred from propagation paths. J Struct Geol 1999;21:–1652.

Miller AD, Foulger GR, Julian BR. Non-double-couple earthquakes 2. Observations. Rev Geophys 1998;36 (4):551–568.

Moos D, Vassilellis G, Cade R, Franquet FA, Lacazette A, Bourtembourg E, Daniel G. Predicting Shale Reservoir Response to Stimulation: the Mallory 145 Multi-well Project. SPE Annual Technical Conference and Exhibition, 30 October–2 November, Denver, CO, SPE 145849; 2011.

Morris AP, Ferrill DA, Henderson DB. Slip tendency and fault reactivation. Geology 1996;24:275–278.

Narr W, Schechter DS, Thompson LB. Naturally Fractured Reservoir Characterization. Soc Petrol Eng 2006:115.

Nelson RA. *Geologic Analysis of Naturally Fractured Reservoirs*. 2nd ed. Boston: Gulf Professional Publishing; 2001. p 352.

Neuhaus CW, Williams-Stroud S, Remington C, Barker WB, Blair K, Neshyba G, McCay T. Integrated microseismic monitoring for field optimization in the Marcellus Shale—a case study. SPE 161965; 2012. 16 p.

Oye V, Ellsworth W. Orientation of three-component geophones in the San Andreas Fault Observatory at depth Pilot Hole, Parkfield, California. Bull Seismol Soc Am 2005;95:751–758. DOI: 10.1785/0120040130.

Pearson C. The relationship between microseismicity and high pore pressures during hydraulic stimulation experiments in low permeability granitic rocks. J Geophys Res 1981; 86:7855–7864.

Pollard DD, Aydin A. Progress in understanding jointing over the past century. Geol Soc Am Bull 1988;100:1181–1204.

Rayleigh CB, Healy JH, Bredehoeff JD. An experiment in earthquake control at Rangely Colorado. Science 1976;191:1230–1237.

Reches Z. Faulting of rocks in three-dimensional strain fields. II. Theoretical analysis. Tectonophysics 1983;95:133–156.

Reches Z, Dieterich JH. Faulting of rocks in three-dimensional strain fields I. Failure of rocks in polyaxial, servo-control experiments. Tectonophysics 1983;95:111–132.

Rogers S, Elmo D, Dunphy R, Bearinger D. Understanding hydraulic fracture geometry and interactions in the Horn River Basin through DFN and numerical modelling. SPE 137488, Canadian Unconventional Resources & International Petroleum Conference held in Calgary, Alberta, Canada, Society of Petroleum Engineering, 19–21 October 2010; 2010. 12p.

Rutledge JT, Phillips WS. Hydraulic stimulation of natural fractures as revealed by induced microearthquakes, Carthage Cotton Valley gas field, East Texas. Geophysics 2003;68:441–452. DOI: 10.1190/1.1567214.

Rutledge JT, Phillips WS, Mayerhofer MJ. Faulting induced by forced fluid injection and fluid flow forced by faulting: An interpretation of hydraulic-fracture microseismicity, Carthage Cotton Valley gas field, Texas. Bull Seismol Soc Am 2004;94:1817–1830.

Seibel M, Baig A, Urbancic T. Single versus multiwell microseismic recording: What effect monitoring configuration has on interpretation? SEG Denver 2010 Annual Meeting; 2010. DOI: http://dx.doi.org/10.1190/segam2013-0654.1.

Shelly DR, Beroza GC, Satoshi I. Non-volcanic tremor and low-frequency earthquake swarms. Nature 2007;446:305–307.

Sibson RH. Fluid involvement in normal faulting. J Geodyn 2000;29:469–499.

Sicking C, Vermilye J, Geiser P, Lacazette A, Thompson L. Permeability field imaging from microseismic. SEG 1383; 2012. 5 p.

Sicking C, Vermilye J, Lacazette A, Yaner A, Klaus A, Bjerke L. Case study comparing microearthquakes, fracture volumes, and seismic attributes. URTeC 1934623;2014. 7 p.

Sondergeld CH, Rai C. Elastic anisotropy of shales. Edge 2011;30:324–331.

Song F, Toksöz N. Full-waveform based complete moment tensor inversion and source parameter estimation from downhole microseismic data for hydrofracture monitoring. Geophysics 2011;76. DOI: 10.1190/GEO2011-0027.1.

Snoke JA. FOCMEC: Focal mechanism determinations. In: Lee WHK, Kanamori H, Jennings PC, Kisslinger C, editors. *International Geophysics*. Volume 81B, Academic Press; 2003. p 1629–1630.

Stein S, Wysession M. *An Introduction to Seismology, Earthquakes, and Earth Structure*. Oxford: Blackwell Publishing; 2002. p 498.

Suckale J. Moderate-to-large seismicity induced by hydrocarbon production. Lead Edge 2010;29:310–314.

Takatoshi I, Kazuo H. Role of stress-controlled flow pathways in HDR geothermal reservoirs. Pure Appl Geophys 2003; 160:1103–1124.

Tamagawa T, Pollard DD. Fractured permeability created by perturbed stress fields around active faults in a fractured basement reservoir. AAPG Bull 2008;92:743–764.

Tary JB, van der Baan M. Potential use of resonance frequencies in microseismic interpretation. Lead Edge 2012;31 (11): 1338–1346.

Tibi R, Estabrook CH, Bock G. The 1996 June 17 Flores Sea and 1994 March 9 Fiji-Tonga deep earthquakes: Sources processes and deep earthquake mechanisms. Geophys J Int 1999; 138:625–642.

Tibi, R, Vermilye J, Lacazette A, Rahak I, Lnh P, Sicking C, Geiser, P. Assessment of hydraulic fracture complexity and stress field variability in an unconventional reservoir from composite moment tensor of double-couple events. 2013 SEG Annual Meeting, 22-27 September, Houston, Texas; 2013. DOI: http://dx.doi.org/10.1190/segam2013-0654.1.

Twiss RJ, Unruh JR. Analysis of fault slip inversions: Do they constrain stress or strain rate? J Geophys Res—Solid Earth 1998;103 (B6):12205–12222.

Urbanic TI, Trifu C-I, Young PR. Microseismicity derived faultplanes and their relationship to focal mechanism, stress inversion, and geologic data. Geophys Res Lett 1993;20 (22):2475–2478.

Vavryčuk V. On the retrieval of moment tensors from borehole data. Geophys Prospecting 2007; 55(3):381–391. DOI: 101111/j.1365-2478.2007.00624.x

Vermilye JM, Scholz CH. The process zone: a microstructural view of fault growth. J Geophys Res—Solid Earth 1998; 103:12223–12237.

Williams-Stroud S, Ozgen C, Billingsley RL. Microseismicity-constrained discrete fracture network models for stimulated reservoir simulation. Geophysics 2013;78(1 [January-February]): B37–B47.

Williams-Stroud SC, Barker WB, Smith KL. Induced hydraulic fractures or reactivated natural fractures? Modeling the response of natural fracture networks to stimulation treatments. Am Rock Mech Assoc 12-667; 2012a. 6 p.

Williams-Stroud S, Neuhaus CW, Telker C, Remington C, Barker W, Neshyba G, Blair G. Temporal evolution of stress states from hydraulic fracturing source mechanisms in the Marcellus Shale. SPE Canadian Unconventional Resources Conference, 30 October–1 November, Calgary, Alberta, Canada, SPE 162786; 2012b. 6 p.

World Stress Map: www.world-stress-map.org.

Xuan R, Sava P. Probabilistic microearthquake location for reservoir monitoring. Geophysics 2010;75 (3):MA9–MA26.

Zoback MD. *Reservoir Geomechanics*. Cambridge: Cambridge University Press; 2010. p 448.

11

GAS TRANSPORT PROCESSES IN SHALE

FARZAM JAVADPOUR[1] AND AMIN ETTEHADTAVAKKOL[2]

[1] Bureau of Economic Geology, Jackson School of Geosciences, The University of Texas at Austin, Austin, TX, USA
[2] Bob L. Herd Department of Petroleum Engineering, Texas Tech University, Lubbock, TX, USA

11.1 INTRODUCTION

Nanoscience is science at tiny scales. For many years, high-tech industries such as microelectronics and bioengineering systems have benefitted and continue to benefit from new advances in nanoscience. The governing physics at the nanoscale are different from those observed at the large or continuum scale. The differences bring new characteristics and, hence, new applications. Recently, nanoscale characteristics of natural systems—in particular, shale gas—have ushered in a new era of nanoscience fossil energy from geological systems. Pores in shale gas strata are at the nanometer scale, and the physics of fluid transport in these pores are different from those described by well-known formulations such as the Darcy equation. The petroleum industry has extensive experience in characterizing pores and modeling fluid flow in the pore networks of hydrocarbon-bearing reservoirs. However, for pores at the nanoscale in shale systems, the characterizing methods need modification and, in many cases, entirely new methods of characterizations are needed. Characterizing pore networks in these systems and developing new formulations for fluid flow in such systems are of great importance and interest.

In addition to having much smaller pore size than conventional gas reservoirs, shale gas also differs in that the source of the initial gas in place (IGIP) in shale is more complicated than the gas source in conventional gas reservoirs. The controlling mechanisms of gas storage and flow in shale gas sediments are a combination of different processes. Gas is stored as compressed gas in pores (free gas), as gas adsorbed to the pore walls, and as dissolved gas in amorphous organic materials (kerogen) and clays (Javadpour et al., 2007). In shale gas reservoirs, gas flows through a network of pores with different diameters, ranging from nanometers ($1\,nm = 10^{-9}\,m$) to micrometers ($1\,\mu m = 10^{-6}\,m$) (Javadpour et al., 2012; Loucks et al., 2012; Milliken et al., 2012). Figure 11.1 shows exemplary pore images of a sandstone sample (left) and a shale sample (right) for comparison of pore distribution in conventional and unconventional (shale) reservoirs. The average size of the pores is much smaller and the number of pores is much higher in the shale sample. In shale gas systems, nanopores play two important roles. First, for the same pore volume, the exposed surface area in nanopores is larger than in micropores. This disparity results because exposed surface area is proportional to $4/d$ (the inner surface of a tube divided by the tube volume, where d is the pore diameter). This large exposed area permits the desorption of large volumes of gas from the surface of the nanopores in kerogen. Diffusion from the kerogen bulk to the nanopores' inner surfaces may also take place. Consequently, high mass transfer of gas molecules occurs inside the bulk kerogen. Second, the applying physics of gas flow in nanopores does not conform to Darcy flow (Akkutlu and Fathi, 2012; Civan, 2010; Darabi et al., 2012; Javadpour, 2009 Singh et al., 2014).

Figure 11.2a–e illustrates the sequence of gas production at different length scales. Gas production from a new hydraulically fractured wellbore (i) takes place through first the larger pores, induced microfractures, and larger fracture conduits (ii); and then through the smaller pores (iii). During reservoir depletion, the thermodynamic equilibrium between kerogen/clays and the gas phase in the pore spaces changes.

Fundamentals of Gas Shale Reservoirs, First Edition. Edited by Reza Rezaee.
© 2015 John Wiley & Sons, Inc. Published 2015 by John Wiley & Sons, Inc.

FIGURE 11.1 Left: Light image of sandstone thin section. Interstices filled with blue resin. Right: AFM topography image revealing nanopores (dark areas). Note scale difference.

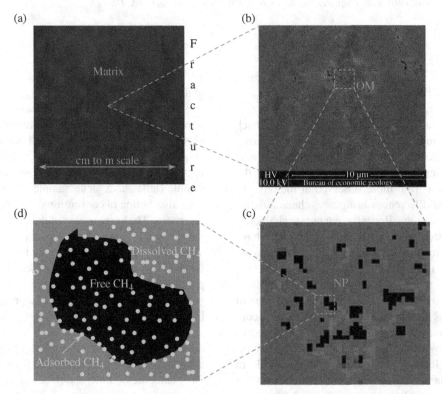

FIGURE 11.2 Multiscale gas transport in shale gas reservoirs: (a) gas transport from shale matrix into fractures (natural or induced); (b) scanning electron microscope (SEM) scale in which organic material (OM) can be identified; (c) higher resolution SEM image to identify nanopore (NP) in organic material; (d) free gas molecules in NP, sorbed gas molecules on NP walls, and dissolved gas molecules in kerogen bulk are shown.

Hence, gas desorbs from the surface of the kerogen/clays (iv). This nonequilibrium process and concentration difference between the kerogen bulk and pore inner surfaces further drives the gas molecules to diffuse from the bulk of the kerogen to the surface of the kerogen (v).

In the shale permeability measurement experiments, accurate measurement of the flow properties in the low-permeability core samples is challenging because (i) the gas-flow rates are extremely small even though the applied pressure difference across the core is large; and (ii) shale permeability is pressure-dependent and therefore multiple experiments should be performed at different mean pressure levels to estimate any permeability model parameters (Blanchard et al., 2007; Rushing et al., 2004). Therefore, these experiments should be efficient in terms of time and cost requirements.

Permeability measurement methods are divided into two categories based on the choice of prevailing gas-flow regime for pressure-flow rate analysis: steady state and unsteady state. Steady-state methods require sufficient time to achieve steady flow across the core sample. The unsteady methods require a more complicated postprocessing because an accurate solution for the transient flow equation is required, and a reliable method for the estimation of the permeability model parameters should be developed.

Permeability measurement experiments typically involve a chronological recording pressure and/or flow rate response under or after the application of a pressure perturbation signal across a core sample. Assuming a one-dimensional flow with constant pressures applied to the upstream and downstream faces of a core sample, the required time to achieve the steady flow condition is proportional to the squared sample length in the flow direction and inversely proportional to the intrinsic permeability. Considering that the shale permeability typically falls in the micro-Darcy down to nano-Darcy range, a relatively long time is required to reach a steady-state flow regime. Therefore, the steady-state flow regime experiments may not be good candidates for shale permeability measurement.

Unsteady-state permeability measurement methods require a relatively short time compared to the steady-state methods; also, they record and monitor the pressure instead of the flow rate, which is less susceptible to the potential measurement errors.

The choice of permeability model is usually independent from the choice of experiment. However, the underlying mathematical model for permeability estimation should be consistent with the prevailing experiment conditions. Unsteady-state methods normally use a larger subset of the recorded pressure data, which often improves the robustness of estimation and tuning of the permeability model. This is especially important for the tuning of multi-parameter and pressure-dependent permeability models.

This chapter presents three common unsteady-state methods for the permeability measurement of ultra-low-permeability shale rocks: the pulse-decay method, the crushed sample test, and the canister desorption test. The pulse-decay method presents more details on the general form of the underlying equations, the fundamental definitions of the physical parameters. The chapter also describes two general workflows for the determination of pressure-dependent permeability models: an analytical and a numerical method. These methods can be applied to the obtained data from the pulse-decay experiment, crushed sample test, or the canister test to estimate the permeability parameters.

11.2 DETECTION OF NANOPORES IN SHALE SAMPLES

There are at least four methods to detect nanopores in shale. Two of these methods are indirect: high-pressure mercury-injection (Javadpour et al., 2007) and nitrogen-adsorption tests (Barrett et al., 1951); the other two are direct methods:

scanning electron microscopy (SEM) (Loucks et al., 2012; Milliken et al., 2012) and atomic force microscopy (AFM) (Javadpour et al., 2012). Mercury-injection tests can reveal pore-throat size distribution in a shale sample, but extremely high pressure (>60,000 psi) are required. Another disadvantage of mercury-injection testing is that, in this process, the sample is destroyed and cannot be used for other tests. Nitrogen-adsorption testing can detect nanopores (Groen et al., 2003). The method is based on condensation of the nitrogen molecules in pores and a material balance of adsorbed nitrogen and nitrogen pressure in the system (Barrett et al., 1951). Pore size is determined by measuring the volume of the nitrogen molecules condensed (adsorbed) versus nitrogen pressure. Other gases such as carbon dioxide (CO_2) can also be used instead of nitrogen gas to reveal sub-nanopores.

In SEM, a high-energy beam of electrons is emitted to the surface and the reflected electrons reveal surface features, such as nanopores. The limit of nanopore detection is about 10 nm (Loucks et al., 2012). SEM can also identify organic material from other minerals on the surface (Fig. 11.3). In most cases, before scanning, the surface of the sample would have previously been well polished using the ion-mill technique (Loucks et al., 2012). Another direct method to detect tiny features is the scanning tunneling microscope (STM), the first member of the atomic force microscope (AFM) family, the introduction of which by Binnig et al. (1982) earned those researchers the 1986 Nobel Prize in physics. In the STM process, a voltage-biased metal tip is brought close to the surface to be scanned, creating a tunneling current inversely proportional to the gap width. Later, Binnig et al. (1986) proposed mounting the tip on a

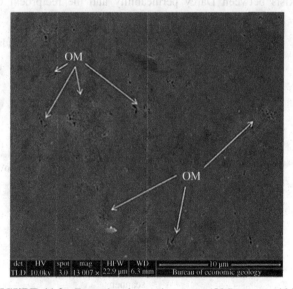

FIGURE 11.3 Example of organic-matter (OM) pores within mudrocks. There are many tiny pores in organic material parts of the sample.

cantilever spring and monitoring the cantilever deflection corresponding to tip-surface forces, inventing AFM in the process. Because the force between tip and surface is a function of gap width, the AFM feedback system can maintain a steady gap between tip and surface by vertical tip displacement compensating for cantilever deflection. This technique allows topographic imaging of any material, either conductive or nonconductive.

The AFM scanner-head system comprises a tip attached to the end of a cantilever, a chip holder, a laser source, a mirror, a quadrant photodiode, and the controlling system (Fig. 11.4). Applications of AFM measurements in shale-reservoir studies are both intriguing and promising and include detection of nanopores in shale samples, identification of different types of organic and inorganic grains in shale samples, and evaluation of elastic properties at small scale (Javadpour et al., 2012). Using sharp tips a few nanometers in diameter, AFM can obtain nanoscale topography of various objects or surfaces. We used topographic images to study nanopores and grain boundaries in shale. An exemplary surface topographic image of a shale gas sample prepared by ion milling is presented in Figure 11.5.

11.3 GAS FLOW IN MICROPORES AND NANOPORES

The Darcy equation (1856) has been used for more than 150 years to linearly relate fluid-flow rate and pressure gradient across a porous system. The linearity of the Darcy equation makes it easy and practical to use in reservoir-engineering analysis and numerical reservoir simulations. Klinkenberg (1941) showed experimentally that a linear relationship exists between Darcy permeability and the reciprocal of mean pressure in the system, that is, between gas-flux reduction and mean pressure increase.

$$k(p_m) = k_D \left(1 + \frac{b}{p_m}\right), \qquad (11.1)$$

where $k(p_m)$ is gas permeability at mean pressure (p_m). The empirical parameters b and k_D are the slope and intercept of the fitted line through the $k(p_m)$ versus $1/p_m$ data. The intercept k_D is the intrinsic permeability or liquid permeability of the sample, that is, $1/p_m \to 0$ as $p_m \to \infty$. The Klinkenberg effect has been used to model gas flow in conventional gas reservoirs (with pores in the range of 10s–100s µm) and

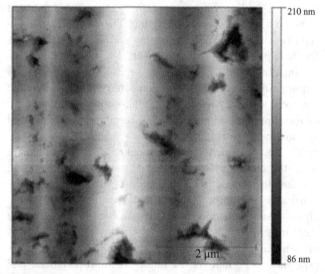

FIGURE 11.5 Atomic force microscope (AFM) topography image of shale sample. Darker areas reveal nanopores.

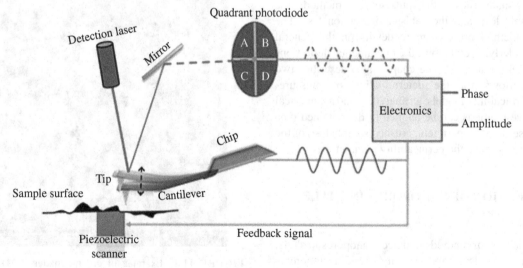

FIGURE 11.4 AFM scanner-head system composed of tip attached to end of cantilever, chip holder, laser source, mirror, quadrant photodiode, and controlling system. Piezoelectric scanner infinitesimally moves the sample up and down with high accuracy.

recently for tight gas systems (with pores of 1–10 µm in size). Brown et al. (1946) added to our understanding of these processes and articulated the concept of slip flow, which provided an explanation for the observed relationship between gas-flow rate and mean pressure.

As explained earlier, the pores in producing shale gas reservoirs are in the range of 1–100s nm; the gas molecules contained in the pores are of comparable size (~0.5 nm), and under certain pressure and temperature conditions the distance between gas molecules (mean free path) exceeds the size of the pores. In such conditions, the gas molecules might move singly through the pores and the concept of continuum and bulk flow may not be applicable. Knudsen number (K_n) is the ratio of mean free path (λ) to pore diameter (d), and can be used to identify different flow regimes.

$$K_n = \frac{\lambda}{d}, \quad (11.2)$$

where

$$\lambda = \frac{k_B T}{\sqrt{2}\pi\delta^2 p}, \quad (11.3)$$

in which k_B is the Boltzmann constant (1.3805×10^{-23} J/K), T is temperature (K), p is pressure (pa), and δ is the collision diameter of the gas molecule (m). Table 11.1 presents flow regimes corresponding to Knudsen number ranges. Continuum no-slip flow or Darcy equation is valid for $K_n < 10^{-3}$. Continuum flow with slip correction (Klinkenberg) is valid for $K_n < 10^{-1}$, which covers most conventional gas reservoirs and many tight gas reservoir conditions as well. However, because of the existence of nanopores in a shale system, K_n could be larger than 0.1, and hence new forms of gas-flow equations are needed. Figure 11.6 presents Knudsen number as a function of pore size and mean reservoir pressure, and shows the validity of different flow equations.

TABLE 11.1 Validity of different flow equations as functions of Knudsen number

	Knudsen number (K_n)			
	Lower bound	Upper bound	Flow regime	
Shale gas reservoirs	0	10^{-3}	Continuum/Darcy flow (no-slip flow)	Navier–Stokes equation
	10^{-3}	10^{-2}	Slip flow (Klinkenberg model)	
	10^{-2}	10^{-1}		
	10^{-1}	10^{0}	Transition flow	
	10^{0}	10^{1}		
	10^{1}	∞	Free-molecule flow	

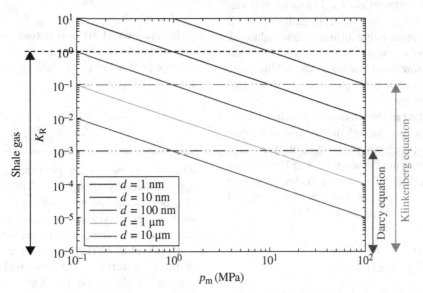

FIGURE 11.6 Knudsen number (K_n) as function of mean reservoir pressure for porous system with different mean pore size. Validity of different gas-flow equations and condition corresponding to shale gas reservoirs are marked.

As shown in Figure 11.6, researchers need an equation that describes flow beyond the limit of slip flow. Molecular dynamic (MD) models (Karniadakis et al., 2005) are powerful models that are capable of modeling interaction of individual gas molecules and molecules at the pore walls. MD models are valid for any range of K_n, but they are computationally inefficient and therefore of limited use for describing large units such as shale systems. Javadpour (2009) proposed a model that includes the two major mechanisms of Knudsen diffusion and slip-flow contribution to the gas flow in a single, straight, cylindrical nanotube. Javadpour also asserted that these two processes exist at any K_n, but their individual contributions to total flux varies.

$$J = \left\{ \frac{2rM}{3\times10^3 RT}\left(\frac{8RT}{\pi M}\right)^{0.5} + F\frac{r^2 \rho_{avg}}{8\mu} \right\}\frac{p_2 - p_1}{L} \quad (11.4)$$

The first and second terms in the right-hand-side bracket in Equation 11.4 refer to Knudsen diffusion and slip flow, respectively. The term F is the slip coefficient and is defined as:

$$F = 1 + \left(\frac{8\pi RT}{M}\right)^{0.5}\frac{\mu}{rp_{avg}}\left(\frac{2}{\alpha} - 1\right) \quad (11.5)$$

where M is molar mass, r is the pore radius, R is the universal gas constant, T is the temperature, ρ_{avg} is the average gas density, μ is the gas viscosity, and p_1 and p_2 are the upstream and downstream pressures, respectively. p_{avg} is the average pressure of the system $p_{avg} = (p_1 + p_2)/2$. The term α is the tangential momentum accommodation coefficient or, simply, the fraction of gas molecules reflected diffusely from the pore wall relative to specular reflection. The value of α varies theoretically in a range from 0 (representing specular accommodation) to 1 (representing diffuse accommodation), depending on wall-surface smoothness, gas type, temperature, and pressure (Agrawal and Prabhu, 2008; Arkilic et al., 2001). Experimental measurements are needed to determine α for specific shale systems.

Javadpour (2009) showed that this model matches data by Roy et al. (2003), from flow through an Anodisc membrane (Whatman Ltd.) with pore sizes of 200 nm, at an average error of 4.5%. By comparing Equation 11.4 to Darcy's law for a single nanotube (Hagen–Poiseuille equation), apparent permeability (k_{app}) for a porous medium containing of straight cylindrical nanotubes can be defined as:

$$k_{app} = \frac{2r\mu}{3\times10^3 p_{avg}}\left(\frac{8RT}{\pi M}\right)^{0.5}$$
$$+ \frac{r^2}{8}\left\{1 + \left(\frac{8\pi RT}{M}\right)^{0.5}\left(\frac{2}{\alpha} - 1\right)\frac{\mu}{rp_{avg}}\right\} \quad (11.6)$$

Equation 11.6 provides an apparent Darcy permeability relationship written in the Klinkenberg form as

$$k_{app} = k_D\left(1 + \frac{b}{p_{avg}}\right) \quad (11.7)$$

$$b = \frac{16\mu}{3\times10^3 r}\left(\frac{8RT}{\pi M}\right)^{0.5} + \left(\frac{8\pi RT}{M}\right)^{0.5}\left(\frac{2}{\alpha} - 1\right)\frac{\mu}{r} \quad (11.8)$$

where k_D is Darcy permeability.

Azom and Javadpour (2012) showed how Equation 11.6 can be corrected for a real gas flowing in a porous medium. The final equation still has the form of Equation 11.7, but with b given a

$$b = \frac{16\mu c_g p_{avg}}{3\times10^3 r}\left(\frac{8zRT}{\pi M}\right)^{0.5} + \left(\frac{8\pi RT}{M}\right)^{0.5}\left(\frac{2}{\alpha} - 1\right)\frac{\mu}{r}, \quad (11.9)$$

where c_g is gas compressibility and z is compressibility factor. Notice that as the real gas becomes ideal, (Eq. 11.9) becomes (Eq. 11.8), because, for an ideal gas, the compressibility $c_g = 1/p_{avg}$ and the compressibility factor $z = 1$.

Darabi et al. (2012) later applied several modifications to adapt the model developed by Javadpour (2009) from being applicable to a single, straight, cylindrical nanotube to being applicable to ultra-tight, natural porous media characterized by a network of inter-connected tortuous micropores and nanopores.

$$k_{app} = \frac{\mu M \phi}{RT\tau\rho_{avg}}(\delta)^{D_f - 2} D_k + k_D\left(1 + \frac{b}{p_{avg}}\right). \quad (11.10)$$

In Equation 11.10, φ is porosity, τ is tortuosity, and δ is normalized molecular radius size (r_m) with respect to local average pore radius (r_{avg}), yielding $\delta = r_m/r_{avg}$. Knudsen diffusion (D_k) is defined as:

$$D_k = \frac{2r_{avg}}{3}\left(\frac{8RT}{\pi M}\right)^{0.5}, \quad (11.11)$$

where r_{avg} is the average pore radius of the porous system, approximated by $r_{avg} = (8k_D)^{0.5}$. The average pore radius can also be determined by laboratory experiments employing such as processes as mercury injection and nitrogen adsorption tests and pore imaging using SEM and AFM.

Darabi et al. (2012) also included the fractal dimension of the pore surface (D_f) to consider the effect of pore-surface roughness on the Knudsen diffusion coefficient (Coppens, 1999; Coppens and Dammers, 2006). Surface roughness is one example of local heterogeneity. Increasing surface roughness leads to an increase in residence time of molecules in porous media and a decrease in Knudsen diffusivity. D_f is a quantitative measure of surface roughness that varies between 2 and 3, representing a smooth surface and a space-filling surface, respectively (Coppens and Dammers, 2006).

Civan (2010) permeability model is based on the Beskok and Karniadakis (1999) approach. The model assumes that permeability is a function of the intrinsic permeability, the Knudsen number (K_n), the rarefication coefficient α_2, and the slip coefficient b,

$$k = k_\infty \left(1 + \alpha_2 K_n\right)\left(1 + \frac{4K_n}{1 - bK_n}\right). \quad (11.12)$$

The dimensionless rarefication coefficient α_2 is given by,

$$\alpha_2 = \alpha_0 \left(\frac{K_n^B}{A + K_n^B}\right). \quad (11.13)$$

The lower limit of α_2 ($\alpha_2 = 0$) corresponds to the slip flow regime and the upper limit α_0 corresponds to the asymptotic limit of α_2 when $K_n \to \infty$, which corresponds to the free molecular flow. A and B serve as the fitting parameters that may be appropriately adjusted based on the dominant flow regime in the shale porous media. Civan (2010) reports the adjusted parameter values, $A = 0.178$, $B = 0.4348$, and $\alpha_0 = 0.1358$ for modeling gas flow in a tight sand example. Civan (2010) assumes $b = -1$ based on the Beskok and Karniadakis (1999) estimate and subsequently estimates the Knudsen number as (Jones and Owens, 1980),

$$K_n = 12.639 k_\infty^{-1/3}. \quad (11.14)$$

With these assumptions, the only unknown parameter remaining in the Civan (2010) model is k_∞, which can be determined from a permeability measurement experiment (e.g., the pulse-decay experiment).

For small Knudsen numbers, that is, $K_n \ll 1$, Civan (2010) estimates the dynamic slippage coefficient b_k as a function of gas viscosity, based on the Florence et al. (2007) study,

$$b_k = \frac{2790\,\mu}{\sqrt{M}} \left(\frac{K_\infty}{\varphi}\right)^{-0.5}. \quad (11.15)$$

Figure 11.7 compares the production performance of an imaginary homogeneous shale gas reservoir when modeled with different gas-flow models. This figure shows the contribution of Knudsen diffusion and underestimation of the Darcy and Klinkenberg models. The effect of Knudsen and slip flow is more pronounced at lower reservoir pressures.

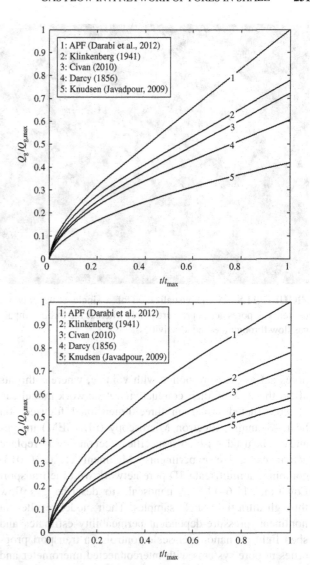

FIGURE 11.7 Comparison of different gas models to predict cumulative gas production from an imaginary homogeneous shale gas reservoir. Input data to the models; Porosity (φ) = 0.05, tortuosity (T) = 5, average pore radius (r_{avg}) = 5 nm, gas: methane (top) initial pressure = 5000 psi, well-flowing pressure = 1000 psi, and (bottom) initial pressure = 1000 psi, well-flowing pressure = 100 psi. The effect of Knudsen and slip on gas flow is more pronounced at lower pressures (bottom).

11.4 GAS FLOW IN A NETWORK OF PORES IN SHALE

Due to high heterogeneity in fluid physics and pore structure in shale, a pore level approach in researching shale petrophysics is imperative. A pore network model serves in capturing the topology of the pore space in a computationally cost-effective manner. A pore network model consists of pores (or nodes) that are connected to each other via throats (or links) (Fig. 11.8). When simulating fluid flow,

FIGURE 11.8 A 3D visualization of a single scale network model. The pores act as pressure points whereas the throats impact gas flow through their conductivities.

pores act as pressure points with volume, whereas throats affect the flow by their conductivity. A network model can be constructed based on three-dimensional focused ion beam–Scanning Electron Microscopy (FIB–SEM) images and can be used to isolate a specific effect on flow or replace actual expensive experiments. Mehmani et al. (2013) combined a multiscale 3D pore network with the transport Equations 11.6–11.8 at nanoscale to describe gas flow through ultra-tight shale samples. Their model resulted in nonlinear, pressure-dependent permeability estimates and shed light on nanometer-scale controls on transport properties in pore systems with interconnected micrometer and nanometer pores.

In order to compute gas flow, one can picture the network model contained in a cubic box. A pressure difference is imposed on the inlet and outlet sides. Conservation of mass is imposed on each pore and given the nature of the constitutive equation (Javadpour, 2009) a nonlinear system of equations is derived. The authors solved the system of equations by using the Newton–Raphson iterative method. One can linearize the equations by assuming an averaged constant density and viscosity, but the larger the pressure difference (larger systems), the more erroneous such a scheme will become.

In order to analyze the interactive effects of multiscale throats on gas flow, Mehmani et al., 2013 proposed 3 two-scale network types as illustrated in Figure 11.9. The authors included nanosize pores by targeting a certain fraction of the pores and shrunk the targeted pores and their adjacent throats. The porosity of the networks was kept constant by enlarging a sufficient number of pores. A description of the network model types are included below:

- Constant cross-section model (CCM) (Fig. 11.9a): The radii of the selected pore-volume compensating pores are enlarged to make up for the loss of volume. The cross-sectional area of the attached throats, however, are kept constant. This manipulation does not affect fluid flow; however, it becomes important in the presence of desorption.
- Enlarged cross-section model (ECM) (Fig. 11.9b): In addition to the sizes of the targeted pores, the sizes of the adjacent throats were enlarged as well. This method allowed for the overlap of the shrunk and enlarged targeted pores and throats. Little difference was observed between this model and CCM.
- Shrunk length model (SLM) (Fig. 11.9c): In this model, the throat lengths are shrunk as well. This procedure deforms the bulk geometry and further amendments are needed to compute gas permeability.

The three network types can be used in constructing a representative network model from FIB–SEM images. Figure 11.10 shows the simulation results of an ECM model with various nanopore fractions to study the effect of multiple-length scales on the apparent permeability. The values are normalized with respect to the maximum Darcy permeability K_D^{max}. Unlike the Darcy permeability, which has a logarithmically linear relationship with the average throat size, there is no single slope that can describe $\log(K_{app})$ versus $\log(r_{avg})$. Furthermore, the spread in a 50% fraction of nanopores could be a key in shedding light on the topology of shale pores.

11.5 GAS SORPTION IN SHALE

Gas desorption from the surface of organic material is a source of gas in shale gas systems (Etminan et al., 2014; Javadpour et al., 2007). The contribution of gas desorption to total gas flux depends on total organic content (TOC), organic type (rank and maturity), and temperature in a shale reservoir (Zhang et al., 2012). The Langmuir isotherm shows the effectiveness of the gas desorption capacity of a shale sample. The simplest theoretical model of monolayer adsorption is the Langmuir model (Ruthven, 1984). Langmuir adsorption isotherm model relates the mass of the sorbed gas to pressure as:

$$\Gamma = \Gamma_{max} \frac{\alpha(T)P}{1+\alpha(T)P}, \qquad (11.16)$$

where Γ is the mass of adsorbed gas (g/unit mass of sample), P is the equilibrium pressure of the gas in the system (atm), T is the absolute temperature (K), Γ_{max} is the maximum

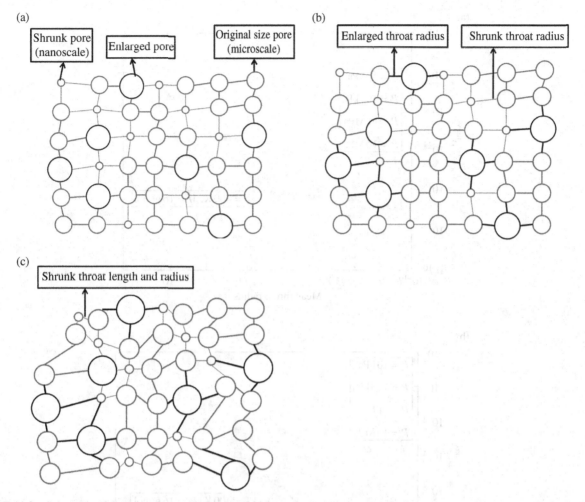

FIGURE 11.9 Schematic of the network models with bimodal pore size distributions. (a) Constant cross-section model (CCM), (b) enlarged cross-section model (ECM), and (c) shrunk length model (SLM).

amount of the sorbed gas at a certain temperature and infinite pressure (per unit mass of sample), and α is the Langmuir parameter (atm^{-1}). Langmuir pressure is the inverse of Langmuir parameter (α). Figure 11.11 is an exemplary Langmuir isotherm of a shale sample.

It is important to distinguish between a source term and a flow term in shale gas analysis. Sorption in a shale gas system is a material balance term, for example, a source term, and does not appear in momentum balance or flow term. Therefore, sorption *per se* does not affect permeability. However, there are two processes involved in sorption that could change permeability. The first is the change in pore size as a result of the release of gas molecules from the inner pore surfaces; the second is the change in pore pressure as a result of sorption. We showed earlier that permeability in a shale system is pressure-dependent. Shabro et al. (2011) and Swami et al. (2013) developed numerical models to link sorption and gas flow in a shale system.

11.6 DIFFUSION IN BULK KEROGEN

As mentioned earlier in this chapter, gas storage in gas shale exists in three major forms: stored as compressed gas in the pore network, sorbed on the surface of organic material and possibly on clay minerals, and dissolved in liquid hydrocarbon and brine (interstitial and clay-bound), and kerogen (Javadpour et al., 2007). Many research studies have addressed the first two storage processes (Chareonsuppanimit et al., 2012; Civan et al., 2012; Darabi et al., 2012; Javadpour, 2009; Zhang et al., 2012), but only limited research has been conducted on the contribution of gas dissolved in organic material in the total gas production from shale reservoirs (Etminan et al., 2014; Moghanloo et al., 2013).

Figure 11.2d shows the three storage processes of gas-in-place in shale gas reservoirs. The compressed gas exists in the micro- and nano-scale pores. Some of the gas molecules are adsorbed on the surface of kerogen and, eventually, some of the gas molecules are dissolved into the kerogen body and

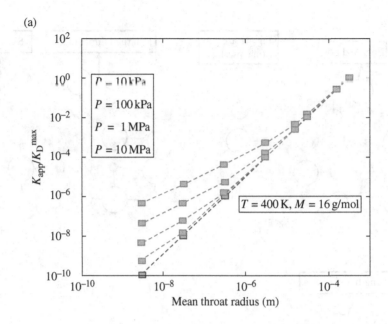

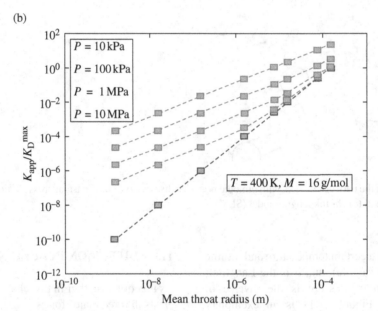

FIGURE 11.10 Effect of throat sizes and pressure on normalized gas permeability K_{app}/K_D^{max} in; (a) a single scale network with a connectivity of 4 and (b) a dual-length scale network with a connectivity of 4 (ECM, $f=0.5$).

become a part of the kerogen in the form of a single phase. The controlling mass transport process of the dissolved gas is molecular diffusion. Depending on the geochemistry of the organic materials (thermal maturity, organic source, etc.), different gas solubility could be expected. The contribution of dissolved gas to gas-in-place and ultimate recovery of a shale reservoir could be significant; hence, evaluation of the gas-diffusion process into kerogen becomes important. In addition to the total contribution of each process, the onset time of each process during production is critical. Once production starts from a reservoir, the compressed gas in interstitial pore spaces expands first; then, adsorbed gas on the surfaces of the pores in kerogen desorbs to the pore network. At this stage, the concentration of gas molecules on the pore inner surface decreases and creates a concentration gradient in the bulk of the kerogen, thereby triggering gas diffusion (Etminan et al., 2014; Javadpour et al., 2007).

Etminan et al. (2014) developed batch pressure decay (BPD) technique to accurately measure the contributions of different storage processes to the total gas-in-place. With the same BPD test, they also measure gas molecular diffusion in kerogen. The method is robust and accurate and can save

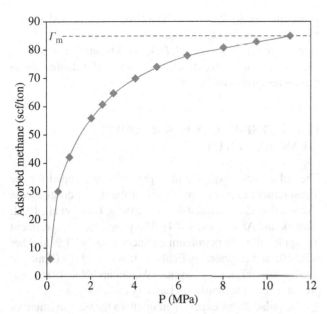

FIGURE 11.11 Methane Langmuir adsorption isotherm of shale sample at 110°C for a shale sample. Maximum adsorption capacity (Γ_{max}) is marked.

time and expense for laboratory-based shale gas characterization and evaluation.

In the BPD method, a thinly-cut plug of a shale sample (e.g., 1 mm thick) was placed in a high-pressure cell. After proper degassing, the cell would be pressurized to a certain pressure usually lower than the critical pressure of the gas. Once gas supply to the cell had ceased, the cell pressure began to decay due to different transport processes, that is, gas expansion in pores, sorption, and diffusion. A record of temporal pressure decay for a few days (or weeks) with a precise pressure transducer reveals the contribution of each transport process. The onset of each transport process changes the slope of decay curve.

11.7 MEASUREMENT OF GAS MOLECULAR DIFFUSION INTO KEROGEN

Etminan et al. (2014) used the Fickian diffusion model in their pulse decay tests to measure molecular diffusion coefficient (D) in kerogen. Assuming that kerogen does not have any volatile components to diffuse into the gas, the unidirectional diffusion of gas molecules into the kerogen body can be modeled by Fick's second law to determine temporal and spatial gas concentration distribution.

$$\frac{\partial^2 C_g}{\partial z^2} = \frac{1}{D}\frac{\partial C_g}{\partial t}, \quad (11.17)$$

where C_g is gas concentration in kerogen bulk, z denotes spatial domain (Fig. 11.12), t is time, and D is molecular

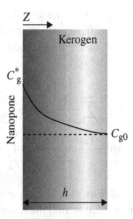

FIGURE 11.12 The process of gas diffusion into kerogen bulk during pressure-decay test.

diffusion in amorphous kerogen. Other assumptions include no volume change in kerogen due to dissolution, constant kerogen density, constant diffusion coefficient, and no chemical reaction between the diffusing gas and the kerogen material.

Initial condition is defined as follows:

$$C_g(z, t=0) = C_{g0}. \quad (11.18)$$

It is easier to set the zero of spatial coordinate at the gas–kerogen interface ($z=0$, Fig. 11.12). At $z=0$, the gas molecules are in adsorbed form. The other boundary at $z=h$ is a no-flow boundary,

$$\left.\frac{\partial C_g}{\partial z}\right|_{z=h} = 0. \quad (11.19)$$

The physical meaning of h is the distance from the surface of a pore into the bulk of the kerogen that feed gas molecules through the diffusion process. An average value of h can be determined based on pore size distribution in kerogen and the average size of the kerogen patches in a sample. Pore size distribution can be obtained from capillary pressure and nitrogen adsorption tests and the distribution of kerogen patches can be determined from analysis of SEM and AFM images.

The interface boundary condition is referred to as the concentration of gas molecules at the interface. This concentration is directly related to the mass of adsorbed gas to the surface of the pores in kerogen. Experimental analysis showed that a constant saturation concentration, C_g^*, equivalent to the equilibrium pressure, can be a valid assumption at the interface boundary condition (Etminan et al., 2014).

$$C_g(z=0, t) = C_g^*(P_{eq}). \quad (11.20)$$

The value of C_g^* could also be estimated from the maximum amount of diffused gas into the kerogen divided by the total volume of TOC in the sample. Equation 11.21 is the solution of partial differential equation (Eq. 11.17) with the defined boundary and initial conditions in Equations 11.18–11.20 (Crank, 1979).

$$C_g(z,t) = C_g^* \left[1 - \frac{4}{\pi} \sum_{n=1}^{\infty} \frac{1}{2n-1} \sin\left(\frac{(2n-1)\pi}{2h} z\right) \exp\left(\frac{-(2n-1)^2 \pi^2 D}{4h^2} t\right) \right]. \quad (11.21)$$

An integration of Equation 11.21 over the volume of kerogen volume is needed to find the mass of gas dissolved.

$$m_{g\text{-Diff}}(t) = \int_{z=0}^{z=h} C_g(z,t) A \, dz. \quad (11.22)$$

In Equation 11.22, A is the area open to molecular diffusion, and h is defined as the depth of diffusion. Evaluation of the area open to diffusion (A) and the average depth of diffusion (h) are important in interpretation of the results. Equation 11.23 is what is obtained by integration over the spatial domain in Equation 11.21.

$$m_{g\text{-Diff}}(t) = \frac{8 A C_g^* h}{\pi^2} \sum_{n=1}^{\infty} \frac{1}{(2n-1)^2} \left[1 - \exp\left(\frac{-(2n-1)^2 \pi^2 D}{4h^2} t\right) \right]. \quad (11.23)$$

Etminan et al. (2014) used Equation 11.23 to model pulse pressure decay data to determine the diffusion coefficient of methane in kerogen material, D. If D is known then, Equation 11.23 can be used to determine the mass of diffusing gas at certain reservoir conditions.

11.8 PULSE-DECAY PERMEABILITY MEASUREMENT TEST

The pulse-decay experiment is primarily developed for the measurement of gas permeability of the tight porous media (Aronofsky, 1954; Aronofsky et al., 1959; Bruce et al., 1952; Wallick and Aronofsky, 1954). The pulse-decay experiment has applications in petroleum engineering as well as in other scientific and engineering fields such as hydrology (Finsterle and Najita, 1997), rock physics (Walder and Nur, 1986), and pressure vessel technology (Lasseux et al., 2011).

The pulse-decay experiment involves the measurement of the pressure-decay response to a pressure perturbation at the upstream face of low-permeability core sample. Figure 11.13 schematically shows the pulse-decay apparatus. The core sample with volume v_p is tightly held in a core holder under a hydrostatic confining pressure p_c. The upstream and downstream core faces are set to communicate with two vessels with finite gas volumes: namely, the upstream vessel with volume V_u and initial pressure p_{ui}, and the downstream vessel with volume V_d and initial pressure p_{di}. Subscripts d and u denote downstream and upstream, and subscript i denotes an initial-value boundary condition. The initial pressure in the upstream is greater than the downstream.

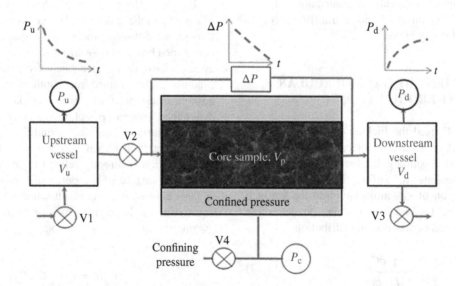

FIGURE 11.13 Schematic diagram of a pulse-decay apparatus showing upstream and downstream reservoir volumes and instantaneous pressures are (V_u, p_u) and (V_d, p_d). Core sample pore volume (V_p) is under a confining pressure of p_c; the temporal pressure difference across the core sample $\Delta p = p_u - p_d$ is measured during the test.

The major steps of a pulse-decay experiment for permeability measurement follow:

1. Valves 1, 2, and 3 are opened to allow the gas flow through the sample and through upstream and downstream vessels.
2. The entire system reaches the equilibrium pressure p_{d0}; valves 2 and 3 are closed.
3. The upstream vessel pressure is increased by a few percent and allowed to reach the new equilibrium state; valve 2 is then opened and the pressure difference (Δp) is measured with respect to time.

The pressure difference $\Delta p = p_u - p_d$ across the core. The pressure-decay response Δp is analyzed to determine the permeability.

Several studies investigated the pressure-decay partial differential equation to determine an analytical solution and subsequently estimate the core permeability. Hsieh et al. (1981) derived the late-transient solution for the pressure difference across the core sample. Jones (1997) used the analytical solution originally derived by Hsieh et al. (1981), and investigated the conditions under which the late-transient solution falls into a single exponential decline. Other studies proposed different solution forms such as series (Brace et al., 1968; Chen and Stagg, 1984; Dicker and Smits, 1988; Haskett et al., 1988; Hsieh et al., 1981; Neuzil et al., 1981; Wang and Hart, 1993), error function (Bourbie and Walls, 1982), or exponential decay (Dana and Skoczylas, 1999; Ivanov et al., 2000). Other studies expand the scope of pulse-decay experiment applications to partially water saturated samples (Homand et al., 2004; Newberg and Arastoopour, 1986) and permeability measurement with incompressible fluids (Amaefule et al., 1986; Trimmer, 1982).

11.8.1 Pulse-Decay Pressure Analysis

The main body of proposed analytical solutions to the pulse-decay equation assumes constant gas density, viscosity, and volumetric compressibility (defined as $c_g = 1/\rho \times \partial \rho / \partial p$). As a result, the gas-flow rate is linearly proportional to the local pressure gradient. Another common assumption is that Darcy's law is valid for the gas flow in the core. With these assumptions the intrinsic permeability is best estimated when the upstream and downstream vessels have equal volumes $V_u = V_d$ (Dicker and Smits, 1988; Jones, 1997). If the core porosity is not known, it's best to have the upstream and downstream volumes set close to a first-order estimate of sample pore volume (Wang and Hart, 1993).

These assumptions of linear proportionality of flow rate and pressure gradient and the Darcy's law may not be valid for gas flow in shales. The Klinkenberg equation (Klinkenberg, 1941) is an alternative pressure-dependent permeability model that may describe the tight gas and shale gas permeability. Kaczmarek (2008) and Wu et al. (1998) proposed analytical interpretations of the pressure-decay response, assuming the Klinkenberg effect and a constant mass flow rate for linear and radial gas flow, respectively. Jannot et al. (2007) investigated the experimental conditions that affect the determination of Klinkenberg parameters and showed that better estimates of k_D and b are achieved with an infinite downstream vessel volume. They also showed that the upstream vessel volume, core diameter, and length do not significantly affect the k_D and b estimations. Darabi et al. (2012) showed that an analytical pseudo-pressure solution may better estimate the permeability, mainly because of the strong sensitivity of the permeability parameter estimates to the variation of fluid properties. In addition, they show the pseudo-pressure partial differential equation may be numerically solved with accurate estimations of gas density, viscosity, and volumetric compressibility.

The following presents the pulse-decay equation along with an analytical solution for the constant permeability and constant μc_g assumptions, and a numerical algorithm with all pressure-dependent parameters. The pulse-decay material balance equation for gas flow in a one-dimensional core sample with gas adsorption is

$$\varphi \frac{\partial \rho}{\partial t} + (1-\varphi)\frac{\partial q}{\partial t} = \frac{1}{r^n}\frac{\partial}{\partial x}\left(r^n \frac{\rho k}{\mu}\frac{\partial p}{\partial x}\right), \quad (11.24)$$

where q is the adsorbate density per unit sample volume, k is permeability (either Darcy permeability or a pressure-dependent permeability function), and $n = 0, 1, 2$, respectively, present one-dimensional flow in Cartesian, Radial, and Spherical coordinates.

Equation 11.24 accounts for the gas desorption. The shale adsorption isotherms are commonly described by the Langmuir isotherms (Cui et al., 2009; Lancaster and Hill, 1993; Ross and Bustin, 2007; Saulsberry et al., 1996)

$$q_a = \frac{q_L p}{p_L + p}, \quad (11.25)$$

$$q = \frac{\rho_s q_a}{V_{std}}, \quad (11.26)$$

where q_L and p_L are the Langmuir volume and pressure, respectively, ρ_s is the bulk density of the core, and V_{std} is the gas molar volume at standard pressure (101,325 Pa) and standard temperature (273.15 K).

The choice of appropriate coordinates depends on the core sample shape and the boundary conditions. We select Cartesian coordinates through the rest of the analysis, that is, $n = 0$. Hsieh et al. (1981) presented the analytical solution assuming Darcy permeability (i.e., $k = k_D$), and constant gas

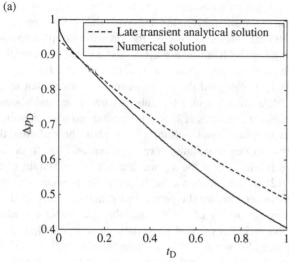
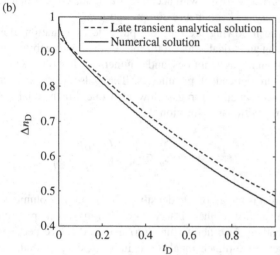
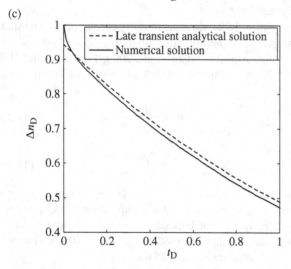

FIGURE 11.14 Comparison of the late-transient analytical solution with numerical solution. The numerical solution assumes that all parameters are pressure dependent. The analytical solution assumes that (a) Darcy permeability is used and ρ, μ, and c_g are constant, (b) Darcy permeability is used, μ and c_g are constant, and ρ is pressure-dependent, (c) APF is used, μ and c_g are constant, and ρ is pressure dependent.

property values for ρ, μ, and c_g. Darabi et al. (2012) showed that the pressure solutions under the typical pulse-decay conditions (i.e., upstream and downstream pressures, standard temperature, large pressure gradient across core sample) with constant fluid properties and pressure-dependent properties are significantly different. Figure 11.14a presents a sensitivity analysis on ρ, μ, and c_g pressure-dependency effect and showed that under the typical pulse-decay conditions, the variations of μ and c_g have a negligible effect on the pressure solution, whereas the pressure dependency of ρ has a significant impact.

The pseudo-pressure function $m(p)$ is defined as follows:

$$m(p) = 2\int_0^p \frac{k}{\mu z} p' dp'. \qquad (11.27)$$

Equation 11.24 is reformulated using Equation 11.27 and the real gas law ($\rho = pM/zRT$) that accounts for pressure dependence of density is

$$\frac{\partial m}{\partial t} = \frac{k}{\mu c_g (\phi + (1-\phi) K_a)} \frac{\partial^2 m}{\partial x^2}, \quad 0 < x \langle L, \ t \rangle 0. \qquad (11.28)$$

Equation 11.28 accounts for the gas desorption in the term K_a, defined as the derivative of adsorbate density with respect to gas density, $K_a = \partial q / \partial \rho$. Equation 11.28 has two boundary conditions and one initial condition,

$$\text{BC-1}: m(0,t) = m_u(t), \quad t > 0, \qquad (11.29)$$

$$\text{BC-2}: m(L,t) = m_d(t), \quad t > 0, \qquad (11.30)$$

$$\text{IC}: m(x,0) = m_{di}, \quad x > 0, \quad m(0,0) = m_{ui}. \qquad (11.31)$$

Three parameters, μ, c_g, and k are assumed to be constant to derive an analytical solution. A constant permeability assumption is a reasonable assumption because the pressure variation across the core is fairly small (e.g., 500 kPa). Darabi et al. (2012) showed that the assumption of constant viscosity-volumetric compressibility product introduce an acceptable error.

Three dimensionless groups, $\Delta m_D = (m_u - m_d)/(m_{u0} - m_{d0})$, $t_D = kt/(\mu c_g \varphi L^2)$, and $x_D = x/L$ are defined and substituted into Equations 11.28–11.31. Using an analogy to the Hsieh et al. (1981) method, the analytical solution to Equation 11.28 is derived for the late-transient time as follows:

$$\ln(\Delta m_D) = \ln(f_0) + s_1 t, \qquad (11.32)$$

$$f_0 = 2\left(a(b^2 + \theta_1^2) + b\sqrt{(a^2 + \theta_1^2)(b^2 + \theta_1^2)}\right) / \left(\theta_1^2(\theta_1^2 + a + a^2 + b + b^2) + ab(a + b + ab)\right), \qquad (11.33)$$

$$s_1 = -\frac{k f_1 A(1/V_u + 1/V_d)}{(\mu L c_g)}, \qquad (11.34)$$

$$f_1 = \frac{\theta_1^2}{(a+b)}, \qquad (11.35)$$

where θ_1 is the first solution to

$$\tan\theta = \frac{((a+b)\theta)}{(\theta^2 - ab)}. \qquad (11.36)$$

The plot of Δm_D on a log scale versus time yields a straight line at late-transient times. Equation 11.34 implies that the slope of the line s_1 is related to permeability,

$$k = \frac{s_1 \mu L c_g}{f_1 A(1/V_u + 1/V_d)}. \qquad (11.37)$$

Equation 11.37 calculates the permeability at average core sample pressure at late-transient time. Figure 11.14b compares the modified late-transient analytical and numerical solutions assuming Darcy permeability (i.e., $k = k_D$). Figure 11.14c compares the same solutions assuming the apparent permeability function (Darabi et al., 2012). Both figures show that the late-transient analytical solution is in good agreement with the numerical solution. Therefore, under typical pulse-decay conditions, that is, high pressure and small pressure difference across core sample, the analytical solution may be used to estimate apparent permeability for shale samples. However, if the pressure difference across the core is large or the initial core pressure is low, the analytical solution may result in a significant error in the permeability estimate.

11.8.2 Estimation of Permeability Parameters with the Pulse-Decay Experiment

Any pressure-dependent permeability model has two or more physical or fitting parameters to be estimated. Using independent experimental methods often help to better estimate the physical parameters that are involved in the permeability model. For example, if we select the apparent permeability function (APF) as the permeability model, then appropriate estimations for tortuosity, tangential momentum accommodation coefficient (TMAC), and the fractal dimension of the pore surface are required before the intrinsic permeability can be determined. Tortuosity may be estimated using SEM and AFM imaging. TMAC can be determined from the oil drop experiments of Millikan, the rotating cylinder method, the spinning rotor gauge method, molecular beam techniques, and flow through microchannels (Agrawal and Prabhu, 2008). Fractal dimensions of surfaces can be determined from small-angle X-ray scattering (Coppens, 1999). The main parameter of the permeability model, the intrinsic permeability, and the remaining unknown parameters may be estimated using the pulse-decay experimental data. Most accurate estimates are achieved by performing several pulse-decay experiments at different mean pressures. To obtain reliable results, the lower and upper bounds of physical parameters should be strictly defined as well. With these considerations, the following optimization problem is formulated and solved to estimate the permeability model parameters,

$$\begin{aligned}\text{Minimize} \quad & f(X) = \sum_{j=1}^{J}(\Delta m_{D,j} - \Delta m_{Ddata,j})^2 \\ \text{s.t.} \quad & X_l \leq X \leq X_u,\end{aligned} \qquad (11.38)$$

where $\Delta m_{D,j}$ is the pseudo-pressure decline estimate, $\Delta m_{Ddata,j}$ is the pseudo-pressure decline experimental data, X is the vector of unknown permeability parameters, and J is the number of data points. Subscripts l and u denote lower and upper parameter bounds, respectively. The dimensionless pseudo-pressure difference in the objective function may be substituted from either numerical or analytical solutions.

Equation 11.38 is classified as constrained and nonlinear optimization problem (Borwein and Lewis, 2000) and can be solved with any appropriate optimization algorithm (e.g., Mehrotra, 1992). One of the following two methods may be used to estimate the permeability parameters:

1. With the analytical pseudo-pressure solution: the objective function in Equation 11.38 is formulated with the analytical solution (Equation 11.32). The optimization problem is then solved to find the best match for the late-transient pressure-decay behavior and subsequently estimate the permeability parameters.

2. With an iterative numerical solution-optimization method: for each optimization step, the numerical solution is first obtained; then the optimization is performed and a new set of fitting parameters is generated, which is used by the numerical simulator in the next step. This procedure is repeated until the gradient of the objective function with respect to all fitting parameters is less than a specified tolerance.

The choice of estimation method is essentially independent from the permeability model; however, the advantage of the numerical method is that it uses all the recorded pressure-decay data to obtain the best parameter estimates, whereas the analytical method only uses the late-transient pressure data. Therefore, the numerical method tends to yield more reliable estimates, especially if the number of available data points is limited. Figure 11.15 presents an example of permeability parameters estimation using the APF based on the analytical and numerical procedures.

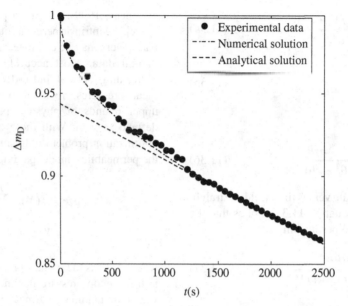

FIGURE 11.15 Dimensionless analytical and numerical solutions for dimensionless pseudo-pressure decay with the APF (Darabi et al., 2012).

11.9 CRUSHED SAMPLE TEST

The crushed sample test is performed on a pack of small rock particles to determine the porosity, adsorption isotherms, and permeability. Figure 11.16 schematically shows the pycnometer apparatus. The apparatus consists of a main vessel and a sample vessel with respective volumes of V_m and V_s. Two pressure transducers measure the main and sample vessel pressures, respectively, p_m and p_s. Three valves control the timing and the stages of the experiment, as described in the following process.

1. The crushed sample is put into the sample vessel, valve 2 is closed. Valve 3 is first opened to fill the sample with the experimental gas (typically He or N_2) and then vacuumed to remove all resident gas in the sample vessel. Valve 3 is then closed and the initial sample vessel pressure (p_{vac}) is recorded.
2. Valve 1 is opened and the main vessel is filled with high-pressure gas. Valve 1 is then closed and the equilibrium pressure in the main vessel (p_{m0}) is recorded.
3. Valve 2 is opened to allow the gas flow from the main vessel to the vacuumed sample vessel. The pressures in both vessels are recorded.

The high-pressure gas in the main vessel occupies the sample vessel void volume almost instantaneously. Thereafter, the gas gradually penetrates into the crushed sample pores until a equilibrium state is reached. The equilibrium pressure of the vessels (p_e) is used to estimate the porosity and the pressure change in the void volume of the two vessels is analyzed to measure the permeability.

Cui et al. (2009) present a straightforward description of the crushed sample test and the procedures for the porosity and permeability measurements. We adapt their description, and with appropriate modifications for a general permeability function (either a constant or pressure-dependent), we develop a methodology as presented in the following sections.

11.9.1 Porosity Measurement

Assuming that no gas adsorption (He or N_2 do not normally adsorb), and known values of sample mass (M) and bulk density (ρ_b), the sample porosity (φ) is calculated:

$$\varphi = \left[\frac{V_m}{V_b}\left(\frac{p_{m0}}{z_{m0}} - \frac{p_e}{z_e}\right) + \left(\frac{V_s}{V_b} - 1\right)\left(\frac{p_{vac}}{z_{vac}} - \frac{p_e}{z_e}\right)\right] / \left(\frac{p_{vac}}{z_{vac}} - \frac{p_e}{z_e}\right), \quad (11.39)$$

where V_b is the sample bulk volume ($V_b = M_s/\rho_b$), and z is the volumetric compressibility factor, V_m and V_s are the main vessel and sample vessel volumes, p_{vac} and p_e are the vacuum and equilibrium pressures in the sample vessel, p_{m0} is the main vessel pressure at the beginning of the gas expansion process, and z_{vac}, z_e, and z_{m0} are the corresponding gas compressibility factors at the vacuum, equilibrium, and initial expansion pressure states, respectively.

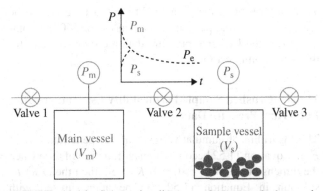

FIGURE 11.16 Schematic diagram of the crushed sample apparatus. The volumes of the main cell (V_m), the sample cell (V_s), and the sample bulk volume (V_b) are known. Using the pressure histories of the main cell and the sample cell (p_m and p_s) and the equilibrium pressure of the two cells (p_e), the porosity and permeability of the sample are measured.

11.9.2 Crushed Sample Pressure Analysis for Permeability Measurement

Once the process of gas expansion into the crushed sample is initiated, the pressure decay is used to determine the permeability (Egermann et al., 2005). The partial differential equation for gas expansion into the crushed spherical particles with radius R_a, density ρ_b, and mass M is,

$$\frac{\partial m}{\partial t} = \frac{k}{\mu c_g \left[\varphi + (1-\varphi) K_a\right]} \frac{1}{r^2}\left(r^2 \frac{\partial m}{\partial r}\right), \quad (11.40)$$

where $m = m(p)$ is the pseudo-pressure potential, defined in Equation 11.27, k is the (either the Darcy permeability or a pressure-dependent permeability), μ is viscosity, c_g is the volumetric compressibility, and φ is porosity. Equation 11.40 accounts for the gas desorption, as defined in Equations 11.25 and 11.26. K_a is the derivative of adsorbate density q with respect to gas density,

$$K_a = \frac{\partial q}{\partial \rho}. \quad (11.41)$$

Equation 11.40 has two boundary conditions and one initial condition,

$$\text{BC-1}: \frac{\partial m}{\partial r} = 0, \quad r = 0, \quad (11.42)$$

$$\text{BC-2}: A_s \frac{k}{\mu c_g} \frac{\partial m}{\partial r} = V_c \frac{\partial m}{\partial t}, \quad r = R_a, \quad (11.43)$$

$$\text{IC}: \begin{matrix} m = m_0 & 0 \leq r < R_a, & t = 0, \\ m = m_{c0}, & r = R_a, & t = 0, \end{matrix} \quad (11.44)$$

where $A_s = 3 M_s/(\rho_b R_a)$ is the total surface area of the spherical crushed particles, m_0 is the initial pseudo pressure of the gas in sample pores, and m_{c0} is the pseudo pressure of the void volume of the main and sample vessels at the beginning of gas penetration into the crushed sample pores.

The analytical solution for pseudo pressure $m(p)$ in the void volume of the main and sample vessels may be found by assuming the permeability k and viscosity-volumetric compressibility μc_g are constant. The constant permeability assumption is reasonable because the pseudo pressure decline after the initial gas expansion into the sample vessel is small. The analytical pseudo-pressure solution under these assumptions is found by an analogy to Carslaw and Jaegar (1947),

$$m = m_{c0} - \frac{m_{c0}}{K_c + 1} + 6 K_c (m_{c0} - m_0)$$
$$\sum_{n=1}^{\infty} \frac{e^{-K \alpha_n^2 t / R_a^2}}{K_c^2 \alpha_n^2 + 9(K_c + 1)}, \quad (11.45)$$

where α_n is the nth solution of,

$$\tan \alpha = \frac{3 \alpha}{3 + K_c \alpha^2}. \quad (11.46)$$

In Equation 11.45, m_{c0} is the average initial pseudo pressure in the main and reference vessels after the equilibrium state in the void volume of the two vessels is reached and before the equilibrium process between the void volume and sample pores starts. m_{c0} is defined as follows:

$$m_{c0} = \frac{m_{m0} V_m + m_0 (V_s - V_b)}{V_m + V_s - V_b}, \quad (11.47)$$

where V_b is the sample bulk volume and V_c is the total void volume of the main and sample vessels,

$$V_c = V_m + V_s - V_b(1 - \varphi). \quad (11.48)$$

We define the cumulative uptake gas penetration ratio F_U as the ratio of the gas mass that penetrated to the sample pores at any time to the ultimate cumulative penetrated gas mass. In the pseudo pressure form, F_U is expressed as follows:

$$F_U = \frac{(K_c + 1)(m_{c0} - m)}{m_0 - m_{c0}}, \quad (11.49)$$

and also the cumulative residual gas penetration ratio F_R, as a complement function of F_U,

$$F_R = 1 - F_U = 1 - \frac{(K_c + 1)(m_{c0} - m)}{m_0 - m_{c0}}. \quad (11.50)$$

In Equations 11.49 and 11.50, K_c is the ratio of the void volume of the main and sample vessels V_c to the crushed sample pore volume,

$$K_c = \frac{\rho_b V_c}{M[\varphi + (1-\varphi)K_a]}, \qquad (11.51)$$

where ρ_b is the sample bulk density and M is the gas molar mass. Substituting the pseudo-pressure solution m in Equation 11.45 into 11.50, F_R is given as follows:

$$F_R = 6K_c(K_c+1)\sum_{n=1}^{\infty} \frac{e^{-K\alpha_n^2 t/R_a^2}}{K_c^2\alpha_n^2 + 9(K_c+1)}. \qquad (11.52)$$

If the ratio of the sample pore volume to the total void volume of the main and sample vessels is small ($K_c \to \infty$), the pseudo-pressure ratio solution F_R simplifies to

$$F_R = \frac{6}{\pi}\sum_{n=1}^{\infty} e^{-\pi^2 n^2 Kt/R_a^2} \frac{1}{n^2}. \qquad (11.53)$$

Cui et al. (2009) show that for $K_c > 50$ Equation 11.53 is an appropriate approximation for the analytical solution, and the early-time and late-time pressure history in the void volume may be used to determine the permeability.

11.9.3 Crushed Sample Permeability Estimation with Early-Time Pressure Data

The early-time solution of Equation 11.53 can be approximated as (Carslaw and Jaeger, 1947; Do, 1998):

$$F_U = 1 - F_R = \frac{6\sqrt{K}}{\sqrt{\pi R_a^2}}\sqrt{t}. \qquad (11.54)$$

Equation 11.42 implies that the early-time cumulative uptake gas penetration ratio F_U versus the square root of time yields a straight line on a linear scale. The line slope s_1 is related to the permeability,

$$k = \frac{\pi s_1^2 R_a^2 [\varphi + (1-\varphi)]\mu c_g}{36}. \qquad (11.55)$$

Equation 11.55 is valid for large values of K_c (>50) and relatively short time after the gas expansion into the sample vessel, specifically when the dimensionless time $\tau = Kt/R_a^2 < 0.0002$ or $F_U < 0.2$ (Cui et al., 2009). The large differential pressure between the high-pressure main vessel and the almost vacuum sample vessel may cause an adiabatic temperature change in the system. Also, pressure measurement at the early times may be affected by the kinetic expansion from the main cell into the sample cell.

This may result in the poor quality and a time lag in the recorded early-time pressure data. Cui et al. (2009) recommend that the late-time pressure data are more reliable for the permeability estimation.

11.9.4 Crushed Sample Permeability Estimation with Late-Time Pressure Data

The logarithm of cumulative residual gas penetration ratio F_R in Equation 11.52 becomes a linear function of time when the dimensionless time $\tau > 0.1$. If $K_{cm} \geq 50$, then the exact F_R solution in Equation 11.52 can be approximated with Equation 11.53. The straight line part of the solution for $\tau > 0.1$ is approximated as follows:

$$\ln(F_R) = f_0 - s_1 t, \qquad (11.56)$$

where s_1 is the slope of the straight line,

$$s_1 = \frac{K\alpha_1}{R_a^2}, \qquad (11.57)$$

and α_1 is the first solution of Equation 11.34, and the y-intercept of the straight line is

$$f_0 = \ln\left(\frac{6K_c(K_c+1)}{K_c^2\alpha_1^2 + 9(K_c+1)}\right). \qquad (11.58)$$

The slope s_1 of the straight-line part of the solution is related to the permeability as

$$k = \frac{R_a^2[\varphi + (1-\varphi)K_a]\mu c_g s_1}{\alpha_1^2}. \qquad (11.59)$$

The full scope of the pressure data, including the early time and the late-time periods, may be used in a numerical estimation method to determine the permeability. The reader may refer to Section 2.2 for the details of the numerical workflow.

11.10 CANISTER DESORPTION TEST

The canister desorption test is performed on the drill cores to estimate the rock permeability and diffusion. Core sample, usually in a cylindrical shape are obtained from the productive zone of the wells and transferred into the canister apparatus, which is schematically shown in Figure 11.17. The drill core is kept at the reservoir temperature and an ambient pressure. The cumulative desorbed gas volumes are recorded and used to determine the permeability and diffusivity.

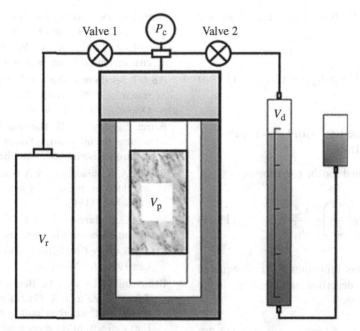

FIGURE 11.17 Schematic model of the canister desorption apparatus. The upper and lower faces of the drill core sample (volume V_p and radius R_a) are sealed to allow for a radial flow. The cumulative volume of desorbed gas (V_d) is chronologically recorded for the permeability estimation.

The theoretical basis of the canister desorption data analysis is similar to the analyses of pulse-decay and crushed sample tests. Assuming that the length of the drill core L is sufficiently larger than its diameter ($L > 2R_a$), the material balance equation is described in cylindrical coordinates as,

$$\frac{\partial m}{\partial t} = \frac{k}{\mu c_g [\varphi + (1-\varphi)K_a]} \frac{1}{r^2} \left(r^2 \frac{\partial m}{\partial r} \right), \quad (11.60)$$

where $m = m(p)$ is the pseudo-pressure potential, defined in Equation 11.27. Equation 11.60 is subject to two boundary conditions and one initial condition,

$$\text{BC-1}: \frac{\partial m}{\partial r} = 0, \quad r = 0, \quad (11.61)$$

$$\text{BC-2}: m = m_e, \quad r = R_a, \quad (11.62)$$

$$\text{IC}: m = m_0, \quad 0 \le r < R_a, \quad t = 0, \quad (11.63)$$

where m_0 is the initial gas pseudo pressure in the drill core, and m_e is the pseudo pressure at the ambient pressure and reservoir temperature conditions.

We define the cumulative desorbed gas mass fraction F_D as the ratio of the cumulative gas desorbed from the drill core to the ultimate cumulative desorbed gas. Assuming a one-dimensional radial flow in an infinitely long cylinder, such that the boundary effects on the top and bottom of the drill core is negligible, the analytical solution of Equation 11.60 in terms F_R is given as,

$$F_D = 1 - 4\sum_{n=1}^{\infty} \frac{1}{\xi_n^2} e^{-\xi_n^2 Kt/R_a^2}, \quad (11.64)$$

where n is the nth root of the Bessel equation, $J_0(\xi_n) = 0$, R_a is the drill core radius, and K is the apparent transport coefficient,

$$K = \frac{k}{\mu c_g [\varphi + (1-\varphi)K_a]}. \quad (11.65)$$

11.10.1 Permeability Estimation with Early Time Cumulative Desorbed Gas Data

Equation 11.64 is approximated for the early time ($\tau = Kt/R_a^2 < 0.0002$) as,

$$F_D = \frac{4\sqrt{K}}{R_a \sqrt{\pi}} \sqrt{t}. \quad (11.66)$$

Equation 11.66 implies the early-time cumulative desorbed gas fraction plotted against the square root of time would yield a straight line,

$$F_D = s_1 \sqrt{t}. \quad (11.67)$$

The slope of the straight line s_1 is related to the permeability as

$$k = \frac{\pi}{16} R_u^2 \left[\varphi + (1-\varphi) K_a \right] \mu c_g s_1^2, \quad (11.68)$$

11.10.2 Permeability Estimation with Late-Time Cumulative Desorbed Gas Data

Equation 11.64 is approximated for the late time as

$$\ln(1 - F_D) = \ln\left(\frac{4}{\xi_1^2}\right) - \frac{K \xi_1^2}{R_a^2} t, \quad (11.69)$$

where ξ_1 is the first root of Bessel function, $J_0(\xi) = 0$, equal to 2.404834. The cumulatively desorbed gas fraction for the late time is described as follows:

$$\ln(1 - F_D) = f_0 - s_1 t, \quad (11.70)$$

and the permeability can be determined with the late-time data as follows:

$$k = \frac{R_a^2 \left[\varphi + (1-\varphi) K_a \right] \mu c_g s_1}{\xi_1^2} \quad (11.71)$$

ACKNOWLEDGMENTS

AFM and SEM images are courtesy of the NanoGeosciences Lab at the Bureau of Economic Geology, The University of Texas at Austin. Ayaz Mehmani wrote the section about network modeling in shale system. Chris Parker edited the manuscript. Publication authorized by the Director Bureau of Economy Geology.

REFERENCES

Agrawal A, Prabhu SV. Survey on measurement of tangential momentum accommodation coefficient. J Vac Sci Technol A 2008;26 (4):634–645.

Akkutlu IY, Fathi E. Multiscale gas transport in shales with local kerogen heterogeneities. SPE J 2012;17(4):1002–1011. SPE-146422-PA.

Amaefule JO, Wolfe K, Walls JD, Ajufo AO, Peterson E. Laboratory determination of effective liquid permeability in low-quality reservoir rocks by the pulse decay technique. SPE 1986;15149:493–502.

Arkilic EB, Kenneth SB, Martin AS. Mass flow and tangential momentum accommodation in silicon micromachined channels. J Fluid Mech 2001;437:29–43.

Aronofsky JS. Effect of gas slip on unsteady flow of gas through porous media. J Appl Phys 1954;25 (1):48–53.

Aronofsky JS, Wallick CG, Reichertz PP. Method of measuring characteristics of porous materials. Patent 2867116. 1959.

Azom P, Javadpour F. Dual-continuum modeling of shale and tight gas reservoirs. SPE Annual Technical Conference and Exhibition; October 8–10, 2012; San Antonio, TX. doi:10.2118/159584-MS.

Barrett EP, Joyner LG, Halenda PP. The determination of pore volume and area distribution in porous media. I. Computations of nitrogen isotherms. J Am Chem Soc 1951;73 (1):373–380.

Beskok A, Karniadakis GE. A model for flows in channels, pipes, and ducts at micro and nano scales. Microscale Thermophys Eng 1999;3 (1):43–77.

Binnig G, Rohrer H, Gerber C, Weibel E. Surface studies by scanning tunneling microscopy. Phys Rev Lett 1982;49:57–61.

Binnig G, Quate CF, Gerber C. Atomic force microscope. Phys Rev Lett 1986;56:930–933.

Blanchard V, Lasseux D, Bertin H, Pichery T, Chauveteau G, Tabary R, Zaitoun A. Gas/water flow in porous media in the presence of adsorbed polymer: experimental study on non-Darcy effects. SPE Reserv Eval Eng J 2007;10 (4):423–431.

Borwein JM, Lewis AS. *Convex Analysis and Nonlinear Optimization, Theory and Examples. Canadian Mathematical Society Books in Mathematics*. New York: Springer; 2000.

Bourbie T, Walls J. Pulse decay permeability: analytical solution and experimental test. SPE J 1982;22:719–721.

Brace WF, Walsh JB, Frangos WT. Permeability of granite under high pressure. J Geophys Res 1968;736:2225–2236.

Brown GP, Dinardo A, Cheng GK, Sherwood TK. The flow of gases in pipes at low pressures. J Appl Phys 1946;17: 802–813.

Bruce GH, Peaceman DW, Hachford HH. Calculation of unsteady-state gas flow through porous media. Trans AIME 1952; 221G:1–16.

Carslaw HS, Jaeger JC. *Conduction of Heat in Solids*. London: Oxford University Press; 1947. p 198–207.

Chareonsuppanimit P, Mohammad SA, Robinson RL Jr, Gasem KAM. High-pressure adsorption of gases on shales: measurements and modeling. Int J Coal Geol 2012;95:34–46.

Chen T, Stagg PW. Semilog analysis of the pulse decay technique of permeability measurement. SPE J 1984;24:639–642.

Civan F. Effective correlation of apparent gas permeability in tight porous media. Transp Porous Media 2010;82 (2):375–384. DOI: 10.1007/s11242-009-9432-z.

Civan F, Rai C, Sondergeld C. Determining shale permeability to gas by simultaneous analysis of various pressure tests. SPE J 2012;17 (3):717–726.

Coppens M-O. The effect of fractal surface roughness on diffusion and reaction in porous catalysts from fundamentals to practical applications. Catal Today 1999;53 (2):225–243.

Coppens M-O, Dammers AJ. Effects of heterogeneity on diffusion in nanopores from inorganic materials to protein crystals and ion channels. Fluid Phase Equilib 2006;246 (1–2):308–316.

Crank J. *The Mathematics of Diffusion*. London: Oxford University Press; 1979.

Cui X, Bustin AM, Bustin R. Measurements of gas permeability and diffusivity of tight reservoir rocks: different approaches and their applications. J Geofluids 2009;9:208–223.

Dana E, Skoczylas F. Gas relative permeability and pore structure of sandstones. Int J Rock Mech Min Sci 1999;36:613–625.

Darabi H, Ettehad A, Javadpour F, Sepehrnoori K. Gas Flow in ultra-tight shale strata. J Fluid Mech FirstView 2012:1–18. DOI: 10.1017/jfm.2012.424.

Darcy H. *Les Fontaines Publiques de la Ville de Dijon*. Paris: Dalmont; 1856.

Dicker AI, Smits RM. A practical approach for determining permeability from laboratory pressure-pulse decay measurements. SPE 1988;15578:285–292.

Do DD. Adsorption analysis: equilibria and kinetics. In: Yang RT, editor. *Series on Chemical Engineering*. Volume 2, London: Imperial College Press; 1998. p 543–545.

Egermann P, Lenormand R, Longeron D, Zarcone C. A fast and direct method of permeability measurements on drill cuttings. SPE Reserv Eval Eng August 2005; 269–275.

Etminan SR, Javadpour F, Maini BB, Chen Z. Measurement of gas storage processes in shale and of the molecular diffusion coefficient in Kerogen. Int J Coal Geol 2014;123:10–19.

Finsterle S, Persoff P. Determining permeability of tight rock samples using inverse modelling. Water Resour Res 1997;33 (8):1803–1811.

Florence FA, Rushing JA, Newsham KE, Blasingame TA. Improved permeability prediction relations for low-permeability sands. Paper SPE 107954-MS Presented at SPE Rocky Mountain Oil and Gas Technology Symposium; April 16–18, 2007; Denver, CO.

Groen JC, Peffer LAA, Pérez-Ramírez J. Pore Size determination in modified micro- and mesoporous materials. Pitfalls and limitations in gas adsorption data analysis. Microporous Mesoporous Mater 2003;60 (1–3):1–17.

Haskett SE, Narahara GM, Holditch SA. A method for the simultaneous determination of permeability and porosity in low permeability cores. SPE 1988;15379:1–11.

Homand F, Giraud A, Escoffier S, Koriche A, Hoxha D. Permeability determination of a deep argillite in saturated and partially saturated conditions. Int J Heat Mass Transf 2004;47:3517–3531.

Hsieh PA, Tracy JV, Neuzil CE, Bredehoeft JD, Silliman SE. A transient laboratory method for determining the hydraulic properties of "tight" rocks—I. Theory. Int J Rock Mech Min Sci Geomech Abstr 1981;18:245–252.

Ivanov AN, Kozlova SN, Pechenov AV. Permeability measurement. Meas Tech 2000;43 (12):1086–1088.

Jannot Y, Lasseux D, Vize G, Hamon G. A detailed analysis of permeability and Klinkenberg coefficient estimation from unsteady-state pulse-decay or draw-down experiments, 2007. International Symposium of the Society of Core Analysts; 2007; Calgary, Canada; Paper SCA2007-08.

Javadpour F. Nanopores and apparent permeability of gas flow in mudrocks (shales and siltstone). J Can Pet Technol 2009;48 (8):16–21. DOI: 10.2118/09-08-16-DA.

Javadpour F, Fisher D, Unsworth M. Nanoscale gas flow in shale gas sediments. J Can Pet Technol 2007;46 (10):55–61. DOI: 10.2118/07-10-06.

Javadpour F, Farshi M, Amrein M. Atomic-force microscopy: a new tool for gas-shale characterization. J Can Pet Technol 2012;51:236–243. DOI: 10.2118/161015-PA.

Jones SC. A technique for faster pulse-decay permeability measurements in tight rocks. SPE Formation Eval; 1997; March issue 19–25. doi:10.2118/28450-PA

Jones FO, Owens WW. A laboratory study of low-permeability gas sands. Paper SPE 7541-PA. J Pet Technol 1980;1631–1640. doi:10.2118/7551-PA

Kaczmarek M. Approximate solutions for non-stationary gas permeability tests. Transp Porous Media 2008. DOI: 10.1007/s11242-008-9216-x.

Karniadakis G, Beskok A, Aluru N. *Microflows and Nanoflows: Fundamentals and Simulation*. New York: Springer Science+Business Media, Inc.; 2005.

Klinkenberg LJ. *The Permeability of Porous Media to Liquids and Gases*. API Drilling and Production Practice, USA; 1941. p 200–213.

Lancaster DE, Hill DG. A multi-laboratory comparison of isotherm measurements of Antrim shale samples. 1993 SCA Conference Paper Number 9303; 1993; Otsego County, MI. p 1–16.

Lasseux D, Jolly P, Jannot Y, Omnes ESB. Permeability measurement of graphite compression packings. Trans ASME J Press Vessel Technol 2011;133:041401.

Loucks RG, Reed RM, Ruppel SC, Hammes U. Spectrum of pore types and networks in mudrocks and a descriptive classification for matrix-related mudrock pores. AAPG Bull 2012;96 (6):1071–1098. DOI: 10.1306/08171111061.

Mehmani A, Prodanovic M, Javadpour F. Multiscale multiphysics network modeling of shale matrix gas flows. Transp Porous Media 2013;99:377–390. DOI: 10.1007/s11242-013-0191-5.

Mehrotra S. On the implementation of a primal-dual interior point method. SIAM J. Optim 1992;2 (4):575–601.

Milliken KL, Esch WL, Reed RM, Zhang T. Grain assemblages and strong diagenetic overprinting in siliceous mudrocks, barnett shale (Mississippian), Fort Worth Basin, Texas. AAPG Bull 2012;96 (8):1553–1578. DOI: 10.1306/12011111129.

Moghanloo RG, Javadpour F, Davudov D. Contribution of methane molecular diffusion in kerogen to gas-in-place and production. SPE 165376, SPE Western Regional & AAPG Pacific Section Meeting, 2013 Joint Technical Conference; April 19–25, 2013; Monterey, CA.

Neuzil CE, Cooley C, Silliman SE, Bredehoeft JD, Hsieh PA. A transient laboratory method for determining the hydraulic properties of 'tight' rocks—II. Application. Int J Rock Mech Min Sci Geomech Abstr 1981;18:253–258.

Newberg MA, Arastoopour H. Analysis of the flow of gas through low-permeability porous media. SPE Reserv Eng 1986: 647–653.

Ross D, Bustin RM. Impact of mass balance calculations on adsorption capacities in microporous shale gas reservoirs. Fuel 2007;86:2696–2706.

Roy S, Raju R, Chuang HF, Cruden BA, Meyyappan M. Modeling gas flow through microchannels and nanopores. J Appl Phys 2003;93:4870–4879.

Rushing JA, Newsham KE, Lasswell PM, Cox JC, Blasingame TA. Klinkenberg-corrected permeability measurements in tight gas sands: steady-state versus unsteady-state techniques. SPE 89867 2004:1–11.

Ruthven DM. *Principles of Adsorption and Adsorption Processes.* New York: John Wiley & Sons, Inc.; 1984.

Saulsberry JL, Schafer PS, Schraufnagel RA. *A Guide to Coalbed Methane Reservoir Engineering. GRI-94/ 0397.* Gas Research Institute: Chicago, IL; 1996.

Shabro V, Torres-Verdin C, Javadpour F. Numerical simulation of shale gas production: from pore-scale modeling of slip flow, Knudsen diffusion and Langmuir desorption to reservoir modeling of compressible flow. SPE 144355, SPE North American Unconventional Gas Conference and Exhibition; June 12–16, 2011; The Woodlands, TX.

Singh H, Javadpour, F, Ettehad, A., Darabi, H., 2014. Non-empirical apparent permeability of shale. SPE Reserv Eval Eng-Reserv Eng 17, (3), 414–424.

Swami V, Settari A, Javadpour F. A numerical model for multimechanism flow in shale gas reservoirs with application to laboratory scale testing. SPE 164840, EAGE Annual Conference & Exhibition incorporating SPE Europec; June 10–13, 2013; London, UK.

Trimmer D. Laboratory measurements of ultralow permeability of geologic materials. Rev Sci Instrum 1982;53 (8):1246–1250.

Walder J, Nur A. Permeability measurement by the pulse-decay method: effect of poroelastic phenomena and nonlinear pore pressure diffusion. Int J Rock Mech Min Sci Geomech Abstr 1986;23 (3):225–232.

Wallick GC, Aronofsky JS. Effect of gas slip on unsteady flow of gas through porous media. Trans AIME 1954;201:32–324.

Wang HF, Hart DJ. Experimental error for permeability and specific storage from pulse decay measurements. Int J Rock Mech Min Sci Geomech Abstr 1993;30 (7):1173–1176.

Wu YS, Pruess K, Persoff P. Gas flow in porous media with Klinkenberg effects. Transp Porous Media 1998;32:19–137.

Zhang TEG, Ruppel S, Milliken K, Yang R. Effect of organic-matter type and thermal maturity on methane adsorption in shale gas systems. Org Geochem 2012;47:120–131.

12

A REVIEW OF THE CRITICAL ISSUES SURROUNDING THE SIMULATION OF TRANSPORT AND STORAGE IN SHALE RESERVOIRS

Richard F. Sigal, Devegowda Deepak and Faruk Civan

Mewbourne School of Petroleum and Geological Engineering, The University of Oklahoma, Norman, OK, USA

12.1 INTRODUCTION

In the past decade, production of hydrocarbons from organic-rich shale reservoirs has exploded onto the world energy landscape. These unconventional reservoirs are now recognized as forming an abundant worldwide resource that has already significantly changed the face of the energy landscape in the United States. Initially, production from these reservoirs was mainly dry gas, but exploration and development activity has largely moved to production of condensates and oils in many parts of the continental United States. The focus of this chapter is on the modeling and simulation of transport and storage in shale reservoirs.

The economic development of shale reservoirs is largely attributed to the introduction and maturation of two technologies: massive hydraulic fracturing and long reach horizontal wells. Because of the very low permeability of shales, these completion and drilling schemes are considered the predominant reasons for the successful development of organic shale reservoirs, and other tight gas and oil resources in the United States. In general, because of the ultra-low permeability of shales, the well architecture and the fracture geometry, including the hydraulic fracture and reactivated natural fracture networks, produced by the stimulations are considered to completely define well drainage volumes within the reservoir.

The very low matrix permeability of the shale reservoir rocks is a consequence of a pore geometry that includes pores less than an order of magnitude larger than a methane molecule. Studies have shown that pore proximity effects in nanopores can potentially alter the behavior of reservoir fluids. Fluid phase behavior and transport deviate significantly from the corresponding formulations for conventional rocks that are characterized by significantly larger pores. Under some conditions, the molecule–wall interactions tend to be more important than the intermolecular interactions within the fluid that have traditionally formed the basis of quantifying fluid PVT properties, transport, and storage. For shales, the small size of the pores restricts the pore to contain only a relatively few molecules whose behavior in this confined environment may not satisfy the assumptions of classical thermodynamics.

Organic shale reservoirs by their nature contain significant volumes of organic material that are matured enough to produce and expel large volumes of hydrocarbons. They therefore have both a complex wettability structure and support significant hydrocarbon storage in an adsorbed state on some pore walls; and at some stage in their evolution, they also have extensive natural fracture systems produced by the process of expelling the hydrocarbons. For a simulator to fully capture the complexity of organic-rich shale reservoirs, it must accommodate potentially two fracture systems with different properties and also at least two pore systems with dramatically different wettability character and pore size distributions. The connectivity of these various reservoir components will differ among reservoirs and often may not be well known, so the connectivity issue must remain as a parameter that can be adjusted in history matching, as demonstrated by the examples given in the upscaling section later. The introduction of a more realistic microgeometry must be accompanied by other modifications. At the least, the equations that define hydrocarbon storage, transport, and the equations of state also must be modified. Adsorption on the pore surface, which for a gas depends on the pore pressure, alters the pore space available for free hydrocarbon storage and transport, so parameters such as permeability and free gas porosity will be explicit functions of pore pressure, not just effective pressure.

12.2 MICROGEOMETRY OF ORGANIC-RICH SHALE RESERVOIRS

The microgeometry of organic-rich shales has been shown to be fairly heterogeneous even on the scale of nanometers (Curtis et al., 2010) and our current understanding of the microstructural features of shale is primarily because of the recent advances in nanoimaging technology done with the help of scanning electron microscopes aided by fixed ion beam milling to smooth and prepare samples for imaging (Ambrose et al., 2010; Curtis et al., 2010, 2011, 2013; Quirein et al., 2012). Although SEM image interpretation tends to be subjective, there is a growing body of work demonstrating the ability of these systems to image 3D connected pore networks within shales. A wide variety of pore types can be found in organic-rich shale reservoir rocks and pores of various types exist in both the organic and inorganic matrix material (Figures 9–12 in Curtis et al., 2010). Within the organic material, irregular cross-sectional pores characterized by some very small pore openings in the order of a few nanometers may be observed. In comparison, a methane molecule is approximately 0.37 nm in diameter. The intrinsic porosity of the organic material can be 30% or greater (Sigal, 2013a). For mature organic shales, the pores in the organic material probably provide most of the gas storage capacity.

Using image analysis methods, SEM images can supply information about the volume weighted pore size distributions, for a volume that is on the order of a cubic micron in size. They typically show pore sizes ranging from a nanometer to hundreds of nanometers, with the majority of the pore volume associated with pores having characteristic sizes larger than 10 nm (Curtis et al., 2013).

Methods such as NMR spectra and gas adsorption measurements can provide pore size distributions on larger scale. For the pores that store methane in a core plug, Sigal (2015) has used NMR measurements combined with standard Langmuir adsorption isotherms to provide volumetric pore size distributions. These samples came from the dry gas zone of the Barnett.

For NMR relaxation curves produced by relaxation from interaction with wall potentials, the curves are modeled as a sum of pore volumes V_i where the pores in each V_i have the same relaxation time as T_{2i}. The V_i sum to the total pore volume $V(t=0)$. That is

$$V(t) = \sum_{1}^{n} V_i e^{-t/T_{2i}} \qquad (12.1)$$

Equation 12.1 is inverted to get the V_i associated with each T_{2i}. The pore sizes are captured in the relaxation times. One has

$$\frac{1}{T_{2i}} = \rho_p \frac{S_i}{V_i} \qquad (12.2)$$

and

$$\frac{S_i}{V_i} = \frac{\alpha}{R_i} \qquad (12.3)$$

S_i is the surface area corresponding to the ith volume, ρ_p is the pore surface relaxivity, R_i is the characteristic size parameter, and α is the pore shape parameter. For a sphere, R_i is its radius and α is 3, for a cylinder R_i is the radius and α is 2, and for a slit pore R_i is its half aperture and α is 1. To get a pore size distribution, an assumption is required for the value of α and a value for ρ_p.

The Langmuir adsorption data provide an estimate for S the total surface area. From Equations 12.2 and 12.3,

$$S = \sum_i S_i = \sum_i \frac{V_i}{\rho_p T_i} \qquad (12.4)$$

Sigal et al. (2013) show that

$$S = \frac{S_{aLmax}}{2 r_m \rho_{maMax}} \qquad (12.5)$$

In Equation 12.5, r_m is the radius of a methane molecule, ρ_{maMax} is the maximum methane density in the adsorbed layer, and S_{aLmax} is the maximum number of moles of methane that can be adsorbed. S_{aLmax} is obtained from the standard Langmuir adsorption measurement. Based on discussion in Sigal et al. (2013), a reasonable estimate for the adsorption layer density is 0.0281 mol/cm³. The methane radius is 1.865×10^{-8} cm.

Sigal and Odusina (2011) reported the NMR methane spectra for several Barnett samples. The NMR samples also had methane adsorption data on companion plugs. Figure 12.1 shows the volumetric pore size distribution data for one of these samples assuming spherical or cylindrical pores. The pore sizes have a minimum value of about 1 nm. This is because the adsorbed methane relaxes too fast to be detected by the NMR measurements reported in Sigal and Odusina (2011), so each NMR pore size estimate has been enlarged by twice the diameter of a methane molecule. The maximum pore size for pores that store methane is a couple of hundred nanometers. Assuming spherical pores, 20% of the pore volume is contained in pores smaller than 10 nm, and assuming cylindrical pores, 30% of the volume is in the small pores. The pore size distributions obtained using NMR and adsorption data are consistent with distributions obtained from SEM studies.

Studies on shale samples of varying organic maturity indicate that porosity development in these shales tends to coincide with the initiation of the formation of hydrocarbon liquids within the organic material although there are no clear trends between organic maturity and porosity. However, the organic material is typically associated with porosities as high as 30% or higher (Sigal, 2013a) and typically most of the hydrocarbon storage is associated with these organic pores (Alfred and Vernik, 2013; Gouth et al., 2007; Sigal and Odusina, 2011). For highly mature shales that are often associated with the gas window, the organic pore walls are probably dominantly gas wetting while the inorganic pores are generally water wetting.

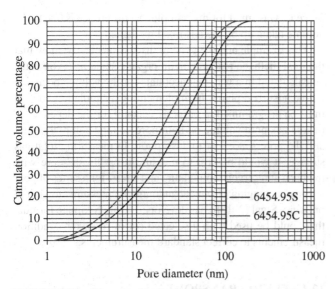

FIGURE 12.1 Pore size distribution from a Barnett core plug for the pores that store methane obtained using NMR and methane adsorption. The distribution 6454.95C (the smaller distribution) represents cylindrical pores and the larger 6454.95S spherical pores.

At earlier maturities in liquid-rich shales, the organic matrix material tends to have a more complex wettability both due to its chemistry and polar molecules that can alter both the wettability of the pores in the organic material and pores in the inorganic material (Hu et al., 2013a, b).

12.3 GAS STORAGE MECHANISMS

There are several potential storage mechanisms for gases in organic shales. They include free gas storage, gas adsorbed on the organic pore walls, gas absorbed/dissolved in the organic material, gas adsorbed on the inorganic pore walls, and gas dissolved in pore water (Civan, 2013). The prevalent practice is to model methane storage as just free and adsorbed gas.

Calculations show that water confined to nanometer-scale pores can dissolve orders of magnitude of more methane than bulk water (Campos et al., 2009, 2011), but these calculations do not apply to clay-bound water, and it is not clear if any suitably small free water containing pores exist at original reservoir conditions. The standard methods to measure the adsorbed gas storage capacity cannot distinguish on which surfaces the methane is adsorbed nor identify if any gas is absorbed rather than adsorbed. A standard Langmuir adsorption model is generally found to fit the adsorption data (Zhang et al., 2012).

The amount of adsorbed gas generally correlates to total organic carbon (TOC) of the sample, but there is a large scatter for any TOC value. This could be due to differences in the chemical or physical nature of the organic material. Another source of the difference is the pore structure of the organic material. The high intrinsic porosity of the organic material implies that most of the adsorption is on pore walls, so an increase in organic porosity would imply an increase in adsorbed gas. Furthermore for the same porosity, a sample with smaller pores will store more adsorbed gas.

In this chapter, the consideration of storage mechanisms to Langmuir adsorption and free gas in the nanopores is restricted. As pore pressure, P increases the volume occupied by the adsorbed methane on the pore walls, so that the pore space available for free gas storage decreases. This implies that total gas storage cannot just be taken as the sum of the adsorbed methane and the amount of methane that could be stored in a pore space determined by a low-pressure helium porosity measurement, both because it is low pressure and helium adsorbs significantly less than methane (Ambrose et al., 2010; Sigal et al., 2010, 2013; Sigal, 2013b). The fact that measurements are usually made on ground up samples only adds to the potential error. In the case of hydrocarbon liquids, the overestimation of reserves could be even worse. This is because liquids are less compressible so that maximum adsorption occurs at lower pressures, and as the nonadsorbed liquids have higher densities than in the case of gas the mass of liquid associated with the adsorbed phase volume is larger than in the case of gas.

Gas adsorption is temperature dependent. For a given absolute temperature T and pore pressure P, the Langmuir adsorption isotherm equation for the moles of gas adsorbed $S_a(P)$ in a volume of porous material V takes the form

$$S_a(P,T) = S(T)_{a\max} \left[P/P_L(T) \right] / \left[1 + \left[P/P_L(T) \right] \right] \quad (12.6)$$

$S_{a\max}$ is the moles of gas stored at infinite pore pressure in V at temperature T, and the Langmuir pressure P_L is the pressure at which S_a is one-half of $S_{a\max}$ (Langmuir, 1918; Rogers, 1994). In deriving the Langmuir equation, the adsorbed gas is assumed to occupy a monolayer, the presence of a molecule at a particular adsorption site does not affect adsorption at an adjacent site, and gas has unobscured access to the adsorbing surface. Molecular dynamic calculations for slit pores with graphene walls show high average methane density in a zone that is one methane molecule in diameter, and a density equal to bulk methane density at the center of the pore. These results suggest the monolayer assumption is at least a reasonable approximation (Sigal et al., 2013).

Equation 12.6 is actually only correct for an ideal gas. Two possible approaches to modify the adsorption equation for real gases are to replace pressure by either gas molar density or fugacity. The use of molar density is just based on the observation that the amount of adsorption should depend on the number of molecules available to adsorb. The density modification is equivalent to what is done for adsorption of solutes (Giles et al., 1974; Sohn and Kim, 2005). Jahediesfanjani and Civan (2007) and Gouth et al. (2007) have formulated Langmuir-type adsorption in terms of fugacity. Despite this, the current industry practice is to fit measured adsorption data to Equation 12.6 (Civan et al., 2012b).

For convenience, we define L as

$$L = \frac{(P/P_L)}{(1+P/P_L)} \quad (12.7)$$

Following the discussion in Sigal et al. (2013), the volume occupied by the adsorbed gas also satisfies a Langmuir pore pressure relationship given by

$$V_a(P) = V_{amax} L \qquad (12.8)$$

V_{amax} is the maximum adsorbed gas volume in moles. The maximum molar density of methane in the adsorbed layer is given by ρ_{amax} and the density at any pressure by ρ_a.

$$r_{amax} = \frac{S_{amax}}{V_{amax}} \qquad (12.9)$$

$$\rho_a = \rho_{amax} L \qquad (12.10)$$

The pore volume available for free methane storage at a given confining pressure and zero pore pressure is V_{p0}. Taking the pore volume compressibility as C_p then $V_p(P)$ the volume occupied by free gas is given by

$$V_p(P) = V_{p0} + C_p V_{p0} P - V_a(P) \qquad (12.11)$$

Using the real gas equation of state, the moles of free gas n_{sp} stored in V_p are given by

$$n_{sp}(P,T) = \frac{PV_p}{(zRT)} \qquad (12.12)$$

where z is the compressibility factor and R is the universal gas constant (McCain, 1990). In nanometer-scale pores, the compressibility factor z can be a function of pore size (Michel et al., 2012). The total number of moles of gas stored in V at pore pressure P and temperature T n_s is then given by

$$n_s = n_{sp} + S_a \qquad (12.13)$$

Based on Equations 12.6–12.13, Sigal et al. (2013) described a methodology for measuring total storage on a core plug. In an example from the paper, the five parameters (S_{amax}, P_L, ρ_{amax}, V_{p0}, and C_p) were determined on a Barnett plug from a fit to measurement of total methane storage as a function of pore pressure as determined from a series of high-pressure pycnometer measurements. For this fit, the bulk fluid value of z in Equation 12.12 was used. Given the five parameters, the conventional methane storage curve for this sample can also be calculated assuming the pore volume is taken for all pressures as V_{p0} and the storage is given by Sigal et al. (2013).

$$n_{sconventional} = PV_{p0}/(zRT) + S_a \qquad (12.14)$$

$n_{sconventional}$ is the total storage calculation used in standard shale gas simulators and most reserve calculations. The conventional calculation overestimates the total methane storage by 24% at initial reservoir pressure (Figure 5 in Sigal et al., 2013).

Since C_p was small for this sample, the difference is due to the volume taken up by the adsorbed methane.

This example and the general discussion clearly show that the total methane resource in an organic shale reservoir must be calculated in a way that accounts for the volume occupied by the adsorbed phase. In implementing a simulation code, it must be programmed to properly account for the dependence of free gas pore volume on both pore volume compressibility and the pore space lost to adsorption. In measuring total gas storage, it would be best to measure it on "fresh" core with methane at reservoir temperature and stress, for a series of pore pressures ranging from abandonment pressure to initial pressure. Sigal (2013b) has developed an approximate formula that could be used with a selected set of reservoir condition measurements to correct standard measurements.

12.4 FLUID TRANSPORT

Current and ongoing research efforts have documented the need for more appropriate physical formulations to describe transport of gas through nanometer-scale porous media, and it is increasingly recognized that Darcy's law with a constant value of permeability may be inadequate in shale nanopores. Various approaches to modifying the gas flow equations have been proposed (Swami et al., 2012). At the time of writing this chapter, none of these have been adequately tested using experimental data from complex porous media. In contrast to the transport equations valid for most conventional reservoirs, the flow in very low permeability shale gas reservoirs undergoes a transition from a Darcy regime, where viscous coupling between molecules controls the flow condition, to other regimes where molecular collisions with the pore walls have a significant effect on transport. Such effects are well known to petrophysicists and are generally referred to as "Klinkenberg effects," after his 1941 paper. These effects cause an increase in the apparent permeability to gas flow as the pore pressure decreases. Although the rock compressibility effects tend to modify the permeability as the pore pressure changes, the phenomenon described in this chapter is purely a result of the transport of fluids confined in nanopores.

Beskok and Karniadakis (1999) (also see Karniadakis et al., 2005), based on theoretical arguments and experimental data, have developed a gas transport equation for a nonadsorbing gas flowing through straight capillary tubes, which is valid for all flow regimes, encompassing no-slip, transition, slip, and free molecular flows. Florence et al. (2007) also considered flow through a single straight tube. Civan (2010a, b) extended the application of their equation to describe transport through a bundle of tortuous flow paths formed in extremely-low permeability porous media with a single size (uniform) tube size. The equation developed by Beskok and Karniadakis (1999) has been used as the starting point in the development of a gas transport equation to be used in more general porous media.

The starting point for this generalization is the direct extension of Beskok and Karniadakis' work to a bundle of tubes (Michel et al., 2011a, b). Consider a bundle of n straight circular cross-section tubes with radii r_i. If the tubes are imbedded in a nonporous matrix they form a simple porous media with total porosity ϕ. The porosity associated with each radius r_i is taken as ϕ_i. For large enough radii fluid flow through the bundle satisfies the Hagen–Poiseuille equation (Bird et al., 2007), that is Darcy's law with a constant permeability k_G. It is then easy to see that

$$k_G = \frac{1}{8}\sum_1^n r_i^2 \phi_i = \frac{1}{8} r_{Geff}^2 \phi, \quad r_{Geff}^2 = \sum_1^n r_i^2 \frac{\phi_i}{\phi} \quad (12.15)$$

where k_G will be referred to as the geometrical permeability as it is the permeability that describes the flow of a viscous liquid that has no chemical interaction with the pore walls, and no slippage at the wall boundary. For porous media with more complex pore geometries that characterize conventional reservoirs, it is generally true that

$$k_G = A r_{Geff}^2 \phi^m \quad (12.16)$$

where m is the Archie cementation exponent (Katz and Thompson, 1986; Sigal, 2002). A is a proportionality constant that depends on rock type, but for a wide range of clastic and carbonate systems its dynamic range is only about 10%. This shows that the difference in formulation for rock permeability between a bundle of tubes and a more general porous media is just the formation factor. The Beskok and Karniadakis modifications to gas flow are a function of the Knudsen number Kn, which essentially quantifies the importance of methane molecule collisions with the pore wall relative to collisions with other methane molecules. It is the ratio of the average distance a molecule goes between collisions with another molecule to the characteristic size scale of the pore. For a circular cross-section tube of radius r, Kn is given by

$$Kn = \frac{\lambda}{r} \quad (12.17)$$

where λ is the mean-free path. For an ideal gas λ is given by (Bird et al., 2007)

$$\lambda = \frac{1}{2^{0.5} n \pi d_m^2} \quad (12.18)$$

In Equation 12.18, d_m is the diameter of the gas molecule and n is the number density. The equation is derived by assuming the gas molecules are spheres that move freely except when they collide. As such it breaks down when molecules cannot be approximated by spheres, and when the density is large enough that the molecules always feel the effects of other molecules. Using Equation 12.12, introducing the Avogadro's number N_A, and using the real gas equation λ can be written as

$$\lambda = \frac{zRT}{2^{0.5} N_A P \pi d_m^2} \quad (12.19)$$

As long as densities are not too high, this should be an adequate extension of the mean-free path of an ideal gas. In terms of Kn, the Beskok and Karniadakis equation for gas flow in a bundle of circular tubes each of radius r is given in terms of v the macroscopic velocity density as

$$v = \left(\frac{1}{\mu}\right)\left(\frac{r^2 \phi}{8}\right) f(Kn) \frac{dP}{dx} = \left(\frac{1}{\mu}\right) k_G f(Kn) \frac{dP}{dx} \quad (12.20)$$

$$f(Kn) = (1+\alpha Kn)\left(1 + \frac{4Kn}{(1+Kn)}\right) \quad (12.21)$$

$$\alpha = \frac{2}{\pi}\alpha_0 \tan^{-1}\left(\alpha_1 Kn^\beta\right) \quad (12.22)$$

$$\frac{2}{\pi}\alpha_0 = \frac{125}{15\pi^2}, \quad \alpha_1 = 4, \text{ and } \beta = 0.4 \quad (12.23)$$

In Equation 12.20, μ is the bulk gas viscosity. Civan (2010a, b, 2011) has proposed an alternative simpler expression for α. The term $k_G f(Kn)$ is the apparent permeability k which describes gas flow in small capillary tubes. When Kn goes to zero at relatively high pressures or in pores associated with larger diameters, the permeability correction goes to unity and the apparent permeability reduces to k_G. It is not only a function of pore size but also a function of pore pressure and temperature. The correction factor $f(Kn)$ accounts for slippage and the modification to viscosity that occurs in nanometer-scale pores. Gouth et al. (2007) has used molecular dynamic simulation to investigate the Beskok and Karniadakis equation. They found that for multicomponent gases viscosity is not correctly treated by the equation. They were not able to find a modification to the viscosity that worked for all possible mixtures of the various components.

Figure 12.2 shows a plot of the correction $f(Kn)$ term as a function of pore pressure for a 2.5 nm effective radius and

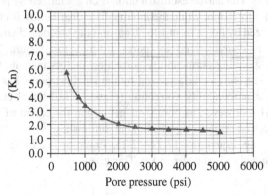

FIGURE 12.2 $f(Kn)$ as a function of pore pressure for radius of 2.5 nm and 350 K.

temperature of 350 K for an ideal gas. For small pore throats, the correction term increases effective permeability k by a factor of 3–5 at abandonment pressures and about 1.5 at initial reservoir pressures. For larger pore throats, the correction term is somewhat smaller.

Measurements of effective pore size based on gas storage measurements and mercury injection measurements suggest that effective pore sizes of 2–3 nm are not unreasonable (Sigal, 2013a, c; Sigal et al., 2013).

For a single size tube k_G and $f(Kn)$ both are functions of a single-scale parameter, the tube radius. Michel et al. (2011a, b) investigated the extension of the Beskok and Karniadakis formulation to bundles of tubes with log normal distributions. It is desirable for the more general case to preserve the form of k as the product of k_G times an apparent permeability correction term $f(Kn)$. For a bundle of tubes, Equation 12.15 or its integral form provides the scale factor r_{Geff} for k_G. The complex functional form of $f(Kn)$ does not allow for a simple analytical expression for r_{Seff}, the effective size parameter that would replace the hydraulic radius, r in Kn when there are multiple tube sizes. For each of the log normal distributions examined a r_{Seff} was found numerically that when substituted into $f(Kn)$ matched the ratio of the calculated value of k/k_G. In general r_{Seff} does not equal r_{Geff}, so that a generalized Beskok and Karniadakis permeability requires two different functions of the pore throat sizes to characterize it. For the case of a general porous media, we have proposed that for simple nonadsorbing gases the complexities of the pore geometry are captured in a k_G term and there exists a r_{Seff} such that the correction term keeps the same form as for the bundle of tube case. That is

$$k = k_G f\left(\text{Kn}\left(r_{Seff}\right)\right) \quad (12.24)$$

For porous media characterized by a pore size distribution, the apparent permeability for a distribution of pore sizes may be quantified through numerical integration of the formulation for apparent permeability for a single capillary tube. Michel et al. (2011a, b) considered three different log-normal pore size distributions (Figure 7 in Michel et al., 2011b). The narrowest distribution is characterized by pores from 1 to 10 nm, while the intermediate distribution is characterized by pores from 1 to 30 nm and the broadest distribution is characterized by pores from 1 to 100 nm. The corresponding permeability correction factors indicate that the distribution characterized by the smallest pores has the largest permeability correction (Figure 8 in Michel et al., 2011b); however, as pore sizes get larger, the extent of this correction remains significant, but diminishes.

Using Equations 12.21–12.23, Equation 12.24 can be expanded to first order in Kn. This gives

$$k = k_G \left(1 + 4\frac{\lambda}{r_{Seff}}\right) \quad (12.25)$$

Equation 12.25 is the Klinkenberg equation (Klinkenberg, 1941) that is used to correct laboratory gas permeability measurements. For an ideal gas from Equations 12.12 and 12.19, the correction term can be written as b/P where b is a constant, so that in the limit of large P k goes to K_G. To first order in Kn, Equation 12.21 for the ideal gas case can also be rewritten as the sum of a Darcy term and a Knudsen diffusion term.

$$\rho v = \frac{\rho}{\mu} k_G \frac{dP}{dx} + 0.589 D_k \frac{d\rho}{dx} \quad (12.26)$$

In Equation 12.26, ρ is the density and D_k is the Knudsen diffusivity (Civan, 2010a, b, 2011; Michel et al., 2011b). The viscosity in the second term is eliminated by using the expression for the viscosity of an ideal gas (Guggenheim, 1960). The apparent diffusion term enters from the gas slippage effect, and does not imply that the transport equation accounts for classical concentration driven diffusion.

The formulation so far discussed has not accounted for gas adsorption on organic pore walls. Adsorption in general will reduce the space available for free gas transport. In addition, adsorption introduces a new transport mode, transport in the adsorbed layer. The effect of adsorption on ideal gas transport in smooth circular cross-section capillary tubes has been investigated in Xiong et al. (2012) (also see Sigal, 2013b). The model assumed the density of the adsorbed layer was in equilibrium with the local pore pressure so that a pore pressure gradient produces a density gradient in the adsorbed layer. The Beskok and Karniadakis equation controlled the transport of the free gas in each tube, and the adsorbed gas transport was calculated by a Fick's law diffusion equation. For this model study, adsorption significantly reduced k_G. The effect on $f(Kn)$ was minimal. Transport in the adsorbed layer in a 2 nm radius tube was only significant for values of diffusivity larger than 0.01 cm²/s.

For a smooth tube with geometrical radius r_0 and with single layer adsorption, in the limit of zero pressure the tube radius for free gas transport is r_0 and its cross-sectional area is πr_0^2. At infinite gas pressure the effective tube radius is $(r_0 - d_m)$ and the cross-sectional area is $\pi(r_0 - d_m)^2$. It is assumed for intermediate pressures that the appropriate tube radius is

$$r_0 - d_m L \quad (12.27)$$

Using Equation 12.27 transport of free methane through a bundle of adsorbing tubes can be computed. Figure 12.3 shows the ratio of k_{Ga}, the geometrical permeability corrected for adsorption to k_G for a tube radius of 2.5 nm, P_L of 1800 psi, at 350 K, with d_m equal to 0.38 nm.

This model would understate the effects of adsorption if the adsorbed zone is not monolayer and if the formation factor is more typical of a conventional reservoir. On the other hand, a rough pore wall surface should act to reduce the permeability loss (Hu et al., 2013a, b).

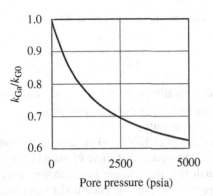

FIGURE 12.3 The geometrical permeability for adsorbing pores at 5000 psia is reduced to 62% of the value it would have without adsorption (modified after Fig. 7 in Sigal, 2013b).

12.5 CAPILLARY PRESSURE, RELAXATION TO EQUILIBRIUM STATE, AND DEPOSITION OF STIMULATION WATER

Because of the nanometer-scale pores in organic shale reservoirs, capillary forces can produce pressure differences between the wetting and nonwetting phases comparable to the in situ reservoir pressure. These small pores also greatly diminish the rate at which fluid saturations can adjust when fluid pressure is changed. Under these conditions, the normal assumption that the wetting and nonwetting phase pressures are related by a capillary pressure curve established for equilibrium saturation states breaks down.

This would not be a significant issue if one could assume immobile water and only simulate gas flow. These reservoirs though are massively hydraulically fractured which could put mobile water into the system. Only a fraction of the stimulation water normally flows back, and much of that is more saline than the initial frac water. Thus, understanding the deposition of this water is important both for reservoir management and to answer environmental concerns.

There are multiple possible reasons for water used in fracturing not being returned to the surface. Among the reservoir conditions that control this are reservoir rock wettability, types and saturation of clays, presence and types of natural fractures, the final connectivity of the stimulation fracture system, reservoir salinity, reservoir gas pressure, and the nature of pores, where the gas is stored in. There is poor quantitative understanding of the effects these conditions have on the amounts of frac fluid returned. Some of the conditions result in the frac water being immobile and some may leave mobile water in the formation.

Qin (2007) studied the deposition of fracture water from a vertical well with a simple multibranched fracture system. The reservoirs simulated were low permeability, low porosity gas reservoirs that were assumed to be water wetting. They had an initial water saturation that allowed limited water mobility. The porosity, permeability, relative permeability, and capillary pressure curves were chosen to be consistent with the then understanding of shale gas reservoirs, and with each other. A range of values for the parameters was investigated. The modeling was performed with a commercial simulator by assuming the stimulation fracture system was filled with water at a prescribed pressure, and then leak off and production was simulated. During the leak off, the well was being produced. Because of the high capillary pressures, and the gas being stored in water wet pores, essentially no water was produced. Countercurrent imbibition drove the frac water into the gas bearing pores and the gas into the fractures. The observation that much more water is produced than would be predicted by the simulation is a strong argument, and the pores where gas is stored in shale gas reservoirs are generally not water wet.

In the formation of the stimulation fracture system, water may be forced into fractures that then become disconnected from the rest of the fracture system trapping water. If this accounted for the large majority of the lost frac water, then most of the created fracture system ends up ineffective for aiding gas production.

Most shale reservoirs have been stimulated with freshwater, but the reservoir salinity is certainly at least that of seawater and may be much higher due to water vapor being removed by the expulsion of oil and gas from the reservoir over geologic time. A regression relationship, the Hill Shirley Kline equation, developed by Shell researchers (Hill et al., 1979) provides a relationship between the amount of clay-bound water and the clay type and the water salinity. The equation reads

$$Ws = \left(0.084\ C_o^{-5} + 0.22\right) CEC \qquad (12.28)$$

where Ws is the clay-bound water in g/100g of dry clay, C_o the salinity in equivalents/l, and CEC the clay cation-exchange capacity in meq/100g.

The equation shows that the amount of clay-bound water held by a clay system increases significantly when the water salinity is reduced as would happen near the fracture face with fresh frac water and a high salinity reservoir. Over long-time periods, diffusion will cause a redistribution of this additional clay-bound water as the salinity in the reservoir comes to equilibrium. The exact amount of frac water immobilized by this mechanism depends on the amount of clay and the clay types, fracture face surface area, the permeability of the clay zones, and the salinity and the effective CEC. Most of these factors are at best poorly known. Clay amounts and types can be reasonably well established from core and/or log measurements. Reservoir salinity may be known. If cation exchange capacity (CRC) values are known for the clay types, they generally vary by factors of two for a given type. Some reservoirs have a significant amount of mixed layer clays. Their CEC values are not found in standard references. Also Equation 12.28, since it is a regression, is controlled by clays that hold large amounts of clay-bound water (Ws) such as montmorillonite. So the quantitative significance of this mechanism is difficult to establish at this time.

The wettability of the organic pores varies as a function of maturity. Even if dominantly gas wet, as one expects for mature organic pores, oxygen sites at the pore walls can hold and trap water (Hu et al., 2013a, b). For less mature organic material, the pore wall can go from mixed wettability to water wet.

Other possible trapping mechanisms are even more poorly understood, so at this time it is impossible to make a good estimate on how much frac water will go into the formation and how much will be mobile. It is, however, clear that current simulators do not properly handle mobile water in the matrix, when there are both low permeability and high capillary pressures.

The assumption that the gas and water are in capillary equilibrium can lead to unrealistic results. Qin (2007) found for gas in water wet pores that under high draw-down pressures and high gas flow rates, this assumption led to negative water pressures.

To overcome the classical assumptions of equilibrium, the variations in phase saturations need to be augmented with a different formulation. One approach to this would be to introduce the capillary force directly into the flow equations. This approach has been applied to model the rise of water in capillary tubes (Hamraoui and Nylander, 2002; Schoelkopf et al., 2000; Zhmud et al., 2000). A second approach is to introduce a relaxation time function that relates the nonequilibrium saturation state to the equilibrium state (Andrade et al., 2010, 2011; Barenblatt et al., 2003; Hanspal and Das, 2011). This is done by relating the effective equilibrium fluid saturation S_e to the instantaneous nonequilibrium fluid saturation S according to the following relationship (Barenblatt et al., 2003; Hassanizadeh and Gray, 1993):

$$S_e - S = X\left(S, \tau, \frac{\partial S}{\partial t}\right) \cong \Psi\left(S, \tau \frac{\partial S}{\partial t}\right) \cong g(S)\tau \frac{\partial S}{\partial t}$$
(12.29)

which is usually approximated simply as

$$S_e = S + \tau \frac{\partial S}{\partial t}$$
(12.30)

where the relaxation time is denoted by τ and simulation time by t, S_e is the effective fluid saturation, and S is the instantaneous fluid saturation under nonequilibrium conditions. Civan (2012) proposed the following formulation for correlation of the relaxation time:

$$p_c^{dyn} - p_c^e = -\tau \frac{dS_w}{dt}$$
(12.31a)

Where the relaxation time τ is given by

$$\tau = \frac{\tau_o}{\left(S_w - S_{wi}\right)^a}, a > 0$$
(12.31b)

Hence, the relaxation time τ varies with the wetting fluid (water) saturation S_w, where the parameters (S_{wi}, τ_o, and a) depend on temperature.

In Barenblatt et al. (2003) where an effective saturation is introduced, it is argued that there exists an effective saturation S_{eff} such that the relative permeability curves and the capillary pressure curve evaluated under equilibrium conditions at this effective saturation are equal to the relative permeability values and capillary pressure value at the actual nonequilibrium saturation S. They recognize that the assumption that the same S_{eff} works for all the functions may not always be true, but assume it for simplicity. In their paper (Barenblatt et al., 2003), they argue that τ for most of its range is a constant. To implement this approach in a simulator requires a better understanding of τ.

For better understanding of the function of τ, Michel et al. (2012, 2013) have explored this for a bundle of capillary tubes where for incompressible fluids of the same viscosity and the transition from one equilibrium state to another can be solved analytically. For this case, the assumption that τ is a constant over most of its range is not satisfied. Though Michel et al. (2012, 2013) were able to find expressions for τ that may be generalizable to the gas/water case, further exploratory work is necessary in order to determine if the capillary nonequilibrium formulation for modeling multiphase gas-water flows in very low permeability systems where the time to reach equilibrium may be significantly longer than the simulation time step is feasible. If not, adding the capillary force directly into the flow equations may be required.

12.6 CHARACTERIZATION OF FLUID BEHAVIOR AND EQUATIONS OF STATE VALID FOR NANOPOROUS MEDIA

Numerous studies have demonstrated that the properties and behavior of fluid systems in nanoporous media deviate from those observed in bulk fluids. This has two primary sources. The first is in nanoporous media where the interaction of the fluid molecules with the pore wall cannot be ignored relative to intermolecular interactions, only the latter of which is important in bulk fluids. The second is that the small number of molecules in the pore may suppress development of multiple phases or the existence of a true liquid phase. In some cases, in hydrocarbon-bearing shale reservoirs, the modification of the fluid properties and behavior may create effects that enhance transport capabilities of the pore fluids. Because the extent of such modifications depends on pore size in addition to other relevant factors, such as heterogeneity, pore connectivity, surface roughness, and wettability, the macro- and micropore geometry of the media is a critical factor in determining the effective porous media modifications. Depending on this, properties such as permeability, pressure, storage capacity, phase behavior, viscosity, interfacial tension, and capillary pressure can be altered significantly. This in turn may have a profound impact on the performance of wells completed in hydrocarbon-bearing shale reservoirs. An accurate understanding of the relevant processes and their roles is instrumental in determining the composition, storage, and transport of fluids. It is required for effective reserves estimation; prediction of variations in fluid composition, condensate dropout, and blockage; and long-term production forecasting and planning of shale reservoirs by means of a model-based approach (Devegowda et al., 2012; Zhang et al., 2013a, b).

In this section, we discuss approaches available for estimating the conditions and properties of fluids contained in nanopores. We start by pointing out the primary factors of interest which include the pore-fluid composition (number density of molecules) and state (gas, liquid, or solid), phase-equilibrium, pressure and temperature, pore geometry and structure (shape, surface roughness, texture, fabric, and pore size distribution), pore-wall proximity (ratio of the pore size to molecule sizes), pore–wall (organic or inorganic) wettability, adsorption, and the potentials that determine the pore–wall interactions with molecules of various types. Different fluid pressures and compositions are attained in different size pores, so the reservoir properties are an average over properties that vary on the nanoscale.

Prediction of the phase behavior and phase-equilibrium in shale is a complicated issue and molecular dynamics (MD) calculations have provided considerable insight into the nanoscale behavior. One of the key advantages of MD simulations is that at the nanoscale it may be prohibitively expensive or even impossible to conduct lab-scale measurements and MD simulations provide a virtual framework for understanding nanoscale phenomena. However, MD calculations currently are only practical on systems with very simple pore geometries, limited fluid complexities, and idealized pore wall potentials. As such they serve very well in exploring the phenomena occurring in nanoscale pores and supplying estimates on changes to parameters such as the critical temperature and pressure, but cannot adequately substitute for laboratory measurements. To use industry-scale reservoir simulation software requires alterations to the algebraic equations of state to account for the above-mentioned unique features of hydrocarbon-bearing shale systems.

The bulk fluid behavior has been satisfactorily described by certain empirical modifications of the van der Waals equation of state (VDW-EoS). The two outstanding candidates are the Soave–Redlich–Kwong (SRK-EoS) and the Peng–Robinson (PR-EoS) EoS. Baled et al. (2012) provide certain improvements on these two equations by incorporating a correlation of an additional volume correction for high-pressure–high-temperature as a function of the reduced temperature, molecular weight, and acentric factor.

Previous studies have made the modifications on such equations to take into account the pore-proximity effect by means of the correlations of the critical properties (density, temperature, and pressure) and the acentric factor against the ratio of the pore size to molecule size, and the reduction of the effective molecular volume of the adsorbate. Proving whether these are the proper and sufficient corrections for pore-proximity is open for future research. Then, the pore-proximity effects observed in various size pores in shale can be averaged over the representative elementary volume of shale porous media considering the pore-size distribution (Civan, 2002b). Theoretically, the conventional bulk fluid properties are attained as the ratio of the pore size to molecule size approaches infinity. Various applications have shown that the fluid properties are close to their bulk values for pore sizes greater than 10 nm. Other types of properties of practical importance, such as the existing real-gas deviation factor and viscosity correlations, such as those given by Dranchuk and Abou-Kassem (1975) and Bergman and Sutton (2007) can also be modified for pore-proximity once the critical properties are correlated against the ratio of the pore size to molecule size.

Examples are elaborated for single- and multiple-component mixtures concerning the gas and gas–condensate applications by Michel et al. (2011a, b), Devegowda et al. (2012), and Zhang et al. (2013a, b). In Michel et al. (2011a, b), the impact of pore proximity on gas z-factors is shown to have an influence on gas formation volume factors and densities (Figure 1 in Michel et al., 2012). Across a wide range of pressures and temperatures, significant differences in the z-factors are observed implying that gas-in-place estimates, gas reservoir material balance calculations, and prediction of gas flow rates are likely to be erroneous if pore proximity effects are not considered.

In Devegowda et al. (2012), the phase behavior of a rich gas–condensate sample was analyzed in different pore sizes, showing significant differences in the phase envelopes and percentages of liquid dropouts (% by volume) in pores of 2, 4, and 5 nm for the gas–condensate mixture (Figures 4 and 5 in Devegowda et al., 2012). The modified phase envelopes were calculated using the Peng-Robinson EOS by inputting the critical point data for the confined fluids (Singh et al., 2009) into a commercially available PVT package (CMG, 2008). Comparisons to the corresponding bulk values are also shown and illustrate dramatic changes in the calculated phase diagrams for the confined fluids. Specifically, the liquid-dropout is significantly lower due to pore-proximity effects in comparison to the corresponding values in bulk (Figure 5 in Devegowda et al., 2012). This has serious implications in the calculations of well productivity and for reservoir management. It is highly likely that reserves calculations based on the traditional method where pore proximity effects are ignored may in fact be underestimating the quantity of recoverable fluids. Additionally, because our calculations indicate fewer issues with condensate dropout in the near wellbore region, very small nanopores may in fact result in very stable production without any drop in productivity for extended periods of time. Presently though, the influence of pore proximity on the acentric factor remains unknown and poorly studied (Cho et al., 1985).

The pore-proximity correction for fluids in a single-pore as described earlier provides different levels of corrections in the various size pores of shale. However, there is another factor of importance when dealing with a network of interconnected pores of prescribed size distributions. When different pore throats and pore bodies are involved in a network having a certain connectivity pattern (coordination number) (Civan, 2001, 2002a), the conditions of pressure and composition of the fluid in different size pores are also likely to be different (Zhang et al., 2013a, b). However, the description of effective fluid phase behavior for relatively larger reservoir or rock volumes continues to remain challenging. At this time, existing knowledge of fluid phase behavior is largely restricted to single pore sizes.

One approach to finding appropriate fluid properties in a connected pore system is to require the chemical potential, and therefore the fugacity of the fluid components are the same in every pore. The Peng–Robinson EOS relating pressure P, temperature T, and molar volume V_m as shown in Equation 12.32 may be used to derive an expression for fugacity for real gases (McCain, 1990).

$$\left(P + \frac{a_T}{V_m(V_m+b)+b(V_m+b)}\right)(V_m-b) = RT \quad (12.32)$$

$$\ln\left(\frac{f}{p}\right) = Z - 1 - \ln(Z-B) - \frac{A}{2^{1.5}B}\ln\left[\frac{Z+\left(2^{0.5}+1\right)B}{Z-\left(2^{0.5}-1\right)B}\right] \quad (12.33)$$

where

$$A = \frac{a_T P}{(RT)^2} \text{ and } B = \frac{bP}{RT}$$

For purposes of demonstrating the challenges associated with describing fluid storage and pore pressure when pore confinement effects become dominant, we examine methane storage for the simple connected pore system, where methane in bulk at 1000 psia and 200F is connected to pores of different sizes of 2, 4, 6, 8, and 10 nm size pores (Figure 13 in Zhang et al., 2013a).

As mentioned earlier, equilibrium conditions may be determined by equating the fugacity of methane in these different pores. In the example earlier, the fugacity of methane in the bulk using bulk properties for methane is 937 psia. The corresponding z-factor is 0.947.

At equilibrium, the chemical potential of the fluid molecules distributed throughout the system should be identical (McCain, 1990). Because the chemical potential is a function of the fugacity for real gases, under these conditions of equilibrium, the fluid distributed throughout the various capillary tubes should also have identical fugacities. By a process of trial and error, the pore pressures in the other pores may be estimated by finding pressure values in Equation 12.32 that result in a fugacity of 937 psia for methane critical pressures and temperatures appropriately chosen for the respective pore sizes (Zhang et al., 2013a, b). The results obtained for the 2, 4, 6, 8, and 10 nm pores are as follows:

1. The pore pressures are estimated as 935, 943, 950, 953, and 957 psia, respectively.
2. The real-gas deviation factors of the confined gases are 1.0, 0.992, 0.987, 0.985, and 0.981, respectively.
3. The densities of methane within each pore can then be estimated as 34.0, 34.5, 34.9, 35.1, and 35.4 g/m^3, respectively.
4. Further, if we assume all cylindrical pores of length 100 nm, then the number of molecules confined in each cylindrical pore are 400, 1626, 3704, 6625, and 10419, respectively.

These results imply that due to pore proximity effects in nanopores resulting in modified critical properties and gas compressibility factors, methane in different pores exist in equilibrium at different densities and at different pore pressures. Although this work presents an exploratory analysis of the implications of pore proximity, to our knowledge this constitutes the first model-based approach to quantifying fluid storage and pore pressure in connected pore systems. The results highlight the complex interplay between pore geometry and fluid properties in nanoporous media and indicate that under equilibrium even a single-component fluid is likely to be characterized by different pore pressures in different pore systems. We are continuing to investigate the full implications of this proposed storage model; however, experimental verification of the results presented earlier are a necessary next step prior to extending the results to large-scale porous media and a discussion of the implications for reservoir performance.

12.6.1 Viscosity Corrections

While in the previous section, the focus was largely centered on phase behavior, Beskok and Karniadakis (1999) also discuss variations of gas viscosity with the Knudsen number Kn previously defined. Due to the effects of rarefaction of the gas, the viscosity is shown to decrease with increasing Knudsen number or decreasing pore size as given by

$$\frac{\mu}{\mu_\infty} = \frac{1}{1+\alpha \text{Kn}} \quad (12.34)$$

where μ_∞ is the unconfined fluid viscosity. The mean-free path in this formulation is taken as the bulk gas value. The gas viscosity representing the resistance to gas transfer under various transport regimes as given by this equation that represents flow beyond true viscous flow should ideally be referred to as the pseudo-viscosity because the concept of viscosity at free molecular and transition flow regimes loses its true meaning. The rarefaction coefficient α is correlated by Civan (2010a, b, 2011).

$$\frac{\alpha_o}{\alpha} - 1 = \frac{A}{Kn^B} \qquad (12.35)$$

where A, B, and α_o are the empirical parameters. The rarefaction coefficient considers all possible regimes in one equation whose value depends on the Knudsen number. In the limit of low Knudsen number flows or under the viscous flow regime, the pseudo-viscosity essentially reduces to the viscosity defined for Darcy law. Gouth et al. (2007) used molecular dynamic simulation to investigate the Beskok and Karniadakis equation. They found for multi-component gases that viscosity is not correctly treated by Equation 12.34. They were not able to find a modification to the viscosity that worked for all percentage mixes of two components.

12.6.2 Corrections for Interfacial Tension

Assume that gas is the nonwetting phase and the liquid film over the pore surface is a wetting phase (Hamada et al., 2007). In a cylindrical pore system containing gas and liquid phases, the diameter of the gas–liquid interface varies with the gas (or liquid) saturation. Therefore, the effect of pore confinement to the apparent interfacial tension (IFT) γ between the gas and liquid phases is given by the following equation (Civan et al., 2012b), modified after Hamada et al. (2007):

$$\gamma = \gamma_\infty - \frac{c}{D_p S_G^{1/2}} \qquad (12.36)$$

where $\gamma = \gamma_\infty$ is the limit value of the apparent gas–condensate IFT when the pore size D_p attains the infinity, D_p is the mean-pore diameter, c is the empirical constant, and S_G is the gas saturation. This correlation can be used to predict the effect of pore proximity on the capillary pressure by using the Leveret J-function. In mature organic pores, gas is relative wetting with respect to water so this approach would need to be modified.

12.7 UPSCALING HETEROGENEOUS SHALE-GAS RESERVOIRS INTO LARGE HOMOGENIZED SIMULATION GRID BLOCKS

As described above, shales are characterized by an organic matrix embedded in an inorganic background. Additionally, there may be microcracks within the matrix and consequently, mature stimulated organic shale reservoirs tend to have a complex pore geometry on the micron scale. There exist pores in the organic and inorganic matrix, and natural and simulation fractures. These four pore systems have different geometries and also different wettabilities. However, in order to perform reservoir-scale modeling, it is necessary to find a set of equivalent grid cell properties such as anisotropic permeabilities, porosity, capillary pressure, and relative permeability curves that accurately represent the flow behavior at the nanoscale. In order to accomplish this, it is necessary to perform some form of multiphase upscaling and this continues to be a topic of active research. Successful standard approaches to upscaling assume the reservoir can on a macroscale be divided into flow units, each of which have capillary pressure curves and relative permeability curves of the same shape, and permeability values that are distributed in some simple way such as a log-normal distribution. It is also assumed that laboratory measurements on core or bulk fluids can be used to provide the input to characterize each flow unit. This is clearly not the case for organic shale reservoirs, and consequently upscaling shale microstructural features may necessitate some new approaches in order to reproduce flow characteristics of shales at the reservoir scale. For the case of nanometer-scale heterogeneity, core measurements are made on samples that are a combination often unspecified of the multiple pore types that exist on the nanoscale. As such to be correctly used they must be modeled to extract the intrinsic properties of each pore type. These can then be input into a nanoscale model and upscaled. It cannot be assumed that the effective value for a quantity obtained from core measurement provides the effective value at any other scale or geometry.

The other issue associated with upscaling is that the connectivity of the different pore systems continues to be poorly understood. Although 3D imaging of shales has revealed some organic pore connectivity, the connectivity of the organics to the inorganic background remains unclear. As shale imaging technology evolves, we are likely to be able to resolve some other challenges, such as the continuity of the organics and whether the organic matrix feeds directly into the high conductivity fractures or whether the hydrocarbons move serially from the organics to the inorganics and eventually to the fracture systems. If gas is almost completely stored in the organic pores than the small amount of water produced during production would seem to imply that the inorganic pores contain no free water and/or the organic pores couple directly into the stimulation fractures.

The accuracy of numerical simulation is better when the grid block sizes are smaller. However, when dealing with simulation of large shale reservoirs, sufficiently large grid blocks need to be used to reduce the number of grid blocks in order to be able to generate simulation results of practical importance with reasonable computational effort and numerical accuracy. This can be accomplished by replacement of intricately detailed models by larger scale models that still correctly capture the production behavior of the detailed models. This process is frequently referred to as upscaling.

The shale formation quad-porosity system contains essentially the subsystems of organic, inorganic, natural fractures, and induced fractures. These subsystems can be connected to each other by six possible paths (Figure 1 in Civan et al., 2012b). However, the nature and degree of connectivity are generally unknown and must be inferred by modeling and history matching. In general, the distribution of these subsystems is also not known a priori but bulk-volume fractions can be inferred by measurements with shale core samples or other means. As demonstrated later, an approach of considering the microdescription as a statistical description and preserving this in the upscaling to a larger scale quad-porosity system in a manner to preserve the total porosity of each system can yield satisfactory results. Under the initial reservoir conditions, the organic and inorganic subsystems can contain gas but the natural and hydraulic fractures contain mostly water. Simulation of gas/water systems also requires the relative permeability data for these subsystems. Adsorption is mostly significant for the organic pores, but usually neglected in the inorganic.

To use commercial simulators, it is sometimes reasonable to group the various flow units of shale into two categories to construct a dual-porosity model. For example, the first can be the inorganics and the fracture network including the microchannels and microfractures. The second can be the organics distributed throughout the inorganics. In another example, the quad-porosity shale system can be separated into a group of the organic subsystem combined with the natural fractures and a second group of the inorganic porosity subsystem combined with the induced fractures. It may sometimes be possible to combine the quad-porosity system into a homogenized single-porosity system. It is important to recognize that there is no general proof that such simplifications will correctly capture production over the life of the reservoir.

The hydraulics of fluid in the heterogonous quad-porosity system can be modeled by a finely detailed reservoir simulation block and the production with time can be predicted by simulation. This result can then be used for upscaling this finely detailed description to a coarse model that matches the production trends of the finely detailed model. For this purpose, two approaches to upscaling are illustrated here. The first is based on upscaling of the finely detailed continuum model of shale to a lumped-parameter leaky tank model of shale, and the second is based on upscaling of the finely detailed continuum model of shale to coarse continuum model.

12.7.1 Upscaling Fine Continuum Model of Shale to Lumped-Parameter Leaky Tank Model of Shale

Extending the approaches by Civan (1993, 1998, 2000), Gupta and Civan (1994), Civan and Rasmussen (2001), and Matejka et al. (2002), the subsystems of the quad-porosity shale are lumped into four tanks and then the material balances of these tanks are written by considering the connectivity of the various tanks and their internal storage and transport mechanisms by appropriate material balance and transfer-rate equations. Obviously, this approach requires various empirical transfer functions whose storage and transport parameters need to be determined to match the results of the finely detailed block simulation. The matching of finely detailed block simulations may not be unique. The leaky tank model approach has the advantage that it does not require a detailed microscopic description once transfer functions are established, but the transfer functions do not seem to have a straightforward interpretation, or tie to core measurements. They though may be obtainable by modeling core measurements with a tank model (Matejka et al., 2002).

Different tank models can be developed for four-porosity systems depending on the nature of connectivity between the subsystems of the quad-porosity shale, such as series, parallel, and parallel with cross flow (Hudson et al., 2012). For example, flow may occur in series through the organic, inorganic, natural fractures, and hydraulic fracture when the organic regions are distributed as isolated pockets in the inorganic. In another example, flow may occur in parallel from organic and inorganic into the fracture subsystem. A more complicated case may involve parallel flow through the organic and inorganic porosity systems feeding into the natural fractures as well as having a cross flow between the organic and inorganic subsystems. The rate equations involving the latter example involving the parallel organic and inorganic flows with cross flow given by Hudson (2011, 2013) can be compiled in a matrix equation form as

$$\frac{d}{dt}\begin{bmatrix} R_O \\ R_I \\ R_N \\ R_F \\ R_W \end{bmatrix} = \begin{bmatrix} -(\lambda_{ON}+\lambda_{OI}) & 0 & 0 & 0 & 0 \\ \lambda_{OI} & -\lambda_{IN} & 0 & 0 & 0 \\ \lambda_{ON} & \lambda_{IN} & -\lambda_{NF} & 0 & 0 \\ 0 & 0 & \lambda_{NF} & -\lambda_{FW} & 0 \\ 0 & 0 & 0 & \lambda_{FW} & 0 \end{bmatrix} \begin{bmatrix} R_O \\ R_I \\ R_N \\ R_F \\ R_W \end{bmatrix}, t>0$$

(12.37)

where λ_{ON}, λ_{OI}, λ_{IN}, λ_{NF}, and λ_{FW} denote the transfer-rate coefficients for mass transfers occurring between the subsystems of the organic O, inorganic I, natural fractures N, induced fractures F, and the well W, and R_O, R_I, R_N, R_F, and R_W denote the mass contained in these subsystems, respectively.

The initial conditions are given by

$$\begin{bmatrix} R_O \\ R_I \\ R_N \\ R_F \\ R_W \end{bmatrix} = \begin{bmatrix} R_{OP}^o \\ R_{IP}^o \\ 0 \\ 0 \\ 0 \end{bmatrix}, t=0 \qquad (12.38)$$

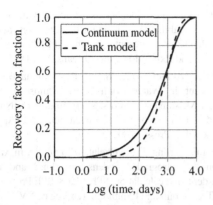

FIGURE 12.4 Recovery factor for quad-porosity shale where organic and inorganic in parallel involving cross flow (modified after Hudson, 2013).

The solution to Equation 12.37 subject to the condition of Equation 12.38 can be obtained analytically and then the overall recovery factor can be determined by

$$RF = 1 - \sum_{i=1}^{N=4} w_i e^{-\lambda_i t}, \ t \geq 0 \quad (12.39)$$

Such that

$$\sum_{i=1}^{N=4} w_i = 1.0, \ t \geq 0 \quad (12.40)$$

where w_i is the weighting coefficient. The best estimate values of the parameters of the tank model are determined to match Equation 12.37 to the recover factor curve obtained by solving the finely gridded reservoir block as described in the next section (Fig. 12.4).

12.7.2 Upscaling Finely Detailed Continuum Model of Shale to Coarse Continuum Model of Shale

Most commercial simulators are based on the continuum approach, which solve the porous media averaged reservoir transport equations over the simulation domain by separating the solution domain into a number of grid blocks using appropriate numerical solution methods such as the method of finite differences and the control volume finite element method. Most commercial simulators can handle the single- and dual-porosity models although attempts are now being made for development of multiporosity versions. The quad-porosity system sometimes can be simulated after some simplifications with dual-porosity model or single-porosity model versions. In the *dual-porosity model* the quad-porosity shale system can be separated, for example, into a group of the organic subsystem combined with the natural fractures and a second subsystem group of the inorganic porosity combined with the induced fractures (Civan et al., 2013; Michel, 2013). In the *single-porosity model*, the whole quad-porosity system is represented by a single homogenized group.

As stated earlier, the actual distribution of and the volume occupied by the quad-porosity subsystems in shale are usually not known. This problem can be alleviated by random distributions in a manner to preserve the total porosity of each subsystem. A second reason for using a random model is for systems that combine very low permeability storage units with high permeability transport units in upscaling is the time it takes fluid to flow into the transport units must in an average sense be preserved. If one just takes a pattern of units and magnifies all the distances without making any other changes, then the production curve will not be preserved. Certain prescribed organic pattern distributions are assumed using information on pore size distributions of organics. For this purpose, for example, the randomly distributed globules and streaks of organic can be considered to simulate the local heterogeneity in shale. Then, the simulated heterogeneous shale block is separated into a fine grid for detailed simulation. Subsequently, an appropriate coarse grid system, such as the Tartan Coarse Grid varying in the grid size from the borders to the center, is used to match the coarse grid simulation results to the detailed heterogeneous shale block simulation results. The volume fraction of the organics and inorganics should be preserved between the finely and coarsely gridded systems (Michel, 2013).

For upscaling purposes, we consider simulating the flow of only the methane gas phase through the shale over a square domain of unit thickness, for example, under isothermal conditions. The initial gas pressure throughout the grid block is prescribed at a constant value. The flow is induced by assuming constant pressures below the initial gas pressure along the edges. First, the fine grid simulation is carried out by randomly distributing the quad-porosity subsystems over a fine grid and the production profiles are determined. Then, the upscaling of shale using the Tartan Grids yields a successful match of the fine-grid production results as illustrated in Figure 12.5.

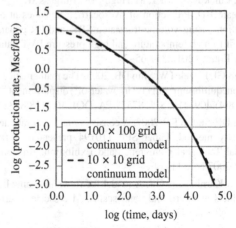

FIGURE 12.5 Upscaling heterogeneous shale with tartan coarse grids (modified after Michel, 2013).

12.8 FINAL REMARKS

This chapter presented an overview of the critical issues that need to be considered in proper simulation of storage and transport processes in organic shale reservoirs. The outstanding features of shale reservoir formations and fluid behavior in their pore structure have been delineated. Modification of the fluid and transport properties in extremely tight porous shale systems containing organic and inorganic constituents and having intricately complicated pore structure has been discussed. The recent advancements made in modeling of the relevant processes have been addressed and the areas needing further work have been emphasized. Upscaling of fine continuum model of shale to a lumped-parameter leaky tank model and coarse continuum model of shale have been developed and implemented for applications providing convenience in large-scale reservoir simulation.

REFERENCES

Alfred D, Vernik L. A new petrophysical model for organic shales. Petrophysics 2013 June;54 (3):240–247.

Ambrose RJ, Hartman RC, Campos MD, Akkutlu IY, Sondergeld C. New pore-scale considerations for shale gas in place calculations. Paper SPE 131772-MS presented at the SPE Unconventional Gas Conference, Pittsburgh, PA, USA; 2010 Feb 23. DOI: 10.2118/131772-MS.

Andrade J, Civan F, Devegowda D, Sigal RF. Accurate simulation of shale-gas reservoirs. Paper SPE 35564-PP, the SPE Annual Technical Conference and Exhibition Florence, Italy; 2010 Sept 19–22.

Andrade J, Civan F, Devegowda D, Sigal RF. Design requirements for a shale gas reservoir simulator and an examination of how the requirements compare to designs of current shale gas simulators. Paper SPE-144401, the SPE Americas Unconventional Gas Conference, The Woodlands, TX; 2011 June 14–16.

Baled H, Enick RM, Wu Y, McHugh MA, Burgess W, Tapriyal D, Morreale BD. Prediction of hydrocarbon densities at extreme conditions using volume-translated SRK and PR equations of state fit to high temperature and high pressure PVT data. Fluid Phase Equilib 2012;317:65–76.

Barenblatt GI, Patzek TW, Silin DB. 2003. The mathematical model of nonequilibrium effects in water-oil displacement. SPE J 2003;8(4):409–416. SPE 87329-PA. DOI: 10.2118/87329-PA.

Bergman DF, Sutton RP. 2007. A consistent and accurate dead-oil-viscosity method. Paper SPE 110194 presented at the SPE Annual Technical Conference and Exhibition, Anaheim, CA, USA; 2007 Nov 11–14.

Beskok A, Karniadakis GE. A model for flows in channels, pipes, and ducts at micro and nano scales. Microscale Thermophys Eng 1999;3 (1):43–77.

Bird RB, Stewart WE, Lightfoot EN. *Transport Phenomena*. New York: John Wiley & Sons, Inc.; 2007.

Campos MD, Akkutlu IY, Sigal RF. A molecular dynamics study on natural gas solubility enhancement in water confined to small pores. SPE 124491 presented at the SPE Annual Technical Conference and Exhibition, New Orleans, LA, USA; 2009.

Campos MD, Akkutlu IY, Sigal RF. Methane Solubility Enhancement in Water Confined to Nanoscale Pores. In: Sushanta KM, Suman C, editors. Microfluidics and Nanofluidics Handbook: Fabrication, Implementation, and Applications. New York: CRC Press; 2011.

Cho S, Civan F, Starling K. A correlation to predict maximum condensation for retrograde condensation fluids and its use in pressure-depletion calculation. SPE 14268, SPE Proceedings 60th Annual Tech. Conf. and Exhibition, Las Vegas, NV, USA; 1985.

Civan F. Waterflooding of naturally fractured reservoirs- an efficient simulation approach, proceedings. Paper SPE 25449-MS, Proceedings of the SPE Production Operations Symposium, Oklahoma City, OK, USA; 1993. pp 395–407.

Civan F. Quadrature solution for waterflooding of naturally fractured reservoirs. SPE Reserv Eval Eng 1998;1 (2):141–147.

Civan F. Leaky-tank reservoir model including the NonDarcy effect. J Pet Sci Eng 2000;28 (3):87–93.

Civan F. Scale effect on porosity and permeability- kinetics, model, and correlation. AIChE J 2001;47 (2):271–287.

Civan F. Relating permeability to pore connectivity using a power-law flow unit equation. Petrophysics 2002a;43 (6):457–476.

Civan F. A triple-mechanism fractal model with hydraulic dispersion for gas permeation in tight reservoirs. Paper 74368 presented at the SPE International Petroleum Conference and Exhibition in Mexico, Villahermosa, Mexico; 2002b.

Civan F. Effective correlation of apparent gas permeability in tight porous media. Transport Porous Med 2010a;82 (2):375–384.

Civan F. A review of approaches for describing gas transfer through extremely tight porous media. Porous Media and Its Applications in Science, Engineering, and Industry, Vafai K. (editor). Proceedings of the Third ECI International Conference on Porous Media and its Applications in Science, Engineering and Industry; 2010 June 20–25, Montecatini Terme, Italy; 2010b. pp 53–58.

Civan F. Porous Media Transport Phenomena. Hoboken, NJ: John Wiley & Sons; 2011. p 463. ISBN: 978-0-470-64995-4.

Civan F. Temperature dependency of dynamic coefficient for non-equilibrium capillary pressure-saturation relationship. R&D note, AIChE J 2012;58 (7):2282–2285. DOI: 10.1002/aic.13817.

Civan F. Modeling gas flow through hydraulically-fractured shale-gas reservoirs involving molecular-to-inertial transport regimes and threshold-pressure gradient. Paper SPE-166324-MS, the 2013 SPE Annual Technical Conference & Exhibition; 2013 Sept 30–Oct 2; New Orleans, LA, USA.

Civan F, Rasmussen ML. Asymptotic analytical solutions for imbibition waterfloods in fractured reservoirs. SPE J 2001;6 (2): 171–181.

Civan F, Devegowda D, Sigal RF. Theoretical fundamentals, critical issues, and adequate formulation of effective shale gas and condensate reservoir simulation. Porous Media and Its Applications in Science, Engineering, and Industry, Vafai K. (editor). Proceedings (CD) of the 4th International Conference on Porous Media and its Applications in Science and Engineering, ICPM4; 2012 June 17–22; Potsdam, Germany, American Institute of Physics; 2012a. pp 155–160.

Civan F, Rai CS, Sondergeld CH. Determining shale permeability to gas by simultaneous analysis of various pressure tests. Paper 144253-PA. SPE J 2012b Sep;17(3):717–726.

Civan F, Devegowda D, Sigal RF. Review of essentials for effective analysis, formulation, and simulation of organic shale reservoirs containing gas, condensate, and/or oil. Proceedings of the 23 ITU Petroleum and Natural Gas Seminar and Exhibition, İstanbul, Turkey, 2013 June 27–28. pp 108–118.

CMG. 2008. http://www.cmgl.ca.

Curtis M, Ambrose R, Sondergeld C, Rai C. Structural characterization of gas shales on the micro- and nano-scales. Paper CUSG/SPE 137693, Canadian Unconventional Resources and International Petroleum Conference, Calgary, Alberta, Canada; 2010 Oct 19–21.

Curtis ME, Ambrose RJ, Sondergeld CH, Rai CS. Transmission and scanning electron microscopy investigation of pore connectivity of gas shales on the nanoscale. Paper SPE144391 presented at the SPE North American Unconventional Gas Conference and Exhibition, The Woodlands, TX, USA; 2011 June 14–16.

Curtis M, Sondergeld C, Rai C. 2013. Investigation of the microstructure of shales in the oil window, URTeC 1581844. Unconventional Resources Technology Conference, Denver, CO, USA; 2013 Aug 12–14.

Devegowda D, Sapmanee, K, Civan F, Sigal RF. Phase behavior of gas condensates in shales due to pore proximity effects: implications for transport, reserves and well productivity. Paper SPE 160099, SPE Annual Technical Conference and Exhibition, San Antonio, TX, USA; 2012 Oct 4–7.

Dranchuk PM, Abou-Kassem JH. Calculation of Z factors for natural gases using equations of state. J Can Pet Tech 1975 July–Sept:34–36.

Florence FA, Rushing JA, Newsham KE, Blasingame TA. Improved permeability prediction relations for low permeability sands. Paper SPE 107954, presented at the 2007 SPE Rocky Mountain Oil and Gas Technology Symposium held in Denver, CO, USA; 2007 April 16–18.

Giles CH, Smith D, Huitson A. 1974, A general treatment and classification of the solute adsorption isotherm. J Colloid Interface Sci 1974 June;47 (3):755–765.

Gouth, F, Collell, J, Galliero G, Wang J. Molecular simulation to determine key shale gas parameters, and their use in a commercial simulator, 2007, EAGE Annual Conference & Exhibition, London, SPE 164790; 2013 June.

Guggenheim EA. *Elements of the Kinetic Theory of Gases*. Oxford: Pergamon Press; 1960.

Gupta A, Civan F. An improved model for laboratory measurement of matrix to fracture transfer function parameters in immiscible displacement. SPE Paper 28929, SPE Annual Technical Conference and Exhibition, New Orleans, LA, USA, 1994 Sept 25–28.

Hamada Y, Koga K, Tanaka H. Phase equilibria and interfacial tension of fluids confined in narrow pores. J Chem Phys 2007;127 (8):084908-1–084908-9.

Hamraoui A, Nylander T. Analytical approach for the Lucas–Washburn equation. J Colloid Interface Sci 2002;250:415–421.

Hanspal NS, Das DB. Dynamic effects on capillary pressure–saturation relationships for two-phase porous flow: implications of temperature. AIChE J 2011;5 (7):1–15.

Hassanizadeh SM, Gray WG. Thermodynamic of capillary pressure in porous media. Water Resour Res 1993;29:3389–3405.

Hill HJ, Klein GE, Shirley OJ, Thomas EC, Waxman WH. 1979. Bound water in Shaly Sands – its relation to Qv and other formation properties. SPWLA 1973 May 1:1979.

Hu Y, Devegowda D, Striolo A, Ho TA, Phan A, Civan F, Sigal RF. 2013a. A pore-scale study describing the dynamics of slickwater distribution in shale gas formations during and after hydraulic fracturing. Paper SPE 164552, presentation at the Unconventional Resources Conference-USA held in, The Woodlands, TX, USA; 2013a April 10–12.

Hu Y, Devegowda D, Striolo A, Civan F, Sigal RF. 2013b. Microscopic dynamics of water and hydrocarbon in shale-kerogen pores of potentially mixed-wettability. Paper SPE 167234, the SPE Unconventional Resources Conference Canada held in, Calgary, Alberta, Canada, 2013b Nov 5–7.

Hudson J. Quad-porosity model for description of gas transport in shale-gas reservoirs. Master's Thesis, Oklahoma University, Norman, OK; 2011.

Hudson J. Unpublished Internal Report for RPSEA, Devegowda D, Civan F, Sigal RF. (Principal Investigators), RFP 2009UN001, Simulation of Shale Gas Reservoirs Incorporating Appropriate Pore Geometry and the Correct Physics of Capillarity and Fluid Transport, funded by RPSEA, University of Oklahoma; 2013.

Hudson J, Civan F, Michel-Villazon GG, Devegowda D, Sigal RF. Modeling multiple-porosity transport in gas-bearing shale formations. Paper SPE 153535 presented at the SPE Latin American and Caribbean Petroleum Engineering Conference, Mexico City, Mexico; 2012 April 16–18.

Jahediesfanjani H, Civan F. Determination of multi-component gas and water equilibrium and nonequilibrium sorption isotherms in carbonaceous solids from early-time measurements. Fuel 2007;86:1601–1613.

Karniadakis G, Beskok A, Aluru N. *Microflows and Nanoflows Fundamentals and Simulation*. New York: Springer Publishers; 2005.

Katz AJ, Thompson AH. 1986. Quantitative prediction of permeability in porous rocks. Phys Rev B 1986;34:8179–8181.

Klinkenberg LJ. *Drilling and Production Practice*. New York: API; 1941. p 200–213.

Langmuir J. The adsorption of gases on plane surfaces of glass, mica and platinum. J Am Chem Soc 1918;40:1361.

Matejka MC, Llanos EM, Civan F. Experimental determination of the matrix-to-fracture transfer functions for oil recovery by water imbibitions. J Pet Sci Eng 2002;33 (4):253–264.

McCain WD. *The Properties of Petroleum Fluids*. 2nd ed. Tulsa, OK: PennWell Books; 1990.

Michel G. Unpublished Internal Report for RPSEA, Devegowda D, Civan F, Sigal RF. (Principal Investigators), RFP 2009UN001, Simulation of Shale Gas Reservoirs Incorporating Appropriate Pore Geometry and the Correct Physics of Capillarity and Fluid Transport, funded by RPSEA, University of Oklahoma; 2013.

Michel G, Sigal RF, Civan F, Devegowda D. Proper modeling of nano-scale real-gas flow through extremely low-permeability porous media under elevated pressure and temperature conditions. 7th International Conference on Computational Heat and Mass Transfer, İstanbul, Turkey; 2011a July 18–22.

Michel G, Civan F, Sigal RF, Devegowda D. Parametric investigation of shale gas production considering nano-scale pore size distribution, formation factor, and NonDarcy flow mechanisms. SPE-147438-PP, the 2011 SPE Annual Technical Conference and Exhibition (ATCE), Denver, CO, USA; 2011b Oct 30–Nov 2.

Michel G, Civan F, Sigal RF, Devegowda D. Effect of capillary relaxation on water entrapment after hydraulic fracturing stimulation. Paper SPE-155787-PP, the 2012 Americas Unconventional Resources Conference held June 5–7 at the David L. Lawrence Convention Center in Pittsburgh, PA, USA; 2012.

Michel G, Civan F, Sigal RF, Devegowda D. Proper simulation of fracturing-fluid flowback from hydraulically-fractured shale-gas wells delayed by NonEquilibrium Capillary Effects. URTeC 1582001 presented at the Unconventional Resources Technology Conference, Denver, CO, USA; 2013 Aug 12–14.

Passey QR, Bohacs K, Esch WL, Klimentidis R, Sinha S. From oil-prone source rock to gas-producing shale reservoir – geologic and petrophysical characterization of unconventional shale gas reservoirs. Paper SPE 131350-MS presented at the SPE International Oil and Gas Conference and Exhibition in China, Beijing, China; 2010 June 8–10.

Qin B. Numerical study of recovery mechanisms in tight gas reservoirs. Masters Thesis, University of Oklahoma, Norman, OK, USA; 2007.

Quirein J, Galford J, Witkowsky J, Buller D, Jerome Truax J. Review and comparison of three different gas shale interpretation approaches. Paper SPWLA-2012-075, Society of Petrophysicists and Well-Log Analysts, SPWLA 53rd Annual Logging Symposium; 2012 June 16–20; Cartagena, Colombia.

Rogers RE. *Coalbed Methane Principles & Practice*. Englewood Cliffs, NJ: Prentice-Hall; 1994.

Schoelkopf J, Ridgway CJ, Gane PAC, Matthews GP, Spielmann DC. Measurement and network modeling of liquid permeation into compacted mineral blocks. J Colloid Interface Sci 2000; 227:119–131.

Sigal RF. Coates and SDR permeability, two variations on the same theme. Petrophysics 2002;43 (1):38–46.

Sigal RF. A note on the intrinsic porosity of organic material in shale gas reservoir rocks. Petrophysics 2013a June, 54(3), 236–239.

Sigal RF. The effects of gas adsorption on storage and transport of methane in organic shales. SPWLA-D-12-00046, SPWLA 54th Annual Logging Symposium, 22–26, New Orleans, LA, USA; 2013b June 22–26; SPWLA-D-12-00046.

Sigal RF. Mercury capillary pressure measurements on Barnett core. SPE Reserv Eval Eng 2013c Nov;16(4), 432–442.

Sigal RF. Pore size distributions for organic shale reservoir rocks from NMR spectra combined with adsorption measurements, submitted for publication, 2015. Accepted for publication in SPEJ.

Sigal RF, Odusina E. Laboratory NMR measurements on methane saturated Barnet shale samples. Petrophysics 2011;52 (1): 32–49.

Sigal RF, Akkutlu IY, Kang SM, Ambrose R. A methodology to measure total gas storage in an organic shale. Hedberg Research Conference "Critical Assessment of Shale Resource Plays," Austin, TX, USA; 2010 Dec 5–10.

Sigal RF, Akkutlu IY, Kang SM, Diaz-Campos M, Ambrose R. The laboratory measurement of the gas storage capacity of organic shales. Petrophysics 2013 June, 54(3), 224–235.

Singh SK, Sinha A, Deo G, Singh JK. Vapor-liquid phase coexistence, critical properties, and surface tension of confined alkanes. J Phys Chem 2009;113 (17):7170–7180.

Sohn S, Kim D. Modification of Langmuir Isotherm in solution systems – definition and utilization of concentration dependent factor. Chemosphere 2005;58:115–123.

Swami V, Clarkson CR, Settari A. NonDarcy flow in shale nanopores: Do we have a final answer? SPE 162665 presented at the SPE Canadian Unconventional Resources Conference; 2012 Oct 30–Nov 1, Calgary, Alberta, Canada.

Xiong X, Devegowda D, Michel GG, Sigal RF, Civan F. A fully-coupled free and adsorptive phase transport model for shale gas reservoirs including NonDarcy flow effects. Paper SPE 159758, SPE Annual Technical Conference and Exhibition, San Antonio, TX, USA; 2012 Oct 4–7.

Zhang T, Ellis GS, Ruppal SC, Milliken K, Yang R. Effect of organic-matter type and thermal maturity on methane adsorption in shale-gas systems. Org Geochem 2012;47:120–131.

Zhang Y, Civan F, Devegowda D, Jamili A, Sigal RF. Critical evaluation of equations of state for multicomponent hydrocarbon fluids in organic rich shale reservoirs. Paper SPE 1581765 the 2013 Unconventional Resources Technology Conference, Denver, CO, USA; 2013a Aug 12–14.

Zhang Y, Civan F, Devegowda D, Sigal RF. Improved prediction of multi-component hydrocarbon fluid properties in organic rich shale reservoirs. Paper SPE 166290 the 2013 SPE Annual Technical Conference & Exhibition in New Orleans, LA, USA; 2013b Sept 30–Oct 2.

Zhmud BV, Tiberg F, Hallstensson K. Dynamics of capillary rise. J Colloid Interface Sci 2000;228:263–269.

13

PERFORMANCE ANALYSIS OF UNCONVENTIONAL SHALE RESERVOIRS

Hossein Kazemi[1], Ilkay Eker[1], Mehmet A. Torcuk[1] and Basak Kurtoglu[2]

[1] Colorado School of Mines, Golden, CO, USA
[2] Marathon Oil Company, Houston, TX, USA

13.1 INTRODUCTION

Producing oil and gas from nano-Darcy shale formations has become possible because of intuitive insights of geoscientists and engineers, ingenuity of well completion engineers, and persistent field tests in the last two decades of the twentieth century. Specifically, in late 1990s, Mitchell Energy began producing commercial gas from Barnett shale using slickwater for hydraulic fracturing instead of the polymer gel fracking fluids. Then, around 2006, EOG Resources, Inc. (EOG) began producing more oil from its North Dakota Bakken leases, using slickwater in the implementation of multistage hydraulic fracturing (Zuckerman, 2013). This achievement spread to Eagle Ford shale and other US shale resources. In 2014, the United States produces nearly 3.5 million barrels/day of new oil from shale reservoirs across the country.

This chapter presents practical approaches for analyzing well performance in shale oil and gas reservoirs. In this regard, the first notable observation is the vast contrast between core-measured permeability versus field-measured permeability from flow tests. Specifically, core-measured permeabilities are two to three orders of magnitude lower than field-measured permeabilities. The prevailing explanation for the larger field-measured permeability points to formation of extensive micro and macro fracture network as byproduct of the multistage well stimulation. The second notable observation is the rapid decline of well flow rates for a short period of time but stabilizing to a gentler decline rate for months and years. For long-term forecasting, engineers use an empirical analysis method, introduced to the industry by Arps (1944). Interestingly enough, this mathematical model is closely related to the pseudo-steady-state flow in high-permeability conventional reservoirs. However, when applied to unconventional shale reservoirs, we will show that it starts with transient flow behavior during early production and converges to boundary-dominated flow (BDF) later.

Five topics are the focus of this chapter: shale reservoir production, flow rate decline analysis, flow rate and pressure transient analysis, reservoir modeling and simulation, and enhanced oil recovery (EOR).

13.2 SHALE RESERVOIR PRODUCTION

Shale is a fissile mudstone consisting of silt, 4–60 µm, and clay-size particles, less than 4 µm, which are largely mineral fragments. Shale hydrocarbon reservoirs, in addition to mineral fragments, include a small amount of organic matter. Under large overburden stress and high temperatures, the organic material slowly converts to hydrocarbon components that create a large internal hydrostatic pressure locally, which could cause creation of microfracture pores because of the fluid expansion force. The pore size in shale could be less than 2 nm and as high as 2 µm. Nanopores create large capillary pressures, lower the critical pressure, and temperature of hydrocarbon components creating a shift in the phase envelope of the resident fluids, and cause capillary condensation and slippage of gas molecules at the pore walls (Knudsen flow). Because of low matrix permeability, Darcy flow (advection) becomes so small that molecular diffusion can play a significant role in the mass transfer of fluids from the

Fundamentals of Gas Shale Reservoirs, First Edition. Edited by Reza Rezaee.
© 2015 John Wiley & Sons, Inc. Published 2015 by John Wiley & Sons, Inc.

matrix to micro and macro fractures. Hydrocarbon-rich shale reservoirs are typically oil wet while their counterparts, tight sandstones, are generally water wet.

Shale reservoirs are classified based on whether the hydrocarbon source is an integral part of the reservoir rock fabric (self-sourced, as in Haynesville), or is adjacent to the reservoir (locally sourced, as in Bakken), or is located at large distances from the reservoir and require significant hydrocarbon migration (externally sourced, as in Austin Chalk) (Tepper et al., 2013). The self-sourcing is the prominent, distinguishing feature of the low-permeability shale reservoirs compared to the externally sourced, low-permeability sandstone reservoirs.

Shale reservoirs have very low permeability and porosity. A typical shale reservoir has a very low permeability matrix of about 10^{-5} to 10^{-2} mD and a porosity of less than 10%. Shale reservoirs must be stimulated to produce commercial amounts of oil and gas. Such reservoirs were considered to be unproductive two decades ago, but persistence led to the development of a new technology known as multistage hydraulic fracturing, which has facilitated oil and gas production from the tight shale matrix. Shale hydraulic fracturing uses slickwater, consisting of 98–99.5% of water plus a relatively small amount of dissolved salt and chemicals, which include friction reducers, acids to remove formation damage, scale and corrosion inhibitors, biocides, and proppants. The hydraulic fracturing process creates local stresses, which break up the shale matrix into smaller fractures that improve oil and gas flow.

To effectively access the reservoir pores, drilling engineers drill long horizontal wells in the formation parallel to the *minimum horizontal stress* direction. Then, completion engineers place a large set of multistage *transverse hydraulic fractures* in each well to stimulate the drainage volume of the well, Figure 13.1. The horizontal well segment is in the range of 4,000–10,000 ft in length (~5,000 ft in Eagle Ford and 9,000 ft in Bakken), consisting of 20–50 transverse hydraulic fractures in the multistage stimulation process. Each horizontal well is usually from 350 to 1200 ft apart (350 and 700 ft in Eagle Ford and 1200 ft in Bakken). The multistage hydraulic fractures create a *dual-porosity environment* in the wellbore drainage area, called the "stimulated reservoir volume (SRV)." The dual-porosity environment makes it easier for hydrocarbons to flow from small pores of the matrix, to micro and macro fractures, and to the wellbore. The inverse of this flow hierarchy is much less effective in fluid injection processes. To confirm the dual-porosity nature of the SRV, reservoir engineers compare the permeability from the rate transient test with that of the cores. If the transient-test permeability is much larger than the core permeability, we conclude that the hydraulic fracturing process has induced macrofractures, which, in turn, has created a larger formation effective permeability than that of the matrix.

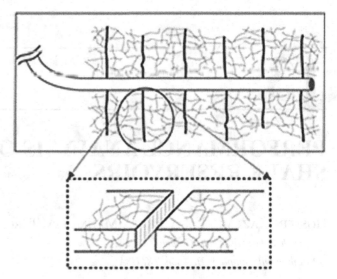

FIGURE 13.1 A conceptual dual-porosity environment created by multistage hydraulic fracturing (Torcuk et al., 2013a).

Hydrocarbon production from unconventional reservoirs is limited to the immediate vicinity of each well because formation permeability is very small and well interference is minimal. Closer well spacing generally improves the inter-well drainage which enhances oil recovery. Salman et al. (2014) explain a tracer flow-back test, which can be used to quantify the inter-well drainage effectiveness. These authors also provide data from a tracer test conducted in two closely spaced Eagle Ford wells.

While well interference testing and analysis is an important technology for quantifying inter-well drainage characteristics, in the following sections, we describe techniques only for analyzing the performance of individual wells in shale reservoirs.

13.3 FLOW RATE DECLINE ANALYSIS

Flow rate decline analysis is a common technique used to forecast future production performance of conventional reservoirs within specific periods (e.g., depletion drive period, waterflood for various in-fill well clusters, EOR with specific well spacing, etc.). In the last decade, decline curve analysis has been also used to forecast individual well performance in unconventional reservoirs. Figure 13.2 presents an example of the decline curve analysis for Elm Coulee field. The field produces oil from the liquid-rich Bakken formation, which is classified as an unconventional resource. The same data when plotted as $\log q$ versus $\log t$ forms a straight line with slope approximately equal to $-1/2$, which is the characteristic of linear flow production into a transverse hydraulic fracture.

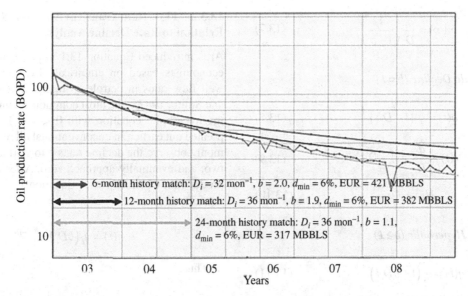

FIGURE 13.2 Hyperbolic decline analysis of an Elm Coulee well (Kurtoglu et al., 2011). The b exponent is 2.0 for the first six months, 1.9 for the first twelve months, and 1.1 for the first 24 months. The b exponent becomes smaller than 1.0 in long-term performance analysis.

13.3.1 Decline Curve Analysis in Unconventional Reservoirs

In what follows, we will show that decline curve analysis (Arps, 1944; Garb and Smith, 1987) has a tractable relationship to the solution of one-dimensional flow in hydraulically fractured reservoirs. Specifically, during reservoir depletion, production flow rate declines in a definite manner if the well's operating conditions remain relatively constant. Equation 13.1 is the classical equation used in decline curve analysis.

$$q(t) = q_i (1 + b D_i t)^{-1/b} \tag{13.1}$$

where

b = rate exponent

D_i = nominal decline rate at time zero, T^{-1}

$q(t)$ = flow rate at time t, $L^3 T^{-1}$

q_i = initial flow rate, $L^3 T^{-1}$

Equation 13.1 is a *three-parameter* performance-matching equation. The initial flow rate q_i is not measureable. Engineers use the rate decline history to calculate the performance parameters by history matching. When b is equal to zero, the flow rate decline is *exponential*, when b is 1.0, the rate decline is called *harmonic*, and when b is between zero and one, the rate decline is *hyperbolic*. However, in shale application, the starting b value is generally 4.0 for a few days, declines and stays at 2.0 for several weeks or months, and eventually becomes zero. Kurtoglu et al. (2011) observed a portion of this behavior in analyzing the flow rate performance of Elm Coulee wells (Fig. 13.2). This behavior is contrary to the decline rate behavior observed in high-permeability, conventional reservoirs. For instance, in conventional reservoirs, the initial b value is zero, resulting in exponential decline at very early times. The numerical value for b is between zero and one after a few days. The latter represents the hyperbolic decline.

Integrating the flow rate equation, Equation 13.1, we obtain the cumulative produced hydrocarbon at time t. Below, we have summarized the integration results for various rate exponents b:

13.3.1.1 Exponential Decline ($b = 0$)

$$q = q_i e^{-D_i t} \tag{13.2}$$

$$N(t) = \frac{q_i}{D_i}\left(1 - e^{-D_i t}\right) \tag{13.3}$$

$$N(\infty) = \frac{q_i}{D_i} \tag{13.4}$$

where $N(t)$ is the cumulative production at time t, and $N(\infty)$ is the ultimate cumulative production.

The time $\tau = 1/D_i$ is the "time constant." At $t = \tau$, $N(\tau) = 1 - e^{-1} = 0.63 \, N(\infty)$

13.3.1.2 Hyperbolic Decline ($0 < b < 1$)

$$q(t) = q_i \left(1 + b D_i t\right)^{-1/b} \tag{13.5}$$

$$N(t) = \frac{q_i}{(b-1)D_i}\left(1 - e^{-D_i t}\right)\left[\left(1 + b D_i t\right)^{\left(1-\frac{1}{b}\right)} - 1\right] \tag{13.6}$$

$$N(\infty) = \frac{q_i}{D_i}\left(\frac{1}{1-b}\right) \quad (13.7)$$

13.3.1.3 Harmonic Decline (b = 1)

$$q(t) = q_i(1+D_i t)^{-1} \quad (13.8)$$

$$N(t) = \frac{q_i}{D_i}\ln(1+D_i t) \quad (13.9)$$

$$N(\infty) = \infty \quad (13.10)$$

13.3.1.4 Beyond Hyperbolic (b ≥ 1)

$$q(t) = q_i(1+bD_i t)^{-1/b} \quad (13.11)$$

$$N(t) = \frac{q_i}{(b-1)D_i}\left(1-e^{-D_i t}\right)\left[\left(1+bD_i t\right)^{\left(1-\frac{1}{b}\right)} - 1\right] \quad (13.12)$$

$$N(\infty) = \infty \quad (13.13)$$

13.3.1.5 Nominal and Effective Decline Rate Terminology

The initial decline rate D_i, known as the *nominal decline rate*, will not remain constant unless the decline is exponential. The definition of decline rate $D(t)$ at time t is

$$D(t) = -\frac{d\ln q}{dt} = \frac{D_i}{1+bD_i t} \quad (13.14)$$

In economic analysis of projects, engineers use the *effective decline rate*, defined below:

$$d = \frac{(q_1 - q_2)}{q_1} = \frac{[q(t_1) - q(t_2)]}{q(t_1)} \quad (13.15)$$

The following equation relates the effective decline rate d to the nominal decline rate $D(t)$:

$$d = 1 - e^{-D} \quad (13.16)$$

Finally, the following equation relates the cumulative production $N(t)$, initial rate q_i, and current flow rate $q(t)$:

$$N(t) = \frac{q_i}{(b-1)D_i}\left[\left(\frac{q_i}{q(t)}\right)^{(b-1)} - 1\right] \quad (13.17)$$

We used Equations 13.6 and 13.12 separately to derive Equation 13.17.

13.3.2 Flow Rate Transient Analysis (RTA) and its Relation to Rate Decline Analysis

Arps introduced Equation 13.1 to petroleum engineers and economists based on empirical observations of declining well flow rates in petroleum reservoirs (Arps, 1944; Garb and Smith, 1987). The most common decline rate equation is the hyperbolic equation with $0 < b < 1$. Nevertheless, as pointed out earlier, in unconventional oil reservoirs, we commonly observe the decline rates b to start as four, decline to two, and eventually approach zero. After a short period of production, when $bD_i t \gg 1$, Equation 13.1 takes the following form:

$$q(t) = q_i(bD_i)^{-1/b} t^{-1/b} \quad (13.18)$$

Thus,

$$\frac{1}{q(t)} = \frac{(bD_i)^{1/b}}{q_i} t^{1/b} \quad (13.19)$$

13.3.2.1 Bilinear Flow Regime
For $b = 4.0$, we obtain the classical bilinear flow equation—a signature of a vertical hydraulic fracture, which is observed in field about a-third of the time (Lacayo and Lee, 2014; Patzek et al., 2013).

$$\frac{1}{q(t)} = \frac{(4D_i)^{1/4}}{q_i} \sqrt[4]{t} \quad (13.20)$$

13.3.2.2 Linear Flow Regime
For $b = 2.0$, we obtain the classical linear flow toward a vertical hydraulic fracture, which is observed in the majority of stimulated wells.

$$\frac{1}{q(t)} = \frac{(2D_i)^{1/2}}{q_i} \sqrt{t} \quad (13.21)$$

Multiplying by Δp, and using formation volume factor to convert surface flow rate to bottom-hole flow rate, we obtain:

$$\frac{\Delta p}{q(t)B} = \frac{\Delta p (2D_i)^{1/2}}{q_i B_i} \sqrt{t} \quad (13.22)$$

Now, we compare the above equation with the RTA equation for multistage hydraulic fractured well, given below:

$$\frac{\Delta p}{qB} = \frac{4.064}{\sqrt{k_{f,\text{eff}}}} \frac{(\pi/2)\mu}{(hn_{\text{hf}} y_{\text{hf}})} \left(\frac{1}{(\varphi c_t)_{f+m} \mu}\right)^{1/2} \sqrt{t} + \frac{141.2 \mu}{k_{f,\text{eff}} h} s_{\text{hf}}^{\text{face}}$$

$$(13.23)$$

Both equations indicate that the rate-normalized pressure plot versus $\sqrt{t}$ is a line with a positive slope. Please note that *when flow rate* is constant, the coefficient ($\pi/2$) disappears in Equation 13.23 (Wattenbarger et. al., 1998).

13.3.2.3 Boundary-Dominated Flow Regime

Now we assess the long-time behavior of the decline rate equation as b approaches 1.0 or becomes smaller than 1.0. For this assessment, we rewrite Equation 13.19 as given below:

$$\frac{\Delta p}{q(t)B} = \frac{\Delta p (bD_i)^{1/b}}{q_i B_i} t^{1/b} = \left[\frac{\Delta p (bD_i)^{1/b}}{q_i B_i} t^{(1/b - 1/2)}\right]\sqrt{t} \quad (13.24)$$

This equation indicates that the slope of the rate-normalized pressure versus $\sqrt{t}$ increases with time, which we observe in field data (Fig. 13.3). The increase in slope is generally a response to the outer boundary effects. The flow behavior is known as *boundary-dominated flow* (BDF).

Numerical simulation indicates that during BDF, the value of b approaches 0.001. With such small b, we approximate the long-term reservoir performance by the exponential decline behavior, whose rate exponent b is zero. Thus, we integrated Equation 13.2 from the onset of the boundary effect t_2 to any future time t to obtain the cumulative production for the $(t - t_2)$. We obtained:

$$N_p(t - t_2) = \frac{q_2}{D_2}\left[e^{(-D_2 t_2)} - e^{(-D_2 t)}\right] \quad (13.25)$$

We also know that the cumulative production $N_p(t - t_0)$ at time t is the sum of cumulative production in the earlier intervals, shown below:

$$N_p(t - t_0) = N_p(t_2 - t_0) + N_p(t - t_2) \quad (13.26)$$

where

$N_p(t_2 - t_0)$ = cumulative production for time interval, $t_2 - t_0$

$N_p(t - t_2)$ = cumulative production for time interval, $t - t_2$

$N_p(t - t_0)$ = cumulative production for total time, $t - t_0$

Equation 13.26 provides a method to estimate the ultimate recovery for the well (as the sum of the measured cumulative production in the first several months plus the late-time extrapolation of the well flow rate). The ultimate recovery is synonymous with the *estimated ultimate recovery* (EUR).

For *black oil* shale reservoirs, the cumulative produced oil recovery per hydraulic fracture stage equals the product of SRV for that stage and the change in Stock tank barrel (STB) of oil per unit pore volume:

$$\frac{N_p(t - t_0)}{n_{hf}} = SRV\left[\left(\phi\frac{S_o}{B_o}\right)^0 - \left(\phi\frac{S_o}{B_o}\right)^n\right] \quad (13.27)$$

The superscripts 0 and n refer to the initial time t_0 and time $t = t_n$.

Similarly, for *dry gas* shale reservoirs, the estimated ultimate recovery (EUR) for gas per hydraulic fracture stage equals the product of SRV and the standard cubic feet (SCF) of gas per unit pore volume:

$$\frac{G_p(t - t_0)}{n_{hf}} = SRV\left[\left(\phi\frac{S_g}{B_g}\right)^0 - \left(\phi\frac{S_g}{B_g}\right)^n\right] \quad (13.28)$$

Equation 13.27 does not account for the natural gas liquid, nor does Equation 13.28 account for the condensate dropping out in the reservoir or in the gas processing plant. Accounting of these details requires compositional material balance. Equation 13.28 is the material balance equation for production of free gas. To account for total gas production, one must add cumulative produced solution gas to the cumulative free gas production. Neither Equation 13.27 nor Equation 13.28 accounts for natural gas liquids.

13.3.3 Field Applications

13.3.3.1 Example 1: Rate Decline Analysis

This example is the production history of an Eagle Ford liquid-rich well. The rate history of the well exhibits three flow regimes: bilinear flow and linear flow followed by BDF, as shown in Figure 13.3. Region 1, the bilinear flow regime, shows the flow rate data from the first practical measurement time t_0 until time t_1. This flow rate region exhibits $-1/4$ slope. Region 2, the linear flow regime, covers the time interval $t_2 - t_1$ and has a slope of $-1/2$. We present the bilinear and linear flow analysis in Section 13.5. However, in this section, we will only focus on estimating the ultimate cumulative hydrocarbon production.

We calculate the cumulative hydrocarbon production for Region 1 and Region 2 using daily production data in the field; however, we calculate the cumulative production for Region 3 using exponential decline equation, Equation 13.25 (Table 13.1). For exponential decline analysis, we used the oil and gas flow rates reported in Figure 13.4a and b. The decline rate D_2 for the oil and gas are 0.0015 and 0.0009 day^{-1}, respectively.

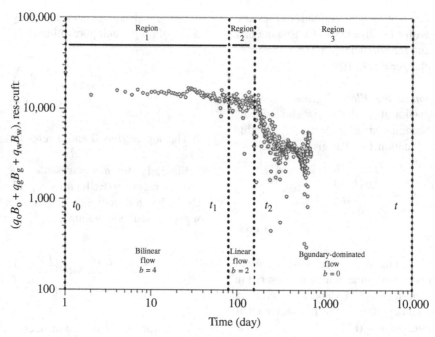

FIGURE 13.3 Field example 1: Rate decline curve analysis plot.

TABLE 13.1 Field example 1: Decline curve analysis

Regions	Cumulative oil produced, STB	Cumulative free gas produced, MMSCF
1	54,516	0
2	48,060	3.3
3	123,470	323.1
Total	226,046	326.4

13.4 FLOW RATE AND PRESSURE TRANSIENT ANALYSIS IN UNCONVENTIONAL RESERVOIRS

In unconventional reservoirs, the formation permeability is in the millidarcy to nanodarcy range even after stimulation. Thus, the well flow rates are generally low and decline continuously. To achieve the highest flow rate for economic viability, wells are operated at low bottom-hole pressure conditions. However, it is difficult to maintain a constant bottom-hole pressure during the well's life because of the ever-changing operating conditions. Thus, engineers use the rate-normalized pressure flow equation to reduce the impact of the varying bottom-hole pressure on the analysis of the well performance. In this section, we will present relevant equations.

After the well rate decline has reached a steady behavior, we can obtain reliable reservoir flow characteristics from a pressure build-up test. The early pressure buildup reflects the bilinear and linear flow regimes, followed by linear flow regime. In the next section, we also present a pressure build-up test from a Bakken well to illustrate the use of pressure build-up data.

13.4.1 Bilinear Flow Regime in Multistage Hydraulic Fracturing

13.4.1.1 Single-Porosity Bilinear Flow In this section, we approximate the dual-porosity flow in the reservoir by a *single-porosity equivalent medium* connected to a finite-conductivity hydraulic fracture. Furthermore, when we move beyond the very early radial and linear flow, the intermediate time flow regime leads to *bilinear flow* presented by Equation 13.29. The hierarchy of flow includes linear flow from macrofractures, surrounding the matrix, to the hydraulic fractures and to the horizontal well.

$$\frac{\Delta p(t)}{q_o B_o + q_w B_w + q_g B_g} = \frac{44.102 \lambda_t^{-1}}{h n_{hf} \sqrt{w_{hf} k_{hf}}} \left[\left(\frac{\lambda_t}{(\phi c_t)_{f+m} k_{f,\text{eff}}}\right)^{1/4}\right] t^{1/4}$$
$$+ \frac{141.2 \lambda_t^{-1}}{k_{f,\text{eff}} h n_{hf}} s_{hf}^{\text{well}}$$

(13.29)

where

$$m_{bl} = \frac{44.102 \lambda_t^{-1}}{h n_{hf} \sqrt{w_{hf} k_{hf}}} \left(\frac{\lambda_t}{(\phi c_t)_{f+m} k_{f,\text{eff}}}\right)^{1/4}$$

(13.30)

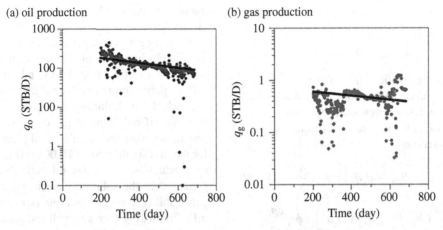

FIGURE 13.4 Field example 1: Exponential decline trend.

$$k_{f,\text{eff}} = k_f \phi_f + k_m \quad (13.31)$$

For pressure buildup, the following equation applies:

$$\frac{p_{ws}(\Delta t) - p_{wf}(\Delta t = 0)}{q_o B_o + q_w B_w + q_g B_g}$$
$$= \frac{44.102 \lambda_t^{-1}}{h n_{\text{hf}} \sqrt{w_{\text{hf}} k_{\text{hf}}}} \left[\left(\frac{\lambda_t}{(\varphi c_t)_{f+m} k_{f,\text{eff}}}\right)^{1/4}\right] \left[(t_p + \Delta t)^{1/4} - (\Delta t)^{1/4}\right]$$

(13.32)

In practice, one obtains more accurate and reliable results from a pressure build-up test—especially if the build-up test is conducted a few weeks after well cleanup.

13.4.1.2 Dual-Porosity Bilinear Flow

The material presented in this section is a more elaborate extension of the previous section. Here, the hierarchy of flow includes flow from matrix to macrofractures, then linear flow from macrofractures to the infinite-conductivity hydraulic fractures and to the horizontal well. Thus, we deal with three media—matrix, macrofractures, and infinite-conductivity hydraulic fractures. Equation 13.33 presents the intermediate bilinear flow conditions. However, its use is more reliable when used in pressure build-up tests, which could yield a numerical value for the dual-porosity shape factor σ (Torcuk et al., 2013a).

$$\frac{\Delta p(t)}{q_o B_o + q_w B_w + q_g B_g}$$
$$= \frac{(45.103)\lambda_t^{-1}}{\sqrt{k_{f,\text{eff}}} h n_{\text{hf}} (2 y_{\text{hf}})} \left\{\left(\frac{1}{1-\omega}\right)^{1/2} \frac{1}{\left[\frac{\sigma}{4} k_m \left(\phi \lambda_t^{-1} c_t\right)_{f+m}\right]^{1/4}}\right\} t^{1/4}$$
$$+ \frac{141.2 \lambda_t^{-1}}{k_{f,\text{eff}} h n_{\text{hf}}} s_{\text{hf}}^{\text{well}}$$

(13.33)

The pressure build-up equation for the dual-porosity bilinear flow regime is

$$\frac{p_{ws}(\Delta t) - p_{wf}(\Delta t = 0)}{q_o B_o + q_w B_w + q_g B_g}$$
$$= \frac{(45.103)\lambda_t^{-1}}{\sqrt{k_{f,\text{eff}}} h n_{\text{hf}} (2 y_{\text{hf}})} \left\{\left(\frac{1}{1-\omega}\right)^{1/2} \frac{1}{\left[\frac{\sigma}{4} k_m \left(\phi \lambda_t^{-1} c_t\right)_{f+m}\right]^{1/4}}\right\}$$
$$\left[(t_p + \Delta t)^{1/4} - (\Delta t)^{1/4}\right]$$

(13.34)

13.4.2 Linear Flow Analysis for Reservoir Permeability

Rate transient analysis is a practical method to calculate the *effective permeability* of the SRV volume. The equation we use for RTA is Equation 13.35, which is the solution of the three-phase, black-oil diffusivity equation for the infinite-acting period (i.e., before the outer boundary effects appear). Similarly, Equation 13.36 is for the gas-condensate reservoirs when the surface oil is the condensate collected in the surface separators. Oil can also condense in the reservoir to form S_o; however, we assume that the condensed oil in the reservoir is not mobile.

$$\frac{\Delta p(t)}{q_o B_o + q_w B_w + q_g B_g}$$
$$= \frac{4.064 \left(\pi/2\right)}{\sqrt{M_t} (h n_{\text{hf}} y_{\text{hf}})} \left[\left(\frac{1}{(\phi c_t)_{f+m}}\right)^{1/2}\right] \sqrt{t} + \frac{141.2}{M_t h n_{\text{hf}}} s_{\text{hf}}^{\text{face}} \quad (13.35)$$

$$\frac{\Delta p(t)}{q_w B_w + (380 \xi_o q_o + q_g) B_g}$$
$$= \frac{4.064 \left(\pi/2\right)}{\sqrt{M_t} (h n_{\text{hf}} y_{\text{hf}})} \left[\left(\frac{1}{(\phi c_t)_{f+m}}\right)^{1/2}\right] \sqrt{t} + \frac{141.2}{M_t h n_{\text{hf}}} s_{\text{hf}}^{\text{face}} \quad (13.36)$$

where

$$M_t = k_{f,\text{eff}} \lambda_t \quad (13.37)$$
$$\lambda_t = \lambda_o + \lambda_w + \lambda_g \quad (13.38)$$

Equation 13.35 is the solution of Equation 13.39 for a horizontal well in oil-producing unconventional reservoirs with the well configuration presented by Figure 13.1. Similarly, Equation 13.36 is the solution of Equation 13.40 for gas-condensate reservoirs.

$$\nabla \cdot (k\lambda_t) \nabla p_o + \left(B_w \hat{q}_w + B_o \hat{q}_o + B_g \hat{q}_g\right) = \phi c_t \frac{\partial p_o}{\partial t} \quad (13.39)$$

$$\nabla \cdot (k\lambda_t) \nabla p_o + \left[B_w \hat{q}_w + B_g \left(380 \xi_o \hat{q}_o\right) + B_g \hat{q}_g\right] = \phi c_t \frac{\partial p_o}{\partial t} \quad (13.40)$$

In Equation 13.40, the coefficient 380 represents the SCF of gas per pound-mole of condensate, ξ_o in molar density in pound-mole per cubic feet of condensate, q_o is the cubic feet of condensate produced, and $\hat{q}_o$ is the specific production rate of condensate per unit rock volume.

For an oil reservoir, the total compressibility is

$$c_t = c_\phi + S_w c_{wa} + S_o c_{oa} + S_g c_g \quad (13.41)$$

where

$$c_\phi = -\frac{1}{\phi} \frac{\partial \phi}{\partial p_o} \quad (13.42)$$

$$c_{wa} = \hat{c}_w + \frac{B_g}{B_w} \frac{\partial R_{sw}}{\partial p_o} \quad (13.43)$$

$$c_{oa} = \hat{c}_o + \frac{B_g}{B_o} \frac{\partial R_{so}}{\partial p_o} \quad (13.44)$$

$$\hat{c}_o = -\frac{1}{B_o} \frac{\partial B_o}{\partial p_o} \quad (13.45)$$

$$c_g = -\frac{1}{B_g} \frac{\partial B_g}{\partial p_o} \quad (13.46)$$

For a *gas-condensate reservoir*, total compressibility is:

$$c_t = c_\phi + S_w c_w + S_o c_o + S_g c_g \quad (13.47)$$

We have provided additional equations in the Appendix to provide more clarity to multi-phase flow equations.

13.4.3 Field Applications

13.4.3.1 Example 1: Rate Transient Analysis Figure 13.5 presents the production history of an oil well in Eagle Ford formation. This production history is a plot of the rate-normalized pressure versus production time. The well was stimulated with 18 fracture stages, and each stage consisted of six clusters. On the figure, we have presented additional data, which are necessary for rate-transient analysis.

Assuming all 18 stages contribute to flow, the analysis of the straight-line slope of the rate-normalized pressure versus square root of time (Eq. 13.35 and Fig. 13.6) yields 0.002 mD for permeability to oil in the first 6 months of production and 0.0015 mD during the second year of production. However, if only nine stages contribute significantly to oil production, then the calculated oil permeability is 0.008 mD for the first six months and 0.006 mD during the second year of production when gas–oil ratio becomes significantly greater than the solution gas–oil ratio. Core plug absolute permeability for this reservoir is of the order of 0.00002 mD. Comparing core and well test permeability reveals that the formation stimulation has created high-permeability channels (or macrofractures) in the drainage volume of the hydraulic fractures. This high-permeability region is designated as SRV.

Analysis of the straight-line slope in Figure 13.6 yields a negative fracture skin factor s_{hf}^{face}, indicating that the hydraulic fracture surface has a larger permeability than the greater SRV region. This is consistent with geomechanical reasoning that macrofractures are wider near the hydraulic fracture surface than far away.

Figure 13.7 presents the plot of the rate-normalized pressure versus $\sqrt[4]{t}$ for *bilinear* flow analysis. The analysis of the straight-line segment on the figure yields $k_{f,\text{eff}} w_{hf} \lambda_t = 52$ mD-ft/cp.

13.4.3.2 Example 2: A Long Pressure Build-up Test Figure 13.8 shows 10 days of pressure buildup in a Bakken open-hole well. This well was stimulated in such a way to create only an axial hydraulic fracture (Kurtoglu, 2013). The analysis of the test produces effective fracture permeability of 0.023 mD if we assume that the well effective length is 50%, Figure 13.9.

13.4.4 Type-Curve Matching

Classic type-curve matching models pertain to radial flow in vertical wells producing from high-permeability conventional reservoirs where the BDF prevails soon after a well begins to produce. However, type-curve matching in unconventional reservoirs must address both transient and BDF in low-permeability reservoirs, which produce from multistage hydraulic fractures in horizontal wells. An appropriate type-curve, covering a broad range of variables, might be too overwhelming. On the other hand, because of the flexibility of reservoir models, we can use these to assess well performance. Reservoir modeling is extremely useful for simulating the primary production and EOR in shale formations under various operating conditions. However, the selection of the appropriate reservoir model and input data is crucial. For instance, for shale reservoir applications, we believe that dual-porosity compositional modeling is most appropriate.

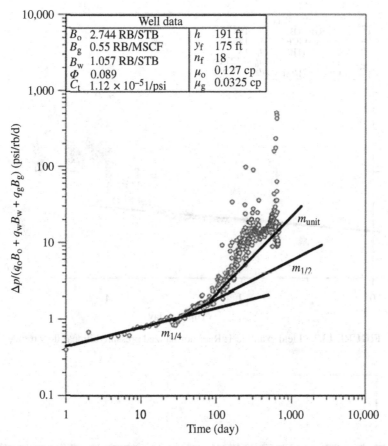

FIGURE 13.5 Field example 1: Diagnostic plot for rate-normalized pressure versus production time.

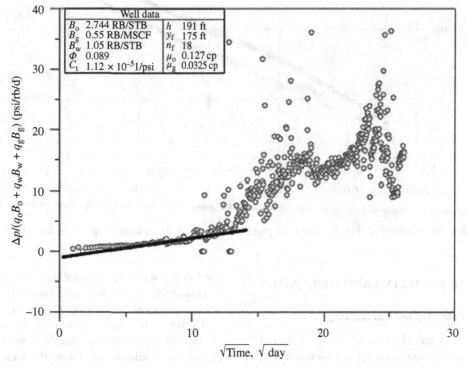

FIGURE 13.6 Field example 1: Rate-normalized pressure linear flow trend.

292 PERFORMANCE ANALYSIS OF UNCONVENTIONAL SHALE RESERVOIRS

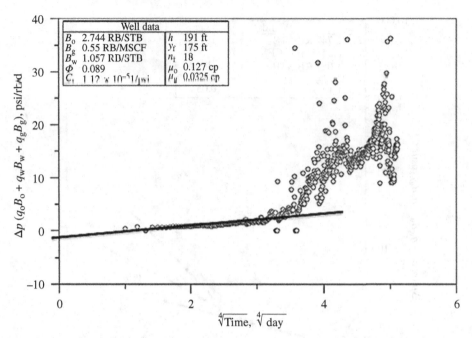

FIGURE 13.7 Field example 1: Rate-normalized pressure bilinear flow trend.

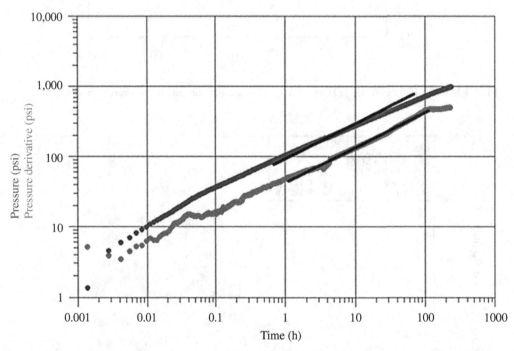

FIGURE 13.8 Field example 2: Ten-day diagnostic plot for pressure buildup in Bakken open-hole well (Kurtoglu, 2013).

13.5 RESERVOIR MODELING AND SIMULATION

13.5.1 History Matching and Forecasting

Core analysis provides matrix permeability k_m while rate-transient analysis yields a numerical value for fracture effective permeability $k_{f,eff}$. The latter permeability is, in reality, the formation enhanced permeability resulting from macro-fracturing because of well stimulation. Both k_m and $k_{f,eff}$ are key reservoir simulation input parameters.

Decline rate analysis provides an empirical estimate of the primary oil production from wells for the specific operating conditions of the wells. Reservoir simulation provides a more reliable estimate of well and reservoir

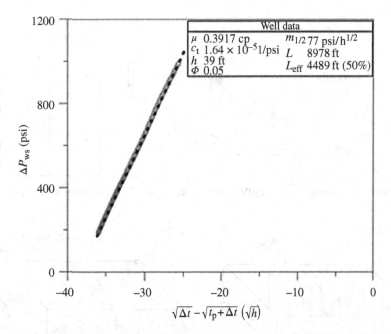

FIGURE 13.9 Field example 2: Ten-day pressure buildup in Bakken open-hole well (Revised from Kurtoglu, 2013).

performance based on a broad range of reservoir and well conditions, such as reservoir re-stimulation and EOR by specific injection–production operations. Two major set of input parameters are *matrix* properties (especially *matrix permeability*), measured in core samples, and the *effective permeability* of the flowing wells derived from pressure and rate transient analyses. The effective reservoir permeability reflects the overwhelming contributions from a network of interconnected macrofractures in the well drainage volume, created because of multistage hydraulic fracturing. We have successfully used these parameters and the relevant characteristics in research and reservoir modeling.

13.5.2 Dual-Porosity Single-Phase Modeling

To acquaint the reader with reservoir modeling approaches used in shale reservoir analysis, in the next section we first present a single-phase, dual-porosity model used in pressure and RTA. Next, we present the formulation for multicomponent gas flow, which accounts for molecular diffusion, Knudsen gas slippage, gas desorption, and gas storage.

The mathematical formulation for the EOR assessment is complex, which involves multiphase, multi-component flow modeling in the nanopore confined space.

The dual-porosity, single-phase model useful for pressure and transient analysis follows:

$$\nabla \cdot \left[\left(0.006328 \, k_{f,\text{eff}} / \mu \right) \nabla \left(\Delta p_f \right) \right] - \tau = \left(\phi c_t \right)_f \frac{\partial \Delta p_f}{\partial t} \quad (13.48)$$

$$\Delta p_f = p_i - p_f$$

The matrix-fracture transfer function τ will be either pseudo-steady state τ_{PSS} or unsteady state τ_{USS} as shown:

$$\begin{aligned} \tau_{\text{PSS}} &= (0.006328)\sigma \frac{k_m}{\mu} \left(p_m - p_f \right) \\ &= (0.006328)\sigma \frac{k_m}{\mu} \left(\Delta p_f - \Delta p_m \right) \end{aligned} \quad (13.49)$$

$$\Delta p_m = p_i - p_m$$

$$\tau_{\text{USS}} = \frac{1}{V_m} \int_0^t \frac{\partial \Delta p_f(x,t)}{\partial \xi} q_{u,m}(t-\xi) d\xi \quad (13.50)$$

$$q_{u,m}(t) = -0.006328 \frac{k_m}{\mu} A_m \frac{2}{r_m} \left[\sum_{l=1}^{\infty} \exp\left(\frac{-\eta_m l^2 \pi^2 t}{r_m^2} \right) \right] \quad (13.51)$$

where, $q_{u,m}$ is the flow rate caused by a unit pressure drop at the matrix surface for spherical matrix blocks; however, similar equations have been developed for other geometries (Torcuk et al., 2013a).

For unconventional reservoirs, because matrix permeability is extremely small (~0.0001–0.01 mD), we use the unsteady state formulation of the transfer function to account for the intricate mass transfer physics. One such issue is how injected gas can extract oil from the matrix. We also use log-distributed grid system to provide more accuracy near hydraulic fractures, which also improves the computational accuracy. We show such a grid system in Figures 13.10 and 13.11.

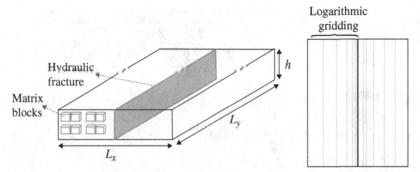

FIGURE 13.10 Conceptual stimulated reservoir volume (SRV) for single-stage hydraulically fractured well (left) and reservoir grid system (right).

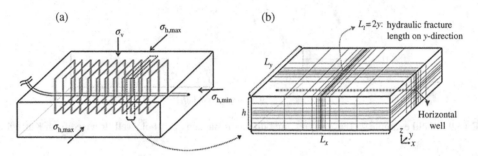

FIGURE 13.11 (a) Conceptual model of multistage hydraulic fractures in horizontal well. (b) Reservoir simulation grid for a single hydraulic fracture stage in horizontal well (Torcuk et al., 2013b).

13.5.3 Dual-Porosity Multicomponent Gas Modeling

In this model, we assume that the natural fracture pore space stores gas without adsorption, but the matrix stores gas both by adsorption on pore surfaces and by volumetric storage within the pore body.

13.5.3.1 Flow Hierarchy and Mathematical Formulation

When a gas well is put on production, gas production at the well causes a pressure drop in the pore body of the matrix pores, which causes gas desorption from the surface of the matrix pores. The desorbed gas mixes with the gas in the matrix pore body, which flows out into the macrofractures and eventually into the hydraulic fracture and the wellbore. Below, we describe the mathematical formulation of this hierarchical flow.

Macrofracture Flow for Component c:

$$-\nabla \cdot \left[-(93.0) D_{c,f}^* \xi_{g,f} \nabla(y_{c,f}) - (0.006328) y_{c,f} \xi_{g,f} k_{f,\text{eff}} \left(1 + \frac{(14.7) b_{c,f}^K}{p_f}\right) \left(\frac{1}{\mu_g}\right) \nabla p_f \right]$$
$$-\sigma \left\{ (93.0) \left(\xi_g S_g D_c^* \right)_m \left[y_{c,f} - y_{c,m} \right] \right\}$$
$$-\sigma \left\{ (0.006328) y_{c,m/f} \xi_{g,m/f} k_m \left(1 + \frac{(14.7) b_{c,m}^K}{p_f}\right) \left(\frac{1}{\mu_g}\right) (p_f - p_m) \right\}$$
$$+ y_{c,f} \xi_{g,f} \hat{q}_g = \frac{\partial}{\partial t} \left(\phi_f y_{c,f} \xi_{g,f} S_{g,f} \right)$$

(13.52)

The first term (divergence term) presents molecular diffusion, Darcy, and Knudsen flow within the interconnected macrofractures for component c. The second term is the diffusion mass transfer between fracture and matrix. The third term presents the Darcy and Knudsen mass transfer between fracture and matrix. The fourth term is the specific mass flux from the well. Finally, the right-hand side is the accumulation term in the macrofractures.

Matrix Flow for Component c:

$$\sigma \left\{ (93.0) \left(\xi_g S_g D_c^* \right)_m \left[y_{c,f} - y_{c,m} \right] \right\}$$
$$+ \sigma \left\{ (0.006328) y_{c,m/f} \xi_{g,m/f} k_m \left(1 + \frac{(14.7) b_{c,m}^K}{p_f}\right) \left(\frac{1}{\mu_g}\right) (p_f - p_m) \right\}$$
$$- \sigma \left\{ (93.0) y_{c,m/s} \xi_{g,m/s} D_{c,m}^{K*} (0.0312 SG_s) B_g \left[a_c(p_m) - a_{c,m} \right] \right\}$$
$$= \frac{\partial}{\partial t} \left(\phi_m y_{c,m} \xi_{g,m} S_{g,m} \right)$$

(13.53)

The first term is the diffusion mass transfer between fracture and matrix. The second term presents the Darcy and Knudsen mass transfer between fracture and matrix. The third term is the mass transfer contribution by gas desorption into the matrix pores. Finally, the right-hand side term is the accumulation term in the matrix.

Matrix Gas Desorption:

$$\sigma (93.0) D_{c,m}^{K*} \left[a_c(p_m) - a_{c,m} \right] = \frac{\partial a_{c,m}}{\partial t} \quad (13.54)$$

The left-hand side represents desorption mass transfer from the pore wall surfaces into the matrix pore body, and the right-hand side presents gas desorption rate.

The following terms are specific to the above formulation, which we present for clarity only.

$$D_c^* = D_c \frac{\phi}{\tau} \qquad (13.55)$$

$$D_c^{K*} = 10^{-3} \frac{b_c^K k^\infty}{\mu} \qquad (13.56)$$

$$b_c^K = 0.795\, k^{-0.4} \quad \text{for } N_2 \qquad (13.57)$$

a_c = Gas adsorption quantity, SCF/ton
D_c^* = Molecular diffusion coefficient for component c, corrected for porosity and tortuosity, cm²/s
D_c = Molecular diffusion coefficient for component c, cm²/s
D^{K*} = Knudsen effective diffusion coefficient, cm²/s
b_c^K = Knudsen slippage coefficient for component c, atm
y_c = Mole fraction of component c in gas phase, fraction
ξ_g = Gas molar density, lb-mol/ft³, $p/(z_g RT)$

13.6 SPECIALTY SHORT-TERM TESTS

In previous sections, we presented the rate-normalized pressure versus square root of time technique for producing wells, which is specifically suitable for unconventional reservoirs. Another technique involves conducting a long pressure build-up test and analyzing the pressure–time data. Both approaches rely on data gathered during long periods. In the specialty short-term tests, the production time duration is short. For instance, a short-duration DST, or mini-DST, requires 1500 seconds of flow period followed by 1 h of pressure buildup. This test produces formation fluids at constant rate from a single perforation into a small chamber, thus the wellbore storage effect is near zero (Kurtoglu, 2013). For engineering applications, we model the flow regime as spherical flow, which we can readily analyze for permeability using appropriate equations in the following section. For greater accuracy, we use numerical modeling to capture various flow regimes more accurately.

13.6.1 Mini-DST

A conventional DST is a short-term production and shut-in test, conducted in the drill stem, in conventional higher permeability reservoirs. Form the analysis of the pressure build-up data of the DST, we calculate formation permeability, formation damage skin factor, and reservoir static pressure. In conventional DST, we can produce several barrels of oil or an equivalent amount of gas depending on the DST configuration. Conventional DST tool is not suitable for use in unconventional shale reservoirs because shale permeability is extremely low, and during shut-in not enough formation fluid enters the drill stem column to create an adequate pressure build-up response. For unconventional shale reservoirs, mini-DST should be used. Mini-DST consists of a 2-l chamber, a constant flow pump, and a pressure gauge (Kurtoglu et al., 2013). The early-time flow regime toward the mini-DST perforation is spherical, which eventually becomes radial flow at larger times. Joseph and Koederitz (1985) derived Equation 13.58, 13.60, and 13.63 of the next section for the spherical flow regime. Kurtoglu et al. (2013) successfully used these equations to analyze a Bakken mini-DST test.

13.6.1.1 Spherical Flow Regime
Joseph and Koederitz (1985) derived the following equations, which are suitable for analyzing spherical flow regime taking place near the mini-DST perforation:

$$\Delta p = \frac{70.6 q B \mu}{k_{sp} r_{sw}}(1+s) - \frac{2453 q B \mu \sqrt{\phi \mu c_t}}{k_{sp}^{3/2}} \frac{1}{\sqrt{t}} \qquad (13.58)$$

where

$$\Delta p = p_i - p_w(t) \qquad (13.59)$$

$$m_{sp} = -\frac{2453 q B \mu \sqrt{\phi \mu c_t}}{k_{sp}^{3/2}} \qquad (13.60)$$

$$k_{sp} = \sqrt[3]{k_x k_y k_z} = \sqrt[3]{k_r^2 k_v} \qquad (13.61)$$

$$r_{sw} = \frac{0.5 h_p}{\ln(h_p/r_w)} \qquad (13.62)$$

r_{sw} = pseudo−spherical wellbore radius in mini−DST
h_p = the open−hole interval between the packers in mini−DST

At sufficiently large flow times, the spherical flow regime becomes radial flow because streamlines become parallel to the formation boundaries. When radial flow regime becomes dominant, we can use the radial flow equations to calculate formation radial permeability for comparison with the early-time spherical flow permeability. Comparing radial and spherical flow permeability is of value in determining whether vertical permeability anisotropy is significant. Kurtoglu et al. (2013) analyzed both the spherical and radial flow pressure transient data for a Bakken well.

For pressure build-up analysis, we use the following approximate equation for spherical flow in mini-DST analysis when pressure buildup is sufficiently long (Stewart and Wittmann, 1979):

$$\Delta p = -\frac{2453 q B \mu \sqrt{\phi \mu c_t}}{k_{sp}^{3/2}} \left(\frac{1}{\sqrt{t_p + \Delta t}} - \frac{1}{\sqrt{\Delta t}} \right) \quad (13.63)$$

where

$$\Delta p \approx p_{ws}(\Delta t) - p_{wf}(\Delta t = 0) \quad (13.64)$$

13.6.1.2 Radial Flow Regime

$$\Delta p = m_r \log t + m_r \left[\log t + \log \left(\frac{k}{\phi \mu c_t r_w^2} \right) - 3.23 + 0.869\, s \right] \quad (13.65)$$

where

$$\Delta p = p_i - p_w(t) \quad (13.66)$$

$$m_r = \frac{162.6 q B \mu}{k_r h_w} \quad (13.67)$$

$$k_r = \sqrt{k_x k_y} \quad (13.68)$$

13.6.2 Mini-Frac Test

In the absence of mini-DST, we can rely on mini-frac pressure–time data to determine matrix permeability. We can also analyze the mini-frac pressure falloff, after injection has ceased, for the *closure pressure* p_c and *leakoff coefficient* C_L. Closure pressure is the pressure at the inflection point on the pressure falloff curve. This pressure is equal also to *minimum horizontal stress* σ_h for mini-frac in *vertical wells*. In this case, the formation breakdown pressure is a function of σ_h and σ_H, given by Equation 13.81. For *horizontal wells*, the formation breakdown pressure, given by Equations 13.82 and 13.83, depends on the direction of the well with respect to the principal stresses in the horizontal plane.

Determination of the in situ matrix permeability is a most crucial parameter for use in reservoir engineering analysis and forecasting. In the absence of a viable pressure or rate transient test, we can estimate matrix permeability from the leakoff coefficient appearing in Equation 13.74. This equation is the slope of fracture-closure pressure falloff in the mini-frac test (Economides and Nolte, 1987). Specifically, the analysis requires plotting $\Delta p(\Delta t_D)$ versus $G(\Delta t_D)$ to yield a straight line. From the slope, we can calculate formation permeability using Equation 13.79.

The mini-frac equations, developed by Nolte in the publication by Economides and Nolte (1987), are

$$\Delta p = \frac{\pi C_L \sqrt{t_p}}{2 c_f} G(\Delta t_D) \quad (13.69)$$

where

$$\Delta p = p_w(\Delta t_D = 0) - p_w(\Delta t_D) \quad (13.70)$$

$$G(\Delta t_D) = \left\{ \frac{16}{3\pi} \left[(1+\Delta t_D)^{3/2} - \Delta t_D^{3/2} - 1 \right] \right\} \quad (13.71)$$

$$\Delta t_D = \frac{t - t_p}{t_p} = \frac{\Delta t}{t_p} \quad (13.72)$$

We rewrite Equation 13.69 as follows:

$$\Delta p = m\, G(\Delta t_D) \quad (13.73)$$

where

$$m = \frac{\pi C_L \sqrt{t_p}}{2 c_f} \quad (13.74)$$

$$c_f = \left[\frac{\pi(1-\nu^2)}{2E} \right] \begin{bmatrix} x_f \\ h_f \\ r_f \end{bmatrix} \quad (13.75)$$

$$\begin{bmatrix} x_f \\ h_f \\ r_f \end{bmatrix} = \frac{q}{2} \frac{\sqrt{t_p}}{\pi h_f C_L} \quad (13.76)$$

where

$$\begin{bmatrix} x_f \\ h_f \\ r_f \end{bmatrix} \Rightarrow \begin{bmatrix} x_f, \text{thin fat fracture} \\ h_f, \text{thin long fracture} \\ r_f, \text{penny-shaped fracture} \end{bmatrix}$$

After calculating C_L from Equation 13.74, we use the following leakoff equation to calculate the formation matrix permeability.

$$C_L = 0.00118\, \Delta p_c \sqrt{\frac{k \phi c_t}{\mu}}; \quad [C_L \text{ in ft}/\sqrt{\min}] \quad (13.77)$$

$$\Delta p_c = p_c - p \quad (13.78)$$

Thus,

$$k_m = \frac{\mu C_L^2}{\phi c_t (0.00118\, \Delta p_c)^2} \quad (13.79)$$

Finally, we estimate the average fracture width during fluid injection from the following equation:

$$\bar{w}_f = c_f (p_{hf} - p_c) \quad (13.80)$$

where

p_{hf} = Fracture propagating pressure within the mini-frac

p_c = Closure pressure

The formation breakdown pressure p_{bd} for vertical well is:

$$p_{bd} = 3\sigma_h - \sigma_H - p + T \text{ (vertical well)} \quad (13.81)$$

From a mini-frac test, we can determine the closure pressure p_c and the formation breakdown pressure p_{bd}. Then, we can calculate σ_H from Equation 13.81 if we replace horizontal stress σ_h with the closure pressure p_c.

The following two equations are useful in estimating the formation breakdown pressure p_{bd} for a well drilled in the σ_H direction (Eq. 13.82) or in the σ_h direction (Eq. 83):

$$p_{bd} = 3\sigma_h - \sigma_v - p + T \text{ (Horizontal well in the direction)} \quad (13.82)$$

$$p_{bd} = 3\sigma_H - \sigma_v - p + T \text{ (Horizontal well in the direction)} \quad (13.83)$$

The formation breakdown pressure in a horizontal well, drilled in the σ_h direction, becomes quite large when $\sigma_H \gg \sigma_h$. Unfortunately, the latter is often the situation in unconventional reservoirs.

13.7 ENHANCED OIL RECOVERY

In the laboratory, we measure permeability, porosity, relative permeability, capillary pressure, and wettability of core samples, then we conduct EOR experiments to determine which EOR process can potentially lead to economic success. Similarly, we conduct well tests in the field to obtain reservoir-scale information on permeability and shape factor to determine macrofracture connectivity and spacing for use in modeling primary production, appraising EOR potential, and reservoir management planning.

The projected maximum oil recovery of shale oil reservoirs, such as the Bakken, is around 10% despite the use of long horizontal wells and reservoir stimulation by multistage hydraulic fracturing. This is because of the ultralow matrix permeability that significantly hinders oil flow from the matrix into the smaller fractures and ultimately into the wellbore. Immiscible displacement of oil by water or gas in such tight formations is not practical because the injected fluids can flow only through the interconnected fractures while having a difficult time entering the tight matrix to displace oil. Miscible gas injection, on the other hand, can potentially mobilize oil from the ultralow permeability shale matrix by solvent extraction and condensing–vaporizing diffusive mixing at the fracture–matrix interface. This extraction process is completely different from oil mobilization in conventional reservoirs, where the injected fluids mobilize oil to form an oil bank ahead of the injected fluid and, then, push the oil bank through the matrix pores to an eventual outlet.

Hawthorne et al. (2013) conducted CO_2 oil extraction experiments in the laboratory at 5000 psi and 230 °F by using millimeter-size Bakken cuttings and centimeter-diameter core plugs. They concluded that oil was extracted from the cuttings or the core plugs because of CO_2 miscibility with reservoir oil, viscosity reduction, and diffusion mass transfer. The exposure time was up to 96 h for the middle Bakken chips (clastic sediment) which resulted in near-complete hydrocarbon recovery. However, the oil extraction experiments required smaller chips and larger exposure time for the upper Bakken (shale). For field applications, solvent extraction is very slow and modest because the specific surface area of reservoir matrix blocks is very small compared to the laboratory samples used by Hawthorne et al (2013). Nevertheless, these experimental results provide the impetus to pursue EOR in unconventional reservoirs, and numerical modeling becomes the tool to scale laboratory results to field. Thus, in unconventional reservoirs, CO_2, natural gas liquids (NGL), liquefied petroleum gas (LPG), and NGL–CO_2 mixture can potentially mobilize matrix oil by miscibility (via condensing–vaporizing gas extraction) which leads to countercurrent oil flow from the matrix. We evaluate these concepts using dual-porosity, multicomponent, numerical simulation of flow in the SRV.

For field implementation of fluid injection to enhance oil recovery, a possible injection-production pattern is the zipper frac pattern, shown in Figure 13.12. This pattern is the counterpart of the conventional five-spot pattern for enriched gas or CO_2 injection. The zipper frac pattern should improve conformance (coverage) of injected fluid and the reservoir

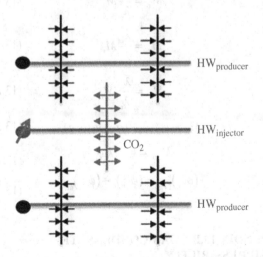

FIGURE 13.12 A proposed inverted five-spot zipper frac injection-production pattern for EOR in unconventional reservoirs.

macrofractures to increase oil extraction from the matrix. To optimize the zipper frac well spacing parameters requires reservoir modeling.

13.8 CONCLUSION

Well performance and transient test analyses are the critical components of reservoir characterization for assessing the ultimate oil recovery from primary production and EOR techniques in unconventional reservoirs. Besides pressure and flow rate analysis, we rely on special core analysis measurements, reservoir fluid PVT characterization, and inflow tracer analysis to determine reservoir stimulation effectiveness.

In this chapter, first we presented practical approaches for analyzing well performance in shale oil and gas reservoirs. Next, we showed how to use the well performance results to estimate the ultimate oil and gas recovery from primary production, and, finally, we shed light on the connection between well test analysis and assessment of EOR for unconventional reservoirs.

ACKNOWLEDGMENTS

The authors gratefully acknowledge the support from Marathon Center of Excellence for Reservoir Studies (MCERS), the Unconventional Oil and Gas Institute (UNGI), and Marathon Oil Corporation, which have provided us the opportunity to conduct research both in conventional and unconventional reservoirs.

APPENDIX 13.A AL BLACK OIL PARAMETERS

$$M_o = \frac{\lambda_o}{\lambda_t} M_t \quad (13.A.1)$$

$$M_g = \frac{\lambda_g}{\lambda_t} M_t \quad (13.A.2)$$

$$M_w = \frac{\lambda_w}{\lambda_t} M_t \quad (13.A.3)$$

$$M_\alpha = k_{f,\text{eff}} \lambda_\alpha \quad (\alpha = w, o, g) \quad (13.A.4)$$

$$k_{f,\text{eff}} = k_m + k_f \phi_f \quad (13.A.5)$$

$$(\phi c_t)_{f+m} = (\phi c_t)_f + (\phi c_t)_m \quad (13.A.6)$$

APPENDIX 13.B GAS-CONDENSATE COMPRESSIBILITY

$$c_t = c_\phi + S_w c_w + S_o c_o + S_g c_g \quad (13.B.1)$$

$$c_\phi = -\frac{1}{\phi} \frac{\partial \phi}{\partial p_o} \quad (13.B.2)$$

$$c_w = -\frac{1}{B_w} \frac{\partial B_w}{\partial p_o} \quad (13.B.3)$$

$$c_g = -\frac{1}{B_g} \frac{\partial B_g}{\partial p_o} \quad (13.B.4)$$

NOMENCLATURE

Basic Symbols

F	Force
L	Length
M	Mass
P	Pressure
T	Time

Field Units

BBL	Barrel
cp	Centipoise
D	Day
ft	Feet
md	Millidarcy
mD	Millidarcy
psi	Pound per square inch
RB	Reservoir barrel
SCF	Standard cubic feet
STB	Stock tank barrel

Symbols

a_c	Gas adsorption quantity, scf/ton
A_m	Surface area of the matrix block, $L^2(\text{ft}^2)$
b	Decline exponent
b_c^K	Knudsen slip coefficient for component c, atm
B	Formation volume factor, L^3/L^3 (RB/STB)
B_g	Gas volume factor, L^3/L^3 (RCF/SCF)
B_o	Oil volume factor, L^3/L^3 (RB/STB)
B_w	Water volume factor, L^3/L^3 (RB/STB)
c_f	Fracture compliance, $L^3 F^{-1}$ (ft/psi)
c_g	Gas compressibility, $L^2 F^{-1} (1/\text{psi})$
$\hat{c}_o$	Oil compressibility, $L^2 F^{-1} (1/\text{psi})$
c_{oa}	Apparent oil compressibility, $L^2 F^{-1} (1/\text{psi})$
c_ϕ	Rock compressibility, $L^2 F^{-1} (1/\text{psi})$
c_w	Water compressibility, $L^2 F^{-1} (1/\text{psi})$
$c_{t,m}$	Total matrix compressibility, $L^2 F^{-1} (1/\text{psi})$

$c_{t,f}$	Total fracture compressibility, L^2F^{-1} (1/psi)	R_{so}	Solution gas–oil ratio, L^3/L^3(SCF/STB)
c_t	Total reservoir compressibility, L^2F^{-1} (1/psi)	R_{sw}	Solution gas–water ratio, L^3/L^3(SCF/STB)
C_L	Leakoff coefficient in mini-frac test, $L/\sqrt{T}$ $\left(\text{ft}/\sqrt{\text{min}}\right)$	r_{sw}	Pseudo-spherical wellbore radius in mini-DST, L(ft)
d	Effective decline rate	s_{hf}^{face}	Skin factor at the face of hydraulic fracture connecting to the reservoir
d_{min}	Minimum decline rate		
D	Decline rate, T^{-1}	s_{hf}^{well}	Skin factor at the well-hydraulic fracture entrance (choked fracture)
D_c	Equivalent binary diffusion coefficient for component c in the mixture, L^2/T (cm²/s)		
		S_w	Water saturation, (fraction)
D_i	Initial decline rate, T^{-1} (1/month)	S_o	Oil saturation, (fraction)
D^{K*}	Knudsen effective diffusion coefficient, L^2/T (cm²/s)	S_g	Gas saturation, (fraction)
		S_{wr}	Irreducible water saturation, (fraction)
E	Young modulus, FL^{-2} (psi)	S_{or}	Residual oil saturation, (fraction)
G_p	Cumulative gas production at time t, L^3(SCF)	$S_{o,rem}$	Remaining oil saturation, (fraction)
h	Formation thickness, L L(ft)	t	Time, T(h, day, month)
h_p	Open-hole interval between two packers in mini-DST, L(ft)	t_D	Dimensionless time
		t_i	Initial production time, T(day)
J	Productivity index, $L^5F^{-1}T^{-1}$ (BBL/D/psi)	t_0	Zero time, T(day)
k	Permeability, L^2(mD)	t_1	End of bilinear flow, T(day)
k_f	Fracture permeability, L^2(mD)	t_2	End of linear flow, T(day)
$k_{f,eff}$	Effective fracture permeability, L^2(mD)	t_p	Production time for well testing, T(h)
k_{hf}	Hydraulic fracture permeability, L^2(mD)	t_p	Pumping time for mini-frac, T(min)
k_m	Matrix permeability, L^2(mD)	T	Temperature, (°R)
k_{sp}	Equivalent spherical permeability, L^2(mD)	T	Tensile strength of the formation, FL^{-2} (psi)
m_{bl}	Slope of bilinear flow, $FL^{-2}T^{-1/4}$ (psi/h$^{1/4}$)	V_m	Matrix control volume, L^3(ft³)
m_r	Slope of semi-log radial flow, FL^{-2}	V_R	Pore volume component, L^3(BBL)
N_p	Cumulative oil production at time t, L^3(STB)	w_{hf}	Hydraulic fracture width, L(ft)
$N(t)$	Cumulative production at time t, L^3(STB)	y_{hf}	Fracture half-length, L(ft)
$N(\infty)$	Ultimate cumulative production, L^3(STB)	y_c	Mole fraction of component c in gas phase, fraction
n_{hf}	Total number of hydraulic fractures		

GREEK

p	Pore pressure, FL^{-2} (psi)
p_c	Closure pressure in mini-frac test, FL^{-2} (psi)
p_{hf}	Hydraulic fracture propagation pressure, FL^{-2} (psi)
p_i	Initial reservoir pressure, FL^{-2} (psi)
p_m	Matrix pressure, FL^{-2} (psi)
p_o	Oil phase pressure, FL^{-2} (psi)
p_r	Reservoir static pressure, FL^{-2} (psi)
p_{wf}	Flowing bottom-hole pressure, FL^{-2} (psi)
$\bar{p}$	Average pore pressure, FL^{-2} (psi)
q	Production rate, L^3T^{-1} (cc/s, BBL/D)
q_g	Gas production rate, L^3T^{-1} (cc/s, SCF/D)
$\hat{q}_g$	Gas production rate per unit rock volume, T^{-1} (1/s, 1/D)
q_i	Initial oil production rate, L^3T^{-1} (cc/s, STB/D)
q_o	Oil production rate, L^3T^{-1} (cc/s, STB/D)
$\hat{q}_o$	Oil production rate per unit rock volume, T^{-1} (1/s, 1/D)
q_w	Water production rate, L^3T^{-1} (cc/s, STB/D)
$\hat{q}_w$	Water production rate per unit rock volume, T^{-1} (1/s, 1/D)
r_m	Radius of spherical matrix block, L(ft)

λ_o	Inverse of oil viscosity, $L^2F^{-1}T^{-1}$(cp^{-1})
λ_w	Inverse of oil viscosity, $L^2F^{-1}T^{-1}$(cp^{-1})
λ_g	Inverse of oil viscosity, $L^2F^{-1}T^{-1}$(cp^{-1})
η_m	Matrix diffusivity coefficient, $k_m/(\phi\mu c_t)_m$, L^2/T (ft²/D)
Δp_{sf}	Skin pressure drop in hydraulic fracture stimulated well, FL^{-2} (psi)
Δp	Well flowing pressure change, FL^{-2} (psi)
ϕ	Porosity, fraction
ϕ_m	Matrix porosity, fraction
ϕ_f	Fracture porosity, fraction
μ	Oil viscosity, $FL^{-2}T$ (cp)
σ	Matrix shape factor for the cubic matrix blocks, $1/L^2$(1/ft²)
σ_h	Minimum horizontal stress, FL^{-2} (psi)
σ_H	Maximum horizontal stress, FL^{-2} (psi)
ν	Poisson ratio
ω	Storativity ratio, fraction

τ Transfer function source term, $T^{-1}\left(day^{-1}\right)$

ξ_g Gas molar density, $M/L^3 (lb\text{-}mol/ft^3)$

ξ_0 Oil molar density, $M/L^3 (lb\text{-}mol/ft^3)$

SUBSCRIPTS

bd   Breakdown
f   Fracture
g   Gas
m   Matrix
o   Oil
w   Water
wf   Well flowing
ws   Well shut in

REFERENCES

Arps JJ. Analysis of decline curves. Technical Paper 1758. AIME; (1944).

Economides MJ, Nolte KM. Reservoir Stimulation. Houston, TX: Schlumberger Educational Services; 1987. p 7.4–7.6.

Garb FA, Smith GL. Estimation of oil and gas reserves. In: Howard B, editor. Petroleum Engineering Handbook. Richardson: SPE; 1987. p 40–27.

Hawthorne SB, Gorecki CD, Sorensen JA, Steadman EN, Harju JA, Melzer S. Hydrocarbon mobilization mechanisms from upper, middle, and lower Bakken Reservoir rocks exposed to CO2. SPE 167200-MS. SPE Unconventional Resources Conference; November 5–7, 2013; Calgary, AB, Canada; 2013. Society of Petroleum Engineers, Richardson.

Joseph JA, Koederitz LF. Unsteady-state spherical flow with storage and skin. SPE J 1985;25(6):804–822.

Kurtoglu B. Integrated reservoir characterization and modeling in support of enhanced oil recovery for Bakken [PhD Thesis]. Petroleum Engineering, Colorado School of Mines; 2013.

Kurtoglu B, Cox SA, Kazemi H. Evaluation of long-term performance of oil wells in Elm Coulee field. CSUG/SPE 149273; November 15–17, 2011; Calgary, AB, Canada; 2011. Society of Petroleum Engineers, Richardson.

Kurtoglu B, Kazemi H, Boratko EC, Tucker J, Daniels R. SPE 159597-PA, Minidrillstem tests to characterize formation deliverability in the Bakken. SPE J 2013;16 (3):317–326.

Lacayo J, Lee WJ. Pressure normalization of production rates improves forecasting results. SPE 168974, SPR URC; April 1–3, 2014; The Woodlands, TX; 2014. Society of Petroleum Engineers, Richardson.

Patzek TW, Male F, Marder M. Gas production in the Barnett shale obeys a simple scaling theory. Proc Natl Acad Sci U S A; December 3, 2013;110(49):19731–19736.

Salman A, Kurtoglu B, Kazemi H. Analysis of chemical tracer flowback in unconventional reservoirs. SPE 171656, SPE/CSUR Unconventional Resources Conference; September 30–October 2, 2014;Calgary, AB, Canada; 2014.

Stewart G, Wittmann M. SPE 8362, Interpretation of the pressure response of the repeat formation tester. SPE Annual Technical Conference and Exhibition, Las Vegas, NV, September 23–26, 1979.

Tepper B, Baechle G, Keller J, Walsh R. Petrophysical evaluation of shale oil and gas opportunities in emerging plays; some examples and learning's from the Americas. IPTC 16926, IPTC; 26–28 March 2013; Beijing, China; 2013. Society of Petroleum Engineers, Richardson.

Torcuk MA, Kurtoglu B, Alharthy N, Kazemi H. Analytical solutions for multiple-matrix in fractured reservoirs: application to conventional and unconventional reservoirs. SPE J 2013a;18 (5):969–981.

Torcuk MA, Kurtoglu B, Fakcharoenphol P, Kazemi H. Theory and application of pressure and rate transient analysis in unconventional reservoirs. SPE 166147, SPE Annual Technical Conference and Exhibition; September 30–October 2, 2013; New Orleans, LA; 2013b. Society of Petroleum Engineers, Richardson.

Wattenbarger RA, El-Banbi AH, Villegas ME, Maggard JB. Production analysis of linear flow into fractured tight gas wells. SPE 3993, SPE rocky Mountain Regional/Low Permeability Reservoirs Symposium and Exhibition; April 5–8, 1998; Denver, CO; 1998. Society of Petroleum Engineers, Richardson.

Zuckerman G. The Frackers. New York: Penguin Publishers; 2013.

14

RESOURCE ESTIMATION FOR SHALE GAS RESERVOIRS

Zhenzhen Dong[1], Stephen A. Holditch[2] and W. John Lee[3]

[1] PTS, Schlumberger, College Station, TX, USA
[2] Petroleum Engineering Department, Texas A&M University, College Station, TX, USA
[3] UH Energy Research Park, University of Houston, Houston, TX, USA

14.1 INTRODUCTION

Many gas shale plays are currently under development in the world oil and gas industry. The use of horizontal drilling in conjunction with hydraulic fracturing has greatly expanded the ability of producers to profitably produce natural gas from low-permeability geologic formations, particularly shale formations.

14.1.1 Unique Properties of Shale

Shale gas refers to natural gas (mainly methane) in fine-grained, organic-rich rocks (gas shales). When talking about shale gas, the word shale does not refer to a specific type of rock. Instead, it describes rocks with more fine-grained particles (smaller than sand) than coarse-grained particles, such as shale (fissile) and mudstone (nonfissile), siltstone, fine-grained sandstone interlaminated with shale or mudstone, and carbonate rocks.

Gas is stored in shales in three ways: (i) adsorbed gas is the gas attached to organic matter or to clays; (ii) free gas is the gas held within the tiny spaces in the rock (pores, porosity, or microporosity) or in spaces created by the rock cracking (fractures or microfractures); and (iii) solution gas is the gas held within other liquids, such as bitumen and oil. Gas shales are source rocks that have not released all of their generated hydrocarbons. In fact, source rocks that are "tight" or "inefficient" at expelling hydrocarbons may be the best prospects for potential shale gas.

Since natural gas occurs both as free (as around the rock structure) and within the rocks, once the wellbore reaches the target zone and has been successfully fracture treated, the free gas flows quickly, causing an initial high production rate. Production then plateaus as the natural gas absorbed in the rock is removed. Thus, for a typical shale gas well, production declines between 70 and 90% in the first year, and an overall average well life may be of the order of 20–30 years.

14.1.2 Petroleum Resources Management System (PRMS)

The terms "resources" and "reserves" have been used in the past and continue to be used to represent different classifications of mineral and/or hydrocarbon deposits. In March 2007, the Society of Petroleum Engineers (SPE), the American Association of Petroleum Geologists (AAPG), the World Petroleum Council (WPC), and the Society of Petroleum Evaluation Engineers (SPEE) jointly adopted and published the PRMS to provide an international standard for classification of oil and gas reserves and resources (Fig. 14.1a). Technically and ERR, commonly used by the Energy Information Administration (EIA), are not formally defined in PRMS.

14.1.3 Energy Information Administration's Classification System

According to the Energy Information Administration (EIA), TRR is the subset of the total resource base that is recoverable with existing technology. The term "resources" represents the total quantity of hydrocarbons that are estimated, at a particular time, to be contained in (i) known accumulations and (ii) accumulations that have yet to be discovered

---

Fundamentals of Gas Shale Reservoirs, First Edition. Edited by Reza Rezaee.
© 2015 John Wiley & Sons, Inc. Published 2015 by John Wiley & Sons, Inc.

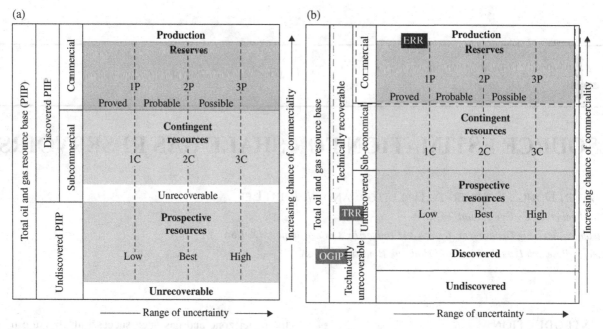

FIGURE 14.1 Flow chart and generalized division of resource and reserve categories from (a) PRMS and (b) EIA definition.

(prospective resources). ERR are those resources for which there are economic incentives for production. It is important to note that economically unrecoverable resources may, at some time in the future, become recoverable, when the technology to produce them becomes less expensive or the characteristics of the market are such that companies can ensure a fair return on their investment by extracting the resources. For our purposes, we considered TRR to be the resources that can be produced within a 25-year time period.

We rearranged categories of PRMS and show how the estimates of technically and ERR are classified (Fig. 14.1b). Commercial resources, which include cumulative production and reserves, are ERR. TRR is the subset of the total resource base that includes commercial resources, contingent resources, and prospective resources. Estimated ultimate recovery (EUR) is not a resources category, but a term that refers to the quantities of petroleum which are estimated to be potentially recoverable from an accumulation, including those quantities that have already been produced.

14.1.4 Reserves Estimate Methodology for Unconventional Gas Reservoirs

The methods that predict well performance can be used to estimate reserves. The principal techniques used for production determination from early stage to mature fields are analogy, volumetric analysis, material balance analysis, decline curve analysis (DCA), and numerical simulation (Table 14.1). The biggest challenge is that methods that we use for conventional reservoirs usually do not work well, without modification, for unconventional reservoirs. For example, DCA is commonly used for future performance prediction and resource estimation when production data are available. However, rate-time production-decline curves have real problems, including (i) a lack of long-term historical production; (ii) optimistic predictions for unconventional reservoirs because boundary dominated flow is not reached within reasonable times in these reservoirs, as required for Arps DCA (e.g., Ilk et al., 2008); and (iii) operating pressure is not constant. In some shale plays, such as the Haynesville Shale, the operating pressure not only has a general continuous declining trend in a wide range but also fluctuates due to choke size changes and frequent shut-ins. In addition, when wells are operated at a restricted rate condition (choked back), they may flow at a nearly constant rate for several months. In such cases, DCA fails because the rate does not decline.

Volumetric analysis coupled with an assumed RF and reservoir simulation with analytical or numerical models have their own challenges. The problems include difficulties in measuring formation properties needed for input into the computational methods.

14.1.5 Monte Carlo Probabilistic Approach

Shale gas plays are generally characterized by low geologic risk and high commercial risk. Uncertainty exists in geologic and engineering data and, consequently, in the results of calculations made with these data. Probabilistic approaches are required to provide an assessment of uncertainty in resource estimates.

Some authors have applied probabilistic approaches to DCA to quantify the uncertainty resulting from the use of these imperfect methods (Cheng et al., 2010; Jochen and

TABLE 14.1 Tools used to estimate reserves

Method	Advantage	Disadvantage	Conventional reservoir	Unconventional reservoir
Analogy	• Best in blanket sands • Best prior to production	The large number of variables and parameters causes high degree of uncertainty	Can be applied in both conventional and unconventional assets	
Volumetric method	• Any stage of depletion • Best prior to production	Has uncertainties of • recovery factor (RF) • actual drainage area	Accurate in blanket reservoir	Used only when no wells have been drilled
Material balance	Best between 10 and 70% depletion	Requires: • accurate average pressure • reservoir fluid properties	Accurate in depletion drive reservoir	• Should never be used • Average pressure cannot be measured accurately
Decline curve analysis	• Simple, easy to apply • Less data required (only rate-time) • Best with long production history • Quick	Requirements difficult to met • Boundary dominated flow • Unchanging drainage area • Fixed skin factor • "b" value constant and should lie between 0 and 1 • Can overestimate reserves	Hyperbolic (small b) or exponential decline usually accurate	Must use hyperbolic decline: • CBM: $b=0–0.5$ • Shale gas and tight gas: b may be larger than 1 • Use best-fit "b" until predetermined minimum decline rate reached, then impose exponential decline • Set "b" to proper "terminal value"
Reservoir simulation	• Best with data rich wells • In conjunction with other methods any time	• Needs good history match • Requires much time, costly	Used to simulate field	Used to simulate individual wells

Spivey, 1996). Reservoir simulation coupled with stochastic methods (e.g., Monte Carlo simulation) has provided an excellent means to predict production profiles for a wide variety of reservoir characteristics and producing conditions. The uncertainty is assessed by generating a large number of simulations, sampling from distributions of uncertain geologic, engineering, and other important parameters. This topic has been an object of study for some time in conventional reservoirs (MacMillan et al., 1999; Nakayama, 2000; Sawyer et al., 1999). However, few applications to unconventional reservoirs can be found in the literature. Oudinot et al. (2005) coupled Monte Carlo simulation with a fractured reservoir simulator, COMET3, to assess the EUR in coalbed methane reservoirs. Schepers et al. (2009) successfully applied this Monte Carlo COMET3 procedure to forecast EUR for the Utica Shale.

14.1.6 Analytical Models

Given the complex nature of hydraulic fracture growth, the extremely low permeability of the matrix rock in many shale gas reservoirs, and the predominance of horizontal completions, reservoir simulation is often the preferred method to predict and evaluate well performance. Analytical solutions for fluid flow in naturally fracture reservoirs were published by Warren and Root (1963) and Kazemi (1969). Semianalytical solutions for hydraulically fractured horizontal wells in fractured reservoirs have been published (Medeiros et al., 2008). PMTx 2.0 (2012), with a number of modeling options, such as a transient dual-porosity reservoir model (Kazemi, 1969), is an analytical unconventional gas reservoir simulator designed to quickly and easily model single-well, single-phase, gas production based on near-wellbore reservoir performance under specified well completion scenarios. One of the important applications of PMTx 2.0 is to estimate ultimate gas recovery for horizontal wells with transverse fractures in a rectangular shale gas reservoir.

14.1.7 Economic Analysis

Almadani (2010) presented a methodology to determine the percentage of TRR that is economically recoverable from the Barnett Shale as a function of gas price and finding and development costs (F&DC). For ERR he applied economic criteria of minimum 20% internal rate of return (IRR) and maximum 5-year payout to recover the initial investment, which are hurdles sometimes used by investors in the oil and gas industry. The author suggested that wells that do not pay out in 5 years are not good investment.

14.1.8 Region-Level World Shale Gas Resource Assessments

The first notable estimate of world shale gas in place was performed by Rogner (1997). In the study, shale-gas OGIP was estimated to be 16,000 Tcf for seven groupings of world countries (Table 14.2). However, Rogner's (1997) world estimate is most likely to be quite conservative, given the recent discoveries of significant shale gas worldwide, such as the Eagle Ford Shale in the United States and the Mikulov Shale in Austria. Actually, a basin-by-basin assessment of shale gas resources in 5 regions containing 32 countries, conducted by EIA (2011a), indicates that shale-gas OGIP (25,840 Tcf) is larger than estimated by Rogner in 1997 (16,112 Tcf), even accounting for the fact that Russia and the Middle East were not included in the EIA study (but were included in Rogner's assessment) (Table 14.2). However, neither Rogner's nor EIA's estimates quantified the considerable uncertainty in shale-gas OGIP. Thus, Dong et al. (2012) presented a probabilistic solution and established the probability distributions of shale-gas OGIP for the seven world regions originally used by Rogner (Fig. 14.2). World shale OGIP was estimated to be between 34,000 (P90) and 73,000 (P10) Tcf, with a P50 value of 50,000 Tcf (Table 14.2). Except for the Middle East and the Commonwealth of Independent States (CIS), the largest and most notable differences between EIA and Dong et al. (2012) estimates are the shale-gas OGIP assessments for Austral-Asia (AAO) and Latin America (LAM).

TRR can be estimated by multiplying the OGIP by a gas RF. For instance, three basic gas RFs, incorporating shale mineralogy, reservoir properties, and geologic complexity, are used in the EIA (2011a) basin-level assessment (Table 14.3). The average RF of shale gas for the basins in the 32 countries is 25%, which is a RF of the gas in place for shale gas basins and formations that have medium clay content, moderate geologic complexity, and average reservoir pressure and properties. North America (NAM) includes United States and Canada in the EIA study. The EIA regional level tabulations of risked gas in place and technically recoverable shale gas resource are provided in Table 14.4. Two specific judgmentally established success/risk factors are used to estimate the risked gas-in-place (GIP) within the prospective area of the

TABLE 14.2 Comparison of regional level shale OGIP assessments, in Tcf

Region	Rogner (1997)	EIA (2011a)	Dong et al (2012) P50
Middle East (MET)	2,547	N/A	15,416
Commonwealth of Independent State (CIS)	627	N/A	15,880
North America (NAM)	3,840	7,140	5,905
Africa (AFR)	274	3,962	3,882
Latin America (LAM)	2,116	4,569	3,742
Austral-Asia (AAO)	6,151	7,042	2,690
Europe (EUP)	549	2,587	2,194
World	16,103	25,300	50,220

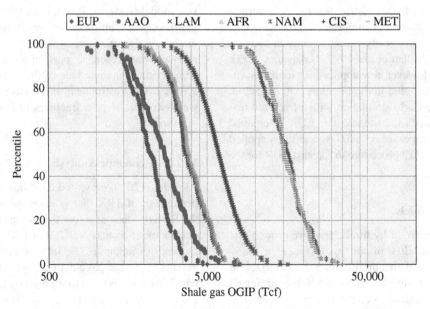

FIGURE 14.2 Probability distributions of shale-gas OGIP for seven world regions (Adapted from Dong et al., 2012).

TABLE 14.3 RFs used in EIA study (EIA, 2011a)

Clay content	Geologic complexity	Reservoir properties	RF (%)
Low	Low to moderate	Favorable	30
Medium	Moderate	Average	25
Medium to high	Moderate to high	Below average	20

TABLE 14.4 Risked gas in place and technically recoverable shale gas resources: five regions (EIA, 2011a)

Region	Risked gas in place, Tcf	Risked technically recoverable, Tcf	Average RF, %
NAM	5,314	1208	23
AAO	7,042	1800	26
LAM	6,935	1906	27
AFR	3,962	1024	26
EUP	2,587	624	24
World	25,840	6562	25

shale gas formations in the EIA study. These two success/risk factors are multiplied by OGIP to derive the risked GIP for the prospective area. The TRR is established by multiplying the risked GIP by a shale gas RF listed in Table 14.3.

14.1.9 Shale Gas OGIP Assessment in North America

Dong et al. (2012) analyzed OGIP in the 16 North American basins listed in Table 14.5. Shale gas resources were evaluated and the results were published.

Smead and Pickering (2008) estimated shale gas in place of 833 Tcf for eight US basins (Table 14.6). Kuuskraa and Stevens (2009) completed in-depth, basin-level assessments for seven gas shales in six North American basins. He estimated that the resource endowment in these six basins is 4789 Tcf. In the same year, DOE (2009) estimated the OGIP in Antrim and New Albany shales at 76 and 160 Tcf, respectively. Formations prospective for shale in the Western Canadian Sedimentary Basin potentially contain 1380–1490 Tcf of shale gas (EIA, 2011a; Kuuskraa and Stevens, 2009).

TABLE 14.5 North American basins assessed

No.	Nomenclature	Full name	No.	Nomenclature	Full name
1	APPB	Appalachian Basin	9	ILLB	Illinois Basin
2	ANAB	Anadarko Basin	10	LMS	Louisiana Mississippi Salt
3	ARKB	Arkoma Basin	11	MICB	Michigan Basin
4	BWB	Black Warrior Basin	12	PERB	Permian Basin
5	DENB	Denver Basin	13	SJB	San Juan Basin
6	ETB	East Texas Basin	14	WGC	Western Gulf Coast
7	FWB	Fort Worth Basin	15	WILB	Williston Basin
8	GGRB	Greater Green River Basin	16	WCSB	Western Canadian Sedimentary Basin

TABLE 14.6 Comparison of original shale gas OGIP assessments by basin, in Tcf

Basins	Smead and Pickering (2008)	Kuuskraa and Stevens (2009)	DOE (2009)		Others	Dong et al. (2012)
ANAB		199				199
APPB		2100	1500	225248	Williams (2006)	1725–2348
ARKB	23	320	52			75–343
BWB	23					23
DENB	13					13
ETB-LMS		790	717			717–790
FWB	168	250	327			168–327
GGRB	265					265
ILLB			160			160
MICB			76			76
PERB	265					265
SJB	61			97	Petzet (2007)	61–97
WGC				950	Hill and Nelson (2000)	950
WILB	15					15
WCSB		1380		1490	EIA (2011a)	1380–1490
TOTAL	833	4789	2832	4774	EIA (2011a)	4774–7341

Table 14.6 presents the shale gas resource estimates compiled for the 16 basins used in the Dong et al. (2012) study. If only one assessment was available for a particular basin, we used that assessment in our study. If multiple assessments were available for a basin, we used the minimum and maximum value among these assessments to generate a GIP range. Shale OGIP in the Marcellus Shale in the Appalachian basin is estimated at 1500 Tcf by DOE (2009) and 2100 Tcf by Kuuskraa (2009). Williams (2006) reported shale OIGP in the Ohio Shale in the Appalachian basin at 225–248 Tcf. The resulting shale OGIP of 1725–2348 Tcf in Appalachian basin was adopted in the Dong et al. (2012) study. In addition, shale OGIP in the Fayetteville Shale in the Arkoma basin is estimated at 52 Tcf by DOE (2009) and 320 Tcf by Kuuskraa (2009). Shale OIGP in the Woodford Shale in the Arkoma basin is reported as 23 Tcf (Smead and Pickering, 2008). The resulting shale OGIP of 75–343 Tcf in Arkoma basin was used in the Dong et al. (2012) study.

The total volume of original shale gas in place for the 16 North American basins was estimated to be 4774–7341 Tcf (Table 14.6). Figures 14.3 and 14.4 show the geographic distribution of shale-gas OGIP in the United States and Canada, respectively. This range obtained from these more recently published assessments significantly exceeds Rogner's (1997) estimate for total North America shale gas resources of 3840 Tcf. The growth in the estimated shale gas resource endowment will likely continue, driven by more intense development of existing shale gas plays as well as the discovery of new plays in North America. Dong et al. (2012) suggest that the range underestimates the uncertainty, so they arbitrarily decided that it represents a 50% confidence interval. In other words, they suggest that there is a 25% probability that the volume of shale-gas OGIP is less than or equal to 4774 Tcf (P25), and a 75% probability that the volume is less than or equal to 7341 Tcf (P75). A lognormal distribution was fit to these two points, which yielded a mean of 6 260 Tcf and standard deviation of 2040 Tcf (Fig. 14.5).

It is clear that there are abundant volumes of natural gas in North America. The question now requiring an answer is this: What portion of the gas resource is technically and economically recoverable? The objective of Dong et al.'s (2013) work was to develop the data sets, methodology, and tools to determine values of OGIP, TRRs, RF, and economic viability in highly uncertain and risky shale gas reservoirs.

14.1.10 Recent Shale Gas Production and Activity Trends

Many gas shale plays are currently under development in the United States. The United States has already experienced the shale revolution, which saw shale gas production increase

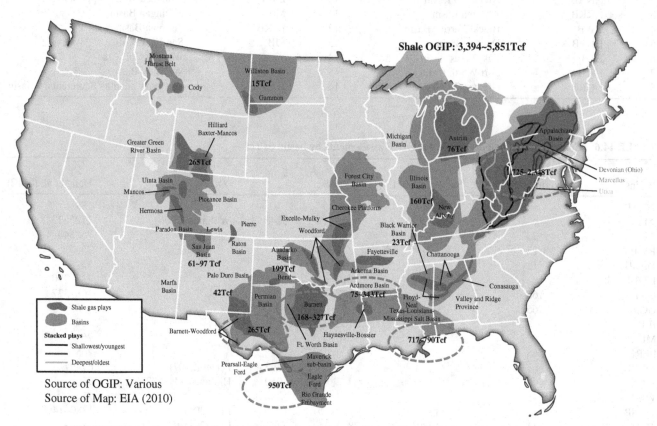

FIGURE 14.3 Graphic distribution of shale-gas OGIP in the United States (Adapted from Dong et al., 2012).

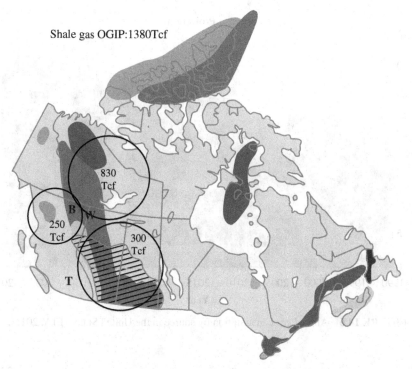

FIGURE 14.4 Graphic distribution of shale-gas OGIP in Canada (Adapted from Dong et al., 2012). Source: PTAC (2006).

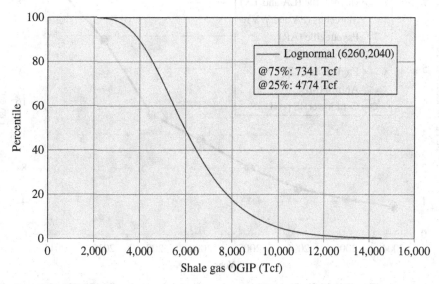

FIGURE 14.5 Probability distribution of original shale gas in place in North America (Adapted from Dong et al., 2012).

from 1% of overall US gas consumption to 32% in 2011, with expectations for it to grow to 56% by 2040 (Fig. 14.6).

As of December 2011, the seven producing shale gas plays were Barnett, Haynesville, Marcellus, Fayetteville, Arkoma Woodford, Eagle Ford, and Antrim shales (Fig. 14.7). The Barnett Shale has been the country's leading shale gas producer during the past decade. Barnett Shale production has grown from 0.06 Bcf in 2004 to 1868 Bcf in 2011 (Fig. 14.7). The Haynesville Shale production in the East Texas–Louisiana–Mississippi salt basins has skyrocketed from nothing in 2006 to produce 2167 Bcf in 2011. Natural gas production from the Louisiana section of the Haynesville overtook the Barnett's volumes in early to mid-February of 2011 (Fig. 14.8).

Unlike most other shale gas plays, the natural gas from the Antrim Shale is biogenic gas generated by the action of bacteria on the organic-rich rock. The Antrim Shale play is winding down as economic limits have been reached.

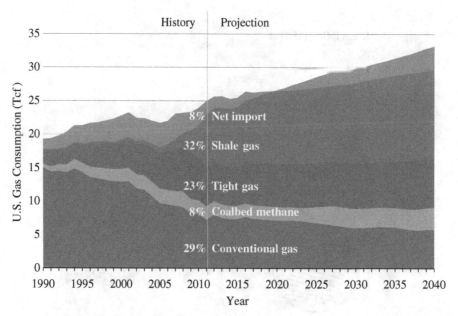

FIGURE 14.6 Annual gas consumption by source in the United States (EIA, 2012).

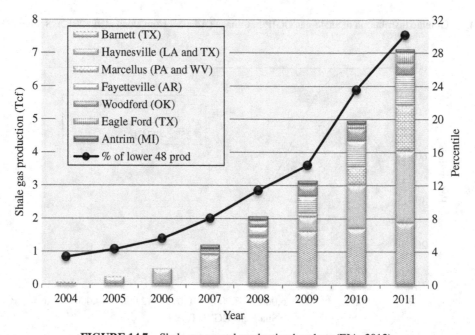

FIGURE 14.7 Shale gas annual production by plays (EIA, 2012).

14.1.11 Drilling, Stimulation, and Completion Methods in Shale Gas Reservoirs

Long horizontal wells (3,000–10,000 ft) are designed to place the gas production well in contact with as much of the shale matrix as technically and economically feasible. Large volume hydraulic fracture treatments, conducted in multiple, closely spaced stages (up to 20 stages), are designed to "fracture" the shale matrix and create permeable flow paths from the reservoir to the wellbore. The production from the hydraulically fracture treated well depends upon the mineralogy of the shale, particularly its relative quartz, carbonate, and clay contents.

- Shale with a high percentage of quartz and carbonate tend to be brittle and will "shatter," leading to a vast array of small-scale induced fractures providing numerous flow paths from the matrix to the wellbore.
- Shale with high clay content tends to be ductile and tends to deform instead of shattering, leading to relatively few induced fractures.

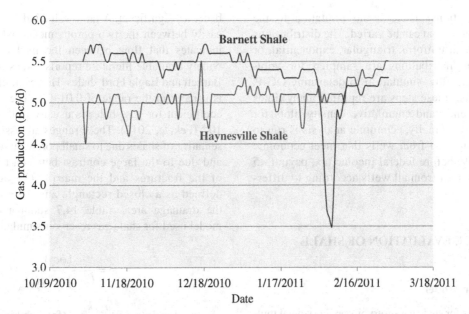

FIGURE 14.8 Haynesville Shale gas production surpasses Barnett Shale as the nation's leading shale gas play (EIA, 2011b).

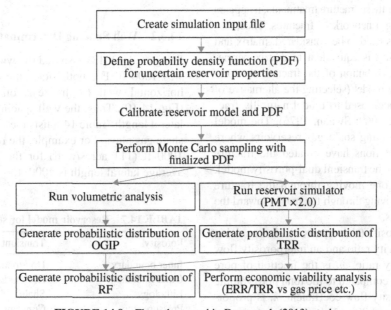

FIGURE 14.9 Flow chart used in Dong et al. (2013) study.

14.2 METHODOLOGY

Shale gas reservoirs are highly heterogeneous, and the well productivity depends on reservoir properties as well as completion and stimulation parameters. Even if finite difference reservoir simulators are available, it can be time-consuming to do a large reservoir simulation study. Dong et al. (2013) developed a computer program, Unconventional Gas Resource Assessment System (UGRAS) to determine the values of ERRs. In the program, they integrated Monte Carlo simulation with an analytical reservoir simulator, PMTx 2.0, to estimate the original volume in place, predict production performance, and estimate the fractions of TRR that are ERRs for a variety of economic situations.

In the study, they applied UGRAS to generate gas production profiles for a variety of reservoir, well, and hydraulic fracture scenarios. Thousands of simulations were run automatically to explore combinations of unknown reservoir and well parameters across their ranges of uncertainty. They used the investment evaluation hurdles IRR >20% and payout time <5 years, applied on an individual-well basis, to determine the fraction of TRR that is ERR for a variety of economic situations. They assumed that if a well does not pay out in 5 years, it is probably not worth drilling at this time. There should be other places to drill and spend capital that are more profitable.

The workflow of the study with the probabilistic reservoir model UGRAS is outlined in Figure 14.9. First, an input file is created and uncertain parameters are assigned probability distributions. There is no limitation to the number of parameters that can be varied. The distributions are typically normal, uniform, triangular, exponential, or lognormal. These distributions are sampled for volumetric analysis and flow simulation to determine OGIP, TRR, and RF. Then, these steps are repeated many times to generate frequency and cumulative density plots for OGIP, TRR, and RF. Finally, economic analysis is run to calculate the production from wells that meet economic criteria (IRR >20% before federal income tax, payout <5 years) over production from all wells according to different F&DC.

14.3 RESOURCE EVALUATION OF SHALE GAS PLAYS

14.3.1 Reservoir Model

Typical completions for shale gas reservoirs are horizontal multistage fractured wells. As more knowledge is gained through microseismic monitoring of these fracture treatments, it appears that they are likely creating a network of fractures. Thus, two permeabilities in gas shales need to be considered: matrix and system. System permeability is equivalent to matrix permeability enhanced by the contribution of the fracture network. The transient dual-porosity model (selecting the alternative of slab matrix blocks) has been used to model naturally fractured reservoirs (Kazemi, 1969; Swaan, 1976). The model can also be used for modeling shale gas reservoirs where multistage fracture completions have created the fracture network (Fekete, 2012). In the transient dual-porosity model, there are two transients: one moving through the fracture system and the second moving through the matrix toward the interior of the matrix blocks.

The transient dual-porosity (slab matrix blocks) model is characterized by a storativity ratio and an interporosity flow coefficient. The storativity ratio, ω, is the fraction of pore volume in the fractures as compared to the total pore volume (Eq. 14.1). The interporosity flow coefficient, λ, is proportional to the ratio of permeabilities between the matrix and the fractures (Eq. 14.2), and it determines the time at which the contribution of flow from the matrix to the fractures becomes significant. A large value indicates that fluids flow easily between the two porous media, while a small value indicates that flow between the media is restricted. No widely available literature reports values of λ and ω for the Barnett and Eagle Ford shales. However, the storativity ratio is usually in the range of 0.01–0.1. The interporosity flow coefficient for gas shales is usually in the range of 10^{-4} to 10^{-8} (Fekete, 2012). These ranges are assumed to be representative of shales due to small pore volume of the fractures, and due to the large contrast between the permeabilities of the fractures and the matrix. The outer boundary is defined as a closed rectangle and the well is centered in the drainage area. Table 14.7 summarizes the reservoir model used for shale gas reservoir simulation.

$$\omega = \frac{(\phi c_t)_f}{(\phi c_t)_f + (\phi c_t)_m} \quad (14.1)$$

$$\lambda = 4n(n+2)\frac{r_w^2}{L^2}\frac{k_m}{k_f} \quad \text{(for slab blocks, } n=1\text{)} \quad (14.2)$$

14.3.2 Well Spacing Determination

Dong et al. (2013) assumed the width of shale gas reservoir was 1000 ft. For both sides, the margin from the end of horizontal well to the reservoir boundary was 400 ft (Fig. 14.10). Thus, the well spacing was determined by the lateral length. Table 14.8 lists the well spacing for the target shale gas plays. For example, the reservoir size is 4800 ft × 1000 ft (111 acres/well) for the Barnett Shale since the average lateral length is 4000 ft.

TABLE 14.7 Reservoir model for shale gas reservoirs

Porosity	Transient dual porosity
Inner boundary	Horizontal with transverse fractures
Outer boundary	Rectangle
Lithology	Shale
Pressure step	Constant
Permeability	Isotropic
Well location	Centered

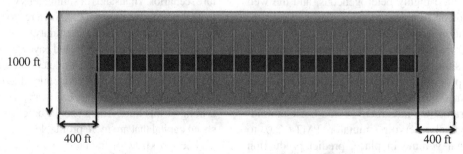

FIGURE 14.10 Well geometry for shale gas reservoirs.

14.3.3 Reservoir Parameters Sensitivity Analysis

Dong et al. (2013) investigated the essential reservoir properties that affect the prediction of OGIP or TRR from shale gas reservoirs. Fifteen different characteristics were analyzed. Table 14.9 lists the primary properties. Besides area, net pay, porosity, and water saturation, shale-gas OGIP is affected by gas content. Gas production from shale gas reservoirs is affected most by completed GIP resources, fracture system permeability, initial reservoir pressure, and gas desorption and diffusion characteristics, relative permeability characteristics, and fracture half-length (Fig. 14.11). Since Langmuir pressure controls the shape of the sorption curve, it impacts how fast gas content changes. Langmuir volume controls the endpoint of the Langmuir curve, but has almost no impact on the rate of change. Horizontal length and well spacing are controllable, so they were not considered to be uncertain parameters. Dong et al. (2013) treated only net pay, initial pressure, system

TABLE 14.8 Well spacing for the four target shale gas plays in the United States

Plays	Average lateral length, ft	Reservoir size	Well spacing, acres
Barnett	4000	4800 ft × 1000 ft	111
Eagle Ford	5600	6400 ft × 1000 ft	147
Marcellus	3700	4500 ft × 1000 ft	104
Haynesville	4700	5500 ft × 1000 ft	

TABLE 14.9 Data source for primary properties of shale gas reservoirs

Primary property	Data source	Controllable	Large uncertainty
Thickness	Log analysis	No	Yes
System permeability	Core analysis, well-test analysis, production analysis	No	Yes
Porosity	MICP, NMR, log analysis	No	Yes
Water saturation	PID, openhole test	No	Yes
Gas content	Desorption canister testing and adsorption isotherms, calibrated log analysis	No	Yes
Reservoir pressure	PID, PITA, openhole test	No	Yes
Effective fracture half-length	Static: Postfracture net-pressure analysis, postfracture flow and buildup Flowing: Rate-transient analysis	No	No
Lateral horizontal length	Well-test analysis	Yes	No
Well spacing	Well-test analysis	Yes	No

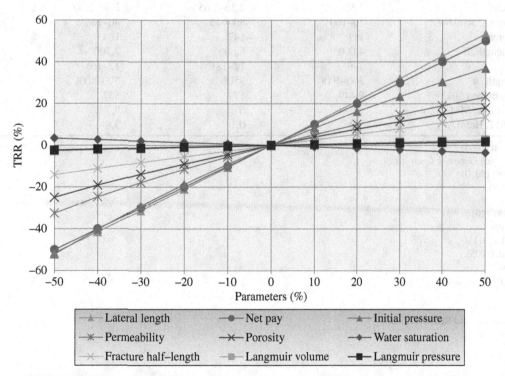

FIGURE 14.11 Sensitivity analysis result of shale gas technically recoverable resources (TRR).

permeability, porosity, water saturation, and gas content as uncertain parameters.

14.3.4 Reservoir Parameters

Table 14.10 summarizes the range of main reservoir parameters for the four target shale gas plays in the United States. Table 14.11 shows the fixed reservoir parameters used for the four shale single-well reservoir simulations in the Dong et al. (2014) study.

14.3.5 Model Verification

Dong et al. (2014) used the HPDI (2011) database as the source for production data. They assumed a uniform density function for net pay, initial reservoir pressure, permeability, porosity, water saturation, and gas content, initially honoring the ranges listed in Table 14.10. They did not consider possible correlations among these parameters. Some or all of these density functions were refined until a reasonable match between the simulated and the actual cumulative distribution of cumulative gas production was obtained (Figs. 14.12, 14.13, 14.14, and 14.15). These cumulative probability distributions in these figures are inverted in contrast to the other decumulative probability distributions shown in the paper. Final density functions of the six uncertain parameters for each play are listed in Table 14.12. For instance, the black curve in Figure 14.12 shows the distribution of 5-year cumulative gas production from 1492 horizontal gas wells in the Barnett shale. The gray curve in Figure 14.12 is the distribution of 5-year cumulative gas production simulated by UGRAS with the reservoir and well parameters in Table 14.11 and the final density functions in Table 14.12.

TABLE 14.10 Summary of the key characteristics for four target shale-gas plays in the United States (Dong et al., 2014)

Parameter	Barnett	Eagle ford	Marcellus	Haynesville
Area, acres	3,200,000	3,000,000	15,000,000	5,760,000
Depth, ft	6,500–8,500[a]	5,500–14,400[b]	3,300–8,800[b]	10,000–14,000[c]
Net Pay, ft	100–600[a,d]	3–326[b]	45–384[b]	200–300[e]
Porosity, %	4–5[a]	3–12[b]	3–13[b]	8–14[e]
System permeability, 10^{-3} md	0.07–5[d]	0.1–0.7[b]	0.2–0.9[b]	0.5–400*
S_w, %	25–43[f]	9–44[b]	6–53[b]	16–41[c]
Average p_i, psia	3,000–5,000[g]	4,300–10,900[b]	2,000–5,100[b]	7,000–10,000[h]
Gas content, scf/ton	60–125[i]	7–120[b]	41–148[b]	100–330
Reservoir temperature, °F	205	170–231[b]	110–160	300–350
TOC, %	2.4–5.1[f]	0.3–5.4[b]	2.0–8.0[b]	0.5–4.0[e]
R_o, %	0.6–1.6[j]	1.5	1.25	2.2
Bulk density, g/cm^3	2.5[k]	2.36–2.65[b]	2.30–2.60[b]	<2.57
Typical well spacing, acres/well	60–160[a]	80–640	40–160	40–560[l]
Well spacing, acres	111	147	104	124
Average lateral length, ft	4,000[m]	5,600[m]	3,700[o]	4,600[p]
Fracture stages	7–9[q]	12–18[q,r]	12–16[o]	12–15[p]
Fracture half-length, ft	300–400[q]	350[q]	300–400[q]	300[q]
Horizontal wells by end 2011	9,449	177	837	1,156
Initial production rate, MMcf/day	1.2–4.7	6	7.7	10
Production in 2011, Bcf/day	5.2	0.1	3.8	6

[a] Hayden and Pursell (2005).
[b] Provided by W.D. Von Gonten & Company.
[c] Wang and Hammes (2010).
[d] Grieser et al. (2008).
[e] Berman (2008).
[f] Bruner and Smosna (2010).
[g] Chong et al. (2010).
[h] Abou-sayed et al. (2011).
[i] Montgomery et al. (2005).
[j] Jarvie et al. (2004).
[k] Kuuskraa et al. (1998).
[l] DOE (2009).
[m] Powell (2010).
[n] Provided by Unconventional Resources, LLC.
[o] Edwards et al. (2011).
[p] Billa et al. (2011).
[q] Kennedy (2010).
[r] Rhine et al. (2011).

RESOURCE EVALUATION OF SHALE GAS PLAYS 313

TABLE 14.11 Fixed reservoir parameters for the shale-gas reservoir simulation (Dong et al., 2014)

Parameters	Barnett	Eagle ford	Marcellus	Haynesville
Reservoir temperature, °F	205	293	144	340
Bottom hole pressure, psia	500	500	500	500
Reservoir length, ft	4800	6400	4500	5400
Reservoir width, ft	1000	1000	1000	1000
Bulk density, g/cc	2.5	2.51	2.53	2.5
Fracture half-length, ft	350	350	300	300
Lateral length of horizontal well, ft	4000	5600	3700	4600
Fracture stages	10	18	12	13
Langmuir pressure, psia	1241	1000	850	1000
Langmuir volume, scf/ton	150	60	100	380
λ (dimensionless)	7×10^{-7}	1×10^{-6}	7×10^{-6}	1×10^{-8}
ω (dimensionless)	0.01	0.01	0.01	0.01

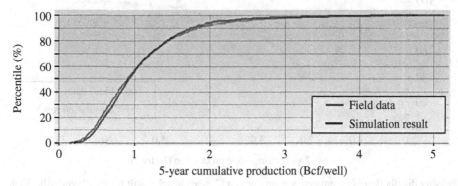

FIGURE 14.12 Probability distribution of cumulative gas production (5-year) match result for the Barnett Shale (Adapted from Dong et al., 2013).

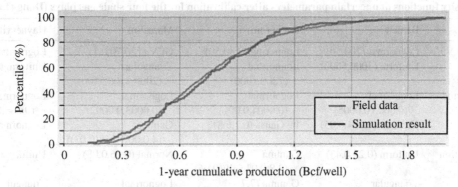

FIGURE 14.13 Probability distribution of cumulative gas production (1-year) match result for the Eagle Ford Shale (Adapted from Dong et al., 2013).

As a further check of the models, Dong et al. (2014) compared simulated and actual production decline trends. They calculated the average monthly gas production from gas wells for each shale gas play (Figs. 14.16, 14.17, 14.18, and 14.19). From the probabilistic model described in Tables 14.11 and 14.12, they plotted well production curves corresponding to the mean, P10, P50, and P90 cumulative production values (Figs. 14.16, 14.17, 14.18, and 14.19). The average gas production of gas wells overlaid the mean TRR simulated by UGRAS for each shale gas play. The good match between the two curves confirmed the reasonableness of the fixed parameters in Table 14.11 and finalized density functions in Table 14.12 to make long-term production prediction.

14.3.6 Resource Assessment

Dong et al. (2014) determined the decumulative probability distributions of OGIP, TRR, and RF for the four target shale gas plays in the United States using the fixed reservoir parameters in Table 14.11 and final density functions in

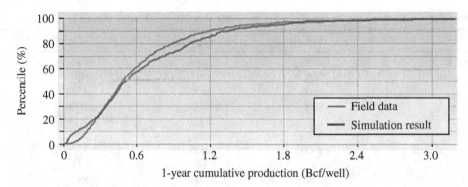

FIGURE 14.14 Probability distribution of cumulative gas production (1-year) match result for the Marcellus Shale (Adapted from Dong et al., 2014).

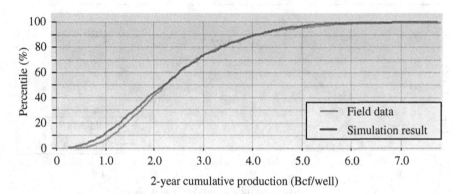

FIGURE 14.15 Probability distribution of cumulative gas production (2-year) match result for the Haynesville Shale (Adapted from Dong et al., 2014).

TABLE 14.12 Density functions of uncertain parameters after calibration for the four shale gas plays (Dong et al., 2014)

Plays	Barnett	Eagle ford	Marcellus	Haynesville
Net pay, ft	Lognormal (200,50)	Lognormal (130,50)	GEV (120,70,0.1)	Lognormal (200,80)
Initial pressure, psi	Uniform (3000,5000)	Lognormal (7200,1000)	Triangular (2000,4100,5100)	Uniform (7000,10000)
System permeability, md	Lognormal (0.0005,0.0005)	Lognormal (0.0004,0.001)	Lognormal (0.0003,0.0002)	Lognormal (0.034,0.032,shift(-0.001))
Porosity, fraction	Uniform (0.004,0.05)	InvGauss (0.1,6.8)	Gamma (4,0.007,shift(0.03))	Lognormal (0.126,0.03)
Water saturation, fraction	Uniform (0.25,0.43)	Gamma (3.8,0.03,shift(0.06))	Normal (0.26,0.08)	Uniform (0.16,0.41)
Gas content, scf/ton	Triangular (60,100,125)	Gamma (7,7)	Lognormal (100,19,shift(-41))	Triangular (100,200,330)

InvGauss (μ, λ): Inverse Gaussian distribution with mean μ and shape parameter λ.
Lognormal (μ, σ): Lognormal distribution with specified mean and standard deviation.
GEV (μ, σ, ξ): Generalized extreme value distribution with mean μ, standard deviation σ, and shape parameter ξ.
Gamma (α, β): Gamma distribution with shape parameter α and scale parameter β.
GEV (μ, σ, ξ)
Generalized extreme value distribution with mean μ, standard deviation σ, and shape parameter ξ.
Triangular (min., most likely, max.): Triangular distribution with defined minimum, most likely, and maximum value.
Uniform (min., max.): Uniform distribution between minimum and maximum.

Table 14.12. Figure 14.20 shows the cumulative distribution of OGIP per section (640 acres) for the four shale gas plays. The Haynesville Shale has the most original shale gas in place per section due to its higher gas content per surface acre.

Figure 14.21 shows the cumulative distribution of TRR per section for the four shale gas plays. TRR is the most sensitive to lateral length and net pay (Fig. 14.11). The Haynesville Shale has the most technically recoverable shale

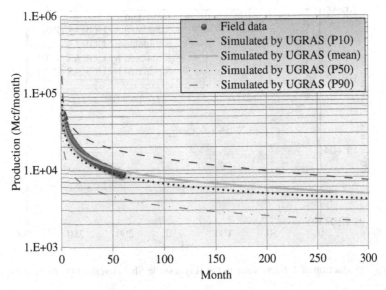

FIGURE 14.16 Average production of 1492 gas wells in the Barnett Shale overlaid the mean TRR simulated by UGRAS.

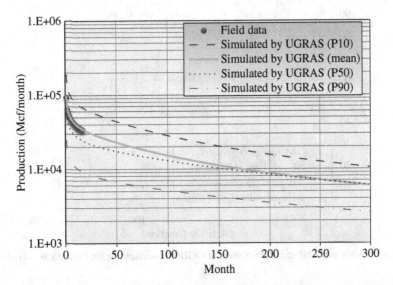

FIGURE 14.17 Average production of 152 dry gas wells in the Eagle Ford Shale overlaid the mean TRR simulated by UGRAS.

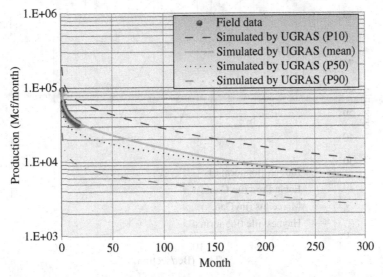

FIGURE 14.18 Average production of 372 gas wells in for the Marcellus Shale overlaid the mean TRR simulated by UGRAS (EIA, 2011b).

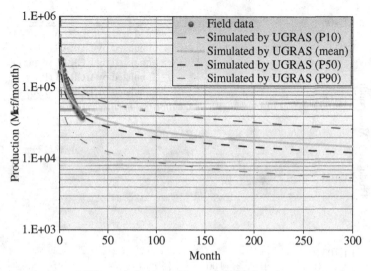

FIGURE 14.19 Average production of 476 gas wells in the Haynesville Shale overlaid the mean TRR simulated by UGRAS.

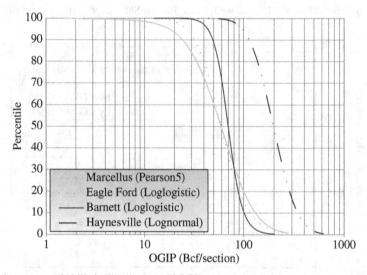

FIGURE 14.20 Comparison between probabilistic distributions of OGIP per section for four shale gas plays in the United States (Adapted from Dong et al., 2014).

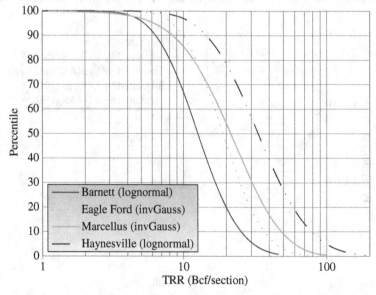

FIGURE 14.21 Comparison between probabilistic distributions of TRR per section for four shale gas plays in the United States (Adapted from Dong et al., 2014).

gas resources per section since it has the largest net pay thickness. The Marcellus Shale has a high RF due to a high value of interporosity flow coefficient ($\lambda = 9 \times 10^{-6}$) (Fig. 14.22). High value of interporosity flow coefficient means that fluids flow easily between the fracture and matrix, while a small value indicates that flow between the media is restricted. The Eagle Ford Shale has a high RF due to its high reservoir pressure.

Table 14.13, constructed for the four major shale gas plays, provides a concise summary of these resource assessments.

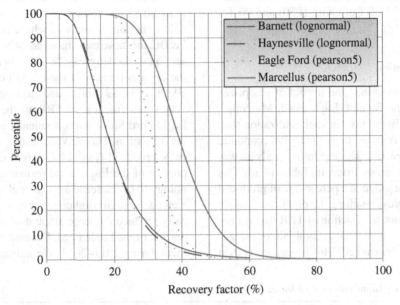

FIGURE 14.22 Comparison between probabilistic distributions of recovery factor (RF) for the four shale gas plays in the United States (Adapted from Dong et al., 2014).

TABLE 14.13 Summary of resource assessment for four target shale gas plays in the United States

Parameter	Barnett	Eagle ford	Marcellus	Haynesville
Area, Acres	3,200,000	3,000,000	15,000,000	5,760,000
Well spacing, acre	111	147	104	124
Number of wells	28,828	20,407	144,230	46,451
Distribution of OGIP	Log-logistic	Log-logistic	Pearson 5	Lognormal
OGIP (P90), Bcf/well	8.4	7.5	3.8	22.0
OGIP (P50), Bcf/well	12.2	13.6	9.6	40.0
OGIP (P10), Bcf/well	17.8	25.3	20.8	68.0
Distribution of technically recoverable resources (TRR)	Lognormal	InvGauss	InvGauss	Lognormal
TRR (P90), Bcf/well	1.1	2.3	1.5	3.1
TRR (P50), Bcf/well	2.2	4.4	3.7	7.1
TRR (P10), Bcf/well	4.5	8.5	8.5	15.2
Distribution of RF	Lognormal	Pearson 5	Pearson 5	Lognormal
RF (P90), %	10	25	29	10
RF (P50), %	18	31	39	18
RF (P10), %	35	40	52	33
OGIP (P90), Bcf/section	48	33	23	114
OGIP (P50), Bcf/section	70	59	59	206
OGIP (P10), Bcf/section	103	110	128	351
TRR (P90), Bcf/section	6	10	9	16
TRR (P50), Bcf/section	13	19	23	37
TRR (P10), Bcf/section	26	37	52	78
OGIP in Dong study (P50), Tcf	352	278	1385	1858
TRR in Dong study (P50), Tcf	63	90	534	330

Using known estimated productive acreage of each shale gas play, Dong et al. (2014) also estimated the OGIP and TRR for the entire plays. For instance, in the Eagle Ford dry gas window, the estimated productive acreage is 3 million acres. Assuming an average well spacing of 147 acres, 20,000 wells could be drilled in the dry gas portion of the Eagle Ford shale. Thus, the resource potential for the entire Eagle Ford dry gas window is 278 Tcf of OGIP (P50) and 90 Tcf of TRR (P50) (Table 14.13).

14.3.7 Reserve Evaluation

Dong et al. (2013) next examined the impact of gas prices and F&DC on ERR in the Barnett, Eagle Ford, Marcellus, and Haynesville shales. To do this, for each realization they determined the gas price required to just meet the economic hurdles for a particular F&DC. The economic analysis was performed for the assumptions listed in Table 14.14. Gas shrinkage results from the usage of a percentage of produced gas for mechanical compression along the pipeline.

Ranking the realizations, the fraction of TRR that is economically recoverable for a particular combination of gas price and F&DC can be determined. This procedure is then repeated to determine the ratio ERR/TRR over a range of gas prices and F&DC (Figs. 14.23, 14.24, 14.25, and 14.26). For instance, with a typical F&DC of US$3 million for the Barnett Shale wells and a gas price of US$4.0/Mcf, 20% of the Barnett Shale gas TRR is economically recoverable (Fig. 14.23).

The Haynesville Shale and dry gas window of the Eagle Ford Shale lie at about the same depth (10,000–14,000 ft) below the land's surface, which results in about the same F&DC for these two shale gas plays. However, the TRR per section of the Haynesville Shale is twice as much as that in the dry gas portion of the Eagle Ford shale. With a typical F&DC of US$9 million and gas price of US$4.0/Mcf, only a very small fraction of TRR in the dry gas portion of the Eagle Ford Shale is economically recoverable (Fig. 14.24), but 37% of the Haynesville Shale gas TRR can be economically produced (Fig. 14.25). It is clear that in the dry gas portion of the Eagle Ford formation, either (i) better technology to (a) increase average well recovery or (b) decrease well costs, or (ii) higher gas prices are required to economically produce the large amount of natural gas in the dry gas portion of the Eagle Ford. Because of the economic environment during 2011–2013, virtually all of the current drilling

TABLE 14.14 Input values used for cash flow statements

	Barnett shale	Eagle ford shale	Marcellus shale	Haynesville shale
Operating cost, US$/Mcf	1.0	1.3	0.7	1.5
Working interest, %	100	100	100	100
Royalty burden, %	25	25	25	22.5
Severance taxes, %	7	7	7	7.5
Gas shrinkage, %	6	6	6	6
Range of F&DC, MMUS$	1–7	6–12	3–9	6–12
Average F&DC, MMUS$	3	9	6	9

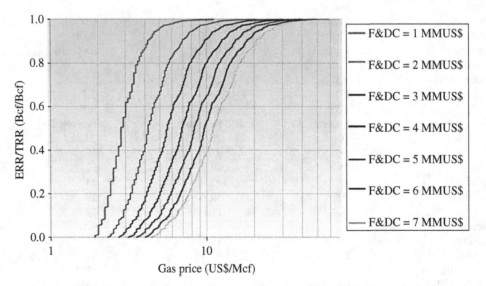

FIGURE 14.23 Ratio of ERR to TRR as a function of gas price and F&DC for the Barnett Shale (Adapted from Dong et al., 2013).

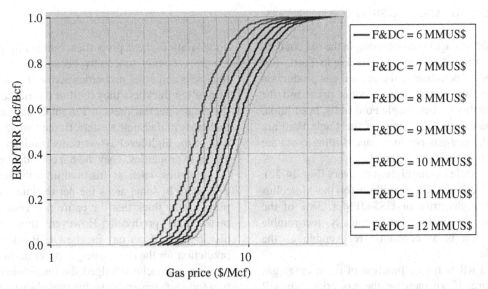

FIGURE 14.24 Ratio of ERR to TRR as a function of gas price and F&DC for the Eagle Ford Shale (Adapted from Dong et al., 2013).

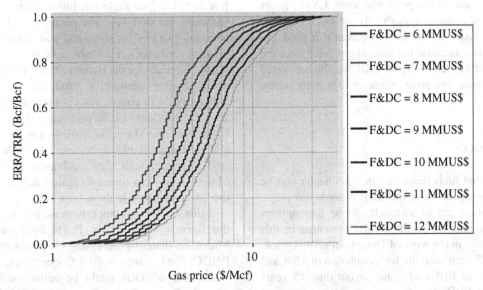

FIGURE 14.25 Ratio of ERR to TRR as a function of gas price and F&DC for the Haynesville Shale.

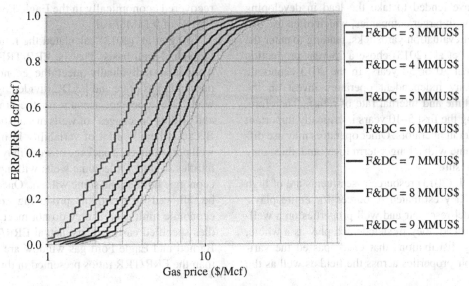

FIGURE 14.26 Ratio of ERR to TRR as a function of gas price and F&DC for the Marcellus Shale.

(other than to hold acreage) was occurring in the gas condensate window or the oil window in the Eagle Ford. The addition of liquids production to the natural gas production significantly improves the average product price and the economic profitability of those Eagle Ford wells. In addition, the gas condensate and oil portions of the Eagle Ford are shallower than the dry gas portion; thus, drilling costs are lower in the shallower portion of the play.

The Marcellus Shale has the highest recovery (Fig. 14.22). With a typical F&DC of US$6 million for the Marcellus Shale wells and a gas price of US$4.0/Mcf, 38% of the Marcellus Shale gas TRR is economically recoverable (Fig. 14.26), which is as economic recoverable as the Haynesville shale.

The value of ERR will be a function of the average gas price in the future. If we increase the gas price, we will increase the fraction of TRR that is economically recoverable. In 2013, the natural gas price was about US$4/Mcf or less, which means many of these wells are not economic. However, the industry is working to increase natural gas demand, which may increase the natural gas price after the year 2013. Therefore, much of the TRR will be recovered whenever the natural gas price increases to a point where more drilling will occur.

14.4 DISCUSSION

The technology and tools described in this chapter can be useful in assessing technically and ERR in shale gas plays. However, it is important to acknowledge the assumptions and uncertainties inherent in the results presented in this chapter, particularly in the work of Dong et al (2013). First, they assumed a 25-year well life for calculation of TRR and economic hurdles of IRR >20% and payout time <5 years for calculation of ERR. These are reasonable criteria, and similar values are used by many smaller companies and investors that have tended to take the lead in developing shale gas plays, with larger companies, which often use different investment evaluation yardsticks, tending to enter the plays later. Dong et al. (2013) chose a 25-year production history rather than 30 or 50 years. In the 2013 economic environment, many independent operators invest on the basis of payout time and internal rate of return. Therefore, production during the first 5–10 years is critically important to many independent operators. Other operators may use different criteria, some with a longer-term focus and, thus, may obtain different results.

The Dong et al. (2013) resource assessments are of high level. Although they estimated resources for entire plays, they did not model reservoir and well properties on a well-by-well basis. Instead, they modeled each play as a whole, using probability distributions that encompassed the variability in reservoir properties across the field as well as the uncertainty in these properties. For example, the OGIP varies from county to county because of differences in net thickness and other properties across the field. The distribution of net thickness they used in the Barnett study covered the greater net thickness in Tarrant County and the lower net thickness in the southwestern Barnett shale. Another limitation of the high-level assessments is related to vertical variability in properties. They did not consider vertical variations in properties, such as fracturability, throughout the zones evaluated. In some areas the net thickness of the shale gas plays are so thick that the entire pay zone cannot be completed and produced. However, they used the same distributions of net pay for the OGIP calculation and TRR prediction for the four shale gas plays in this study.

They also acknowledged the uncertainty in the production forecasts generated by the probabilistic analytical simulator. The input parameters used to generate production forecasts for four shale gas plays were obtained from the literature and well data. The parameter values and forecasts were reviewed by operators and reserves evaluators in these plays to verify their reasonableness. For instance, the probabilistic forecasts for the Barnett Shale were calibrated against actual 5-year cumulative production data (Fig. 14.2), although this, of course, does not guarantee the accuracy of 25-year forecasts. Little performance data exists for the Eagle Ford and Marcellus Shale—only one year production history was available at the time of the Dong et al. (2012) study. Even though they calibrated the Eagle Ford and Marcellus dry gas forecasts against actual production data, there is uncertainty in these forecasts.

Table 14.15 lists opportunities for increasing ERR in the Barnett, Haynesville, Eagle Ford gas window, and Marcellus shales by increasing gas prices and/or decreasing F&DC. For instance, if F&DC were decreased to 2 MM$/well, 50% of TRR could be economically recoverable from the Barnett Shale at a gas price of US$4.0/Mcf. If gas price increased to US$6.0/Mcf, 25% of TRR could be recovered economically in the Eagle Ford gas window at a F&DC of 9 MM$/well.

Dong et al. (2013) calculated the ratio ERR/TRR on an individual-well basis. That is, ERR/TRR is the TRR from wells that individually meet the economic hurdles at a particular gas price and F&DC divided by the TRR from all wells. This does not account for the practice of budgeting and drilling "packages" of wells in statistical shale gas plays. Because there is a lot of variability in individual shale gas well performance, a package of wells may meet the economic hurdles overall while some wells will individually meet the economic hurdles and some will not. Once drilled, wells will be allowed to continue producing to a net-cash-flow economic limit even if they do not meet the economic hurdles specified earlier. Thus, actual ERR/TRR ratios for the Barnett and Eagle Ford gas window are potentially greater than the ERR/TRR ratios presented in this chapter.

TABLE 14.15 Opportunities for increasing the ERR/TRR ratio

ERR/TRR, %	Barnett Shale		Eagle Ford gas window		Marcellus		Haynesville	
	F&DC, MMUS$/well	Gas price, US$/Mcf	F&DC, MMUS$/well	Gas price, US$/Mcf	F&DC, MMUS$/well	Gas price, US$/Mcf	F&DC, MMUS$/well	Gas price, US$/Mcf
75	1	3.1	6	6.5	3	4.0	6	4.7
	2	5.1	7	7.2	4	5.0	7	5.4
	3	7.1	8	8.3	5	6.0	8	6.1
	4	9.0	9	9.0	6	7.0	9	7.0
	5	10.5	10	10.0	7	8.3	10	7.5
	6	12.0	11	10.6	8	9.8	11	8.2
	7	16.0	12	11.0	9	10.5	12	9.0
50	1	2.8	6	5.2	3	2.8	6	3.5
	2	4.0	7	6.0	4	3.4	7	4.0
	3	5.1	8	7.0	5	4.0	8	4.6
	4	7.0	9	7.2	6	5.0	9	5.0
	5	8.0	10	8.0	7	6.0	10	5.6
	6	9.8	11	9.0	8	6.4	11	6.0
	7	10.5	12	8.0	9	7.0	12	6.8
25	1	2.1	6	4.2	3	1.8	6	2.8
	2	3.1	7	5.0	4	2.1	7	3.0
	3	4.1	8	5.5	5	2.8	8	3.4
	4	5.0	9	6.0	6	3.0	9	3.9
	5	6.0	10	6.2	7	3.8	10	4.1
	6	7.0	11	7.0	8	4.0	11	4.7
	7	8.0	12	7.5	9	4.4	12	5.0

NOMENCLATURE

c_t	Total compressibility, 1/psi
InvGauss (μ, λ)	Inverse Gaussian distribution with mean μ and shape parameter λ.
k_m	Matrix permeability, md
k_f	Fracture permeability, md
L	Characteristic length of a matrix block
Logistic (α, β)	Logistic distribution with location parameter α and scale parameter β.
Loglogistic (γ, β, α)	Log-logistic distribution with location parameter γ, scale parameter β, and shape parameter α.
n	Number of flow dimensions, dimensionless
Pearson5 (α, β)	Pearson type V (or inverse gamma) distribution with shape parameter α and scale parameter β.
P90	Value for which there is at least a 90% probability that the value will equal or exceed the estimate, indicated by the 90th percentile on a cumulative probability distribution plot. Similarly for P50 and P10.
r_w	Wellbore radius
φ	Wellbore raporosity
ω	Storativity ratio, dimensionless

REFERENCES

Abou-sayed IS, Sorrell MA, Foster RA, Atweed EL, Youngblood DR. Haynesville Shale Development Program—From Vertical to Horizontal. Paper SPE 144425-MS presented at North American Unconventional Gas Conference and Exhibition; January 1, 2011; Woodlands, TX; 2011.

Almadani HS. A methodology to determine both the technically recoverable resource and the economically recoverable resource in an unconventional gas play [MS Thesis]. Texas A&M University, College Station, TX; 2010.

Berman A. The Haynesville Shale Sizzles with the Barnett Cools. World Oil Mag 2008;229 (9):18–22.

Billa RJ, Mota JF, Schneider B. et al. Drilling Performance Improvement in the Haynesville Shale Play. Paper SPE 139842-MS presented at SPE/IADC Drilling Conference and Exhibition; March 1, 2011; Amsterdam, the Netherlands; 2011.

Bruner KR, Smosna R. *A Comparative Study of the Mississippian Barnett Shale, Fort Worth Basin, and Devonian Marcellus Shale, Appalachian Basin*. U.S. Department of Energy: Washington, DC; 2010.

Cheng Y, Wang Y, McVay DA, et al. Practical application of a probabilistic approach to estimate reserves using production decline data. SPE Econ Manage 2010;2 (1):19–31.

Chong KK, Grieser WV, Passman A. et al. A Completions Guide Book to Shale-Play Development: A Review of Successful Approaches Towards Shale-Play Stimulation in the Last Two Decades. Paper SPE-133874-MS presented at

Canadian Unconventional Resources and International Petroleum Conference; October 19, 2010; Calgary, AB, Canada; 2010.

DOE. Modern Shale Gas Development in the United States: A Primer, Washington, DC: U.S. Department of Energy; 2009.

Dong Z, Holditch SA, McVay DA. et al. Global unconventional gas resource assessment. SPE Econ Manage 2012;4(4): 222–234. SPE-148365-PA.

Dong Z, Holditch SA, McVay DA. Resource evaluation for shale gas reservoirs. SPE Econ Manage 2013;5(1): 5–16. SPE-152066-PA.

Dong Z, Holditch SA, McVay DA. et al. Probabilistic Assessment of World Recoverable Shale Gas Resources. Paper SPE-1667768-MS presented at 2014 SPE/EAGE European Unconventional Conference and Exhibition; February 2, 2014; Vienna, Austria; 2014.

Edwards KL, Weissert S, Jackson JB. et al. Marcellus Shale Hydraulic Fracturing and Optimal Well Spacing to Maximize Recovery and Control Costs. Paper SPE 140463-MS presented at SPE Hydraulic Fracturing Technology Conference; January 1, 2011; Woodlands, TX; 2011.

EIA. World Shale Gas Resources: An Initial Assessment of 14 Regions Outside the United States. Washington, DC: Energy Information Administration; 2011a.

EIA. 2011b. Haynesville Surpasses Barnett as the Nation's Leading Shale Play. Energy Information Administration. Available at http://www.eia.gov/todayinenergy/detail.cfm?id=570. Accessed January 17, 2014.

EIA. Annual Energy Outlook 2012. Washington, DC: Energy Information Administration; 2012.

Fekete. 2012. Available at http://www.fekete.com/SAN/TheoryAndEquations/WellTestTheoryEquations/Dual_Porosity.htm. Accessed January 17, 2014.

Grieser W., Shelley RF, Johnson BJ, et al. Data analysis of Barnett shale completions. SPE J 2008;13(3): 366–374. SPE-100674-PA.

Hayden J, Pursell D. The Barnett Shale: Visitor's Guide to the Hottest Gas Play in the US. Houston, TX: Pickering Energy Partners, Inc.; 2005.

Hill DG, Nelson CR. Gas productive fractured shales: An overview and update. GasTIPS 2000;6 (2):4–13.

HPDI. 2011. HPDI Production Data Applications, Version 6.0.1.2. Available at http://hpdi.com//index.jsp. Accessed January 17, 2014.

Ilk D, Rushing JA, Perego AD. et al. Exponential vs. Hyperbolic Decline in Tight Gas Sands—Understanding the Origin and Implications for Reserve Estimates Using Arps' Decline Curves. Paper SPE 116731-MS presented at SPE Annual Technical Conference and Exhibition; September 21, 2008; Denver, CO; 2008.

Jarvie DM, Pollastro RM, Hill RJ, et al. Geochemical Characterization of Thermogenic Gas and Oil in the Ft. Worth Basin, Texas. Paper presented at AAPG National Convention; April 18, 2004; Dallas, TX; 2004.

Jochen VA, Spivey JP. Probabilistic Reserves Estimation Using Decline Curve Analysis with the Bootstrap Method. Paper SPE 36633-MS presented at the SPE Annual Technical Conference and Exhibition; October 6, 1996; Denver, CO; 1996.

Kazemi H. Pressure transient analysis of naturally fractured reservoirs with uniform fracture distribution. SPE J 1969;9(4):451–462. SPE-2156-PA.

Kennedy RB. Shale Gas Challenges/Technologies Over the Asses Life Cycle. Paper present at U.S.-China Oil and Gas Industry Forum; September 1, 2010; Fort Worth, TX; 2010.

Kuuskraa VA, Stevens SH. Worldwide Gas Shales and Unconventional Gas: A Status Report. Paper present at United Nations Climate Change Conference; December 9, 2009;Copenhagen, Denmark; 2009.

Kuuskraa VA, Koperna G, Schmoker J, et al. Barnett shale rising star in Fort Worth basin. Oil Gas J 1998;96 (21):71–76.

MacMillan DJ, Pletcher JL, Bourgeois SA. Practical Tools to Assist History Matching. Paper SPE 51888-MS presented at the SPE Reservoir Simulation Symposium; February 14, 1999; Houston, TX; 1999.

Medeiros F, Ozkan E, Kazemi H. Productivity and drainage area of fractured horizontal wells in tight gas reservoirs. SPE Reserv Eval Eng 2008;11 (5):902–911.

Montgomery SL, Jarvie DM, Bowker KA, et al. Mississippian Barnett shale, Fort Worth basin, North-Central Texas: Gas-shale pay with multidashtrillion cubic foot potential. AAPG Bull 2005;89 (2):155–175.

Nakayama K. Estimation of Reservoir Properties by Monte Carlo Simulation. Paper SPE 59408-MS presented at the SPE Asia Pacific Conference on Integrated Modelling for Asset Management; April 25, 2000; Yokohama, Japan; 2000.

Oudinot AY, Koperna GJ, Reeves SR. Development of a Probabilistic Forecasting and History Matching Model for Coalbed Methane Reservoirs. Paper presented at 2005 International Coalbed Methane Symposium May 5, 2005; Tuscaloosa, AL; 2005.

Petzet A. More operator eye maverick shale gas, tar sand potential. Oil Gas J 2007;105(30):38–40.

PMTx 2.0. 2012. Available at http://www.phoenix-sw.com/. Accessed January 17, 2014.

Powell G. Shale energy: Developing the Barnett—Lateral lengths increasing in Barnett shale. World Oil 2010;231 (8).

Rhine T, Loayza MP, Kirkham B. et al. Channel Fracturing in Horizontal Wellbores: The New Edge of Stimulation Techniques in the Eagle Ford Formation. Paper SPE 145403-MS presented at SPE Annual Technical Conference and Exhibition; October 2, 2011; Denver, CO; 2011.

Rogner HH. An assessment of world hydrocarbon resource. Annu Rev Energy Environ 1997;22:217–262.

Sawyer WK, Zuber MD, Williamson JR. A Simulation-Based Spreadsheet Program for History Matching and Forecasting Shale Gas Production. Paper SPE 57439-MS presented at the SPE Eastern Regional Conference and Exhibition; October 21, 1999; Charleston, WV; 1999.

Schepers KC, Gonzalez RJ, Koperna GJ. et al. Reservoir Modeling in Support of Shale Gas Exploration. Paper SPE 123057-MS presented at the Latin American and Caribbean

Petroleum Engineering Conference; May 31, 2009; Cartagena de Indias, Colombia; 2009.

Smead RG, Pickering GB. *North American Natural Gas Supply Assessment*. Chicago: Navigant Consulting, Inc.; 2008.

Swaan AD. Analytic solutions for determining naturally fractured reservoir properties by well testing. Soc Petrol Engineers J 1976;16(3):117–122. SPE-5346-PA.

Wang FP, Hammes U. Effects of petrophysical factors on Haynesville fluid flow and production. World Oil Mag 2010;COMP: Place section title as recto RH.231(6).

Warren JE, Root PJ. The behavior of naturally fractured reservoirs. SPE J 1963;3(3):245–255. SPE-426-PA.

Williams P. The grande dame of tight gas. A Supplement to Oil and Gas Investor 2006; March: 5–7, 2006.

15

MOLECULAR SIMULATION OF GAS ADSORPTION IN MINERALS AND COAL: IMPLICATIONS FOR GAS OCCURRENCE IN SHALE GAS RESERVOIRS

Keyu Liu[1,2], Shuichang Zhang[1], Shaobo Liu[1] and Hua Tian[1]

[1] PetroChina Research Institute of Petroleum Exploration and Development, Beijing, China
[2] CSIRO Division of Earth Science and Resource Engineering, Perth, WA, Australia

15.1 INTRODUCTION

15.1.1 Molecular Dynamics Simulation

Molecular simulation (MS) includes molecular dynamics (MD) and Monte Carlo (MC) simulations. MD simulation (Allen and Tildesley, 1989; Frenkel and Smit, 1996; Sadus, 1999) is based on the principles of Newtonian mechanics. It simulates physical movements of interacting atoms and molecules in a complicated and complex system. The trajectories of atoms and molecules are determined by numerically solving the Newton's equations of motion, where forces between the atoms and molecules are defined by molecular mechanics force fields. Thermodynamic, structural, and transport properties, such as potential energy, radial distribution function (RDF), and diffusion coefficient are derived from the trajectories of a system according to the principle of statistical mechanics. MS can also be performed in the grand canonical statistical ensemble (GCMC) (Zhang et al., 2014a).

The basic procedure of MD simulation includes (i) creating a system including hundreds and thousands of atoms and molecules, (ii) applying the law of classical Newtonian mechanics to all the particles in the system, (iii) solving the coupled Newtonian's equations of motion numerically and obtaining coordinates and velocities and each particle by integration, (iv) updating the positions and velocities of the particles and recording the trajectories and velocities of the particles for every instant of time, and (v) computing properties of interest. The MD simulation workflow is shown in Figure 15.1.

The natural systems usually consist of a vast number of particles. It is impossible to find properties of such complex systems analytically. MD circumvents this problem by using numerical methods. It has been widely used as a useful predictive tool for new materials without synthesizing them and for virtual experiments to provide an insight into relationship between microstructures and macrothermodynamic, elastic, dielectric and transport properties under some extreme conditions, where experiments are impractical or impossible. MD serves as a bridge between the microscopic and the macroscopic realms and a bridge between theoretical and experimental work. Testing agreement between theory and simulations allows theory to be refined and comparing simulation experiment results to be extrapolated.

Advantages of MD over experiments lie in the possibility of obtaining detailed time evolution of a system and allowing the information on the dynamics of the system and the atomic processes to be fully presented, which otherwise might be difficult or impossible to realize in the laboratory. It can solve problems with many degrees of freedom and under a variety of initial and external conditions.

The disadvantage of MD is that the reliability of the calculations depends on the quality of the employed interaction model, as the construction of the model always involves compromises between accuracy and speed of calculation. It has its intrinsic time and length scale problem, where simulation size is at nanoscale level with simulation durations being typically no more than hundreds of nanoseconds. Nowadays, MD has been developed into an extremely useful,

---

Fundamentals of Gas Shale Reservoirs, First Edition. Edited by Reza Rezaee.
© 2015 John Wiley & Sons, Inc. Published 2015 by John Wiley & Sons, Inc.

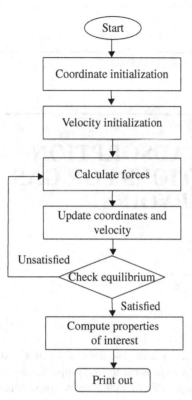

FIGURE 15.1 Workflow and procedure of performing molecular dynamic simulations.

but computationally very intensive tool for studies of chemical and biomolecular systems and has been widely applied to various fields including geochemistry, chemical engineering, materials science, medical science, petroleum science, and so on, to understand and characterize the properties of molecular systems.

15.1.2 Major Challenges in Shale Gas Research

Shale gas is presently a research focus in the world largely sparked from the recently shale gas revolution in the United States. Shale gas was first discovered in the Devonian Ohio Shale of the Appalachian Basin, USA, in 1821. However, it was not produced commercially on a large scale until a decade ago owing to technology breakthrough in horizontal drilling and hydraulic fracturing, as well as the demand of cleaner fossil fuels. So far, over 20 shale gas fields have been discovered in 11 basins in the United States (Energy Information Administration (EIA), 2013). Many countries have now embarked on shale gas R&D and exploration attempting to replicate the successful story in the United States.

Because of the complexity of geological conditions of shale gas reservoirs, the distribution of the shale gas and its abundance vary greatly from basin to basin and between different shale formations. This makes the search for commercially viable shale gas plays and their development extremely difficult despite the huge reserves predicted by EIA (2013) in 137 shale formations of over 40 countries. Shale gas is known to store primarily in pores of nanometer scale (1–100 nm; e.g., Curtis et al., 2012; Loucks et al., 2012; Slatt and O'Brien, 2011; Zou et al., 2010) and in adsorbed phase on organics and minerals. This poses enormous challenges for the traditional macroscopic characterization and conventional laboratory experiments regarding the occurrence and phase behavior under reservoir conditions.

15.1.3 MS of Gas Adsorption

MS has been used as a powerful tool for solving scientific and engineering problems and has found extensive applications in the studies of equilibrium thermodynamics of fluids involving phase transition of bulk fluids (Harris and Yung, 1995; Wei and Sadus, 2000), nanofluid in confined geometry (Todd and Daivis, 2000; Zhang and Todd, 2004; Zhang et al., 2007), gas adsorption on porous materials (Aukett et al., 1992; Sweatman and Quirke, 2001a, b; Zhang et al., 2012), fluid interfaces with clays (Chang et al., 1998; Skipper et al., 1995a, b), gas adsorption in coal (Brochard et al., 2012; Hu et al., 2010; Tambach et al., 2009), hydrate formation (Duffy et al., 2004; Moon et al., 2003; Rodger et al., 1996; Zhang et al., 2008), multicomponent gas separation (Ghoufi et al., 2009; Hamon et al., 2009), and so on.

To adequately evaluate shale gas storage in shale and organic matter, equilibrium adsorption methods can be used to characterize those nanoporous materials and quantify the storage capacity of gas (Gensterblum et al., 2014). In laboratory experiments, volumetric methods and gravimetric methods are most commonly used for high pressure adsorption. The amount of adsorbed gas is the difference between that admitted to the sample container and that remaining in the dead space (or isolated pores). The dead space refers to the void space and the volume of the connecting tubing (pore throat). In analyzing a shale reservoir rock, a distinction is made between the stored free gas, which satisfies as a function of pressure using an equation of state such as the compressibility equation of state, and the adsorbed gas, which satisfies as a function of pressure using an adsorption equation such as the Langmuir equation. The dead space is the volume that is occupied by the free gas. It is a combination of volumes contained in the equipment such as pipes and the part of the sample pore volume not occupied by adsorbed gas. Isolated pores by their definition are not part of the dead space as no gas can enter them.

To accurately determine the amount of the adsorbed gas, the dead space needs to be experimentally obtained as a slight change in the dead space value might shift the high pressure part of the adsorption isotherms upward or downward. The void volume is usually estimated by the helium expansion method prior to the introduction of gas into the sample cell. During the adsorption isotherm measurement, at each pressure step in the experiment, the gas adsorbed reduces the void volume. As a result, the initially determined void volume must be corrected at the beginning and the end of each pressure step (Ambrose et al., 2012; Menon, 1968).

Measurement of the adsorbed phase density or its volume is not a trivial matter. The local density for the adsorbed phase is expected to vary across the pore, different from its average bulk density due to the added interactions between the adsorbed phase and the solid matrix. Further, in gas shale where pressure and temperature may be significantly greater than the supercritical thresholds of some gases, it makes experimental measurements extremely difficult. There are several estimates of the adsorbed phase density such as densities at the critical point (Tsai et al., 1985) or densities calculated from the "b" term in the cubic equation of state (Dubinin, 1960; Haydel and Kobayash, 1967), and liquid densities at their boiling points at one atmosphere (Menon, 1968). Ozawa et al. (1976) considered the adsorbed phase as a superheated liquid with a density dependent upon the thermal expansion of the liquid. However, the density calculated from the "b" term in the cubic equation of state does not take into account of the temperature and pressure effects. Thus the liquid densities at their boiling points at one atmosphere cannot always be used since gas often would not become liquid state at one atmosphere. Li et al. (2003) compared all of the above-mentioned methods to a Langmuir–Freundlich adsorption model, and found that the adsorbed phase density is temperature dependent.

To circumvent the problems facing these estimates, Ambrose et al. (2012) used equilibrium MS involving methane in small carbon slit-pores of varying sizes and temperatures to address the fundamental issues related to the adsorbed phase density and phase transition. They performed MD simulations with constant numbers of molecules, constant volume, and constant temperature (NVT ensemble) with the total number of methane molecules being the same for two different channel widths, but by changing the dimension in one direction to obtain the same volume for the two slit pores. They used the density profiles to analyze the adsorbate density and further determine the adsorbed phase and bulk phase.

For adsorption studies, a more suitable ensemble to use is the grand-canonical ensemble. In this ensemble, the temperature, volume, and chemical potential are fixed. It simulates the experimental setup allowing the adsorbed gas to equilibrate with the gas reservoir. Heat exchange is allowed between the system studied and a heat bath, and heat transfer takes place until a thermal equilibrium is reached. The gas in the system and a gas reservoir can be exchanged to allow equilibrium of chemical potential. This makes the grand canonical simulations different from other ensembles, where the numbers of molecules are fixed. We describe how the MS in the grand canonical ensemble can be successfully applied to shale gas research with a number of simulation examples in the following sections.

15.1.4 Methodology and Workflow of Molecular Simulation

Molecular simulation of gas adsorption in shale was carried out by combining MC and MD simulations. The objectives were to determine the absolute and excess adsorption isotherms, adsorbed phase density, RDFs between the minerals and gases such as CH_4, C_2H_6, C_3H_8, and CO_2 and their diffusion coefficients. The workflow consists of setting-up molecular structures and their force fields based on experimental and empirical data.

15.1.5 Simulation Algorithms and Software

The molecular simulation procedure consists of MD simulations in the constant pressure and temperature ensemble and MC simulations performed in the GCMC (Zhang et al., 2014a). The former is performed using the Gromacs software (Bae and Bhatia, 2006; Busch et al., 2003; Fitzgerald et al., 2005), and the latter is performed using an open-source package, RASPA 1.0, developed by Dubbeldam et al. (2008). The MD and MC methods were previously applied to study hydrate formation (Zhang et al., 2008), gas adsorption in porous material (Zhang et al., 2012), and coal (Zhang et al., 2014a). In our simulations, the temperature was fixed by using the Berendsen thermostat (Berendsen et al., 1984). As with the temperature coupling, the system can also be connected to a pressure bath. We use the Berendsen pressure coupling scheme to reach the target pressure, and then switch to the Parrinello–Rahman coupling (Parrinello and Rahman, 1981) to keep the pressure constant once the system is in equilibrium as the Berendsen pressure coupling allows the system to reach equilibrium quickly, whereas the Parrinello–Rahman coupling can adjust the pressure finely and avoid unexpected deformation and fluctuation of the simulation box. Adsorption isotherms are obtained from the GCMC simulations (Siepmann and Frenkel, 1992). The GCMC simulation can be switched to the MD simulation to allow the volume to swell or shrink when simulate CH_4 adsorption in coal (Zhang et al., 2014a). The GCMC algorithm allows the system density to fluctuate with insertion and deletion of adsorbate molecules. Equilibrium is attained when the numbers of successful insertion and deletion attempts balance each other. The GCMC method is detailed in Dubbeldam et al. (2004a, b).

15.2 MS OF GAS ADSORPTION ON MINERALS

Most of the gas shales comprise silty fractions of detrital quartz and feldspar minerals, biogenic silica, carbonate, clays, and organics (kerogen). The mineralogical compositions,

TABLE 15.1 Summary of various simulations performed

Adsorbent phase	Surface area (Å)	Layer interval (Å)	Temperature (K)	Pressure (MPa)	Moist content	Gas composition
Quartz	24.55 × 27.01	10, 15, 20	310	0–15	Dry	Natural gas[b]
Wyoming-type[a]	31.68 × 27.12	20	353.5	5, 30	Dry, 7 wt%	Natural gas[b]
Zeolite (FAU)	—	—	288, 308, 328	1–100	Dry	CH_4, CO_2
Coal	—	—	308, 370	0–10	Dry, 1.2%, 3%	CH_4

[a] Montmorillonite.
[b] 92% CH_4, 6% C_2H_6, and 2% C_3H_8.
Zeolite (FAU) simulation cell: 25.099 × 25.099 × 25.099 Å³.

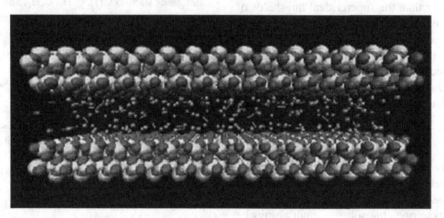

FIGURE 15.2 Structural model of a quartz nanochannel used in the MD simulation.

sedimentary texture and organic content, and maturity collectively determine whether a shale can become an effective gas reservoir with adequate storage space, adsorption capacity, and hydraulic frackability (Loucks and Ruppel, 2007; Slatt and O'Brien, 2011; Zou et al., 2010). The majority of shale gas plays currently produced in North America are between 1000 and 3000 m at present with reservoir temperatures ranging from 50 to 100°C and a pressure regime of 10–30 MPa (EIA, 2013), although there are rare cases of deep, hot, and overpressured shale gas plays reported (e.g. Zou et al., 2010; EIA, 2013). As summarized in Table 15.1, we conducted a series of simulations of gas adsorption on quartz, montmorillonite (Wyoming-type), zeolite (FAU), and coal at median reservoir conditions with temperatures up to 80°C and pressure up to 100 MPa. A typical natural gas composition with 92% CH_4, 6% C_2H_6, and 2% C_3H_8 was used, in addition to pure CH_4 and CO_2 for comparison. The nanoscale pore models used in the simulations are in accordance with analytical measurements and experimental observation of typical gas shale structures. The effect of moist (water) on gas adsorption has also been simulated by varying the moist contents in some simulations. We used MS to address some myths relating to the occurrence and phase behavior of shale gas under reservoir conditions that cannot be routinely observed or measured by laboratory experiments due to physical constraints.

15.2.1 MD Simulation of Gas Adsorption on Quartz

Quartz or silica (SiO_2), either of detrital or of biogenic origin, is a major component in gas shales (Loucks and Ruppel, 2007). Its contribution to gas adsorption in gas shales under reservoir conditions has not been studied explicitly. We simulate gas adsorption on SiO_2 via nanochannels (10–20 Å) under reservoir conditions of 310 K at 0–15 MPa pressure range with the aim to understand the relative adsorption capacities of natural gas on SiO_2, and the effect of nanopore throat sizes and pressures on the gas adsorption.

15.2.1.1 Model Construction Nanochannel models of SiO_2 were constructed with a surface area of 24.55 × 27.01 Å² and with varying channel widths of 10, 15, and 20 Å (Fig. 15.2). The simulations were carried out at a temperature of 310 K, with changing pressures of 0, 5, 10, and 15 MPa. A typical natural gas composition with 92% CH_4, 6% C_2H_6, and 2% C_3H_8 was used in the simulation as the absorbate. The simulation box is connected to a reservoir containing the natural gas with the same pressure and temperature condition to allow exchange of molecules between the system and the reservoir at each simulated pressure.

Natural gas adsorption on quartz mineral was simulated by performing GCMC simulations. Computer models of two opposing (1, 0, 0) crystal surfaces of α-quartz (dimensions

TABLE 15.2 Different types of atoms and their LJ force-field

Atom type	Label	$C^{(6)}$ kJ nm^6 mol^{-1}	$C^{(12)}$ kJ nm^{12} mol^{-1}	$Q(e)$
Hydrogen (surface)	H			+0.40
Oxygen (crystal)	OS	0.22617×10^{-2}	0.74158×10^{-6}	
Oxygen (surface)	OA	0.22617×10^{-2}	0.15062×10^{-5}	−0.71
Silica (crystal)	SI	0.22617×10^{-2}	0.22191×10^{-4}	
Silica (surface)	SI	0.22617×10^{-2}	0.22191×10^{-4}	+0.31
Methane	CH$_4$	1.32400×10^{-2}	0.35651×10^{-4}	
CH$_3$ in propane	CH$_3$(P)	1.03800×10^{-2}	0.36789×10^{-4}	
CH$_2$ group	CH$_2$	0.70000×10^{-2}	0.24806×10^{-4}	
CH$_3$ in ethane	CH$_3$(E)	1.00233×10^{-2}	0.28987×10^{-4}	

TABLE 15.3 Energy terms and parameters for the bonded potential of ethane and propane

Component	Parameter value	Energy
Harmonic bond	$k_1 = 801.96$ (kJ mol^{-1}/Å^2), $r_0 = 1.54$ Å	$U^{bond} = \frac{1}{2} k_1 (r - r_0)^2$
Harmonic bond	$k_2 = 519.37$ (kJ mol^{-1}/rad^2), $\theta_0 = 114°$	$U^{bond} = \frac{1}{2} k_2 (\cos\theta - \cos\theta_0)^2$

24.55×27.01 Å^2) were constructed with varying layer spacing. Silanol groups were created by adding hydrogen atoms to the surface oxygen atoms and generating the necessary bonds. The methane, ethane, and propane are described with a united atom model, in which each CH$_4$, CH$_3$, and CH$_2$ group is treated as a single interaction centre (a united atom) with their own effective potentials. The united atoms are connected by harmonic bond length and angle potentials. The nonbonded interactions between methane, ethane, propane, and quartz are modeled by using the Lennard–Jones (LJ) potentials and the Coulomb potentials. The Coulomb interactions are calculated using the Ewald summation (Allen, 1987). The LJ potential is expressed as follows:

$$V_{LJ}(r_{ij}) = \frac{C_{ij}^{(12)}}{r_{ij}^{12}} - \frac{C_{ij}^{(6)}}{r_{ij}^6} \quad (15.1)$$

Cross interactions are calculated by the Jorgensen mixing rules:

$$C_{ij}^{(6)} = \left(C_{ii}^{(6)} C_{jj}^{(6)} \right)^{1/2} \quad (15.2)$$

$$C_{ij}^{(12)} = \left(C_{ii}^{(12)} C_{jj}^{(12)} \right)^{1/2} \quad (15.3)$$

The interaction energy between the adsorbed gas and the α-quartz mineral, the gas adsorption isotherm, and its structure properties were studied. The separation between the crystal surfaces was varied, monitoring the layer separation dependence of the interaction energy and the natural gas adsorption amount.

The LJ parameters for α-quartz, methane, ethane, and propane are taken from Wensink et al. (2000) and listed in Tables 15.2 and 15.3.

15.2.1.2 Result and Discussion Our simulation box has x, y, z-dimensions of $24.55 \times 27.01 \times 32.6315$ Å^3 with channel width of 10 Å (thickness of the quartz layer: 22.63 Å). The z-dimension depends on the channel width. Obviously, the pore is at nanoscale and the gas adsorbed in the pore is regarded as adsorbed gas. When 1.8 wt% of water is present, the adsorption amount of methane, ethane, and propane reduces about 3.8, 2.4, and 2.2%, respectively.

Figures 15.3 and 15.4 display the simulation outputs for natural gas adsorption on SiO$_2$. It can be observed that the adsorption of CH$_4$ in SiO$_2$ nanochannels (1–2 nm) increase from 0.75 mole fraction to 0.87 mole fraction from 0 to 15 MPa (Fig. 15.3a), while that of C$_2$H$_6$ decrease from 0.15 mole fraction to 0.095 mole fraction (Fig. 15.3b) for the same pressure increase. Comparatively, C$_2$H$_6$ appears to be more easily absorbed in the nanochannels with reference to its bulk phase mole fraction (0.06), whereas CH$_4$ adsorbed relatively less with reference to its bulk phase mole fraction (0.92) (Fig. 15.3a and b). In absolute term, CH$_4$ adsorption appears to be increased strongly with pressure from near zero at 0 MPa to 2.5–4.0 mole/kg at 15 MPa (Fig. 15.3a). CH$_4$ adsorption increases with increasing nano SiO$_2$ channels from 1 to 2 nm. By contrast, the adsorbed mole fraction

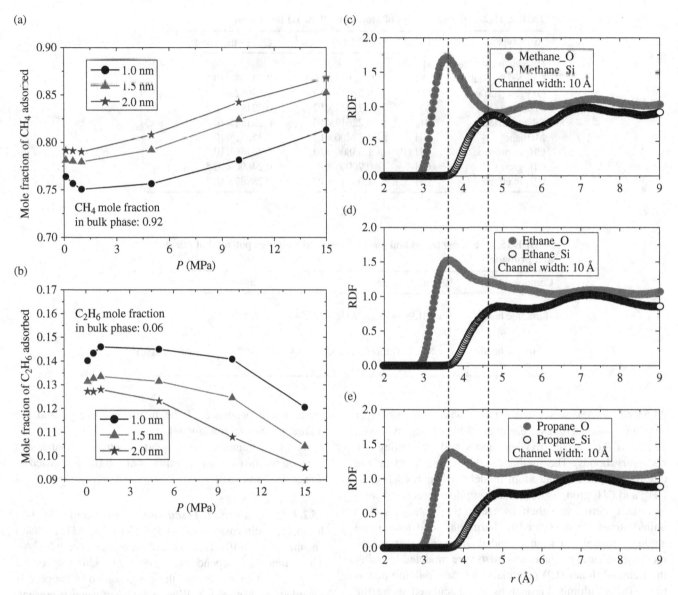

FIGURE 15.3 MD simulation results: (a) Mole percentage of CH_4 adsorption in quartz nano channels showing the adsorption increases with increasing pressure and channel widths, (b) mole percentage of C_2H_6 in the adsorbed gas phase showing that the C_2H_6 molar percentage decreases with pressure and channel widths, (c) RDF for CH_4 with reference to O- and Si- atoms, (d) RDF for C_2H_6 with reference to O- and Si- atoms, and (e) RDF for C_3H_8 with reference to O- and Si- atoms.

C_2H_6 appears to be inversely correlated with pressures and channel widths (Fig. 15.3b) with the least adsorption for the large (2 nm) channel at high pressure (15 MPa).

On the basis of the simulated isotherms and RDF for the natural gas adsorption in SiO_2 nanochannels (Figs. 15.3 and 15.4), it can be interpreted that (i) both pressures and SiO_2 channel widths have positive effects on CH_4 adsorption, but negative effects on that of C_2H_6 and C_3H_8; (ii) CH_4, C_2H_6, and C_3H_8 are adsorbed on SiO_2 in a single-layer mode with a closer contact to the O atom than the Si atom in the SiO_2 molecular as indicated by the RDF (Fig. 15.3c–e). The RDF for CH_4–O and CH_3–O of SiO_2 show sharp peaks at 0.36 nm for CH_4 and CH_3 unit in C_2H_6 and C_3H_8. The RDF of CH_2–O has a broad peak around 0.48 nm (Fig. 15.4).

15.2.2 Molecular Dynamic Simulation of Gas Adsorption on Wyoming-Type Montmorillonite

15.2.2.1 Model Construction Shale contains a high fraction of compacted clay minerals, such as smectite, which comprise negatively charged mica-like sheets, and held together by charge balancing interlayer cations, such

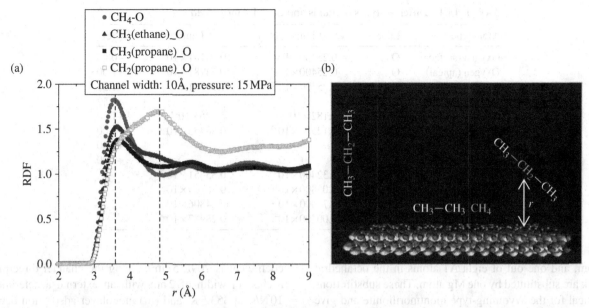

FIGURE 15.4 MD simulation results: (a) RDF of CH_4, CH_3- and CH_2- with references to the O- atom in quartz showing that the adsorption of CH_4, C_2H_6, and C_3H_8 are all in single layer mode and (b) schematic illustration showing the attachment mode of CH_4, C_2H_6, and C_3H_8 SiO_2. r: RDF range.

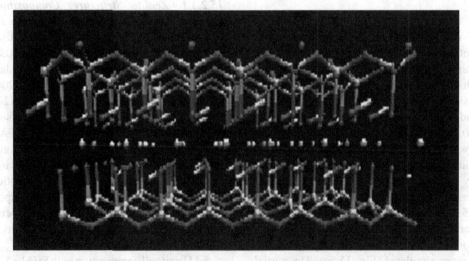

FIGURE 15.5 Molecular structural model of the Wyoming-type montmorillonite: Layered structure of the Wyoming-type montmorillonite.

as potassium or sodium. Montmorillonite (Wyoming-type) was chosen as a typical clay mineral in gas shale as the molecular structure of montmorillonite has been well established and it is sensitive to moist.

The composition of a typical nature gas was used in the simulation with 92% CH_4, 6% C_2H_6, and 2% C_3H_8. The modeling temperature was set to 353.5 K or 80.5°C. Five pressure points were simulated including 5, 10, 15, 20, and 25 MPa. The simulations were performed for both dry and moist scenarios with a water content of 7.15 wt%.

MC simulations of natural gas adsorption in the interlayers of 2:1 Na-saturated clay minerals were performed. Clay surface with the dimensions $31.68 \times 27.42 \times 6.56 Å^3$ was constructed with a layer spacing of 20Å, which contains 36 clay unit cells details are in the literature (Pusch and Madsen, 1995). The clay mineral studied is a Wyoming-type montmorillonite, with a unit cell formula as M_x^+ [$Si_a Al_{8-a}$] $(Al_b Mg_{4-b})O_{20}(OH)_4$,

where M represents a monovalent cation; in the present work Na^+ is considered, $x=0.75$, $a=7.74$, and $b=3.5$. The Wyoming-type montmorillonite is a member of the smectite family and consists of two tetrahedral layers, sandwiching an octahedral layer (2:1), shown in Figure 15.5. One out of thirty-two Si atoms in the tetrahedral sheet is replaced by an

TABLE 15.4 Different types of atoms and their LJ force-field

Atom type	Label	$C^{(6)}$ (kJ nm^6 mol^{-1})	$C^{(12)}$ (kJ nm^{12} mol^{-1})	$Q(e)$
Oxygen (surface)	O	0.25400×10^{-2}	0.24781×10^{-5}	−0.80
Oxygen (apical)	O	0.25400×10^{-2}	0.24781×10^{-5}	−1.00
Oxygen (OH)	O	0.25400×10^{-2}	1.24781×10^{-5}	−1.52
Hydrogen	H			+0.52
Silica	Si	0.182×10^{-2}	0.626×10^{-7}	+1.20
Sodium	Na	0.12500×10^{-2}	0.58453×10^{-7}	+1.00
Aluminum	Al			+3.00
Magnesium	Mg			+2.00
Methane	CH$_4$	1.32400×10^{-2}	0.35651×10^{-4}	
CH$_3$ in propane	CH$_3$(P)	1.03800×10^{-2}	0.36789×10^{-4}	
CH$_2$ group	CH$_2$	0.70000×10^{-2}	0.24806×10^{-4}	
CH$_3$ in ethane	CH$_3$(E)	1.00200×10^{-2}	0.28987×10^{-4}	

Al atom, and one out of eight Al atoms in the octahedral network are substituted by one Mg atom. These substitutions are typical for the Wyoming-type montmorillonite and give rise to an overall negative charge on the clay framework. This negative charge is balanced by the presence of monovalent interlayer counter ions M (Na$^+$). The LJ potential parameters and partial charges are taken from Jorgensen et al. (1984) (Table 15.4).

15.2.2.2 Results and Discussion

The simulations of natural gas adsorption on the Wyoming-type montmorillonite results are summarized in Figure 15.6. The simulated isotherm for CH$_4$ fits to the classic Langmuir curve extremely well with an R^2 of 0.999 (Fig. 15.6a). The fitting parameters for the maximum adsorption capacity and adsorption rate are 5.744 mol/kg and 0.112 MPa^{-1}, respectively. The adsorption of CH$_4$ increases with pressure from less than 2 mol/kg to over 4.0 mol/kg from 5 to 25 MPa, while that of C$_2$H$_6$ appears to be unaffected by pressure (Fig. 15.6b). The adsorption of C$_3$H$_8$ decreases from 0.3 to 0.2 mol/kg with pressure from 5 to 25 MPa (Fig. 15.6b). The adsorption of CH$_4$ on dry the Wyoming-type montmorillonite is consistently higher than that for the moist one with 7.15 wt% of water by up to 40% over the pressure range simulated (Fig. 15.6c). The partial density of the CH$_4$ in the z-direction of the simulation box attains a maximum of 570 kg/m^3 at a distance of 0.15 nm from Na$^+$ and 0.38 nm from the clay surface (Fig. 15.6d). RDF indicates that CH$_4$ has a close contact of 3.2 Å with Na$^+$ in the Wyoming-type montmorillonite followed by CH$_4$–O (3.9 Å) and CH$_4$–Si (4.6 Å). The mole fraction of methane in the adsorbed phase is lower than that in the bulk phase. However, both mole fractions of ethane and propane in the adsorbed phase are higher than those in the bulk phase.

It is noticed that the presence of water in the Wyoming-type montmorillonite reduces the adsorption capacity of methane. During the nonequilibrium MS on methane diffusion in montmorillonite, we have found that (i) the diffusion coefficient increases with external force; (ii) the diffusion coefficient is 209E-5 cm^2/s along the channel direction for a channel width of 2 nm with an external acceleration of 10 N/s^2 at 353.5 K; and (iii) monolayer adsorption behavior of methane along montmorillonite channel was observed.

15.2.3 MD Simulation of Gas Adsorption on Zeolite

15.2.3.1 Zeolite Structure, Compositions, and Model Construction

MS has been played an important role in understanding the fundamentals of gas adsorption in zeolite at the molecular level (Beerdsen et al., 2003; Demontis et al., 1992; Dubbeldam and Smit, 2003; Dubbeldam et al., 2004a, b; García-Pérez et al., 2006; Granato et al., 2007; Wender et al., 2007; Zhang et al., 2012). Calero et al. (2004) developed a force field to accurately describe the adsorption properties of n-alkanes in the sodium form of FAU-type zeolites. García-Pérez et al. (2006) extended the force field of Calero et al. (2004) by including calcium-type ions and the force field was applied to study the adsorption properties of n-alkanes in LTA 5A over a range of temperatures and pressures. In our simulation, we primarily focused on the effect of the number density of the nonframework cations, the ratio of Si/Al and Na$^+$/Ca^{2+} on gas adsorption.

FAU unit cell composition with Si/Al = 1.18 corresponding to 88 aluminum atoms per unit cell is used in our simulation (Fig. 15.7). The negative charges introduced by Al replacing Si are compensated by Na$^+$ (NaX). Atomic charges are chosen as q_{Na} = +1, q_{Si} = +2.05, q_{Al} = +1.75 (Zhang et al., 2012, 2014b). The crystal structures of Na$_{88}$Al$_{88}$Si$_{104}$O$_{384}$ is used for FAU zeolite X, with a lattice parameter of 25.099 Å. Cations present in the zeolite framework are Na$^+$. The zeolite is assumed to be rigid. We studied the influence of temperature on methane and carbon dioxide adsorption in FAU zeolite.

The interactions of adsorbate molecules with the zeolite host framework are dominated by the forces between the adsorbate and the oxygen atoms of zeolite. The contribution of Si and Al are taken into account through an effective

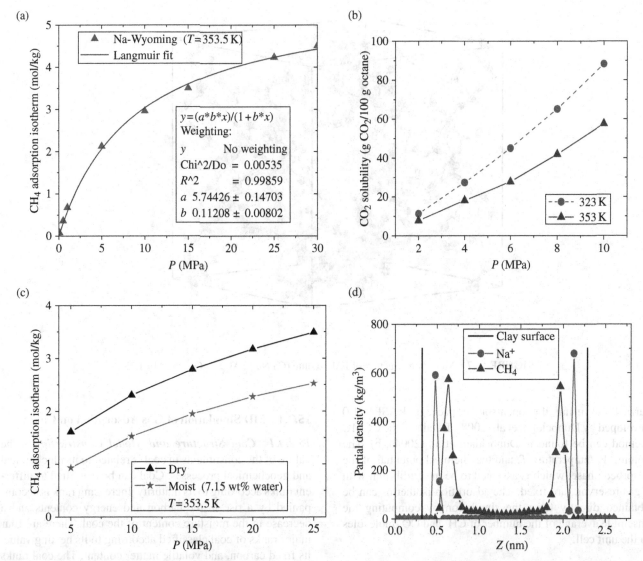

FIGURE 15.6 MD simulation results: (a) Adsorption isotherm of CH_4 in the Wyoming-type montmorillonite showing an excellent fit to the Langmuir isotherm curve; (b) adsorption isotherms of CH_4, C_2H_6, and C_3H_8 showing that the CH_4 adsorption increases with pressure, and that of C_2H_6 remains largely unaffected by pressure, while that of C_3H_8 decreases with pressure; (c) CH_4 adsorption isotherms on dry and wet (7.15% water) Wyoming-type montmorillonite showing that moist reduced the adsorption capacity of CH_4 in the Wyoming-type montmorillonite; (d) spatial distribution of CH_4 in referencing with Na^+ in the Wyoming-type montmorillonite showing that CH_4 has affinity to the Na^+ and the adsorption is in single layer.

potential with only the oxygen atoms, as the polarizability of Si and Al is much lower than those of oxygen atoms. Therefore, the Si atoms are randomly selected to be replaced by Al atoms. The nonframework cations are allowed to move freely within the system, but the movement is controlled by the force field applied.

The initial setup would not affect the results (Calero et al., 2004; García-Perez et al., 2006). The interactions between cations and the zeolite host framework were modeled by Lennard-Jones and Coulomb potentials. The Coulomb interactions were calculated using the Ewald summation (Frenkel and Smit, 1996). The LJ parameters for the oxygen of the zeolite host framework, cations and methane and partial charges are taken from García-Pérez et al. (2006). We used a truncated and shifted potential with a cutoff radius of 12 Å. The partial charges on Si (+2.05e), Al (+1.75e), O_{Al} (−1.20e), and O_{Si} (−1.025e) of the zeolite host framework system are fixed. It should be noted that O_{Al} are oxygens bridging one Si and one Al, while O_{Si} are oxygens bridging two Si atoms. GCMC simulations were carried out for a pressure range of 10^2–10^6 Pa and at temperatures of 288, 298, 308, and 328 K using one orthorhombic unit cell of dimensions $25.099 \times 25.099 \times 25.099$ Å^3 with a typical number of MC steps of five millions. The simulations were

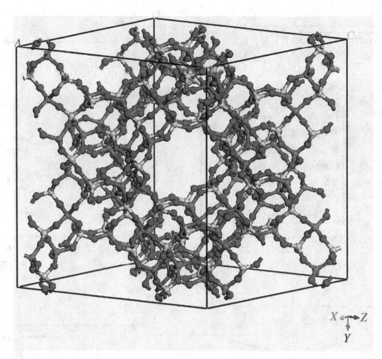

FIGURE 15.7 Molecular structure of FAU zeolite ($Ca_xNa_{88-2x}Al_{88}Si_{104}O_{384}$, Si/Al = 1.18).

carried out using the open-source package RASPA 1.0 developed by Dubbeldam et al. (2008). Details of the GCMC method can be found in Dubbeldam et al. (2004a, b). The volume V, temperature T, and the chemical potential of the adsorbed phase, which is assumed to be in equilibrium with a gas reservoir, are fixed. The adsorption isotherms can be obtained directly from the simulation by computing the ensemble average of the number of CH_4 and CO_2 molecules in the unit cell.

15.2.3.2 Result and Discussion Figure 15.8 shows the isotherms of CH_4 and CO_2 on FAU–zeolite for four temperatures (Fig. 15.8a) and variations of CH_4 densities for both the adsorbed phase and bulk with pressure (Fig. 15.8b). It can be observed that for both CO_2 and CH_4 adsorption increases with pressure. The simulation results have been compared with both the Langmuir and Toth models. The results demonstrate that the Toth model can better describe the adsorption data, suggesting that NaX presents a heterogeneous surface in the adsorption especially for CO_2. The effect of the temperature on the adsorption for both CO_2 and CH_4 at high pressures appears negligible. From the simulation trend, we infer that the adsorption capacity (the maximum adsorption amount) is independent of the temperature. Both CO_2 and CH_4 would have the same adsorption capacity. At low pressure, the adsorbed phase density linearly relates to the bulk density. As the bulk density increases, the adsorbed phase densities at different temperatures converge and reach a plateau as pores being eventually filled (Fig. 15.8b).

15.2.4 MD Simulation of Gas Adsorption on Coal

15.2.4.1 Coal Structure and Model Construction The nature of the constituents in coal is related to its biochemical and geochemical processes. Coal can be classified as different ranks according to its maturity. Increasing rank is accompanied by a rise in the carbon and energy contents and a decrease in the moisture content of the coal. There are four major ranks of coal classified according to its heating value, its fixed carbon, and volatile matter content. The coal ranks from lowest to highest in heating value are lignite, sub-bituminous, bituminous, and anthracite. Of the four ranks, bituminous coal is an intermediate-rank coal and is the most common coal. There are a large number of molecular representations for coals with different ranks. Generating realistic molecular models of coal is an essence of coal simulations applied in coal related research. Here we focus on a model representation of a bituminous coal (Spiro and Kosky, 1982). The development of the molecular coal model started with an intermediate-rank bituminous coal building block ($C_{100}H_{82}O_5N_2S_2$) of 191 atoms, shown in Figure 15.9 (Zhang et al., 2014a). The model was constructed by using the PRODRG server (Schuttelkopf and Van Aalten, 2004). In this model, carbon, hydrogen, oxygen, nitrogen, and sulfur cover about 82.53, 5.64, 5.5, 1.93, and 4.4% of the total mass of the coal, respectively. Constituents and their ratios in this model are similar to that observed in natural coal and account for the amorphous and chemically heterogeneous structure of the natural coal, although they vary widely from one coal

MS OF GAS ADSORPTION ON MINERALS 335

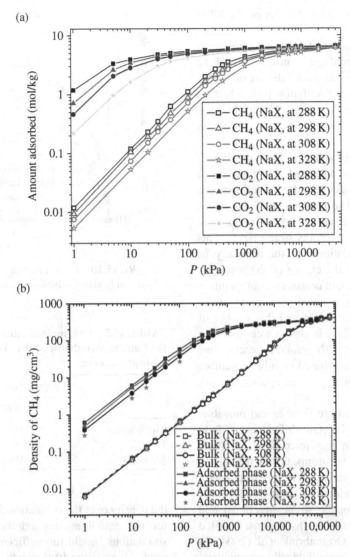

FIGURE 15.8 MD simulation results of CH_4 and CO_2 adsorption on FAU zeolite: (a) Adsorption isotherm of CH_4 and CO_2 at different temperatures (288, 298, 308, and 328 K) and (b) variations of densities of the adsorbed phase and bulk gas with pressure.

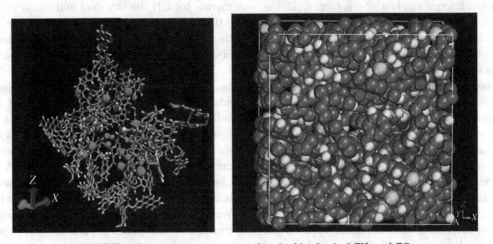

FIGURE 15.9 Molecular structure of coal with adsorbed CH_4 and CO_2.

sample to the other (Day et al., 2008; Ottiger et al., 2008; St George and Barakat, 2001).

The system simulated consists of coal and pure component of CH_4. The initial configuration consists of 12 randomly placed coal molecules. All molecules are randomly placed in an empty space of a simulation box, which is large enough to accommodate the coal molecules and has x, y, and z-dimensions of $3.2 \times 3.2 \times 3.2\,nm^3$, shown in Figure 15.9.

In MD simulations, the behavior of a large number of atoms in the simulation box is followed as a function of time by Newton's laws. Periodic boundary conditions are applied in three directions. The movement of the atoms within the simulation box is governed by an interatomic interaction potential between all atoms (force field). This is the core of the MD simulations, as the accuracy of the results mainly depends on the choice of the potential. The most accurate potential would originate from quantum mechanical calculations, but in practice, it is computationally expensive, limiting the timescale and the number of atoms that can be simulated. There is always a compromise between accuracy and feasibility. Therefore, in practice, we make use of empirical potentials, based on a large number of parameters, which are typically fitted against density functional theory calculations.

The coal molecules modeled are flexible and movable. They are described using the GROMOS force field (Oostenbrink et al., 2004). In this force field, aliphatic carbon atoms are treated as united atoms, that is, the carbon and the hydrogens that are bonded to it are treated as a single atom, reducing computational effort up to a factor of 9 at the expense of neglecting the slight directional and volume effects of the presence of these hydrogens. Detailed parameter sets can be found in Oostenbrink et al. (2004). In contrast to other biomolecular force fields, this parameterization of the GROMOS force field is based primarily on reproducing the free enthalpies for a range of compounds. The relative free enthalpy is a key property in many biomolecular processes of interest and is why this force field was selected. Hydrocarbon molecules are also modeled using GROMOS force field treating the carbon and the hydrogens that are bonded to it as a single atom. The nonbonded interactions between atoms which are separated by more than three bonds, or belong to different molecules, are described by pair wise-additive LJ.

The equations of motion were integrated with a time step of 0.001 ps (peco-second). Each MD run was done in two steps. The first step consisted of a 5 ps simulation using the steepest-descent method to perform energy minimization to reduce the thermal noise in the structures and potential energies, which can prevent the crash of the simulation due to bad contact (extremely large force) between molecules. The second step of 50 ns (nano-second) included equilibrium run (10 ns) and production run (40 ns). The evolution of

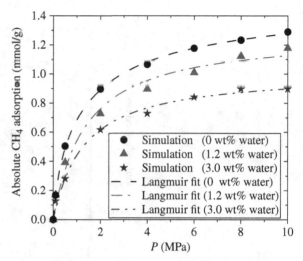

FIGURE 15.10 Comparison of experimental and simulated CH_4 adsorption isotherm (absolute adsorption).

TABLE 15.5 Langmuir constants extracted by fitting the Langmuir isotherm model to our simulated absolute adsorption curve

	Langmuir constant L (mmol/g)	Langmuir constant B (MPa^{-1})
0 wt% water	1.396	1.052
1.2 wt% water	1.275	0.746
3.0 wt% water	1.036	0.700

the density over the simulation time was monitored to check for the equilibration conditions. To get good statistics, simulations should run sufficiently long enough. A typical production run was 40 ns after the system reaches equilibrium.

15.2.4.2 Result and Discussion Absolute adsorption (the actual amount of adsorbate present in coal matrix) isotherms for CH_4 on dry coal and moist coal with 1.2 and 3.0 wt% water at 308 K are shown in Figure 15.10. To investigate the adsorption rate, we fit the simulation data of the absolute adsorption using the Langmuir equation.

The Langmuir constants obtained by the fitting are listed in Table 15.5. Figure 15.10 indicates that the absolute adsorption of CH_4 on both the dry and the moist coal follows the Langmuir isotherm. Sitprasert et al. (2011) has justified the Langmuir isotherm as a good approximation to the results from microscopic simulations being based on curve fitting with simulation results. We obtained the Langmuir sorption constant of 1.396 mmol/g for CH_4 on intermediate rank coal at 35°C on a dry and ash-free basis. Weniger et al. (2010) reported the Langmuir sorption constant of 0.99 mmol/g, obtained from approximation of their experimental data on a coal sample with a total organic carbon (TOC) of 43.9% at 45°C on a dry and ash-free basis, and of 0.43 and 0.46 mmol/g on two other coal samples with undetermined TOC at 35°C.

Coal with 3 wt% moist has the lowest adsorption capacity as compared to that of 1.2% moist and dry coal; and at pressure of 10 MPa the reduction of adsorption capacity in reference to dry coal to coal with 1.2 and 3.0% moist are 10 and 30%, respectively.

15.3 CONCLUSIONS

MS comprising MD and MC simulations can mimic physical movements of interacting atoms and molecules in a complex system. In MS, the trajectories of atoms and molecules are determined by numerically solving the Newton's equations of motion, where forces between atoms and molecules are defined by fields of molecular mechanics force. MS can deal with a vast number of particles and associated properties in complex systems numerically and make it ideal for simulating gas adsorption in porous medium. The MS of the adsorption of natural gas, CH_4 and CO_2 on quartz, montmorillonite (Wyoming-type), zeolite (FAU), and coal under various reservoir conditions has shown that MS can generate adsorption isotherms for both minerals and coal that complements the laboratory measurements and provide the following insight for understanding gas occurrence and phase behavior in shales:

1. Gas (C_1–C_3) adsorbed on minerals (e.g., SiO_2 and clay) in single layer;
2. C_2H_6 and C_3H_8 have comparatively stronger adsorption capacity than CH_4 at low pressure;
3. CH_4 adsorption increases with pore/channel sizes between 1 and 2 nm;
4. At high pressure (>30 MPa), temperature has little effect on the CH_4 adsorption capacity;
5. Both CH_4 and CO_2 have similar adsorption capacity at high pressure;
6. Water has an adverse effect on gas adsorption in minerals and coal.

In relation to the shale gas resource assessment (i.e., adsorption capacity), the simulation indicates that pressure has strong effect on the gas adsorption capacity as compared with temperature. In fact at high pressure, temperature does not appear to have any effect on the gas adsorption capacity. At high pressure, the contribution of the free gas in the pore space becomes significant due to the increased bulk density. Natural gas can enter into nanochannels of 1 nm in width. Water-wet channels are conducive to gas migration and transport due to the adverse effect of water on gas adsorption. Gas can move freely even in nanometer channels because gases are shown to be primarily adsorbed on the minerals in single layers.

ACKNOWLEDGMENTS

This work is financially funded by the Research Institute of Petroleum Exploration and Development, PetroChina through the "Element and process constraint petroleum system modeling" project (No. 2011A-0207) under the PetroChina Science Innovation program. We are grateful to many colleagues for their support and encouragement for the research work including Dr Junfang Zhang of CSIRO Division of Earth Science and Resource Engineering and Prof. Xiancai Lu of Nanjing University.

NOMENCLATURE

RDF, also called pair correlation function, is defined as the ratio of the number of atoms at a distance r from a given atom compared with the number of atoms at the same distance in an ideal gas with the same density.

RASPA 1.0 is a computer program for MC molecular simulations developed by Dubbeldam and coauthors from the Computational Chemistry Group, the University of Amsterdam. The software is described in Dubbeldam, S. Calero, D. Ellis, R. Snurr, "RASPA, 1.0: Molecular Software Package for Adsorption and Diffusion in Nanoporous Materials," in Northwestern University, Evanston 2008.

GROMACS is a computer software for MD simulation developed by Eril Lindahl and coauthors (Lindahl et al., 2001) and is described by in the widely cited paper "GROMACS 3.0: a package for MS and trajectory analysis," in Molecular Modelling Annual 7 (8) 306–317.

PRODRG is a server can take a description of a small molecule (as PDB coordinates/MDL Molfile/SYBYL Mol2 file/text drawing) and from it generate a variety of topologies for use with GROMACS, WHAT IF, Autodock, HEX, CNS, REFMAC5, SHELX, O and other programs, as well as energy-minimized coordinates in a variety of formats. (http://davapc1.bioch.dundee.ac.uk/cgi-bin/prodrg)

MS: Molecular simulation
MD: Molecular dynamics
MC: Monte Carlo
GCMC: Grand canonical statistical ensemble

REFERENCES

Allen MP. *Computer Simulation of Liquids*. Oxford: Clarendon Press; 1987.

Allen MP, Tildesley DJ. *Computer Simulation of Liquids*. Oxford: Clarendon Press; 1989.

Ambrose RJ, Hartman RC, Diaz-Campos M, Akkutlu IY, Sondergeld CH. Shale gas-in-place calculations part I: new pore-scale considerations. SPE J 2012;17:219–229.

Aukett PN, Quirke N, Riddiford S, Tennison SR. Methane adsorption on microporous carbons—a comparison of experiment, theory, and simulation. Carbon 1992;30:913–924.

Bae JS, Bhatia SK. High-pressure adsorption of methane and carbon dioxide on coal. Energy Fuels 2006;20:2599–2607.

Beerdsen E, Dubbeldam D, Smit B, Vlugt TJH, Calero S. Simulating the effect of nonframework cations on the adsorption of alkanes in MFI-type zeolites. J Phys Chem B 2003;107:12088–12095.

Berendsen HJC, Postma JPM, Vangunsteren WF, Dinola A, Haak JR. Molecular-dynamics with coupling to an external bath. J Chem Phys 1984;81:3684–3690.

Brochard L, Vandamme M, Pelenq RJM, Fen-Chong T. Adsorption-induced deformation of microporous materials: coal swelling induced by CO_2–CH_4 competitive adsorption. Langmuir 2012;28:2659–2670.

Busch A, Gensterblum Y, Krooss BM. Methane and CO_2 sorption and desorption measurements on dry Argonne premium coals: pure components and mixtures. Int J Coal Geol 2003;55:205–224.

Calero S, Dubbeldam D, Krishna R, Smit B, Vlugt TJ, Denayer JF, Martens JA, Maesen TL. J Am Chem Soc 2004;126(36): 11377–11386.

Chang FRC, Skipper NT, Sposito G. Monte Carlo and molecular dynamics simulations of electrical double-layer structure in potassium–montmorillonite hydrates. Langmuir 1998;14: 1201–1207.

Curtis ME, Sondergeld CH, Ambrose RJ, Rai CS. Microstructural investigation of gas shales in two and three dimensions using nanometer-scale resolution imaging. AAPG Bull 2012;96: 665–677.

Day S, Fry R, Sakurovs R. Swelling of Australian coals in supercritical CO_2. Int J Coal Geol 2008;74:41–52.

Demontis P, Suffritti GB, Fois ES, Quartieri S. Molecular dynamics studies on zeolites. 6. Temperature dependence of diffusion of methane in silicalite. J Phys Chem 1992;96:1482–1490.

Dubbeldam D, Smit B. Computer simulation of incommensurate diffusion in zeolites: understanding window effects. J Phys Chem B 2003;107:12138–12152.

Dubbeldam D, Calero S, Vlugt TJH, Krishna R, Maesen TLM, Beerdsen E, Smit B. Force field parametrization through fitting on inflection points in isotherms. Phys Rev Lett 2004a;93: 88302–88306.

Dubbeldam D, Calero S, Vlugt TJH, Krishna R, Maesen TLM, Smit B. United atom force field for alkanes in nanoporous materials. J Phys Chem B 2004b;108:12301–12313.

Dubbeldam D, Calero S, Ellis D, Snurr R. RASPA, 1.0: Molecular Software Package for Adsorption and Diffusion in Nanoporous Materials. Evanston: Northwestern University; 2008.

Dubinin MM. The potential theory of adsorption of gases and vapors for adsorbents with energetically nonuniform surfaces. Chem Rev 1960;60:235–241.

Duffy DM, Moon C, Rodger PM. Computer-assisted design of oil additives: hydrate and wax inhibitors. Mol Phys 2004;102: 203–210.

Energy Information Administration (EIA). Technically recoverable shale oil and shale gas resources: an assessment of 137 shale formations in 41 countries outside the United States. Washington, DC; June 2013.

Fitzgerald JE, Pan Z, Sudibandriyo M, Robinson RL, Gasem KAM, Reeves S. Adsorption of methane, nitrogen, carbon dioxide and their mixtures on wet Tiffany coal. Fuel 2005;84: 2351–2363.

Frenkel D, Smit B. Understanding Molecular Simulation: From Algorithms to Applications. Boston: Academic Press; 1996. p 572.

García-Perez E, Dubbeldam D, Maesen TLM, Calero S. Influence of cation Na/Ca ratio on adsorption in LTA 5A: a systematic molecular simulation study of alkane chain length. J Phys Chem B 2006;110:23968–23976.

Gensterblum Y, Busch A, Krooss BM. Molecular concept and experimental evidence of competitive adsorption of H_2O, CO_2 and CH_4 on organic material. Fuel 2014;115:581–588.

Ghoufi A, Gaberova L, Rouquerol J, Vincent D, Llewellyn PL, Maurin G. Adsorption of CO_2, CH_4 and their binary mixture in Faujasite NaY: a combination of molecular simulations with gravimetry–manometry and microcalorimetry measurements. Microporous Mesoporous Mater 2009;119:117–128.

Granato MA, Vlugt TJH, Rodrigues AE. Molecular simulation of propane–propylene binary adsorption equilibrium in zeolite 4A. Ind Eng Chem Res 2007;46:321–328.

Hamon L, Llewellyn PL, Devic T, Ghoufi A, Clet G, Guillerm V, Pirngruber GD, Maurin G, Serre C, Driver G, van Beek W, Jolimaitre E, Vimont A, Daturi M, Ferey G. Co-adsorption and separation of CO_2–CH_4 mixtures in the highly flexible MIL-53(Cr) MOF. J Am Chem Soc 2009;131:17490–17499.

Harris JG, Yung KH. Carbon dioxide's liquid-vapor coexistence curve and critical properties as predicted by a simple molecular model. J Phys Chem 1995;99:12021–12024.

Haydel JJ, Kobayash R. Adsorption equilibria in methane-propane-silica gel system at high pressures. Ind Eng Chem Fundam 1967;6:546–554.

Hu HX, Li XC, Fang ZM, Wei N, Li QS. Small-molecule gas sorption and diffusion in coal: molecular simulation. Energy 2010;35:2939–2944.

Jorgensen WL, Madura JD, Swenson CJ. Optimized intermolecular potential functions for liquid hydrocarbons. J Am Chem Soc 1984;106:6638–6646.

Li M, Gu AZ, Lu XS, Wang RS. Determination of the adsorbate density from supercritical gas adsorption equilibrium data. Carbon 2003;41:585–588.

Lindahl E, Hess B, van der Spoel D. GROMACS 3.0: a package for molecular simulation and trajectory analysis. J Mol Model 2001;7:306–317.

Loucks RG, Ruppel SC. Mississippian Barnett Shale: lithofacies and depositional setting of a deep-water shale-gas succession in the Fort Worth Basin, Texas. AAPG Bull 2007;91:579–601.

Loucks RG, Reed RM, Ruppel SC, Hammes U. Spectrum of pore types and networks in mudrocks and a descriptive classification for matrix-related mudrock pores. AAPG Bull 2012;96: 1071–1098.

Menon PG. Adsorption at high pressures. Chem Rev 1968;68: 277–294.

Moon C, Taylor PC, Rodger PM. Molecular dynamics study of gas hydrate formation. J Am Chem Soc 2003;125:4706–4707.

Oostenbrink C, Villa A, Mark AE, van Gunsteren WF. A biomolecular force field based on the free enthalpy of hydration and solvation: the GROMOS force-field parameter sets 53A5 and 53A6. J Comput Chem 2004;25:1656–1676.

Ottiger S, Pini R, Storti G, Mazzotti M. Competitive adsorption equilibria of CO2 and CH4 on a dry coal. Adsorption 2008;14: 539–556.

Ozawa S, Kusumi S, Ogino Y. Physical adsorption of gases at high pressure. IV. An improvement of the Dubinin—Astakhov adsorption equation. J Colloid Interface Sci 1976;56:83–91.

Parrinello M, Rahman A. Polymorphic transitions in single crystals: a new molecular dynamics method. J Appl Phys 1981;52: 7182–7190.

Pusch R, Madsen FT. Aspects on the illitization of the Kinnekulle bentonites. Clays Clay Miner 1995;43:261–270.

Rodger PM, Forester TR, Smith W. Simulations of the methane hydrate/methane gas interface near hydrate forming conditions. Fluid Phase Equilib 1996;116:326–332.

Sadus RJ. Molecular Simulation of Liquids: Theory, Algorithms, and Object Orientation. Amsterdam: Elsevier; 1999.

Schuttelkopf AW, van Aalten DMF. PRODRG: a tool for high-throughput crystallography of protein-ligand complexes. Acta Crystallogr D 2004;60:1355–1363.

Siepmann JI, Frenkel D. Configurational bias Monte-Carlo—a new sampling scheme for flexible chains. Mol Phys 1992;75: 59–70.

Sitprasert C, Zhu ZH, Wang FY, Rudolph V. A multi-scale approach to the physical adsorption in slit pores. Chem Eng Sci 2011; 66:5447–5458.

Skipper NT, Chang FRC, Sposito G. Monte Carlo simulation of interlayer molecular structure in swelling clay minerals. 1. Methodology. Clays Clay Miner 1995a;43:285–293.

Skipper NT, Sposito G, Chang FRC. Monte Carlo simulation of interlayer molecular structure in swelling clay minerals. 2. Monolayer hydrates. Clays Clay Miner 1995b;43:294–303.

Slatt MS, O'Brien NR. Pore types in the Barnett and Woodford gas shales: contribution to understanding gas storage and migration pathways in fine-grained rocks. AAPG Bull 2011;95: 2017–2030.

Spiro CL, Kosky PG. Space-filling models for coal 2: extension to coals of various ranks. Fuel 1982;61:1080–1084.

St George JD, Barakat MA. The change in effective stress associated with shrinkage from gas desorption in coal. Int J Coal Geol 2001;45:105–113.

Sweatman MB, Quirke N. Characterization of porous materials by gas adsorption: comparison of nitrogen at 77 K and carbon dioxide at 298 K for activated carbon. Langmuir 2001a;17: 5011–5020.

Sweatman MB, Quirke N. Modelling gas adsorption in slit-pores using Monte Carlo simulation. Mol Simul 2001b;27: 295–321.

Tambach TJ, Mathews JP, van Bergen F. Molecular exchange of CH4 and CO2 in coal: enhanced coalbed methane on a nanoscale. Energy Fuels 2009;23:4845–4847.

Todd BD, Daivis PJ. The stability of non-equilibrium molecular dynamics simulations of elongational flows. J Chem Phys 2000;112:40–46.

Tsai MC, Chen WN, Cen PL, Yang RT, Kornosky RM, Holcombe NT, Strakey JP. Adsorption of gas mixture on activated carbon. Carbon 1985;23:167–173.

Wei YS, Sadus RJ. Equations of state for the calculation of fluid-phase equilibria. AlChE J 2000;46:169–196.

Wender A, Barreau A, Lefebvre C, Di Lella A, Boutin A, Ungerer P, Fuchs AH. Adsorption of n-alkanes in faujasite zeolites: molecular simulation study and experimental measurements. Adsorption 2007;13:439–451.

Weniger P, Kalkreuth W, Busch A, Krooss BM. High-pressure methane and carbon dioxide sorption on coal and shale samples from the Parana Basin, Brazil. Int J Coal Geol 2010;84: 190–205.

Wensink EWJ, Hoffmann AC, Apol MEF, Berendsen HJC. Properties of adsorbed water layers and the effect of adsorbed layers on interparticle forces by liquid bridging. Langmuir 2000;16:7392–7400.

Zhang JF, Todd BD. Pressure tensor and heat flux vector for inhomogeneous non-equilibrium fluids under the influence of three-body forces. Phys Rev E 2004;69:31111–31122.

Zhang JF, Hansen JS, Todd BD, Daivis PJ. Structural and dynamical properties for confined polymers undergoing planar Poiseuille flow. J Chem Phys 2007;126 (1–14):144907.

Zhang JF, Hawtin RW, Yang Y, Nakagava E, Rivero M, Choi SK, Rodger PM. Molecular dynamics study of methane hydrate formation at a water/methane interface. J Phys Chem B 2008; 112:10608–10618.

Zhang JF, Burke N, Yang YX. Molecular simulation of propane adsorption in FAU zeolites. J Phys Chem C 2012;116: 9666–9674.

Zhang J, Burke N, Zhang S, Liu K, Pervukhina M. Thermodynamic analysis of molecular simulations of CO2 and CH4 adsorption in FAU Zeolites. Chem Eng Sci 2014a;113:54–61.

Zhang J, Clennell MB, Dewhurst D, Liu K. Combined Monte Carlo and molecular dynamics simulation of methane adsorption on dry and moist coal. Fuel 2014b;122:186–197.

Zou C, Dong D, Wang S. Geological characteristics and resource potential of shale gas in China. Petrol Explor Develop 2010;37 (6):641–653.

16

WETTABILITY OF GAS SHALE RESERVOIRS

Hassan Dehghanpour, Mingxiang Xu and Ali Habibi
School of Mining and Petroleum Engineering, University of Alberta, Edmonton, Alberta, Canada

16.1 INTRODUCTION

The rapid increase of energy demand has shifted the focus of petroleum industry toward abundant unconventional resources worldwide. Organic shales have become an important hydrocarbon source in North America, and are being explored as a resource in other continents as well. From 2000 to 2012, the contribution of shale gas to the total natural gas production increased from 1% in the United States and Canada, to 39% in the United States and 15% in Canada (Stevens, 2012; US EIA, 2013). The abundant hydrocarbon resources in tight formations are now technically accessible due to advances in horizontal drilling and multistage hydraulic fracturing. However, measurement and modeling of petrophysical properties required for reserve estimation and reservoir-engineering calculations are among the remaining challenges for the development of tight formations. In particular, characterizing wettability (wetting affinity) of tight rocks and shales is challenging due to their complex pore structure, which can be either in hydrophobic organic materials or in hydrophilic inorganic materials.

16.2 WETTABILITY

Wettability is the measure of the preferential tendency of a fluid to wet the rock surface in the presence of the other fluid(s) (Agbalaka et al., 2008). Four general states of wettability have been recognized as (i) water-wet, (ii) fractional-wettability, (iii) mixed-wettability, and (iv) oil-wet (Donaldson and Alam, 2008). The pore-scale distribution of reservoir fluids strongly depends on the wettability state, which in turn depends on various factors such as rock mineralogy and the properties of the materials coating the rock surface (Anderson, 1987; Hamon, 2000; Mohammadlou and Mork, 2012; Rao and Girard, 1996). Rock wettability strongly influences the capillary pressure and relative permeability, which depend on pore-scale positioning of reservoir fluids. Therefore, selection of relevant capillary pressure and relative permeability curves for reservoir engineering calculations of unconventional reservoirs requires characterizing the wetting state of the reservoir rock. Furthermore, the interaction of fracturing and treatment fluids with the shale matrix strongly depends on shale wettability, which is poorly understood. In general, wettability characterization of organic shales is important for (i) selecting fracturing and treatment fluids, (ii) investigating residual phase saturation and its pore-scale topology, (iii) investigating the occurrence of water blockage at fracture face, and (iv) selecting relevant capillary pressure and relative permeability models for reservoir engineering calculations.

The wetting state of a reservoir rock can be identified by measuring the equilibrium contact angle (Johnson and Dettre, 1964), the Amott wettability index (Amott, 1959), the United States Bureau of Mines (USBM) wettability index (Donaldson et al., 1969), spontaneous imbibition rate/volume (Kathel and Mohanty, 2013; Morrow, 1990; Ma et al., 1999; Olafuyi et al., 2007; Zhou et al., 2000), hysteresis of the relative permeability curves (Jones and Roszelle, 1978), and nuclear magnetic relaxation (Brown and Fatt, 1956).

Fundamentals of Gas Shale Reservoirs, First Edition. Edited by Reza Rezaee.
© 2015 John Wiley & Sons, Inc. Published 2015 by John Wiley & Sons, Inc.

The conventional methods such as Amott and USBM can hardly be applied for measuring wettability of tight rocks primarily due to their extremely low permeability and porosity, complex pore structure, presence of organic materials and their mixed-wet characteristics (Odusina et al., 2011; Sulucarnain et al., 2012). For example, Amott and USBM techniques require a forced displacement which may not be practical for shales due to their ultra-low permeability. Recently, a technique has been applied based on interpretation of nuclear magnetic resonance (NMR) signals to study the wettability of different shale formations such as Eagle Ford, Barnett, Floyd, and Woodford (Odusina et al., 2011; Sulucarnain et al., 2012). These studies show that both brine and oil can wet the shale samples, which indicates their mixed-wet characteristics due to the presence of organic and inorganic materials. However, representative interpretation of NMR signals from unconventional rocks is more challenging than those from conventional rocks (Washburn and Birdwell, 2013). In many situations, the NMR signal from organic materials and clay-bound water can influence the resulting response of saturating fluids. Moreover, as pore size decreases the error associated with NMR measurement increases (Prince et al., 2009). Finally, invaded water-based or oil-based mud can influence NMR results (Looyestijn and Hofman, 2006). In general, accurate measurement of liquid-phase saturation in ultra-low permeability rocks using NMR is challenging due to the challenges involved in NMR signal processing and also in calibration of the system.

Spontaneous imbibition has been used as a reliable technique to quantify the wettability of reservoir rocks such as sandstones and carbonates (Akin et al., 2000; Ma et al., 1999; Morrow and Mason, 2001; Ma et al., 1999; Takahashi and Kovscek, 2010; Zhou et al., 2002). This technique is specifically attractive for tight rocks such as shales since a forced displacement in such low-permeability rocks requires a significant pressure drop, which may induce artificial cracks. Therefore, measuring and interpreting spontaneous imbibition of oleic and aqueous phases can be an alternative approach to quantify wettability of tight rocks such as shales. However, interpretation of imbibition data for characterizing the wetting state is challenging due to the adsorption of water by clay minerals and oil by organic materials, and also due to the complexity of the rock pore structure (Clarkson et al., 2012, 2013). In general, liquid imbibition in gas shales is controlled by many factors in addition to wettability such as sample expansion due to clay swelling (Dehghanpour et al., 2012, 2013), depositional lamination (Chalmers et al., 2012; Makhanov et al., 2014), osmosis effect (Bai et al., 2008; Chen et al., 2010; Chenevert, 1989; Neuzil, 2000; Xu and Dehghanpour, 2014), water adsorption by clay minerals (Chenevert, 1970; Hensen and Smit, 2002), and the connectivity of pore network (Xu and Dehghanpour, 2014).

16.3 IMBIBITION IN GAS SHALES

During spontaneous imbibition, the nonwetting phase initially saturating the porous medium is naturally displaced by a wetting phase (Brownscombe and Dyes, 1952; Engelberts and Klinkenberg, 1951; Leverett, 1940). This process can also occur at the field scale and may lead to significant consequences. During a hydraulic fracturing operation, fracturing fluid is pumped into the well to create fractures or fissures in the rock formation. The induced fracture network produces a pathway for hydrocarbon flow toward the wellbore (Novlesky et al., 2011). However, a significant fraction of injected fracturing fluid can imbibe into natural fractures and shale matrix, during and after fracking operations (Dutta et al., 2012; Holditch, 1979; Odusina et al., 2011; Roychaudhuri et al., 2011). Field data show that more than 50% of fracking water can remain in the reservoir even several months after opening the wells for flowback (Makhanov et al., 2013). The imbibed water can reduce fracture conductivity, fracture effective length, and fracture face permeability (Bahrami et al., 2012; Paktinat et al., 2006).

On the other hand, recent simulation (Cheng, 2012), experimental (Dehghanpour et al., 2012, 2013), and field studies (Adefidipe et al., 2014; Ghanbari et al., 2013) show that effective imbibition during shut-in periods can accelerate early-time gas production rate. Such observations has encouraged the industry to understand the physics of spontaneous imbibition in gas shales during shut-in periods, and to estimate the water loss versus soaking time (Lan et al., 2014a). Moreover, spontaneous imbibition of water, brine, or surfactants has also been considered as an enhanced oil recovery method in shale reservoirs (Wang et al., 2011, 2012a). Therefore, understanding water–oil countercurrent imbibition is critical for optimizing fracture and treatment fluids for enhancing the transport of oil from the tight rock matrix into the fracture network of oil shales.

Recently, extensive imbibition experiments have been conducted on various shale samples from the Horn River Basin, which is located in Canada, and this has been identified as the third largest North American natural gas accumulation discovered prior to 2010, with 110 Tcf (trillion cubic foot) recoverable gas in place (Johnson et al., 2011). The Horn River Basin consists of several subsurface producing members including Evie (E), Otter Park (OP) and Muskwa (M), overlaid by Fort Simpson (FS) non-producing shale. The total organic carbon (TOC) of Horn River shales has been reported to be up to 5.9 wt.% (Chalmers et al., 2012; Reynolds and Munn, 2010).

Dehghanpour et al. (2012) characterized several shale samples from the Horn River Basin by measuring porosity, contact angle, and mineralogy; then interpreting the well log data and scanning electron microscopy (SEM) images. Table 16.1 summarizes the mineralogy of example samples from the Fort Simpson, Muskwa, and Otter Park shale

TABLE 16.1 Average mineral concentration (wt.%) of the shale sections determined by X-ray diffraction

Formation	Calcite	Quartz	Dolomite	Chlorite IIb2	Illite 1 Mt	Plagioclase albite	Pyrite	Matrix density
FS	0.5 ± 0.4	29 ± 1.3	2.7 ± 0.3	6.5 ± 0.8	55.4 ± 1.7	4.1 ± 0.5	1.7 ± 0.2	2.747
M	0	36.7 ± 1.2	5.2 ± 0.4	4.4 ± 0.4	48.3 ± 1.5	3.6 ± 0.5	1.7 ± 0.2	2.744
OP	12.9 ± 0.4	43.6 ± 1.1	2.2 ± 0.5	0	33.8 ± 1.2	4.4 ± 0.4	3.2 ± 0.2	2.772

members. The samples are mainly composed of illite (clay mineral) and quartz (nonclay mineral). With increasing the depth, the illite concentration decreases and the quartz concentration increases. The Fort Simpson samples have the highest clay content while the Otter Park samples have the least clay content.

Figure 16.1 shows the gamma ray response, density porosity, and neutron porosity of Muskwa, Otter Park, and Evie members, and highlights the approximate location of the samples selected for the imbibition study. The gamma ray responses of all the shale members are relatively high, which indicates the high concentration of clay minerals and organic materials. The large separation between the neutron porosity and density porosity indicates that Fort Simpson is a water-saturated and relatively young shale. The increasing of density porosity log and the decreasing of neutron porosity log with respect to depth indicate that gas saturation increases from top to bottom of the Muskwa member. Similarly, the overlap of density porosity and neutron porosity indicates the presence of gas in the Otter Park member. In general, the electrical resistivity values, shown by AIT curves in Figure 16.1a and b, increases by increasing the depth. This increasing trend, which is more pronounced when moving from Fort Simpson to Muskwa, is another indication of increasing gas saturation with depth. The overlap of resistivity curves with different depth of investigation indicates negligible drilling fluid invasion due to low permeability of these shales.

Figure 16.2 compares the oil and water droplets equilibrated on the surface of Fort Simpson, Muskwa, and Otter Park samples at room temperature and atmospheric pressure. The needle shown in the pictures is used to support the weight of droplet and to reduce the impact of gravity on the measured contact angle. The results suggest that all samples are preferentially oil-wet as oil completely spreads on their surface while water partially wets their surface.

Figure 16.3 shows that water and brine uptake of Fort Simpson, Muskwa, and Otter Park samples is significantly higher than their oil uptake. This observation indicates that the connected pore network of the samples is strongly water-wet. Surprisingly, in contrast to water, oil hardy imbibes into the shale samples, and this observation contradicts the contact angle results, which indicate the samples are strongly oil-wet.

A similar trend was observed in other studies (Dehghanpour et al., 2013; Makhanov et al., 2012, 2013, 2014), which measured and compared the imbibition rate of various aqueous and oleic phases into the similar shale samples. Interestingly, the water uptake decreases by increasing the salt concentration, and this observation indicates the role of osmotic potential, which will be discussed later. Furthermore, the water uptake of Fort Simpson samples with the maximum clay content is considerably higher than that of Otter Park samples with the minimum clay content. This observation indicates the role of clay–water interaction, which will be detailed later.

The earlier experimental results show the strong water uptake of gas shales, which are strongly oil-wet, based on contact angle measurements. Now the challenging question is: why does oil, which completely spreads on the surface of these samples, hardly imbibes into their pore network, while water can easily and spontaneously imbibe into the samples. To answer this question, it is necessary to consider the possible mechanisms, which control liquid imbibition in gas shales. In general, clay hydration, microfracture induction, lamination, and osmotic effect are collectively responsible for the excess water uptake, which will be discussed in detail later. For instance, the significant water uptake of the Fort Simpson samples can be due to adsorption of water by clay minerals and imbibition-induced microcracks which can increase the sample porosity and permeability.

16.4 FACTORS INFLUENCING WATER IMBIBITION IN SHALES

This section presents various experimental evidences to discuss the parameters that control water flow in gas shales.

16.4.1 Sample Expansion

The imbibition experiments (Dehghanpour et al., 2012, 2013; Morsy et al., 2014a) on unconfined and intact core samples show that water uptake can induce microfractures in some of the shale samples, especially those which are clay-rich and naturally laminated. The clay hydration leads to sample expansion, which increases the porosity and permeability of the samples and, in turn, results in higher water imbibition rate. For example, Figure 16.4 shows a Muskwa sample before and after water imbibition. It is observed that significant cracks are induced by water imbibition into the clay-rich and laminated shale sample. In general, negatively

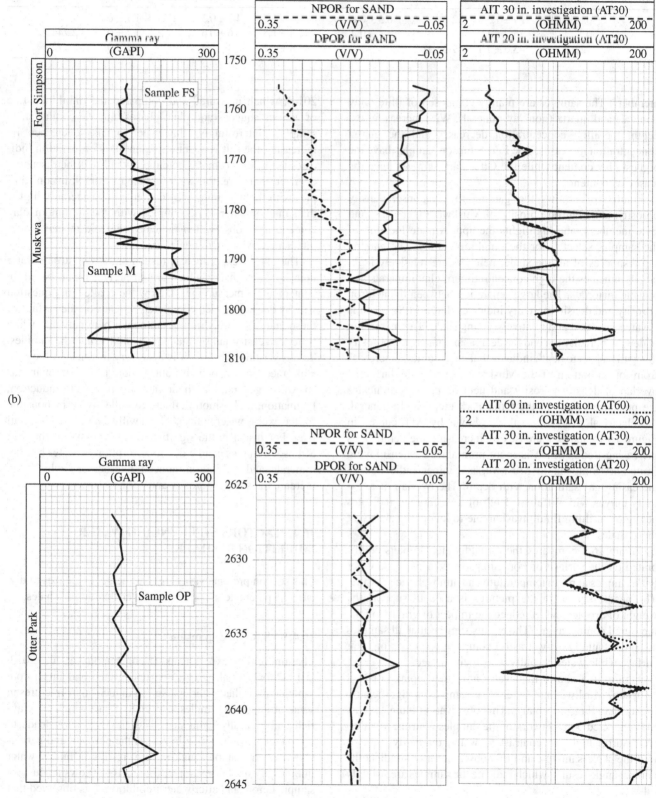

FIGURE 16.1 Gamma ray, neutron porosity, density porosity, and resistivity logs and the approximate location of the samples selected for the imbibition experiments for FS (a), M (a), and OP (b) formations.

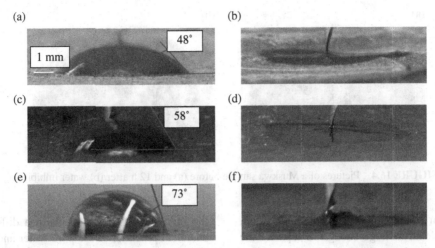

FIGURE 16.2 The picture of water (left) and oil (right) contact angles of FS (a, b), M (c, d), and OP (e, f).

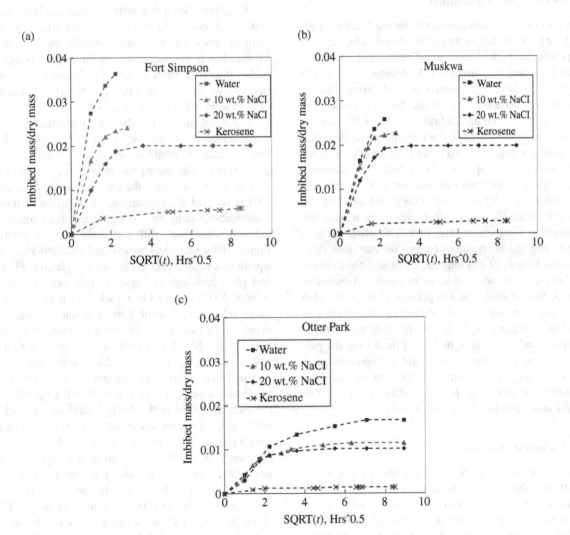

FIGURE 16.3 Normalized mass of NaCl brine, oil (kerosene), and deionized water imbibed into (a) FS, (b) M, and (c) OP samples. The curves are incomplete for FS and M samples because of their reaction with water, which leads to sample expansion, and thus the imbibition mass cannot be measured accurately.

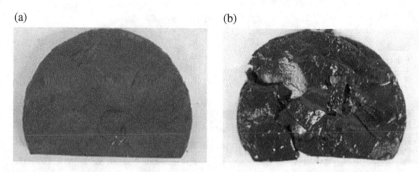

FIGURE 16.4 Pictures of a Muskwa sample before (a) and 12 h after (b) water imbibition.

charged clay layers can adsorb a significant volume of water, which induces a tensile stress large enough to separate the clay layers.

16.4.2 Depositional Lamination

It is well known that shales commonly have a layered structure, and the permeability along the lamination is higher than that perpendicular to the lamination (Chalmers et al., 2012). Consistently, recent measurements (Makhanov et al., 2012) show that water imbibition parallel to the bedding plane is faster than that perpendicular to the bedding plane. For example, Figure 16.5 compares brine (2 wt.% KCl) imbibition parallel and orthogonal to the bedding plane direction for Fort Simpson, Muskwa, and Otter Park samples, respectively. Evidently, brine uptake in both directions decreases by increasing the depth from Fort Simpson to Otter Park.

Furthermore, the degree of anisotropy, indicated by the separation between the curves, decreases by increasing the depth from Fort Simpson to Otter Park. This observation is consistent with the decreasing of clay concentration from the top to the bottom of this interval. Evidently, the samples with higher clay concentration are more laminated and show a higher degree of directional dependency. It should also be noted that clay swelling during water imbibition can enhance the anisotropy by increasing the distance between the clay platelets, and in turn, by further increasing the permeability along the lamination. This phenomenon can explain why imbibition anisotropy is more pronounced for water imbibition than that for oil imbibition (Li, 2006; Makhanov et al., 2014).

16.4.3 Chemical Osmosis

The other mechanism responsible for the excess water uptake of gas shales is the higher chemical potential of freshwater compared with pore water, which provides an additional force for water imbibition (Bai et al., 2008; Chen et al., 2010; Chenevert, 1989; Neuzil and Provost, 2009; Zhang et al., 2004, 2006). During water imbibition tests, the salts initially existing in the pore network dissolve into the imbibed water and result in a difference between the chemical potential of the pore water and the external water. This chemical potential difference acts as an additional driving force for the transport of water molecules into the sample. Xu and Dehghanpour (2014) measured and compared the imbibition rate of freshwater and NaCl brine with concentrations of 10 and 20 wt.% to study the effect of osmotic pressure on water imbibition. As shown in Figure 16.3, freshwater uptake of all samples is significantly higher than their brine uptake. Furthermore, the imbibition rate of low salinity brine (10 wt.% NaCl) is higher than that of high salinity brine (20 wt.% NaCl).

In general, increasing the salt concentration of external water reduces the osmotic effect and in turn reduces the liquid uptake. It should be noted that the salt concentration gradient decreases during the imbibition process, as ions diffuse from the rock into the external water. Ghanbari et al. (2013) verified the countercurrent diffusion of ions during water imbibition by measuring the electrical conductivity of external water using different shale samples. Interestingly, Figure 16.6a that plots normalized imbibed volume versus square root of time (SQRT) is well correlated to Figure 16.6b that plots conductivity versus square root of time. In both figures, Muskwa and Otter Park data points show a good linear relationship, while Fort Simpson data points can be divided into two relatively linear periods. The linear relationship in a SQRT plot indicates that the transport process can be described by a one-dimensional linear diffusion equation. Therefore, the transport of pressure during the imbibition process and that of ions during the diffusion process follow the linear diffusivity equation. Interestingly, we observe two different linear parts for the imbibition and diffusion profiles of the Fort Simpson sample. The higher slope (Region 1) of the Fort Simpson sample represents a higher imbibition rate and indicates that water imbibes through microfractures which have a higher permeability. This region continues until water fills all the microfractures. The lower slope (Region 2) shows a lower water imbibition rate and indicates that water imbibes into the matrix with ultra-low permeability. Fewer microfractures in the Muskwa and Otter Park samples explain the absence of Region 1 in Figure 16.6a.

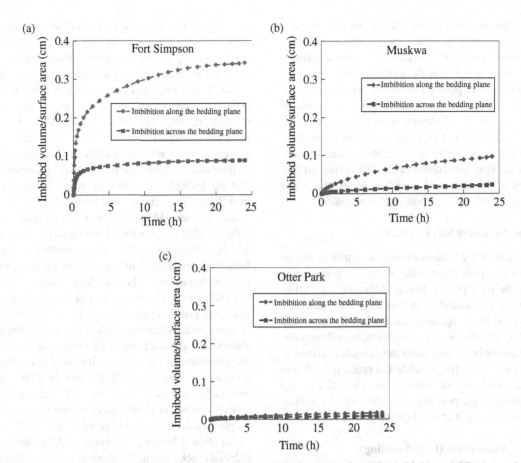

FIGURE 16.5 The effect of anisotropy on brine (2 wt.% KCl) imbibition into FS (a), M (b), and OP (c) samples.

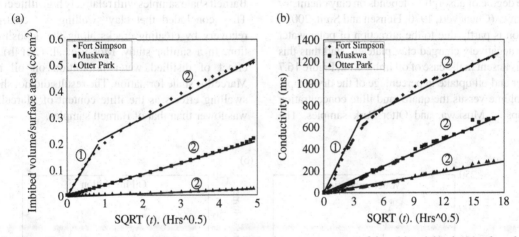

FIGURE 16.6 Normalized volume data for deionized water imbibition versus square root of time (a) and the change of conductivity versus square root of time (b) for FS, M, and OP shale samples (Ghanbari et al., 2013).

The shape of conductivity profile of the Fort Simpson sample can also be explained in a similar way. In simple, the effective diffusion coefficient for the ion transport in microfractures is higher than that in the tight rock matrix.

In general, as imbibition proceeds, the electrical conductivity of the external water increases for the three samples. This observation indicates the transport of salt out of the samples. Interestingly, the ion transport rate is lithology-dependent and decreases by increasing the depth from Fort Simpson to Muskwa and Otter Park. Fort Simpson with the highest clay concentration is more soluble in water which leads to a higher diffusion rate. Furthermore, diffusion rate strongly depends on porosity. Water uptake of Muskwa samples creates microfractures and enhances porosity as it is shown for example in Figure 16.4. The increase in porosity results in a lower tortuosity value, and in turn, in a higher effective diffusion coefficient. In addition, when the porosity is very low, surface interaction between the ions and shale minerals dominates. As porosity increases, surface interaction becomes less important and diffusion rate increases. Furthermore, the results of diffusion experiments can be used to explain the gradual increase in salt concentration of produced water during flowback operations (Sharak et al., 2014). Moreover, Fakcharoenphol et al. (2014) showed that osmotic potential can also lead to excess water imbibition and countercurrent oil release in oil-saturated shales.

16.4.4 Water Film and Salt Crystals

The strong affinity of dry shales to water can partly be due to the presence of water (or brine) film and/or salt crystals initially coating the rock pore network. Although the earlier experiments were conducted on dehydrated shale samples, parts of the pore space, originally saturated with water, may still be coated by a film of water. In addition, the salt initially dissolved in the pore brine may form salt crystals coating the surface of pore network. The possible salt precipitates in the pore space can stabilize the water film (Hematfar et al., 2013). The affinity of the pore surface coated by a brine film to water is stronger than that to oil (Schenk et al., 2006).

16.4.5 Water Adsorption (Clay Swelling)

Clay minerals of shales can adsorb a considerable amount of water, and the degree of adsorption depends on clay chemistry and water salinity (Chenevert, 1970; Hensen and Smit, 2002). This adsorption is partly due to the attraction of polar water molecules by negatively charged clay platelets, and thus this driving force is absent in the case of oil imbibition. Figure 16.7 plots the water and oil uptake in percentage of the dry sample initial pore volume versus the quartz and illite concentration for Fort Simpson, Muskwa, and Otter Park samples. The water uptake is strongly correlated to the concentration of illite, which is the dominant clay mineral. Although, illite is conventionally known as a nonswelling clay, this correlation indicates the effect of water adsorption on water uptake of gas shales. This is not a new observation, as previous experiments show that water adsorption can even alter illitic shales (Chenevert, 1989; Hensen and Smit, 2002). In addition, the presence of a small amount of mixed layer clay (interlayered or interstratified mixtures of illite and montmorillonite) may be the possible reason for the observed alteration. It is known that the hydration tendency of mixed layer clays is greater than that of illite (Hensen and Smit, 2002).

Furthermore, Makhanov et al. (2014) observed that imbibition of KCl brine is less than that of DI (deionized) water, which is another sign of water adsorption by clay minerals. Imbibition in shales is affected by the presence of potassium ions, which act as a clay swelling inhibitor. The potassium ion (K^+) can replace cations already present in the shale structure. Low hydration energy of (K^+) inhibits the reaction of clays with water molecules, hence it helps to keep the shale structure intact. Since KCl brine inhibits clay swelling and microfracture generation, it results in a reduced imbibition rate as compared to freshwater. In a similar study, Xu and Dehghanpour (2014) showed that increasing NaCl concentration can significantly reduce the alteration degree of gas shales, as shown in Figure 16.8.

Morsy and Sheng (2014) observed that clay swelling and induced cracks due to the imbibition of distilled water and a low pH aqueous solution can help oil production from Barnett shale samples with relatively high illite concentration. They concluded that clay swelling can contribute to oil recovery by creating cracks along the depositional limitation. In a similar study, Morsy et al. (2014b) studied the effect of distilled water imbibition on oil recovery in Marcellus shale formation. The results did not show any clay swelling effects as the illite content of Marcellus samples was lower than that of Barnett samples.

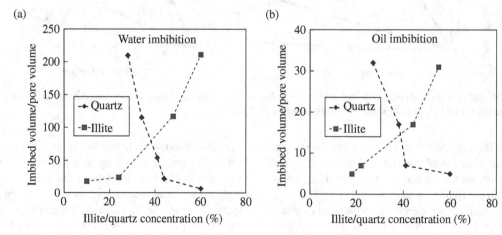

FIGURE 16.7 Normalized volume of water (a) and oil (b) gained by the Horn River samples versus illite and quartz concentration.

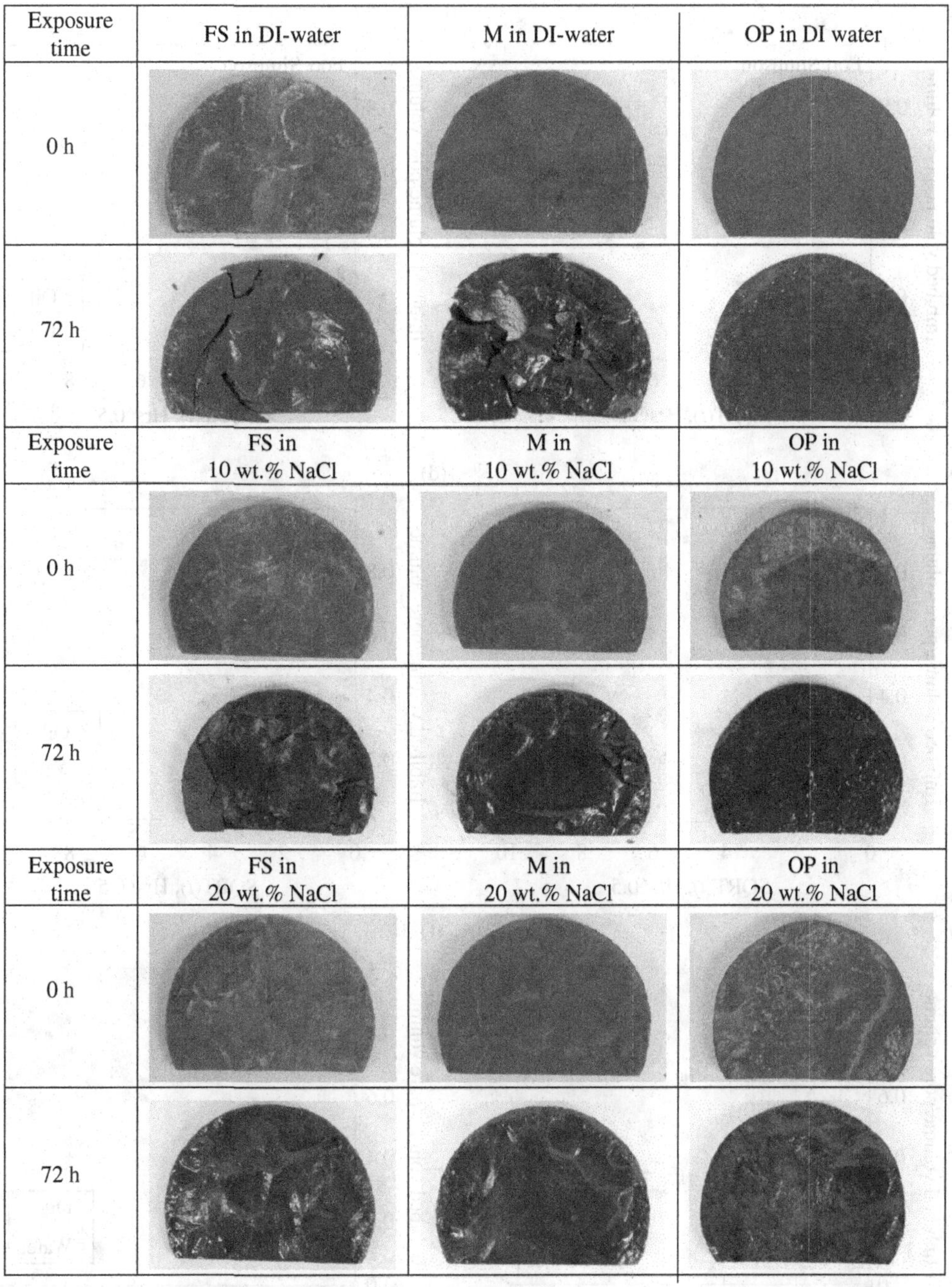

FIGURE 16.8 Pictures of FS, M, and OP shale samples used in the spontaneous imbibition experiments using 20 wt.% NaCl brine, 10 wt.% NaCl brine, and DI water (Xu and Dehghanpour, 2014).

16.4.6 Connectivity of Hydrophobic and Hydrophilic Pore Networks

It is well known that the shale pores can either be in organic or inorganic materials (Sondergeld et al., 2010). The organic part of the rock is hydrophobic (Mitchell et al., 1990), while the inorganic part can be hydrophilic, especially in the presence of clay minerals. Therefore, organic shales are usually a mixture of hydrophilic and hydrophobic materials. Significant water uptake of gas shales may indicate that the hydrophilic pore network is relatively well connected. Furthermore, this network may be coated by water film and salt which increase its affinity to water, as discussed earlier. On the other hand, insignificant oil uptake of gas shales may indicate that the hydrophobic pore space, partly coated by organic carbon, is poorly connected.

350 WETTABILITY OF GAS SHALE RESERVOIRS

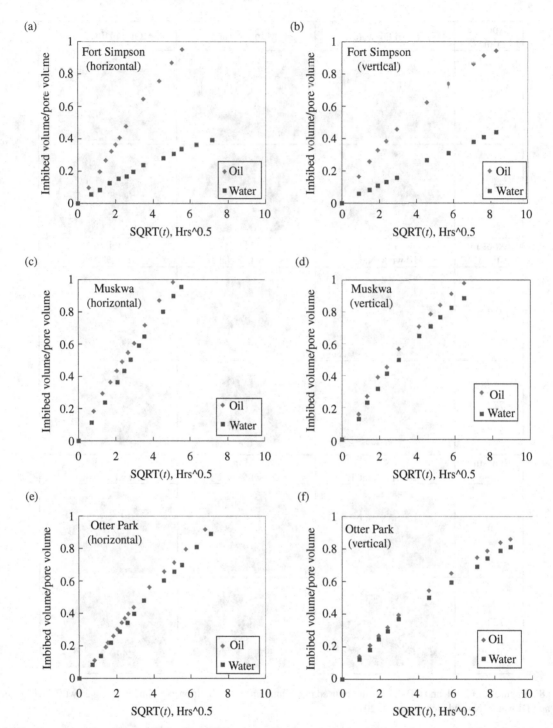

FIGURE 16.9 Comparison between imbibition rate of DI water and kerosene in horizontal (a, c, e) and vertical (b, d, f) crushed-shale packs from FS (a, b), M (c, d), and OP (e, f) sections.

To investigate the effect of pore network connectivity on the imbibition behavior, Xu and Dehghanpour (2014) conducted several horizontal and vertical imbibition experiments on crushed shale packs produced from the Horn River shale members. The objective was to investigate water and oil imbibition behaviors in a synthesized porous medium, in which both hydrophobic and hydrophilic pore networks are artificially well connected. In addition, a crushed-shale pack is relatively isotropic compared with intact samples that are significantly anisotropic. The shale samples were first crushed into around 1-μm diameter particles, and then packed into 1-in. diameter and 10-in long plastic tubes. The shale packs were confined to prevent the possible expansion during the imbibition tests. Interestingly, Figure 16.9 shows that oil (kerosene), which hardly imbibes into intact samples of the similar shales, imbibes into the crushed samples faster than water. In simple, the crushed samples are more oil-wet than the intact samples based on the comparative imbibition behavior. This comparative study indicates that the hydrophobic pore network of the intact samples is poorly connected. The strong affinity of crushed samples to oil is in agreement with the complete spreading of oil on the fresh break of intact samples, as demonstrated in Figure 16.2. The impact of organic pore connectivity on oil uptake of tight rocks is also indicated from recent imbibition experiments conducted on the Montney tight gas samples (Lan et al., 2014b). Analysis of high-resolution images and the results of organic petrology indicate that oil preferentially flows through the porous pyrobitumen and the pores coated by degraded bitumen which is strongly hydrophobic.

16.4.7 Effect of Polymer and Surfactant

Makhanov et al. (2014) investigated the role of various additives on water imbibition in the Horn River shales. The imbibition rate of various polymer and surfactant solutions in different shale members is compared in Figure 16.10. Xanthan Gum (XG) polymer solution with concentrations of 0.28 and 0.56 wt.% shows the lowest imbibition rates, primarily due to its high viscosity. However, the rate and magnitude of XG solution imbibition into the shale samples is still considerable, which is surprising when considering its high viscosity compared with water viscosity (almost 600 times higher than that of the other aqueous solutions at the representative shear rate). The XG molecules can hardly enter the shale pore network due to their large size compared with shale pore size and also due to the high solution viscosity. However, the observed weight gain of the shale

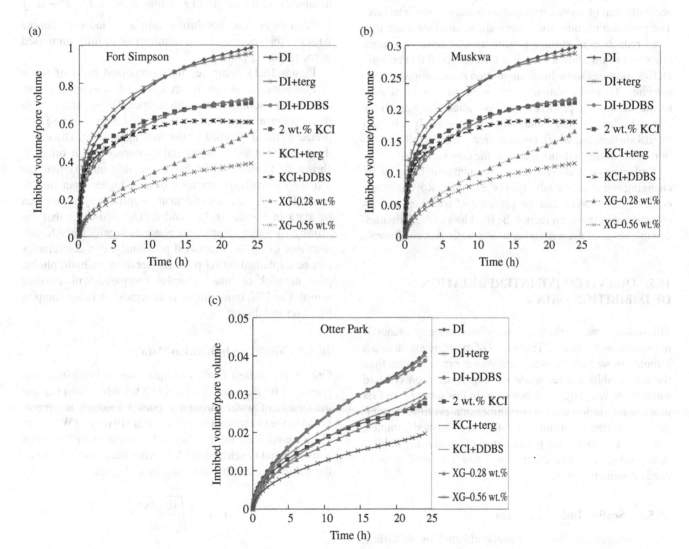

FIGURE 16.10 Plots of normalized imbibed volume versus of time. Imbibition data of various aqueous fluids imbibing into the FS (a), M (b), and OP (c) formation samples.

samples exposed to XG solutions indicates that water uptake is partly controlled through preferential adsorption of water molecules by the clay particles, and the high viscosity of polymer solution can only partly reduce the imbibition rate.

Figure 16.10 also compares the imbibition behavior of an anionic surfactant in freshwater (DI + sodium dodecylbenzenesulfonate (DDBS)), a nonionic surfactant in freshwater (DI + Tergitol (Terg)), an anionic surfactant in 2 wt.% KCl brine (KCl + DDBS), and a nonionic surfactant in 2 wt.% KCl brine (KCl + Terg). Expectedly, the imbibition rate of all surfactant solutions is lower than that of DI water, which is primarily due to their lower surface tension. The imbibition rate of KCl + DDBS is lower than that of KCl + Terg. This difference can be explained by the change of surface properties of anionic surfactants in the presence of solution ions (Lowe et al., 1999). Takaya et al. (2005) speculated that the salt particles in water decrease the repulsion forces between anionic surfactant molecules, which, in turn, allow accumulation of more surfactant molecules at the interface. The presence of more surfactants at the interface leads to a more reduction of surface tension and, in turn, to a more reduction of imbibition rate. It is also observed that anionic DDBS solutions show lower imbibition rates compared with nonionic Tergitol solutions. This behavior can be related to the adsorption properties of clay minerals. Negatively charged clay particles may repulse negatively charged DDBS surfactants, which could decrease the overall adsorption of the surfactant molecules on the clay surface.

Surfactants can also influence the imbibition behavior by changing the rock wettability. For example, Roychaudhuri et al. (2013) showed that the presence of additives such as cationic surfactants can reduce the fluid loss to the formation due to change of the wettability to less water-wet conditions.

16.5 QUANTITATIVE INTERPRETATION OF IMBIBITION DATA

This section presents the methods to characterize spontaneous imbibition in gas shales. The results of experiments on intact samples show that the water imbibition rate is higher than the oil imbibition rate, while the experiments on crushed samples show the opposite behavior. In this section, we first plot the imbibition data versus dimensionless time to compare the water/oil affinity of intact and crushed samples. Then, we investigate the performance of the existing imbibition models for predicting water and oil imbibition in crushed shale packs.

16.5.1 Scaling Imbibition Data

The imbibition results are mainly affected by rock/fluid properties and geometrical parameters. These factors include porosity and permeability of porous media, fluid viscosity, interfacial tension, wettability, boundary conditions, and sample shape. The objective of this section is to compare the wettability of intact and crushed samples, using the concept of dimensional analysis. The basic model for scaling laboratory imbibition data was proposed by Rapoport (1955). For scaling the imbibition results of oil/water/rock systems, Mattax and Kyte (1962) proposed the most frequently used dimensionless time (t_D):

$$t_D = t\sqrt{\frac{k}{\varnothing}}\frac{\sigma}{\mu_{gm}}\frac{1}{L_c^2} \qquad (16.1)$$

where μ_{gm} is the geometric mean of water and oil viscosities (Shouxiang et al., 1997). L_c is the characteristic length that depends on samples' shape and boundary condition (Zhang et al., 1996). For example, $L_c = (L/2)$ for cocurrent imbibition of a liquid phase into a linear porous medium with the length of L, while $L_c = \left(dL/2\sqrt{d^2 + 2L^2}\right)$ for countercurrent imbibition into a cylindrical sample with the thickness of L and diameter of d, fully immersed in the imbibing phase.

Figure 16.11 compares the normalized mass of water and oil imbibed into the intact and crushed samples versus the corresponding dimensionless time. The separation between water and oil data on the scaled plots can be interpreted as the difference in the wetting affinity. Therefore, the higher values of water data compared with oil data, shown in Figure 16.11a, b, c, and d, confirms that the affinity of crushed samples to oil is higher than that to water. However, the water data are significantly higher than oil data in Figure 16.11e and f. This indicates that the affinity of intact samples to water is significantly higher than that to oil. As discussed previously, this discrepancy can be explained by (i) poor connectivity of hydrophobic pore network of intact samples compared with crushed samples and (ii) induction of microcracks in intact samples by water imbibition.

16.5.2 Modeling Imbibition Data

One of the earliest models of spontaneous imbibition was presented by Bell and Cameron (1906), who found that the movement of water through a porous medium is proportional to the square root of time. Lucas (1918) and Washburn (1921) combined the Laplace relationship with Poiseuille equation, and developed the following relationship between the imbibition length and square root of time:

$$L_s(t) = \sqrt{\frac{\lambda \sigma \cos\theta}{4\mu}} t^{\frac{1}{2}} \qquad (16.2)$$

where L_s is the distance between the inlet and the imbibition front at time t. λ is the effective pore diameter. σ and μ are

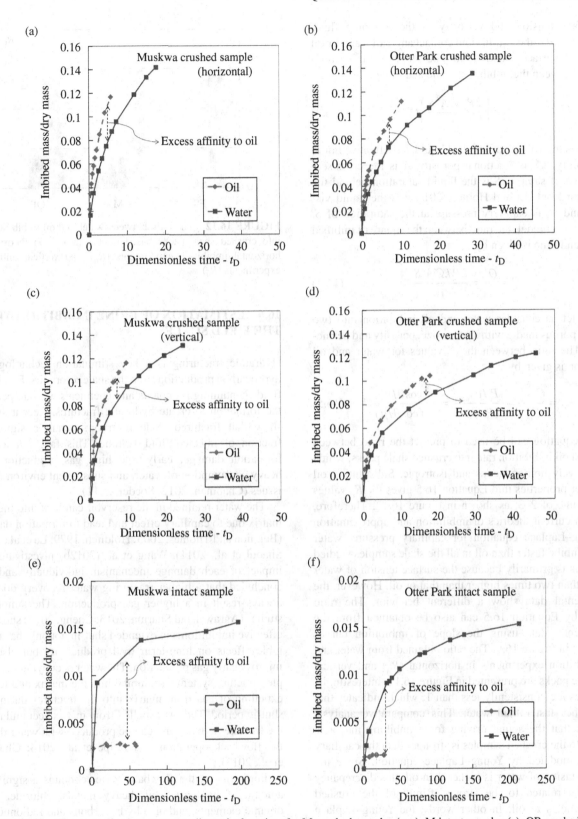

FIGURE 16.11 The normalized mass versus dimensionless time for M crushed samples (a, c), M intact samples (e), OP crushed samples (b, d), and OP intact samples (f). FS samples are not presented here due to the significant alteration of the samples during water imbibition tests.

the surface tension and viscosity of the imbibing fluid, respectively. θ is the equilibrium contact angle of water or oil on the rock surface. Handy (1960) developed a similar relationship between the imbibed volume and time:

$$Q^2 = \left(\frac{2P_c k \emptyset A^2 S}{\mu}\right) t \qquad (16.3)$$

where Q is the volume of imbibed liquid, k is effective liquid permeability, $\emptyset$ is fractional porosity, A is the cross-sectional area of sample, S is the liquid saturation behind the imbibition front (Li and Horne, 2001), μ is the liquid viscosity, and P_c is capillary pressure at the saturation of S. Based on this model, the ratio between the square of imbibed volume and time is given by

$$C = \frac{Q^2}{t} = \frac{2P_c k \emptyset A^2 S}{\mu} \qquad (16.4)$$

Now let us consider water and oil imbibition into two separate porous media with similar size, porosity and permeability. The ratio between the C values for water and oil imbibition is given by

$$R_C = \frac{C_{\text{water}}}{C_{\text{oil}}} = \frac{(P_c/\mu)_{\text{water}}}{(P_c/\mu)_{\text{oil}}} = \frac{(\sigma\cos\theta/\mu)_{\text{water}}}{(\sigma\cos\theta/\mu)_{\text{oil}}} \qquad (16.5)$$

This equation can be used to predict the ratio between water and oil imbibition rate into crushed shale packs, which are relatively homogeneous and isotropic. Substituting oil and water properties into Equation 16.5 gives the R_C values in the range of 2–3, as shown in Figure 16.12. Therefore, based on current theories of imbibition and approximation of Young–Laplace equation for capillary pressure, water should imbibe faster than oil in all the shale samples studied here. This is primarily because the surface tension of water is more than two times higher than that of oil. However, the experimental data show a different behavior. The ratio defined by Equation 16.5 can also be obtained from the experimental data using the slope of imbibition curves presented in Figure 16.9. The ratio obtained from water and oil imbibition experiments in horizontal (R_H) and vertical (R_V) shale packs are presented in Figure 16.12. Interestingly, the values are consistently less than 1, which indicates that oil imbibes faster than water. This comparative analysis indicates that the actual driving force imbibing the oleic phase into the crushed samples is stronger than the capillary pressure modeled by Young–Laplace equation. Since the surface tension of water is higher than oil, this discrepancy should be related to the strong affinity of the crushed shale particles to oil. In other words, the Young–Laplace equation, even with the contact angle of zero (due to the complete spreading of oil on the shale surface), underestimates the strong suction or adsorption of oil by the crushed shale particles.

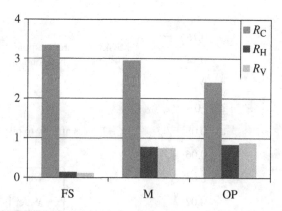

FIGURE 16.12 The ratio between water and oil imbibition rate into crushed shale packs based on the imbibition theory (R_C), horizontal imbibition experiments (R_H), and vertical imbibition experiments (R_V).

16.6 ESTIMATION OF BRINE IMBIBITION AT THE FIELD SCALE

Hydraulic fracturing is a key stimulation technology for hydrocarbon production from the shale reservoirs. Fracturing fluids containing proppants and chemicals are pumped into the formation to create hydraulic fractures. Recent studies show that fractured shale reservoirs retain a significant fraction of injected fluid volume. This fact can lead to formation damage, early-time high gas production rate, heavy consumption of water, and subsequent environmental issues (Chapman, 2012; Soeder, 2011).

The water retained in the reservoir can leak into the rock matrix due to capillary effect and lead to formation damage (Bennion and Thomas, 2005; Holditch, 1979; Le et al., 2009; Shaoul et al., 2011). Wang et al. (2012b) investigated the impact of each damage mechanism individually and also concluded that a higher fracturing water recovery does not always result in a higher gas production. The simulation studies (Agrawal and Sharma, 2013; Cheng, 2012) show that effective imbibition and extended shut-in not only has negligible effects on long-term well productivity but also can improve early gas production. The water occupying the complex fracture system can imbibe into the matrix and lead to expelling of gas from matrix into the fractures during the shut-in period. The gas expelled from matrix accumulates in the fracture network and can be produced with water during the flowback operation (Adefidipe et al., 2014; Ghanbari et al., 2013).

Furthermore, the flowback water contains a significant amount of hardness, soil, heavy metals, chloride, salts, organic elements, and also hydrocarbons and radionuclides from shale formations (Jiang, 2013). The recent strict environmental regulations prohibit flowback disposal especially in the North America (Jiang, 2013). Another motive to manage flowback water comes from the freshwater shortage in

TABLE 16.2 Parameters used for a sensitivity analysis of the total fracture–matrix interface in Muskwa shale

Injected volume (m³)	Number of stages (n_f)	Leak-off (%)	Aperture (w)	Interface (A_{cm})
51,000	20	10	2 prop.	1,910,000
51,000	20	10	3 prop.	1,275,000
51,000	20	10	4 prop.	478,000
51,000	20	15	2 prop.	1,810,000
51,000	20	15	3 prop.	1,200,000
51,000	20	15	4 prop.	452,000
51,000	20	30	2 prop.	1,490,000
51,000	20	30	3 prop.	992,000
51,000	20	30	4 prop.	372,000

the arid areas, especially in the United States. For example, the data from Natural Resources Conservation Service shows that the majority of Colorado is an arid place with the range of rainfall less than 10 in. By considering the development of fracturing operations, the demand for water is rising dramatically (Jiang, 2013). Goodwin et al. (2014) assessed the water intensity of shale gas resources in Northeastern Colorado and compared it with the consumption of water for extraction of other fuels from other energy sources. They concluded that the consumptive water intensity (the ratio between the consumption water and the estimated ultimate energy recovered) for shale gas development is estimated to be between 1.8 and 2.7 gal/MMBtu and is similar to that for surface coal mining. They concluded that flowback and produced water must be managed to minimize the environmental and public health risks. Therefore, maximizing fracturing water recovery is critical for minimizing the overall water consumption for shale gas development. The recovered water can be used for subsequent fracturing operations.

A study of 18 multifractured horizontal wells completed in the Horn River Basin shows that on average only 25% of injected water is recovered after nearly 40 days of flowback operations (Ghanbari et al., 2013). Fracturing fluid loss into the shale matrix at the field scale is a strong function of the soaking time (time period between the end of fracturing fluid injection and the beginning of the flowback operation), fracture–matrix interface, and fluid/rock properties.

Makhanov et al. (2014) estimated the volume of brine imbibed into the Muskwa shale using the material balance approach. This method assumes that the fracture volume equals to the difference between the total injected volume and fluid leak-off volume during the fracturing process. Therefore, assuming the rectangular slab geometry of fractures, the total fracture–matrix interface (A_{cm}) can be approximated by

$$A_{cm} = 2 \frac{V_{inj} - V_{leak}}{w} \qquad (16.6)$$

where A_{cm} is the total fracture–matrix interface, V_{inj} is the total injected volume, V_{leak} is the fluid leak-off volume during the fracturing operation, and w is the average fracture aperture, which is an uncertain parameter. Fluid leak-off volume represents how much of the injected fracturing fluid goes into the matrix and naturally existing fractures due to the high injection pressure. Rogers et al. (2010) anticipated up to 30% of leak-off in the Horn River Basin. The actual injected volume and the number of fracture stages related to the wells completed in the Muskwa formation are listed in Table 16.2. By doing a sensitivity analysis, Makhanov et al. (2014) investigated how the leak-off volume and fracture aperture influence the total fracture–matrix interface and in turn the total imbibed volume.

Figure 16.13 shows that the amount of brine imbibed into the formation increases with increasing the fracture–matrix interface. The left vertical axis represents the actual volume of imbibed brine into the formation. The right vertical axis represents the percentage of the injected volume that is imbibed into the formation. The horizontal axis represents the soaking time. Fracture aperture (w) and fracture–matrix interface (A_{cm}) are inversely proportional as described by Equation 16.6. As the fracture aperture decreases, the fracture–matrix interface increases and accordingly the imbibed volume of fracturing fluid increases. Moreover, a lower leak-off (during the fracturing operation due to differential pressure) percentage results in a greater imbibition volume during the shut-in period. High leak-off into the formation hinders generation of large fractures. The results show that fluid loss in Muskwa formation can reach up to 40% of injected volume after 90 days of shut-in period.

In a similar study, Roychaudhuri et al. (2013) estimated the fluid loss into the Marcellus gas shales. They observed the effect of microfracture network embedded in the samples on water imbibition. They found that water imbibition increases drastically as a result of microfracture network which accelerates the imbibition process. They concluded that the fluid loss during hydraulic fracturing can be explained by imbibition process, and showed that decreasing

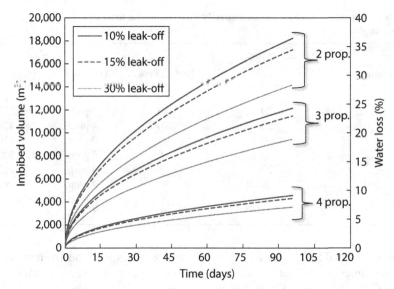

FIGURE 16.13 Imbibed volume of brine versus the soaking time for the Muskwa shale with different values of fracture aperture.

fracture aperture size increases the imbibed volume rapidly, mainly because fracture–matrix interface increases.

16.7 INITIAL WATER SATURATION IN GAS SHALES

The imbibition experiments reported in this chapter were conducted on dry shale samples while in situ shales may have some initial water saturation. The initial hydration state can influence the spontaneous imbibition rate in gas/water/rock systems (Li et al., 2006). However, the initial water saturation of some tight gas (Newsham and Rushing, 2002) and gas shale (Wang and Reed, 2009) reservoirs is abnormally low. Such reservoirs are in a state known as "sub-irreducible initial water saturation," created by the excessive drying at high paleo temperatures and pressures, and the lack of sufficient water for increasing the irreducible water saturation (Wang and Reed, 2009). Furthermore, thermal decomposition of hydrocarbons during deep burial in gas shales can consume the pore water (Seewald, 1994; Siskin and Katritzky, 1991; Wang et al., 2013).

16.8 CONCLUSIONS

The experimental data presented in this chapter demonstrate the complexity of liquid flow in gas shales. Although the studied samples are strongly oil-wet based on contact angle measurements, water uptake of all intact samples is considerably higher than their oil uptake. This observation indicates that in contrast to the fresh break of the rock samples, the connected pore network, which imbibes a considerable volume of water, is strongly water-wet. The results show that the water imbibition rate (i) is correlated to clay concentration, (ii) is faster along the depositional lamination, and (iii) decreases with increasing the salt concentration. The data suggest that in addition to capillarity, adsorption effect of clay minerals and osmotic potential influence water uptake of gas shales. Furthermore, in contrast to the intact samples, the crushed shale packs imbibe more oil than water. In a crushed sample, both hydrophobic and hydrophilic pores are well connected. Therefore, the observed difference between the oil uptake of crushed and intact samples is primarily due to the difference in connectivity of pore network in crushed and intact samples. In simple, the poorly connected hydrophobic pore network of intact samples becomes artificially well connected by crushing the samples. This interpretation is backed by the complete spreading of oil on fresh breaks of all the shale samples studied here.

REFERENCES

Adefidipe OA, Dehghanpour H, Virues CJ. Immediate gas production from shale gas wells: a two-phase flowback model. Paper presented at the Society of Petroleum Engineers – SPE USA Unconventional Resources Conference 2014, 1–3 April, The Woodlands, TX, USA, p 247–262.

Agbalaka C, Dandekar AY, Patil SL, Khataniar S, Hemsath JR. The effect of wettability on oil recovery: a review. Paper presented at the SPE Asia Pacific Oil and Gas Conference and Exhibition 2008 – "Gas Now: Delivering on Expectations;" 2008; Perth, Western Australia, Australia, 1, p 73–85.

Agrawal S, Sharma MM. Impact of liquid loading in hydraulic fractures on well productivity. Paper presented at the Society of Petroleum Engineers – SPE Hydraulic Fracturing Technology Conference 2013, 4–6 February, The Woodlands, TX, USA, p 253–268.

Agrawal S, Sharma M. Impact of liquid loading in hydraulic fractures on well productivity. SPE Hydraulic Fracturing Technology Conference; Society of Petroleum Engineers; 2013.

Akin S, Schembre JM, Bhat SK, Kovscek AR. Spontaneous imbibition characteristics of diatomite. J Petrol Sci Eng 2000;25(3–4): 149–165.

Amott E. Observations relating to the wettability of porous media. Trans AIME 1959;216:156–162.

Anderson WG. Wettability literature survey-part 6: the effects of wettability on waterflooding. J Petrol Technol 1987;39 (12): 1605–1622.

Bahrami H, Rezaee R, Clennell B. Water blocking damage in hydraulically fractured tight sand gas reservoirs: an example from Perth Basin, Western Australia. J Petrol Sci Eng 2012;88:100–106.

Bai M, Guo Q, Jin ZH. Study of wellbore stability due to drilling fluid/shale interactions. The 42nd US Rock Mechanics Symposium (USRMS); June 29–July 2, 2008; San Francisco, CA, USA.

Bell JM, Cameron FK. The flow of liquids through capillary spaces. J Phys Chem 1906;10:658–674.

Bennion DB, Thomas FB. Formation damage issues impacting the productivity of low permeability, low initial water saturation gas producing formations. J Energy Resour Technol 2005;127: 240–247.

Brown RJS, Fatt I. Measurements of fractional wettability of oil fields' rocks by the nuclear magnetic relaxation method. Fall Meeting of the Petroleum Branch of AIME; Society of Petroleum Engineers, 1956, 14–17 October, Los Angeles, CA, USA.

Brownscombe ER, Dyes AB. Water-imbibition displacement—can it release reluctant spraberry oil? Oil Gas J 1952;50:264–265.

Chalmers GR, Ross DJ, Bustin RM. Geological controls on matrix permeability of Devonian Gas Shales in the Horn River and Liard basins, northeastern British Columbia, Canada. Int J Coal Geol 2012;103:120–131.

Chapman A. Managing surface water use for hydraulic fracturing, BC oil and gas commission. 2012. Available at https://www.bcwwa.org/index.php?option=com_bcwwaresourcelibrary&view=resource&Itemid=74&id=1041. Accessed January 17, 2015.

Chen G, Ewy RT, Yu M. Analytic solutions with ionic flow for a pressure transmission test on shale. J Pet Sci Eng 2010;72: 158–165.

Chenevert M. Shale alteration by water adsorption. J Pet Technol 1970;22:1141–1148.

Chenevert ME. Lecture: diffusion of water and ions into shales. ISRM International Symposium; August 30–September 2, 1989, Pau, France.

Cheng Y. Impact of water dynamics in fractures on the performance of hydraulically fractured wells in gas-shale reservoirs. J Can Pet Technol 2012;51 (02):143–151.

Clarkson CR, Jensen JL, Pedersen PK, Freeman M. Innovative methods for flow-unit and pore-structure analyses in a tight siltstone and shale gas reservoir. AAPG Bull 2012;96 (2):355–374.

Clarkson CR, Solano N, Bustin RM, Bustin AMM, Chalmers GRL, He L, Melnichenko YB, Radliński AP, Blach TP. Pore structure characterization of North American shale gas reservoirs using USANS/SANS, gas adsorption, and mercury intrusion. Fuel 2013;103:606–616.

Dehghanpour H, Zubair HA, Chhabra A, Ullah A. Liquid intake of organic shales. Energy Fuels 2012;26(9):5750–5758.

Dehghanpour H, Lan Q, Saeed Y, Fei H, Qi Z. Spontaneous imbibition of brine and oil in gas shales: effect of water adsorption and resulting microfractures. Energy Fuels 2013;27(6): 3039–3049.

Donaldson EC, Alam W. *Wettability*. Houston, TX: Gulf Publishing; 2008. p 1–336.

Donaldson EC, Thomas RD, Lorenz PB. Wettability determination and its effect on recovery efficiency. Soc Pet Eng J 1969;9:13–20.

Dutta R, Lee C, Odumabo S, Ye P, Walker SC, Karpyn ZT, Ayala LF. Quantification of fracturing fluid migration due to spontaneous imbibition in fractured tight formations. Paper presented at the Society of Petroleum Engineers – SPE Americas Unconventional Resources Conference 2012, 5–7 June, Pittsburg, PA, USA, p 265–275.

Engelberts WF, Klinkenberg LJ. Laboratory experiments on the displacement of oil by water from packs of granular material. Proceedings of the 3rd World Petroleum Congress; 1951; The Hague, Netherlands, vol 2, p 544–554.

Fakcharoenphol P, Kurtoglu B, Kazemi H, Charoenwongsa S, Wu Y. The effect of osmotic pressure on improve oil recovery from fractured shale formations. Paper presented at the Society of Petroleum Engineers – SPE USA Unconventional Resources Conference 2014, 1–3 April, The Woodlands, TX, USA, p 456–467.

Ghanbari E, Abbasi MA, Dehghanpour H, Bearinger D. Flowback volumetric and chemical analysis for evaluating load recovery and its impact on early-time production. Paper presented at the Society of Petroleum Engineers – SPE Canadian Unconventional Resources Conference 2013, 5–7 November, Calgary, Alberta, Canada, 1, p 533–548.

Goodwin S, Carlson K, Knox K, Douglas C, Rein L. Water intensity assessment of shale gas resources in the Wattenberg field in northeastern Colorado. Environ Sci Technol 2014;48(10):5991–5995.

Hamon G. Field-wide variations of wettability. Paper presented at the Proceedings – SPE Annual Technical Conference and Exhibition, 2000, 1–4 October, Dallas, TX, USA, SIGMA, p 447–455.

Handy LL. Determination of effective capillary pressures for porous media from imbibition data. Trans AIME 1960;219:75–80.

Hematfar V, Maini B, Chen ZJ. Influence of clay minerals and water film properties on in-situ adsorption of asphaltene. 2013 SPE Heavy Oil Conference; June 11–13, 2013; Calgary, Alberta, Canada.

Hensen EJM, Smit B. Why clays swell. J Phys Chem B 2002; 106:12664–12667.

Holditch S. Factors affecting water blocking and gas flow from hydraulically fractured gas wells. J Pet Technol 1979;31:1515–1524.

Jiang X. *Flowback quality characterization for horizontal wells in Wattenberg field* [thesis]. Fort Collins, CO: Colorado State University; 2013.

Johnson RE, Dettre RH. Contact angle, wettability, and adhesion. Adv Chem Ser 1964;43:112.

Johnson MF, Walsh W, Budgell PA, Davidson JA. The ultimate potential for unconventional gas in the Horn River Basin: integrating geological mapping with Monte Carlo simulations. Paper presented at the Society of Petroleum Engineers – Canadian Unconventional Resources Conference 2011, 15–17 November, Calgary, Alberta, Canada, CURC 2011, 2, p 1399–1415.

Jones SC, Roszelle WO. Graphical techniques for determining relative permeability from displacement experiments. J Pet Technol 1978;30:807–817.

Kathel P, Mohanty KK. Wettability alteration in a tight oil reservoir. Energy Fuels 2013;27:6460–6468.

Lan Q, Ghanbari E, Dehghanpour H, Bearinger D. Water loss versus soaking time: spontaneous imbibition in tight rocks. Energy Technol 2014a;2:1033–1039.

Lan Q, Dehghanpour H, Wood J, Sanei H. Wettability of the Montney tight gas play. SPE Canadian Unconventional Resources Conference; Society of Petroleum Engineers, 2014b, 30 September–2 October, Calgary, Alberta, Canada.

Le D, Hoang H, Mahadevan J. Impact of capillary suction on fracture face skin evolution in water blocked wells. Paper presented at the Society of Petroleum Engineers – SPE Hydraulic Fracturing Technology Conference 2009, 19–21 January, The Woodlands, TX, USA, p 574–594.

Leverett MC. Capillary behavior in porous solids. Trans Am Inst 1940;142 (1):152–169.

Li Y. An empirical method for estimation of anisotropic parameters in clastic rocks. Leading Edge 2006;25 (6):706–711.

Li K, Horne R. Characterization of spontaneous water imbibition into gas-saturated rocks. SPE J. 2001;6:375–384.

Li K, Chow K, Horne RN. Influence of initial water saturation on recovery by spontaneous imbibition in gas/water/rock systems and the calculation of relative permeability. SPE Reserv Eval Eng 2006;9 (4):295–301.

Looyestijn WJ, Hofman JP. Wettability-index determination by nuclear magnetic resonance. SPE Reserv Eval Eng 2006;9 (2):146–153.

Lowe DF, Oubre CL, Ward CH, editors. Surfactant/cosolvent enhanced recovery of NAPL. In: Surfactants and Cosolvents for NAPL Remediation: A Technology Practices Manual. Volume 1, Boca Raton, FL: CRC Press; 1999. p 43.

Lucas R. Rate of capillary ascension of liquids. Kolloid Z 1918; 23:15–22.

Ma S, Zhang X, Morrow NR, Zhou X. Characterization of wettability from spontaneous imbibition measurements. J Can Petrol Technol 1999;38:1–8.

Makhanov K, Dehghanpour H, Kuru E. An experimental study of spontaneous imbibition in Horn River shales. Paper presented at the Society of Petroleum Engineers – SPE Canadian Unconventional Resources Conference 2012, CURC 2012, 30 October–1 November, Calgary, Alberta, Canada, 1, p 732–745.

Makhanov K, Kuru E, Dehghanpour H. Measuring liquid uptake of organic shales: a workflow to estimate water loss during shut-in periods. SPE Unconventional Resources Conference Canada; Society of Petroleum Engineers; 2013.

Makhanov K, Dehghanpour H, Kuru E. Measuring liquid uptake of organic shales: a workflow to estimate water loss during shut-in periods. Paper presented at the Society of Petroleum Engineers – SPE Canadian Unconventional Resources Conference 2013, 5–7 November, Calgary, Alberta, Canada, 1, p 433–450.

Makhanov K, Habibi A, Dehghanpour H, Kuru E. Liquid uptake of gas shales: a workflow to estimate water loss during shut-in periods after fracturing operations. J Unconventional Oil Gas Resour 2014;7:22–32.

Mattax CC, Kyte JR. Imbibition oil recovery from fractured, water-drive reservoir. Old SPE J 1962;2:177–184.

Mitchell AG, Hazell LB, Webb KJ. Wettability determination: pore surface analysis. Paper presented at the Proceedings – SPE Annual Technical Conference and Exhibition, 1990, 23–26 September, New Orleans, LA, USA, Gamma, p 351–360.

Mohammadlou M, Mørk MB. Complexity of wettability analysis in heterogeneous carbonate rocks a case study. SPE Europec/EAGE Annual Conference; Society of Petroleum Engineers, 2012, 4–7 June, Copenhagen, Denmark.

Morrow NR. Wettability and its effect on oil recovery. J Pet Technol 1990;42:1476–1484.

Morrow NR, Mason G. Recovery of oil by spontaneous imbibition. Curr Opin Coll Interf Sci 2001;6:321–337.

Morsy S, Sheng JJ. Imbibition characteristics of the Barnett shale formation. Paper presented at the Society of Petroleum Engineers – SPE USA Unconventional Resources Conference, 2014, 1–3 April, The Woodlands, TX, USA, p. 283–290.

Morsy S, Gomma A, Hughes B, Sheng JJ. Imbibition characteristics of Marcellus shale formation. Paper presented at the Proceedings – SPE Symposium on Improved Oil Recovery, 2014a, 12–16 April, Tulsa, OK, USA, 1, p 83–91.

Morsy S, Sheng JJ, Gomaa AM. Improvement of Mancos shale oil recovery by wettability alteration and mineral dissolution. SPE Improved Oil Recovery Symposium; Society of Petroleum Engineers; 2014b.

Neuzil CE. Osmotic generation of "anomalous" fluid pressures in geological environments. Nature 2000;403:182–184.

Neuzil CE, Provost AM. Recent experimental data may point to a greater role for osmotic pressures in the subsurface. Water Resour Res 2009;45(3):1–14.

Newsham E, Rushing JA, Chaouche A, Bennion DB. Laboratory and field observations of an apparent sub capillary-equilibrium water saturation distribution in a tight gas sand reservoir. SPE 75710. Presented at the Gas Technology Symposium, Calgary, Canada, 2002, 30 April–2 May.

Novlesky A, Kumar A, Merkle S. Shale gas modeling workflow: from microseismic to simulation—a horn river case study. Canadian Unconventional Resources Conference; November 15–17, 2011; Calgary, Alberta, Canada.

Odusina E, Sondergeld C, Rai C. An NMR study on shale wettability. Paper presented at the Society of Petroleum Engineers – Canadian Unconventional Resources Conference 2011, CURC 2011, 15–17 November, Calgary, Alberta, Canada, 1, p 515–529.

Olafuyi OA, Cinar Y, Knackstedt MA, Pinczewski WV. Spontaneous imbibition in small cores. Paper presented at the SPE – Asia Pacific Oil and Gas Conference, 2007, 30 October–1 November, Jakarta, Indonesia, 2, p 713–722.

Paktinat J, Pinkhouse J, Johnson N, et al. Case studies: optimizing hydraulic fracturing performance in northeastern fractured shale formations. SPE Eastern Regional Meeting; October 11–13, 2006; Revitalizing Appalachia, Canton, OH, USA.

Prince CM, Steele DD, Devier CA. Permeability estimation in tight gas sands and shales using NMR – a new interpretive methodology. 9th AAPG ICE Meeting in Rio de Janeiro, Brazil, 2009, November 15–18

Rao DN, Girard MG. A new technique for reservoir wettability characterization. J Can Pet Technol 1996;35(1):31–39.

Rapoport LA. Scaling laws for use in design and operation of water-oil flow models. Trans AIME 1955;204:143–150.

Reynolds M, Munn D. development update for an emerging shale gas giant field—Horn River Basin, British Columbia, Canada. SPE Unconventional Gas Conference; February 23–25, 2010; Pittsburgh, PA, USA.

Rogers S, Elmo D, Dunphy R, Bearinger D. Understanding hydraulic fracture geometry and interactions in the Horn River Basin through DFN and numerical modeling. Paper presented at the Society of Petroleum Engineers – Canadian Unconventional Resources and International Petroleum Conference 2010, 19–21 October, Calgary, Alberta, Canada, 2, p 1426–1437.

Roychaudhuri B, Tsotsis T, Jessen K. An experimental investigation of spontaneous imbibition in gas shales. Paper presented at the Proceedings – SPE Annual Technical Conference and Exhibition; October 30–November 2, 2011; Denver, CO, USA, 6, p 4934–4944.

Roychaudhuri B, Tsotsis TT, Jessen K. An experimental investigation of spontaneous imbibition in gas shales. J Petroleum Sci Eng 2013;111:87–97.

Schenk O, Urai JL, Piazolo S. Structure of grain boundaries in wet, synthetic polycrystalline, statically recrystallizing halite–evidence from cryo-SEM observations. Geofluids 2006;6: 93–104.

Seewald JS. Evidence for metastable equilibrium between hydrocarbons under hydrothermal conditions. Nature 1994;370: 285–287.

Shaoul J, Van Zelm L, De Pater CJ. Damage mechanisms in unconventional-gas-well stimulation—a new look at an old problem. SPE Prod Oper 2011;26(4):388–400.

Sharak AZ, Ghanbari E, Dehghanpour H, Bearinger D. Fracture architecture from flowback signature: a model for salt concentration transient. Paper presented at the Society of Petroleum Engineers – SPE Hydraulic Fracturing Technology Conference 2014, 4–6 February, The Woodlands, TX, USA, p 253–268.

Shouxiang M, Morrow NR, Zhang X. Generalized scaling of spontaneous imbibition data for strongly water-wet systems. J Pet Sci Eng 1997;18:165–178.

Siskin M, Katritzky AR. Reactivity of organic compounds in hot water: geochemical and technological implications. Science 1991;254(5029):231–237. Available at 10.1126/science.254.5029.231. Accessed January 17, 2015.

Soeder JD. Environmental impacts of shale-gas production. American Institute of Physics. Phys Today 2011;64(11):8.

Sondergeld CH, Ambrose RJ, Rai CS, Moncrieff J. Micro-structural studies of gas shales. Paper presented at the SPE Unconventional Gas Conference 2010, 23–25 February, Pittsburg, PA, USA, p 150–166.

Stevens P. The "shale gas revolution": developments and changes. Chatham House Briefing Paper; 2012, August 2012.

Sulucarnain ID, Sondergeld CH, Rai CS. An NMR study of shale wettability and effective surface relaxivity. SPE Canadian Unconventional Resources Conference; Society of Petroleum Engineers; 2012.

Sulucarnain I, Sondergeld CH, Rai CS. An NMR study of shale wettability and effective surface relaxivity. Paper presented at the Society of Petroleum Engineers – SPE Canadian Unconventional Resources Conference 2012, 30 October–1 November, Calgary, Alberta, Canada, 1, p 356–366.

Takahashi S, Kovscek AR. Spontaneous countercurrent imbibition and forced displacement characteristics of low-permeability, siliceous shale rocks. J Pet Sci Eng 2010;71:47–55.

Takaya H, Nii S, Kawaizumi F, Takahashi K. Enrichment of surfactant from its aqueous solution using ultrasonic atomization. Ultrason Sonochem 2005;12:483–487.

US Energy Information Administration. October 23, 2013. North America leads the world in production of shale gas. Available at http://www.eia.gov/todayinenergy/detail.cfm?id=13491. Accessed February 18, 2014.

Wang FP, Reed RM, John A, Katherine G. Pore networks and fluid flow in gas shales. Paper presented at the Proceedings – SPE Annual Technical Conference and Exhibition, 2009, 4–7 October, New Orleans, LA, USA, 3, p 1550–1557.

Wang D, Butler R, Liu H, et al. Flow-rate behavior and imbibition in shale. SPE Reserv Eval Eng 2011;14:485–492.

Wang D, Butler R, Zhang J, Seright R. Wettability survey in Bakken shale with surfactant-formulation imbibition. SPE Reserv Eval Eng 2012a;15(6):695–705.

Wang Q, Guo B, Gao D. Is formation damage an issue in shale gas development? SPE International Symposium and Exhibition on Formation Damage Control; Society of Petroleum Engineers, 2012b, 15–17 February, Lafayette, LA, USA.

Wang Y, Zhang S, Zhu R. Water consumption in hydrocarbon generation and its significance to reservoir formation. Pet Explor Dev 2013;40 (2):259–267.

Washburn EW. The dynamics of capillary flow. Phys Rev 1921;17:273.

Washburn KE, Birdwell JE. A new laboratory approach to shale analysis using NMR relaxometry. The SPE conference at Unconventional Resources Technology Conference; August 12–14, 2013; Denver, CO, USA.

Xu M, Dehghanpour H. Advances in understanding wettability of gas shales. Energy Fuels 2014;28(7), 4362–4375.

Zhang X, Morrow NR, Ma S. Experimental verification of a modified scaling group for spontaneous imbibition. SPE Reserv Eng 1996;11(4):280–285.

Zhang J, Chenevert ME, AL-Bazali T, Sharma M. A new gravimetric-swelling test for evaluating water and ion uptake in shales. SPE Annual Technical Conference and Exhibition; Society of Petroleum Engineers, 2004, 26–29 September, Houston, TX, USA.

Zhang J, Clark DE, Al-Bazali TM, et al. Ion movement and laboratory technique to control wellbore stability. Fluids Conference; April 11–12, 2006; the Greenspoint Hotel, Houston, TX.

Zhou X, Morrow NR, Ma S. Interrelationship of wettability, initial water saturation, aging time, and oil recovery by spontaneous imbibition and waterflooding. SPE J 2000;5 (2):199–207.

Zhou D, Jia L, Kamath J, Kovscek AR. Scaling of counter-current imbibition processes in low-permeability porous media. J Pet Sci Eng 2002;33(1–3):61–74.

17

GAS SHALE CHALLENGES OVER THE ASSET LIFE CYCLE

ROBERT "BOBBY" KENNEDY
Petroleum Engineering, Unconventional Resources Team, Baker Hughes, Inc., Tomball, TX, USA

17.1 INTRODUCTION

The asset life cycle is a data-driven, integrated, multidiscipline approach for gas shale plays, which includes recommended technologies/solutions for operators to use when analyzing, developing, and producing these unconventional resource reservoirs. In essence, recommended practices are offered that will address the operator's challenges over the complete life cycle of a shale gas reservoir. The earlier chapters of this book have discussed, in detail, shale characteristics and the evaluation, reservoir analysis, drilling, completion, and hydraulic fracturing (stimulation) of shale gas reservoirs. It is not the intent of this author to duplicate that, but only to discuss the integration of these various analysis techniques operational technologies with respect to the complete asset life cycle. During this process, however, certain aspects of the various technologies and processes will have to be repeated.

The asset life cycle includes five phases—exploration, appraisal, development, production, and rejuvenation. Most of these terms for the life cycle phases have been around the industry for a number of years, and are not new with the exception of possibly the rejuvenation phase. What is new, however, are the technologies and solutions that are employed to address operator challenges and proposed objectives during each phase of the shale gas life cycle.

The three key points to the asset life cycle data-driven approach which are designed to maximize the net value of an operator's asset are

- Begin with a complete understanding of the reservoir
- Use a multidiscipline and integrated approach across each phase of the life cycle
- Effectively use fit-for-purpose modern technology

17.2 THE ASSET LIFE CYCLE

The asset life cycle includes five phases: (i) exploration, (ii) appraisal, (iii) development, (iv) production, and (v) rejuvenation (see Fig. 17.1). The discussion under each phase will be divided into current common practices and recommended practices. Choices (technology and application) must be made at every phase of the life cycle, and these choices can affect ultimate recovery. Not all shale gas reservoirs are the same, and each may require different choices. Also, each choice can affect later options. As this discussion ensues, it will become obvious that the description of the objectives and challenges of the operator, the technologies, and data required to implement each phase of the life cycle do address the uniqueness of shale gas. The following is an outline of the discussion of the five life cycle phases.

17.2.1 Exploration Phase Objectives—Recommended Practices

- Conduct a basin/area screening study to identify core areas (sweet spots) and to determine an initial estimate of gas in place

---

Fundamentals of Gas Shale Reservoirs, First Edition. Edited by Reza Rezaee.
© 2015 John Wiley & Sons, Inc. Published 2015 by John Wiley & Sons, Inc.

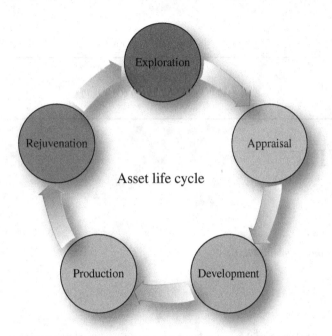

FIGURE 17.1 The shale gas asset life cycle (Source: Baker Hughes).

- Begin to characterize the reservoir with exploration wells
- Determine the initial economic value and reservoir potential/evaluate

17.2.2 Appraisal Phase Objectives—Recommended Practices

- Drill the appraisal wells and continue to characterize the reservoir
- Build reservoir model(s) for simulation
- Generate a field development plan
- Validate the economics of the play/evaluate
- Consider a pilot project/evaluate

17.2.3 Development Phase Objectives—Recommended Practices

- Implement the field development plan or pilot
- Install surface production and export facilities, including compression and pipelines
- Design wells and optimize drilling costs
- Refine and optimize the hydraulic fracturing and wellbore completion designs (characterize the horizontal laterals)

17.2.4 Production Phase Objectives—Recommended Practices

- Monitor and optimize producing rates
- Manage the water cycle—sourcing for drilling and fracturing water; lifting, treating, handling, storage, and disposal of well flowback water
- Reduce corrosion, scaling, and bacterial contamination in wells and facilities
- Install artificial lift if required
- Protect the environment

17.2.5 Rejuvenation Phase Objectives—Recommended Practices

- Evaluate wells for Re-frac candidates and then implement Refracs
- Analyze field for redevelopment potential—infill drilling

17.3 EXPLORATION PHASE DISCUSSION

The life cycle starts with the exploration and then appraisal phases (which are similar in many respects) where the operator begins to obtain an understanding of the reservoir aspects of these unique shale formations. It has been said by other authors that there is not really an exploration phase, because shales are continuous formations and their locations are known. The objective is to locate the part of the formation that is more prolific (sweet spots) than the others.

A screening study is particularly important when entering a new basin or area. The primary purpose of the study is to identify the core areas, that is, locate the sweet spots. Well-by-well production data indicate that shale formations have small spots of very productive wells (sweet spots), surrounded by large areas of wells that produce far less gas or oil. Sweet spots are a function of total organic carbon (TOC), thermal maturity, thickness, gas in place (GIP), natural fractures, mineralogy, and field stress in the area. Sweet geologic spots may not necessarily be sweet economic spots. Also, if an area possesses most of these attributes, but is not a favorable area in which to frac (mineralogy or stresses), it is not a sweet spot. It may sound trite, but develop the sweet spots first, then go back to the less attractive areas (Kennedy et al., 2012a).

17.3.1 Screening Study—Current Practice

Most small operators in the United States do not conduct basin screening studies, especially in the established plays. There is no doubt, however, that the major operators and large independents have conducted such studies prior to entering a new play area in the United States. From this author's experience with National Oil Companies (NOCs) and International Oil Companies (IOCs), they are all conducting or have conducted basin screening studies prior to entering international virgin areas that do not have any shale wells or existing production. Studies are known to exist in China, Saudi Arabia, Argentina, Australia, and Mexico.

17.3.2 Screening Study Recommended Practices

Conduct a basin screening study prior to entering a new basin or area that involves gathering and analyzing data including:

- Geology—sedimentology, stratigraphy, and depositional environment
- Geochemistry—TOC (initial reserve estimate), thermal maturity (type of hydrocarbon)
- Geomechanics—stress regime for well drilling and fracturing design and placement
- Petrophysics—rock type, lithology/mineralogy, porosity (from cores and logs)
- Existing well data

To begin initial characterization of the reservoir, it is recommended that 3D seismic be conducted over the potential play area. From 3D seismic the typical information on faults, formation thickness, depth, and lateral continuity can be obtained. However, 3D seismic can also be used to

- Identify areas of highest TOC using acoustic impedance
- Increase understanding of natural fractures using seismic attributes
- Provide azimuthal anisotropy data related to natural fracture orientation and horizontal stress anisotropy
- Assist in identification of sweet spots using seismic cross-plots

An optimal reservoir characterization workflow for stress analysis/seismic interpretation is described by Sena et al. (2011), and the primary steps are listed in the following:

- Seismic rock properties (advanced rock properties analysis)
- Azimuthal analysis
  - Azimuthal migration
  - Reservoir oriented gather conditioning
  - Azimuthal velocity/AVO analysis
- Seismic analysis
  - Pre stack inversion
  - Pore pressure prediction
- Stress estimation
- Multiattribute/integration
  - Rock property prediction
  - Interpretive correlation of seismic attributes with geologic and engineering data

Data from exploration and appraisal wells are used to pinpoint the location of sweet spots through further formation assessment consisting of detailed mineralogical, structural, and geomechanical characterization (pulsed neutron spectroscopy, nuclear magnetic resonance, and acoustic logs). These data are more location specific, while information gathered through seismic is broad, spread out, and certainly not as accurate as that obtained from an actual wellbore penetrating the formation. The detailed discussion of this topic is included later in this chapter. Seismic information is relevant through the exploration, appraisal, development, and rejuvenation phases of the life cycle (Kennedy et al., 2012a). Also, during the exploration phase the operator should begin to characterize the reservoir from the exploratory wells.

17.3.3 Reservoir Characterization—Current Practice

Many US operators will log and core vertical exploratory and appraisal wells to gather data for overall reservoir characterization; however, very little data are collected on the lateral of horizontal appraisal wells. Industry data indicate that in the United States less than 9% of horizontal laterals of development wells are logged or any data gathered about the reservoir along these 5000–6000 ft horizontals. Such data could be used to refine the characterization of the reservoir, and lead to more intelligent placement of future development wells.

17.3.4 Reservoir Characterization—Recommended Practices

Open-hole logs (conventional, pulsed-neutron, and spectroscopy) and cores from exploratory wells provide data for petrophysical analysis for initial reservoir characterization for shale reservoirs. Wellbore image logs and nuclear magnetic resonance (NMR) logs provide necessary information for shale. An example of one of these special log and analysis techniques is shown in Figure 17.2. This shale gas facies expert system provides operators with a quick and accurate method of classifying shale gas reservoirs, identifying favorable zones for hydraulically fracturing, identifying frac barriers, and locating zones from which to drill horizontal laterals (Jacobi et al., 2009; LeCompte et al., 2009; Pemper et al., 2009). It should be noted that cores are a must, either whole cores or sidewall (rotary) cores for analysis and direct measurement of certain parameters which are also used to calibrate logs.

These logs assist in locating where to locate the hydraulic fractures. A fracture should be initiated in the most "fracable" (most brittle mineralogy) locations, so more energy can be used to propagate fractures and create more complex fractures. In addition, these locations should also have the higher TOCs in the well; thus, they are targeted at the sweet spots. To determine this, lithofacies data are examined in context with TOC content, geomechanical properties, and porosity to identify specific intervals within the most

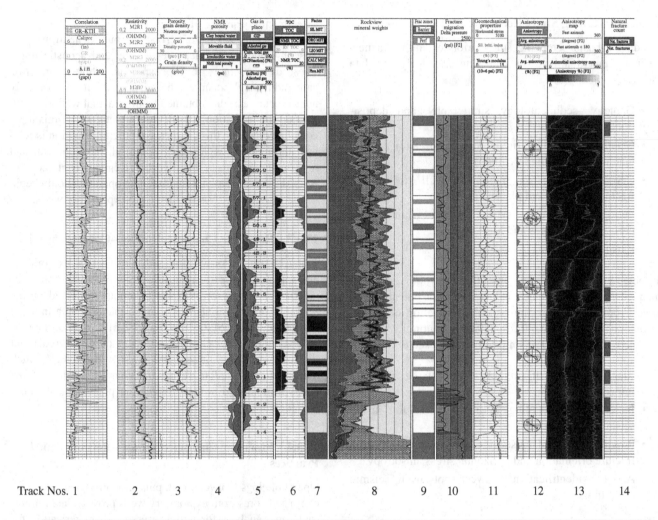

FIGURE 17.2 Integrated shale analysis plot with mineralogy/lithology legend (Source: Baker Hughes).

favorable facies as the best intervals to initiate a fracture. Likewise, the most unfavorable sections of the most unfavorable facies are also identified. The method classifies zones for unfavorable frac intervals and for favorable frac intervals. A similar criterion is applied for identifying the optimum location to place the horizontal lateral.

These types of logging surveys and cores are normally only run or taken in vertical wells. From the numerous mineral spectroscopy logs that have been run in shale wells, it is known that geochemistry (primarily TOC), lithology, and mineralogy can vary foot by foot both vertically and laterally in shale reservoirs. This author recommends that reservoir characteristics should be determined along the horizontal lateral, and this topic is discussed later in this chapter. It is recommended that the logs and cores also be run and obtained from vertical appraisal wells and that the characterization of the reservoir continue through the appraisal phase.

Conventional whole cores or sidewall cores and analysis are required for calibrating logs and obtaining true measured properties. Recommended analyses should include porosity, permeability, saturations, gas analysis to include adsorption/desorption isotherms and isotopes, triaxial compression testing for static elastic properties, and thin section studies. A wide variety of technologies are used to examine the lithology and mineral composition of shale. "Sample" for this analysis comes from cuttings, rotary sidewall cores, conventional core plugs, and whole core. These different types of analyses can often be combined and performed on the same sample.

17.3.5 Determining Initial Economic Value and Reservoir Potential

An initial assessment of reservoir potential and economic value can be determined from all the above data. Individual operators have different drivers and specific financial and leasehold situations in the United States. Gas price, regulatory, and infrastructure are all different for countries outside the United States. Martin and Eid (2011) cover these topics at length in their paper. The final step of the exploration phase for the operator is to evaluate to determine whether or not to proceed with appraisal and development phases.

17.4 APPRAISAL PHASE DISCUSSION

17.4.1 Drill Appraisal Wells—Current Practice

More wells are drilled during the appraisal phase than in the exploration phase; thus, data from these additional wells should continue to be used to further characterize the reservoir.

17.4.2 Drill Appraisal Wells—Recommended Practices

Both vertical and horizontal "appraisal" wells should be drilled. Vertical wells are required to collect data, and some horizontal appraisal wells should be drilled to test the hydraulic fracturing and mechanical well completion designs to obtain estimates of initial production potential of the designs. Horizontal wells will also provide information to assist in initially determining optimum lateral length, and to begin early drilling optimization.

17.4.3 Build Reservoir Models for Simulation—Current Practice

Applying classic reservoir engineering techniques to unique shale reservoirs is problematic due to the length of time to reach pseudo-steady-state flow and/or establish a constant drainage area. This leads to the inability to accurately estimate the recoverable reserves in a timely and consistent manner. Both decline curve and material balance methods have been found to have serious drawbacks when applied to shale gas reservoirs that had not established a constant drainage area. Kupchenko et al. (2008), upon recognizing that production performance from tight gas reservoirs (similar to shale reservoirs) displays steep initial decline rates and long periods of transient flow, realized that inaccurate forecasts would result from using this transient production data. Also in 2008, Ilk et al. (2008) introduced the "power law exponential decline" (form of power law loss ratio) concluding that it offered a better match to production rate than hyperbolic decline. Others, including Duong (2010), have also developed and proposed decline curve analysis (DCA) methods, and some have offered new techniques for using the material balance approach (Engler, 2000; Payne, 1996).

The industry has taken a traditional approach to developing shale gas/oil, looking at these unconventional shale plays in a statistical manner. Since it is simpler to apply the classic decline curve analysis and type curve approach, it is being applied. Expected ultimate recoveries (EURs) are also easily determined by calculating the cumulative production from the type curves for 20–30 years. However, the "average or type curves" that have been developed are not truly representative of the physics of shale gas flow. Actual performance has been found to be quite dissimilar from these "average or type curves." Since operators do not understand the exact reasons for the deviation, they have been limited in their ability to optimize the development and properly prioritize operations based on sound engineering and geological information. A more reliable analysis and predictive approach is needed.

17.4.4 Build Reservoir Models for Simulation—Recommended Practices

According to Vassilellis et al. (2010), conventional reservoir engineering tools have been found to be inadequate for use with the change in reservoir characteristics after hydraulically fracturing a "shale" well. This complex newly altered reservoir (after fracturing) must be described and properly modeled in order to reliably predict long-term production and recovery. Vassilellis and his coauthors introduced a multidisciplinary integrated approach called "shale engineering." Shale engineering involves building three models—reservoir, well, and fracturing models—and tuning the models for reliable long-term prediction and recovery. Data and analysis techniques involve the disciplines of geology, petrophysics, geomechanics, geochemistry, seismology, and, of course, reservoir engineering. Application of the shale engineering techniques has been documented by Vassilellis et al. (2011) and Moos et al. (2011). The workflow developed for this process is shown in Figure 17.3. Cipolla et al. (2009a, b) also have introduced a new approach to more comprehensive modeling of complex shale. Holditch (2006) concludes that the most accurate reservoir analysis technique for tight gas (and shale gas reservoirs) is to build a reservoir model that includes layers, and it is also suggested that a dual porosity model be used.

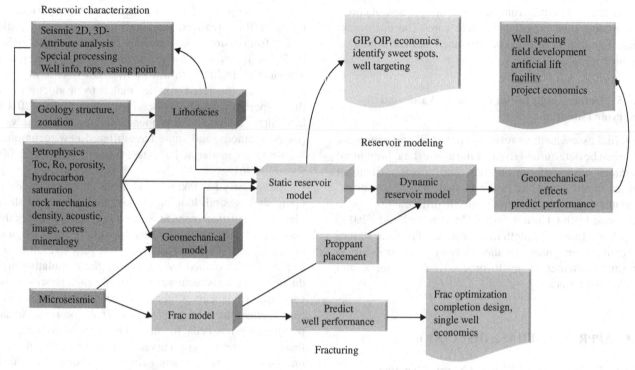

FIGURE 17.3 Shale engineering life cycle workflow (Source: Baker Hughes).

17.4.5 Generate a Field Development Plan—Current Practice

Most US operators have utilized the classic approach for field development plans (FDP), having been successful in generating FDPs in the past for conventional reservoirs. However, shale gas has introduced uncertainty in the traditional approach. The US operators have certainly realized that there are other considerations for development with horizontal wells in shale, especially including lease line considerations.

17.4.6 Generate a Field Development Plan—Recommended Practices

Field development plans should include well type, placement, attitude, direction (azimuth), and spacing (drainage area considerations). For shales, wells should be drilled in the direction of minimum principal stress, which maximizes access to existing natural fractures when transverse-trending hydraulic fractures intersect the natural fractures near the wellbore. Therefore, it is important to understand the stress regime in the field. Most development wells in the United States are horizontals, this is partly due to the "relatively thin" shale formations ranging from 20 to 600 ft thick. Horizontal wells also maximize reservoir contact resulting in a cost advantage over drilling a larger number of vertical wells. The literature says that one horizontal well can replace four vertical wells (number can be higher depending on the particular shale play). Usually, a full plan also includes completion and fracturing designs. It has been determined that a large number of wells are required to develop either a shale gas or shale oil play. Typical shale gas well spacing in the United States is approximately 116 acres (Kuuskraa et al. (EIA), 2011); however, the continuous shale formations extend over large geographical areas. Figure 17.4 shows the number of existing wells in the six major US shale gas plays and the total number of wells required to develop the technically recoverable resources (TRR) from EUR per well (Kuuskraa et al. (EIA), 2011) for each play using the typical number of 200–300 wells required to recover 1 tcf of gas. Most plays have not yet even approached the required number of wells.

17.4.7 Validate Economics of the Play or Pilot Project

Now, with all of the data collected from the drilling and analysis of the appraisal wells, and with an understanding of the unique aspects of these unconventional reservoirs and characterization data for the particular play, operators can complete the final step of the appraisal phase—validating the economics of the play. Operators can then evaluate prior to taking the decision whether or not to proceed with play development. This has been the typical sequence of events in the United States.

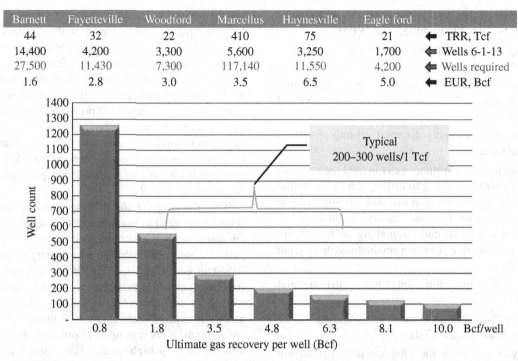

FIGURE 17.4 Number of wells required to develop 1 tcf of gas (Updated from Kennedy et al., 2012b).

Internationally, it has been observed that NOC operators plan to take the approach of conducting a pilot program prior to full field development. In these virgin areas with very few wells or any offset wells to observe, this approach certainly makes economic sense. Pilot programs are conducted to prove commerciality, and/or to define best technologies and development plan. After conducting the pilot program, international operators will stop to evaluate prior to taking the decision to proceed with full field development.

17.5 DEVELOPMENT PHASE DISCUSSION

17.5.1 Implement the Field Development Plan

In implementing a field development plan that includes a large number of wells, an operator is cautioned not to become complacent and continue to allow the rig schedule to totally drive the plan. Interim analyses should be undertaken to ensure that drilling and completion programs are delivering wells with the expected initial potential (IP)-producing rates.

17.5.2 Surface Facilities

Surface facilities will not be discussed here. They are only included in the life cycle for completeness and to ensure construction schedules match the timing of well completion and availability.

17.5.3 Design Wells and Optimize Drilling Costs—Current Practice

The early shale gas wells in the Barnett (discovery in 1981) were mostly verticals. It was not until 2003 that there was a total shift to horizontal wells for developing shale in the Barnett as well as all other US shale basins that followed. Now that the template has been set for shale gas drilling in the United States, service companies have been successfully reducing drilling costs through optimized drilling and new technology. To date there have been approximately 150,000 shale wells drilled in the United States. (It should be noted that some 10,000 of these wells were drilled in the old biogenic Antrim Basin, Michigan, and do not behave like the major shale thermogenic basins where today's activity is centered.)

Listed in the following are some of the current typical practices used for drilling horizontal wells:

- Curve Section—Build rate is typically 6–8/100 ft.
- Optimum (and shorter) lateral lengths are now preferred over "longer lateral lengths." Length of horizontal sections varies from play to play, and ranges from 3000 to 6000 ft for shale gas.
- Most operators convert to some form of oil-based mud (OBM) prior to drilling the curve and lateral.
- Only a few wells in the United States are being drilled with environmentally friendly water-based mud (WBM).

It is understood that most international horizontal shale wells will be probably be drilled with WBM.
- Mud weights depend on formation, which ranges from normal to overpressured.
- In a cost-sensitive environment, motors (positive displacement motors, PDM) are used to drill the curve and the lateral sections.
- Most operators use poly-diamond crystalline (PDC) bits in the shale.
- Early preference for drilling wells in the "toe up" attitude is gradually changing to drilling the lateral as flat as possible and perfectly horizontal (toe up attitude selection based on fact that gravity would facilitate drainage of any fluids collecting along the lateral into the mother hole, thus allowing production/lifting from the well).
- Wells are drilled in the direction of minimum horizontal stress.
- Typical drilling time in Barnett and Marcellus = 12 days, and Eagle Ford = 17 days.
- Typically, wells using the plug-and-perf method are completed with a 4½" or 5½" production string.
- US operators are quickly adopting the practice of pad drilling (4–10 wells per pad), and currently over 60% of US shale wells are being drilled from pads. Savings are achieved from logistics and the operation protects the environment.
- The "drilling cost" constitutes 40–50% of the total well cost (hydraulic fracture treatment cost is the major part of a completed well cost).

Due to the unique nature of shale, every basin, play, well, and pay zone may require a different approach. Complex reservoir characteristics pose different drilling challenges related to well placement, wellbore stability, higher torque and drag, inconsistent buildup rates, geological uncertainties, lost circulation zones, and other issues.

17.5.4 Design Wells and Optimize Drilling Costs— Recommended Practice

Operators have already seen the benefits of horizontal wells in US shale plays. With the available modern horizontal drilling technology, it is possible to efficiently drill and complete these wells. This author supports following a strict regimen of careful planning and execution that is included in drilling optimization process. Drilling optimization involves a proven and highly disciplined process that includes the following steps: (i) pre-project analysis, (ii) planning, (iii) implementation, (iv) detailed post well analysis, and (v) Knowledge capture. The effective application of this process has delivered substantial drilling improvements in many drilling environments worldwide. Drilling optimization is based on the continuous improvement cycle. Each well has slightly different challenges, but successive well captures additional drilling efficiency. Drilling shale wells presents technical challenges that impact the overarching goal of optimizing drilling costs and reducing days on wells. The quality of pre-well planning impacts both drilling spend and ultimate recovery. Therefore, an integrated approach is required to ensure that all the available reservoir data, geoscience, and geomechanical knowledge and downhole conditions are well understood prior to planning the details of shale wells.

The planning for the drilling pathway must take into account the need to maximize the contact of the well with the reservoir. It is recommended that the curve and lateral part of shale wells be drilled using rotary steerable systems (RSS) in order to reduce drilling risk, promote wellbore stability, increase rig efficiency, increase the contact of the well with the reservoir, and deliver holes of superior quality at a price comparable to motors. These new tools are becoming increasingly popular in shale plays. The importance of a high-quality hole cannot be emphasized too much. Poor quality boreholes will, in many cases, lead to a poorly completed well or one where significant amount of repair must be undertaken prior to well completion. An RSS is typically more efficient than using motors because of the continuous near-bit information and automatic steering capability. Today with the advancement of modern drilling technology, it is possible to drill the build/curve section and the lateral section in "one single run" using high build rate RSS.

Some of the other drilling technological advances include Janwadkar et al. (2006), bottom assembly hole (BHA) and drilling string modeling to optimize Barnett drilling performance; Janwadkar et al. (2007), advanced (LWD) and directional drilling technologies to overcome completion challenges in Barnett horizontals; Janwadkar et al. (2009), innovative rotary steerable system to overcome challenges of the Woodford complex well profiles; Isbell et al. (2010), new use of PDC bits to improve performance in shale plays; and Janwadkar et al. (2010), using electromagnetic MWD to improve Fayetteville drilling performance.

Once the target zone has been identified, it is important to place the well bore in the optimal part of the reservoir following the natural formation dip and moving to follow faults. With today's modern MWD (measurement while drilling) and LWD (logging while drilling) tools it is possible to use reservoir navigation services (Fig. 17.5) where engineers use the shale analysis with available seismic and offset well data to plan the trajectory of the well bore. During drilling, the LWD data are monitored by experts located in remote drilling centers and modifications are made to the trajectory to maintain the well bore in the optimal zone and in the sweet spots.

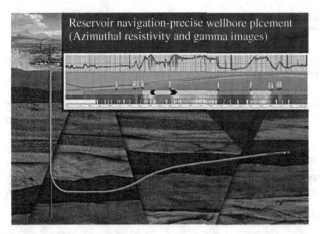

FIGURE 17.5 Reservoir navigation to stay in the zone and the sweet spot (Source: Baker Hughes).

Drilling fluid system design and selection should involve a fluid that achieves (i) proper hole cleaning, (ii) stabilization of clays in the formation, (iii) overall wellbore stability, (iv) reduction of drag and stuck pipe, and (v) improvement in rate of penetration (ROP). Drilling fluids are often considered as commodity items, because operators are not always aware of the positive effects of utilizing a high-performance drilling fluid. Because of the high environmental sensitivity of many shale wells, WBMs have been designed to provide superior performance while at the same time lowering environmental impact and reducing disposal costs.

Since so many wells have been drilled in the same general area of a shale play, there are numerous opportunities during post well analysis (of the drilling optimization process) for constructive follow-up to assess successes and failures of the various components of the drilling system. Of particular interest will be drilling runs where one just one parameter has been altered to assess its effect on the ROP as well as the quality, stability, and accuracy of placement of the well.

Knowledge capture during the drilling optimization process involves documenting lessons learned from the field, so that subsequent drilling runs will benefit from previous ones. This continuous improvement becomes a cycle as knowledge of products and processes is recycled so that time and cost-saving measures are adopted as standard practices and costly mistakes are avoided.

17.5.5 Refine and Optimize Hydraulic Fracturing and Wellbore Completion Design—Current Practices (Characterize the Lateral)

The last step of the development phase of the life cycle involves optimizing the hydraulic fracturing and wellbore completion designs and specifically includes characterizing the horizontal lateral.

17.5.6 Current Hydraulic Fracturing Practices

Because permeability is so low in shale reservoirs, little to no production from a well can occur without first breaking up the rock by some kind of fracturing process. Fracturing the rock and propping open the induced fractures creates high permeability pathways, which allows the reservoir formation to produce at much higher flow rates than it could naturally. This process dramatically increases hydrocarbon production and makes the well profitable, where it would not have been without fracturing. The importance of the "customized/optimized" hydraulic fracture design for each individual well is paramount to success, although the "cookie-cutter" approach is still quite common. This design approach/methodology may be partly justified by the lack of the information required for designing the frac and/or by the economics. The fact that reservoir properties vary significantly, both vertically and aerially along the lateral, creates additional challenges in designing the "optimum" treatment. It is difficult to model the fracture complexity in these naturally fractured shale formations even with the use of the modern fracture simulators. The variations and the uncertainty in the geomechanical and petrophysical properties complicate this challenging task even further. Therefore, some operators choose for the completion and stimulation design, which was previously proven to be successful in an offset well, in a different part of the field or even in a different formation.

Due to the great variability of rock properties and significant variations in stimulation treatments of different shale plays, it is "unfair" to summarize the trends, common practices, and rules of thumb. However, Table 17.1 has been included. The table shows typical fracturing treatments of some of the major shale plays. A completion strategy and a hydraulic fracture design should be localized and limited to a certain area within a particular shale play. For example, the variations for the stage spacing and number of clusters are quite significant: from 150 to 750 ft spacing (250 to 350 ft is used most frequently) between stages, and from three to eight clusters per stage, in the cemented (cased hole) plug-and-perf type of completion. The same applies to open-hole completions with the packers and frac sleeves. Spacing can differ based on the experience in a certain area. Unfortunately, the industry has tended to use the arbitrary geometric method of placing frac stages along the horizontal laterals of shale wells. It is not the intent here to discuss hydraulic fracturing in gas shale, as this has been covered in detail other chapter and can be found in two other very comprehensive papers by King (2010, 2012). Starting with "No two shale alike," King also states "There are no optimum, one-size-fits-all completion or stimulation designs for shale wells."

Quite often, poor production from some shale wells is blamed on the completion and the fracture treatment. Several technical publications have presented data demonstrating

TABLE 17.1 Typical fracturing treatments of some of the major shale plays

	Bakken	Barnett	Eagle Ford	Haynesville	Marcellus
Average MD, ft	17,535	10,873	14,643	16,566	10,722
Average TVD, ft	10,207	7,331	9,392	11,941	6,937
Horiz Perfed, ft	7,401	2,788	4,311	4,355	3,331
Average ft /stage	550	450	270	325	275
Average BHP, psi	5,310	4,213	7,550	10,870	7,650
Average rate BPM	24.8	73.3	81.6	71.2	83.5
Average number of stage	13	6	16	13.3	12
Average number of stages/day	3.4	1.9	2.6	1.9	1.5
Amount of proppant per stage, lbm	160,300	286,000	292,600	357,800	399,500
Amount of proppant per well, lbm	1,998,000	1,515,000	4,304,000	4,675,500	4,425,600

BHP, borehole pressure; BPM, barrel per minute; lbm, pound-mass; MD, measured depth; TVD, true vertical depth.

that production logging tools (PLT) have verified that 30–40% of the perforations of a typical well are not contributing to any production at all. In a very few cases similar numbers have been confirmed with distributed temperature sensing (DTS). The question begs, "What is the cause?" In many cases the best fracturing job possible has been designed, or the design of a successful offset well used runs the best mechanical wellbore completion design; and yet many wells experience production rates that are below those that were predicted and certainly not optimum. There are two possible reasons: (i) frac placement did not intersect the natural fractures in the well, and/or (ii) reservoir quality (i.e., TOC levels, thermal maturity—formation either immature or overcooked, or not true organic source rocks) was low or nonexistent at the locations where fracture stages were placed. Also, mineralogy and stress regime may not have been ideal to initiate and propagate the induced fractures. Reservoir quality is probably a significant part of the reason, which brings us to, "Was the well placed in the specific area of the 'sweet spots' in the play?"

The first fracs in the Barnett were gelled fracs, until the successful slickwater fracs became the default design not only in the Barnett, but also other US shale plays. Slickwater fracs are typically composed of 94% water (no polymer gelling agents) and 6% sand proppant and chemicals (friction reducers, surfactants, biocides, and clay stabilizers). These fracs are pumped at very high rates. Slickwater fracs are less expensive than polymer gel fracs. Most shale wells today use what are called geometric fracs, that is, a frac stage every 250–350 ft with four to eight perforation clusters per stage. This approach totally ignores the changing reservoir characteristics along the 3000–6000 ft long laterals of shale gas wells. Geometric fracs are used, because those changing characteristics along the lateral are not known, quantitatively at least. Only about 9% of the US shale wells logs or any characterization is being done for the laterals, which could provide information as to where to place stages (and where to avoid, i.e., faults and geohazards) and perf clusters.

Only some operators are now opting to monitor fracturing treatments, shale and tight gas (Warpinski et al., 2010), in real time using microseismic. Monitoring does require a nearby offset well in which to run sondes for recording the data. Microseismic monitors the treatment as to the direction (azimuth) and height, and whether the treatment is going out of zone, into a water zone, or being lost to a fault. This provides the operator with the ability to stop or alter the treatment if not going as planned.

17.5.7 Hydraulic Fracturing—Recommended Practices

The more the fracturing process can penetrate the rock, the more successful the frac will be in allowing hydrocarbons to flow from the reservoir and into the wellbore. A successful fracture stimulation is the one that increases productivity index as well as ultimate recovery of a producing well in an economic, safe, and environmentally friendly manner. Technology to fracture these reservoirs must be efficient in order to bring acceptable economic returns to the owners of these assets. This need for efficient fracture technology and processes has driven innovation in the oil service industry to understand shale reservoirs and develop tools and techniques to exploit them more efficiently. Overcoming all of the hydraulic fracturing challenges of (i) achieving designed fracture geometry, (ii) transporting proppant to the right location, (iii) achieving final conductivity, (iv) encountering expected reservoir properties, and (v) avoiding geohazards can certainly lead to the "optimum fracturing treatment."

To design an optimum fracture treatment, a number of factors that relate to the variability of the well must be considered. Every shale play is different and, this makes a certain amount of customization of the shale play inevitable. The shale frac design process cannot be one size fits all. Engineers and geoscientists must strive to understand the idiosyncrasies of each shale play and be willing to utilize a number of technologies and process designs in order to most effectively exploit an operator's assets. The design of the

TABLE 17.2 Suggested fracture treatment types for dry gas, wet gas, and oil

Fracfluid type	Formation	Pump rate	Conductivity	Play application
Slickwater/linear gel	Dry gas or low liquid	High rates, 100+bpm	Infinite to gas	Barnett, Marcellus, Fayetteville
Hybrid frac	Gas condensate	Low, 60–80 Bpm	More conductive Frac	Eagle Ford, Utica
Crosslinked frac	Oil bearing	Low, 40–60 Bpm	Highly conductive frac	Bakken, Niobrara, Eagle Ford

frac begins with collecting a number of key parameters (especially the stress profile along the lateral) about the reservoir and probable frac plan, which provide the input to a fracturing model (Simulator). A geomechanical model, reservoir model, and fracture stimulation model, along with multiple diagnostics such as surface tilt meters and downhole microseismic monitoring, can provide the basis for a sound engineering solution. This solution should result in optimized spacing between the stages, optimized number of clusters required, and the overall optimum treatment design. Fracture modeling, design, and posttreatment diagnostics (history matching) are also the important factors in the optimization of the stimulation treatment.

The understanding of the local stress regime will determine the optimum horizontal wellbore azimuth, which will facilitate the control of the fracture orientation relative to the wellbore axis. The relative deviation between the well and stress field may also cause tortuosity (as the fracture propagates), resulting in higher initiation and treating pressures, and lower near wellbore conductivity. The knowledge of the in situ stress is critical for the successful fracture placement. The complexity of the fracture network may be achieved by multistage fracturing with low viscosity fluid, provided that the reservoir properties "allow" the fracture complexity to develop.

High-rate slickwater fracturing creates tensile fractures as well as shears the existing fractures in brittle shale formations with low horizontal stress anisotropy. Slickwater fracturing has become the norm in the Barnett and Marcellus shale plays. Fracture complexity of the hydraulic stimulation is highly desired. The Barnett Shale is one of the best examples of a successful application of slickwater fracturing. No two shales are alike (King, 2010), and there is no other shale exactly like the Barnet that is, with the identical rock properties. Other shale reservoirs require a unique completion and stimulation strategy. Although slickwater fracturing has proved itself in a number of US shale plays, there are many cases where slickwater fracturing has not provided sufficient propped flow capacity to develop a gas or oil productive shale. Unfortunately, slickwater fluid is an inherently poor proppant carrier, necessitating high pump rates to achieve flow velocities sufficient to overcome the tendency of the proppants to settle. Advanced fluid technologies combine the best attributes of slickwater and conventional cross-linked fluids systems to maximize proppant transport through the surface equipment and long laterals, before breaking to create a desirably complex fracture network (Brannon and Bell, 2011). To enhance the proppant placement in a low viscosity fluid (like slickwater fracs), lightweight and ultra-lightweight (ULW) proppants can be considered to provide more improved effective fracture length than can be achieved with conventional proppants. In cases where the natural fracture network is deemed of secondary importance to productivity, drilling the well in the direction of minimum principal stress is preferred in order to favor the creation of longitudinally trending hydraulic fractures. Longitudinal fractures reduce radial convergence by maximizing exposure of the wellbore to the hydraulically created fracture and usually eliminate the need for high-conductivity proppants (Cramer, 2008). The Niobrara shale play is an example of this, that is, low numbers and presence of natural fractures.

Selection of the fracturing fluids is very important for the successful hydraulic fracture stimulation treatment. There is a wide variety of fracturing fluids, and the optimum one must be chosen depending on the shale reservoir type, that is, dry gas, wet gas, or oil. Table 17.2 compares suggested fracture treatment types for dry gas, wet gas, and oil. The industry continues to optimize fracturing fluids, improving the performance of the fluids and addressing environmental concerns.

There are several methods of fracture treatment monitoring, including but not limited to tilt meters, microseismic monitoring, DTS, radioactive and nonradioactive tracers, pressure monitoring, and production logging. Currently, the most frequently used/effective monitoring method is microseismic monitoring. With microseismic, the fracturing operation can be monitored near real time by stage and changes made to existing or subsequent stages. Microseismic can monitor the treatment as to the direction (azimuth) and height, and provides the operator with the ability to stop the treatment if not going as planned (i.e., whether the treatment is going out of zone, into a water zone, or being lost to a fault, all in near real time). It can validate the stress profile and fracture geometry, providing an estimate of the stimulated reservoir volume (SRV). Figure 17.6 shows an analysis of where the fracturing fluids have interacted with the rock enough to create very small seismic disturbances (displayed as a 3D microseismic cloud). As shown, seismicity can be pinpointed very precisely with the right array of sensors along with good interpretation expertise and the appropriate representational software. Microseismic does not indicate where the proppant or fluid actually goes, but where the rock has slipped or cracked. Some US operators

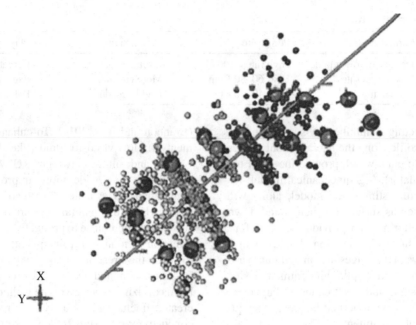

FIGURE 17.6 Microseismic cloud for multizone fractured well.

are running microseismic with every frac job (at least those where a nearby observation well is available).

Multistage fracturing requires significant amount of water. Water availability has become a worldwide concern. To address the water shortage issue, the industry has identified the alternatives to freshwater. These include recycled water (treated flowback and produced water), formation water (water source wells), and/or seawater. Many companies are reevaluating these resources as a cheaper and better base fluid for frac fluid. In areas with developed oil and gas infrastructure, the cost of processing and using this fluid for fracs is both economically and technically attractive to scarce freshwater (King, 2012).

Stage size and cluster spacing are important design variables to create contributing transverse fractures along the horizontal wellbore, but often fluid loss and mechanical interaction of the created fractures are overlooked. There are two factors to consider when deciding on stage size and cluster spacing: (i) the physics of creating multiple transverse fractures and (ii) the longer-term production interference (interaction) of those fractures. Bazan et al. (2012) provide some information on this. Several other publications discuss the above topic (Bhattacharya and Nikolaou, 2011; Cheng, 2010; Jo, 2012; Meyer and Bazan, 2011; Song et al., 2011).

17.5.8 Characterize the Lateral

It is strongly recommended that the lateral should be characterized to (i) gain critical information about location of natural and conductive fractures and identify faults, and (ii) collect reservoir information, that is, mineralogy for brittleness, TOC for level of potential hydrocarbon, porosity for determining GIP, and stress profile for frac initiation and propagation. This information allows the frac/completion engineer to effectively plan the completion design. An informed decision can be made that uses science to determine the optimum placement and spacing of hydraulic fractures and perforation clusters (in plug-and-perf completions), and ultimately to maximize production and recovery. For collecting reservoir data in horizontal wells, there are some tools available for use in a logging while drilling (LWD) mode; currently, however, LWD or its wireline equivalents (which must be deployed with tractors or coiled tubing) are infrequently used or not always available. Probably few operators have opted for this costly and possibly problematic logging method. However, high definition resistivity imaging LWD tools that can provide information on natural fractures along the lateral are available. Although costs are relatively inexpensive, and the process is transparent to well drilling, only a limited number of operators are running these logs. Imaging tools can identify location/prevalence of natural fractures, locate faults, bedding planes, and even induced fractures from nearby offset wells (see Fig. 17.7). LWD resistivity imaging is a significant part of characterizing the horizontal lateral as opposed to using the arbitrary geometric fracture stage placement. It has been shown that imaging provides important information about fractures and hazards along the lateral, resulting in increased production rates where stages have been moved or altered. This has been documented in a case history described in Kennedy et al. (2012b). There are wireline tools available to measure reservoir data in only vertical wells; however, similar data can be collected along the horizontal lateral by analyzing drill cuttings for TOC, mineralogy, porosity, and some rock

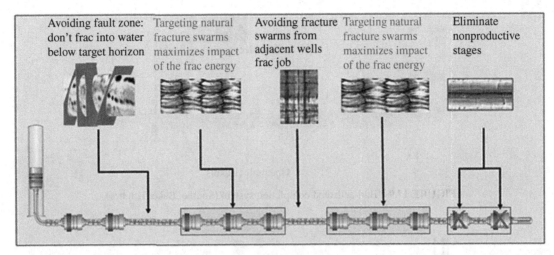

FIGURE 17.7 LWD resistivity imaging tool provides valuable information on frac stage placement (Source: Baker Hughes).

mechanical properties and by employing advanced mud logging (to include gas chromatography). Utilizing the two tool/processes can provide valuable information necessary for the optimum placement of frac stages and perforation clusters (in plug-and-perf) for well completion.

It should be noted that resistivity-imaging tools require drilling the well with a conductive fluid (WBM) in order to obtain reliable high-quality image data. It has been observed for most shale wells in the United States, the drilling fluid is converted to some form of OBM prior to drilling the curve and lateral. With the advancement of WBM fluid technology and its environmental friendly characteristics, in the United States, it has been observed that operators are beginning to accept use of WBM to drill the lateral section. WBM has proven to be more cost effective than using OBM.

17.5.9 Current Wellbore Completion Practice

The three types of completions that have proven to be the most effective and efficient in the North American shale plays are plug-and-perf, ball-activated systems, and coiled tubing–activated systems.

Plug-and-perf (Fig. 17.8) uses perforations to divert the frac fluid, composite bridge plugs to isolate the fracture through the tubing, and cement to isolate the annulus of the open hole and liner string. Ball-activated systems (Fig. 17.9) use frac sleeves containing ball seats that correspond to different size frac balls. When the frac balls are pumped onto the seats, pressure opens the sleeve to divert the frac, and the ball also provides through-tubing isolation from the previously fractured stage. Annular isolation is achieved by using either hydraulic-set open-hole packers or swell element open-hole packers. The third completion type system (Fig. 17.10) use frac sleeves that are opened by means of coiled tubing. Through-tubing isolation is achieved with a coiled tubing packer and annular isolation is accomplished

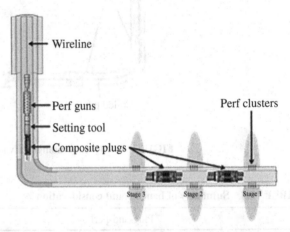

FIGURE 17.8 Plug-and-perf completion system (Source: Baker Hughes).

with cement. The recent technology advances for these types of completions have resulted in a variety of hybrid systems as well. These hybrid completions can use either cement or open-hole packers for annular isolation (Burton, 2013).

17.5.10 Wellbore Completion—Recommended Practices

Each of the three types of completion techniques has its own unique benefits, and there is no "silver bullet" when it comes to unconventional completions. There are a number of benefits and considerations for each technique (see Table 17.3; Burton, 2013). Historically, the majority of shale gas wells have been completed using plug-and-perf, and for liquids plays the choice of operators is the ball-activated completion type. Burton (2013) states that the selection of the appropriate completion type is application specific, depending on the phase of development of the particular shale play, and includes answering the questions

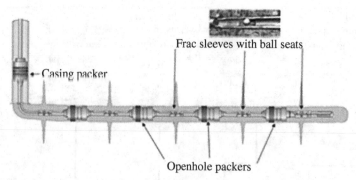

FIGURE 17.9 Ball-activated completion system (Source: Baker Hughes).

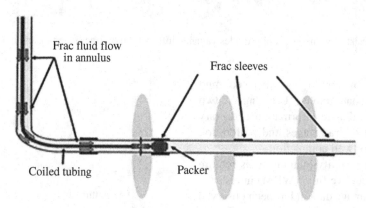

FIGURE 17.10 Coil tubing–activated completion system (Source: Baker Hughes).

TABLE 17.3 Summary of benefits and consideration

	Plug-and-perf	Ball-activated	Coil tubing activated
Number of stages	Virtually unlimited	Limited	Virtually unlimited
Stage placement	Flexible	Fixed	Fixed with frac sleeves flexible with SJP
Contingency options	Full diameter	Diameter restrictions	Full diameter and CT in hole
Fracturing logistics	Pressure pumping, wireline, CT	Pressure pumping	Pressure pumping, CT
Fracturing ops efficiency	Rig up/down between stages	Nonstop	Brief frac shut down between stages
Post fracture	Mill out plugs	Restricted diameter	Full production diameter

concerning operational efficiency, cost, logistics, flexibility, and application.

Plug-and-perf has a distinct advantage during the appraisal phase. The stage placement has not been locked into place until the perforation gun has been fired into the liner string. If it is determined that a stage needs to be moved to another location along the lateral, it is simply a matter of moving the perforations to that depth. This allows time for data gathering and changing the stage placement on the fly, if required. There is, however, a lack of efficiency during the fracturing process, because of rigging up and rigging down of wireline and pressure pumping equipment between each stage. When operators move into the development phase, it is the time to focus on more efficient completion techniques, depending on the economics and availability of services. Simultaneous operations can be utilized in some applications to optimize plug-and-perf completions. If two wells are spaced close enough together, wireline can be operating on one well while pressure pumping is operating on the other. When both operations are complete, the services switch wells and minimize the nonproductive time for both wireline and pressure pumping (a simultaneous frac operation).

Ball-activated systems are primarily used for completion optimization, because of the efficiency and logistical benefits during the fracturing process. These systems are simpler from a logistical point of view, because pressure pumping is the only service that is required on location. Also, the frac job does not have to be shut down between stages, because the frac balls can be dropped while the fluid is being pumped. The total time of the fracturing process is dependent on the frac plan, but in some instances 40 stages have been fractured in only 24 h with a ball-activated system. The ball seats, however, do create an ID restriction. In most scenarios (as shale wells tend to be low flow rate) this is not an issue unless a screenout occurs. Therefore, this technique may not be the best choice during the appraisal phase. Once operators reach the development phase and are more familiar with the formation/reservoir, screenouts are less likely to happen. Ball-activated systems can help drive efficiency during the fracturing process, and allows for wells to come on production faster than with plug-and-perf (Burton, 2013).

Coil tubing activated systems offer accuracy and contingency options during the frac process. The frac maintains efficiency, requiring only brief shutdowns between stages to move the coiled tubing assembly to the next stage. Shutting down the fracturing allows for the most accurate fluid placement into each stage, which is significant in formations where over displacement of fluids is an issue. Also, the coiled tubing is already in the wellbore. If a screenout occurs, this would keep nonproductive time to a minimum. This system can actually allow a more aggressive frac design. The frac is performed through the annulus of the coiled tubing and the liner string; therefore, the fracturing ID is significantly reduced. This completion system is dependent on coiled tubing; therefore, depth and availability of coiled tubing at the same time as a pressure pumping crew can result in limitations on this type of completion. Because the coil tubing–activated system uses frac sleeves, the placement of the stage cannot be adjusted on the fly. If there were ample data to determine the placement of the stage before the well is completed, this system would be a viable option during the appraisal phase. Primarily, because the system offers the flexibility, should a screenout occur. If additional data gathering that could impact the placement of the stages is performed after the completion is set, this may be a better option for completion optimization during the development phase (Burton, 2013).

17.5.11 Drilling Considerations for Completion Methods

- The quality and gauge of the wellbore affects which type of completion can reach the intended depth and whether or not cement or open-hole packers are required to achieve isolation for all stages of the well.
- If deviation exceeds 15°/100 ft, it is recommended to run a torque and drag simulation to determine if the completion will physically go in the wellbore.
- A reamer run is recommended if there is a small clearance between the wellbore ID and the completion tool OD.

17.5.12 Fracturing Considerations for Completion Method

- The cost of services required for the completion must be considered for the operator's economic concerns.
- Availability of services and supplies must be considered.
- If there is a shortage of frac fluids or proppants, how much of a benefit is nonstop fracturing?
- The size and type of completion directly affects the fracturing operation parameters including fracture ID, pump rates, slurry concentrations, pressure ratings, and a number of other factors.

17.6 PRODUCTION PHASE DISCUSSION

The production phase includes (i) monitoring/optimizing production; (ii) managing the complete water cycle; (iii) reducing corrosion, scaling, and bacterial contamination in wells and facilities; (iv) installing artificial lift if required; and (v) protecting the environment.

17.6.1 Monitor and Optimize Producing Rates—Current Practice

Managing and controlling well flowback rates are the first steps in optimizing production and ultimate recovery. Many operators open wells on full choke in order to obtain maximum production rates immediately. This may possibly be an acceptable practice for shale oil wells, but not for shale gas wells.

17.6.2 Monitor and Optimize Producing Rates—Recommended Practices

Multistage hydraulically fractured wells require a poststimulation flow period to prepare the well for long-term production. This is one of the most critical times in life of well, more so for shale gas wells as opposed to tight gas wells and shale oil wells. Excessive flowback rates are known to have caused proppant flowback or fracture collapse. Intensive management of flowback can yield significant improvement in well's long-term performance (Crafton, 2010; Crafton and Gunderson, 2007). An operator in the Haynesville reported in 2010 that "Haynesville wells … have been produced using restricted rate production practices. Additionally, initial decline rates appear more gradual as a result of restricting production." This operator also stated that the decline curves modeled higher EURs from the restricted well rates. It appears that this is one instance of a technique that could slow down the dramatic initial decline rates characteristic of shale gas wells.

Total production from a multistage hydraulically fractured well can be monitored, but there has not been a truly effective method of determining production rates coming from individual perforated stages. Some operators have run production-logging tools (PLT) in horizontal wells to monitor production (Heddleston, 2009). However, the question begs as to whether any significant remedial work might be attempted (or be economically justified) aside from a cleanout, chemical treatment, or possible refrac. One other method of monitoring production from individual stages has been employed, that is, DTS. A few wells have been equipped with DTS, but the fiber-optic cable and equipment must be installed as part of the original completion and the cost is difficult to justify based on as yet unproven benefit.

17.6.3 Manage the Water Cycle—Recommended Practices

Most shale gas wells do not produce any significant amounts of water, and the water produced by tight gas wells is handled with deliquification techniques—plunger lift, foam sticks/continuous foam injection, gas lift, beam lift pumps, and jet pumps. Water that is of concern with shale wells is frac flowback water. Although not all of the frac water comes back (typically about 30%), the amount that does brings with it formation salts, scale, and sometimes low-level radiation (NORM). Frac flowback water must be treated, whether it is to be reused or disposed. With the water situation in many parts of the country and the world, reuse is strongly recommended. Service companies provide water-treating services for flowback and produced water. Currently, the most popular and effective equipment uses electrocoagulation technology to remove suspended solids and heavy metals from flowback and produced water. Freshwater is not required for fracturing wells. Formation brine water and seawater are other alternatives. Additional frac additive chemicals are required due to salt content of these waters. There is more to managing the water cycle than treating and or disposal. Water sourcing for both drilling and fracturing has become significant. In the Eagle Ford, water is being sourced from shallow salt water formations and being lifted from wells using large-volume electric submersible pumps (ESPs). On the surface, produced and treated water must be handled and transported to central processing/treating facilities or removed for disposal. This requires piping and surface transfer pumps.

17.6.4 Preventing Corrosion, Scaling, and Bacterial Contamination in Wells and Facilities

Corrosion, scaling, and bacterial contamination in wells and facilities is handled much the same way as in traditional oil and gas fields. Components of a production chemical program include the chemical treatment program, monitoring of program effectiveness, and design and installation of necessary equipment. Chemical automation systems can be used, especially for remote locations, widespread operations, and in low-winter temperature operations.

17.6.5 Protecting the Environment

Protection of the environment should be included in every phase of the life cycle; however, it is particularly critical during the production phase when hydrocarbons and water are brought to the surface.

17.7 REJUVENATION PHASE DISCUSSION

It is the opinion of this author that the most significant opportunity to accomplish rejuvenation lies with refracs. As has been discussed, unconventional wells decline rapidly reaching low unacceptable rates after only a few years of production. It has not been proven that any form of production management or enhancement has been successful in arresting the rapid decline or restoring original production rates. In his database of 100 published studies on refracs, Vincent (2010) attributed re-frac success to a number of mechanisms as listed in the following:

- Enlarged fracture geometry, enhancing reservoir contact
- More thorough lateral coverage in horizontal wells or initiation of more transverse fractures
- Increased fracture conductivity compared to initial frac
- Restoration of fracture conductivity loss due to embedment, cyclic stress, proppant degradation, gel damage, scale, asphaltene precipitation, fines plugging, and so on
- Increased conductivity in previously unpropped or inadequately propped portions of fracture
- Improved production profile in well, preferentially stimulating lower permeability intervals (reservoir management)
- Use of more suitable fracture fluid
- Reenergize or reinflate natural fractures
- Reorientation due to stress field alterations, leading to contact of "new" rock

Production rates from refracs have matched, or sometimes exceeded those from the original frac. Examples in the literature show increased production rates, especially in the Barnett, where wells were restimulated with slickwater fracs versus the original gel fracs.

Redevelopment of a shale or tight gas field will more than likely involve infill drilling to accomplish downspacing. This has already been seen in the Pinedale and Piceance tight gas plays, where original well spacing of 160 acres has now gone down to 5–10 acres. Redevelopment should be considered when the primary development in the sweet spots of the field has been or are nearing depletion.

17.8 CONCLUSIONS—RECOMMENDED PRACTICES

A listing of the pertinent recommended practices (included in this chapter's discussion) to overcome an operator's challenges in each phase of the asset life cycle of a shale gas well are shown in the following:

Exploration Phase

1. Conduct basin/area screening study to identify core area/sweet spots and for initial estimate of G/OIP. Recommended for new basins or new areas within a producing basin.
2. Conduct 3D seismic in new basins/areas to (i) Identify areas of highest TOC using seismic impedance, (ii) Increase understanding of natural fractures using seismic attributes, (iii) provide azimuthal anisotropy data related to natural fracture orientation and/or horizontal stress anisotropy, and (iv) assist in the identification of sweet spots using seismic cross-plots.
3. Geophysical evaluation optimal workflow includes
   a. Seismic rock properties
   b. Azimuthal analysis
   c. Seismic analysis
   d. Stress estimate
   e. Multiattribute/integration
4. Begin to characterize reservoir by running specific logs and taking/analyzing cores in exploration and appraisal vertical wells
5. Recommended logs include: conventional suite plus—spectroscopy, pulsed neutron, and NMR for determining mineralogy, porosity, O/GIP, optimum frac and lateral placement locations, and log-derived mechanical properties to assist in frac design.
6. Obtain cores conventional whole cores or sidewall cores and analyze for calibrating logs and obtaining true measured properties. Recommended analyses should include porosity, permeability, saturations; gas analysis to include adsorption/desorption isotherms and isotopes; and triaxial compression testing for static elastic properties. A wide variety of technologies are used to examine the lithology and mineral composition of shale. "Sample" for these analyses come from cuttings, rotary (SW) cores, conventional core plugs, and whole core. These can often be combined and performed on the same sample.
7. Determine initial economic value and reservoir potential (should be accomplished by the operator).

Appraisal Phase

1. Drill appraisal wells. Number depends on a number of considerations, but recommendation is to drill both vertical (to continue to gather data to characterize the reservoir) and horizontal wells for testing the completion and hydraulic fracturing design to determine potential producing rates from the frac design/implementation.
2. Build reservoir models for simulation—start with the geocellular (static) model and reservoir analysis techniques for shale.
3. Use the integrated multidisciplinary approach including reservoir engineering ("shale engineering"), geology, petrophysics, geomechanics, geochemistry, and seismology for building models.
4. Generate a field development plan to include well types, placement, attitude, direction (azimuth), and spacing (drainage area considerations).
5. Shale wells should be drilled in the direction of minimum horizontal stress, which maximizes access to existing natural fractures so that the induced hydraulic fractures intersect the natural fractures in transverse manner near the wellbore creating more complex fractures.
6. Based on the US experience and the characteristics of shale, a large number of development wells are required to develop shale gas reservoirs.
7. Validate the economics of the play or consider conducting a pilot project to determine whether or not to continue into the development phase of the play.
8. Pilot projects are conducted to prove commerciality and to determine the best technologies and optimum development plan.

Development Phase

1. Implement the field develop plan or pilot project.
2. Install surface production and export, compression, pipelines facilities.
3. Design and drill wells; optimize drilling costs.
   - Drill the lateral as flat as possible and perfectly horizontal (not undulating), which can cause difficulty running casing, and result in low spots that can collect liquids causing hydrocarbon flow obstruction or liquid loading of gas wells.
   - High-build rate RSS can achieve "one-run curve and lateral" wells (one BHA, one bit, one trip);

several drilling days. A rotary steerable system (RSS) is typically more efficient due to its automatic steering capability and results in a much better hole condition than provided by motors.

- Drilling optimization increases drilling efficiency and reduces days on wells (cost).
- Effective well placement using reservoir navigation maximizes reservoir exposure and locates the well in the sweet spots.
- Environmentally friendly WBM, which is more economic and provides the preferred characteristics of an OBM, is recommended for drilling the curve and lateral.

4. A successful fracture stimulation is the one that increases productivity index as well as ultimate recovery of a producing well in an economic, safe, and environmentally friendly manner. This involves not only executing a proper treatment design but also the placement of the frac stages at the proper place along the horizontal lateral.

5. Overcoming all of the hydraulic fracturing challenges of (a) achieving designed fracture geometry, (b) transporting proppant to the right location, (c) achieving final conductivity, (d) encountering expected reservoir properties, and (e) avoiding geohazards will certainly lead to the "optimum fracturing treatment."

6. The design of the frac begins with collecting a number of key parameters about the reservoir and the well. These data are used as input in a fracture simulator (recommended) to design the treatment with proper frac fluid, proppant, additives, and pumping schedule. There is no optimum, one-size-fits-all stimulation design for shale wells; each frac design is dependent on the parameters of a specific well in a specific formation, at a specific location, and for the type of produced fluid expected.

7. Use microseismic while implementing the frac in order to obtain near real-time information to control frac direction and limit loss. Microseismic also provides an approximate stimulated reservoir volume for input into a reservoir simulator to determine predicted recovery.

8. In order to place the frac stages and perforation clusters in the optimum location along the lateral, the lateral should be characterized in order to obtain critical information about natural and conductive fractures and reservoir characteristics.

9. Characterizing the lateral to obtain these two different data aspects can be accomplished by running an LWD high-definition image log (for fracture information) and running advanced mud logging with cuttings analysis (mineralogy and reservoir data).

10. The type of wellbore completion used directly affects efficiency during the fracturing process.

11. Each completion technique has its own benefits and considerations, dependent on application and the current phase of the asset life cycle.

12. Use plug-and-perf during appraisal phase to provide flexibility of stage placement, resulting in more time to gather data.

13. Ball-activated systems provide efficiency during development phase (when more information is known about the reservoir). This type of completion system is preferred by operators in liquids plays.

14. Coiled tubing–activated systems offer accuracy and contingency options and can result in optimized fracs during the development phase. These systems can be run in both open-hole and cased hole completions.

Production Phase

1. Monitor and control the flowback rate of the completed well in order to optimize total recovery and prevent well damage.
2. Periodically, run PLTs to assess production problems and consider possible remedial well work.
3. Manage the complete water cycle—sourcing for drilling and fracture water; lifting, treating, handling, storage, and disposal of well flowback water.
4. Install artificial lift if required.
5. Implement production chemical monitoring and treatment equipment and programs.
6. Protect the environment throughout all operations, especially when hydrocarbons and water are brought to the surface.

Rejuvenation Phase

1. Evaluate wells as re-frac candidates. Select the appropriate wells and conduct the re-frac operation. Refracs have been determined to be the one significant method to arrest the rapid decline in shale wells and increase production rate.
2. Redevelopment (probably infill drilling) can be considered when the primary development in the sweet spots of the field have been or are nearing depletion.

ACKNOWLEDGMENTS

This author would like to thank the management of Baker Hughes for allowing him to write and contribute this chapter to the book. I would also like to thank and show my appreciation to the members of the Baker Hughes Unconventional

Resources Team and the US Land Geomarket and Product Lines for their support and contribution to the contents of this chapter.

REFERENCES

Bazan LW, Lattibeaudiere MG, Palisch TT. Hydraulic Fracture Design and Well Production Results in the Eagle Ford Shale: One Operator's Perspective. Paper SPE 155779 presented at the SPE Americas Unconventional Resource Conference; June 5–7, 2012; Pittsburgh, PA; 2012.

Bhattacharya S, Nikolaou M. Optimal Fracture Spacing and Stimulation Design for Horizontal Wells in Unconventional Gas Reservoirs. Paper SPE 147622 presented at the SPE Annual Technical Conference an Exhibition; October 30–November 2; Denver, CO; 2011.

Brannon HD, Bell CE. Eliminating Slickwater Compromises for Improved Shale Stimulation. Paper SPE 147485 presented at the SPE Annual Technical Conference and Exhibition; October 30–November 2; Denver, CO; 2011.

Burton WA. Unconventional Completions: Which One is Right for Your Application. Paper SPE 166431-MS presented at the SPE Annual Technical Conference and Exhibition; New September 30–October 2; Orleans, LA; 2013.

Cheng Y. Impact of the Number of Perforation Clusters and Cluster Spacing on Production Performance of Horizontal Shale-Gas Wells. Paper SPE 138843 presented at the SPE Eastern Regional Meeting; October 12–14; Morgantown, WV; 2010.

Cipolla CL, Lolon EP, Erdle JC, Rubin B. Reservoir Modeling in Shale-Gas Reservoirs. Paper SPE 125530 presented at the SPE Eastern Regional Meeting; September 23–25; Charleston, WV; 2009a.

Cipolla CL, Lolon EP, Erdle JC, Tathed V. Modeling Well Performance in Shale Gas Reservoirs. Paper SPE 125532 presented at the SPE/EAGE Reservoir Characterization and Simulation Conference; October 19–21; Abu Dhabi, UAE; 2009b.

Crafton JW. Flowback Performance in Intensely Naturally Fractured Shale Gas Reservoirs. Paper SPE 131785 presented at the 2010 SPE Unconventional Gas Conference; February 23–25; Pittsburgh, PA; 2010.

Crafton JW, Gunderson D. Stimulation Flowback Management: Keeping a Good Completion Good. Paper SPE 110851 presented at the 2007 SPE Annual Technical Conference and Exhibition; November 11–14; Anaheim, CA; 2007.

Cramer DD. Stimulating Unconventional Reservoirs: Lessons Learned, Successful Practices, Areas For Improvement. Paper SPE 114172 presented at the 2008 SPE Unconventional Resources Conference; February 10–12; Keystone, CO; 2008.

Duong AN. An Unconventional Rate Decline Approach for Tight and Fracture-Dominated Gas Wells. Paper SPE 137748 presented at the Canadian Unconventional Resources & International petroleum Conference; October 19–21; Calgary, AB, Canada; 2010.

Engler TW. A New Approach to Gas Material Balance in Tight Gas Reservoirs. Paper SPE 62883 presented at the 2000 SPE Annual Technical Conference and Exhibition; October 1–4; Dallas, TX; 2000.

Heddleston D. Horizontal Well Production Logging Deployment and Measurement Techniques for US Land Shale Hydrocarbon Plays. Paper SPE 120591 present at the 2009 SPE Production and Operations Symposium; April 4–8; Oklahoma City, OK; 2009.

Holditch SA. Tight Gas Sands. Paper SPE 103356, Distinguished Author Series, JPT; June, 2006.

Ilk D, Rushing JA, Perego JD, Blasingame TA. Exponential vs. Hyperbolic Decline in Tight Gas Sands—Understanding the Origin and Implications for Reserve Estimates Using Arps' Decline Curves. Paper SPE 116731 presented at the SPE Annual Technical Conference and Exhibition; September 21–24; Denver, CO; 2008.

Isbell S, Scott D, Freeman M. Application-Specific Bit Technology leads to Improved Performance in Unconventional Gas Shale Plays. Paper SPE 128950 presented at the 2010 IADC/SPE Drilling Conference and Exhibition; February 2–4; New Orleans, LA; 2010.

Jacobi D, Breig J, LeCompte B, Kopal M, Hursan G, Mendez F, Bliven S, Longo J. Effective Geochemical and Geomechanical Characteristics of Shale Gas Reservoirs from the Wellbore Environment: Caney and Woodford Shale. Paper SPE 124231 presented at the 2009 SPE Annual Technical Conference and Exhibition; October 4–7; New Orleans, LA; 2009.

Janwadkar S, Fortenberry DG, Roberts GK, Kramer M, Devon S, Trichel DK, Rogers T, Privott SA, Welch B, Isbell MR. BHA and Drillstring Modeling Maximizes Drilling Performance of Lateral Wells of Barnett Shale Gas Field of N. Texas. Paper SPE 100589 presented at the 2006 SPE Gas Technology Symposium; May 15–17; Calgary, AB, Canada; 2006.

Janwadkar S, Morris S, Potts M, Kelley J, Fortenberry D, Roberts G, Kramer M, Devon S, Privott S, Roger T. Advanced LWD and Directional Drilling Technologies Overcome Completion Challenges of Lateral Wells in the Barnett Shale. Paper SPE 110837 presented at the 2007 SPE Annual Technical Conference and Exhibition; November 11–14; Anaheim, CA; 2007.

Janwadkar S, Hummes O, Fryer C, Rogers T, Simonton S, Black D. Innovative Design Rotary Steerable Technologies Overcome Challenges of Complex Well Profiles in Fast-Growing Unconventional Resource—Woodford Shale. Paper SPE 119959 presented at the SPE/IADC Drilling Conference and Exhibition; March 17–19; Amsterdam, the Netherlands; 2009.

Janwadkar S, Klotz C, Welch B, Finegan S. Electromagnetic MWD Technology Improves Drilling Performance in Fayetteville Shale of North America. Paper SPE 128905 presented at the 2010 IADC/SPE Drilling Conference and Exhibition; February 2–4; New Orleans, LA; 2010.

Jo H. Optimizing Facture Spacing to Induce Complex Fractures in a Hydraulically Fractured Horizontal Wellbore. Paper SPE-154930-MS presented at the Americas Unconventional Resources Conference; June 5–12; Pittsburgh, PA; 2012.

Kennedy RL, Knecht WN, Georgi DT. Comparisons and Contrasts of Shale Gas and Tight Gas Developments, North American Experience and Trends. Paper SPE 160855 presented at the SPE Saudi Arabia Section Technical Symposium and Exhibition; April 8–1; Al-Khobar, Saudi Arabia; 2012a.

Kennedy RL, Gupta R, Kotov S, Burton WA, Knecht WN, Ahmed U. Optimized Shale Resource Development: Proper Placement of Wells and Hydraulic Fracture Stages. Paper SPE 162534 presented at the Abu Dhabi International Petroleum Exhibition and Conference; November 11–14; Abu Dhabi, UAE; 2012b.

King GE. Thirty years of Gas Shale Fracturing: What Have We learned? Paper SPE 133456 presented at the SPE Annual Technical Conference and Exhibition; September 19–22; Florence, Italy; 2010.

King GE. Hydraulic Fracturing 101: What Every Representative, Environmentalist, Regulator, Reporter, Investor, University Researcher, Neighbor and Engineer Should Know About Estimating Frac Risk and Improving Frac Performance In Unconventional Gas and Oil Wells. Paper SPE 152596 presented at the SPE Hydraulic Fracturing Technology Conference; February 6–8; The Woodlands, TX; 2012.

Kupchenko CL, Gault BW, Mattar L. Tight Gas Production Performance Using Decline Curves. Paper SPE 114991 presented at the CIPC/SPE Gas Technology Symposium 2008 Joint Conference; June 16–19; Calgary, AB, Canada; 2008.

Kuuskraa V, Stevens S, Van Leeuwen T, Moodhe K. World Shale Gas Resources: An Initial Assessment of 14 Regions Outside the United States. Prepared by Advanced Resources International Inc.; February 17, 2011; for the U.S. Energy Information Administration, U.S. Department of Energy; April 2011; Washington, DC.

LeCompte B, Franquet JA, Jacobi D. Evaluation of Haynesville Shale Vertical Well Completions with a Mineralogy Based Approach to Reservoir Geomechanics. Paper SPE 124227 presented at SPE Annual Technical Conference and Exhibition; October 4–7; New Orleans, LA; 2009.

Martin AN, Eid R. The Potential Pitfalls of Using North American Tight and Shale Gas Development Techniques in the North African and Middle Eastern Environments. Paper SPE 141104 presented at the SPE Middle East Oil and Gas Show and Conference; September 25–28; Manama, Bahrain; 2011.

Meyer BR, Bazan LW. A Discrete Fracture Network Model for Hydraulically Induced Fractures: Theory, Parametric, and Case Studies. Paper SPE 140514 presented at the SPE Hydraulic Fracturing Technology Conference; January 24–26; The Woodlands, TX; 2011.

Mitra A, Warington D, Sommer A. Application of Lithofacies Models to Characterize Unconventional Shale Gas Reservoirs and Identify Optimal Completion Intervals. Paper SPE 132513 presented at the SPE Western Regional Meeting; May 27–29; Anaheim, CA; 2010.

Moos D, Vassilellis GD, Cade R, Franquet J, Lacazette A, Bourtenbourg E, Daniel G. Predicting Shale Reservoir Response to Stimulation in the Upper Devonian of West Virginia. Paper SPE 145849 presented at the SPE Annual Technical Conference and Exhibition; October 30–November 2; Denver, CO; 2011.

Payne DA. Material-Balance Calculations in Tight-Gas Reservoirs: The Pitfalls of p/z Plots and a More Accurate Technique. Paper SPE 36702 presented at the SPE Annual Technical conference and Exhibition; October 6–9; Denver, CO; 1996.

Pemper R, Han X, Mendez F, Jacobi D, LeCompte B, Bratovich M, Feuerbacher G, Bruner M, Bliven S. The Direct Measurement of Carbon in Wells Containing Oil and Natural Gas Using Pulsed Neutron Mineralogy Tool. Paper SPE 124234 presented at 2009 SPE Annual Technical Conference and Exhibition; October 4–7; New Orleans, LA; 2009.

Sena A, Castillo G, Chesser K, Voisey S, Estrada J, Carcuz J, Carmona E, Hodgkins P, Hamson-Russell Software & Services, a CGGVeritas Company. Seismic reservoir characterization in resource shale plays: stress analysis and sweet spot discrimination. The Leading Edge, July 2011.

Song B, Economides MJ, Ehlig-Economides C. Design of Multiple Transverse Fracture of Horizontal Wells in Shale Gas Reservoirs. Paper SPE 140555 presented at the SPE Hydraulic Fracturing Technology Conference; January 24–26; The Woodlands, TX; 2011.

Vassilellis GD, Li C, Seager R, Moos D. Investigating the Expected Long-Term Production Performance of Shale Reservoirs. Paper SPE 138134 presented at the Canadian Unconventional Resources & International Petroleum Conference; October 19–21; Calgary, AB, Canada; 2010.

Vassilellis GD, Li C, Bust VK, Moos D, Cade R. Shale Engineering Application: The MAL-145 Project in West Virginia. Paper SPE 146912 presented at the Canadian Unconventional Resources Conference; November 15–17; Calgary, AB, Canada; 2011.

Vincent MC. Refracs—Why Do They Work, and Why Do They Fail? In 100 Published Field Studies? Paper SPE 134330 presented at the SPE Annual Technical Conference and Exhibition; September 19–22; Florence, Italy; 2010.

Warpinski NR, Waltman CK, Weijers L. An Evaluation of Microseismic Monitoring of Lenticular Tight Sandstone Stimulations. Paper SPE 131776 presented at the SPE Unconventional Gas Conference; February 23–25; Pittsburgh, PA; 2010.

18

GAS SHALE ENVIRONMENTAL ISSUES AND CHALLENGES

TERENCE H. THORN
JKM 2E Consulting, Houston, TX, USA

18.1 OVERVIEW

Experts agree that the storage and treatment of liquid drilling wastes, methane emissions, water usage, and site construction are among the biggest environment issues surrounding shale gas development. These environmental priorities were confirmed in Resources For the Future's (RFF) Center for Energy Economics and Policy, *Pathways to Dialogue: What the Experts Say about the Environmental Risks of Development*, released in early 2013 (Krupnick et al., 2013). The report was the first survey-based, statistical analysis of experts from government agencies, industry, academia, and nongovernmental organizations to identify the priority environmental risks related to shale gas development. What was most interesting about the survey was a high degree of consensus as to the risks that should be identified as priorities. Only two of the consensus risks identified by the experts are unique to the shale gas development process, and both have potential impacts on surface water. The remaining 10 consensus risks relate to practices common to gas and oil development in general, such as the construction of roads, well pads and facilities related to the actual drilling of the well, and pipelines, gathering and processing facilities, and the potential for leaks in well casing and cementing.

The environmental concerns about the effects of shale gas development can be grouped into five general categories:

(i) Water: the overall water consumption requirements, the improper disposal of wastewater, and the possibility that underground fracking fluids can migrate into aquifers;

(ii) Climate: Emissions of CO_2, methane, and other greenhouse gases during the production, processing, and transportation of shale gas;

(iii) Air Quality: the local impact of emissions from drilling rigs, trucks, and compressors;

(iv) The social impacts on shale gas communities including demands on local infrastructure and social services and environmental degradation; and

(v) Earthquakes caused by wastewater disposal in deep injection wells.

18.2 WATER USE

Access to water is critical to shale gas development. The amount of water needed to hydraulically fracture a well varies greatly, depending on how hard it is to extract oil and gas from each geological formation, but on average it can take 2–5 million gallons (7–19 million liters) of water to frack a well, and a well may be fracked multiple times. Depending on state and local water laws, producers may draw their water for free from underground aquifers or rivers, or may buy and lease supplies belonging to water districts, cities, and farmers. Even if some of the water can be recycled, the process requires a major withdrawal from an aquifer or other water resources during the fracking process.

The U.S. Environmental Protection Agency (EPA) in the 2011 draft study on the impact of fracking on drinking water estimated that that the 35,000 oil and gas wells used for fracking consume between 70 billion and 140 billion gallons of water every year (EPA, 2011a). That's about equal, EPA

Fundamentals of Gas Shale Reservoirs, First Edition. Edited by Reza Rezaee.
© 2015 John Wiley & Sons, Inc. Published 2015 by John Wiley & Sons, Inc.

says, to the water use in 40–80 cities with populations of 50,000 people, or one to two cities with a population of 2.5 million each. In the Marcellus formation in Pennsylvania, New York, and West Virginia, water usage for well fracking could reach 650 million barrels per year according to a report done by All Consulting for the U.S. Department of Energy (DOE) and state authorities (Arthur et al., 2010).

When compared to the other water uses in the Marcellus Basin, shale development is a fraction of total water usage for agricultural, industrial, and recreational purposes. For example, the total volume of water needed to meet estimated peak shale gas development represents about 0.8% of the 85 billion barrels (there are 42 gallons in a barrel) per year that are currently consumed in the region.

A University of Texas at Austin Bureau of Economic Geology (BEG) study found that total water use for fracking in Texas has risen by about 125%, from 36,000 acre-ft in 2008 to about 81,500 acre-ft in 2011 (Nicot and Scanlon, 2012). For a comparison, the 800,000-person city of Austin used about 107,000 acre-ft of water in 2011. About one-fifth of the current total water used in Texas comes from recycled or brackish water, a category of water supply that has been growing (Kalaswad et al., 2012). Water use for fracking is not expected to exceed 2% of the statewide total water use.

But drilling can send water use percentages much higher in rural areas and aggregate numbers for use hide important conflicts in places where farmers compete with natural gas developers for their water supply, especially in drought stricken areas or areas traditionally with low rainfall. In these areas, fracking is driving up the price of water and burdening already depleted aquifers and rivers.

A comprehensive May 21, 2013, report by Ceres, a Boston-based nonprofit organization, stated that 47% of oil and gas fracking sites are in high or extremely high water-stressed water basins (Freyman and Salmon, 2013). The study was based on water consumption by 25,450 fracking wells that drillers voluntarily reported to the FracFocus, an industry run database, between January 2011 and September 2012. The data was then laid on top of water risk maps developed by the World Resources Institute (WRI). According to the report, during the study period, U.S. fracking operations consumed 68.5 billion gallons of water—equivalent to the amount 2.5 million people would use in a year. Ceres researchers believe this figure is most likely an underestimate—oil and gas companies are not required to report how much water they are using. A second Ceres report in February 2014, *Hydraulic Fracturing & Water Stress: Water Demand by the Numbers* (Freyman, 2014), provides first-of-its-kind data on the various water sourcing risks facing oil and gas companies in eight regions of intense shale development in the United States and Canada. An April 2013 report by the Western Organization of Resource Councils, a regional network of grassroots community organizations, found that fracking is using 7 billion gallons of water a year in four western states: Wyoming, Colorado, Montana, and North Dakota (Western Organization of Research Councils, 2013).

Water use for fracking will remain a serious issue in the parts of the country where significant oil and gas shale development is taking place in drought stressed areas or areas traditionally facing water restraints. The February 2014 Ceres report found that more than half of the wells—55%—were in drought-stricken areas, and nearly half were in regions under high or extremely high water stress. Even in areas with adequate water supplies, better water management will be necessary to reduce amounts of fracking wastewater, the disposal of which creates its own set of environmental problems.

18.3 THE DISPOSAL AND REUSE OF FRACKING WASTEWATER

With each round of fracking, about half of the fracking fluid, now called wastewater or flowback, returns to the surface along with the gas, via the collection pipes. Although fracking fluids are more than 99% water and sand, they also contain a number of chemicals, including some that are toxic at the parts-per-billion level. These include benzene, antimicrobial agents, and corrosion inhibitors. The U.S. House Energy and Commerce Committee released a report in April 2011 that identified 29 chemicals that are either known or possible carcinogens that would be normally subject to EPA regulation under the Clean Water Act. Oil and gas fracking, however, was exempted from the act in 2005 (Chemicals, 2011).

Many environmentalists have severely criticized the handling of wastewater, claiming it results in toxic waste and surface water contamination. A key problem is the disposal of the increasing amounts of fracking fluid. As described in the Vaughn and Purcell study *Frac Attack: Risks, Hype, and Financial Reality of Hydraulic Fracturing in the Shale Plays*, fracking chemicals and drilling waste are more hazardous above ground than several miles underground and pose a more serious environmental hazard than potential contamination of groundwater from fracking (Vaughan and Pursell, 2010).

For a producer, there are four options to handle wastewater: storage in tanks or tarp lined pits where the wastewater is allowed to evaporate or be recycled, disposal in deep underground injection wells, trucking to publicly owned treatment plants that may or may not be designed to handle fracking chemicals, or trucking to centralized waste treatment facilities some of which can safely recycle the water. Closed tanks are also sometimes used for collection of produced water during the flowback period, sometimes with secondary containment. Open impoundments, also called pits, are typically subject to requirements designed to minimize the risk of contamination. As with any liquid material in storage, accidental spills and mismanagement can cause releases to the environment that could contaminate nearby waters and soils.

In November 2010, the Pennsylvania Department of Environmental Protection (PADEP) discovered a discharge during an inspection of the Penn Township facility, where a PADEP inspector observed wastewater spilling from an open valve from a series of interconnected tanks. At the time, XTO Energy Inc., a subsidiary of Exxon Mobil Corporation, stored wastewater generated from energy extraction activities conducted throughout Pennsylvania at its Penn Township facility and, at the time of the release, stored produced fluid from its operations in the area. Pollutants from the release were found in a tributary of the Susquehanna River basin.

EPA, in consultation with PADEP, conducted an investigation and determined that wastewater stored in the tanks at the Penn Township facility contained the same variety of pollutants, including chlorides, barium, strontium, and total dissolved solids, that were observed in those surface waters (EPA, 2013a). The federal settlement between parties required that XTO implement a comprehensive plan to improve wastewater management practices to recycle, properly dispose of, and prevent spills of wastewater generated from natural gas exploration and production activities in Pennsylvania and West Virginia (XTO Energy, 2013).

Historically, the practice of disposing of waste in deep injection wells has not gotten much attention. However, in many regions of the country, injection wells have become the preferred method for disposing of the liquid waste (primarily oil-field brine) produced during the hydraulic fracturing process. In the Southwest United States, producers reinject the wastewater into abandoned wells. Injection wells are not impacted by waste type or contaminants during disposal. The 1974 Clean Water Act, among other things, requires EPA to protect underground sources of drinking water and granted EPA the power to regulate injection wells. Injection wells are classified into six classes according to the type of fluid they inject and where the fluid is injected. Class 2 wells inject fluids associated with oil and natural gas production operations (EPA Classes of Injection Wells). In the United States, there are more than 151,000 waste-Class 2 injection wells. More than 2 billion gallons of waste, mostly brine, from oil and gas drilling and production are injected into those wells every day.

States like Texas have many deep underground injection wells where companies dispose of the salty and chemical- and mineral-laden shale wastewater. The state has more than 8000 active disposal wells, about 850 of which are large commercial operations, according to the Texas Railroad Commission (TRC), the regulator of oil and gas production. Texas has another 25,000 wells that accept waste fluids and use them to retrieve additional oil and gas (Texas Railroad Commission, 2013).

The amount of wastewater being disposed of in Texas wells has skyrocketed with the spread of fracking, to nearly 3.5 billion barrels in 2011 from 46 million barrels in 2005, according to data from the TRC. On average, companies in Texas dispose of 290 million barrels of wastewater—equivalent to about 18,500 Olympic-size swimming pools every month, The Colorado Oil and Gas Conservation Commission, charged with both promoting and regulating the oil and gas industry, has issued over 3000 permits letting companies dispose of liquid waste in evaporative ponds, shallow pits, and in 300 plus super-deep injection wells.

In states like Pennsylvania where, according to PADEP data, over 8418 Marcellus wells have been permitted as of August 2013, there are few Class 2 wells. New York State has no disposal wells. The lack of injection wells has led to a significant increase in the amount of wastewater trucked to Ohio for disposal via underground injection, from roughly 26 million gallons in 2010 to 106 million gallons in 2011. In 2012, Ohio injection wells handled 588 million gallons of wastewater, the majority of which was received from Pennsylvania (Ohio and Fracking, 2013). After a series of small earthquakes near Ohio disposal sites in 2011, Ohio regulators now require far lengthier and more thorough review of geological records which will make underground disposal there much more expensive.

According to a study by researchers at Kent State and Duke University, despite producing less wastewater per unit of gas produced than a conventional gas well, developing the Marcellus shale has increased the total wastewater generated in the region by about 570% since 2004, overwhelming existing wastewater disposal infrastructure capacity (Lutz et al., 2013). Wastewater was initially shipped to publicly owned wastewater treatment plants. These treatment plants were designed to treat municipal or county wastewater, and can remove biological contaminants and some heavy metals but are far less capable of removing radioactive contaminants. When wastewater is sent to municipal sewage facilities, harmful chemicals and other pollutants are merely diluted, rather than removed, and then released into the state's rivers, lakes and streams potentially affecting drinking water supplies.

On April 19, 2011, the PADEP asked all Marcellus Shale natural gas drilling operators to stop delivering wastewater from shale gas extraction to 15 different wastewater treatment plants around the Pittsburgh region by May 19. This request was greatly influenced by research conducted by a Carnegie Mellon river monitoring project that showed elevated levels of bromide in the Monongahela River, a source of drinking water for over 800,000 people in southwestern Pennsylvania. Carnegie Mellon University Professor Jeanne VanBreisen was first to discover high levels of bromides in drinking water sourced from rivers that have received treated fracking wastewater (VanBriesen, 2012). Bromide itself is nontoxic, except when it reacts with the chlorine that is used during treatment. It forms brominated by-products that have been found to be carcinogenic to humans. Since these problems were highlighted, most drilling companies in Pennsylvania have stopped sending their wastewater through treatment plants that were unable to remove many of the contaminants before the water was discharged into rivers. State regulators and drinking water operators are also now testing more regularly for radioactive and other toxic elements in the drilling wastewater.

The final option for disposal is shipping the wastewater to centralized waste treatment facilities, which are private commercial wastewater treatment operations that handle industrial waste. These facilities handle all the types of waste fluids produced from oil and gas operations and release it into waterways or send it for reuse after it is processed. Even with waste treatment facilities that have been designed to specifically treat the wastewater from the fracking process, radiological components, chemicals, and toxins have been released and later detected in freshwater sources. Proper sampling methodology needs to be put into place and strictly enforced to ensure that water quality is minimally affected by the treatment and release of this wastewater.

In May 2013, environmental regulators in Pennsylvania discovered high levels of radium around the Josephine Treatment Plant discharge pipe in Indiana County (Ferrar et al., 2013). The levels of radioactivity found at the Josephine plant were not high enough to cause any health threat to passersby or to workers. However, radium can also accumulate in fish, meaning that fish in the creek ingesting the radioactive metal could carry higher levels than found in the water.

In response to tougher state requirements for wastewater disposal, new facilities are being built that can handle wastewater treatment. In April 2013, the wastewater treatment company Aquatech opened a new centralized shale wastewater treatment plant in Tioga County, Pennsylvania (Aquatech, 2013). The new plant is equipped to process up to 200 gallons per minute and sits in an area central to a huge amount of shale drilling. The Aquatech site can recycle backflow and other drilling-generated liquids to filter out solid materials, disinfect, and distill the water to two different levels. At the first level, or clean level, filtration clears water for reuse in the industry. At the second level, or ultra-clean level, processed water meets state standards for general use.

Fully cleaning waste comes down to economics and technical issues. Cost estimates for various methods of wastewater disposal and recycling vary, depending on both the reservoir in question and the information source. Technology exists to clean liquid waste right up to drinking water standards, but it is expensive, and far more costly compared to buying freshwater for drilling and fracking. Industry estimates that the cost to dispose of Marcellus shale fracking fluids at a proper wastewater management facility are roughly $3.00 per barrel to dispose of it, and $7.00–$10.00 for it to be hauled away. That equals between $10.00 and $13.00 for the disposal of a single barrel, which holds 42 gallons of wastewater. In 2012, new energy policy proposals were put forth in Ohio that would raise brine disposal fees from 5 to 10 cents on in-state waste, and from 20 cents to $1.00 on out-of-state waste. Under new proposed costs, disposing of liquid waste from a single well in this manner would cost $5,700 for in-state waste and $57,000 for out-of-state disposal.

As the cost of wastewater treatment is increased, recycling, where wastewater is blended with freshwater and hydraulic fracturing chemicals for use in subsequent hydraulic fracture treatments, is growing in popularity. However, there are limitations on the amounts of total dissolved solids (TDS), barium, and other contaminants that can be present for the waste to acceptable for reuse in hydraulic fracturing fluids (Wilson et al., 2014). While the recycling of fracking waste is conceptually a good thing, there is concern that the residual waste coming out of that process could be toxic and is not governed under waste regulations.

As local and federal regulators raise questions about water consumption and its disposal from drilling operations, U.S. oil and natural gas producers are asking service companies to improve their handling of the millions of gallons of fluids involved in fracking an average well. Halliburton has a stated goal for the entire oil and gas industries to use an average of 25% less freshwater in fracking jobs by the end of 2014. Producers are implementing comprehensive water management plans which incorporate strategies for reducing consumption of water, reducing the loss or waste of water, improving or maintaining efficiency in the use of water, and increasing recycling and reuse of water.

18.4 GROUNDWATER CONTAMINATION

There are concerns that the potentially carcinogenic chemicals used in the fracking process can find their way underground into drinking water sources. Water and sand make up 98–99.5% of the fluid used in hydraulic fracturing. In addition, chemical additives are used. The exact formulation varies depending on the well. Overall the concentration of additives in most slickwater fracturing fluids is a relatively consistent 0.5–2% with water making up 98–99.5%.

Typical shale gas deposits are located several thousand feet below the deepest potential sources of underground drinking water. Further, the low permeability of shale rock and other intervening formation horizons present additional impediments to the flow of fracking chemicals from target zones upward into aquifers. The likelihood of water contamination as a consequence of fluids migration up through several thousand feet of strata has proven to be extremely unlikely. Most agree that more likely candidates as sources of possible water contamination include improper well design, inadequate surface casing and substandard or improper cementing, improper handling of surface chemicals, improper design/performance of holding ponds, and improper storage and disposal of wastes and produced water.

A study of the Marcellus Shale by Duke University published in June 2013 found that drinking water wells within 1 km of natural gas fracking wells (horizontal drilling/fracturing) were contaminated with stray petroleum gases including methane, ethane, and propane, with methane concentrations an average of six times higher than those wells farther away (Jackson et al., 2013). The researchers said there was no evidence of any chemical contamination from drilling–neither chemicals used for fracking nor from naturally occurring brine water that comes along later in the process. The study states that the two simplest explanations for the contamination in drinking water are faulty or inadequate steel casings and imperfections in cement sealings.

A controversial study conducted by the Energy Institute at the University of Texas at Austin, *Fact-based Regulation for Environmental Protection in Shale Gas Development*, February 2012, found that many problems attributed to hydraulic fracturing were related to processes common to all oil and gas drilling operations, such as drilling pipe inadequately cased in concrete (Groat and Gromshaw, 2012). Many reports of contamination can be traced to above ground spills or other mishandling of wastewater produced from shale drilling and not from hydraulic fracturing.

In July 2013, the U.S. Department of Energy's National Energy Technology Laboratory (NETL) in Pittsburgh, Pennsylvania, after a year-long study at a western Pennsylvania drilling site in Washington County, concluded that there was no evidence that chemicals used during the hydraulic fracturing process have contaminated drinking water aquifers adjacent to the drilling site (NETL, 2013). The study found that hydraulic fracturing fluids remained thousands of feet below the shallower portions of the aquifer that supply drinking water.

Eight wells in the Marcellus Shale formation were monitored seismically and one was injected with four different "tracers" at different stages in the hydraulic fracturing process. The depth of the injection of the tracers was at approximately 8000 ft below the surface of the well bore. None of the tracers were detected in a monitoring zone at a depth of 5000 ft. The study also tracked the maximum extent of the man-made fractures, and all were at least 6000 ft below the surface of the well bore. Finally, the study also monitored a separate series of older gas wells that are about 3000 ft above the Marcellus Shale formation to determine the impacts, if any, of hydraulic fracturing process on these wells. Ultimately, the DOE study did not detect the tracers in these older gas wells.

The DOE cautioned that the results of the study are preliminary, as the study is ongoing, but that the results are the first independent examination of whether the chemicals utilized during the hydraulic fracturing process pose a threat to local drinking water supplies. DOE's initial conclusion suggests that hydraulic fracturing does not impact drinking water supplies. Samples from 100 private drinking water wells revealed that arsenic, selenium, strontium, and TDS exceeded the Environmental Protection Agency's Drinking Water Maximum Contaminant Limit (MCL) in some samples from private water wells located within 3 km of active natural gas wells. Lower levels of arsenic, selenium, strontium, and barium were detected at reference sites outside the Barnett Shale region as well as sites within the Barnett Shale region located more than 3 km from active natural gas wells. Methanol and ethanol were also detected in 29% of samples.

In July 2013, a study by researchers at the University of Texas at Arlington, *An Evaluation of Water Quality in Private Drinking Water Wells Near Natural Gas Extraction Sites in the Barnett Shale Formation* (Fontenot et al., 2013), found elevated levels of arsenic and other heavy metals in groundwater near natural gas fracking sites in Texas' Barnett Shale where there are 16,000 active wells. Historical databases from the same areas sampled never had these issues until the onset of all the fracking.

In March 2013, Resources for the Future, in a paper published in the *Proceedings of the National Academy of Sciences* relied upon more than 20,000 surface water quality observations taken over 11 years in Pennsylvania to estimate the effects of shale gas development on downstream water quality through 2011 (Olmstead et al., 2013). The authors were unable to find any statistical evidence of water contamination due to leaks at the actual well sites. The authors did find some evidence of pollution downstream from wastewater treatment facilities in the form of raised chlorine levels. Fracking was not identified as the cause of the contamination, but the study shows these issues do occur in close relation, geographically to natural gas extraction. The results indicated statistically significant water quality impacts from wastewater sent to treatment plants and runoff from well pad development. The study found no systematic statistical evidence of spills or leaks of flowback and produced water from shale gas wells into waterways.

Vaughan and Pursell in their 2010 paper summarize the available studies and information on hydraulic fracturing and provide an objective look at the debate. The authors confirm that water-supply contamination from the so-called stray gas occurs more often from failures in well design and construction, breaches in spent hydraulic-fracturing water-containment ponds, and spills of leftover natural gas liquids used in drilling. In this respect, waste disposal and safe materials handling are the biggest challenges to producers.

The authors analyze incidents of contamination cited by environmental advocates as evidence of contamination caused by fracking and conclude that most of those incidents are either naturally occurring gas in water sands or problems caused by mistakes in well design—improper cementing—not related to fracking.

No one is saying that there are no potential risks to water resources from hydraulic fracturing. However, available scientific literature appears to support a cautiously optimistic view of the dangers posed by fracking to freshwater supplies. Studies have thus far failed to establish any systematic relationship between drilling activity and water pollution. Others caution that although the studies do not confirm any cases of drinking water contamination caused by fracking, which does not mean that such contamination is impossible or that hydraulic fracturing chemicals cannot get loose in the environment in other ways such as through spills of produced water.

Definitive evidence may have to wait until the EPA completes in 2014 its study of hydraulic fracturing and its potential impact on drinking water resources (EPA, December 2012a). Nonetheless two things are clear: in areas of fracking, in order to trace the source of contamination, there is a need to know what chemicals are being used in the fracking process before fracking starts. Secondly, proper baseline water sampling needs to be conducted prior to the start of drilling.

18.5 METHANE EMISSIONS

Methane emissions from natural gas extraction, especially shale gas, has been getting a lot of attention as environmentalists focus on the full life cycle of energy production and use and the amount of natural gas that is released into the atmosphere unburned as part of the hydraulic fracturing process, transportation by pipelines, and from compressors and processing units. The issue of methane leaks has caused a major split between environmental groups. Since power plants that burn natural gas emit about half the amount of the greenhouse gases as coal-fired power plants, the recent shift to gas-fired generation has allowed the United States to become the only major industrialized country to significantly reduce greenhouse emissions. But others believe the methane leaks during the production of shale gas negate any benefits over coal substitution, since methane is a highly potent greenhouse gas.

A controversial paper by Cornell's Robert Howarth jump-started this debate (Howarth et al., 2011). He argued that natural gas from fracking operations can be worse for the atmosphere than coal because of methane seepage into the atmosphere. Another Cornell study also suggested that life cycle greenhouse gas emissions from shale gas are 20–100% higher than coal on a 20-year timeframe basis. Studies by DOE's National Energy Technology Laboratory (NETL), The Worldwatch Institute, Carnegie Mellon, and Deutsche Bank challenged these conclusions and showed that shale gas production has a GHG footprint that is 20–50% lower than for coal.

In the 2012 Massachusetts Institute of Technology (MIT) study, researchers argue that the amount of methane emissions caused by shale gas production has been largely exaggerated (O'Sullivan and Paltsev, 2012). Their analysis was based on data from each of the approximately 4000 wells drilled in the five main U.S. shale-drilling sites during 2010. Wells in two of those sites, Texas' Barnett shale and the Haynesville shale on the Texas-Louisiana boarder, had been studied by Robert Howarth from Cornell University last year when he looked at potential emissions released by the industry.

In studying potential emissions, where Howarth found 252 mg of methane emissions per well in the Barnett site and 4638 mg per well in the Haynesville site, the MIT researchers, using their comprehensive well dataset, found that the potential emissions per well in the Barnett and Haynesville sites were in fact 147 mg of methane (273 thousand cubic meters of natural gas) and 633 mg (1,177 thousand cubic meters of gas), respectively. When accounting for actual gas handling field practices, these emission estimates were reduced to about 35 mg per well of methane from an average Barnett well and 151 mg from an average Haynesville well. The MIT study found that companies are already capturing about 70% of potential "fugitive" emissions.

In April 2013, EPA dramatically lowered its estimate of how much of a potent heat-trapping gas leaks during natural gas production in its latest GHG inventory to less than half of what had been previously estimated (EPA, April 2013). Among the key findings of the analysis:

(i) The effective natural gas emissions rate per unit of natural gas production is 1.5%. This emissions rate is lower than earlier estimates of 2.2–2.4% using data from prior EPA inventories for 2009 and 2010.

(ii) The EPA revised its estimates of natural gas system methane emissions downward 33% for 2010, from 215.4 million metric tons of carbon dioxide equivalent in the 2012 inventory to 143.6 MMTe in the 2013 inventory.

(iii) The long-term trend for methane emissions from natural gas systems is downward. Annual methane emissions have dropped 10% since 1990 and are 17% below the all-time peak set in 2007.

(iv) Distribution system methane emissions have dropped 16% since 1990, even as the industry added nearly 300,000 miles of distribution mains to serve 17 million more customers, an increase of 30% in both cases.

In August 2013, scientists with the National Oceanic and Atmospheric Administration (NOAA) and the University of

Colorado (UC) at Boulder published a new paper on methane leakage in the journal *Geophysical Research Letters*. It reports an alarmingly high level of methane emissions in the Uintah Basin of Utah—6.2–11.7% of total production for an area about 1000 square miles. The scientists have 3 h of observations, and no direct way of knowing whether those observations are representative of methane emissions over longer periods of time. Most emissions in the basin are due to gathering, processing, and transmission, rather than the fracking process itself. A GAO study in 2010 estimated that a full 93% of fugitive methane emissions came from pneumatic devices and glycol dehydrators—equipment used in gathering, processing, and transmission, not in fracking itself (GAO, 2010). Indeed, the GAO estimates only 4% of emissions come from well completions.

Two other recent studies help frame the debate over methane emissions. A study conducted by the University of Texas and sponsored by the Environmental Defense Fund and nine petroleum companies takes a comprehensive look at the extent to which methane leaks during drilling and production offset the environmental benefits of the clean-burning natural gas the wells produce and was the first to conduct detailed examinations of individual drilling sites (Allen et al., 2013a). The study concluded that emissions of methane, a greenhouse gas and primary component of natural gas, were lower than previously estimated during well completion but are higher than previously estimated from other aspects of production, including from pneumatic controllers and equipment leaks. The study's finding that 99% of methane emissions during completion are captured by containment measures is significantly greater than had been previously estimated using engineering estimates and emissions factors developed in the early 1990s. The report estimates the national methane leakage rate associated with the phase of natural gas extraction to be equivalent in line with EPA's current emission inventory estimate for the production segment of the supply chain. A second study coauthored by the Colorado NREL and Joint Institute for Strategic Analysis and published in the journal *Science* in February 2014 (Brandt et al., 2014), reviewed more than 200 earlier studies. The study concluded that methane leakage was not great enough to negate the climate benefits of switching from coal to natural gas as a fuel for electricity although methane emissions are 50% higher than the estimates by the EPA. The authors said that fracking likely accounts only for a small portion of the excess methane emissions. Natural gas production and processing, leaks from distribution systems, and abandoned oil and gas wells are all likely to be larger sources of fugitive emissions.

Finally, an analysis released in March 2014 by ICF International shows that the U.S. oil and gas industries can significantly and cost-effectively reduce emissions of methane using currently available technologies and operating practices (Economic, 2014). Total methane emissions from U.S. oil and gas are projected to increase 4.5% by 2018 as emissions from industry growth—particularly in oil production—outpace reductions from regulations already on the books. However, the industry could cut methane emissions by 40% below projected 2018 levels at an average annual cost of less than 1 cent per thousand cubic feet of produced natural gas by adopting available emissions-control technologies and operating practices. This would require a capital investment of $2.2 billion, which *Oil & Gas Journal* data shows to be less than 1% of annual industry capital expenditure.

The full life cycle impact of natural gas production is attracting increased interest as studies such as Horwarth's surface and energy policies include an expanded role for natural gas. All sources of the so-called greenhouse gases are important and every effort to reducing those methane emissions should be a priority for the natural gas industry. Any methane emissions are unwanted. But it is important to separate emissions from above ground operations versus fracking itself. Fortunately, there are straightforward technological requirements that can be imposed on gathering and processing systems that would greatly cut down on this leakage. The Howarth study is an important reminder that the whole life cycle impact is what matters, not just the immediate emissions.

18.6 OTHER AIR EMISSIONS

Other air quality impacts from shale gas operations include emissions of carbon dioxide stripped from the gas, volatile organic compounds, sulfur dioxide and/or hydrogen sulfide from treating sour water for use as hydraulic fracture fluid, and NOX and other emissions from compressors, pollution from diesel engines, and ground level ozone. EPA has identified these emissions as one of the largest sources of air pollution from the energy industry and is working on new emission regulations for all oil and gas field productions.

In response to tightening emissions standards and to reduce operational costs, companies such as Cabot Oil & Gas, a leading independent natural gas producer with significant operations in the Marcellus, recently announced that it is using natural gas from the Marcellus to fracture wells via an innovative dual-fuel technology. The use of this technology, which uses engines that operate on a mixture of both natural gas and diesel, may help reduce the use of diesel—the traditional fuel of choice to operate hydraulic fracturing equipment—by as much as 70%.

According to a statement released by the company, dual-fuel technology offers several benefits, including (1) reduced air emissions for a cleaner environment, due to a reduction in diesel usage, (2) reduced truck traffic when field gas at or near the well site is used due to a reduction in the transportation of diesel fuel to site, and (3) reduced costs, as natural gas can be a less expensive fuel option than diesel, providing potential cost savings for the industry and for energy consumers. With its foray into dual-fuel technology, Cabot joins the likes of Apache Corporation, which in January became the first energy exploration and production company to power a full hydraulic fracturing operation using natural gas-burning engines at its Granite Wash operations in Oklahoma. By switching to natural gas, Apache said it expects to reduce fuel costs by roughly 60%, while also lowering emissions.

18.7 SOCIAL IMPACTS ON SHALE GAS COMMUNITIES

While many communities have embraced shale gas development for its economic benefits, there is recognition of the need to understand the potential negative impacts shale gas development can have on communities in production areas. These impacts can include:

(i) increased demands on local infrastructure and utilities including a severe strain on roads due to increased truck traffic (including accidents and congestion);

(ii) the "Boom Town" effects of rapid population growth such as inadequate educational and medical facilities and other social services, housing shortages or lack of affordable housing, and increased crime;

(iii) potential effects on human health due to water contamination, noise level increase, and poor air quality from the emissions associated with fracking operations;

(iv) the physical footprint of drill pads, roads, storage sites, truck traffic, compressors, and rigs in fracking areas; and

(v) habitat fragmentation, particularly in and around recreational areas, and the disturbance of vegetation and soils that are disturbed where gas wells require new roads, clearing, and leveling.

Community concerns about these impacts are reflected in survey studies in the areas of Pennsylvania that are in the early Marcellus boom stages (Brasier et al., 2011). Respondents were most concerned about the impact on the local economy; social relations; and aesthetics, amenities, and environmental quality. The potential for road damage as a result of the heavy traffic from drilling trucks carrying equipment was another impact that was mentioned repeatedly.

The degree of surface impacts can be affected by many factors, such as the location and the rate of development; geological characteristics; climatic conditions; the use by companies of new technologies and best practices; and regulatory and enforcement activities. In a report issued in December 2012, the GAO noted that because shale development is relatively new in some areas, the long term effects—after operators are to have restored portions of the land to pre-development conditions—have not been evaluated (GAO, 2012). Without this data, the cumulative effects of shale oil and gas development on habitat and wildlife are largely unknown. GAO also found that, in general, shale gas development impacts can vary significantly even within specific shale basins.

In response to growing community concerns about the impact of shale gas development, on July 11, 2014, the American Petroleum Institute has issued ANSI-API Bulletin 100-3, which sets forth detailed recommendations for oil and gas companies seeking to engage with the communities affected by exploration and development activities (American Petroleum Institute, 2014). The standards came 1 week after New York ruled its towns and cities could ban fracking. Other states including California and Colorado have also pushed for bans on drilling and hydraulic fracturing.

18.8 INDUCED SEISMICITY: WASTEWATER INJECTION AND EARTHQUAKES

As part of its ongoing effort to study a variety of potential impacts of U.S. energy production, United States Geological Survey (USGS) scientists have been investigating the recent increase in the number of magnitude 3 and greater earthquakes in the midcontinent of the United States (Ellsworth et al., 2012). The largest of these was a magnitude 5.6 event in central Oklahoma that destroyed 14 homes and injured two people. The mechanism responsible for inducing these events appears to be the well-understood process of weakening a preexisting fault by elevating the fluid pressure. The fact that the disposal (injection) of wastewater produced while extracting resources has the potential to cause earthquakes has long been known. One of the earliest documented case histories with a scientific consensus of wastewater inducing earthquakes is at the Rocky Mountain Arsenal well, near Denver. There, a large volume of wastewater was injected from 1962 to 1966, inducing a series of earthquakes below magnitude 5.

Beginning in 2001, the average number of earthquakes occurring per year of magnitude 3 or greater increased significantly, culminating in a six-fold increase in 2011 over twentieth century levels. The scientists then began taking a closer look at earthquake rates in regions where energy production activities have changed in recent years. The lead researcher in the paper, Mr. Ellsworth, believes the increased number of earthquakes is not associated with hydraulic fracturing, but instead with the disposal of drilling waste fluids in deep well injection sites, of which there are 144,000 in the United States. None of what government researchers consider being man-made earthquakes has caused significant damage.

The USGS is coordinating with other federal agencies, including the EPA and Department of Energy, to better understand the occurrence of induced seismicity through both internal research and by funding university-based research with a focus on injection-induced earthquakes from wastewater disposal technologies. For instance, USGS and its university partners have deployed seismometers at sites of known or possible injection-induced earthquakes in Arkansas, southern Colorado, Oklahoma, and Ohio.

Currently, there are no methods available to anticipate whether a planned wastewater disposal activity will trigger earthquakes that are large enough to be of concern. Evidence from some case histories suggests that the magnitude of the largest earthquake tends to increase as the total volume of injected wastewater increases. Injection pressure and rate of injection may also be factors. More research is needed to determine answers to these important questions. Industry will need to investigate all areas at risk in which seismicity may be induced through current activity of shale gas production. More needs to be known about how these activities interact with in situ stresses and possibly affect seismic activity.

18.9 REGULATORY DEVELOPMENTS

Most everyone agrees that fracking needs some level of regulation. State and federal regulators are scrambling to adapt their existing regulatory programs to the boom in shale gas production and adjust those regulations to address new or newly recognized risks. We will continue to see the EPA propose changes to its Clean Air Act and Clean Water Act regulations to address specific pollution issues associated with hydraulic fracturing.

The natural gas industry, environmentalists, the general public, and state and federal regulators continue to fiercely debate just what level of regulation is required and whether the regulations should be enforced at the state or federal level. (The federal government does maintain control of production on federal and Indian lands.) The industry and several state governors maintain that oversight should remain at the state level, where regulation of oil and gas production has traditionally taken place. They note that many states have passed new regulations and rules governing drilling and some have even banned shale gas fracturing. Many environmental groups are advocating for uniform federal regulation of gas drilling and more stringent environmental protections for water resources. Environmental groups point to the uneven enforcement by state agencies as evidence for the need for federal regulation.

States argue that local control is important because it accounts for differences in geology and geography. Even with the states continuing to regulate the drilling process, the EPA is currently considering rules to regulate air pollution and wastewater from fracking and also investigating the impacts of fracking on drinking water, which the agency says is necessary to address broad public concerns. A common focus of industry and regulators has been the importance of a continuous improvement in the various aspects of shale gas production that relies on best practices and is tied to measurement and disclosure.

Regulators and legislators at both the federal and state levels are taking steps to address the environmental concerns about shale gas development by initiating studies to analyze the risks of fracking and imposing additional regulatory requirements on hydraulic fracturing operations. In 2013, the Resources for the Future (RF) catalogued the range of state regulations relevant to the shale gas development process across 31 states that have significant shale gas reserves (Richardson et al., 2013). State legislatures have reacted to environmental concerns with a patchwork of different regulations RFF identified 20 different regulatory elements in state fracking regulations, ranging from freeboard requirements to fluid storage options to underground injection regulations. Only the regulations in New York and West Virginia contained all 20 elements, but the average state did possess 15.6 elements. Only four states regulated less than 10 elements, each of which have yet to see any significant amount of shale development.

18.10 DISCLOSURE OF FRACKING CHEMICALS

Those states that have seen a dramatic spike in exploration and production activity have been quick but deliberate in adjusting their regulations accordingly, and many of the stated goals of the environmental groups on hydraulic fracturing, such as chemical disclosure and management of water resources, have been or are being addressed by state regulators such in programs taking effect in Texas, Wyoming, Colorado, Oklahoma, Illinois, New York, and Pennsylvania. At least 15 of the 29 states with confirmed hydraulic fracturing activity have laws requiring disclosure of information about the chemicals and additives in hydraulic fracturing fluid. Legislation requiring disclosure is pending in at least an additional seven states. Organizations such as the Environmental Defense Fund believe that chemical disclosure requirements at either the state or federal level should extend beyond the chemicals used in hydraulic fracturing to all of the chemicals used on a production site, such as those used in drilling muds.

In most of the states, chemicals are listed on the industry web site FracFocus.org. Environmental groups argue the site limits provide less transparency and accountability than standard government disclosure. Especially contentious is trade secret protection that the industry says is critical in order to reward development of unique products in the marketplace (Elgin, 2012). Well operators have sole discretion to determine when to assert trade secrets. As a result, inconsistent trade secret assertions are made throughout the registry.

According to a 24-page report released on February 24, 2014, the DOE Secretary's Energy Advisory Board (SEAB) Task Force said 84% of the wells registered in FracFocus invoked a trade secret exemption for at least one chemical since June 1, 2013 (SEAB, 2014). Another issue the task force discovered was errors in the registry. It said that after examining a limited sample of FracFocus 2.0 records, it found "a variety of errors, partly due to many different companies contributing data to an individual FracFocus record, besides the operator of the well." In Texas alone, 5509 of the 6406 disclosures made in the same time period invoked a trade secret exemption. The trade secret claims made in Texas hid information that included descriptions of ingredients as well as identification numbers and concentrations of the chemicals used. In Wyoming and Pennsylvania, companies must submit trade secret information so that the validity of companies' claims that information is proprietary can be evaluated.

FracFocus is revamping its system to let regulators for the first time search and aggregate the information. This change resolves the criticism from environmental groups by converting the online information into a database of chemicals used in individual wells. Of 20 states that require companies to disclose their chemicals, 12 require or allow the reporting to be on FracFocus.

18.11 AT THE FEDERAL GOVERNMENT LEVEL

EPA has begun a new study of hydraulic fracturing at the direction of Congress and is in the early stages of collecting information on the potential environmental impact of fracking on drinking water resources (EPA, February 2011). The study is a welcomed first step in a scientific analysis of the risks of fracking and in potentially developing industry best management practices. In December 2012 (EPA, December 2012a), the agency released a proposed Study Plan that outlined several basic questions with the promise that a dedicated team of scientists will begin providing answers:

(i) Water Acquisition: What are the potential impacts of large volume water withdrawals from ground and surface waters on drinking water resources?

(ii) Chemical Mixing: What are the possible impacts of surface spills on or near well pads of hydraulic fracturing fluids on drinking water resources?

(iii) Well Injection: What are the possible impacts of the injection and fracturing process on drinking water resources?

(iv) Flowback and Produced Water: What are the possible impacts of surface spills on or near well pads of flowback and produced water on drinking water resources?

(v) Wastewater Treatment and Waste Disposal: What are the possible impacts of inadequate treatment of hydraulic fracturing wastewaters on drinking water resources?

On December 9, 2013, EPA reconvened the study's Technical Roundtable. Subject-matter experts discussed the outcomes of the 2013 Technical Workshops, stakeholder engagement, and plans for the draft assessment report expected to be released in December 2014 (EPA, 2013a).

On July 28, 2011, EPA announced the release of a 604-page suite of proposed air emission regulations for oil and gas production, processing, transmission, and storage. The proposed rules were finalized on April 17, 2012. Companies now have until January 1, 2015 (rather than the 60 days in the original proposal) to begin using "green completion" equipment that can pare emissions at natural gas wells. A green completion is where gas and liquid hydrocarbons are separated from the wastewater using tanks, gas–liquid–sand separator traps, and gas dehydration equipment. If gathering lines are not available to collect the gas, it can be flared. The most notable new source of air emissions that EPA has targeted is that of well completions and recompletions, that is, the process of preparing gas wells for production. This means that hydraulically fractured wells are subject to federal air regulations for the first time.

During the transition period, companies can use both green completions and flaring. After January 1, 2015, companies cannot use flaring. Covered operations and equipment would include completions and recompletions of hydraulically fractured natural gas wells, compressors, pneumatic controllers, various storage tanks, and gas processing plants. Older pipelines and processing plants must also be retrofitted with new gear to reduce leaks, something that can be easily done for low cost and with existing technology.

Many drilling companies already use green-completion systems. Southwestern Energy Co. and Devon Energy Corp. say that they already use systems to capture methane and other fumes at wells, a key requirement of the new rule. Drilling has not slowed in Colorado or Wyoming where technology to capture emissions has been required by the state since 2009 and 2010. Of wells drilled in 2011 by eight members of America's Natural Gas Alliance, an association of natural gas producers, 93% used systems to capture stray gas, according to Sara Banaszak, chief economist, with the Washington-based group.

On October 19, 2011, EPA unveiled plans to set national standards for wastewater discharges from natural gas drilling amid growing concern over water pollution from fracking.

The study will examine the full cycle of water in hydraulic fracturing, from the acquisition of the water, through the mixing of chemicals and actual fracturing, to the postfracturing stage, including its ultimate treatment and disposal. EPA has selected locations for five retrospective and two prospective case studies. Separately, EPA will develop the first national standards for wastewater produced along with natural gas. It said any water pretreatment standards would be based on economically achievable technologies. A draft report will be released in late 2014.

On May 13, 2014, EPA posted on its web site a prepublication version of an Advanced Notice of Proposed Rulemaking (ANPRM) to regulate chemicals used in hydraulic fracturing under Sections 8(a) and 8(d) of the Toxic Substances Control Act. This ANPRM does not propose any actual regulation at present but instead seeks "comment on the information that should be reported or disclosed for hydraulic fracturing chemical substances and mixtures and the mechanism for obtaining this information" from the public and interested parties. EPA also indicates it is considering whether to require such reporting through regulatory action under TSCA authority, to implement a voluntary program, or to have a combination of both. The ANPRM asks for comments within 90 days of publication in the Federal Register. This time frame appears somewhat limited given the broad scope of questions on which EPA has requested comments. Operators, well services companies, and chemical manufacturers may all be subject to portions of any new rule that results from this process.

The U.S. Department of the Interior (DOI, 2013) has been working on fracturing regulations for the 3400 wells drilled every year on public lands, 90% of which use fracking and horizontal drilling. The draft rules focus on the disclosure of chemical identities, well-bore integrity, and management of wastewater disposal. On May 4, 2012, the Bureau of Land management (BLM) issued the proposed rule for oil and gas on public lands that will for the first time require disclosure of the chemicals used in the process (DOI/BLM, 2012). Companies will have to reveal the composition of fluids only after they have completed drilling—a sharp change from the government's original proposal, which would have required disclosure of the chemicals 30 days before a well could be started. The draft rule affects drilling operations on the 700 million acres of public land administered by BLM.

On May 25, 2013, the Bureau of Land Management (BLM) issued new draft regulations (DOI/BLM, 2013). Environmental groups say the new draft provides weaker water protections than a version DOI proposed a year earlier, while oil industry groups said they wanted regulation left in the hands of states and were opposed to any federal rules. Environmentalists expressed disappointment that the regulations do not include a ban on the storage of waste fluids in open, lined pits. They also wanted complete disclosure of chemicals used in fracking, which the regulations would not require.

In the last draft of the rules, the BLM required companies to test the integrity and strength of each well bore to ensure they do not leak fracking chemicals and hydrocarbons—and submit those results to the agency for review—before the well could be fracked. Under the new rules, companies can avoid obtaining or submitting this information for wells if they are similar to one that has been shown to have a strong enough cement job. Companies do not have to submit this information until after fracking has occurred. In the last draft of the rules, the BLM required companies to test the integrity and strength of each well bore to ensure they do not leak fracking chemicals and hydrocarbons—and submit those results to the agency for review—before the well could be fracked. The new rule still requires companies to disclose the chemical composition of their fracking fluids after the well is fracked. Now, though, it allows them to do it through FracFocus.org.

The American Petroleum Institute (API), oil and gas trade group, has developed a series of shale development guidance documents that encompass well integrity and production operations. Historically, API standards have been integrated into state regulatory frameworks. Such an approach benefits all parties in shale gas production: regulators will have more complete and accurate information; industry will achieve more efficient operations; and the public will see continuous, measurable improvement in shale gas activities. The Interstate Oil and Gas Compact Commission, the Marcellus Shale Coalition, the State Review of Oil and Natural Gas Environmental Regulation (STRONGER), the Groundwater Protection Council, and the Intermountain Oil and Gas Project are all working to identify best practices.

18.12 CONCLUSION

No energy is produced without risk and without some environmental cost. The extraction, processing, and transportation of natural gas all affect the environment. However, expansion of the supply of natural gas permits the displacement of more polluting forms of energy. With the shale gas boom continuing to gather steam, hydraulic fracturing will likely remain a focus for environmental and citizen groups concerned about its potential environmental impacts. Since October 2010, more than 100 bills across 19 states have been introduced relating to hydraulic fracturing for natural gas. The most active states include New York and Pennsylvania. This explosion of recent attention is largely attributed to fracking in regions where it is not as familiar to the affected communities, such as in the Marcellus Shale region. The most prominent recent trend in state legislation is an attempt to require chemical disclosure, along with other fluid regulation such as proper disposal and additive stipulations.

Industry is well aware that failure to manage some of the attendant impacts surrounding the development of this resource such as water use and contamination concerns, the public disclosure of the composition of fracking fluids, and fugitive emissions will seriously hamper efforts to fully develop this resource.

The scientific analysis investigating potential environmental threats of fracking has not yet concluded and testing will indefinitely continue. Working with industry, federal and state regulators and legislators will continue to monitor developments and develop regulations and protocols that will minimize shale gas developments' environmental footprint and any long-term impacts that it might have. More needs to be done by way of air emission monitoring, baseline water quality testing, and assaying of wastewater produced. If the environmental and social impacts of shale gas development are not addressed properly, there is the real possibility that public opposition will slow and perhaps stop new shale gas development.

Recognizing this reality, producers agreed on March 20, 2013, to a set of 15 standards focused on some of the most pressing problems with shale gas development (Brownstein, 2013). The companies also agreed to a certification process by the new Pittsburgh-based Center for Sustainable Shale Development. Companies will held accountable for complying with those standards in the Marcellus Shale. The new standards include limits on emissions of methane and the flaring, or burning off, of unwanted gas; reductions in engine emissions; groundwater monitoring and protection; improved well designs; stricter wastewater disposal; the use of less-toxic fracking fluids; and seismic monitoring before drilling begins (Center for Sustainable Shale Gas Development, 2013).

The project will cover Pennsylvania, West Virginia and Ohio as well as New York and other states in the East that currently have moratoriums on new drilling. Participants in the agreement include Shell, Chevron, the Environmental Defense Fund, the Clean Air Task Force, the Heinz Endowments, EQT Corp., Consol Energy, and the Pennsylvania Environmental Council.

Over time, the technology, best practices, and regulatory frameworks necessary to achieve shale gas' potential in a safe and environmentally acceptable manner will become the industry standard. The U.S. experience and technology innovations can then be carried to the rest of the world.

REFERENCES

Allen D, Torres V, et al. Measurements of methane emissions at natural gas production sits in the United States. Center for Energy and Environmental Resources. Austin: University of Texas; 2013a.

Allen D, Torres V, Thomas J, Sullivan D, et al. Measurements of methane emissions at natural gas production sites in the United States. Proceedings of National Academy of Science USA September 16, 2013b;110:17768–17773.

American Petroleum Institute. Community engagement guidelines. ANSA/API Bulletin 100-3. July 2014. http://www.api.org/~/media/Files/Policy/Exploration/100-3_e1.pdf.

Aquatech Formally Opens Central Water Treatment Facility in Tioga County, Pennsylvania for Shale Gas Producers. April 2013. Available at http://www.aquatech.com/MediaandLiterature/PressReleases/News.aspx?NewsID=116. Accessed January 17, 2015.

Arthur J, Uretsky M, Wilson P. Water resources and use for hydraulic fracturing in the marcellus shale region. All Consulting, LLC 2010. Available at http://www.all-llc.com/publicdownloads/WaterResourcePaperALLConsulting. Accessed January 17, 2015.

Brandt A, Heath G, et al. Methane leaks from North American natural gas systems. Science February 2014;343(6172):733–735.

Brasier K, Filteau M, McLaughlin Jacquet J, Stedman R, Kelsey T, Goetz S. Residents' perceptions of community and environmental impacts from development of natural gas in the Marcellus Shale: a comparison of Pennsylvania and New York cases. J Rural Social Sci 2011;26 (1):32–61.

Brownstein M. Industry and environmentalists make progress on fracking. Energy EDF March 28, 2013. Available at http://www.edf.org/blog/2013/03/28/industry-and-environmentalists-make-progress-fracking. Accessed January 17, 2015.

Center for Sustainable Shale Gas Development. Performance Standards. Geographic scope and applicability of CSSD performance standards March 2013. Available at http://037186e.netsolhost.com/site/wp-content/uploads/2013/03/CSSD-Performance-Standards-3-13R.pdf. Accessed January 17, 2015.

Chemicals Used in Hydraulic Fracturing. United States House of Representatives. Committee on Energy and Commerce. Minority Staff. April 2011. Available at http://democrats.energycommerce.house.gov/sites/default/files/documents/Hydraulic-Fracturing-Chemicals-2011-4-18.pdf. Accessed January 17, 2015.

DOI. Oil and gas; hydraulic fracturing on Federal and Indian lands, Supplemental Notice of Proposed Rulemaking, 43 CFR Part 3160; 2013.

DOI, BLM. Proposed rule oil and gas; well stimulation, including hydraulic fracturing, on Federal and Indian lands. May 11, 2012.

DOI, BLM. Revised proposed rule, oil and gas; hydraulic fracturing on Federal and Indian Lands. May 24, 2013.

Economic Analysis of Methane Emissions Reduction Opportunities in the U.S. Onshore Oil and Natural Gas Industries. ICF International. March 2014. Available at http://www.edf.org/sites/default/files/methane_cost_curve_report.pdf. Accessed January 17, 2015.

Elgin B, Haas B, Kuntz P. Pivot Upstream Group. Analyzed by Bloomberg News. Fracking secrets by thousands keep U.S. clueless on wells. November 2012.

Ellsworth W. US Geological Survey scientist. (December 2012). Injection-induced earthquakes. Science 12 July 2013a;341(6142). DOI: 10.1126/science.1225942.

Ellsworth W, Hickman S, Leons A, McGarr A, Michael A, Rubinstein J. USGS. Are seismicity rate changes in the midcontinent natural or man made? April 2012.

EPA. Company to upgrade treatment and pay penalties after discharge violations at western Pa. Oil and Gas Wastewater Treatment Facilities May 2013a. Available at http://yosemite.epa.gov/opa/admpress.nsf/0/9FC2BD02936B253785257B73006C68F7. Accessed January 17, 2015.

EPA. Draft plan to study the potential impacts of hydraulic fracturing on drinking water resources. Office of Research and Development. February 7, 2011a.

EPA. Inventory of U.S. greenhouse gas emissions and sinks: 1990–2011. EPA 430-R-13-001. April 12, 2013b.

EPA. Study of the potential impacts of hydraulic fracturing on drinking water resources. Progress report. EPA 601/R-12/011, December 2012a. Available at www.epa.gov/hfstudy.

EPA. Potential impacts of hydraulic fracturing on drinking water resources: study update. December 2013c. Available at http://www2.epa.gov/sites/production/files/2013-12/documents/study_update-. Accessed January 17, 2015.

Ferrar K, Michanowicz D, Christen C, Mulcahy N, Malone S, Sharma R. Assessment of effluent contaminants from three facilities discharging Marcellus shale wastewater to surface waters in Pennsylvania. Environ Sci Technol May 2013;47(7): 3472–3481.

Fontenot B, Hunt L, Hildenbrand Z. et al. (July 25, 2013). An Evaluation of water quality in private drinking water wells near natural gas extraction sites in the Barnett shale formation. Environ Sci Technol 2013;47(17):10032–10040.

Freyman M. Ceres. Hydraulic fracturing & water stress: water demand by the numbers. February 2014.

Freyman M, Salmon R. Ceres. Hydraulic fracturing & water stress: growing competitive pressures for water. May 2013. Available at http://www.ceres.org/resources/reports/hydraulic-fracturing-water-stress-growing-competitive-pressures-for-water. Accessed January 17, 2015.

GAO. Opportunities exist to capture vented and flared natural gas, which would increase royalty payments and reduce greenhouse gases. October 2010. Available at http://www.gao.gov/new.items/d1134.pdf. Accessed January 17, 2015.

GAO. Report to congressional requesters, oil and gas: information on shale resources, development and environmental and public health Risks. GAO-12-732. September 2012.

Groat C, Gromshaw T. Fact-based regulation for environmental protection in shale gas development. The Energy Institute. The University of Texas at Austin. February 2012. Available at http://www.velaw.com/UploadedFiles/VEsite/Resources/ei_shale_gas_reg_summary1202[1].pdf. Accessed January 17, 2015.

Howarth R, Santoro R, Ingraffea A. Methane and the greenhouse-gas footprint of natural gas from shale formations. Climatic Change March 13, 2011. DOI 10.1007/s10584-011-0061-5 – cornell.edu.

Jackson R, Vengosh A, Darrah T, Warner N, Down A, Poreda R, Osborn S, Zhao K, Karr J. Increased stray gas abundance in a subset of drinking water wells near Marcellus shale gas extraction. Proceedings of the National Academy of Sciences of the United States of America June 24, 2013.

Kalaswad S, Christian B, Petrossian R. Texas Water Development Board. Brackish groundwater in Texas. 2012.

Krupnick A, Gordon H, Olmstead S. Resources for the Future. Pathways to dialoque. What the experts say about the environmental risks of shale gas development. February 2013. Available at http://www.rff.org/Documents/RFF-Rpt-PathwaystoDialogue_Overview.pdf. Accessed January 17, 2015.

Lutz B, Lewis L, Doyle M. Generation, transport, and disposal of wastewater associated with Marcellus shale gas development. Water Resourc Res February 2013;49(2):647–656.

NETL. Monitoring of Air, Land, and Water resources during Shale Gas Production. 2013. Available at http://www.netl.doe.gov/publications/factsheets/rd/R&D167.pdf. Accessed January 17, 2015.

Nicot J, Scanlon B. Water use for shale-gas production in Texas, U.S. Bureau of Economic Geology, Jackson School of Geosciences, University of Texas at Austin. Environ Sci Technol March 2012. dx.doi.org/10.1021/es204602t | Environ Sci Technol 2012;46:3580–3586. Accessed January 17, 2015.

"Ohio and Fracking." Source Watch. Source Watch, n.d. Web. May 28, 2013. Available at http://www.sourcewatch.org/index.php?title=Ohio_and_fracking. Accessed January 17, 2015.

Olmstead S, Muehlenbachs L, Shih J. Resources for the Future. Impacts of shale gas developments on rivers and streams. 2013.

O'Sullivan F, Paltsev S. Shale gas production: potential versus actual greenhouse gas emissions. MIT Joint program on the Science and Policy of Global Change. November 2012.

Richardson N, Gottlirb M, Krupnick A, Wiseman H. The state of shale gas regulation by state. Resources for the Future. Center for Energy Economics and Policy May 2013. Available at http://www.rff.org/rff/documents/RFF-Rpt-StateofStateRegs_Report.pdf. Accessed January 17, 2015.

SEAB, Task Force Report on FracFocus 2.0. U.S. Department of Energy February 24, 2014. Available at http://energy.gov/sites/prod/files/2014/03/f8/FracFocus%20TF%20Report%20Final%20Draft.pdf. Accessed January 17, 2015.

Texas Railroad Commission. H-10 report by district or county. Available at http://webapps.rrc.state.tx.us/H10/h10PublicMain.do. Accessed January 17, 2015.

VanBriesen J. Center for Water Quality in Urban Environmental Systems (Water QUEST). Carnegie Institute of Technology. Bromide levels in the Monongahela River. 2012.

Vaughan A, Pursell D. Frac attack: risks, hype, and financial reality of hydraulic fracturing in the shale plays. Reservoir Research Partners; and Tudor, Pickering, Holt & Co. 2010.

Western Organization of Resource Councils. Gone or good, fracking water in the West. 2013. Available at http://www.worc.org/userfiles/file/Oil%20Gas%20Coalbed%20Methane/Hydraulic%20Fracturing/Gone_for_Good.pdf. Accessed January 17, 2015.

Wilson J, Wang Y, VanBriesen J. Sources of high total dissolved solids to drinking water supply in southwestern Pennsylvania. J Environ Eng 2014; 140(5):B4014003.

XTO Energy, Inc. Settlement. July 18, 2013 Available at http://www2.epa.gov/enforcement/xto-energy-inc-settlement. Accessed January 17, 2015.

FURTHER READING

ALL Consulting, LLC, for U.S. Department of Energy. Estimates of peak drilling activity in New York, Pennsylvania, and West Virginia. July 2010.

American Petroleum Institute. (2012). Overview of industry/guidance/best practices supporting hydraulic fracturing (HF). 2012.

Argonne National Laboratory. Consumptive water use in the production of ethanol and petroleum gasoline. ANL/ESD 09-1 update. Updated July 2011.

Belcher M, Resnikoff M. Hydraulic fracturing radiological concerns for Ohio. Radioactive Waste Management Associates. Fact sheet prepared for the FreshWater Accountability Project Ohio. June 13, 2013.

Department of Energy & Climate Change. UK Government. Government response to Royal Academy of Engineering and Royal Society report on shale gas extraction in the UK: a review of hydraulic fracturing. December 2012.

DOE/NETL. Life cycle analysis: natural gas combined cycle (NGCC) power plant 2010;403/110509:127. Available at http://www.netl.doe.gov/energy-analyses/pubs/NGCC_LCA_Report_093010.pdf. Accessed January 17, 2015.

DOE, Office of Fossil Fuel. National Energy Technology Laboratory, Modern shale gas development in the United States: a primer. April 2009.

Ellsworth W. (July 2013b). Injection-induced earthquakes. USGS. Science ;341(6142).

Engelder T. Department of Geosciences. The Pennsylvania State University, The mechanisms by which the Marcellus gas shale sequesters residual treatment water. Review of National Academy of Sciences. Attachment #3 (Engelder Commentary), May 2012. Available at http://s3.documentcloud.org/documents/395440/attachment-3-engelder-commentary-may-29-2012.pdf. Accessed January 17, 2015.

EPA. Key Documents about Mid-Atlantic Oil and Gas Extraction. Letter to Pennsylvania Department of Environmental Protection. May 2011b. Available at http://www.epa.gov/region3/marcellus_shale/. Accessed January 17, 2015.

EPA. Study of the potential impacts of hydraulic fracturing on drinking water resources. Progress report. EPA 601/R-12/011, December 2012b. Available at www.epa.gov/hfstudy.

Frohlich C. Two-year survey comparing earthquake activity and injection-well locations in the Barnett Shale, Texas. Proc Natl Acad Sci USA 2012; 109(35):13934–13938. http://dx.doi.org/10.1073/pnas.1207728109.

Fulton M, Mellquist N. Worldwatch Institute. Comparing life-cycle greenhouse gas emissions from natural gas and coal. August 25, 2011.

Hammer R, VanBriesen J. In fracking's wake: new rules are needed to protect our health and environment from contaminated wastewater. NRDC Document. D:12-05-A. May 2012.

Howarth R, Santoro R, Ingraffea A, Cathles L, Brown L, Taam M, Hunter A. The greenhouse-gas footprint of natural gas in shale formations. July 2012. Available at http://www.springerlink.com/content/x001g12t2332462p/. Accessed January 17, 2015.

Konschnik K, Holden M, Shasteen A. Harvard Law School. Legal fractures in chemical disclosure laws. Why the voluntary chemical disclosure registry FracFocus fails as a regulatory compliance tool. April 2013. Available at http://blogs.law.harvard.edu/environmentallawprogram/files/2013/04/4-23-2013-LEGAL-FRACTURES.pdf. Accessed January 17, 2015.

Maloney K, Yoxtheimer D. Production and disposal of waste materials from gas and oil extraction from the Marcellus shale play in Pennsylvania. US Geological Survey–Leetown Science Center. Northern Appalachian Research Laboratory, Wellsboro, Pennsylvania. Penn State Marcellus Center for Outreach and Research, University Park, Pennsylvania. September 2012.

Massachusetts Institute of Technology. The future of natural gas. An interdisciplinary MIT study. June 2011. Available at http://mitei.mit.edu/system/files/NaturalGas_Report.pdf. Accessed January 17, 2015.

McKenzie L, Witter R, Newman J, Adgate J. Colorado School of Public Health, University of Colorado, Anschutz Medical Campus, Aurora, Colorado, USA. Human health risk assessment of air emissions from the development of unconventional natural gas resources. February 2012.

Nicot J, Reedy R. Oil and gas water use in Texas: update to the 2011 mining water use report, Bureau of Economic geology, The University of Texas Austin. September 2012.

NRDC. Leaking Profits. March 2012. Available at http://www.nrdc.org/energy/files/Leaking-Profits-Report.pdf. Accessed January 17, 2015.

Olmstead S, Muehlenbachs L, Shih J, Krupnick J. Resources for the Future. Shale gas development impacts on surface water quality in Pennsylvania. August 2012.

Pennsylvania Department of Environmental Protection. Utilization of mine influenced water for natural gas extraction activities. January 2013.

Resources for the Future. Managing the risks of shale gas. June 2013. Available at http://www.rff.org/rff/documents/RFF-Rpt-ManagingRisksofShaleGas-KeyFindings.pdf. Accessed January 17, 2015.

Schaefer K. Fracking and water: a new way to profit from the industry's biggest problem. *Oil Price* February 2012. Available at http://oilprice.com/Energy/Energy-General/Fracking-and-Water-A-New-Way-To-Profit-from-the-Industrys-Biggest-Problem.html. Accessed January 17, 2015.

SEAB, Shale Gas Production Subcommittee Second Ninety Day Report. November 18, 2011. Available at http://www.shalegas.energy.gov/resources/111811_final_report.pdf. Accessed January 17, 2015.

Styles P, Brian Baptie B. Induced seismicity in the UK and its relevance to hydraulic stimulation for exploration for shale gas. April 2012.

VanBriesen J. Letter to Pennsylvania Department of Environmental Protection, Center for Water Quality in Urban Environmental Systems (Water QUEST). Carnegie Institute of Technology. April 2011.

Warner N, Christie C, Jackson R, Vengosh A. Impacts of shale gas wastewater disposal on water quality in Western Pennsylvania. Environ Sci Technol October 2013;47:11849–11857. dx.doi.org/10.1021/es402165b

INDEX

acoustic impedance (AI), 201
adsorption index (AI), 58
adsorption isotherm, 124, 326
    Langmuir equation, 124, 269
    Langmuir isotherm, 124, 252
amplitude versus offset and azimuth
    (AVOAz), 192
anisotropic differential effective medium
    approach, 195
apparent permeability function (APF), 259
Archie equation, 130
aspect ratio, 143, 195
asphaltenes, 49, 57
asset life cycle, 361
    appraisal phase, 365, 377
    development phase, 367, 377
    exploration phase, 361, 377
    production phase, 375, 378
    rejuvenation phase, 376, 378
atomic force microscopy (AFM), 247
AVO response, 201–3

Backus model, 193–5
ball-activated systems, 373–4
Barnett Shale, 74
basin type(s), 11–13
batch pressure decay (BPD) technique, 254
biogenic aggregation, 23
black shale, 21–2, 27, 29, 30, 32, 34
boundary-dominated flow (BDF), 287
Bowers's method, 150
brittle mineral(s), 3
brittleness index, 132
Brunauer–Emmett–Teller (BET) method, 92

canister desorption test, 262
cap rock, 91, 117
capillary force, 274
capillary pressure, 91, 105, 273
carbon isotopic measurements, 59, 63
clay content, 13, 118, 145, 343
clay diagenesis, 142
clay hydration, 343
clay minerals, 21, 89
clay swelling inhibitor, 348
clay-bound water, 104, 125, 273
coal-bed methane (CBM), 57
Coates equation, 107
coherent surface wave noise, 222
coiled tubing, 373–4
compaction disequilibrium, 140
connecting pores, 143
constant cross-section model
    (CCM), 252
contact angle, 342
contained joints, 212
continental margins, 30
contractional fractures, 211
crosshole continuity logging, 175
cutting mechanisms, 174
    brittle regime, 174
    ductile regime, 174

the dead space, 326
Dean Stark method, 122
decline curve analysis (DCA), 302
dehydration, 143
deliquification techniques, 376
density log, 126, 148, 159

depositional processes, 23, 38
diagenetic processes, 23, 39
discrete fracture network (DFN), 240
drill stem tests (DSTs), 295
    mini-DST, 295
drilling mud, 184
dry gas ratio (DGR), 59, 60
dual-porosity, 284, 289

Eaton's method, 147
effective field theory, 195
effective porosity, 121, 143
effective saturation, 274
effective stress method, 149
elastic properties, 178, 192
    bulk moduli, 178
    Poisson's ratio, 132, 178
    shear moduli, 178
    Young's modulus, 132, 178
enlarged cross-section model
    (ECM), 252
estimated ultimate recovery
    (EUR), 287
export production, 24, 33

Fick's second law, 255
field development plans (FDP), 366
flame ionization detector (FID), 51, 52
flow rate decline analysis, 284
fluid extraction, 122
fluid mud, 23
focal mechanism solutions, 213
Fourier transform infrared transmission
    spectroscopy (FTIR), 120

Fundamentals of Gas Shale Reservoirs, First Edition. Edited by Reza Rezaee.
© 2015 John Wiley & Sons, Inc. Published 2015 by John Wiley & Sons, Inc.

fracture imaging methods, 232
fracture treatment monitoring, 371
  microseismic, 216, 370–371
fracturing fluids, 354, 371

gamma ray log, 72, 125
  NGT, 125
  SGR, 125
gas content, 57
gas desorption, 252
gas in place, 119, 304–7
  initial gas in place, 245
  OGIP, 304–7
gas shale, 1, 89, 117, 301
gas slippage factor, 95
Gassmann substitution, 195
geological properties
  depositional environment, 1, 2, 13, 34–9, 53
  depositional setting, 30
  organic geochemistry, 2
  thermal maturity, 54–5, 62, 118, 126–7
geomechanical properties, 120
geometric fracs, 370
geophone, 222
  buried geophone, 222
  surface geophone, 222
grain size, 30, 121
grand canonical statistical ensemble (GCMC), 325, 327
gravimetric method, 124
grid design, 226
groundwater contamination, 384

Haynesville/Bossier shales, 80
the Hill Shirley Kline equation, 273
hydraulic fracturing, 3, 211, 370, 385, 390–394
hydrogen content, 50
hydrogen indices, 49
hydrostatic pressure, 144

illite/smectite (I/S), 90, 108
initial water saturation, 356
interfacial tension (IFT), 277

joints, 211

kerogen
  cracking, 49
  density, 130
  type I, 2, 124
  type II, 2, 49, 118, 124
  type III, 2, 118, 124
  type IV, 52
kicks, 144
Klinkenberg effects, 270
Klinkenberg equation, 272
Klinkenberg permeability, 95
Knudsen diffusion, 250
Knudsen number, 249

level of maturity (LOM), 130
long time aperture (LTA), 227
long-period long-duration (LPLD) seismic, 208
low pressure adsorption measurement, 92
low pressure nitrogen adsorption, 92

macrofracture flow, 294
Marcellus Shale, 78
material balance equation, 263
matrix flow, 294
matrix gas desorption, 294
mechanical stratigraphy, 211
mercury injection capillary pressure (MICP), 89
mercury porosimetry, 121
methane emissions, 386
microearthquake (MEQ), 208
microfossils, 73
microgeometry, 268
microstructure, 193
mineralogy, 3, 120, 127, 130, 364
mini-frac equations, 296
modular formation dynamic tester (MDT), 146
Mohr–Coulomb
  (strength) criterion, 181–3
  envelope, 177
molecular dynamics (MD), 325
molecular simulation (MS), 325
moment tensor, 213
monitoring well, 216
Monte Carlo probabilistic approach, 302
Montney Formation, 80
mud, 21–3, 34–7, 73
  biogenous mud, 21
  terrigenous mud, 21–2, 31
mud log, 144, 151, 154
mud weight window (MWW), 178, 181
mudstone, 47, 64, 83
multiregression analysis, 108

nanoindentation testing (NIT), 173
nanoporous media, 276
nanoscience, 245
nanoscratch test, 174
narrow azimuth (NA), 202
natural fracture reactivation, 212
neutron log, 126
neutron-density cross-plot, 145
New Albany Shale, 78
normal compaction trends (NCTs), 144
nuclear magnetic resonance (NMR), 89, 123, 268
nuclear magnetic resonance log, 128

oil window, 49, 53
organic matter, 21
organic richness, 49
organic-rich shales (ORSs), 191

organoclasts, 52
organoporosity, 55, 56
osmotic effect, 346

paleogeography, 28, 38
partial saturation, 195
Passey methodology, 130
passive seismic methods, 207
pay zone, 14
Peng–Robinson EoS (PR-EoS), 275
permeability, 95, 107–8, 123, 247, 257, 262, 272
photic zone, 24, 29, 37
photoelectric factor log, 127
physiography, 28–30, 33
plug-and-perf, 373–4
polar anisotropy, 191
pore diameter, 89, 103–4
pore geometry, 89, 267
pore network model, 251
pore pressure, 139, 146
  abnormal pressure, 139
  overpressure, 139
pore size classification, 90
  macropores, 90
  mesopores, 90
  micropores, 90
pore size distribution, 89, 118, 272
pore-proximity effect, 275
porosity
  density porosity, 145, 159, 160, 164, 343
  effective porosity, 121, 143
  MICP porosity, 103, 113
  neutron porosity, 145, 159, 160, 164, 343
  NMR porosity, 104, 128
pressure gradient, 139
principal stresses, 171
pseudo-pressure function, 258
pulsed neutron mineralogy log, 127
pulsed neutron mineralogy tool, 130
pyrolysis, 50–52, 55–6

quad-porosity shale, 278

radial distribution function (RDF), 337
rarefaction coefficient, 277
rate of penetration (ROP), 144
recover factor curve, 279
refractory kerogen, 49
repeated formation tests (RFTs), 146
reservoir characterization, 363
resins, 49, 57
resistivity log, 125
retort distillation method, 122
reverse time imaging, 218
rocks radioactivity, 125–6
  potassium, 125
  thorium, 126
  uranium, 125

sample preservation, 172, 191
scanning electron microscopy (SEM), 92, 247, 268
scratch tests, 174
sedimentation rates, 25, 33, 37–8
seismic consistency, 237
seismic emission tomography (SET), 223
sequence stratigraphic model, 71
    HST, 71
    relative sea level, 27, 72, 75
    SB, 71
    TSE, 71
    TST, 71
shale engineering, 365
shale maps, 11
short time aperture (STA), 227
shrunk length model (SLM), 252
slick water, 47
slip flow, 250
sonic log, 127
source rock, 2, 49, 201
spherical flow, 295
spontaneous imbibition, 342
static corrections, 225
stimulated reservoir volume (SRV), 210, 240, 284
stimulation, 48, 220, 371
storage pores, 143
strain, 209
stress, 209
stress field, 152
subhorizontal cracks, 191
sub-irreducible initial water saturation, 356
subvertical microcracks, 191
surface area, 92, 119
surface monitoring, 222
surface-to-volume (S/V) ratio, 110
surfactants, 352
Swanson method, 123
sweet spot, 85, 362

T_2 relaxation time, 93
tangential momentum accommodation coefficient (TMAC), 259
technically recoverable resource (TRR), 1
tectonic
    activity, 151, 164, 165
    loading, 141, 156
    processes, 27
    setting, 76
tessellated surface, 232
thermogravimetric analyzer (TGA), 122
Thomsen's parameters, 176
tortuosity, 90, 250
total organic carbon (TOC) (content), 50, 118, 201–2
total porosity, 128, 143
triaxial test, 170, 171
Tributary Drainage Volume (TDV), 240

ultrasonic tests, 172
    P-wave, 172
    S-wave, 172
unconventional
    hydrocarbon resource, 1
    resource shales, 71, 72
    shale gas resource systems, 47
    shale reservoirs, 283
Unconventional Gas Resource Assessment System (UGRAS), 309
under-compaction, 140
uniaxial compressive strength (UCS), 174
unsteady-state techniques, 123
    pressure decay, 123
    pulse decay, 123
upscaling, 278
upwelling currents, 73
US Bureau of Mines (USBM), 57

velocity anisotropy, 127, 133
vertical effective stress, 141
vertical well(s), 61, 196, 297
vitrinite, 54, 129
vitrinite reflectance, 3, 54–5, 129
volumetric gas adsorption, 124
VTI behaviour, 176

water column, 24, 32, 33, 36
water resources, 13, 381, 386, 389–90
water use, 381
well instability, 193
well spacing, 207, 310
wettability, 267, 274, 341
wide azimuth (WA), 202
Woodford Shale, 74

Xanthan Gum (XG) polymer, 351
X-ray diffraction (XRD), 90, 160

Printed in the United States
By Bookmasters